Synthetic Membranes:
Science, Engineering and Applications

NATO ASI Series

Advanced Science Institutes Series

A series presenting the results of activities sponsored by the NATO Science Committee, which aims at the dissemination of advanced scientific and technological knowledge, with a view to strengthening links between scientific communities.

The series is published by an international board of publishers in conjunction with the NATO Scientific Affairs Division

A Life Sciences B Physics	Plenum Publishing Corporation London and New York
C Mathematical and Physical Sciences	D. Reidel Publishing Company Dordrecht, Boston, Lancaster and Tokyo
D Behavioural and Social Sciences E Engineering and Materials Sciences	Martinus Nijhoff Publishers The Hague, Boston and Lancaster
F Computer and Systems Sciences G Ecological Sciences	Springer-Verlag Berlin, Heidelberg, New York and Tokyo

Series C: Mathematical and Physical Sciences Vol. 181

Synthetic Membranes: Science, Engineering and Applications

edited by

P. M. Bungay

Chemical Engineering Section, Biomedical Engineering & Instrumentation Branch,
National Institutes of Health, U.S. Department of Health and Human Services,
Bethesda, Maryland, U.S.A.

H. K. Lonsdale

Bend Research, Inc., Bend, Oregon, U.S.A.

and

M. N. de Pinho

Chemical Engineering Department,
Instituto Superior Técnico,
Technical University of Lisbon, Portugal

Technical editor:

Janet K. Bungay

National Heart, Lung and Blood Institute,
National Institutes of Health,
Bethesda, Maryland, U.S.A.

D. Reidel Publishing Company

Dordrecht / Boston / Lancaster / Tokyo

Published in cooperation with NATO Scientific Affairs Division

Proceedings of the NATO Advanced Study Institute on
Synthetic Membranes: Science, Engineering and Applications
Alcabideche, Portugal
June 26-July 8, 1983

Library of Congress Cataloging in Publication Data

NATO Advanced Study Institute on Synthetic Membranes: Science, Engineering, and Applications (1983: Alcabideche, Portugal)
Synthetic Membranes—Science, engineering, and applications.

(NATO ASI series. Series C, Mathematical and physical sciences; vol. 181)
"Proceedings of the NATO Advanced Study Institute on Synthetic Membranes: Science, Engineering, and Applications, Alcabideche, Portugal, June 26–July 8, 1983"—T.p. verso.
"Published in cooperation with NATO Scientific Affairs Division."
Includes index.
1. Membranes—Congresses. I. Bungay, Peter M., 1941– . II. Lonsdale, Harold K. (Harold Kenneth), 1932– . III. Pinho, Maria Norberta de, 1945– . IV. North Atlantic Treaty Organization. Scientific Affairs Division. V. Title. VI. Series: NATO ASI series. Series C, Mathematical and physical sciences; no. 181.
TP159.M4N38 1983 660.2′842 86–13108
ISBN 90–277–2293–5

Published by D. Reidel Publishing Company
P.O. Box 17, 3300 AA Dordrecht, Holland

Sold and distributed in the U.S.A. and Canada
by Kluwer Academic Publishers,
101 Philip Drive, Assinippi Park, Norwell, MA 02061, U.S.A.

In all other countries, sold and distributed
by Kluwer Academic Publishers Group,
P.O. Box 322, 3300 AH Dordrecht, Holland

D. Reidel Publishing Company is a member of the Kluwer Academic Publishers Group

---

Printed in The Netherlands

CONTENTS

PREFACE

The chapters in this book are based upon lectures given at the NATO Advanced Study Institute on Synthetic Membranes (June 26–July 8, 1983, Alcabideche, Portugal), which provided an integrated presentation of synthetic membrane science and technology in three broad areas.

Currently available *membrane formation* mechanisms are reviewed, as well as the manner in which synthesis conditions can be controlled to achieve desired membrane structures. Membrane performance in a specific separation process involves complex phenomena, the understanding of which requires a multidisciplinary approach encompassing polymer chemistry, physical chemistry, and chemical engineering. Progress toward a global understanding of membrane phenomena is described in chapters on the *principles of membrane transport*. The chapters on *membrane processes and applications* highlight both established and emerging membrane processes, and elucidate their myriad applications.

It is our hope that this book will be an enduring, comprehensive compendium of the state of knowledge in the field of synthetic membranes. We have been encouraged in that hope by numerous expressions of interest in the book, coming from a variety of potential users.

At this time, we wish to acknowledge several individuals and groups whose contributions were essential to the success of this venture: Dr. Craig Sinclair and the NATO Science Committee; Ms. Aida de Sousa, Ms. Judite Fialho and their colleagues at SERV-INTERNACIONAL, Lisbon, Portugal; Dr. Murray Eden, Dr. Robert Dedrick, Ms. Leisa Lyles, and Ms. Inyoung Kwon, of the Biomedical Engineering and Instrumentation Branch, National Institutes of Health; the Department of Chemical Engineering, Instituto Superior Técnico, Lisbon, Portugal; and in particular, the members of the Advanced Study Institute faculty. Corporate sponsors of the ASI in 1983, as well as sponsors of the book, are listed separately; their valued assistance is gratefully acknowledged.

Peter M. Bungay
Harold K. Lonsdale
Maria Norberta de Pinho

ACKNOWLEDGMENTS

Book Sponsors

The costs of typesetting this volume were met in part by the following corporate sponsors, whose generosity is gratefully acknowledged.

FilmTec Corporation
Minneapolis, Minnesota, U.S.A.

The Millipore Foundation
Millipore Corporation
Bedford, Massachusetts, U.S.A.

W.L. Gore & Associates, Inc.
Membrane Products Division
Elkton, Maryland, U.S.A.

De Martini S.p.A.
Biella, Italy

Merck Sharp & Dohme Research Laboratories
Division of Merck & Co., Inc.
Rahway, New Jersey, U.S.A.

Celanese Separations Products
Charlotte, North Carolina, U.S.A.

Bend Research, Inc.
Bend, Oregon, U.S.A.

Journal of Membrane Science
Elsevier Science Publishers BV
Amsterdam, The Netherlands

Gambro Dialysatoren KG
Hechingen, Federal Republic of Germany

Syntex Research
Division of Syntex (U.S.A.) Inc.
Palo Alto, California, U.S.A.

OSMO Membrane Division
Osmonics, Inc.
Minnetonka, Minnesota, U.S.A.

Solvay & Cie.
Direction Centrale des Recherches
Brussels, Belgium

Dow Chemical U.S.A.
Membranes Department
Midland, Michigan, U.S.A.

Advanced Study Institute Sponsors

The following organizations contributed financial support to the Advanced Study Institute on Synthetic Membranes. Their assistance is gratefully acknowledged.

National Institutes of Health
Biomedical Engineering and Instrumentation Branch
Division of Research Services
Bethesda, Maryland, U.S.A.

National Science Foundation
Fellowships Section
Washington, D.C., U.S.A.

E.I. du Pont de Nemours & Company
Polymer Products Department
Wilmington, Delaware, U.S.A.

Merck, Sharp & Dohme Research Laboratories
Rahway, New Jersey, U.S.A.

INTER Research Corporation
Lawrence, Kansas, U.S.A.

Allied Corporation
Corporate Technology
Morristown, New Jersey, U.S.A.

Consiglio Nazionale delle Richerche
Progetto Finalizzato "Chimica Fine e Secondaria"
Roma, Italy

Electricité de France
Membrane Club (D.E.R.)
Clamart, France

Société Nationale Elf Aquitaine (Production)
Paris - La Défense, France

Société C.E.C.A., S.A.
Velizy - Villacoublay, France

Gambro Dialysatoren KG
Research & Development
Hechingen, Federal Republic of Germany

Seitz-Filter-Werke
Theo & Geo Seitz GmbH und Co.
Bad Kreuznach, Federal Republic of Germany

Solvay & Cie
Laboratoire Central
Bruxelles, Belgium

Acknowledgment is extended to the following publishing houses for contributions to the library of the Advanced Study Institute.

Elsevier Science Publishers
Science & Technology Division
Amsterdam, The Netherlands

D. Reidel Publishing Company
Dordrecht, Holland

Verlag Moritz Diesterweg
Frankfurt, Federal Republic of Germany

Typesetting for this book was done by Photo Data, Inc., 419 Seventh St., N.W., Suite 500, Washington, D.C. 20004. Special acknowledgment is extended to Ms. Sheila Duncan, Ms. Becky Saunders, Mr. Reginald Saunders, Ms. Mickie Swann and Ms. Vera Thompson.

SYMBOLS

The following symbols constitute a common nomenclature for the book. Symbols given specialized meanings in a particular chapter may be found at the end of that chapter. Units are first indicated in terms of base physical quantities: length (L), mass (M), time (t), temperature (T), amount of substance (mol) and electric current (A, ampere). Alternative expressions for units contain derived quantities: energy ($E=ML^2/t^2$), pressure ($p=M/Lt^2$) and electric potential (V, volt).

a	activity, mol/L^3
A	area, L^2
A_V	solvent permeation coefficient (volume), L^2t/M or L/tp
A_w	solvent permeation coefficient (mass), t/L or M/L^2tp
b	radius of spherical particle, L
B_i	solute permeation coefficient, L/t
c	concentration, M/L^3 or mol/L^3
C	heat capacity, L^2/t^2T or E/MT
d	differential operator
d	diameter (with subscript), L
d_h	hydraulic diameter, L
d_i	molecular diameter of species i, L
d_p	pore diameter, L
d_t	tube diameter, L
D	diffusion coefficient, L^2/t
D_T	thermodynamic diffusion coefficient, L^2/t
e	base of natural logarithms
E	energy, ML^2/t^2 or E
$\Delta\tilde{E}_a$	activation energy, ML^2/t^2 mol or E/mol
$\dot{E}$	rate of energy consumption, ML^2/t^3 or E/t
f	number of degrees of freedom
f_{ij}	friction coefficient, M/t mol or Et/L^2mol
F	Faraday constant, At/mol
g	gravitational acceleration, L/t^2
g_{ij}	Flory-Huggins interaction parameter of species i and j, ML^2/t^2 mol or E/mol
G	Gibbs free energy, ML^2/t^2 or E
h	membrane thickness, L

H	enthalpy, ML^2/t^2 or E
$\Delta\hat{H}_v$	latent heat of vaporization, ML^2/t^2 or E
i	electric current density, A/L^2
i_{lim}	limiting current density, A/L^2
I	electric current, A
J_i	flux of species i, M/L^2t or mol/L^2t
J_V	volumetric flux, L/t
k	thermal conductivity, ML/t^3T or E/LtT
κ	Boltzmann constant, ML^2/t^2T or E/T
k_f	forward reaction rate constant, $(mol/L^3)^{1-n}/t$, n = order of reaction
k_i	mass transfer coefficient for species i, L/t
k_r	reverse reaction rate constant, $(mol/L^3)^{1-n}/t$, n = order of reaction
K_{eq}	equilibrium coefficient, various units
K_i	distribution coefficient for species i, a_i/a_i', dimensionless
K_s	solubility product, various units
L	membrane length, L
L_p	hydraulic permeability, hydraulic conductivity or filtration coefficient, L^2t/M or L/tp, L^3t/M or L^2/tp
L_{ij}	phenomenological conductance coefficient, various units
m	molal concentration, mol/M
M	mass, M
$\tilde{M}$	molecular weight, M/mol
n	number of moles
$\tilde{N}$	Avogadro constant, $(mol)^{-1}$
Δp	pressure drop or loss, M/Lt^2 or p
p	pressure or partial pressure (with subscript), M/Lt^2 or p
P	permeability coefficient, various units
Pe	Peclet number, vd/D, dimensionless
q	heat flux, M/t^3 or E/L^2t
$\mathcal{Q}$	heat, ML^2/t^2 or E
$\dot{\mathcal{Q}}$	heat consumption rate, ML^2/t^3 or E/t
Q	volumetric flow rate, L^3/t
r	radial coordinate, radial position, or radius (with subscript), L
r_p	pore radius, L
r_t	tube radius, L
r_{ij}	phenomenological resistance coefficient, various units
$\mathcal{R}$	rejection or retention, dimensionless
R	gas constant, ML^2/t^2Tmol
R	resistance coefficient (with subscript), various units
R_c	cake resistance, M/L^2t or pt/L
R_e	electrical resistance, V/A
R_m	membrane resistance, M/L^2t or pt/L
Re	Reynolds number, vd/ν, dimensionless
s_i	solubility coefficient of gaseous species i, various units

S	entropy, ML^2/t^2T or E/T
Sc	Schmidt number, ν/D, dimensionless
Sh	Sherwood number, k_iL/D, dimensionless
t	time, t
t	transference number (with subscript), dimensionless
$t_{1/2}$	half-life, t
T	temperature, T
T_b	normal boiling point, T
T_c	critical temperature, T
T_g	glass transition temperature, T
u_i	absolute mobility of species i, t mol/M or L^2mol/Et
v	velocity, L/t
v_m	membrane permeation rate, L/t
V	volume, L^3
$\tilde{V}_i$	partial molar volume of species i, L^3/mol
w	channel width, L
w_i	mass fraction of species i, dimensionless
x,y,z	rectilinear coordinates, L
x_i	mole fraction of species i, dimensionless
X	thermodynamic force, E/mol
Y	yield, dimensionless
z_i	valency of species i (with sign)
α_{ij}	separation factor, $y_i''y_j'/y_i'y_j''$, $y = c, p, w$, etc., dimensionless
β_i	enrichment factor, y_i''/y_i', $y = c, p, w$, etc., dimensionless
γ	surface tension, M/t^2 or E/L^2
γ_i	activity coefficient of species i, a_i/c_i, dimensionless
$\gamma_\pm$	mean activity coefficient, $\sqrt{\gamma_+\gamma_-}$, dimensionless
Γ_{ij}	selectivity coefficient, D_iK_i/D_jK_j or D_is_i/D_js_j, dimensionless
δ	polarization layer thickness, L
δ	solubility parameter (with subscript), $(E/L^3)^{1/2}$
∂	partial differential operator
Δ	difference operator
ϵ	porosity, dimensionless
η	efficiency (with subscript), dimensionless
η	viscosity, M/Lt
θ	tortuosity factor, dimensionless
λ	channel height, L
μ	chemical potential, ML^2/t^2 mol or E/mol
ν	kinematic viscosity, L^2/t
ν	vibrational frequency, t^{-1}
ν_i	number of ions per molecule of electrolyte i
ξ	current utilization, dimensionless
π, Π	osmotic pressure, M/Lt^2 or p

ρ	density or mass concentration (with subscript), M/L^3
σ	reflection coefficient, dimensionless
Σ	summation operator
τ	time constant or transition time, t
ϕ	volume fraction, dimensionless
ϕ_f	fraction free volume, dimensionless
χ	Flory-Huggins interaction parameter, dimensionless
ψ	electric potential, V
ω	solute permeability coefficient, $t\,\text{mol}/LM$, $t\,\text{mol}/M$

Diacritical Marks

~	per mole
^	per unit mass
–	average value
·	time rate of change

Superscripts

	value in the membrane indicated by absence of superscript
′	value in feed stream or on high pressure side of membrane; value in phase external to the membrane
″	value in extract, permeate, product or on low pressure side of membrane
°	standard reference state

Subscripts

A,B	particular components
bulk	interior of stream outside of boundary layer
i	general species index or solute species i
int	membrane-solution interface
In	inlet
j	species j
l	liquid
m	membrane or confined to membrane
obs	observed or apparent
Out	outlet
p	product, permeate, or permeant
r	retentate or reject
s	solution
t	tube
v	vapor
w	water or solvent
0	initial value

ABBREVIATIONS

Chemical Names:

CA	cellulose acetate
CE	cellulose ester
CN	cellulose nitrate
CTA	cellulose triacetate
DMAc	dimethyl acetamide
DMF	dimethyl formamide
DMSO	dimethyl sulfoxide
EDTA	ethylenediaminetetraacetic acid
NMP	N-methy-2-pyrrolidone
NR	natural rubber
PA	polyamide
PC	polycarbonate
PE	polyethylene
PI	polyimide
PS	polystyrene
PO	polyolefin
PAN	polyacrylonitrile
PBD	cis-polybutadiene
PEA	polyethylacrylate
PEI	polyethyleneimine
PEM	polyethylmethacrylate
PMA	polymethylacrylate
PPN	polyphosphonate
PPr	polypropylene
PSF	polysulfone
PVA	polyvinyl alcohol
PVC	polyvinyl chloride
PVP	polyvinylpyrrolidone
PDMS	polydimethylsiloxane
PTFE	polytetrafluoroethylene
PVDF	polyvinylidene flouride
SR	silicone rubber
TDI	toluene diisocyanate
THF	tetrahydrofuran
VAMA	vinylacetate/methylacrylate copolymer

Processes:

DO	direct osmosis
ED	electrodialysis
IE/IX	ion exchange
ME	multiple effect
MF	microfiltration
MSF	multistage flash evaporation
PRO	pressure-retarded osmosis
RED	reverse electrodialysis
RO	reverse osmosis
UF	ultrafiltration
VC	vapor compression

Other:

GPD	gallons per day
GSFD	gallons/ft^2/day
TDS	total dissolved solids

SYNTHETIC MEMBRANES AND THEIR PREPARATION

H. Strathmann

Fraunhofer-Institut für
Grenzflächen- und Bioverfahrenstechnik
Nobelstrasse 12
7000 Stuttgart 80
Federal Republic of Germany

In recent years, separations with synthetic membranes have become increasingly important processes in the chemical industry, in food and waste-water processing, and in medical treatment. Synthetic membranes made from a variety of polymers are used in processes such as microfiltration, ultrafiltration, reverse osmosis, electrodialysis, and gas separations. In this paper, the different membrane structures, their function and application in various separation processes are described. The preparation procedures of the various membrane types are discussed with special emphasis on symmetric and asymmetric membranes obtained by the phase inversion process.

P. M. Bungay et al. (eds.), Synthetic Membranes: Science, Engineering and Applications, 1–37.

1. INTRODUCTION

In recent years conventional mass separation techniques such as distillation, crystallization, solvent extraction, etc., have been supplemented by a class of processes that utilize semipermeable membranes as the essential element for the separation of molecular or particulate mixtures. Membrane processes can be very different in their application, in the structures used as the separating barrier, and in the driving forces used for the transport of the different chemical components. They include processes such as reverse osmosis or ultrafiltration, which utilize asymmetric microporous polymer structures as membranes and a hydrostatic pressure difference as the driving force; or electrodialysis, which utilizes ion-exchange membranes and an electrical potential difference for the separation of charged components; or the controlled release of active agents, where usually homogeneous polymer films are used as the separating barriers and concentration differences as driving forces. A summary of technically relevant membrane separation processes, including their operating principles and their main areas of application, is presented in Table 1. Although membrane separation processes are very different in their mode of operation and their application, several common features make them particularly attractive. In many cases membrane processes are faster, more efficient and economical than conventional separation techniques. With membranes the separation is usually performed at ambient temperature; thus a temperature-sensitive solution can be treated without the constituents being damaged or chemically altered. This is important for mass separation problems in the food and drug industry and in biotechnology, which often require processing of temperature-sensitive products.

TABLE 1. Technically relevant membrane separation processes, their operating principles, and their application

Separation process	Membrane type	Driving force	Method of separation	Range of application
Microfiltration	Symmetric microporous membrane 0.1 to 10 μm pore radius	Hydrostatic pressure difference 0.1 to 1 bar	Sieving mechanism due to pore radius and absorption	Sterile filtration clarification
Ultrafiltration	Asymmetric microporous membrane 1 to 10 nm pore radius	Hydrostatic pressure difference 0.5 to 5 bar	Sieving mechanism	Separation of macromolecular solutions
Reverse osmosis	Asymmetric "skin type" membrane	Hydrostatic pressure 20 to 100 bar	Solution-diffusion mechanism	Separation of salt and microsolutes from solutions
Dialysis	Symmetric microporous membrane 0.1 to 10 nm pore radius	Concentration gradient	Diffusion in convection free layer	Separation of salts and microsolutes from macromolecular solutions
Electrodialysis	Cation- and anion-exchange membranes	Electrical potential gradient	Electrical charge of particle and size	Desalting of ionic solutions
Gas separation	Homogeneous or porous polymer	Hydrostatic pressure concentration gradient	Solubility, diffusion	Separation from gas mixture

Consequently, membranes are used today on a large scale to produce potable water from the sea; to clean industrial effluents and recover valuable constituents; to concentrate, purify, or fractionate macromolecular solutions in the food and drug industry, to remove urea and other toxins from the blood stream in an artificial kidney, or to release certain drugs such as scopolamine, nitroglycerin and pilorcarpine at a predetermined rate in medical treatment (1). Membranes can to a certain extent be "tailormade" so that their separation properties can be adjusted to a specific separation task. The field of membrane science and technology is interdisciplinary, involving polymer chemists to develop new membrane structures, physical chemists and mathematicians to describe the transport properties of different membranes using mathematical models to predict their separation characteristics, and chemical engineers to design separation processes for large-scale industrial utilization.

The most important element in membrane separation processes is the membrane itself.

To gain an understanding of the significance of the various structures used in different separation processes, a brief discussion of the basic properties and functions of membranes and the driving forces and fluxes involved is essential.

1.1. Definition of a Membrane

A precise and complete definition of a membrane that covers all its aspects is rather elusive, even when the discussion is limited to synthetic structures as in this outline. In the most general sense, a synthetic membrane is an interphase that separates two phases and restricts the transport of various chemical species in a rather specific manner. A membrane can be homogeneous or heterogeneous, symmetric or asymmetric in structure; it may be solid or liquid and it may be either neutral, or may carry positive or negative charges, or both.

Its thickness may vary between less than 100 nm to more than a centimeter. The electrical resistance may vary from thousands of megohms to a fraction of an ohm. Mass transport through a membrane may be caused by diffusion of individual molecules or by convection induced by gradients in electrical potential, concentration, pressure or temperature.

The term "membrane", therefore, includes a great variety of materials and structures, and a membrane can often be better described in terms of what it does rather than what it is. Some materials, though not meant to be membranes, show typical membrane properties, and in fact are membranes, e.g., protective coatings, or packaging materials. All materials functioning as membranes have one characteristic property in common: They restict the passage of various chemical species in a very specific manner.

1.2. Fluxes and Driving Forces in Membrane Separation Processes

Separation in membrane processes is the result of differences in the transport rates of chemical species through the membrane interphase. The transport rate is determined by the driving force or forces acting on the individual components and their mobility and concentration within the interphase. The mobility and concentration of the solute within the interphase determine how large a flux is produced by a given driving force. The mobility is primarily determined by the solute's molecular size and the physical structure of the interphase material, whereas the concentration of the solute in the interphase is primarily determined by chemical compatibility of the solute and the interphase material.

In membrane separation processes there are three basic transport forms of mass through the membrane interphase, as indicated in Figure 1. The simplest form is the so-called "passive transport". Here the membrane acts as a physical barrier through which all components are transported under the

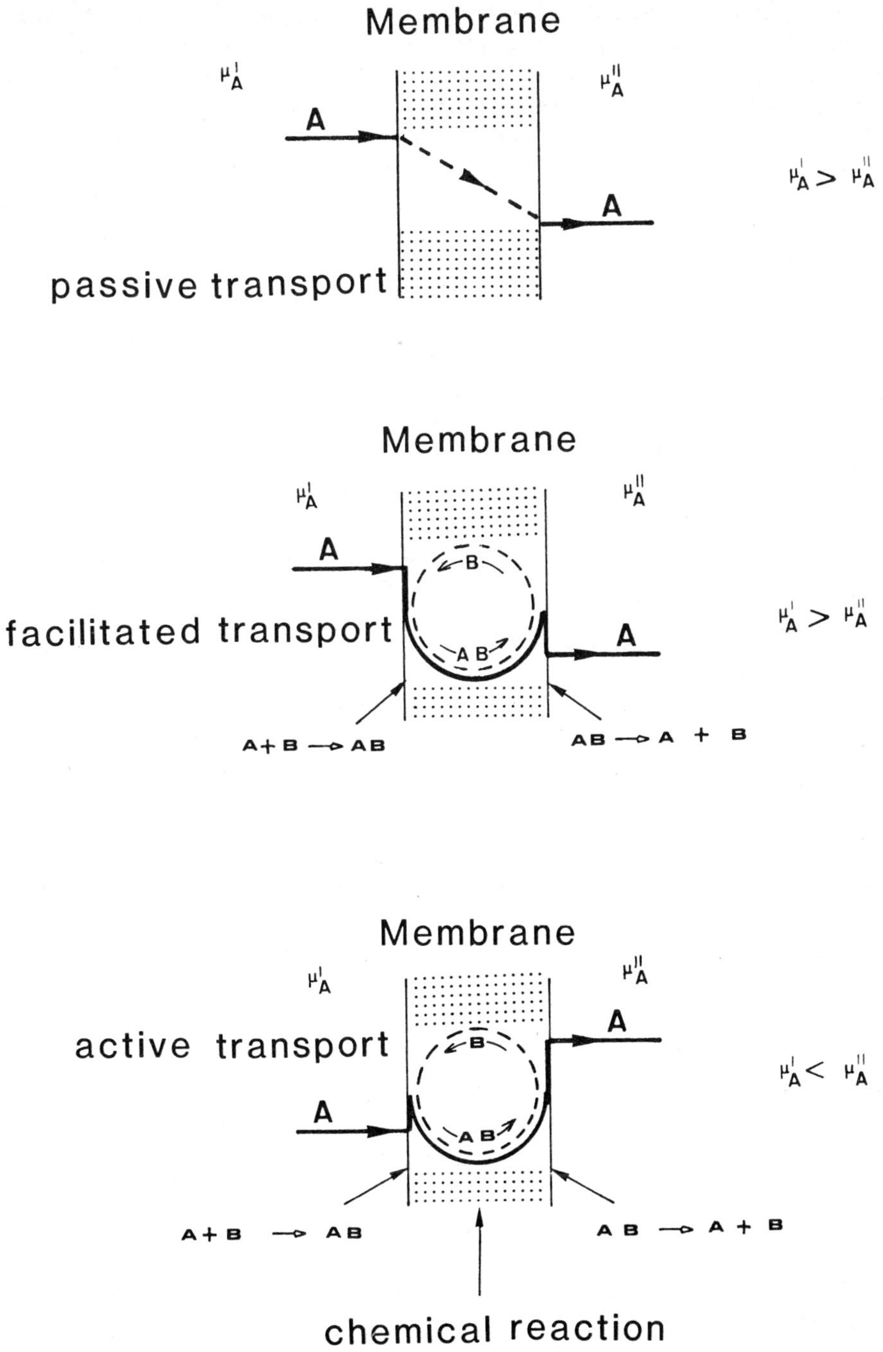

FIGURE 1.—Schematic diagram of mass transport processes in synthetic membranes.

driving force of a gradient in their electro-chemical potential. Gradients in the electro-chemical potential of a component in the membrane phase may be caused by differences in hydrostatic pressure, concentration, temperature or electrical potential between the two phases separated by the membrane. In the so-called "facilitated transport", the driving force for the transport of the various components is again the gradient in their electro-chemical potential across the membrane. The different components, however, are coupled to a specific carrier in the membrane phase. Facilitated transport, therefore, is just a special form of the passive transport, which generally is highly selective. Completely different, however, is the so-called "active transport". Here various components may be transported against the gradient of their electrochemical potential. The driving force for the transport is provided by a chemical reaction within the membrane phase. Active transport, usually coupled with a carrier in the membrane interphase, is found mainly in the membranes of the living cells. To date, it has had no significance in technically relevant membrane separation processes.

The transport process itself is a nonequilibrium process and is conventionally described by a phenomenological equation that relates the flows to the corresponding driving forces in the form of proportionalities. Fick's law, for example, describes the relation between flow of matter and a concentration gradient and the constant of proportionality is the diffusion coefficient. Ohm's law describes the relation between an electrical current and an electrical potential gradient, while Fourier's law describes the relation between heat transport and a temperature gradient. Some of the more familiar phenomenological equations are given in Table 2.

TABLE 2. Phenomenological relationships between various fluxes and the corresponding driving forces

Phenomenological relationship	Flux	Driving force	Constant of proportionality
Fick's law	Mass	Concentration difference	Diffusion coefficient
$J = -D\ \Delta c/\Delta x$	J	Δc	D
Ohm's law	Electricity	Electrical potential difference	Electrical resistance
$I = \Delta\psi/R_e$	I	$\Delta\psi$	R_e
Fourier's law	Heat	Temperature differences	Heat conductivity
$q = -k\ \Delta T/\Delta x$	q	ΔT	k
Hagen-Poiseuille's law	Volume	Pressure differences	Hydrodynamic permeability
$J_V = h_d\ \Delta p/A$	J_V	Δp	h_d

Driving forces in some membrane processes may be interdependent, giving rise to new effects. Thus, a concentration gradient across a membrane may not only result in a flow of matter but, under certain conditions, can also cause the build-up of a hydrostatic pressure difference, the phenomenon called osmosis. Similarly, a gradient in hydrostatic pressure may not only lead to a volume flow but may also result in the formation of a concentration gradient; this phenomenon is called reverse osmosis. Another example is a temperature gradient across a membrane which may result not only in a flow of heat, but may also lead to a transport of matter. This process is then called thermodiffusion or thermoosmosis. The reverse process, when a mass flow causes a temperature gradient, is known as the Dufour effect. Frequently, fluxes as well as driving forces are coupled and the flow of one component causes a flow of another. One example of the coupling of fluxes is the transport of bound water with an ion which is driven across a membrane by an electrical potential gradient.

For membrane separation processes, only driving forces that can lead to a significant flux of matter are of practical importance. These driving forces are hydrostatic pressure, concentration, and electrical potential differences.

a) A hydrostatic pressure difference between two phases separated by a membrane can lead to a volume flux and to a separation of chemical species when the hydrodynamic permeability of the membrane is different for different components.
b) A concentration difference between two phases separated by a membrane can lead to a transport of matter and to a separation of various chemical species when the diffusivity and the concentration of the various chemical species in the membrane are different for different components.
c) A difference in the electrical potential between two phases separated by a membrane can lead to a transport of matter and to a separation of various chemical species when the different charged particles show different mobilities in the membrane.

Note, however, that the overall driving force for the transport of a chemical component through a membrane is the gradient in its chemical potential, which may consist of additive terms of the gradients in the hydrostatic pressure, in the concentration, and in the electrical potential. Therefore, a hydrostatic pressure gradient, for example, will not necessarily result in a flux of mass when it is counterbalanced by a concentration gradient, as the phenomenon of osmotic equilibrium shows.

The driving forces and corresponding mass fluxes through a membrane are related by permeability coefficients, which strongly depend on the chemical nature and physical structure of the membrane and the components to be transported. For example, in homogeneous polymer membranes, the various chemical species are transported under a concentration or pressure gradient by diffusion; the permeability of these membranes is determined by the diffusivities and solubilities of the various components in the membrane matrix. The transport rates are, in general, relatively slow. In porous membrane structures, however, mass is transported under the driving force of a hydrostatic pressure difference via viscous flow and, in general, permeabilities are significantly higher than in diffusion-controlled membrane transport. In electrically charged membranes, usually referred to as ion-exchange membranes, ions carrying the same charge as the membrane material are more or less completely excluded from the membrane and therefore unable to penetrate the membrane. The type of membrane and driving force required for a certain mass separation will depend on the specific properties of the chemical species in the mixture.

2. CLASSIFICATION OF MEMBRANES: STRUCTURES AND METHODS OF PREPARATION

Although synthetic membranes show a large variety in their physical structure and chemical nature they can conveniently be classified in four basic groups: (1) microporous media, (2) homogeneous films, (3) asymmetric structures and (4) electrically charged barriers. This classification, however, is rather arbitrary and many structures would perfectly fit more than one of the above mentioned classes, e.g., a membrane may be microporous, asymmetric in structure, and electrically charged. Other schemes, such as a phenomenological categorization, would serve the same purpose.

2.1. Microporous Media

The microporous media represent a very simple form of a membrane as far as mass transport properties and the separation mode are concerned. Microporous membranes consist of a solid matrix with defined holes or pores which have diameters ranging from less than 5 nm to more than 50 μm. Separation of the various chemical components is achieved strictly by a sieving mechanism with the pore diameters and the particle sizes being the determining parameter. Microporous membranes can be made from various materials, such as ceramics, graphite, metal or metal-oxides and various polymers. In Table 3, the properties and applications of various microporous filter media of technical relevance are summarized and the scanning electron micrograph of Figure 2 shows the structure of typical microporous membranes.

Sintered membranes Sintered membranes are most simple in their function and in the way they are prepared. The structure of a typical sintered membrane is shown in the scanning electron micrograph of Figure 2a. This photograph shows a microporous membrane made of polytetrafluoroethylene by pressing a fine powder into a film or plate of 100 to 500 μm thickness and then sintering the film at a temperature that is just below the melting point of the polymer (2). This process yields a microporous structure of relatively low porosity in the range of 10 to 40% and a rather irregular pore structure with a very wide pore size distribution.

TABLE 3. Microporous membranes: properties, preparation, and application

Membrane material	Pore size	Manufacturing process	Application
Ceramic, metal or polymer powder	1–20 μm	Pressing and sintering of powder	Microfiltration
Homogeneous polymer sheets (PE, PTFE)	0.5–10 μm	Stretching of extruded polymer sheet	Microfiltration, burn dressings, artificial blood vessels
Homogeneous polymer sheets (PC)	0.02–10 μm	Track-etching	Microfiltration
Polymer solution (CN, CA)	0.01–5μm	Phase inversion	Microfiltration, sterilization

CA = cellulose acetate; CN = cellulose nitrate; PC = polycarbonate; PE = polyethylene; PTFE = polytetrafluoroethylene

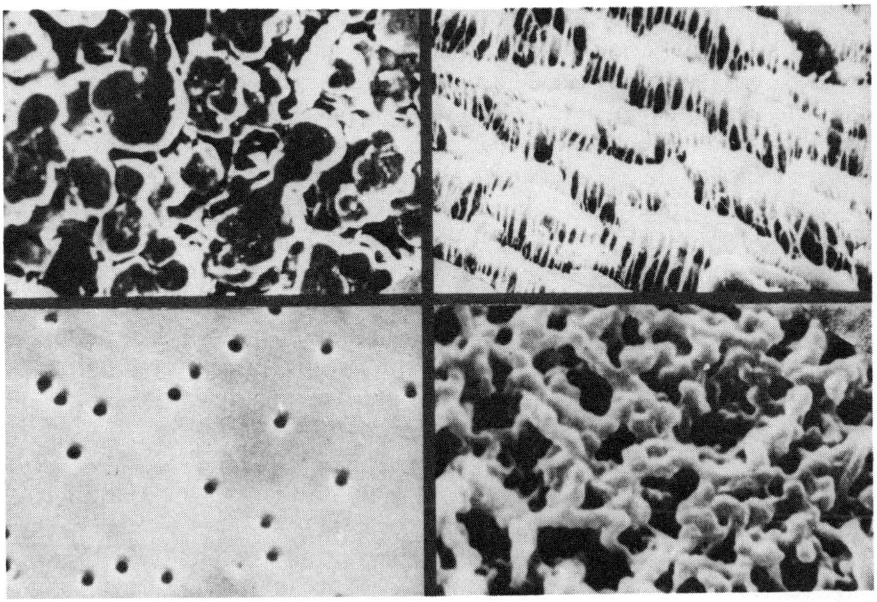

FIGURE 2.—Scanning electron micrograph of typical microporous membranes: a) sintered membranes, b) stretched membranes, c) capillary pore membranes, d) phase inversion membranes.

Sintered membranes are made on a fairly large scale from ceramic materials and also from graphite and metal powders such as stainless steel, tungsten, etc. The particle size of the powder is the main parameter determining the pore sizes of the final membrane, which is made in the form of discs, cartridges, and fine-bore tubes. Sintered membranes are used for filtration of colloidal solutions and suspensions and when manufactured from inorganic materials for the filtration of corrosive solutions, such as acids and bases at elevated temperatures. Membranes of this type have been made with sufficiently fine pores that they are marginally suitable for gas separation. They have gained some significance in the separation of radioactive isotopes (3).

Stretched membranes Another relatively simple procedure for preparing microporous media is based on stretching a homogeneous polymer film of partial crystallinity. This technique is mainly employed with films of polyolefins or polytetrafluoroethylene, which have been extruded from a powder and are then stretched perpendicular to the direction of extrusion (4). This leads to a partial fracture of the film and relatively uniform pores with diameters of 0.1 to 20 μm. A typical stretched membrane prepared from tetrafluoroethylene is shown in the scanning electron micrograph of Figure 2b.

These membranes, which have a very high porosity, up to 90%, and a fairly regular pore size, are now widely used for microfiltration of acid and caustic solutions, organic solvents, and hot gases. They have now largely replaced the previously used sintered materials in this application.

The stretched membrane made of polytetrafluoroethylene, e.g., that manufactured by the Gore Corporation under the trade name Gore-Tex, has gained a special significance in its use in rain-protecting clothing for parkas, tents, sleeping bags, etc. This membrane type, because of its very high porosity, has a high permeability for gases and vapors, but, up to a certain hydrostatic pressure, is completely impermeable to aqueous solutions because of the hydrophobic nature of the basic polymer. This is why this membrane provides complete protection in rain or water, but permits the water vapor of sweat to permeate and thus keeps the protected person completely dry. More recently this membrane has also been used for a novel process generally referred to as membrane distillation.

Capillary pore membranes Microporous membranes with very uniform, almost perfectly round cylindrical pores are obtained by a process generally referred to as track-etching (5). The membranes are made by a two-step process. During the first step, a homogeneous 10- to 20-μm thick polymer film is exposed to collimated, charged particles from a nuclear reactor. As particles pass through the film, they leave sensitized tracks where the chemical bonds in the polymer backbone are broken. In the second step, the irradiated film is placed in an etching bath. In this bath, the damaged material along the tracks is preferentially etched forming uniform cylindrical pores. The entire process is schematically shown in Figure 3. The pore density of a track-etched

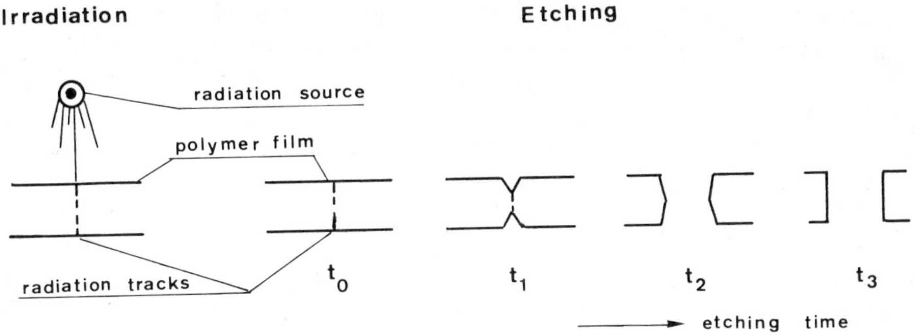

FIGURE 3.—Schematic diagram of the "track-etching" membrane preparation procedure.

membrane is determined by the residence time in the irradiator, while the pore diameter is controlled by the residence time in the etching bath. The scanning electron micrograph in Fig. 2c shows a typical track-etched polycarbonate membrane. Capillary pore membranes are prepared today mainly from polycarbonate and polyester films. One of the reasons this process is limited presently to these polymers is that they are commercially available as uniform films of 10–20 μm thickness. This thickness corresponds to about the maximum penetration depth that can be achieved by the collimated nuclear particles, which have an energy of about 0.8 to 1 MeV. Particles with higher energy up to 10 MeV may be obtained in an accelerator. They are used today to irradiate thicker polymer films up to 50 and more μm thickness or inorganic materials such as mica. But these membranes are not yet available on a commercial basis. Because of their narrow pore size distribution and low tendency of plugging, capillary pore membranes made from polycarbonate and polyester have found application on a large scale in analytical chemistry, microbiological laboratories, and in medical diagnosis and treatment. On an industrial scale capillary pore membranes are used for the production of ultrapure water for the electronic industry. Here they show certain advantages over other membrane products because of their short "rinse down" time (6).

Microporous phase inversion membranes The most important commercial microporous membranes are usually prepared from cellulosic polymers by the so-called phase inversion process, in which a polymer is dissolved in an appropriate solvent and cast as a 20 to 200 μm thick film. A non-solvent is added to this liquid film from the vapor phase, causing precipitation and separation into a solid polymer and a liquid solvent phase. The precipitated polymer forms a porous structure containing a network of more or less uniform pores. A microporous cellulosic membrane is shown in the scanning electron micrograph of Figure 2d.

Although this type of membrane was prepared first from cellulosic polymers it can now be made from almost any polymer that is soluble in an appropriate solvent and can be precipitated in a non-solvent (7).

By varying the polymer, the polymer concentration, the precipitation medium and the precipitation temperature, microporous phase inversion membranes can be made in a very large variety of pore sizes, with different chemical, thermal, and mechanical properties.

This type of membrane is utilized for various separation tasks on a laboratory or industrial scale ranging from the clarification of turbid solutions, to the removal of bacteria or enzymes, the detection of pathological components, and the detoxification of blood in an artificial kidney. The separation mechanism is that of a typical depth filter, which traps the particles somewhere within the structure. In addition to the simple "sieving" effect, microporous phase inversion membranes often show, because of their extremely large internal surface, a high tendency of adsorption. They are therefore particularly suitable when a complete removal of components such as viruses is desired. They are suited for immobilization of enzymes or even microorganisms to be used in modern biotechnology.

The phase inversion process is today the most important procedure of making membranes. It is not only used for the preparation of microporous structures but also for making asymmetric skin-type ultrafiltration and reverse osmosis membranes. It is, therefore, of great significance and will be discussed in more detail in connection with asymmetric membrane structures.

Other microporous membrane structures Microporous membranes made from glass or metal alloys have gained some significance in mass separation. The preparation of glass membranes is relatively simple: Two different types of glass are homogeneously mixed, then one glass type is removed by acid or base leaching. Thus, a microporous structure is obtained with well-defined pore sizes (8). Porous glass membranes can be made in various configurations, such as flat sheets, tubes, and hollow fibers, which are of particular importance because of their high surface area. Microporous metal membranes can be prepared from metal alloys such as Ni-Al-Cr by subsequent leaching of one component (9). These membranes have found their main application in gas separation processes.

2.2. Homogeneous Membranes

A homogeneous membrane merely consists of a dense film through which a mixture of chemical species is transported under the driving force of a pressure, concentration, or electrical potential gradient. The separation of various components in a solution is directly related to their transport rate within the membrane phase,which is determined by their diffusivity and concentration in the membrane matrix. An important property of homogeneous membranes is that chemical species of similar size, and hence identical diffusivities, may be separated when their concentration, or more appropriately their solubility in the film, differs significantly. The membrane phase itself may be solid or liquid. The mass transport in homogeneous membranes is always strictly by diffusion; thus permeabilities are rather low. Homogeneous membranes should therefore always be as thin as possible. Technically relevant homogeneous membranes are summarized in Table 4.

TABLE 4. Homogeneous membranes: properties, preparation and application

Membrane type	Membrane material	Manufacturing	Application
Homogeneous membrane	Polymer (SR)	Extrusion of films, casting of solutions	Gas separation
Ion-exchange membrane	Polymer Ion-exchange resin	Casting of solutions	Electrodialysis
Mosaic membrane	Polymer Ion-exchange resin	Casting of solutions	Piezodialysis

SR = silicone rubber

Homogeneous polymer membranes Although there are a number of homogeneous membranes made from inorganic materials such as glass or certain metals, the technically more important structures are of polymeric origin. Modern polymer chemistry is highly proficient in tailoring polymers to specific aims in terms of mechanical or thermal stability as well as chemical compatibility, to satisfy the needs in various separation processes. In general, mass transfer will be greater in amorphous polymers than in highly crystalline or cross-linked polymers. Thus crystallization and orientation are to be avoided as much as possible. However, the strength and physical properties of the polymer may then be adversely affected, and the final product will represent a compromise between necessary strength properties and desirable mass-transfer rates. The principal aim is to create as thin a barrier as possible, consistent with the requested minimum strength properties and without pinholes or defects. There are two basic membrane configurations: flat sheeting or hollow fiber shapes. Film shapes can be prepared by casting from a solution or from a polymer melted by extrusion, blow and press molding. Hollow fibers are generally made by extrusion with central gas injection.

Homogeneous membranes are used in various applications, the most important of which are gas separation and pervaporation. As basic material silicon rubber, because of its relatively high permeability, is the more widely used polymer, particularly for preparation of gas separation membranes. (10).

Homogeneous metal and glass membranes The only homogeneous metal membranes of technical importance are palladium, palladium-silver or palladium-yttrium alloy membranes, which are used for the separation and purification of hydrogen. Likewise, homogeneous glass membranes are used nearly exclusively as pH-electrodes. Their preparation is described in some detail in the literature.

Liquid membranes Liquid phase membranes have gained increasing significance in recent years in separation processes. When used in combination with "carriers" capable of transporting certain components, such as metal ions, these membranes can achieve high selectivity and relatively high transport rates (11).

Liquid membranes are used today in two different configurations. In the first, the liquid membrane is composed as a thin film stabilized by a surfactant in an emulsion-type mixture. The second configuration is generally referred to as supported liquid membranes. Here, the separating medium is a microporous polymer structure the pores of which are filled with the liquid membrane phase. In this configuration the microporous structure provides the mechanical strength while the liquid-filled pores function as the selective separation barrier. Both types of membranes are being tested in pilot plants for the selective removal of heavy metal-ions or certain organic solvents from industrial waste streams. Also, recently a liquid membrane system has demonstrated rather efficient separation of oxygen and nitrogen (12). The preparation techniques and the use of liquid membranes especially in combination with selective carriers will be discussed in more detail in the chapter on carrier-mediated transport.

Ion-exchange membranes Ion-exchange membranes, another type of homogeneous membrane, consist of highly swollen gels carrying fixed positive or negative charges. A membrane with fixed positive charges is referred to as an anion-exchange membrane since it binds anions from the surrounding fluid, whereas a membrane containing fixed negative charges is called a cation-exchange membrane. Separation in charged membranes is achieved by exclusion of co-ions, i.e., ions that bear the same charges as the fixed ions of the membrane structure. Separation properties of these membranes are determined by the charge and concentration of the ions in the surrounding solution and in the membrane structure. Ion-exchange membranes are generally prepared by dispersing a conventional ion-exchange material in a polymer matrix, or by polymerization of film-forming ionic monomers, or by introducing ionic groups into a film-forming polymer. The most common fixed-ion groups in cation-exchange membranes are sulfonic and to a lesser extent carboxylic groups. But phosphoric, phosphinic and selenoic groups are also sometimes used in cation-exchange membranes. In anion-exchange membranes, the fixed groups are usually quaternary ammonium and to a lesser extent quaternary phosphonium, and tertiary sulfonium groups. The basic groups are generally inherently less stable than the acid groups, and therefore cation-exchange membranes are usually more stable than anion-exchange membranes. The reaction schemes for preparing cation- and anion-exchange membranes are shown in Figures 4 and 5.

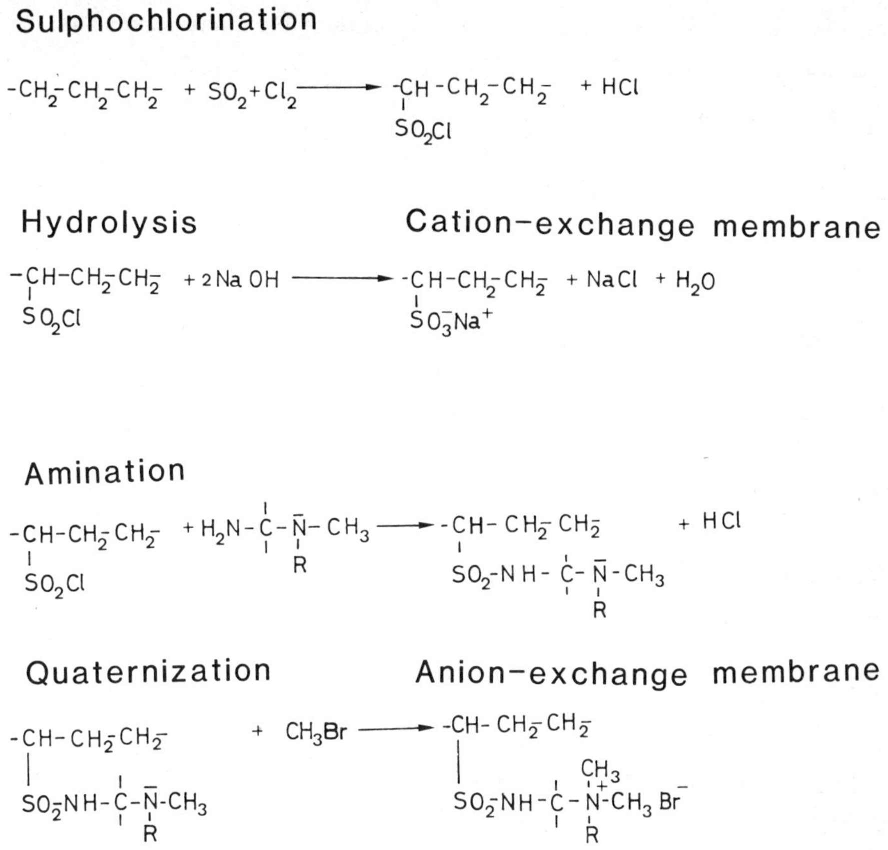

Sulphochlorination

$-CH_2-CH_2-CH_2-\ +\ SO_2+Cl_2 \longrightarrow\ \begin{array}{c}-CH-CH_2-CH_2-\\ |\\ SO_2Cl\end{array}\ +\ HCl$

Hydrolysis

$\begin{array}{c}-CH-CH_2-CH_2-\\ |\\ SO_2Cl\end{array}\ +\ 2\,NaOH \longrightarrow$

Cation-exchange membrane

$\begin{array}{c}-CH-CH_2-CH_2-\\ |\\ SO_3^-Na^+\end{array}\ +\ NaCl\ +\ H_2O$

Amination

$\begin{array}{c}-CH-CH_2\,CH_2-\\ |\\ SO_2Cl\end{array}\ +\ H_2N-\overset{|}{\underset{R}{C}}-\bar{N}-CH_3 \longrightarrow\ \begin{array}{c}-CH-CH_2\,CH_2-\\ |\\ SO_2-NH-\overset{|}{\underset{R}{C}}-\bar{N}-CH_3\end{array}\ +\ HCl$

Quaternization

$\begin{array}{c}-CH-CH_2\,CH_2-\\ |\\ SO_2NH-\overset{|}{\underset{R}{C}}-\bar{N}-CH_3\end{array}\ +\ CH_3Br \longrightarrow$

Anion-exchange membrane

$\begin{array}{c}-CH-CH_2\,CH_2-\\ |\\ SO_2NH-\overset{|}{\underset{R}{C}}-\overset{CH_3}{\underset{|}{\overset{+}{N}}}-CH_3\ Br^-\end{array}$

FIGURE 4.—Reaction scheme for the preparation of cation- and anion-exchange resins for the preparation of ion-exchange membranes.

The performance of an ion-exchange membrane is determined mainly by its selectivity to ions of opposite charges, its electrical resistance and its mechanical and chemical stability. Such properties depend largely on the nature and density of the ionic groups and on the matrix polymer, i.e., the balance of the hydrophilic and hydrophobic content, degree of cross-linking and crystallinity. In highly cross-linked membranes the Donnan-exclusion is effective, and mechanical stability is excellent, but these membranes are not easily swollen and the mobility of the counter-ions is small. On the other hand, non-cross-linked membranes are easily swollen and ionic mobility is high but the mechanical stability is generally poor, to such an extent that these membranes often dissolve in deionized water. Thus, the density of the charged groups and the degree of cross-linking has to be well balanced in an efficient ion-exchange membrane. The fixed charge density of a typical ion-

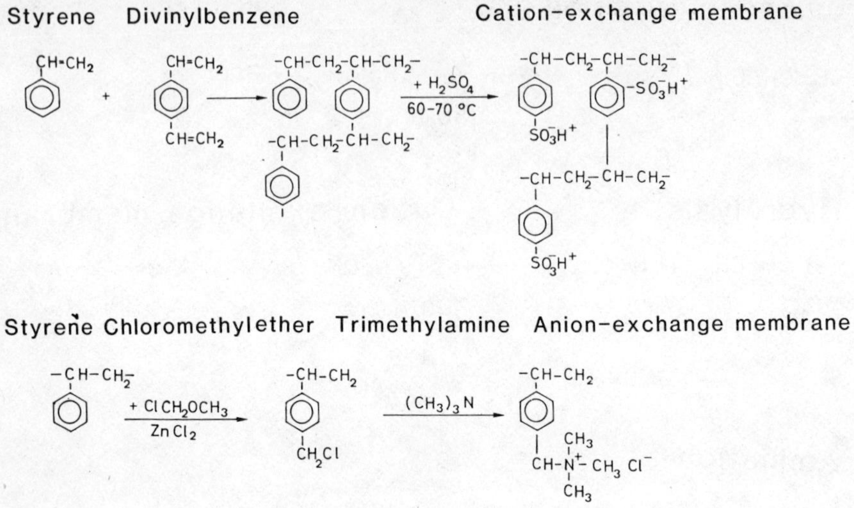

FIGURE 5.—Reaction scheme for the preparation of homogeneous cation- and anion-exchange membranes.

exchange membrane as used in electrodialysis is 1–2 mequiv/cm^3 and its electrical area resistance in a 0.1 N solution of a univalent electrolyte is less than 10 Ωcm^2. The main application of ion-exchange membranes is in electrodialysis and related processes, which will be discussed in detail in a subsequent chapter.

2.3 Asymmetric Membranes

The most important membrane used today in separation processes is composed of a rather sophisticated asymmetric structure. In this membrane two basic properties requested of any membrane used in mass separation processes, i.e., high mass transport rates for certain components and good mechanical strength, are physically separated. An asymmetric membrane consists of a very thin (0.1 to 1μm) polymer layer on a highly porous 100- to 200-μm-thick sublayer, as indicated in the schematic drawing of Figure 6, which shows the cross-sections of a) a symmetric and b) an asymmetric membrane. The very thin skin represents the actual membrane. The separation characteristics are determined by the nature of the skin polymer or pore size and the mass transport rate mainly by the thickness, since the mass transport rate is inversely proportional to the thickness of the actual barrier layer. The highly porous sublayer serves only as a support for the very thin and fragile skin and has no, or very little, effect on separation characteristics and the mass transfer rate of the membrane.

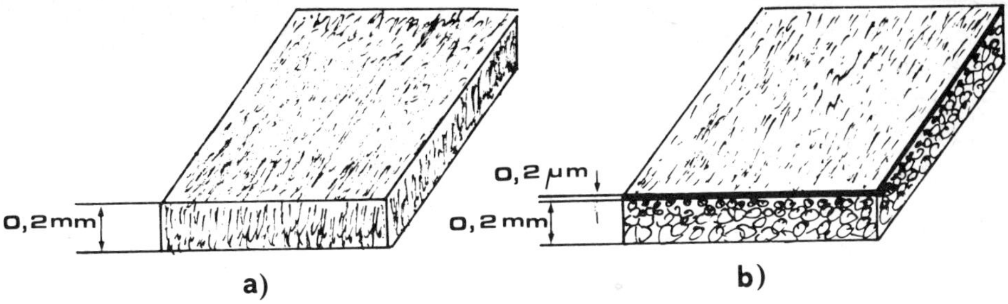

FIGURE 6.—Schematic diagram of a) a symmetric and b) an asymmetric membrane.

Asymmetric membranes are used primarily in pressure-driven membrane processes, such as reverse osmosis, ultrafiltration, or gas separation, where their unique properties in terms of high mass transfer rates and good mechanical stability can be utilized best.

In addition to high filtration rates, asymmetric membranes have another significant advantage. Conventional symmetric structures act as depth filters and retain most particles within their internal structure. These trapped particles plug the membrane and the flux declines during use. Asymmetric membranes are surface filters retaining all rejected materials at the surface, where they can be removed by shear forces applied by the feed solution moving parallel to the membrane surface. The difference in the filtration behavior between a symmetric and an asymmetric membrane is shown schematically in Figure 7. In the preparation of asymmetric membranes, two techniques have been developed: one utilizes the phase inversion process and the other leads to a composite structure by depositing an extremely thin polymer film on a microporous substructure.

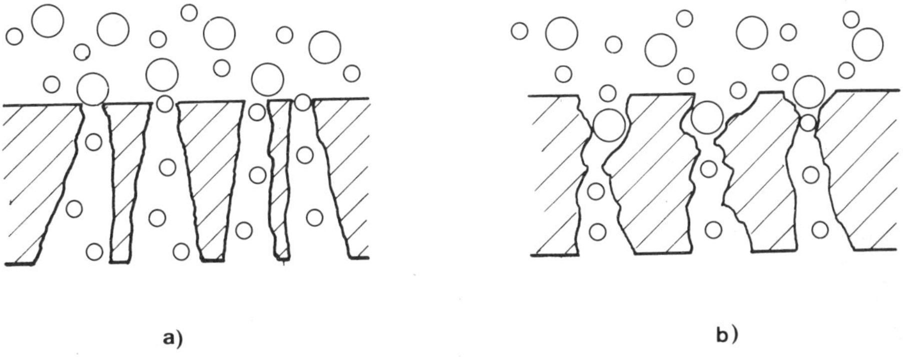

FIGURE 7.—Schematic diagram of the filtration behavior of a) an asymmetric and b) a symmetric membrane.

Integral asymmetric membranes The development of the integral asymmetric membrane was a major breakthrough in the application of ultrafiltration and particularly reverse osmosis (13). These membranes, which were first prepared from cellulosic polymers, yield 10 to 100 times higher fluxes than symmetric structures with comparable separation characteristics and at about the same hydrostatic pressure-driving forces. Soon it also became apparent that asymmetric membranes could be prepared not only from cellulose acetate but from a wide variety of different polymers, using the same general procedure:

1. A polymer is dissolved in an appropriate solvent to form a solution containing 10 to 30 wt.% polymer.
2. The solution is cast into a film of typically 100 to 500 μm thickness.
3. The film is quenched in a nonsolvent which for most polymers is typically water or an aqueous solution.

During the quenching process, the homogeneous polymer solution is separated into two phases: a polymer-rich solid phase, which forms the membrane structure, and a solvent-rich liquid phase, which forms the liquid-filled membrane pores. Generally, the pores at the film surface where precipitation occurs first and most rapidly are much smaller than those in the interior or the bottom side of the film, and this leads to the asymmetric membrane structure. The actual membrane formation procedure is easily rationalized with the aid of a three-component phase diagram, as shown in Figure 8. There are different variations to this general preparation procedure described in the literature; e.g., Loeb and Sourirajan (13) used an evaporation step to increase the polymer concentration in the surface of the cast polymer solution and an annealing step during which the precipitated polymer film is exposed for a certain time period to 70 to 80°C water.

The original recipes and subsequent modifications for preparing asymmetric membranes are deeply rooted in empiricism; detailed descriptions of the membrane preparation techniques are given in the literature (14, 15). Only after the extensive use of the scanning electron microscope, which provided the necessary structural information, was it possible to rationalize the various membrane preparation process parameters in a comprehensive theory about the formation mechanism of asymmetric membranes. At first they were limited mainly to membranes from cellulosic polymers. But later it became apparent that the process of making skin-type asymmetric membranes by precipitating a polymer solution is just a special case of a general procedure, which Kesting has called phase inversion (14).

Phase inversion can be achieved in several ways. In one technique inversion is achieved by controlled evaporation of a volatile solvent from a three-component mixture of solvent-precipitant-polymer, causing precipitation as the system becomes enriched in precipitant. This technique was used by Zsigmondy (16) and more recently by Kesting (17). Alternatively, precipitation of a simple two-component polymer-solvent casting solution can be brought about by imbibing precipitant from the vapour phase. This technique was the basis of the original microporous membranes and is still used com-

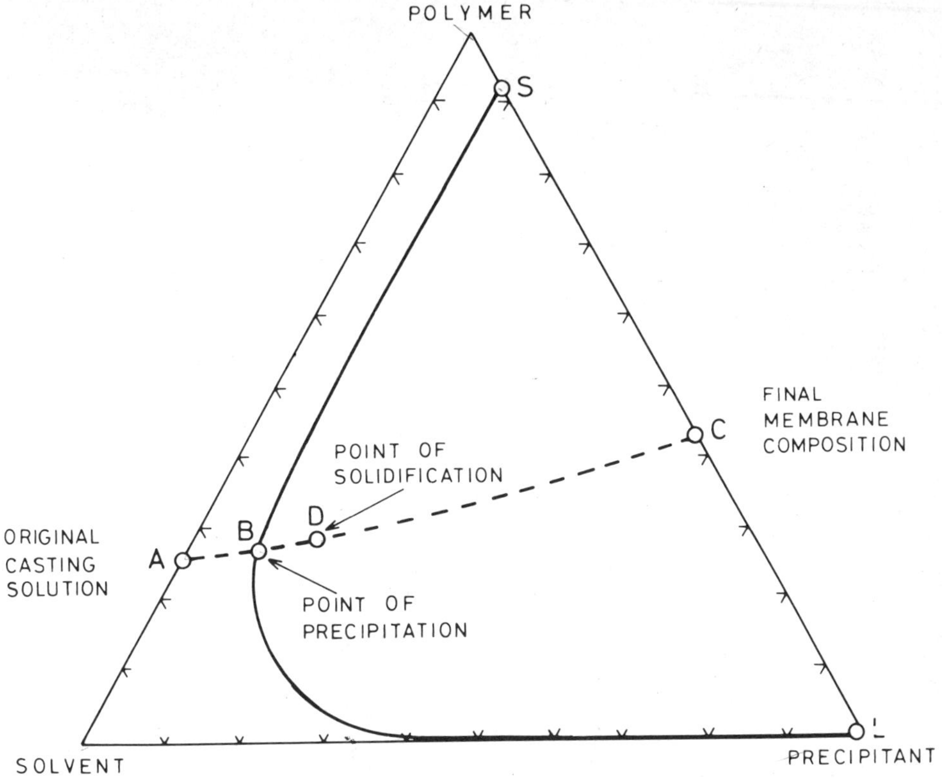

FIGURE 8.—The phase diagram of the system polymer-solvent-precipitant showing the precipitation pathway of the casting solution during membrane formation.

mercially by several companies. Yet another technique, thermal gelation, brings about precipitation by cooling a casting solution that has a composition close to the point of precipitation.

Phase inversion membranes can be made from almost any polymer in a very large variety of structures and separation properties. By variation of different preparation parameters membranes can be "tailor-made" for a given mass-separation task. The effect of the various preparation parameters is well understood and can be correlated with the different membrane structures and transport properties.

Structural differences of integral asymmetric membranes.—Use of optical and electron microscope techniques permits observation of three typical membrane structures, depending on the preparation procedure. These typical structures, shown in the scanning electron micrographs of Figure 9a-c, are related to typical membrane properties. Photographs of the cross-sections of all three membranes show a thin and dense skin on the surface of the microporous support structure. However, the pores of the three membranes

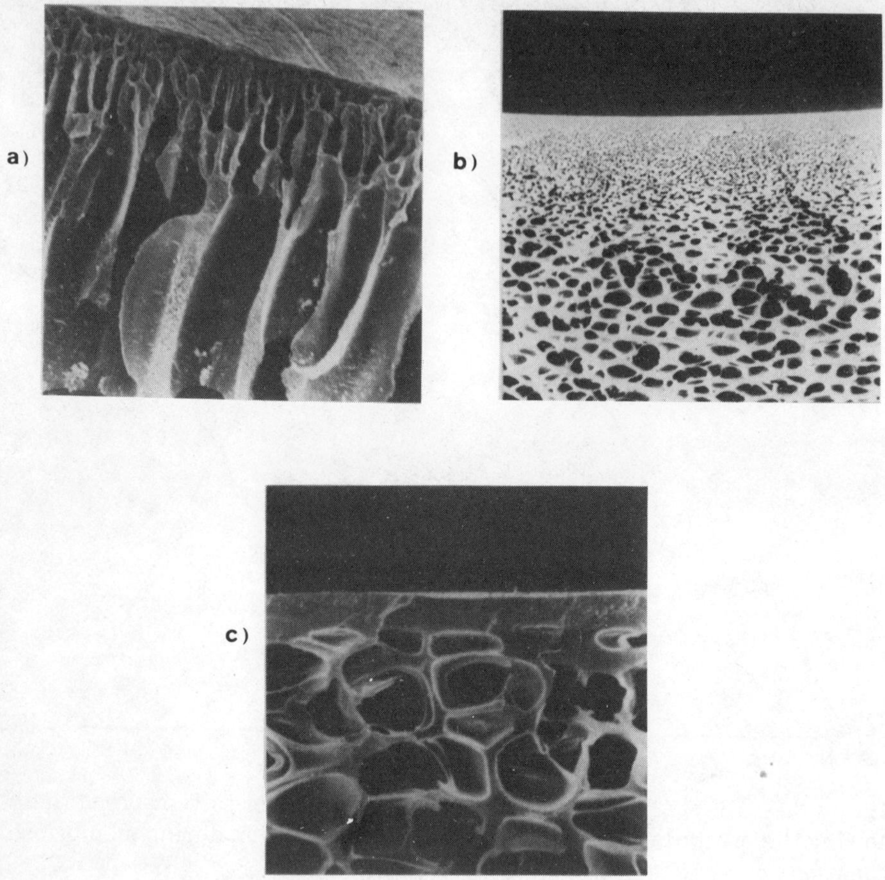

FIGURE 9.—Scanning electron micrographs of integral asymmetric membranes:

a) ultrafiltration membrane with a "finger" structure prepared from polysulfone dissolved in dimethyl formamide and precipitated in water
b) reverse osmosis membrane with a "sponge" structure and a pore size that increases from the top to the bottom surface, prepared from Nomex® dissolved in dimethyl acetamide (DMAc) and precipitated in water
c) reverse osmosis membrane with a "sponge" structure and a uniform pore size over the entire cross-section, prepared from polyimid dissolved in DMAc precipitated in water and thermally cross-linked.

are rather different in their sizes and shapes. Figure 9a shows pores in the substructure that penetrate the entire cross-section of the membrane like fingers or capillaries, with their diameter increasing from the top to the bottom surface of the membrane.

The inside surfaces of these pores are also coated with a dense polymer layer. Membranes with the "finger" structure generally have no salt rejection capability. Their hydrodynamic permeability is very high. Their mechanical strength is rather poor and hydrostatic pressures in excess of 20 bar (2000 KPa) result in very high compaction rates due to a collapse of the porous substructure under pressure. These membranes are used in ultrafiltration with hydrostatic pressure driving forces of 2 to 10 hours.

The structure shown in Figure 9b is generally found in reverse osmosis membranes. It is mechanically quite strong and can support hydrostatic pressures of up to 100 bars without significant deformation of the substructure or changes in the hydrodynamic permeability. These membranes have good salt rejection properties, but fairly low hydrodynamic permeabilities, which is only 5 to 10% of typical ultrafiltration membranes. The porous support shows a typical sponge structure with pore diameters that increase from the top to the bottom surface. The membrane generally has to be kept wet. When dried completely without a surfactant it will undergo irreversible shrinkage and its permeability will be significantly decreased.

The structure shown in Figure 9c is again found in reverse osmosis membranes. It is mechanically quite strong and can support hydrostatic pressures up to 100 bars without changes in the hydrodynamic permeability. It differs primarily from the structure in Figure 9b in the pore size of the support structure, which is more or less the same over the entire membrane cross-section. This membrane type has excellent salt retention qualities and very often it may be dried completely without irreversibly losing its water permeability. This property is very useful for storing, handling and processing the membrane.

The three structures shown in Figure 9 represent extreme forms and in practice, many different intermediate membrane structures can be found. Moreover, the different structures are not necessarily a function of the polymer used in the casting solution. In general, all three structures can be made from the same polymer and with different polymers exactly the same structures may be obtained when all other preparation parameters are selected accordingly. This is indicated in Figure 10a-c, which shows three membranes with virtually the same structure and filtration properties made from three entirely different polymers.

Figure 11a-c shows three different membranes made from the same polymer–solvent–precipitant system. Photograph 11a shows the cross-section of a membrane made from a 10% solution of Nomex® in dimethyl acetamide (DMAc) precipitated by water vapor. This membrane has virtually no continuous skin on the surface and a very uniform pore structure over the entire cross-section of the membrane. It has rather high water fluxes but virtually

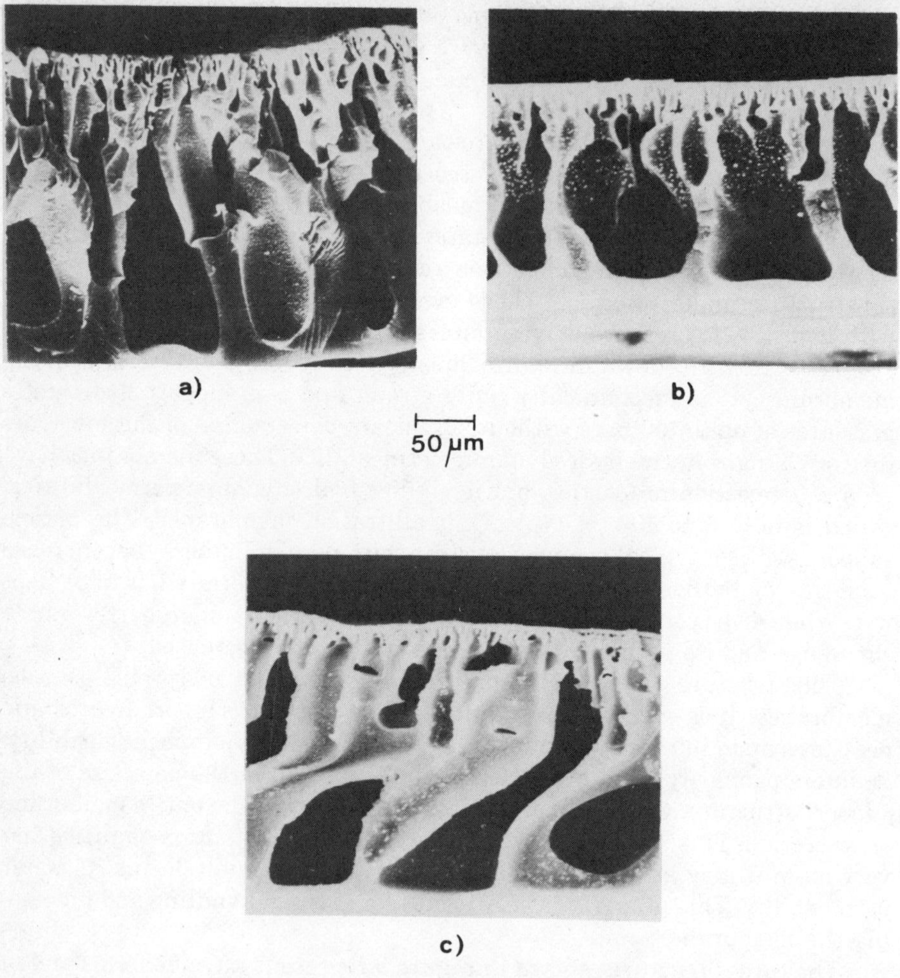

FIGURE 10.—Scanning electron micrographs of cross-sections of membranes made from a solution of

a) 12% cellulose acetate in dimethyl acetamide
b) 12% Nomex® in dimethyl sulfoxide
c) 12% polysulfone in dimethyl formamide
by precipitation in water.

no rejection for molecules with a molecular weight of less than 5 million Dalton. Photograph 11b shows the cross-section of a membrane made from a 10% solution of Nomex® in DMAc precipitated by quenching in a water bath at 25° C. This is a typical ultrafiltration membrane with dense skin on the surface and a molecular weight cut-off of 50000 Dalton. Photograph 11c shows the cross-section of a membrane made from a solution of 22% Nomex® in DMAc precipitated in a water bath at 0°C. This is a typical reverse osmosis

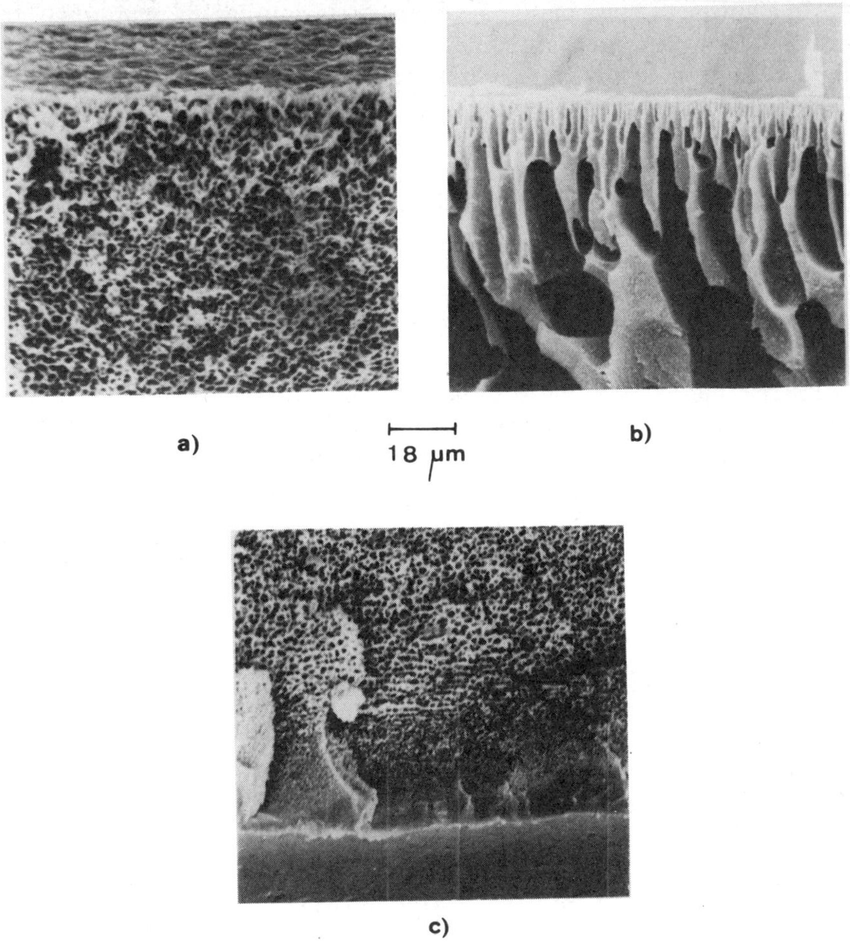

FIGURE 11.—Scanning electron micrographs of cross-sections of membranes made from a solution of

a) 10% Nomex® in DMAc precipitated by water
b) 10% Nomex® in DMAc precipitated at 25°C in a water bath
c) 22% Nomex® in DMAc precipitated at 0°C in a water bath.

membrane, which displays better than 99% NaCl rejection from a 1% solution at 100 bar hydrostatic pressure.

The effect of different preparation parameters on membrane structures and filtration properties.—Parameters determining the structure and properties of phase inversion membranes are:

- the polymer and its concentration in the casting solution
- the solvent or solvent system
- the precipitant or precipitant system
- the form of the precipitant (vapor or liquid)
- the temperature of precipitation

In addition to these main parameters determining the membrane structure, there are pre- and post-precipitation procedures, such as an evaporation step or an annealing step, which have some effect on the membrane properties.

a) Precipitation rate and membrane structure. Before discussing the various preparation parameters and their effect on membrane structure and filtration properties, some general observations can be made about the preparation of all phase inversion membranes. For instance, certain membrane structures can always be correlated with the rate of precipitation during the structure formation stage. High precipitation rates always lead to a "finger" structure, while slow precipitation rates lead to asymmetric membranes with a "sponge" structure, and very slow precipitation rates very often lead to symmetric membranes with no defined skin at the surface and a sponge structure with a very uniform pore size distribution over the entire cross-section of the membrane. The rate of precipitation and its relation to the membrane structure can be observed rather impressively under an optical microscope.

Figure 12 shows two series of photographs taken during the precipitation process. The magnification and the time intervals at which the photographs were taken are the same in both cases. The rate of precipitation slows down as the precipitation front moves further into the casting solution. The photographs also show that the finger structure (series I) precipitates much faster than the sponge structure (series II). From this type of photograph, the rate of precipitation can be followed quantitatively and Figure 13 shows some typical results for a 20% Nomex® solution precipitated in water to give a finger structure and in glycerin to give a sponge structure. From a plot of the time versus the distance the precipitation front moves, the diffusivity of the precipitant can be calculated. These calculations indicate that the transport rate of the precipitant into the casting solution is too fast to be a strictly diffusive process (18).

b) The effect of the polymer–solvent–precipitant system on membrane structure and properties. The effect of the polymer on membrane structure and properties is closely related to the solvent used in the casting solution and the precipitant. The solvent and the precipitant used in membrane precipitation determine both the activity coefficient of the polymer in the solvent–precipitant mixture and the concentration of the polymer at the point of precipitation and solidification. Unfortunately, values for the activity coefficients of the polymer, the solvent, or the precipitant, and the dependence of these activity coefficients on the composition, are difficult to obtain. A quantitative treatment of the actual membrane formation process is therefore rather difficult. However, the polymer–solvent interaction can be approximately expressed in terms of the disparity of the solubility parameter of polymer and solvent. The smaller the solubility parameter disparity of solvent and polymer, the better is the compatibility of solvent and polymer, the more time it takes to remove the solvent from the polymer structure, and the

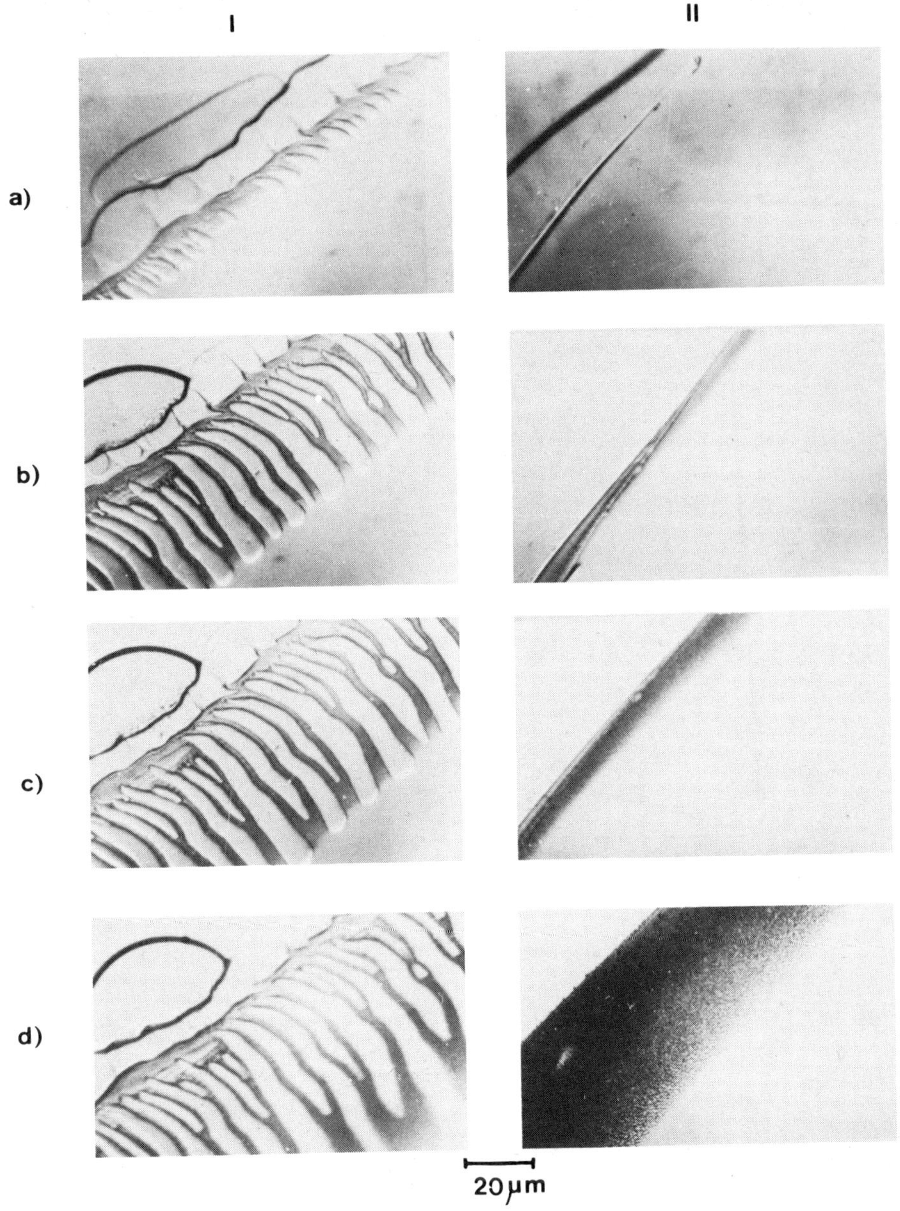

FIGURE 12.—Photomicrographs of the casting solution-precipitant interface at (a) the beginning of the precipitation and after (b) 12 sec., (c) 24 sec. and (d) 5 min. Series I is precipitated from a casting solution of 15% Nomex® in DMAc and series II from a casting solution of 15% Nomex® in DMAc + 6% formic acid.

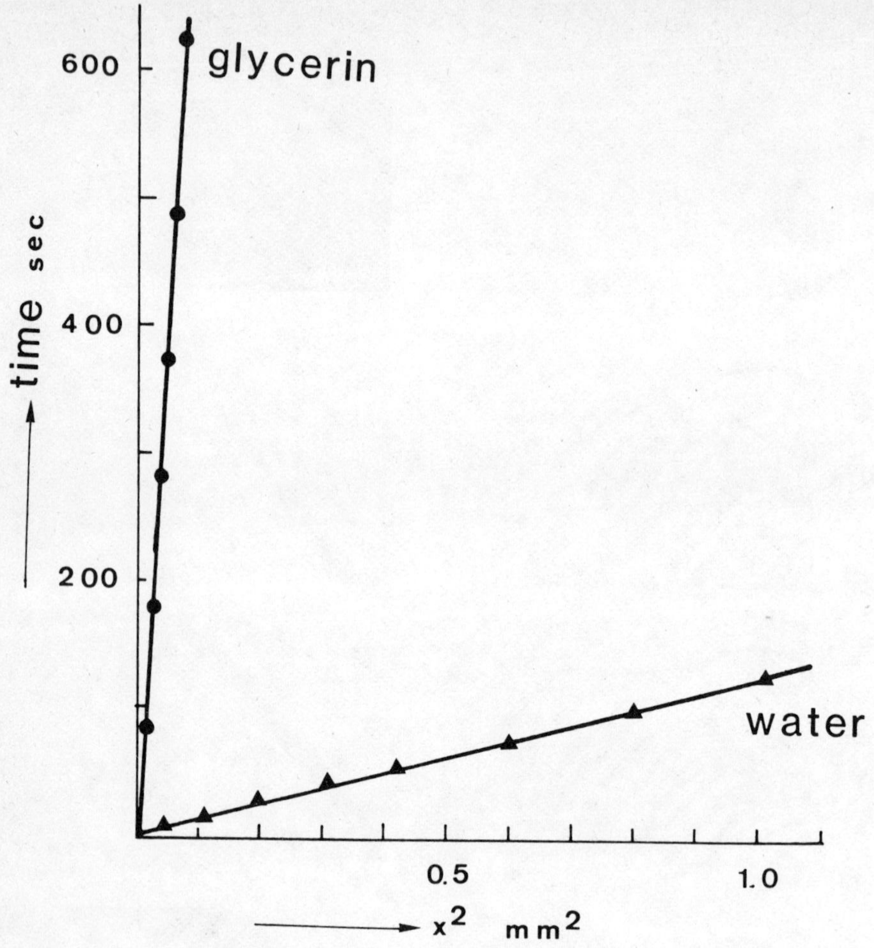

FIGURE 13.—The rate of precipitation of a 20% Nomex® solution precipitated in water, yielding a finger structure, and in glycerin, yielding a sponge structure.

slower is the precipitation of the polymer. Therefore, when all other parameters are kept constant, the tendency for a change from a sponge to a finger structure membrane increases with decreasing compatibility of solvent and polymer. The compatibility of polymer and precipitant can also be expressed in terms of the solubility parameter disparity. The higher this disparity, the less compatible are polymer and precipitant, the higher the activity coefficient of the polymer in the solvent-precipitant mixture will be, and the faster the precipitation will be. The tendency to change from a sponge to a finger structure will increase with decreasing compatibility of polymer and precipitant.

This is demonstrated in the scanning electron micrographs of Figure 14, which shows the cross-sections of membranes prepared from a solution of 15% Nomex® in DMAc and precipitated in water–glycerin mixtures of different composition. Since the compatibility of the polymer with glycerin is slightly better than that with the water, i.e., the disparity in the solubility parameter of polymer–water is larger than that of polymer–glycerin, the tendency of the membrane structure to go from a finger to a sponge structure will increase with increasing glycerin content in the precipitation bath.

Additives to the casting solution have a similar effect as shown in the scanning electron micrographs of Figure 15. This figure shows the cross-sections of membranes precipitated from a solution of 15% Nomex® in a) DMAc, b) 75% DMAc + 25% benzene and c) 60% DMAc + 40% benzene. Here a drastic change from a typical finger to sponge structure can be observed. Membrane properties do change in a similar pattern as membrane structures, i.e., sponge structures generally have lower water fluxes and lower molecular weight cut-offs and better resistance to compaction. The selection of the polymer–solvent–precipitant system is thus a very effective tool for tailoring phase inversion membranes for a certain mass separation problem.

c) The effect of the polymer concentration on membrane structure and properties. The polymer concentration is a very significant parameter for tailoring a membrane in terms of its structure and separation properties. This is demonstrated in Figure 16, where the scanning electron micrographs show membranes that have been prepared from casting solutions of different Nomex® concentrations in N-methyl pyrrolidone (NMP). With increasing polymer concentration, the structure changes from a typical finger structure (5% Nomex® in the casting solution) to a typical sponge structure (22% Nomex® in the casting solution).

The flux and retention properties, the membrane porosity and the rate of precipitation change in a corresponding pattern as shown in Table 5. With increasing polymer concentration in the casting solution the retention increases, while the permeability, the overall porosity and the rate of precipitation decrease.

d) The effect of pre- and post-precipitation procedures. In the original recipes for making phase inversion membranes, the evaporation step prior to the precipitation of the film was considered essential for the formation of the asymmetric membrane. Subsequently, it has become apparent that "skin type" asymmetric membranes can be prepared without an evaporation step. The effect of the evaporation is an increase of the polymer concentration at the surface of the cast film due to a loss of solvent. This does not necessarily lead to membranes with a higher salt rejection, however, as is easily demonstrated by the procedure first suggested by Manjikian (15), whereby cellulose–acetate membranes are prepared using a casting solution consisting of 25% cellulose–diacetate in a solvent mixture of 60% acetone and 40% formamide. When the polymer film is exposed to air at room temperature for increasing

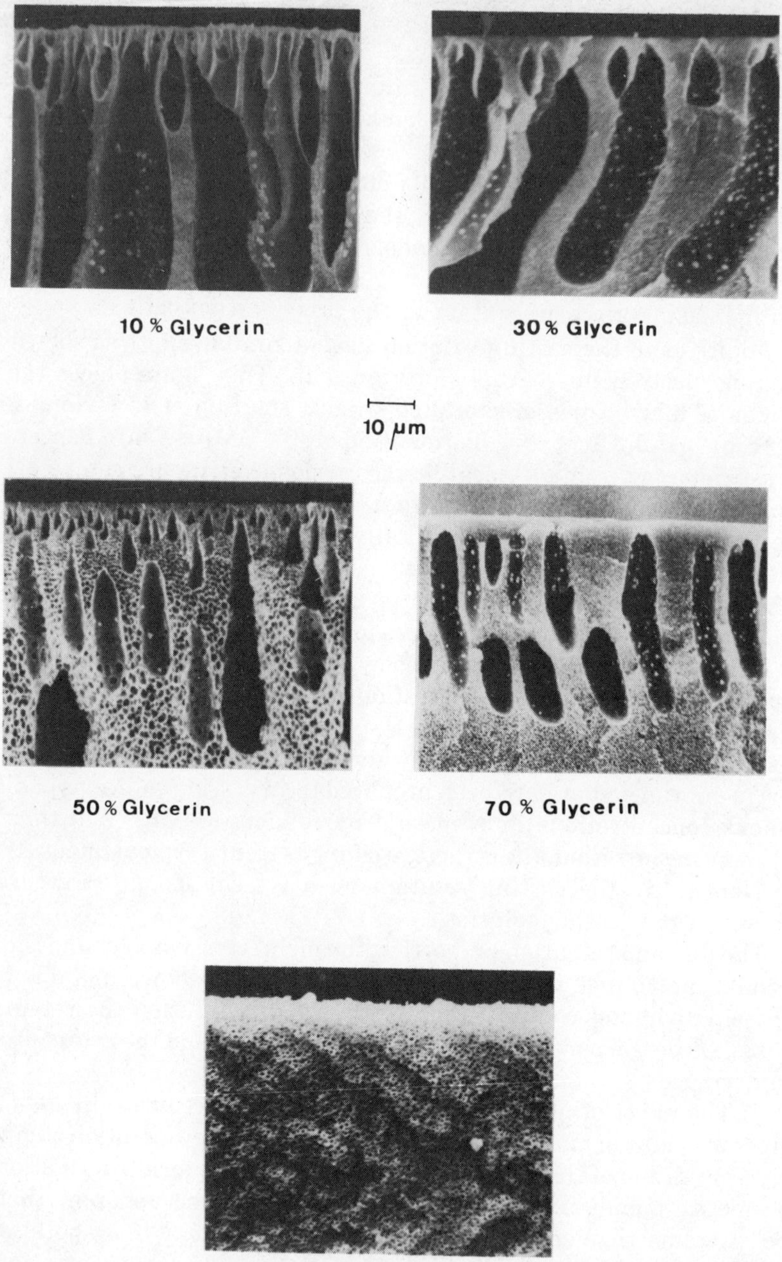

FIGURE 14.—Scanning electron micrographs of membranes prepared from a solution of 15% Nomex® in DMAc and precipitated at room temperature in water–glycerin mixtures of different glycerin content.

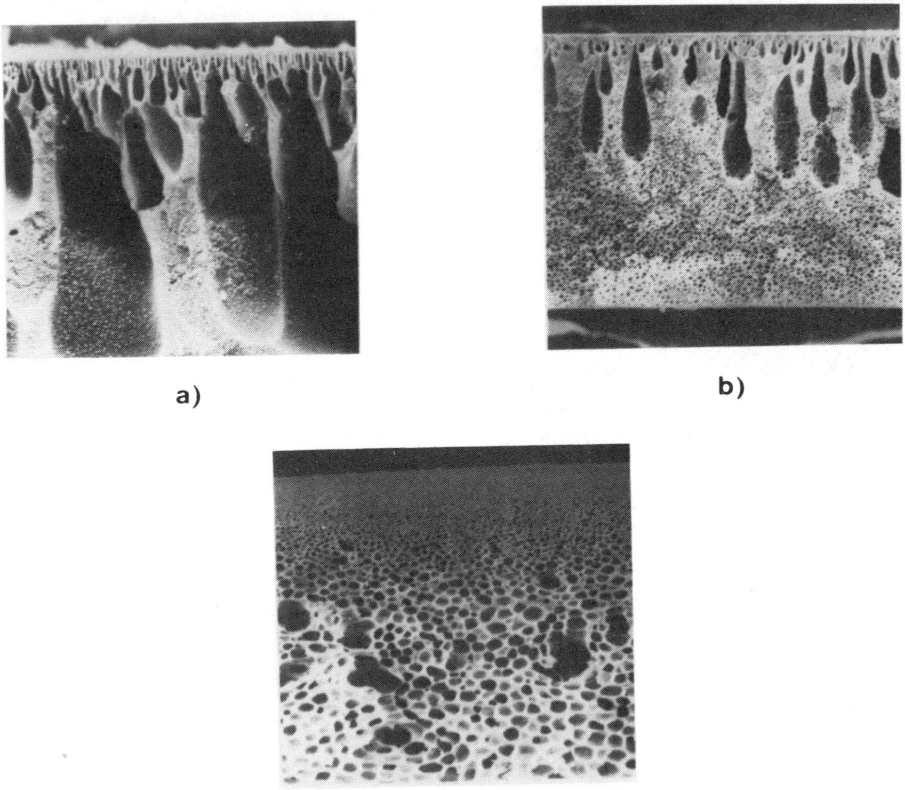

a) b) c)

FIGURE 15.—Scanning electron micrographs of membranes prepared from casting solutions of 15% Nomex® in a) DMAc, b) 75% DMAc + 25% benzene and c) 60% DMAc + 40% benzene by precipitation in water at room temperature.

time periods prior to precipitation in the water bath at very short exposure periods (< 1 minute), a slight increase in rejection is observed. Longer exposure periods (> 2 minutes), however, lead to a drastic decrease in rejection. This is shown in Figure 17, where the salt rejection of cellulose acetate membranes prepared from a casting solution containing 25% cellulose–acetate, 45% acetone, and 30% formamide, precipitated in water at 0°C and annealed at 75°C for 2 minutes, is shown as a function of the evaporation time. The tests were carried out at 100-bar hydrostatic pressure with a 1% NaCl solution.

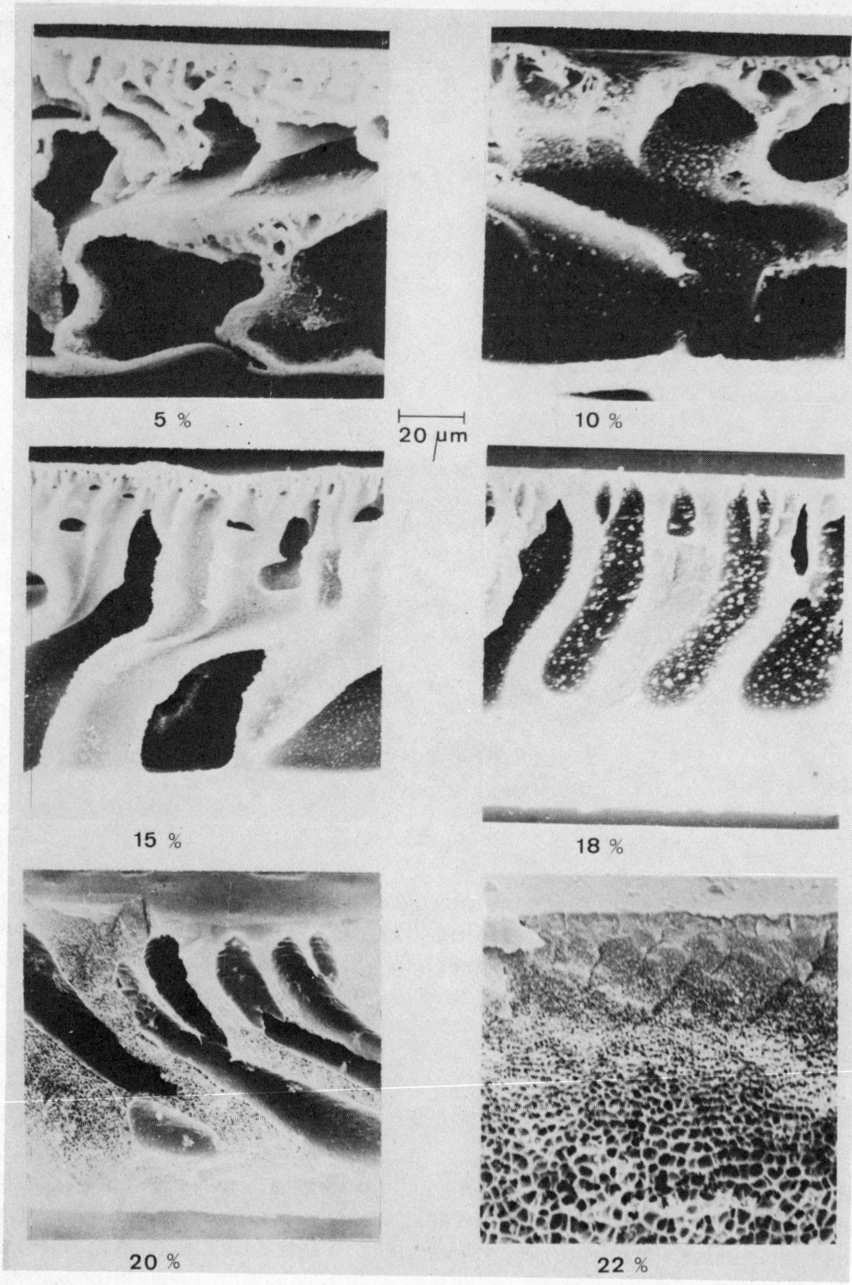

FIGURE 16.—Scanning electron micrographs of membranes prepared from casting solutions of Nomex® in NMP with different polymer concentrations by precipitation in water at room temperature.

TABLE 5. The effect of the polymer concentration in the casting solution on the porosity, the precipitation rate, and the filtration properties of membranes prepared from Nomex®-NMP-casting solutions of different polymer content by precipitation in water at room temperature

Polymer concentration (%)	Rejection+ (%)			Water flux (cm/s × 10^4)	Porosity (Vol. %)	Precipitation rate (s)
	$MgSO_4$	Cytochrome C	Bovine serum albumin			
5 x	0	0	10	56	91	32
10 x	0	43	84	32	85	40
15 x	8	92	100	9	81	52
18 xx	75	100	100	18	79	83
20 xx	90	100	100	4	77	142
22 xx	98	100	100	1.6	76	212

+ Determined using solutions of 1 wt.% solids
x Operating pressure 5 bar
xx Operating pressure 100 bar

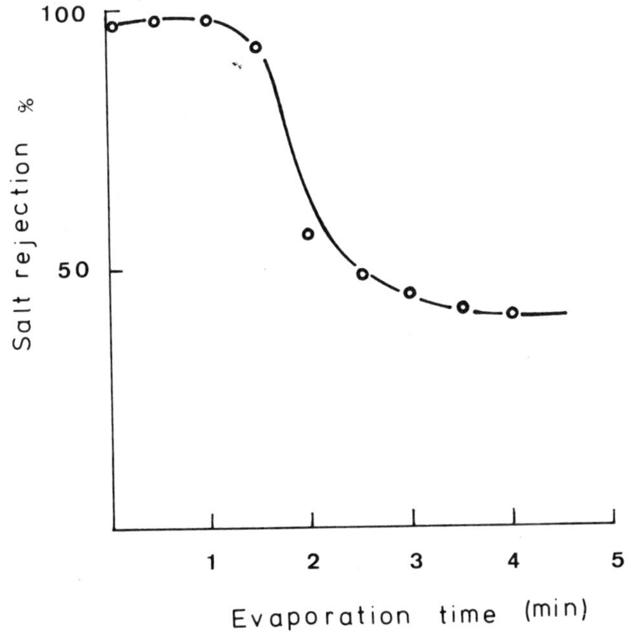

FIGURE 17.—Effect of evaporation time at 25°C on salt rejection of membranes cast from a solution of 25% cellulose acetate, 30% formamide, and 45% acetone. The membranes were annealed for 2 minutes at 75°C and then tested at 100 bar with a 1% NaCl solution.

The drastic loss in rejection with increasing evaporation time can easily be explained by the fact that during the air exposure time acetone, the more volatile solvent, will be evaporated preferentially, while the formamide concentration will increase in the film surface. Thus, the original acetone–formamide ratio of the solvent system will be changed. In an independent set of experiments, however, it was shown that cellulose acetate membranes, made from casting solutions in which the formamide concentration exceeded 35%, had no rejection capability for NaCl.

This is demonstrated in Figure 18, where the rejection and trans-membrane fluxes of membranes prepared from casting solutions containing varying portions of acetone and formamide are shown as a function of the formamide concentration in the casting solution. All membranes were precipitated with 1-minute evaporation time in ice water, annealed for 2 minutes at 75°C and then tested at 100 bar with 1% NaCl solution.

The results shown in Figures 17 and 18 demonstrate quite clearly that an evaporation step can lead not only to an increase in the polymer concentration at the surface of a cast film, it might also lead to a significant change in the

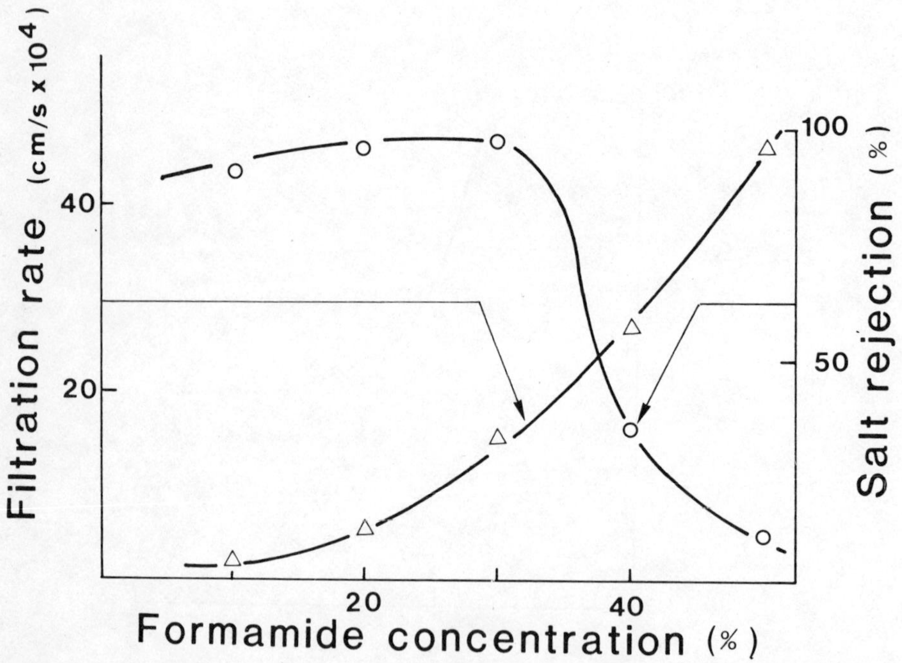

FIGURE 18.—Rejections and trans-membrane fluxes of membranes prepared from casting solutions containing 25% cellulose acetate and varying proportions of acetone and formamide. The cast polymer films were evaporated for 1 minute before precipitation in ice water. All membranes were annealed at 75°C for 2 minutes and then tested at 100 bar with a 1% NaCl solution.

casting solution composition when solvent mixtures with significant differences in their boiling temperatures are used, as in the case of acetone (b.p. 56.2°C) and formamide (b.p. 105°C).

As indicated already, membranes made from a casting solution containing 25% cellulose acetate, 45% acetone and 30% formamide do not show any NaCl rejection unless they are annealed in hot water; thus for this type of membrane the annealing is an essential post-precipitation membrane treatment step when high salt rejections are desired. The effect of the annealing post-treatment procedure on membrane flux and salt rejection is demonstrated in Figure 19, where the flux and salt rejections of membranes precipitated in ice water from a casting solution of 25% cellulose acetate, 45% acetone and 30% formamide are shown as a function of the annealing temperature. The annealing time was kept constant at 2 minutes and the membranes were tested at 100 bars with a 1% NaCl solution.

Figure 19 shows that with increasing annealing temperature, the transmembrane flux decreases drastically and the rejection correspondingly. The

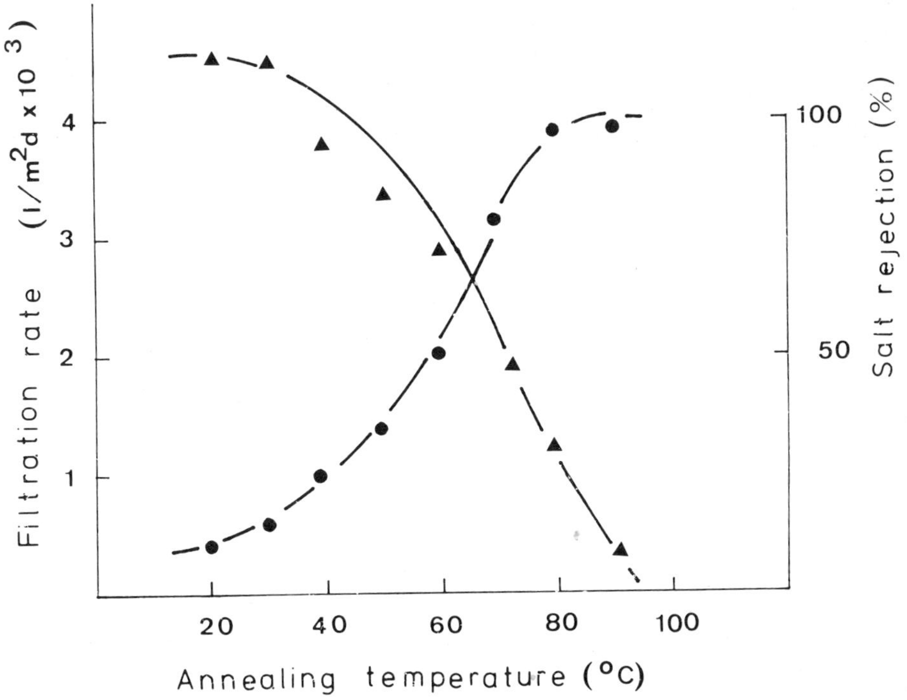

FIGURE 19.—Effect of annealing temperature at a constant annealing time of 2 minutes on rejections (●) and fluxes (▲) of membranes prepared from a solution of 25% cellulose acetate, 45% acetone and 30% formamide. The membranes were tested at 100 bars with a 1% NaCl solution.

reduction in pore size accompanying the shrinkage that takes place on annealing is probably partially responsible for these effects. An additional cause is the restricted polymer mobility accompanying the increased degree of crystallinity. Thus, while dry cellulose acetate has a glass transition temperature of over 100°C, in the plasticized wet form it may have a considerably lower value. The X-ray photographs of Figure 20 confirm the increased crystallinity upon annealing. The membrane (b) annealed for 5 minutes at 100°C shows a significantly higher degree of crystallinity than the unannealed membrane (a). The increased crystallinity reduces the freedom of movement of the polymer chains, causing the hydrated salt to be excluded to a greater degree than the smaller water molecules.

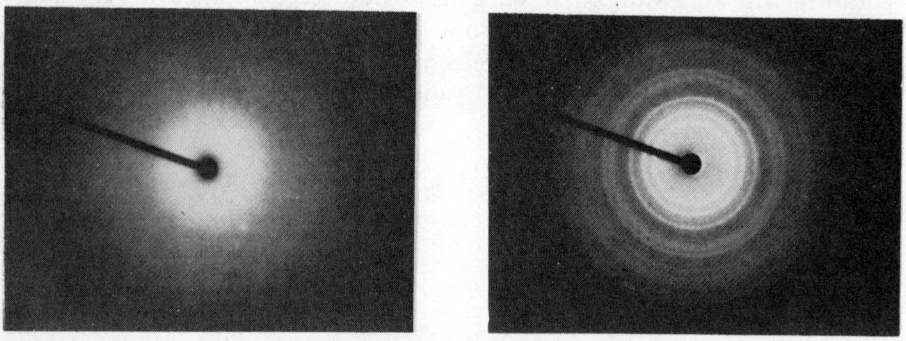

(a) (b)

FIGURE 20.—X-ray photograph of an unannealed membrane (a) and a membrane annealed for 5 minutes at 100°C (b).

Composite membranes The composite or "thin-film composite" membrane is mainly used for reverse osmosis applications. In this type of membrane the actual selective membrane barrier is deposited onto the surface of a suitable finely porous substrate. The first composite membranes were made by casting a thin film of cellulose esters on a water surface from a very dilute solution. After the solvent evaporated the thin film was lifted off onto a substrate (19). These techniques of making composite membranes, however, were rather impractical for large-scale production. The same problem was experienced when plasma polymerization was applied for cross-linking monomers into a homogeneous film on the surface of a porous substructure. The real breakthrough was achieved when interfacial polymerization techniques were utilized to form the actual polymer film at the membrane surface (20). The process is illustrated in Figure 21, which shows the formation of the polymer film at the interface of an aqueous solution of, for example, polyethylenimine and a toluene solution of diisocyanate, which are not miscible. This cross-linked structure is shown in Figure 22. The membranes can be made very thin, i.e., less than 50 nm, and virtually free of imperfections by this method.

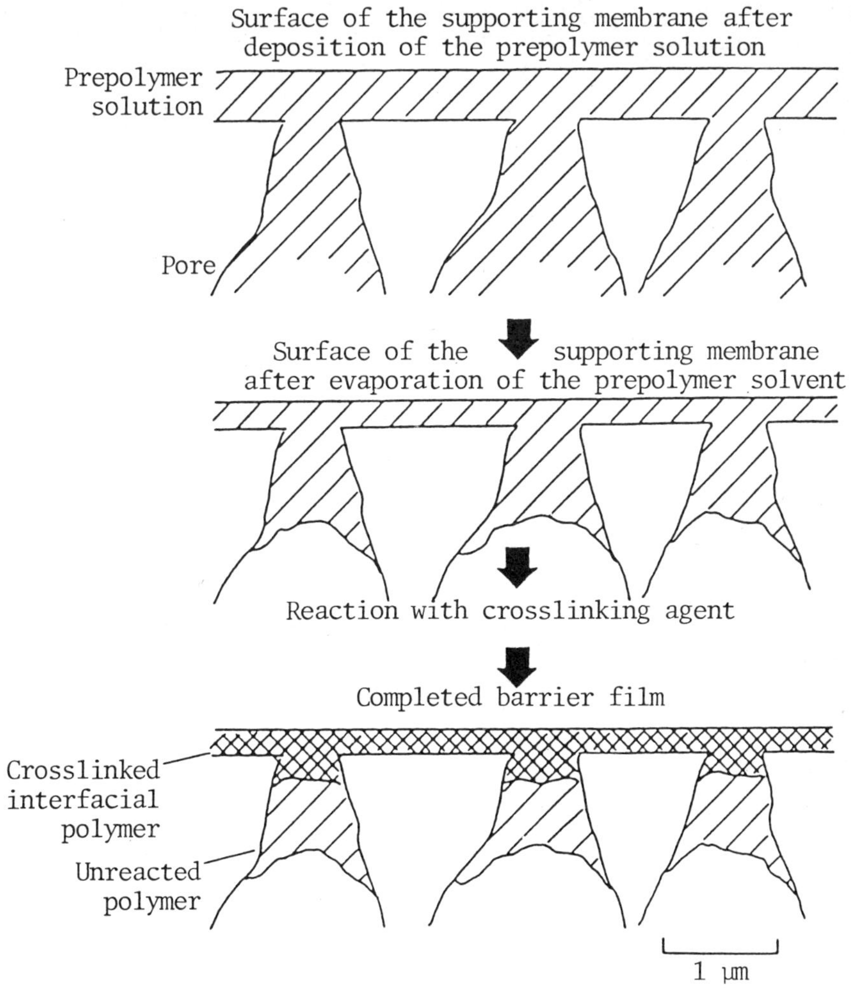

FIGURE 21.—Schematic diagram illustrating the formation of a composite membrane by interfacial polymerization of polyethyleneimine with diisocyanate.

If the reactants are sufficiently mutually reactive, a film is formed very quickly and, once it is formed, it does not grow in thickness appreciably, since the film is resistant to further permeation of the reactants. Furthermore, the film is essentially self-sealing. In actual practice cross-linking is achieved by a heat treatment procedure.

The performance of a composite membrane is not determined only by the selective surface film. The microporous support structure, or rather the pore

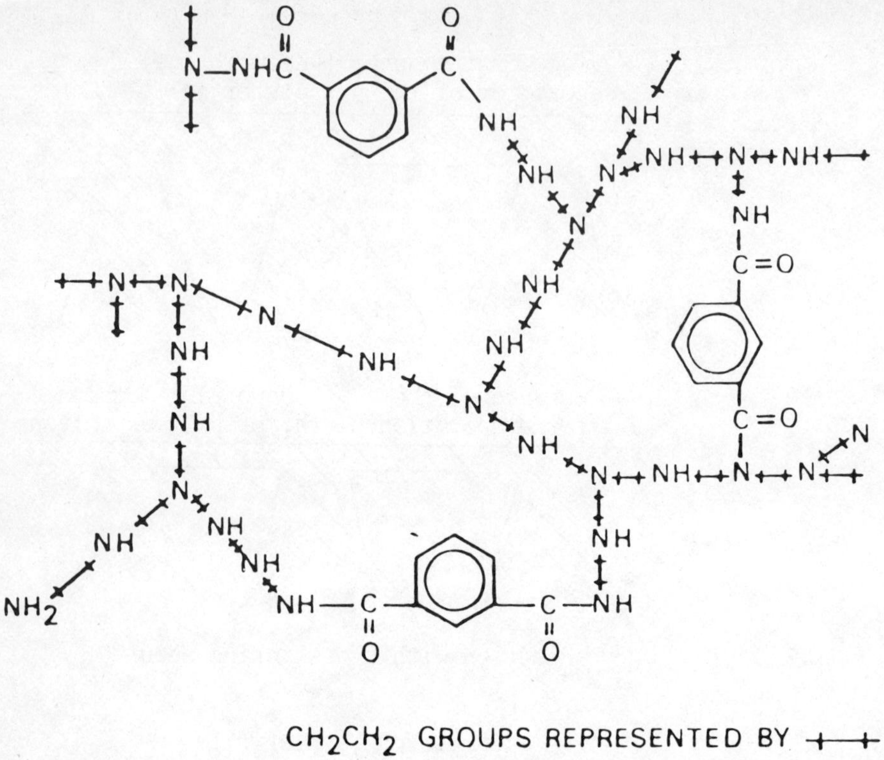

FIGURE 22.—The reaction scheme of cross-linking polyethyleneimine with diisocyanate.

size, the pore distribution and the overall porosity, is also of importance. The porosity of the microporous substructure should be as high as possible to minimize the proportion of the surface film that is in contact with the support and, hence, less available for transport. The pore diameter, however, should be as small as possible to minimize the distance between unsupported points of the polymer layer. In addition, the penetration of cross-linked and unreacted polymer into the support can be an important factor in determining overall transport characteristics of the composite.

REFERENCES

1. Lonsdale, H.K., 1982: "The growth of membrane technology." J. Membrane Sci. *10,* 81.
2. Metzger, H., Kock, K., Dt. Pat. 1769595.
3. Kammermeyer, K, 1968: "Gas and vapor separations by means of membranes," in "Progress in Separation and Purification," Perry E.S. (ed.), New York, Wiley-Interscience, Volume 1.
4. Gore, R.W., 1967: "Very highly stretched polytetrafluoroethylene and process therefor." U.S. Patent 3962153.

5. Fleischner, R.L., Price, P.B., and Walker, R.M., 1969: "Nuclear tracks in solids." Sci. Amer. *220,* 30.
6. Porter, M.C., 1979: "Membrane filitration," in Handbook of Separation Techniques for Chemical Engineers, Schweitzer, Ph.A., ed. New York, McGraw Hill, Section 2.1.
7. Strathmann, H., 1979: "Trennung von molekularen Mischungen mit Hilfe synthetischer Membranen." Darmstadt, Steinkopff.
8. Haller, W., 1965: "Chromatography on glass of controlled pore size." Nature *206,* 693.
9. McBride, McKinley, D.L., 1965: "A new hydrogen recovery route." Chem. Engr. Progr. *61,* 81.
10. Ward, W.J., Browall, W.R. and Salemme, R.M., 1976: "Ultrathin silicone/polycarbonate membranes for gas separation processes." J. Membrane Sci. *1,* 99.
11. Li, N.N., 1971: "Permeation through liquid surfactant membranes." AIChE J. *17,* 459.
12. Baker, R.W., Roman, I.C., Smith, K.L., and Lonsdale, H.K., 1982: "Liquid membranes for the production of oxygen-enriched air," Industrial Heating (July issue).
13. Loeb, S., Sourirajan, S., 1962: "Sea water demineralization by means of an osmotic membrane." Advan. Chem. Ser. *38,* 117–132.
14. Kesting, R.E., 1971: Synthetic Polymer Membranes. New York, McGraw-Hill.
15 Manjikian, S., 1967: "Desalination membranes from organic casting solutions," Ind. Engr. Chem. Prod. Res. Develop. *6,* 23.
16. Zsigmondy, R., Carius, C., 1927: Chem. Ber. *60* B, 1074.
17. Kesting, R.E., 1973: "Concerning the microstructure of dry-RO membranes." J. Appl. Polymer Sci. *17,* 1771.
18. Strathmann, H., Kock, K., Amar, P. and Baker, R.W., 1975: "The formation mechanism of asymmetric membranes." Desalination *16,* 179.
19. Riley, R.L., Lonsdale, H.K., Lyons, C.R., and Merten, U., 1967: "The preparation of ultrathin reverse osmosis membranes and the attainment of 'theoretical' salt rejection." J. Appl. Polymer Soc. *11,* 2143.
20. Cadotte, J.E., King, R.S., Majerle, R.J., and Petersen, R.J., 1981: "Interfacial synthesis in the preparation of reverse osmosis membranes." J. Macromol. Sci.-Chem. A *15,* 727.

PREPARATION OF ASYMMETRIC MEMBRANES BY THE PHASE INVERSION PROCESS

J.G. Wijmans and C.A. Smolders

Department of Chemical Technology
Twente University of Technology
P.O. Box 217, 7500 AE Enschede
The Netherlands

The formation of membranes by the phase inversion process is discussed. In this process a polymer solution is brought to phase separation by an exchange of solvent and nonsolvent. The structure of the membrane is the result of an interplay of phase separation and mass transfer. Typical morphological features of the membrane (skin, sponge structure, conical voids) are discussed in relation to the preparation procedure. It is shown that the skin layer is formed by gelation at increased polymer concentration, while liquid-liquid phase separation is responsible for the porous sublayer.

P. M. Bungay et al. (eds.), Synthetic Membranes: Science, Engineering and Applications, 39–56.

1. INTRODUCTION

Since the beginning of this century, when the first microporous membranes were prepared by Bechold (1), the potential of synthetic membranes for the separation of dissolved and suspended solutes from the solvent has been recognized. Industrial-scale applications remained limited until Loeb and Sourirajan (2) developed the first asymmetric skinned membranes in 1962. These membranes owe their practical value to the presence of a skin: a very thin (0.1 to 0.5 μm) and very dense layer that possesses selective properties. The skin is supported by a porous sublayer of 0.1 to 0.2 mm thickness, which gives the membrane mechanical stability. Due to the thin skin layer the hydrodynamic resistance is low compared with symmetric membranes having the same selectivity.

Synthetic membranes can be prepared in various ways. In this chapter we will focus on the so-called phase inversion process, by which the majority of the commercially available membranes are produced. The concept of phase inversion, introduced by Kesting (3), can be defined in the following way: a homogeneous polymer solution is transformed into a two-phase system in which a solidified polymer phase forms the porous membrane structure, while a liquid phase, poor in polymer, fills the pores.

The next sections include a discussion of the various preparation procedures of phase inversion membranes and the phenomena that are characteristic of the phase inversion process; these phenomena are then correlated with the resulting membrane structure in a discussion of the mechanism of membrane formation.

2. PHASE INVERSION PROCESS: PREPARATION PROCEDURES

The membranes produced by phase inversion are polymeric. To obtain these membranes the polymer is dissolved in a solvent which can be a single component solvent or a mixture of solvents and nonsolvents. Generally this solution is cast on a support, i.e., a glass or metal plate or a nonwoven textile fabric, and then treated in a specific way [see (a) to (d) below] in order to precipitate the polymer and thus obtain flat or tubular membranes. In another procedure the polymer solution is spun through a spinneret which has an extra outlet for pressurized air or liquid in the centre of the opening, resulting in hollow fiber membranes. For all these membranes the structure formation phenomena are basically the same.

Within the phase inversion process four different techniques can be distinguished:

(a) Precipitation from the vapour phase (4,5). In this very early developed technique, membrane formation is accomplished by penetration of a precipitant for the polymer into the solution film from the vapour phase, which is saturated with the solvent used. A porous membrane is produced without a skin and with an even distribution of pores over the membrane thickness.

(b) Precipitation by controlled evaporation (6–8). The polymer is dissolved in a mixture of a good and a poor solvent, of which the good solvent is more volatile. The polymer precipitates when the solvent mixture shifts in composition during evaporation to a higher nonsolvent content. A skinned membrane can be the result.

(c) Immersion precipitation (5, 9–12). This technique, which was first used successfully by Loeb and Sourirajan for the preparation of a reverse osmosis membrane, has been studied and exploited most for the production of skinned membranes. The characteristic feature is the immersion of the cast polymer film in a nonsolvent bath. The polymer precipitates as a result of solvent loss and nonsolvent penetration.

(d) Thermal precipitation (12). Here a solution of polymer in a mixed solvent, which is on the verge of precipitation, is brought to separation by a cooling step. When evaporation of the solvent has not been prevented the membrane can have a skin.

From these procedures we can infer that for the phase inversion process at least the following features are characteristic:

(*i*) A ternary system. The process involves at least one polymer component, one solvent and one nonsolvent. The latter two must be miscible.

(*ii*) Mass transfer. The polymer solution is subject to a transfer of solvent and nonsolvent in such a way that the nonsolvent concentration in the film increases. Mass transfer starts at the interface between the polymer film and the coagulation medium (vapour or liquid). The changes in composition in the film are governed by diffusion. No mass transfer takes place in thermal precipitation without evaporation.

(*iii*) Precipitation. As a result of the increase of nonsolvent content the polymer solution becomes thermodynamically unstable and phase separation will occur. So an important aspect of the phase inversion process is associated with the demixing phenomena possible in ternary polymer systems. These phenomena include not only the phase equilibria but also the kinetics of phase separation, as the formation of membranes is a dynamic process.

3. PHASE SEPARATION IN POLYMER SYSTEMS

3.1. Thermodynamics

Information on the stability of a solution can be obtained from the Gibbs free energy of mixing. For the system of our interest, a mixture of polymer, solvent and nonsolvent, we will use the Flory–Huggins (13) expression as derived by the lattice model:

$$\Delta G_m/RT = n_1\ln(\phi_1) + n_2\ln(\phi_2) + n_3\ln(\phi_3) + g_{12}n_1\phi_2 + g_{13}n_1\phi_3 + g_{23}n_2\phi_3 \quad (1)$$

ΔG_m is the Gibbs free energy of mixing; R is the gas constant and T the temperature in Kelvin. The subscripts refer to nonsolvent [1], solvent [2] and polymer [3]. The number of moles and the volume fraction of component i are n_i and ϕ_i, respectively. The g_{ij} parameter is the component i–component j interaction parameter.

The first three terms on the right hand side of equation 1 represent the ideal entropy of mixing of the solution, divided by R. The last three terms describe the enthalpy of mixing of the solution (divided by RT) when only binary interactions have been taken into account. If equation 1 is to describe the free enthalpy of mixing accurately, then the interaction terms will also have to incorporate the inevitable deviations from the ideal entropy term. This is one of the reasons why the interaction parameters are assumed to be concentration dependent (14).

In the last two decades other expressions for ΔG_m have been derived using equation-of-state theories (15). These formulations are certainly improvements in the theoretical sense but require the knowledge of properties of the components which are available only for a limited number of compounds. On the other hand, information on the interaction parameters can be deduced from ΔG_m measurements easily. These measurements have been compiled for binary low molecular weight mixtures (16). For mixtures containing a polymer component some data are available, but relatively simple techniques exist which yield interaction parameter values: vapour pressure osmometry, high pressure membrane osmometry and swelling experiments.

From equation 1 the expressions for the chemical potentials of the components can be derived using

$$\Delta\mu_i = \partial\Delta G_m/\partial n_i \quad (2)$$

In the differentiation procedure the eventual concentration dependence of the g parameters has to be accounted for. Every system will try to minimize its Gibbs free energy. The entropy terms of equation 1 will always be negative, whereas the enthalpy terms are positive in the case of positive interaction parameters. If the g parameters become too large demixing will occur. The equilibrium state is reached when the chemical potentials in each phase formed are equal for all components.

3.2. Liquid–Liquid Phase Separation

If a solution becomes thermodynamically unstable it is possible that it can lower its free enthalpy of mixing by separating into two liquid phases. These two phases have different compositions but are in equilibrium with each other. The reason for the solution to become unstable could be a decrease in temperature, a loss of solvent or an increase in nonsolvent content.

In order to make the mechanism of liquid–liquid (L–L) phase separation clear, we will first consider a binary system: a polymer and a solvent. Figure 1 shows the dependence of ΔG_m on concentration in such a system for two temperatures. At temperature T_a the system is completely miscible over the whole concentration range. If one lowers the temperature the interaction parameter g will increase and for a certain temperature T_b the ΔG_m vs concentration curve is displayed in Figure 1b. The upward bend in the curve is the result of the increased enthalpy term. From this figure it can be seen that all compositions between ϕ' and ϕ'' can lower their free enthalpy by demixing into two phases. The compositions of those two phases are ϕ' and ϕ'', respectively, and the phases are in equilibrium with each other since they lie on the same tangent to the ΔG_m curve.

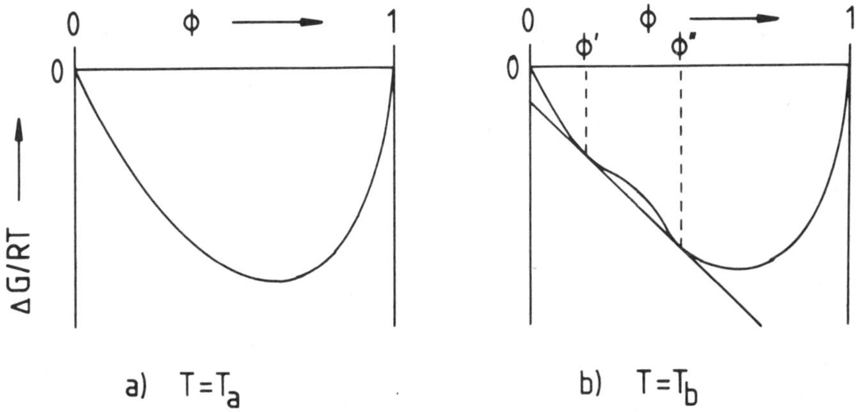

FIGURE 1.—ΔG_m vs composition curves for a binary system. $T_a > T_b$. ϕ: polymer volume fractions.

There are two different kinetic ways for L–L separation to occur (see Figure 2). Within the composition range $\phi^1 - \phi^2$ the curvature is such that

$$\partial^2 \Delta G_m / \partial \phi^2 < 0 \qquad (\phi^1 < \phi < \phi^2) \qquad (3)$$

Condition 3 implies that the solution is unstable with respect to even the smallest amplitude of certain concentration fluctuations. The solution will then separate spontaneously (17, 18) into very small, interconnected regions with compositions ϕ' and ϕ'': the so-called spinodal decomposition. In the composition regions $\phi' < \phi < \phi^1$ and $\phi^2 < \phi < \phi''$ the second derivative of ΔG_m with respect to ϕ is positive, which means that there are no spontaneously growing concentration fluctuations, so there is no spontaneous phase separation. Demixing can only start if the concentration fluctuations have generated at least one stable nucleus, i.e., the decomposition regions are metastable. A nucleus is stable if it lowers the total free enthalpy of the system; hence in the range $\phi' < \phi < \phi^1$ the nucleus must have a composition near ϕ'' and in the range $\phi^2 < \phi < \phi''$ it must have a composition near ϕ'. After the nucleation of the "second phase" the nuclei will grow while the surrounding phase gradually moves toward the composition of the other equilibrium phase. The two possibilities indicated above are illustrated in Figure 2. In a relatively diluted solution the dispersed phase will have composition ϕ'' and in a more concentrated solution the dispersed phase will have composition ϕ'.

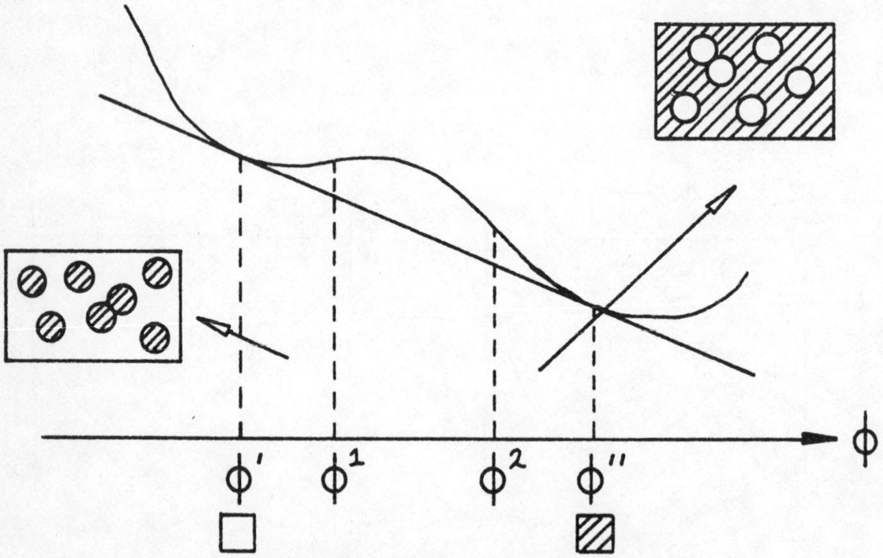

FIGURE 2.—Part of the ΔG_m vs ϕ graph of figure 1b.
○ and □: composition ϕ'
⊘ and ▨: composition ϕ''

In a three-component system the phenomena are basically the same, but here a decrease in temperature is not necessary to induce L–L phase separation; a change in composition is sufficient. In Figure 3 the ΔG_m surface for a ternary system is schematically drawn. All pairs of compositions with a common tangent plane to the ΔG_m surface together constitute the solid line in the phase diagram at the bottom of Figure 3. This line is the L–L boundary, the binodal. The dotted line is the spinodal, inside which all compositions are unstable. The point in which the spinodal and binodal touch is the critical composition. It is the location of the critical point in the phase diagram that determines whether the nuclei formed will have a composition high or low in polymer concentration. If one wants to reach the spinodal area by a change in composition, the composition has to travel first through the metastable region where nucleation and growth take place. In our opinion the latter process is fast compared with the rate of mass transfer and in that case the spinodal region is not reached. Therefore, we think it is highly improbable that spinodal decomposition has a role in membrane formation.

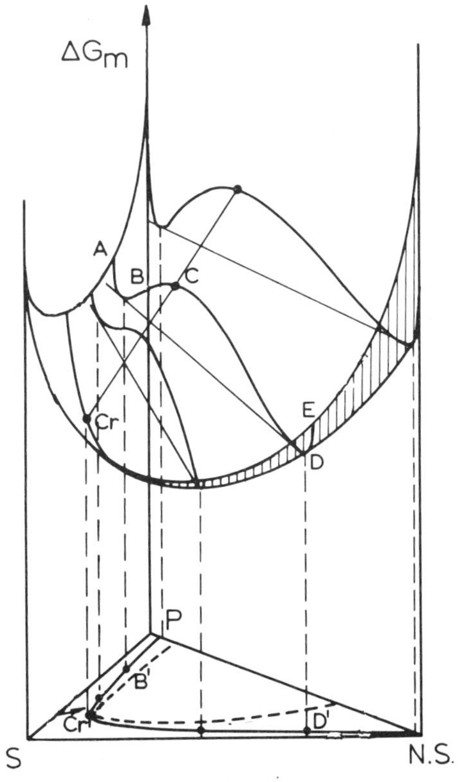

FIGURE 3.—Sketch of ΔG_m surface and miscibility gap for the system polymer (P), solvent (S), nonsolvent (NS). Cr: critical composition

3.3. Crystallization and Gelation

When the thermodynamic quality of a polymer solution is decreasing, which may occur by loss of solvent, by lowering of the temperature or by the introduction of a nonsolvent, most polymers are able to form ordered agglomerates. In very dilute solutions the polymer molecules can form single crystals of lamellar type, being only a few hundred Angströms thick and often many microns in the lateral direction. From solutions of medium concentration, more complex morphologies occur, i.e., dendrites or spherulites. These latter structures may contain, except for the ordered regions, appreciable amounts of amorphous polymeric material.

In the preparation of membranes polymer solutions of medium and high concentrations are used. In these systems the time available for the crystallization becomes important since this process takes place through nucleation and growth. Especially if crystallization is induced by a change in composition, i.e., by mass transfer, there will be a competition between the rate of the mass transfer and the rate of nucleation. A transfer, slow compared with the nucleation process, will induce crystallization at a low level of supersaturation and there will be a limited number of nuclei, which will grow considerably. A rapid mass transfer will lead to crystallization at high polymer concentrations: many nuclei are present because of the high supersaturation, but their chance to grow will be limited. At high concentrations the numerous microcrystalline regions act as physical crosslinks in the solution and a thermoreversible gel is formed.

In Figure 4 we have given a schematic representation of the free enthalpy behaviour in the region of high polymer/low nonsolvent content of the three-component system. One sees that at higher concentrations the free enthalpy of mixing of the solution can be lowered by the formation of solid, crystalline polymer in equilibrium with a solution of a certain lower polymer content. At a lower (or higher) temperature the ΔG_m surface will change its shape and the crystalline polymer will be in equilibrium with a solution of lower (or higher) polymer concentration.

As can be seen from Figure 4, at high polymer and high nonsolvent concentrations both L–L phase separation and crystallization are possible thermodynamically. In these regions the kinctics of the two phenomena will determine which one actually takes place.

Some polymers are completely amorphous, i.e., there is no ordered structure that lowers the free enthalpy with respect to the amorphous state. In these systems there is no liquid–solid transition in the thermodynamic sense, but nevertheless at very high polymer concentrations the fluidity of the solution becomes zero. The polymer molecules are so densely packed and there are so many entanglements that in effect a gel has formed. The solution–gel boundary is a viscosity boundary in this case.

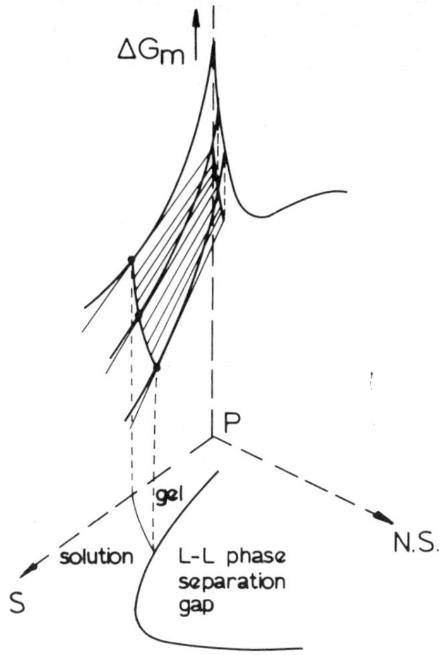

FIGURE 4.—Free enthalpy behavior at high polymer/low nonsolvent content, explaining the solution–crystal transition.

3.4. Experimental Data on Phase Separation

Liquid–liquid phase separation is readily detected by turbidity measurements (19) and light scattering methods (20). If the interaction parameters of a system are available, one can calculate the L–L demixing gap, using the thermodynamic equilibrium expressions. Altena (21) has shown that for most systems consisting of a polymer, a solvent and a nonsolvent, L–L phase separation should be expected.

Crystallinity in gels is hard to demonstrate when diffraction techniques are used, since the crystalline entities are of submicroscopic order. If surface effects are not too large, calorimetric measurements can be used (22). Data on the kinetics of crystallization can be obtained from Pulse Induced Critical Scattering (23).

4. FORMATION OF ASYMMETRIC MEMBRANES

The origin of the asymmetric structure of membranes, i.e., the presence of the skin, has received much attention in membrane research. There are three views on the formation of the skin:

a) The asymmetry is already present in the cast film of the concentrated polymer solution before precipitation takes place, due to surface tension effects. Further steps in the preparation process, such as coagulation and heat treatment, will only fix the already existing asymmetry. Representatives of this approach are Panar (24) (nodular morphology) and Tanny (12). Since it is possible to obtain both skinned and nonskinned membranes from one and the same polymer solution (25), we think this view cannot be true.

b) The skin is formed through evaporation from the upper layer of the cast film. The duration and the conditions of the evaporation step determine to a great extent the resulting properties of the membrane. This approach is followed by Sourirajan (26) and Kunst (29) (solution structure-evaporation rate concept), Kesting (3) and Anderson (28). We do not agree that solvent evaporation in general is a necessary step to induce the formation of the asymmetric structure. The following objections can be made:

—One can prepare excellent asymmetric membranes from polymer/solvent/nonsolvent systems in which evaporation of the solvent is negligible, e.g., polysulfone/DMF/water and cellulose acetate/dioxane/water.

—Also systems with a volatile solvent (CA/acetone-formamide/water) cast and coagulated under circumstances that the atmosphere is saturated with the solvent acetone give excellent asymmetric membranes (Sarbolouki (29)).

—Asymmetric hollow fibres with a skin at the interior surface can be made which have had no contact with air during the preparation (Strathmann (30)).

c) The coagulation process is responsible for the formation of the asymmetry. Skin formation and formation of the porous sublayer are the result of a complex interplay of phase separation and diffusional processes. Authors who have made important contributions are Frommer (31), Strathmann (5) and Koenhen (11).

The formation of asymmetric membranes according to approach c) is described below.

4.1. Mechanism of Formation

In the immersion precipitation process a polymer solution, cast on a support, is immersed in a bath containing a nonsolvent. As was first suggested by Koenhen (11), we assume that the skin is formed by gelation and that the porous sublayer is the result of liquid–liquid phase separation by nucleation and growth. The factor determining the type of phase separation at any point in the cast film is the local polymer concentration at the moment of precipitation. In the first split second after immersion there is a rapid depletion of solvent from the film and a relatively small penetration of nonsolvent. This means that the polymer concentration at the film/bath interface increases and that the gel boundary is crossed (see Figure 4). The thin and dense gel layer that is formed in this way, the skin, will act as a resistance to the outdiffusion of solvent and at the positions beneath the top layer demixing will occur at lower polymer and higher nonsolvent concentrations. So here the type of demixing will be liquid–liquid phase separation (see Figure 4). The demixing gap is entered at the polymer-rich side of the critical point, so the nuclei consist of the polymer-poor phase and the result is a porous structure, the pores of which are filled with the dilute solvent/ nonsolvent phase. The porous structure will be fixed by gelation of the concentrated, continuous polymer phase.

The approximate changes of composition for the toplayer and the substructure, the so-called "coagulation paths", are given in Figure 5. A more detailed description of the concentration profiles in a precipitating casting solution can be found in the paper by Bokhorst *et al.* (32).

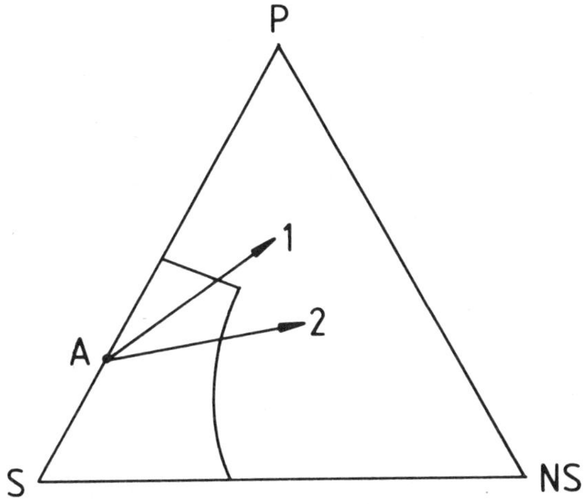

FIGURE 5.—Schematic course of the composition for the skin layer (1) and the bottom layer (2) of a polymer film with initial composition A upon immersion in a nonsolvent bath.

5. MEMBRANE STRUCTURES IN RELATION TO PREPARATION PROCEDURES

5.1. Precipitation from the Vapour Phase

The polymer solution is subject to a nonsolvent inflow, while the solvent saturated vapour phase prevents the outflow of solvent: the polymer concentration decreases gradually. The only possible demixing mechanism is L–L phase separation and a microporous membrane without a skin is the result. The mass transfer at the solution/vapour interface is slow compared with the diffusion in the polymer solution, so the concentration profile inside the film is very flat. This implies that in the whole polymer film demixing takes place at approximately the same time and at the same polymer concentration. The result is an even pore distribution over the cross-section of the membrane: they are essentially symmetric.

The size of the pores can be controlled by the preparation variables. In general the pore radius will increase when the temperature is increased (gelation at higher concentrations) or when the polymer concentration of the casting solution is decreased. The polymer content should not be too small, however, since for a porous structure it is necessary that the L–L demixing gap is entered at the polymer-rich side of the critical point.

5.2. Precipitation by Controlled Evaporation

When a polymer-solvent solution, without nonsolvent, loses solvent through evaporation, the polymer concentration gradually increases and gelation will take place. In this manner dense polymer films are obtained, which are useful in pervaporation (33). If gelation occurs through crystallization, the properties of the film depend on the rate of evaporation and diffusion, i.e., the degree of supersaturation.

Porous structures are formed when one or more nonsolvents are added to the solution. In this case the precipitation is determined by the volatility of the compounds used, the initial concentrations in the solution and the phase boundaries of the system. Kesting (34) has utilized controlled evaporation to produce asymmetric membranes with reverse osmosis properties: the "Dry-RO" membranes. In this procedure the solvent must have a boiling temperature which is at least 30°C under the boiling temperatures of the nonsolvents used. In order to obtain high flux membranes more than one nonsolvent must be present in the casting solution (34). The asymmetry of the membranes is the result of the fast evaporation of the solvent from the upper layer of the polymer solution.

5.3. Immersion Precipitation

In the section on the formation of asymmetric membranes the phenomenon of immersion precipitation has been discussed. Because this process is used in the production of the majority of the asymmetric membranes we will focus here to some extent on the variables determining the membrane morphology.

The ultimate determining factor for the skin formation is the local polymer concentration in the top layer in the polymer solution at the moment of precipitation. In the immersion technique the solvent depletion from the top layer of the solution film is extremely fast. An increase in polymer concentration in the top layer is the result. This increase improves the conditions for gelation to occur. The gelation will be favoured by the penetration of nonsolvent. The higher the polymer concentration has become before nucleation in the skin sets in, the more numerous and the smaller will be the nuclei because of higher supersaturation.

It will be clear now which factors favour the formation of a more finely structured, i.e., denser skin and therefore, a more selective membrane:

—a higher initial polymer concentration of the solution will favour the conditions for a large supersaturation in the top layer before nucleation sets in;

—a lower tendency of the nonsolvent to induce L–L separation in the system or to penetrate the cast film will delay the onset of gelation until sufficient solvent depletion has been obtained and again supersaturation is favoured. A proper choice of nonsolvent type and the use of certain additives in the coagulation bath (salt, glycerin, etc.) will serve these purposes;

—lowering the temperature of the coagulation bath will increase supersaturation while decreasing growth kinetics for the nuclei.

The skin is formed due to a high outflow of the solvent. When the solvent used is added to the nonsolvent bath this outflow is decreased as well as the rate of penetration of the nonsolvent. It appears that in every ternary system, which yields skinned membranes when the coagulation bath is pure nonsolvent, microporous membranes without a skin are obtained if sufficient solvent is added to the coagulation bath (25). Each system has its own minimum solvent concentration, which can be more than 80 percent by weight of solvent. On the basis of thermodynamic evaluations it seems plausible that the addition of solvent to the bath relatively favours the penetration of nonsolvent into the polymer solution (25) and thus favours L–L phase separation in the top layer.

Under the skin we find a porous substructure (see Figure 6). It will be clear that the pores in this sponge structure are the grown-out nuclei of the dilute phase dispersed within the matrix of the polymer solution, and that the matrix has solidified by gelation at a certain stage after L–L separation. If the concentration of the polymer at the locus where L–L phase separation sets in does not increase too much with depth into the solution layer, the nucleation density will also not vary very much across the film and a uniform pore structure will be the result. In order to obtain an "open pore" sponge structure, a certain amount of coalescence of the drops should occur before the walls of concentrated polymer solution between the pores solidify by gelation.

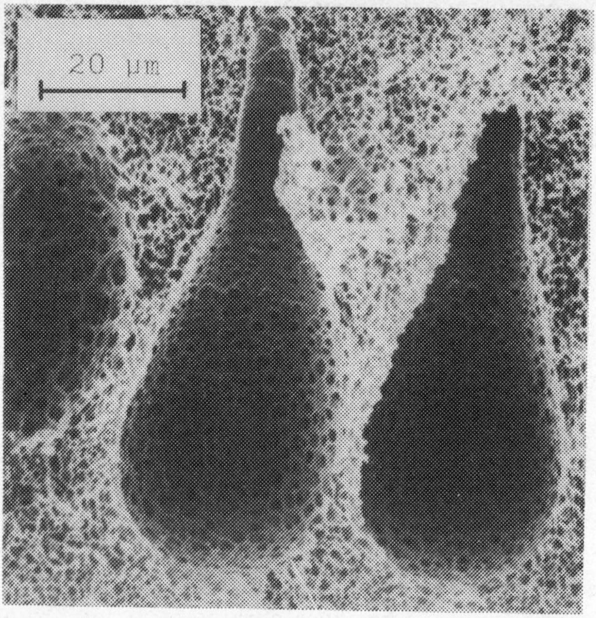

FIGURE 6.—Cross-section of a PSF membrane, showing voids with open walls.

This can be monitored by choosing the proper initial polymer concentration (not too high), or by adding a nonsolvent to the polymer solution. Bokhorst (32) obtained membranes with an open, very regular sponge structure by adding maleic acid to the cellulose acetate–dioxane casting solution. The absence of very small pores directly under the skin layer makes these membranes wet–dry reversible (in small pores the capillary forces are large upon drying).

A very important feature in immersion-coagulated membranes (in UF membranes as well as in RO types) is the presence of the large voids with a length from several microns to, sometimes, the total thickness of the membrane. The cavities were first observed in wet-spun fibres (35–39). The absence of the voids in dry-spun fibres (38) and in dry RO membranes (34) indicates that their formation only occurs in the case of immersion precipitation.

At first it was assumed that the formation was associated with volume changes in the precipitated polymer phase (36). Later Craig (35) proposed the mechanism of penetration of nonsolvent through defects (cracks) on the surface of the spinning filament. The same mechanism in the case of immersion-coagulated membranes was proposed by Strathmann (30). In a systematic study of the kinetics of void formation in PAN fibre spinning, Gröbe (37) came to the conclusion that diffusion of both solvent and nonsolvent to certain areas was the basis for void formation. In our opinion the available experimental data are consistent with this mechanism. Some typical facts remain interesting for further consideration:

—through optical microscopy one observes that the voids move inward faster than the diffusion controlled coagulation front;

—the boundary of the voids does not solidify by gelation during its growth, since coalescence with small pores remains possible (Figure 6);

—often the growth of the voids slows down and the coagulation front for L–L separation passes beyond the lower end of the cavity.

Smolders (40) has proposed the following mechanism for the appearance of the conical voids:

—When the skin is formed, nonsolvent penetrates into the underlying polymer solution faster at certain spots in the skin, e.g., a thinner part of the skin or a local loose arrangement of the structural units in the skin, giving a more favourable pathway for diffusion. Only in systems with a large driving force for solvent/nonsolvent mixing is this heterogeneous type of nucleus formed.

—Solvent is expelled from the surrounding polymer solution (syneresis) to these statistically spread loci and a gradient in nonsolvent concentration is set up in the void, ranging from a rather low value near the interface with the polymer solution to a high value near the skin surface. There is a fluid interface between the polymer solution and the void.

—Because of the syneresis effect and the more rapid diffusion in the void in comparison with the polymer solution phase, the void may grow faster initially than the coagulation front proceeds. When the syneresis becomes less effective the growth of the voids depends solely on the diffusion of solvent over larger distances to the voids, so that the growth of the voids may slow down and the coagulation front may proceed beyond the void.

Up to this point, we have focussed on the preparation procedures rather than on the compounds used in the procedure. There are countless combinations of a polymer, a solvent and a nonsolvent which constitute a membrane-forming system, but only some will yield membranes with commercially interesting properties. For a typical polymer widely used in membrane fabrication, cellulose acetate, the influence of the choice of the solvent on membrane properties is illustrated in Table 1.

TABLE 1. Influence of solvent used in membrane preparation on membrane properties. Polymer: cellulose acetate, nonsolvent: water.

Solvent	Membrane structure and properties
Dioxane	Skinned, Reverse Osmosis
Acetone	Homogeneous, Dense
Acetone/Formamide	Skinned, Reverse Osmosis
Triethylphosphate	Homogeneous, Finely Porous
N,N–Dimethylformamide	Skinned, Ultrafiltration

5.4. Thermal Precipitation

The polymer is dissolved at a high temperature in a mixture of a solvent and a nonsolvent. Upon lowering the temperature the solution becomes unstable with respect to L–L phase separation and a porous structure is formed. If, at the solution/vapour interface, concentration of the polymer through evaporation takes place, a skin layer might be the result. A mixture of a polymer and low molecular weight compounds is not always employed. Polymer blends, miscible at high temperatures, will also yield porous structures if the temperature is decreased. In that case one polymer component, which is dispersed in the matrix, is washed away. This procedure is used in the dry-spinning of porous hollow fibers.

6. CONCLUSIONS

It has been shown that in membrane-forming systems the skin is formed by gelation of the top layer, at increased polymer concentration due to solvent loss, whereas liquid–liquid phase separation is responsible for the formation of the porous sublayer. The diffusive exchange of solvent and nonsolvent determines at which place in the polymer solution and at what time these demixing processes take place. Although in different systems the membrane structure and properties may differ, the basic phenomena responsible for the membrane formation are the same.

REFERENCES

1. Bechold, H.: 1907, "Investigations on colloids by filtration method." Biochem. Z. 6, pp. 379–391
2. Loeb, S. and Sourirajan, S.: 1962, "Sea water demineralization by means of an osmotic membrane." Advan. Chem. Ser., pp. 117–132
3. Kesting, R.E.: 1971, "Synthetic Polymeric Membranes." McGraw Hill, New York.
4. Zsigmondy, R. and Bachmann, W.: 1918, "Ueber neue filter." Z. Anorg. Allgem. Chem. 103, pp. 119–128
5. Strathmann, H. and Kock, K.: 1977, "The formation mechanism of phase inversion membranes." Desalination 21, pp. 241–255
6. Ferry, J.D.: 1936, "Ultrafilter membranes and ultrafiltration." Chem. Rev. 18, pp. 373–455
7. Maier, K. and Scheuermann, E.: 1960, "Ueber die Bildungsweise teildurchlässiger Membranen." Kolloid. Z. 171, pp. 122–135
8. Kesting, R.E.: 1973, "Concerning the microstructure of Dry–RO membranes." J. Appl. Polym. Sci. 17, pp. 1771–1785
9. a. Sourirajan, S.: 1970, "Reverse Osmosis," Acad. Press, New York
   b. Kesting, R.E.: 1971, "Synthetic Polymeric Membranes," McGraw Hill, New York
   c. Lonsdale, H.K. and Podall, H.E., eds.: 1972, "Reverse Osmosis Membrane Research," Plenum Press, New York
10. Guillotin, M., Lemoyne, C., Noel, C. and Monnerie, L.: 1977, "Physicochemical processes occurring during the formation of cellulose diacetate membranes. Research of criteria for optimizing membrane performance. IV. Cellulose diacetate - acetone organic additive casting solutions." Desalination 21, pp. 165–181

11. Koenhen, D.M., Mulder, M.H.V. and Smolders, C.A.: 1977, "Phase separation phenomena during the formation of asymmetric membranes." J. Appl. Polym. Sci. 21, pp. 199–215
12. Tanny, G.B.: 1974, "The surface tension of polymer solutions and asymmetric membrane formation." J. Appl. Polym. Sci. 18, pp. 2149–2163
13. Flory, P.J.: 1953, "Principles of Polymer Chemistry." Cornell University Press, Ithaca
14. Koningsveld, R.: 1967, Thesis, Leiden University
15. Flory, P.J.: 1965, "Statistical thermodynamics of liquid mixtures." J. Amer. Chem. Soc. 87, pp. 1833–1838
16. Wisniak, J. and Tamir, A.: 1978, "Mixing and Excess Thermodynamic Properties," Elsevier, Amsterdam
17. Cahn, J.W.: 1965, "Phase separation by spinodal decomposition." J. Chem. Phys. 42, pp. 93–99
18. Smolders, C.A., Van Aartsen, J.J. and Steenbergen, A.: 1971, "Liquid–liquid phase separation in concentrated solutions of noncrystallizable polymers by spinodal decomposition." Kolloid Z. Z. Polym. 243, pp. 14–20
19. van Emmerik, P.T. and Smolders, C.A.: 1973, "Differential scanning calorimetry of poly (2,6 dimethyl–1, 4 phenylene oxide)-toluene solutions." Eur. Polym. J. 9, pp. 293–300
20. Derham, K.W., Goldsbrough, J. and Gordon, M.: 1974, "Pulse Induced Critical Scattering (PICS) from polymer solutions." J. Pure and Appl. Chem. 38, pp. 97–107
21. Altena, F.W. and Smolders, C.A.: 1982, "Calculation of liquid–liquid phase separation in a ternary system of a polymer in a mixture of a solvent and a nonsolvent." Macromolecules 15, pp. 1491–1497
22. Altena, F.W. and Smolders, C.A.: 1981, "Phase separation phenomena in solutions of cellulose acetate. I. Differential scanning calorimetry of cellulose acetate in mixtures of dioxane and water." J. Polym. Sci., Polym. Symp. 69, pp. 1–10.
23. Koenhen, D.M., Smolders, C.A. and Gordon, M.: 1977, "Phase separation phenomena in solutions of poly (2,6 dimethyl–1, 4 phenylene oxide). III. Pulse induced critical scattering in solutions of toluene." J. Polym. Sci., Polym. Symp. 61, pp. 93–100
24. Panar, M., Hoehn, H.H. and Herbert, R.R.: 1973, "The nature of asymmetry in reverse osmosis membranes." Macromolecules 6, pp. 777–780
25. Wijmans, J.G., Baaij, J.P.B. and Smolders, C.A.: 1983, "The mechanism of formation of microporous or skinned membranes produced by immersion precipitation." J. Membr. Sci. 14, pp. 263–274
26. Sourirajan, S. and Kunst, B.: 1977, in "Synthetic Membranes," Sourirajan, S., ed., National Research Council, Canada, Ottawa, pp. 129–152
27. Kunst, B. and Vajnaht, Z.: 1974, "On the structure of concentrated cellulose acetate solutions." J. Appl. Polym. Sci. 21, pp. 2505–2514
28. Anderson, J.E. and Ullman, R.: 1973, "Mathematical analysis of factors influencing the skin thickness of asymmetric reverse osmosis membranes." J. Appl. Phys. 44, pp. 4303–4311
29. Sarbolouki, M.N.: 1973, "Preparation of the skinned membranes without evaporation step." J. Polym. Sci., Polym. Lett. 11, pp. 753–754
30. Strathmann, H., Kock, K., Amar, P. and Baker, R.W.: 1975, "The formation mechanism of asymmetric membranes." Desalination 16, pp. 179–203
31. Frommer, M.A. and Lancet, D.: 1972, in "Reverse Osmosis Membrane Research," Lonsdale, H.K. and Podall, H.E., eds., Plenum Press, New York, pp. 349–360
32. Bokhorst, H., Altena, F.W. and Smolders, C.A.: 1981, "Formation of asymmetric cellulose acetate membranes." Desalination 38, pp. 349–360
33. Mulder, M.H.V., Kruitz, F. and Smolders, C.A.: 1982, "Separation of isomeric xylenes by pervaporation through cellulose ester membranes." J. Membr. Sci. 11, pp. 349–363
34. Kesting, R.E.: 1975, United States Patent 3 884 801, May 20
35. Craig, J.P., Knudsen, J.P. and Holland, V.F.: 1962, "Characterization of acrylic fiber structure." Textile Res. J. 32, pp. 435–448
36. Gröbe, V. and Meyer, K.: 1959, "Acrylic fibers. XXII. Influence of precipitant on fiber formation in the wet spinning of poly-acrylo-nitrile." Faserforsch. Textiltechn. 10, pp. 214–224

37. Gröbe, V., Mann, G. and Duve, G.: 1966, "Structure formation in coagulating polyacrylonitrile." Faserforsch. Textiltechn. 17, pp. 142–147
38. Ziabicki, A.: 1976, "Fundamentals of Fibre Formation," John Wiley, London
39. Knudsen, J.P.: 1963, "The influence of coagulation variables on the structure and physical properties of an acrylic fibre." Textile Res. J. 33, pp. 13–20
40. Smolders, C.A.: 1980, in "Ultrafiltration Membranes and Applications," Cooper, A.R., ed., Plenum Press, New York

SYMBOLS

ΔG_m	Gibbs free energy of mixing
ϕ	polymer volume fraction

Subscripts

1	nonsolvent
2	solvent
3	polymer

TRANSPORT PRINCIPLES—SOLUTION, DIFFUSION AND PERMEATION IN POLYMER MEMBRANES

Geoffrey S. Park

Department of Applied Chemistry
University of Wales Institute of Science and Technology
P.O. Box 13, Cardiff CF1 3XF, Wales, United Kingdom

The concept of the diffusion coefficient, D, and Fick's Laws are considered. Measurements of D and of solubility, s, of gases in polymers are outlined. The factors influencing D and s of gases in polymers are discussed. Methods for obtaining D and s of organic vapours in elastomeric polymers are given. The effects of concentration, temperature, diffusant size and polymer properties on D are correlated using free volume ideas. Equilibrium vapour uptake is discussed. Vapour sorption anomalies in glassy polymers and the reasons for these are outlined. D, s, and permeability, P, of water in both hydrophobic and hydrophilic polymers are considered and interpreted in terms of hydrogen bonding to polymer sites and water clustering by hydrogen bonding. D, s, and P in non-uniform polymers are considered and discussed in terms of the effects of fillers, crystallites and layer structures.

P. M. Bungay et al. (eds.), Synthetic Membranes: Science, Engineering and Applications, 57–107.

1. BASIC IDEAS ON DIFFUSION

1.1. Membrane Function

Membrane action is concerned with the exchange of matter or energy between two environments that are separated by the membrane, or between the membrane and the environment in which it is immersed. It is mainly the exchange of matter that we will be considering here. This might, for instance, be the release of a drug from a drug release membrane, it could be the loss of solute from a membrane to the environment, or, in separation processes, the transmission of one solute and the non-transmission of another from one environment to a second one.

1.2. Equilibrium and Dynamic Aspects

The interchange between a membrane and the environment can be considered under two headings. The first of these is concerned with the equilibrium set up between the material in the environment (gas, vapour, solvent, solute, ions, drugs, herbicides, etc.) and the membrane surface. This, then, will be a thermodynamic consideration and in the first instance will be expected to lead more or less instantaneously to a constant surface condition. The second consideration is that of kinetic transport, which is concerned with the transmission of material from the surface into the bulk of the membrane or in the reverse direction. Alternatively, this might also lead to the transmission of material from one surface to the other and, hence, from one environment to the other separated by the membrane. In this chapter it will be assumed that transport within the membrane occurs by diffusion. Diffusion can be roughly defined as the process that tends to remove differences of concentration by means of random molecular motions.

1.3. The Diffusion Coefficient

Both equilibrium distribution and diffusion are important but it is the diffusion process that is most varied in membrane behaviour, so as a first instance it is necessary to establish a rate constant for diffusion and to consider how this can be used in considering simple solutes in polymeric membranes. The simplest law of diffusion is Fick's law (1)

$$J_i = -D_i \mathrm{d}c_i/\mathrm{d}x \tag{1}$$

Here, J_i is the flux of species i in terms of the amount passing in unit time through the unit area of section in the direction of the gradient of concentration, $\mathrm{d}c_i/\mathrm{d}x$. D_i will usually have the dimensions of length^2 × time^{-1} (m^2 s^{-1} in SI, or cm^{2}s^{-1} in CGS) and the flux and concentration must involve the same measure of the amounts of material (mass, moles, number of molecules, etc.).

The quantity, D_i, is the diffusion coefficient and can be regarded as the rate constant for diffusion. The concentration gradient, $\mathrm{d}c_i/\mathrm{d}x$, could be thought of as the driving force for diffusion but it is not, in fact, a force; a simple derivation of a relationship for diffusive flow in terms of a driving force is as follows. Consider two planes at a distance, $\mathrm{d}x$, apart. If the mean energy per molecule in one plane is q_i, and in the other is $q_i + \mathrm{d}q_i$, then the work involved in moving a molecule the distance, $\mathrm{d}x$, between one plane and the next will be $\mathrm{d}q_i$. It follows that the external driving force involved in the transfer must be $\mathrm{d}q_i/\mathrm{d}x$. In terms of simple thermodynamic ideas, the mean energy per mole in a uniform system is the partial molar Gibbs function, $\tilde{G}_i$, or chemical potential, μ_i. It follows that the mean energy per molecule is $\mu_i/\tilde{N}$, where $\tilde{N}$ is Avogadro's number. If this energy were dissipated in moving a molecule down the chemical potential gradient, the driving force, X_i, per molecule of species, i, would be:

$$X_i = -(1/\tilde{N})\, \mathrm{d}\mu_i/\mathrm{d}x \tag{2}$$

Assuming a velocity-dependent frictional resistance, f_i, to motion (having the nature of a viscous resistance), the mean molecular velocity, v_i, from one plane to the next will be

$$v_i = X_i/f_i \tag{3}$$

So,

$$v_i = -(\mathrm{d}\mu_i/\mathrm{d}x)/(\tilde{N}f_i) \tag{4}$$

If the molar concentration of species, i, is c_i, then the number of moles passing through unit area in unit time, J_i, will be

$$J_i = v_i c_i \tag{5}$$

and so

$$J_i = -(1/\tilde{N}f_i)\, c_i\, (d\mu_i/dx) \tag{6}$$

For an ideal solution,

$$\mu_i = \mu_i^o + RT \ln c_i \tag{7}$$

and so

$$J_i = -(RT/\tilde{N}f_i)\, c_i (d \ln c_i/dx) \tag{8}$$

and hence,

$$J_i = -(RT/\tilde{N}f_i)\, dc_i/dx \tag{9}$$

Comparing this with Fick's law, the value for the diffusion coefficient is given by

$$D_i = RT/\tilde{N}f_i \tag{10}$$

The driving force per molecule for diffusion in an ideal solution is thus given by

$$X_i = -(RT/\tilde{N}c_i)\, dc_i/dx \tag{11}$$

and not by dc_i/dx on its own. If the solution in the membrane is not ideal, then

$$\mu_i = \mu_i^o + RT \ln a_i \tag{12}$$

where a_i is the thermodynamic activity. It then follows that

$$J_i = -(RT/\tilde{N}f_i)\, (d\ln a_i/d\ln c_i) \cdot dc_i/dx \tag{13}$$

so that the diffusion coefficient is given by

$$D_i = (RT/\tilde{N}f_i)\, (d\ln a_i/d\ln c_i) \tag{14}$$

or

$$D_i = (RT/\tilde{N}f_i)\, [1 + (c_i/\gamma_i)\, (d\gamma_i/dc_i)] \tag{15}$$

where γ_i is the activity coefficient of species i.

In the non-ideal situation, the quantity $RT/\tilde{N}f_i$ is sometimes called the thermodynamic diffusion coefficient, D_T. Causes other than concentration changes can produce variations in μ_i and so equation 6 can also be used to predict relations between mass transport and temperature gradients, between mass transport and hydrostatic pressure, and so on, but at this stage we are concerned only with transport resulting from concentration differences.

1.4. Fick's Second Law

Fick's first law can easily be used to derive the second order differential equation between concentration, time and distance that goes under the name of Fick's second law (2)

$$\partial c/\partial t = \partial (D\partial c/\partial x)/\partial x \tag{16}$$

(For simplicity the subscript i is now omitted.)

Consider the slice of thickness $2\mathrm{d}x$ between two unit area planes. Then the rate at which diffusing substance enters the slice through the first plane at $x-\mathrm{d}x$ is given by $J-(\partial J/\partial x)\mathrm{d}x$, where J is the rate of transfer through unit area of plane at the centre of the slice. Similarly, the rate of loss of diffusing substance through the other face is given by $J+(\partial J/\partial x)\mathrm{d}x$. It is clear that the contribution to the rate of increase of diffusing substance in the slice from these two faces is equal to $-2\mathrm{d}x(\partial J/\partial x)$, but in terms of the change of concentration with time in the slice, the rate of increase of diffusing substance is $2\mathrm{d}x(\partial c/\partial t)$ and hence

$$\partial c/\mathrm{d}t + \partial J/\partial x = 0 \tag{17}$$

It is clear that J is given by Fick's first law, equation 1, and so on substitution, equation 16 is obtained. This one-dimensional expression of Fick's second law can be integrated with any appropriate boundary conditions to derive flux rates into and out of membranes, to obtain concentration/distance relationships in membranes, and so on, in many different situations (2,3). In heterogeneous membranes, fluxes parallel to the membrane faces might also be important and then the three-dimensional form of equation 16 is useful:

$$\partial c/\partial t = \partial(D\partial c/\partial x)/\partial x + \partial(D\partial c/\partial y)/\partial y + \partial(D\partial c/\mathrm{d}z)/\mathrm{d}z \tag{18}$$

1.5. Measurement of Diffusion Coefficients from Permeation through Membranes

Fick's first law gives the simplest technique for obtaining the diffusion coefficient of small molecules in membranes (4). This is most simply considered in terms of the diffusion of a gas in a membrane. One side of the membrane, of thickness h, is contacted with the diffusant at constant activity, a', so that eventually a constant concentration, c_1, is established in the surface of the membrane. The other surface of the membrane is maintained at zero activity and hence zero concentration and the number of moles, n, of issuing diffusant is measured from the zero concentration face as a function of time. When steady conditions have been reached, Fick's first law shows the issuing flux, J, of diffusant is given by

$$J = (1/A)\mathrm{d}n/\mathrm{d}t = \mathrm{D}c_1/h \tag{19}$$

Knowing the membrane area, A, the thickness h, and the surface concentration c_1, one can easily calculate the diffusion coefficient. This technique is particularly useful for studying the diffusion of gases in membranes; here the activity and the concentration in the membrane surface is controlled by the pressure p' of gas contacting the membrane surface. c_1 cannot be measured directly and in the steady state, the permeability coefficient, P, is obtained when the pressure is used instead of the concentration.

$$P = Jh/p' \tag{20}$$

It is easy to see that the solubility coefficient, $s = c/p'$, of the gas enables the diffusion coefficient to be obtained from P,

$$P = Ds \tag{21}$$

and so in determining diffusion coefficients from steady state permeation, the solubility coefficient, s, has to be obtained in a separate experiment. It can be shown, however, that permeation rate experiments enable D to be obtained from the rate of approach to steady state conditions. For an initially gas-free membrane and with one surface maintained at zero concentration of gas throughout while the other is raised to concentration $c = c_1$ at time $t = 0$, solution of equation 16 yields the relationship

$$n/(Ahc_1) = Dt/h^2 - 1/6 + (2/\pi^2)\sum_{\mathrm{m}=1}^{\mathrm{m}=\infty} [(-1)^2/\mathrm{m}^2] \exp(-D\mathrm{m}^2\pi^2 t/h^2) \tag{22}$$

where m is an integer.

This has the form shown in Figure 1 and at large values of t, it reduces to

$$n/A = (Dc_1/h)(t - h^2/6D) \tag{23}$$

This linear relationship extrapolates back to cross the n/A axis at a time t', the time lag, given by

$$t' = h^2/6D \tag{24}$$

This time lag is independent of the solubility, s. It enables D to be obtained directly from permeation experiments (5) and then c_1 and hence s can be

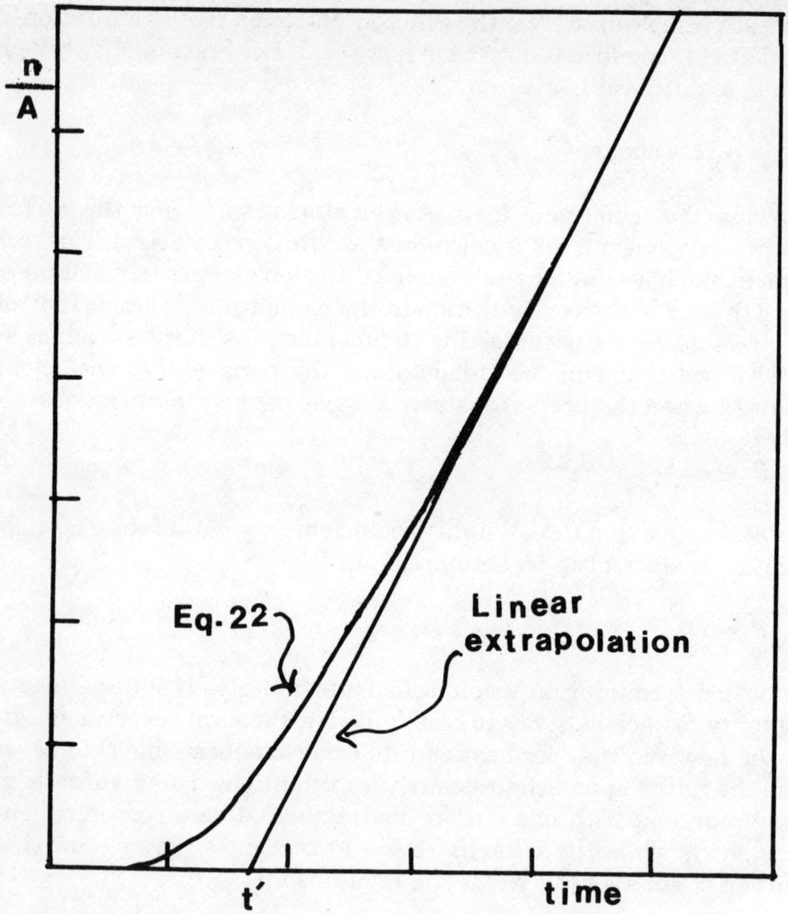

FIGURE 1.—Time lag for the establishment of steady state.

obtained from the steady state using equations 20 and 21. Another solution of equation 16 gives

$$J = (1/A)\mathrm{d}n/\mathrm{d}t = 2c_1(D/\pi t)^{1/2} \sum_{\mathrm{m}=1}^{\mathrm{m}=\infty} \exp\left[-(2\mathrm{m}+1)^2 h^2/(4Dt)\right] \quad (25)$$

At low values of t this can be written as

$$\ln(Jt^{1/2}) = \ln[2c_1(D/\pi)^{1/2}] - h^2/4Dt \quad (26)$$

and hence values for D can be obtained from the slope of plots of ln $Jt^{1/2}$ against $1/t$ (6).

To obtain the diffusion of gases in membranes from the principles given above, two kinds of experimental assemblies have been used. One is illustrated diagrammatically in Figure 2 (7). This is a vacuum system in which the only gas is the one being investigated. Here, the constant pressure, p', is measured by a manometer or pressure transducer and the quantity of gas flowing out at the zero concentration face of the membrane is obtained from the small pressure increase, δp, in an evacuated known volume. Problems associated with this technique include the provision of mechanical support for the membrane, which leads to uncertainty of membrane area and provision of vacuum tight seals. For this and other reasons, techniques involving carrièr gases (helium) have been used (8). This enables a situation with no total pressure differential to be used and the kind of detection methods commonly used in gas chromatography can be employed on the outgoing side of the membrane. In this technique, both the steady state and the approach to the steady state can be investigated, but problems brought about by the finite

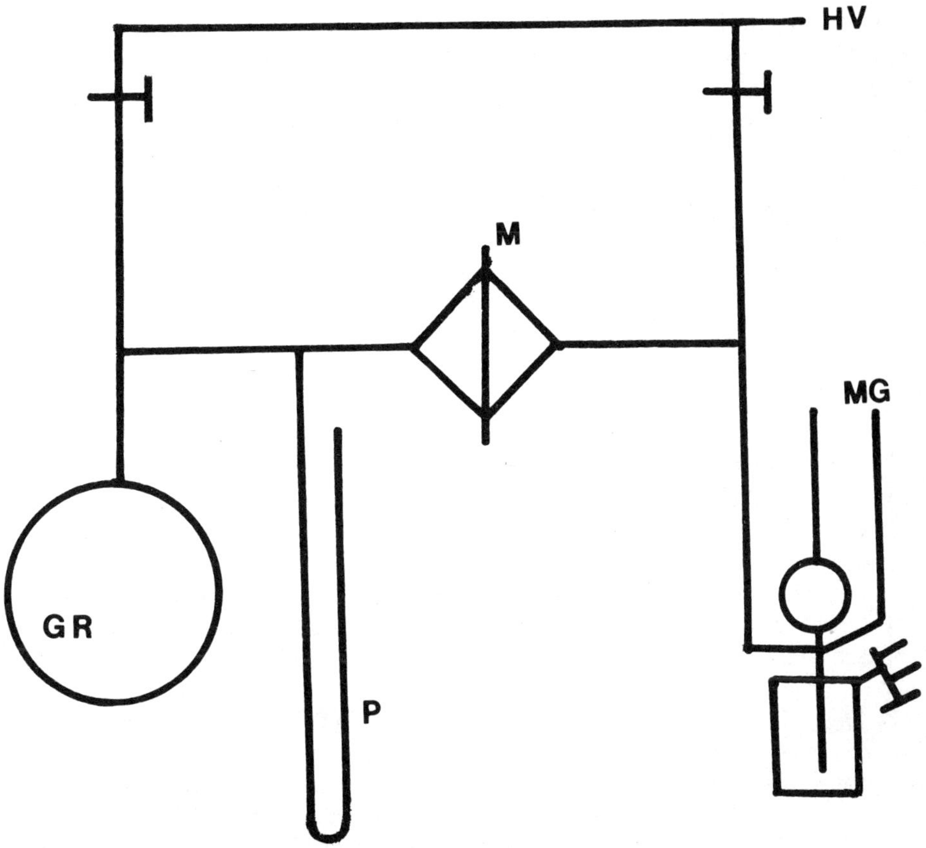

FIGURE 2.—Gas permeability apparatus. GR, gas reservoir; HV to high vacuum pumps; M, membrane in holder; MG, Macleod gauge (or Pirani); P, manometer or pressure transducer.

flow rate of carrier gas and diffusion through the carrier gas layer at the surface of the membrane have to be dealt with. Since the flux, J, is measured directly, the time t'' required for J to reach half its steady state value enables D to be estimated from (9)

$$t'' = h^2/(7.199D) \tag{27}$$

When the permeability principle is applied to diffusion of substances that are solutes in condensed phases in contact with the membrane surfaces, diffusion through the solvent layer may cause problems and time lags tend to be unreliable unless special precautions are taken (10).

2. NOBLE GAS DIFFUSANTS AS MODELS FOR DIFFUSION AND PERMEATION IN MEMBRANES

The monatomic noble gas molecules, helium, neon, argon, krypton, xenon, are probably the simplest substances to permit a few general principles to be established for diffusion in polymeric membranes.

2.1. Dependence of Diffusion Coefficient on Diffusant Size

For a spherical body of radius, b, moving through a continuum of viscosity, η, Stoke's law indicates that the frictional resistance of equation 3 is

$$f_i = 6\pi\eta b \tag{28}$$

and so

$$D_T = RT/6\pi\eta\tilde{N}b \tag{29}$$

This relationship is moderately successful for large rigid molecules diffusing in low-molecular-weight solvents. In some of these systems an inverse relationship between D_T and b holds with reasonable accuracy. For the simple noble gases diffusing in polymers, however, the diffusion coefficient is far more sensitive to the molecular radius than is predicted by equation 29. Figure 3 gives the logarithmic plot of the diffusion coefficients of four noble gases against ln b for diffusion in a polyethylmethacrylate membrane at 25°C. Here, the relationship obtained is $D \propto b^{-17}$. Furthermore, viscosity values calculated from equation 29 fall between 0.0052 poise for helium and 18.6 poise for krypton, whereas the bulk viscosity of the polymer is about 10^{10} poise. This is because gas diffusion is essentially measuring the microviscosity involved in the small local motion round the diffusing molecule instead of overall flow of the polymer system. This leads to the idea of disturbance in a small local zone of the polymer with the zone size depending upon the size of the diffusing molecule. It fits in well with the zone theory of

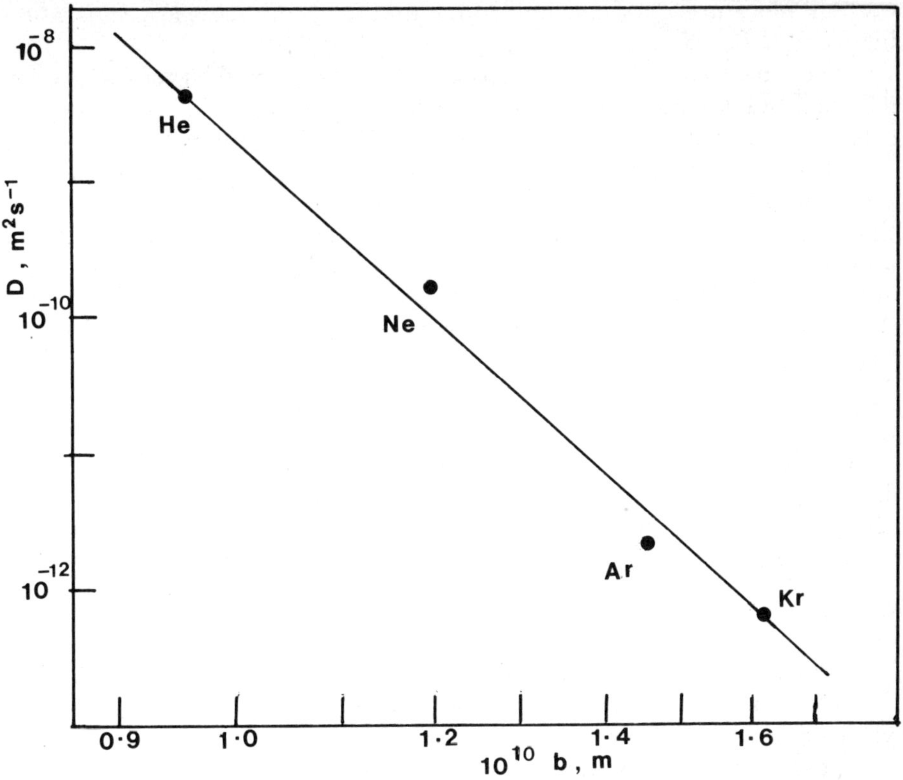

FIGURE 3.—Diffusion coefficient, D, in polyethylmethacrylate at 25°C as a function of atomic radius, b.

diffusion (13) that has been extensively used in discussing the effects of temperature on the diffusion coefficient.

2.2. Effect of Temperature

Diffusion is an activated process so that the Arrhenius activation energy, $\Delta\tilde{E}_a$, is defined by

$$\Delta\tilde{E}_a = -R \, d \ln D/d(1/T) \tag{30a}$$

which for $\Delta\tilde{E}_a$ independent of T gives the linear relationship

$$D = D_o \exp(-\Delta\tilde{E}_a/RT) \tag{30b}$$

This relationship works well for many simple gases in polymers, and $\Delta\tilde{E}_a$ values fall between about 5 and 200 kJ mol^{-1}. In some cases, curved Arrhenius plots are obtained indicating that $\Delta\tilde{E}_a$ falls as the temperature increases (12).

The zone theory of diffusion accounts for this. From the previous section it is clear that the microviscosity needed for equation 29 involves a zone of the polymer that increases rapidly as the size of the diffusing molecule goes up. Barrer (13), following Wheeler's ideas (14) on viscosity, assumed that a large zone of polymer was involved in each unit step of the gas diffusion process so that vibrational motion of several polymeric atoms was involved in each unit diffusion step. This implies that the activation energy, $\Delta\tilde{E}_z$, will be distributed over several f vibrational degrees of freedom. In unimolecular gas reaction theory for activation with several degrees of freedom, the Arrhenius expression is replaced by

$$D = [\nu\lambda^2\sigma/(f-1)!]\,(\Delta\tilde{E}_z/RT)^{(f-1)}\exp(-\Delta\tilde{E}_z/RT) \tag{31}$$

Here, ν is the vibrational frequency of the diffusant, λ is the average jump distance of the diffusant, and σ is the probability that the f degrees of vibrational freedom will collaborate in the diffusional step. It follows from equation 31 that the Arrhenius activation energy is given by

$$\Delta\tilde{E}_a = -R\,\mathrm{dln}\,D/\mathrm{d}(1/T) = \Delta\tilde{E}_z - (f-1)RT \tag{32}$$

This accounts well for the curved Arrhenius plots and from the variations in Arrhenius activation energy obtained, f values of 10–20 or even greater are involved in the diffusion zone.

There are other ways of thinking about the variation of Arrhenius activation with temperature, including the free volume approach to diffusion, which will be discussed when the diffusion of larger organic diffusants is considered.

2.3. A Useful Empirical Relationship Between Diffusion Coefficient and Activation Energy for Diffusion

For many activated processes, an empirical correlation between the pre-exponential term and the activation energy exists. For gas diffusion in polymer systems the relationship

$$\ln(D_o/\mathrm{m^2s^{-1}}) = -18.9 + 0.626\Delta\tilde{E}_a/RT \tag{33}$$

was suggested by Barrer and Chio (15). This is reasonably accurate, not only for simple gases, but also for the diffusion of more complicated molecules. It is particularly useful since combination with equation 30 gives

$$\ln(D/\mathrm{m^2s^{-1}}) = -18.9 - 0.374\Delta\tilde{E}_a/RT \tag{34}$$

which enables activation energies to be estimated when the diffusion coefficient is known at only one temperature. Equally the diffusion coefficient over a range of temperatures can be estimated from one observation at one temperature.

2.4. Glass Transition Effects

Non-crystalline polymers exist in two forms: (a) a hard glassy form in which molecular motions are extremely limited and microvoids are frozen into the system, and (b) at higher temperatures above a critical glass transition temperature, T_g, a leathery, elastomeric or liquid-like state, in which long-range molecular motions are possible and permanent voids do not exist. Although in many cases reasonable linear Arrhenius plots are obtained continuously through the transition temperature, T_g, between these two zones, in some cases a break occurs in the region of T_g with a lower activation energy in the glassy state than in the region above T_g [Figure 4; (16)]. Although there are variations of detail from one system to another, it appears that the lower activation energy below T_g arises from diffusion through pre-existing cavities, whereas the energy of cavity formation could account for the higher activation energies at higher temperatures. One of the most thorough investigations of this phenomenon is that of Meares (17), who showed that the transition from one state to another spreads over an appreciable temperature range. For the diffusion of noble gases in poly(vinyl acetate), he suggested that for a few degrees below T_g, islands of glassy material are being formed as the temperature drops and it is not until about 10° below T_g that a completely glassy state is obtained. This leads to a linear Arrhenius plot above T_g and a different linear plot starting about 10° below T_g joined by a somewhat indefinite region. As will be seen later, the diffusion of organic molecules in glassy polymers is subject to time effects, which produce considerable complication. Diffusion coefficients of gases obtained purely from steady state measurements can be quite different from those obtained from time lag measurements. One explanation of this may also be that the glassy state leads to time effects and the slow establishment of surface equilibrium or slow changes of the diffusion coefficient.

2.5. Nature of the Polymer

The magnitude of the diffusion coefficient for a given diffusant varies very much from one polymer to another. Thus, the diffusion coefficient of krypton in silicone rubber at 25°C is 1.3×10^{-9} m^2s^{-1} (15), whereas in poly(vinyl acetate) at the same temperature, the value is 1.9×10^{-13} m^2s^{-1} (17) and the activation energy increases from 12 kJ mol^{-1} in the silicone to 61 kJ mol^{-1} in the poly(vinyl acetate). This effect, which can be correlated with a considerable drop in the value of T_g, is much greater for larger diffusants and will be considered when the diffusion of organic molecules in polymer membranes is discussed.

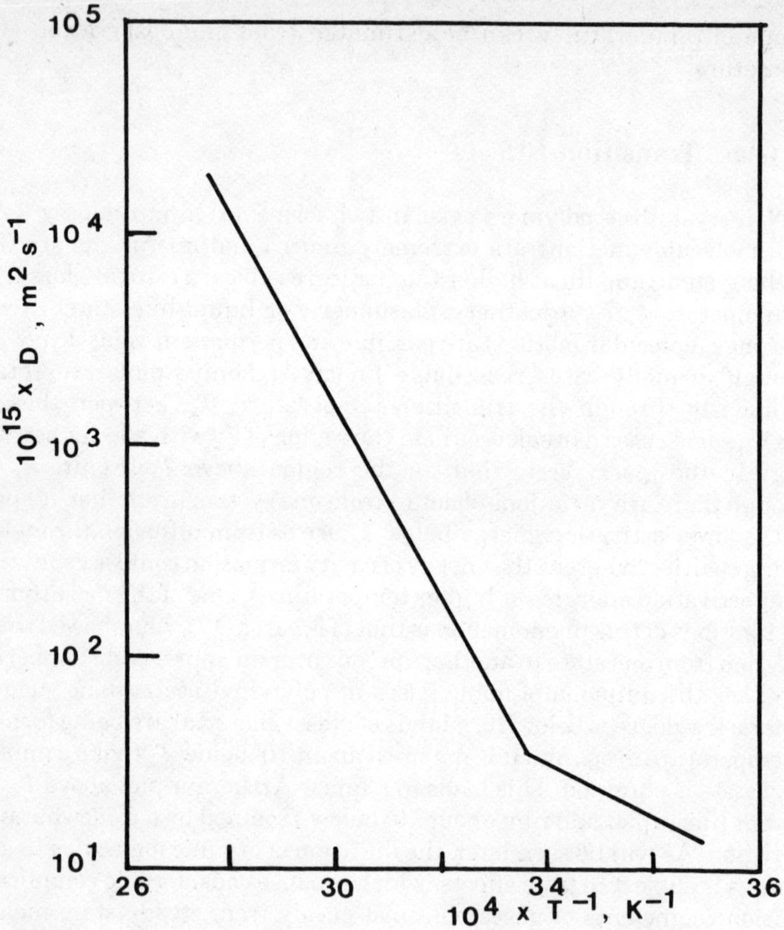

FIGURE 4.—Break in Arrhenius plot for CO_2 in a vinyl acetate/vinyl chloride copolymer. Reproduced with permission from Kumins, C.A. and Roteman, J.: 1961, "Diffusion of gases and vapors through polyvinyl chloride - polyvinyl acetate copolymer films. I. Glass transition effect." J. Polymer Sci. 55, p. 683. John Wiley and Sons Inc., Copyright owners.

3. THE SOLUBILITY AND PERMEABILITY COEFFICIENTS OF GASES IN POLYMER MEMBRANES

The solubility coefficient, s, is the equilibrium constant expressing the distribution of the gas molecules between the gas phase and the polymer phase.

$$s = c/p' \tag{35}$$

Both the diffusion coefficient, D, and the solubility coefficient, s, govern the permeability of a membrane to the gas (equation 21). Although variations in D are much greater than the variations in s, it is interesting to see how s varies with different conditions. The solubility can be measured directly from equilibrium measurements in which a polymer sample is allowed to equilibrate with gas at a known pressure and then the volume of gas that desorbs into a vacuum is obtained. Another much used method is by combination of steady state and time lag or some other early time transient permeation rate (equations 24, 26, and 27).

3.1. Effect of Temperature

Because s is an equilibrium constant, it follows that

$$s = \exp(\Delta\tilde{S}^{\circ}/R)\exp(-\Delta\tilde{H}^{\circ}/RT) \tag{36}$$

Here, $\Delta\tilde{S}^{\circ}$ and $\Delta\tilde{H}^{\circ}$ are the standard entropies and enthalpies of solution for the gas in the polymer. $\Delta\tilde{H}^{\circ}$ is often negative and generally falls between ± 10 kJ mol^{-1}. This is a small range compared with the range of the activation energy, $\Delta\tilde{E}_a$, for diffusion and, since from equations 21, 30 and 36

$$P \propto \exp[(-\Delta\tilde{H}^{\circ} - \Delta\tilde{E}_a)/RT] \tag{37}$$

the temperature dependence of the permeability, P, is usually mainly governed by the temperature dependence of the diffusion coefficient and P increases with temperature. It is, however, possible, when $\Delta\tilde{E}_a$ is small, that $-\Delta\tilde{H}^{\circ} > \Delta\tilde{E}_a$ and so in the case of some diffusants in polydimethylsiloxane membranes, the permeability decreases with increasing temperature (15).

3.2. Dependence of Solubility on Molecular Size

Because interaction of simple gas molecules with a polymer matrix tends to be rather small, the main factor governing s is the ease of condensation of the gas and so linear relations between ln s and the critical temperature of the gas, T_c, the normal boiling point, T_b, and the Lennard Jones 6–12 potential force constant, ε/K, have been proposed (18–20).

In general, these parameters all increase with molecular size and thus, solubilities increase with molecular size (Table 1). Usually, the increase of s with increasing size is much less than the decrease of D and so permeabilities decrease with increasing size of diffusant almost as much as the diffusion coefficient.

3.3. Dependence of Solubility on the Polymer

Relations proposed by both Gee (19) and by Michaels and collaborators (20) express variations in gas solubility on passing from one polymer to

TABLE 1. Condensation parameters and gas solubility in polymers

Gas	10^{10} *b*, m	T_b, K	T_c, K	εK^{-1}, K (15)	$10^2 s$, cm^3 (STP) $cm^{-3}atm^{-1}$	
					PEM (11)	PDMS (15)
He	0.95	4	5	10	0.12	4.3
Ne	1.18	27	44	35	1.5	9.0
Ar	1.44	87	151	121	21.5	34
Kr	1.60	121	209	165	44.0	98
Xe	1.76	166	290	221	—	400

b, atomic radius; T_b, normal boiling point; T_c, critical temperature, εK^{-1}, Lennard-Jones 6–12 potential force constant; *s*, gas solubility at 25°C; KPEM, poly(ethylmethacrylate); PDMS, poly(dimethylsiloxane).

another solely in terms of the Flory–Huggins interaction parameter, χ, so that

$$s \propto \exp(-\chi) \tag{38}$$

The changes in χ from one amorphous polymer membrane to another tend to be rather small and so *s* is far less dependent on the nature of the polymer than is the case for *D*. Thus, from a random selection of non-crystalline elastomeric polymers at 25°C (21), *D* for nitrogen increases by about 500-fold from polyethylmethacrylate (11) to polydimethylsiloxane (15), whereas the maximum range in *s* is only about sixfold, with the extremes being an isoprene/methacrylonitrile copolymer (12) and polydimethylsiloxane (15).

To summarize, it appears that for a wide variety of temperatures, gases or polymers, the changes in the solubility coefficient are much less than the changes in the diffusion coefficient and it can be concluded that variations in the diffusion coefficient are the main changes affecting changes in gas permeability.

4. ORGANIC MOLECULES IN ELASTOMERIC POLYMERS

4.1. Theory and Technique for Obtaining the Diffusion Coefficient

The methods for simple gases can be used with organic vapours but these larger diffusant molecules have low diffusion coefficients, which lead to low permeation rates and long lag times. These problems can be overcome by using thin membranes but these are fragile and clamping errors such as membrane leakage are liable to occur. For these reasons, the diffusion coefficients of organic molecules in polymer membranes have usually been obtained by other methods.

The most frequent technique depends upon measurement of vapour sorption kinetics by thin membranes immersed in the organic vapour at constant pressure. If it is assumed that the constant activity of the vapour leads to a constant surface concentration at both faces of the membrane, then the solution of Fick's second law for an initially diffusant-free membrane (or for a membrane containing a constant concentration of diffusant) enables concentration–distance relationships to be obtained and hence the two relationships

$$M_t/M_\infty = (16Dt/\pi h^2)^{1/2} \{1 + \pi^{1/2} \sum_{n=1}^{n=\infty} (-1)^n \, \text{ierfc}[n(h^2/4Dt)^{1/2}]\} \qquad (39)$$

$$M_t/M_\infty = 1 - (8/\pi^2) \sum_{n=0}^{n=\infty} [1/(2n+1)^2] \exp[-(2n+1)^2\pi^2 Dt/h^2] \qquad (40)$$

can be derived for the mass, M_t, of vapour absorbed by a membrane of thickness, h, at time, t (2). Here, M_∞ is the mass sorbed by the membrane at equilibrium, n is an index with integer values, and the integrated error function complement, ierfc x, is given by

$$\begin{aligned} \text{ierfc}\, x &= (4/\pi)^{1/2} \int_x^\infty \int_z^\infty \exp(-y^2) dy\, dz \qquad (41) \\ &= (1/\pi)^{1/2} [\exp(-x^2) - 2x \int_x^\infty \exp(-y^2) dy] \end{aligned}$$

When $M_t/M_\infty < 1/2$, equation 42 gives a very good approximation to equation 39, and when $M_t/M_\infty > 1/2$, equation 43 is a very good approximation to equation 40.

$$M_t/M_\infty = (16\, Dt/\pi\, h^2)^{1/2} \qquad (42)$$

$$\ln(M_\infty - M_t) = \ln(8\, M_\infty/\pi^2) - \pi^2 Dt/h^2 \qquad (43)$$

Equation 42 enables the diffusion coefficient to be obtained from plots of M_t/M_∞ against $t^{1/2}$ for the initial period of vapour sorption, whereas at the end of the sorption process, logarithmic plots of $M_\infty - M_t$ against time also yield the diffusion coefficient.

A great difference between the sorption of organic vapours by polymer membranes and the sorption of permanent gases is that the vapour sorptions lead to much larger concentrations of diffusant in the membrane. The consequent swelling that occurs during the vapour sorption means that the thickness of the membrane increases and relations for the diffusion coefficient that neglect this could be in error.

The problem can be dealt with by careful definition of the frame of reference through which diffusion occurs. It has been comprehensively dealt with by Crank (2). For swelling occurring only at right-angles to the membrane surface, the sorption technique yields a diffusion coefficient, D_v^m, of the vapour, v, in a frame of reference fixed with respect to the membrane, m. This diffusion coefficient is related to the more normally defined diffusion coefficient, D^V, by the relationship

$$D^V = D_v^m/(1 - \phi_v)^2 \tag{44}$$

Here, ϕ_v is the volume fraction of the penetrant vapour. Modifications of this relationship have been proposed to relate the mutual diffusion coefficient to that given by the sorption technique when some of the swelling occurs in the plane of the membrane as well as at right-angles to it (22).

$$D^V = D^1/(1 - \phi_v)^\lambda \tag{45}$$

In this relationship, D^1 is the measured diffusion coefficient and λ is a power $<$ 2. In practice, the contribution of the concentration dependence of the diffusion coefficient from the volume fraction term in these relationships is small compared with the total concentration dependence. At zero concentration it disappears, and when the volume fraction is only a few percent, the correction can usually be ignored.

Several experimental techniques have been used for obtaining M_t as a function of t. Systems in which all permanent gas can be removed using a vacuum line are normal. The vapour pressure can be made constant using thermostatted liquid diffusant or using a very large thermostatted buffer volume of diffusant vapour. The vapour pressure can be measured using a mercury manometer or a pressure transducer and the extent of vapour sorption, M_t, can be followed from the extension of a quartz helix or using a vacuum microbalance. A diagrammatic representation of the equipment used is shown in Figure 5.

4.2. Complications Due to Concentration Dependence of the Diffusion Coefficient

Although equation 42 is not obeyed in glassy polymers, good linear plots of M_t/M_∞ against $t^{1/2}$ and the linear logarithmic plots suggested by equation 44, are obtained for organic vapours in non-glassy polymer membranes. Equation 43 is valid, however, at much longer times than would be suggested by equation 39 and the linear logarithmic plots cover a much shorter period than that predicted by equation 40. These deviations occur because the diffusion coefficient values increase with increasing penetrant concentration in the membrane. The concentration dependence of D can be determined in

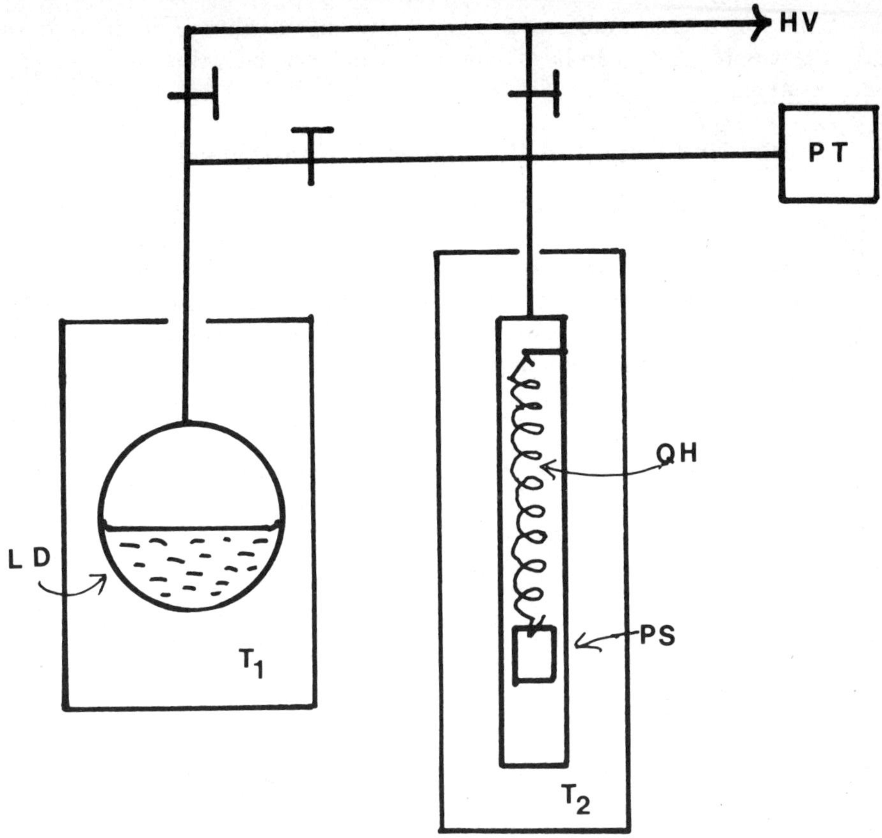

FIGURE 5.—Vapour sorption assembly. HV to high vacuum pumps; LD, liquid diffusant at temperature T_1; PT, pressure transducer or manometer; PS, polymer specimen at temperature T_2; QH, quartz helix or vacuum microbalance.

several ways. It can be shown, for instance (2), that for sorption from zero concentration of penetrant to an equilibrium concentration, c_1, in the membrane, equation 42 gives an average value of the diffusion coefficient, $\bar{D}$, which is approximated by the relationship

$$\bar{D} = (1/c_1) \int_0^{c_1} D\mathrm{d}c \tag{46}$$

and the value of the diffusion coefficient from equation 44 is the value at concentration, c_1 (23). By carrying out vapour sorption at several different vapour pressures, a range of c_1 values can be investigated, and so the D versus c relationship can be obtained as well as equilibrium solubility data.

Such concentration effects are generally absent when permanent gases are investigated. This can be attributed to the very low values of c_1 found in such systems.

4.3. Nature and Extent of the Concentration Dependence

The magnitude of the change of the diffusion coefficient for a relatively small change in penetrant concentration is the most startling feature of the concentration dependence in many polymer membranes. Figure 6 shows that for benzene in poly(vinyl acetate) at 45°C, there is an almost exponential dependence on concentration and practically a 1000-fold increase in D, going from 0 to 0.1 volume fraction of penetrant (22). Linear logarithmic plots are only obtained over a relatively small concentration range and at higher

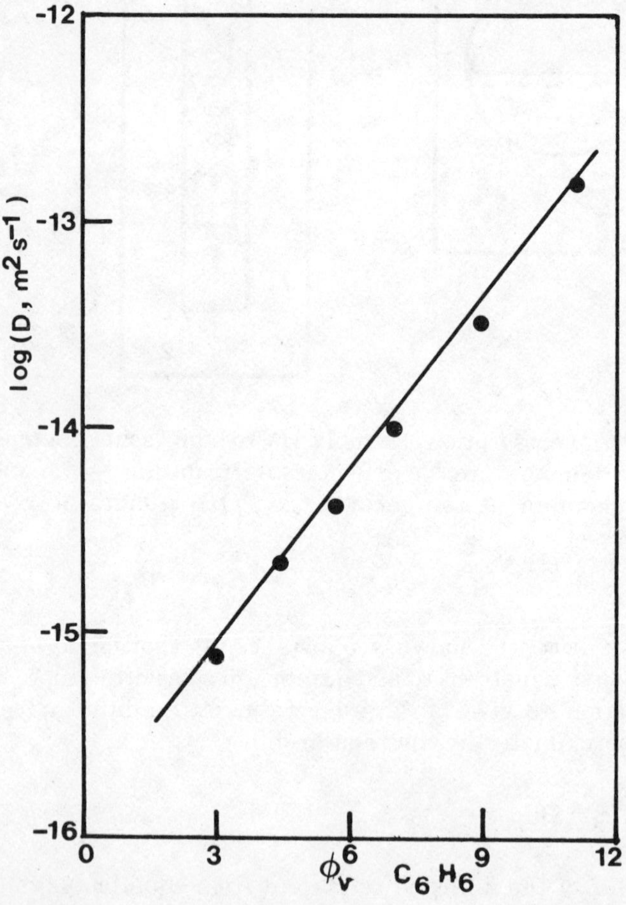

FIGURE 6.—Diffusion coefficient, D, of benzene in polyvinylacetate at 45°C. Reproduced with permission from Brown, W.R. and Park, G.S.: 1970, "Diffusion of solvents and swellers in polymers." J. Paint Technol. 42, p. 16. Federation of Societies for Coatings Technology, Copyright owners.

concentrations the dependence on concentration is less marked. The concentration dependence varies from polymer to polymer and for benzene in polydimethylsiloxane the diffusion coefficient is almost independent of concentration (24), whereas in natural rubber there is only about a threefold increase (25) compared with the 1000-fold increase in poly(vinyl acetate). It appears that when the diffusion coefficient at zero concentration is small, the dependence on concentration is great.

The free volume theory of diffusion (discussed later) leads to curved plots of log D against the volume fraction of penetrant but at low volume fractions, ϕ_v, it gives a constant α in

$$D = D_0 \exp(\alpha \, \phi_v) \tag{47}$$

in agreement with Figure 6.

4.4. Effect of Diffusant Size and Shape

The variation of the diffusion coefficient with the geometry of the diffusant is best discussed in terms of the diffusion coefficient at zero concentration, D_0. As with permanent gases, D_0 decreases with increasing penetrant size. This is clearly shown by the data in Figure 7 (26) for four alkylacetates in poly(methylacrylate), but more important than actual diffusant size is diffusant shape. This was recognised 30 years ago (27) for the diffusion of three isomeric pentanes in polyisobutene. n-Pentane, 1-methylbutane, and 2,2-dimethylpropane have values for $10^{14}\, D_0/\mathrm{m^2 s^{-1}}$ at 25°C of 10.8, 4.7 and 2.0. Recent studies by a radiotracer technique of the large relatively involatile molecules, n-hexadecane and DDT [1,1-bis(4-chlorophenyl)-2,2,2-trichloroethane] in polybutadiene (28) and in plasticized poly(vinyl chloride) (29) have shown that the sinuous flexible n-hexadecane molecule has a diffusion coefficient up to 30 times as great as the more rigid DDT molecule, even though the molar volume of the DDT is less than that of the n-hexadecane. Decrease of D_0 with increasing size is understandable in terms of the increased disturbance to the polymer needed to enable motion of the larger molecule. The increase in D_0 with increasing sinuousness of the diffusant suggests that such molecules can move through narrow gaps in a number of coordinated unit steps involving much less disturbance to the polymer chains than for large inflexible molecules.

4.5. Variation of Diffusion Coefficient with Temperature

As with permanent gases, large organic molecules show a rapid increase of the diffusion coefficient with increasing temperature (Figure 7). Since for small temperature changes, reasonable Arrhenius plots can be obtained, it is possible to interpret the temperature dependence in terms of the activation energy, $\Delta\tilde{E}_a$. Tables 2, 3 and 4 show that the temperature dependence and, hence, $\Delta\tilde{E}_a$, decrease with increasing penetrant concentration, increase with

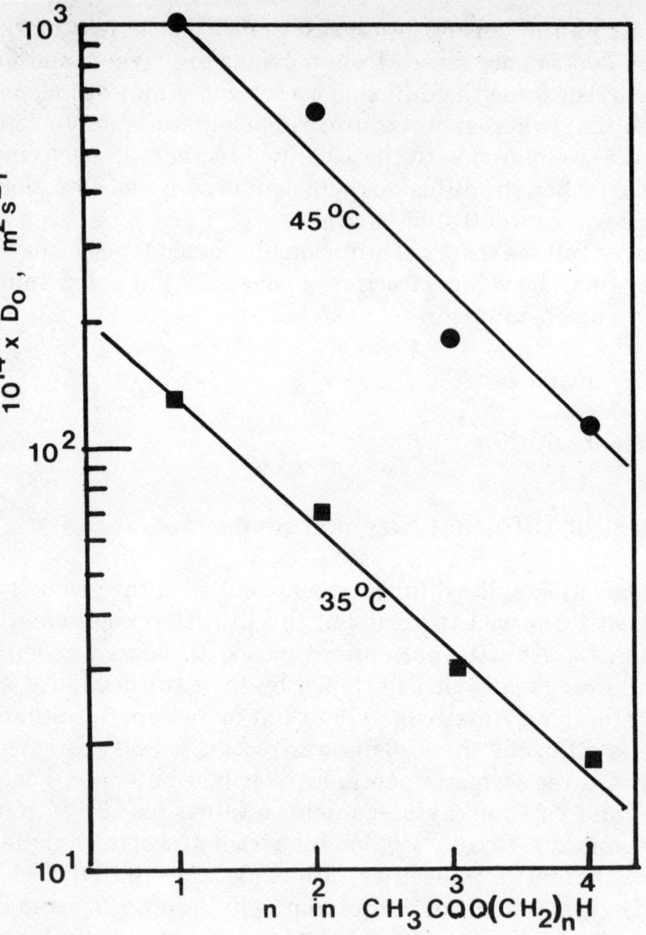

FIGURE 7.—Diffusion coefficient, D_0, of various alkyl acetates in polymethylacrylate (26). Reproduced with permission from Brown, W.R. and Park, G.S.: 1970, "Diffusion of solvents and swellers in polymers," J. Paint Technol. 42, p. 16. Federation of Societies for Coatings Technology, Copyright owners.

increasing penetrant size, and show considerable change from one polymer to another. As illustrated in Tables 2–4, whenever the diffusion coefficient is low, the dependence on temperature is generally high. This is part of a more general phenomenon that low diffusion coefficients of organic materials in polymers show considerable dependence on any change of condition (of concentration, penetrant size, temperature, etc.), whereas for high diffusion coefficient values there is little dependence on these factors.

A noteworthy feature of Table 2 is that although $\Delta\tilde{E}_a$ increases with penetrant size, it tends towards a constant value for large penetrants. This

TABLE 2. Effect of penetrant size on the activation energy, $\Delta\tilde{E}_a$, for diffusion at zero concentration in poly(vinyl acetate) at 40°C (30)

Diffusant	$10^6 \times$ Molar volume $\tilde{V}$, $m^3\ mol^{-1}$	D_0, m_2s^{-1}	$\Delta\tilde{E}_a$, kJ mol^{-1}
Water	18	1.2×10^{-11}	60
Methanol	41	1.5×10^{-13}	90
Acetone	76	1.3×10^{-15}	160
n-Propanol	76	1.1×10^{-16}	170
Benzene	91	4.8×10^{-17}	150

TABLE 3. Variation of the activation energy, $\Delta\tilde{E}_a$, for diffusion in poly(vinyl formate) with volume fraction, ϕ_v, of dichloromethane at 25°C (31)

$100\ \phi_v$	0	2	4	6	8
$\Delta\tilde{E}_a$, kJ mol^{-1}	173	163	152	142	109

TABLE 4. Dependence of the activation energy, $\Delta\tilde{E}_a$, for the diffusion of benzene on polymer type at 35°C

Polymer	PBD	NR	PEA	PMA	VAMA
T_g, K	171	200	249	278	~293
$\Delta\tilde{E}_a$, kJ mol^{-1}	23	39	62	166	290

PBD:—cis-polybutadiene (32) NR:—natural rubber (25) PEA:—poly(ethylacrylate) (33) PMA:—poly(methylacrylate) (33) VAMA:—46:54 vinyl acetate/methylacrylate copolymer (34)

value has been identified as equivalent to the activation energy for viscous flow of the polymer and implies that the same molecular motion is involved in the diffusion of large molecules and the viscous flow of the whole polymer matrix (30).

4.6. Dependence on the Glass Transition of the Polymer

The very large effect of the nature of the polymer on the diffusion coefficient in a membrane is most clearly shown by examining the D values at zero penetrant concentration. It is found that D_0 depends amongst other things on tacticity, cis/trans ratio, copolymer ratio, main chain flexibility, and bulkiness of side groups. For a given penetrant, D_0 can vary by as much as ten orders of magnitude from the most permeable polymer to the least. Probably the best correlation of D_0 with polymer properties is with the glass transition temperature, T_g [Figure 8; (35)].

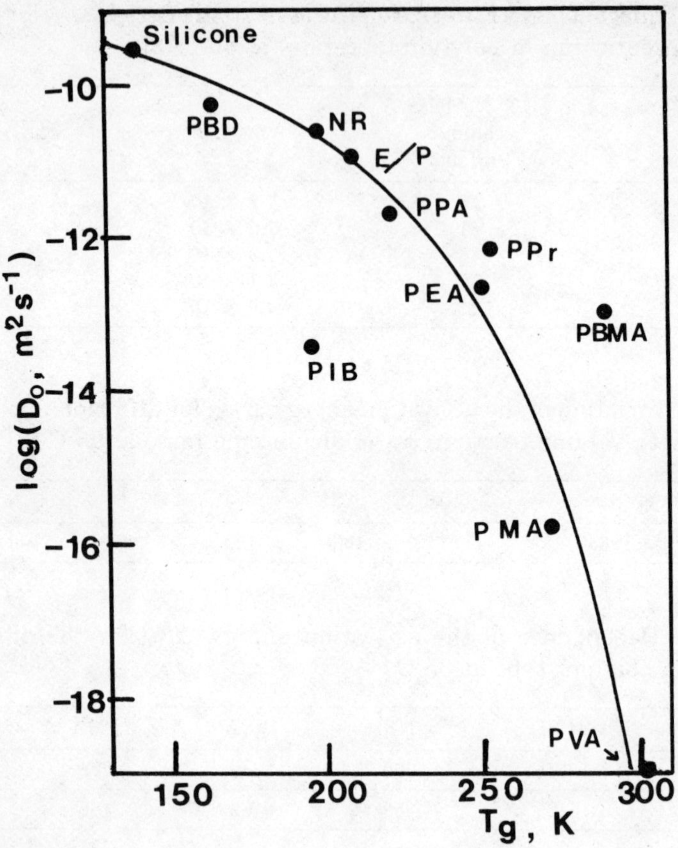

FIGURE 8.—Correlation of diffusion coefficient, D_0, for benzene with the glass transition, T_g, of the diffusion medium. Reproduced with permission from Brown, W.R. and Park, G.S.: 1970, "Diffusion of solvents and swellers in polymers." J. Paint Technol. 42, p. 16. Federation of Societies for Coatings Technology, Copyright owners.
PBD = cis polybutadiene, NR = natural rubber, PEA = polyethylacrylate, PMA = polymethylacrylate, PVA = polyvinylalcohol, E/P = ethylene/propylene copolymer, PPA = polypropylacrylate, PPr = polypropylene, PBMA = polybutylmethacrylate, PIB = polyisobutene.

As might be expected, the very considerable increase of D_0 with decreasing value of T_g is accompanied by a considerable decrease in the concentration dependence of the diffusion coefficient. Thus, for benzene at 25°C in natural rubber, in polyethylacrylate, and in polymethylacrylate, with T_g of 200K,

250K, and 280K, the diffusion coefficient increases by 2.9-, 20-, and 340-fold, going from zero penetrant concentration to 10% by volume.

4.7. Plasticization

Plasticizers lower the glass transition temperature of a polymer and, as would be expected, this increases the diffusion coefficient, decreases the dependence on concentration, decreases $\Delta\bar{E}_a$ except for small diffusants, and it tends to decrease the dependence on diffusant size and shape (36).

4.8. Free Volume Theory of Diffusion

The variation of the diffusion coefficient with concentration, temperature, T_g, and penetrant size can easily be rationalized in terms of the free volume theory of diffusion. This theory is based on the assumption that a diffusing molecule can only move from one place to another when the local specific volume, and hence the local amount of empty space (free volume) around the diffusing molecule, exceeds a certain critical value. Because the probability of finding such sufficient local free volume is proportional to $\exp(-B/\phi_f)$ where B expresses the amount of local free volume needed and ϕ_f is the fractional free volume (the volume fraction of 'empty space'), Fujita and others (37) have shown that

$$D_T = RTA_f \exp(-B/\phi_f) \tag{48}$$

Here, A_f is a proportionality factor.

The fractional free volume of any system increases with increasing temperature, and decreases with increasing glass transition temperature in polymer systems; because penetrants have more free volume than polymers, it increases with increasing concentration of penetrant. It follows then from equation 48 that the diffusion coefficient will increase with increasing concentration, with increasing temperature, and with decreasing glass transition temperature. It will also increase with decreasing value of B and since the critical local free volume needed for diffusion will be larger when the diffusant is larger, equation 48 predicts that the diffusion coefficient will increase with decreasing size of diffusant.

4.9. Free Volume Theory for Concentration Dependence

Assuming additivity of free volume, the overall fractional free volume of a polymer penetrant system, ϕ_f, will be given by

$$\phi_f = \phi_{f0} + \beta_f \phi_v \tag{49}$$

where ϕ_{f0} is the fractional free volume of pure polymer, β_f is the difference between the fractional free volume of pure polymer and pure organic penetrant, and ϕ_v is the volume fraction of penetrant. Combining this with equation 48 gives

$$\ln(D_T/D_0) = \beta_f B\phi_v/(\phi^2_{f0} + \phi_{f0}\beta_f\phi_v) \tag{50}$$

This predicts a linear relationship between $1/\ln(D_T/D_0)$ and $1/\phi_v$. The slope and intercept enable values of B and β_f to be obtained (37). In general, the B values are less than unity and realistic figures for β_f have been found using this relationship.

Discrepancies arise, however, at high diffusant concentrations and when the size of the diffusing molecule is very different from the size of a polymer segment. A more comprehensive theory has been given by Vrentas and Duda (38), but the treatment of Fujita given here is reasonably satisfactory for most organic diffusants in membranes.

4.10. Free Volume Theory for Temperature Dependence Above the Glass Transition

The equation

$$\phi_{f0} = \phi_{fg} + \alpha_f(T - T_g) \tag{51}$$

relates the fractional free volume of a polymer at temperature, T, to the fractional free volume, ϕ_{fg}, at the glass transition, and α_f, the expansion coefficient of free volume. Combination of this relationship with equation 48 leads to

$$T - T_g = -\phi_{fg}/\alpha_f - (B/\alpha_f)/(\ln D + \ln A_f/RT) \tag{52}$$

Equation 52 clearly shows the inverse relationship between the logarithm of the diffusion coefficient and the distance apart of the glass transition temperature and the temperature of measurement. Equation 52 fits the data for the diffusion of benzene in a variety of polymers, and yields a reasonable value for B/α_f (24).

5. EQUILIBRIUM BETWEEN ORGANIC VAPOURS AND POLYMER MEMBRANES

5.1. Flory–Huggins Relationship

The very low solubility of permanent gases in polymer membranes ensures that Henry's law is obeyed, i.e., the concentration of gas in the polymer is directly proportional to the gas pressure. The much greater solubility of organic vapours results in deviations from this ideal law and a very common method of representing the relationship between the volume fraction, ϕ_v, of the organic vapour, and the vapour pressure, p', is given by the Flory–Huggins relationship. This is derived in terms of the statistics of placing penetrant molecules and polymer segments on a lattice structure. For very high molecular weight polymers, it has the form

$$\ln p'/p^0 = \ln \phi_v + (1 - \phi_v) + \chi(1 - \phi_v)^2 \tag{53}$$

Here, p^0 is the saturation vapour pressure and χ is the Flory–Huggins interaction parameter. This parameter is small when the polymer and organic penetrant liquid are completely miscible and is large when the organic vapour and the polymer are incompatible.

χ can be estimated from a knowledge of the solubility parameters, δ_v and δ_m, using equation 54.

$$\chi = z + (\delta_v - \delta_m)^2 \tilde{V}_1/RT \tag{54}$$

Here, z is the coordination number in the lattice structure and $\tilde{V}_1$ is the molar volume of the organic liquid. It follows from this that high solubility of penetrant in the membrane occurs when $\delta_v \simeq \delta_m$ and solubilities are low when δ_v and δ_m are far apart. In practice, χ usually varies with concentration as well as temperature, and so equation 54 is mainly used as an empirical relationship.

Departure from Henry's law means that the measured diffusion coefficient, D, is no longer the same as the thermodynamic diffusion coefficient, D_T, and

$$D = D_T \, \partial \ln a/\partial \ln c = D_T \partial \ln p'/\partial \ln \phi_v \tag{55}$$

From equation 53

$$\partial \ln p'/\partial \ln \phi_v = 1 - (2\chi + 1)\phi_v + 2\chi\phi_v^2 \tag{56}$$

This non-ideality term is unity when $\phi_v = 0$ and it decreases with increasing ϕ_v.

For volume fractions up to about 10%, the decrease is not very large; thus, for $\chi = 0.5$ and $\phi_v = 0.1$, $\partial \ln p'/\partial \ln \phi_v = 0.8$. When $\phi_v = 0.5$, however, this drops to 0.25. It follows that D_T, which gives the best measure of penetrant mobility in the membrane, increases with concentration more rapidly than the measured diffusion coefficient, D.

5.2. The Zimm and Lundberg Cluster Theory

Although the Flory–Huggins relationship is very useful for summarizing the equilibrium behaviour of organic vapours in polymer membranes, the ideas of Zimm and Lundberg (39, 40) enable vapour sorption characteristics to be interpreted in terms of the degree of clustering of the sorbed organic molecules. This makes use of a function that we can call G_c (G_{11}/v_1 in Zimm's and Lundberg's nomenclature). According to Zimm and Lundberg, $\phi_v G_c$ expresses "the mean number of molecules of type 1 in excess of the mean concentration of type 1 molecules amongst the neighbours of a given type 1 molecule when the type 1 molecules are at low concentrations." G_c is -1 for a purely random distribution of type 1 molecules. Taking organic penetrant as consisting of type 1 molecules, one can calculate the values of G_c from the vapour pressure, p', against volume fraction, ϕ_v, data. It can be shown, neglecting the compressibility of the system, that

$$G_c = (1/\phi_v - 1)(\partial \ln \phi_v/\partial \ln p') - 1/\phi_v \tag{57}$$

For organic molecules in many polymers, the slope of the ϕ_v against p' isotherm leads to $\phi_v G_c$ values of $+1$ or more, showing that penetrant clustering occurs in the membrane. The behaviour of the function, G_c, is much more complicated in water/polymer systems and further discussion of this theory occurs later.

6. ORGANIC MOLECULES IN GLASSY POLYMERS

6.1. Vapour Sorption Anomalies

Early studies of vapour sorption in cellulose esters, polystyrene and other hard polymers showed that plots of sorption against $t^{1/2}$ were not linear, as predicted by equation 42, but instead sorption and desorption curves similar to those shown in Figure 9 were obtained. It is not possible to explain these results in terms of concentration dependence of the diffusion coefficients, or by assuming that the barrier to diffusion varies with the distance from the membrane surface (41).

Equations 39–43 predict that the time required to achieve any one value of M_t/M_∞ for sheets of different thickness should be proportional to h^{-2}. In these anomalous systems, however, it turns out that t is proportional to h^a where a can have a variety of values between -2 and 0. Amongst other anomalies, diffusion coefficients obtained from steady state studies (permeation rates) and transient state studies do not agree with each other. It is also found that boundaries of advancing vapour concentration into polymer sheets move at rates that are not proportional to $t^{1/2}$ as would be predicted by simple theory. These anomalies are often called non-Fickian and are due to a variety of causes.

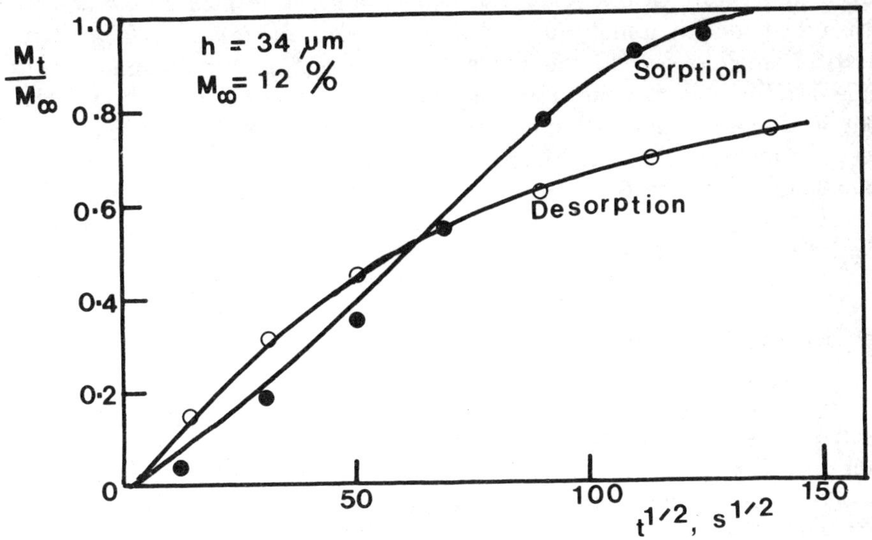

FIGURE 9.—"Anomalous" sorption and desorption kinetics for dichloromethane in polystyrene sheet. Comparison of the experimental points with the curves calculated from a history dependent diffusion coefficient. Reproduced with permission from Crank, J.: 1953, "A theoretical investigation of the influence of molecular relaxation and internal stress on diffusion in polymers." J. Polymer Sci. 11, p. 151. John Wiley and Sons Inc., Copyright owners.

6.2. Correlation with Glassy State and Time Effects for the Diffusion Coefficients and Vapour Solubilities

An examination of the systems in which non-Fickian vapour sorption occurs, and comparison with those that show Fickian kinetics, lead to the tentative conclusion that non-Fickian vapour sorption is a characteristic of the glassy state. This has been confirmed by experiments on poly(vinyl acetate), which show typical non-Fickian behaviour at temperatures below the glass transition and normal behaviour at temperatures above (42). Two characteristics of the glassy state are that 1) this has an elastic modulus that is about 10^6 times as great as for a non-glassy polymer, and 2) the response of a glass to any external change, such as the expansion in volume following a change in temperature, involves a very slow component (43).

From this second characteristic of the glassy state, it is possible that the non-Fickian behaviour results from slow changes of surface equilibrium at fixed vapour activity or that after the establishment of a given concentration of diffusant, *c*, the diffusion coefficient, *D*, takes some time to reach an equilibrium value. Crank (41,44) developed a sorption model in which the diffusion coefficient is not only a function of concentration but also of the length of time for which the concentration has been established in any element of the polymer. Suitable choice of parameters permits a good fit for both sorption and desorption measurements, as shown in Figure 9, where the lines are from the theory and the points are the experimental values. Delay in the setting up of surface equilibrium can also reproduce the main characteristics of non-Fickian vapour sorption (2,41) but this is best discussed in terms of two-stage sorption.

6.3. Differential Swelling Stress

The high elastic modulus of glassy polymers leads to the possibility that the stresses between the outer swollen parts of a polymer membrane and the inner unswollen parts can affect both the concentration–activity relationship and the penetrant mobility in the polymer. Sorption of the outer layers is inhibited by the unswollen core of the membrane and this lowers the vapour uptake. It is also possible that the compressive forces exerted by the unattacked core of the membrane densify the surface layers and the consequent lowering of the amount of free volume would be expected to decrease the diffusion coefficient.

Lowering of the diffusion coefficient and restriction on vapour sorption decrease as time progresses because of the decreasing extent of the unattacked membrane core. Crank (44) developed a simple model for this situation, in which only effects on the diffusion coefficient were considered. A reasonable fit with the experiments was obtained and this has led the way to the more detailed treatments given by Petropolous and Roussis (45).

A noteworthy consequence of the differential stress mechanisms is that in the earlier stages of vapour sorption the flux through the membrane surface of a glassy polymer is often greater for a thin membrane than for a thick one. Figure 10 clearly shows this for dichloromethane in cellulose acetate (46).

6.4. Two-stage Vapour Sorption

If, in a given system, the change of surface concentration with time is very slow, and the diffusional transport of material from the surface to the centre of the membrane is very rapid, it should be possible for the vapour uptake to be completely independent of the diffusion coefficient. This would give rise to vapour sorption kinetics that are independent of membrane thickness. This effect has been called Case III transport by Lewis (47) and until recently (48) it was only observed in certain interval sorption experiments.

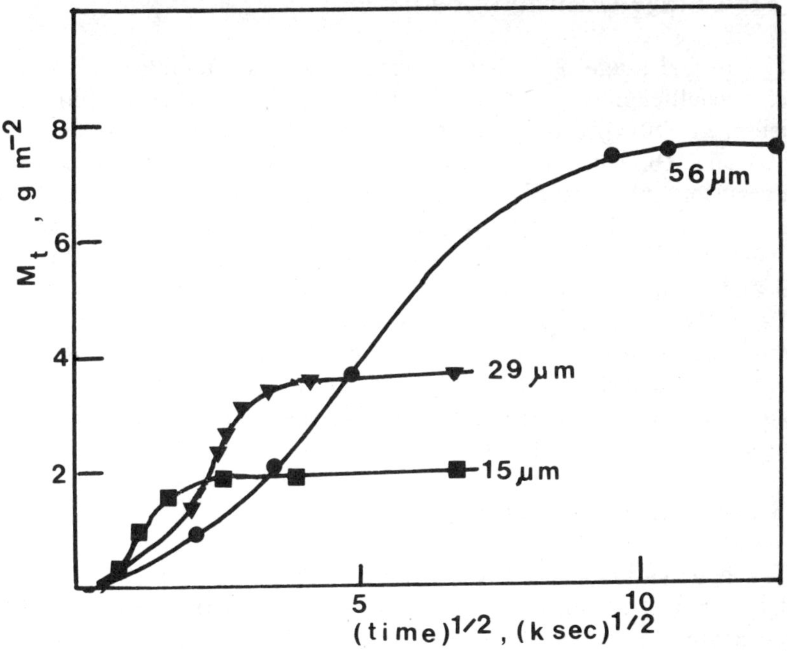

FIGURE 10.—Sorption of dichloromethane for unit area of cellulose acetate sheets of various thicknesses. Reproduced with permission from Park, G.S.: 1953, "An experimental study of the influence of various factors on the time dependent nature of diffusion in polymers." J. Polymer Sci. 11, p. 97. John Wiley and Sons Inc., Copyright owners.

Many glassy polymers that have reached equilibrium with vapour at one vapour pressure show two-stage vapour sorption kinetics when the vapour pressure is increased (49). It is now well established that on increasing the vapour pressure, there is an immediate increase of surface concentration in these experiments and this leads to rapid Fickian sorption governed by a fairly high diffusion coefficient. At the end of this stage, the surface concentration is beginning to increase further with time due to the relaxation of internal stress, and this change in surface concentration is transmitted rapidly throughout the whole of the membrane by very rapid diffusion. This leads to a second stage of sorption in which the relaxation process is rate-controlling and so in this stage, there is no dependence on membrane thickness. The second stage relaxation rate is very dependent on penetrant concentration and varies considerably from one polymer to another. It appears that the concentration dependence is much greater for non-crystalline glassy polymers than for those having crystalline or dipolar order (50).

6.5. First Stage Diffusion Coefficient

The initial stage of two-stage vapour sorption enables values for the diffusion coefficient to be obtained in the polymer. Use of this diffusion coefficient in interpreting one-stage non-Fickian sorption requires some care, because the concentration activity relationships involved in the two-stage sorption are unusual. The initial Fickian stage occurs when there is a large increment in the vapour pressure (thermodynamic activity) and a relatively small increment in the concentration. High values of $\partial \ln p'/\partial \ln \phi_v$ result and so the effective diffusion coefficient, D, is very much larger than the thermodynamic diffusion coefficient, D_T (51). The dependence of D on this thermodynamic term as well as on D_T means that another time-dependence has to be considered in glassy polymers, because the activity concentration relationship varies with time, not only at the surface of the polymer, but throughout the whole of the membrane.

6.6. Case I, II and III Transport

Transport of low molecular weight penetrants into polymers was classified by Alfrey et al. (52) in terms of the motion of the advancing front of penetrant into the polymer. Sorption in which this advance was proportional to $t^{1/2}$ was called Case I, and the other extreme at which the front advanced at a rate proportional to t, was termed Case II. Whereas Case I is normal diffusion-controlled transport with a constant surface concentration, Case II, in which the advancing front consists of a sharp concentration discontinuity, is thought to be related to the stresses and strains developed at this discontinuity.

For Case I sorption, the time needed for M_t/M_∞ to reach 0.5 is inversely proportional to the square of the thickness. For Case II sorption, it is inversely proportional to the thickness. Lewis (47) has proposed the classification Case III for sorption in which this time is independent of the thickness. He has shown that this occurs for the sorption of vapours by cellulose acetate films plasticized with nitroglycerine, even though M_t/M_∞ is approximately proportional to $t^{1/2}$. This behaviour, which is the same as in the second stage of two-stage sorption, must be completely independent of diffusional transport and the sorption rate must result purely from a relaxation mechanism occurring uniformly throughout the membrane system. The reasons for Case I and Case III absorption are thus fairly clear but various explanations have been put forward for Case II.

Early theories of Case II transport contained mathematical factors that were not obviously related to physical properties (53) or relied on events that only occurred in a limited number of the systems in which Case II behaviour was observed (54,55). Perhaps the most recent and successful explanation of Case II transport is that of Thomas and Windle (56). The most obvious distinguishing feature that accompanies Case II transport is the large discontinuity in penetrant concentration that constitutes the advancing front of the penetrating organic material. All theories focus on this aspect and use it to explain the linear advance of the concentration front.

The three-part theory of Thomas and Windle considers first the equilibrium relationship between pressure, concentration and activity in the absence of other factors in the polymer penetrant system. It then calculates the kinetics of swelling for an element that is so thin that diffusional transport is rapid enough to be considered instantaneous. The viscous resistance to change of shape and size is then the controlling factor and the effect of plasticizing penetrant on this is the important feature. Finally, the complete diffusion process for the whole bulk of the system is obtained in terms of equation 6, where the relationship between μ_i and c_i depends on the pressure and the viscous flow in the system. The theory involves the calculation of the change in concentration profile caused by creep at constant activity and combines this with the change in activity profile at a constant activity/concentration ratio to obtain the concentration distance relationship as a function of time in the system. This treatment combines all the suggestions that have been proposed to explain non-Fickian sorption in the previous sections. It places these, however, on a firm footing in relation to the known rheological behaviour of plasticized polymers and shows that the sharp concentration front is due to concentration dependence of viscous flow.

6.7. Diffusion Coefficient at Zero Concentration

Vapour sorption in glassy polymers at very low vapour activities is usually Fickian. It is relatively slow, but by using polymer powders consisting of uniform microspheres, one can conduct quite rapid vapour sorption experiments. Berens and Hopfenberg (84) have shown the very considerable dependence of the diffusion coefficients obtained on both the size and shape of the diffusant in glassy polyvinylchloride, polymethylmethacrylate and polystyrene. Going from neopentane to helium, the diffusion coefficients increase by about 10^{10}-fold and going from neopentane to n-pentane a 300-fold increase occurs.

The microsphere technique developed by Berens has also proved very useful for investigating the effect of history on vapour sorption equilibria in glassy systems.

7. WATER IN POLYMER MEMBRANES

7.1. Importance of Hydrogen Bonding

The energy of interaction between diffusing molecules and the substrate, or between one diffusing molecule and another, is usually relatively small and does not affect the diffusion behaviour to any great extent. One of the main reasons that makes water a unique substance is the extensive hydrogen bonding that occurs between one water molecule and another and the possibility of hydrogen bonding between water molecules and other polar substances. The hydrogen bond energies produce sorption and diffusion phenomena not exhibited by other diffusants. Simple alcohols show this behaviour to some extent, but here the greater molecule size lowers the interaction energy per unit volume so that, for instance, whereas water has a cohesive energy density of 2120 kJ dm^{-3}, this drops to 810 in the case of methanol, which is thus halfway between water and benzene, which has a cohesive density of 350 kJ dm^{-3}. It is these hydrogen bonding effects, as much as the technological importance of water, that necessitate special consideration of water as a diffusant and as a sorbate in membranes.

7.2. Techniques for Measurement of Diffusion, Permeability and Solubility of Water

A number of simple routine techniques for measuring the permeability of polymeric membranes to water vapour are used; the cup technique is perhaps the simplest. In this, the polymer layer makes a barrier across the mouth of a cup in which water, an aqueous solution, or a dessicant is kept. The apparatus is placed in a constant humidity region or in a dessicator and the rate of permeation is obtained by direct weighing of the cup and its contents. This very simple method is subject to error from the air barriers on both sides of the membrane and if the permeability is low, membrane sealing problems arise. When more accurate permeabilities are needed, the techniques described for permanent gases can be used but then problems may arise from the rather low vapour pressures used and from the sorption of water on glass surfaces. With reasonable precautions these effects do not prevent the measurement of steady state permeation rates, but permeation lag times are unreliable (57).

Polymer membranes can be roughly divided into two categories—hydrophobic and hydrophilic. With hydrophilic polymers, permeation rates can be relatively high but when the water sorption values are very low, as in hydrophobic polymers, or with hydrophilic polymers at low vapour pressure, vapour sorption techniques may be preferable. These methods usually involve the use of a vacuum microbalance. The high latent heat of water can cause very large temperature changes during sorption into hydrophilic polymers when a vacuum technique is used. In some cases, this even leads to vapour sorption kinetics that are controlled entirely by the rate of loss of heat from the sorbent (58) and then convective cooling in a carrier gas stream is necessary.

7.3. Concentration Dependence of Diffusion in Hydrophobic Polymers

Polymer membranes in which hydrogen bonding sites are absent are usually hydrophobic and the water content, even at saturation vapour pressure, is often less than 1%. It is not surprising that over this range of water concentration the diffusion coefficient does not increase appreciably. However, in many cases the diffusion coefficient decreases as the concentration increases! As early as 1936, this phenomenon was demonstrated from the measurements of the concentration distance relationship in natural rubber [(59) Figure 11]. The diffusion coefficient of water at the maximum concentration of 2% is only one-tenth of that in the dry polymer.

The diffusion coefficient apparently decreases with increasing water content in polyurethanes (57) and in polyalkylmethacrylates and ethyl cellulose. In polyethylene early reports suggested the diffusion coefficient decreased with increasing water concentration but later work suggested a constant diffusion coefficient. Constant diffusion coefficients have also been obtained in polypropylene, poly(vinyl acetate), and polyethyleneteraphthalate. In some cases, such as silicone rubbers (60), the behaviour is very dependent on impurities. The decrease of the diffusion coefficient has been attributed to clustering to give relatively immobile hydrogen bonded aggregates in equilibrium with free monomeric water in the polymer. The negative concentration dependence arises from the decreasing proportion of mobile

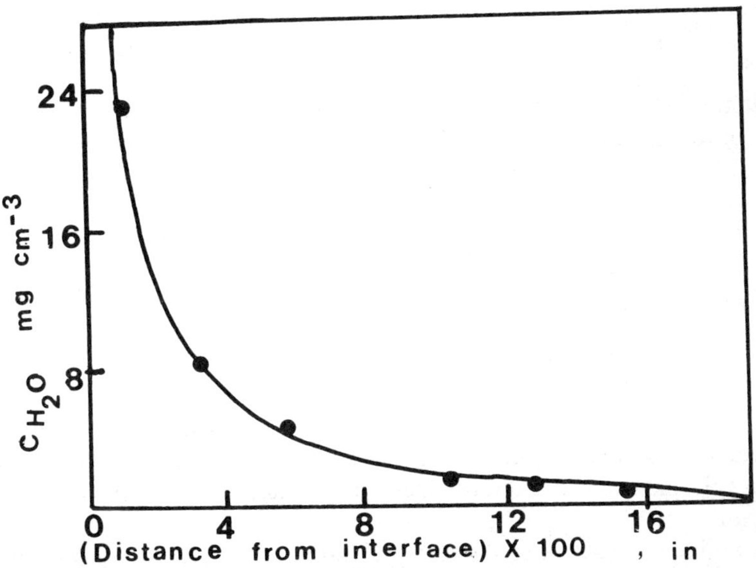

FIGURE 11.—Concentration-distance plot for steady state permeation of water through natural rubber. Reproduced with permission from Taylor, R.L., Herrmann, D.B. and Kemp, A.R.: 1936, "Diffusion of water through insulating materials. Rubber, synthetic resins and other organics." Ind. Eng. Chem. 28, p. 1255. American Chemical Society, Copyright owners.

monomeric water as the equilibrium shifts in favour of the aggregates at high water contents. The measured overall diffusion coefficient is given by

$$J = -D\frac{\partial c}{\partial x} \tag{1}$$

where c is the total concentration of water in the system. If it is assumed that only a portion of this water, concentration c_1, is mobile, then the diffusion coefficient, D_1, of this mobile portion is given by

$$J = -D_1\frac{\partial c_1}{\partial x} \tag{58}$$

It follows that

$$J = -D_1\frac{\partial c}{\partial x}\frac{\partial c_1}{\partial c} \tag{59}$$

and so the overall diffusion coefficient is related to the diffusion coefficient of mobile monomeric water by

$$D = D_1\frac{\partial c_1}{\partial c} \tag{60}$$

A very simple, but rather unrealistic, model of immobilisation by clustering is that of dimerization of the diffusant. It is easy to show that in this case

$$\frac{\partial c_1}{\partial c} = (1 + 8K_\mathrm{d}c)^{-1/2} \tag{61}$$

and hence,

$$D = D_1/(1 + 8K_\mathrm{d}c)^{1/2} \tag{62}$$

where K_d is the equilibrium constant for the dimerization. This clearly shows the diffusion coefficient decreasing steadily with increasing concentration; similar relationships are obtained for more complicated forms of diffusant clustering.

In silicone elastomers containing sodium chloride (60), the diffusion coefficient is constant up to a certain concentration and then it drops rapidly with increasing concentration. The reason for this is probably the variation of $\partial c_1/\partial c$ with increasing concentration. It is likely that the concentration, c_1, is equal to c at low activities of vapour, and this equality remains until a concentration is reached at which the activity of the water is equal to that of a saturated sodium chloride solution. From this point onwards most of the water will go to form minute droplets round the tiny sodium chloride crystals in the polymer, and so $\partial c_1/\partial c$ falls rapidly with the consequent dropping of the diffusion coefficient, D, according to equation 60. A resultant cloudiness of the polymer membrane is a fairly general phenomenon in many moderately hydrophobic polymers.

7.4. Hydrophobic Glassy Polymers

Diffusion coefficient values and diffusion coefficient concentration relationships for hydrophobic polymers reported by different investigators often show considerable divergences. This in part is caused by the very low solubilities being affected by traces of impurities but in non-elastomeric polymers, there is a systematic difference depending upon the technique used for measurement. When measurement of permeability time lag is used, it is often found that the diffusion coefficient is independent of concentration, while steady state measurements frequently lead to diffusion coefficients decreasing with concentration (61), as is shown in Figure 12. This could be readily explained if clustering were a slow process, so that during the transient state all the water present would be in the mobile monomeric form, whereas during steady state measurements immobile associated water would be in equilibrium with monomeric water. Clustering cannot easily account for the similarity of diffusivities obtained from steady state and sorption desorption measurements, but Petropolous and Roussis propose an explanation based on relaxation theory (62).

Non-ideal *gas* sorption in glassy polymers and concentration dependence of the diffusion coefficient at quite low concentrations have been explained in terms of a dual sorption theory (63). This assumes that sorbed gas molecules are distributed throughout the polymer in ideal solution obeying Henry's law, but they are also adsorbed in microcavities of the polymer according to a Langmuir relationship. When the absorbed material is water, this dual sorption model is further complicated by clustering. The net result is that $\partial c_1/\partial c$ increases with concentration at low activities in the Langmuir sorption region and then decreases as clustering becomes important. Water in polyacrylonitrile shows this behaviour (64) so that the diffusion coefficient first increases with concentration and then decreases. A recent mathematical analysis of this system has been given by Mauze and Stern (65).

7.5. Sorption Equilibria

In the absence of surface adsorption and specific interactions the solubility of water in a hydrophobic membrane would obey Henry's law but clustering leads to increased non-linear sorption at higher concentrations. If water clustering can be represented by the equilibrium

$$\mathrm{n\ H_2O \rightleftharpoons (H_2O)_n} \tag{63}$$

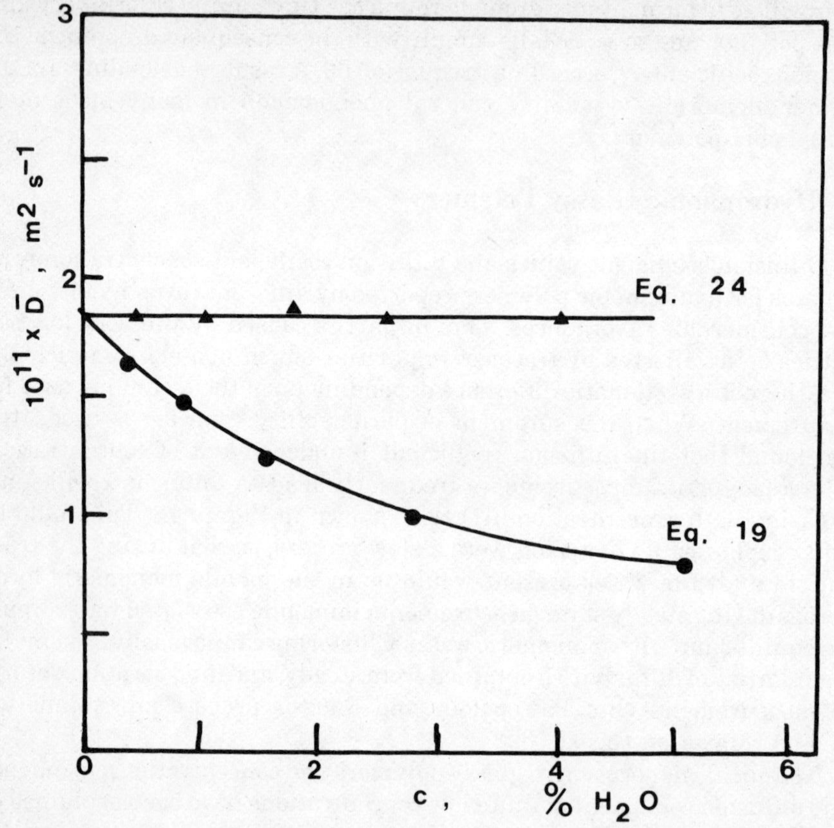

FIGURE 12.—Integral diffusion coefficient ($\int_0^1 Ddc/c_0$), from steady state (equation 19) and time lag (equation 24) permeation of water in ethyl cellulose. Reproduced with permission from Wellons, J.D. and Stannett, V.: 1966, "Permeation, sorption, and diffusion of water in ethyl cellulose." J. Polymer Sci. A1, 4, p. 593. John Wiley and Sons Inc., Copyright owners.

then the total concentration of water, c, is related to the activity, a, by

$$c = K_c s^n a^n / n + sa \tag{64}$$

Here, K_c is the equilibrium constant for the clustering reaction, and s is the Henry's law solubility coefficient of monomeric water. This kind of behaviour is clearly shown by the interpolated data in Table 5 for water in PVC (66). Water sorption in polyacrylonitrile, however, gives the sigmoid plots shown in Figure 13. This is explained by a Henry's law or Flory–Huggins-type sorption complicated by a Langmuir-type sorption on internal pore surfaces at low

TABLE 5. Difference, c_δ, between total membrane water concentration, c_T (from equilibrium measurements) and the Henry's law concentration, c_H (from permeation lag time measurements) as a function of relative humidity, p'/p^0, by PVC at 25°C from the measurements of Williams, Hopfenberg and Stannett (66)

$$[c_\delta \propto a^{\approx 1.8}]$$

Activity, $a = p'/p^0$		0.2	0.4	0.6	0.8	0.9
Water uptake	c_T	0.96	2.47	4.15	6.85	9.10
cm^3 H_2O (STP)·	c_H	0.56	1.12	1.69	2.25	2.53
$(cm^3\ PVC)^{-1}$	c_δ	0.40	1.35	2.46	4.40	6.57

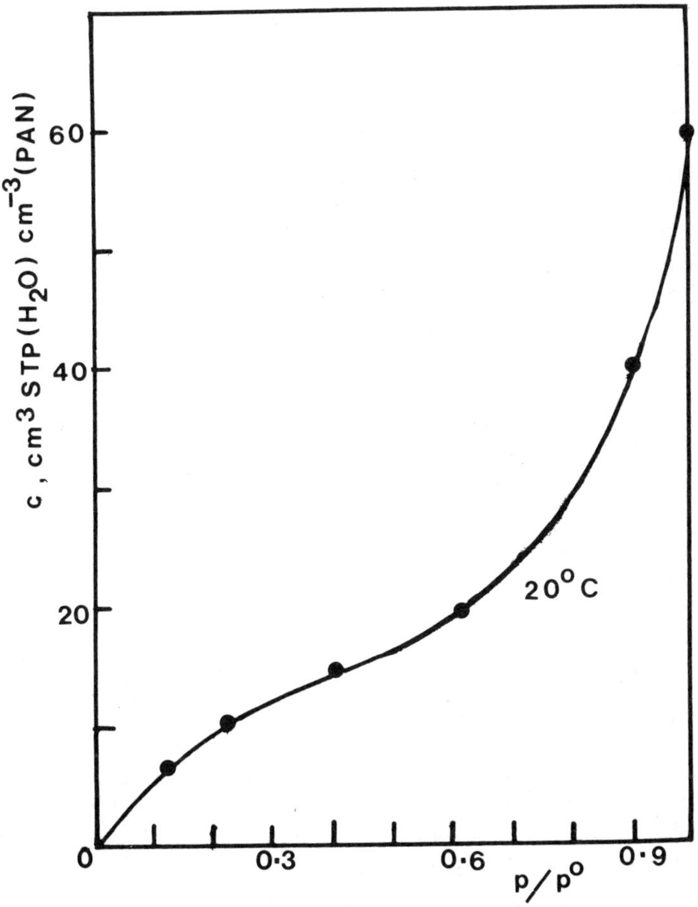

FIGURE 13.—Equilibrium isotherm for water in polyacrylonitrile. Reproduced with permission from Stannett, V., Haider, M., Koros, W.J. and Hopfenberg, H.B.: 1980, "Sorption and transport of water vapor in glassy poly(acrylonitrile)." Pol. Eng. Sci. 20, p. 300. Society of Plastics Engineers, Inc., Copyright owners.

activities, and at higher activities by clustering. An oversimplified relationship taking account of the three modes of sorption would be

$$c = Ab_{\mathrm{L}}a/(1 + b_{\mathrm{L}}a) + K_{\mathrm{c}}s^{\mathrm{n}}a^{\mathrm{n}}/\mathrm{n} + sa \tag{65}$$

Here, A is a measure of the extent of the internal pore surfaces per unit volume of membrane, and b_{L} is the Langmuir sorption parameter.

The Zimm and Lundberg cluster theory treatment, equation 57, applied to Figure 13, gives negative values of G_{c} at low pressures and it is not until $a \simeq 0.5$ that reasonable positive values indicating clustering are found. Calculation of G_{c} values from water vapor sorption isotherms gives a very useful indication of the polymeric state of the sorbed water and helps considerably with the interpretation of diffusion coefficient variations with concentration.

The relative immobilisation of water molecules by surface adsorption and by clustering leads to the concentration dependence of the diffusivities in hydrophobic polymers. Because the concentration, c_1, of mobile water is linearly related to the vapour pressure, the permeability coefficient, P, would be expected to be almost independent of concentration, which is the case in many hydrophobic polymer/water systems. In some cases, however, the permeability does vary with concentration but the variation is not as marked as for the diffusion coefficient.

Application of equation 57 to the water vapour sorption isotherms for hydrophilic systems leads to quite large negative values of G_{c} at low concentrations, as low as -5 for 10% volume fraction of water in collagen (40). G_{c} increases with concentration but in the collagen system it does not reach -1 until a volume fraction of 25% water is obtained. This low value of G_{c} is attributable to hydrogen bonding of the water molecules to the polar groups in the polymer in low concentrations and it is only at higher concentrations when most of these sorption sites have been covered that a moderate degree of clustering is possible.

7.6. Hydrophilic Polymers

Hydrogen bonding sites in such polymers as poly(vinyl alcohol), polyacrylic acid, proteins and cellulose, result in very high water sorption or even complete miscibility of water and polymer. In these polymers, the water molecules are strongly attracted to sorption sites at low vapour pressures, as is indicated by the large negative values of G_{c}. This means that c_1 is very much smaller than c, and hence $\partial c_1/\partial c$ is often very small, but tends to increase quite rapidly with increasing c. It follows from equation 60 that the diffusivity of water in these systems is quite low at low activities. The value increases, however, quite rapidly with increasing water content. This increase would occur even if D_1 in equation 60 were constant, but because water contributes free volume to the system, D_1 also increases with the water content, which results, in cellulose for instance, in a very considerable increase of diffusion coefficient with concentration (Table 6). In glassy polymers the high water

TABLE 6. Diffusion coefficient, D, of water in cellulose calculated from mean diffusion coefficient, $\bar{D}$, data obtained at various concentrations, c, by Gillespie and Williams (76)

$10^3\ c$, g cm^{-3}	0	1	2	3	4	5	6	7	8
$10^{14}\ D$, m^2 s^{-1}	1.5	3.5	6	10	15	22	31	45	52

uptakes of hydrophilic polymers result in considerable stress in the system and lead in some cases to the production of massive fissures, as was shown by P.H. Hermans (67) for liquid water diffusing into cellulose model filaments.

These stresses can also lead to quite marked non-Fickian sorption behaviour of the type described for organic penetrants in glassy polymers and so effects such as two-stage vapour sorption, and Case II transport, occur in these systems.

8. DIFFUSION SOLUBILITY AND PERMEATION IN NON-UNIFORM POLYMERS

8.1. Theoretical Predictions About the Effect of Fillers

The most general kind of heterogeneous polymer system is one in which small islands of a disperse phase, A, are distributed throughout a continuous phase, B. Even though such systems are heterogeneous on a microscopic scale, if a scale that is large compared with the size of the islands is chosen, the polymer can be considered homogeneous and it is possible to discuss the system in terms of the overall diffusion coefficients D, solubility s, and permeability coefficient P. The theoretical problem is to relate these values of D, s, and P to the values D_A, s_A, P_A of the disperse phase, and D_B, s_B, P_B of the continuous phase.

Because the solubilities in the two phases are additive, it is clear that

$$s = s_A \phi_A + s_B (1 - \phi_A) \tag{66}$$

Here, ϕ_A is the volume fraction of the disperse phase, A.

Although in some composites and, for instance, in some block and graft copolymers the disperse phase is permeable, in many cases of polymer/filler systems it can be assumed that D_A, s_A and P_A are zero and then

$$s = s_B (1 - \phi_A) \tag{67}$$

The relationships between the permeabilities and the diffusion coefficients are not easy to derive but steady state permeabilities obey the same mathematical relationships as both thermal conductivity and electrical conductivity; perhaps the first published work on this was that of Clarke Maxwell (68), who derived the effect of uniform spheres of A distributed at high dilution and at random in a continuum of B. In permeability terms his result is

$$(P - P_B) / (P + 2P_B) = \phi_A (P_A - P_B) / (P_A + 2P_B) \tag{68}$$

For an impermeable disperse phase ($P_A = 0$) this reduces to

$$P = P_B (1 - \phi_A) / (1 + \phi_A/2) \simeq P_B (1 - 3\phi_A/2) \tag{69}$$

Extensions of these treatments to ellipsoids and to particles of different sizes have been made. The results are not too different and as an example, equation 70 gives the relationship obtained by Fricke (69) for a dispersion of spheroids

$$P = P_B (1 - \phi_A) / (1 + \phi_A/x) \simeq P_B [1 - \phi_A (1 + 1/x)] \tag{70}$$

Here, the parameter, x, is a function solely of the spheroid geometry and it increases with the asymmetry of the spheroid.

Equations 68–70 are only valid for dilute systems. More complicated relationships have been obtained for regular arrays of particles and for particles having different geometries but up to a volume fraction of disperse phase = ~ 1/2, the different relationships give rather similar results and at this volume fraction the value of P/P_A is about 0.4. It is rather surprising that the drop in permeability is so small but, of course, in the case of a regular array of very asymmetric particles, such as parallel thin plates, the permeability can drop by much more.

Numerical solutions for specific geometries using finite difference methods have been discussed by Crank (2). Here, problems arise with angular particles because of the sharp differences at the corners. Another technique makes use of an electrical analogue in which conductivities are measured of a continuous phase conductor in which a dispersion of insulating particles is contained.

The steady state diffusion coefficients are given by equation 21 and so the diffusion coefficient is less dependent than the permeability on the concentration of dispersed phase. Equation 69, for instance, gives P/P_B as 0.4 for a volume fraction of disperse phase of 0.5, whereas

$$D/D_B = (P/P_B) / (1 - \phi_A) \tag{71}$$

which results from equations 21 and 67, gives a value of D/D_B of about 0.8.

8.2. Observed Effects in Filled Systems

When the theoretical expressions for solubility and permeability are compared with measured values, reasonable agreements are obtained for large filler particles.

Hamilton and Crosser (70) examined the effect of aluminium particles of various shapes on the permeability of rubber. They found that equation 70 predicts the behavior quite well if x is calculated from the relationship

$$x + 1 = 3/\psi \quad (72)$$

where ψ is the ratio of the surface area of a sphere to the surface area of the particle having the same volume. This value of $x + 1$ differs from that proposed by Fricke, which should be $3/\psi^2$ for prolate ellipsoids and $3/\psi^{3/2}$ for oblate ellipsoids.

With finely dispersed fillers, adsorption at the filler surface can sometimes occur so that a filled polymer actually absorbs more gas than an unfilled one. Thus, whereas at low volume fractions of filler, propane is absorbed in a rubber/zinc oxide system according to equation 67, at higher concentrations of filler an extra term proportional to the volume fraction of filler has to be inserted. This suggests that at low concentrations of filler, the filler surface is completely wetted by the polymer, at higher concentrations some of the filler has a partially free surface, and this enables adsorption of the gas to take place. Strangely enough, with benzene in the same filled system (71) and with butane in silica-filled polydimethylsiloxane, equation 67 appears quite adequate. Barrie and Machin (72) found that silicone-filled polydimethylsiloxane absorbed more than twice as much water as the unfilled polymer but the amount absorbed appeared to be independent of the volume fraction of the filler. Investigation of water vapour permeability in this system shows the diffusion coefficient increasing and then decreasing again as the water content is increased. This is easily explicable in terms of equation 60 with immobilization of the water on the silica surface at low concentrations, and immobilization by clustering at high concentrations.

In discussing permeation in filled systems, there are two other peculiarities to be considered. The first is concerned with what in paint technology is called 'critical pigment volume'. For a simple cubic lattice of touching spheres the interstitial volume is almost one half of the total volume. Tighter packing is possible but the minimum interstitial volume will still be about 0.4. It follows that for uniform spherical particles, volume fractions above about 0.5 can only occur when the filler particles project from the surfaces and vacuoles are produced in the bulk of the polymer. The empty space introduced at the critical pigment volume leads to a dramatic increase in the permeability. For non-uniform particles and for some non-spherical ones, this critical pigment volume can be much higher.

The final point that must be mentioned is the effect of the interaction between filler particles and the polymer on the properties of the polymer. Typical of this is the finding of Kwei and Kumins (73) that the sorption of chloroform by an epoxy resin was lowered by about 70% when 5% of filler was incorporated, in contrast to the 5% reduction in sorption predicted by equation 66. There are several examples of this behavior in filled systems and it is clear that it must be attributed to very strong interactions between the filler surfaces and the polymer which, in the case of very small filler particles, is equivalent to a change of the 'chemistry' of the polymer system.

8.3. Polycrystalline Membranes

In a first analysis, crystallites, like filler particles, can be regarded as discrete, impermeable islands in the polymer system and then equation 67, in which ϕ_A is the volume fraction of crystallinity, should describe solubility and D/D_B is given by a structure factor which depends on the crystallite morphology and on ϕ_A. Changes in ϕ_A occur on going through the melting region, on annealing, on altering the degree of chain branching in the polymer, and on changing its stereoregularity. An interesting example (74) shows the effect on the permeability coefficient in polyethylene due to increased crystallinity which, at the same time, increases the density. Figure 14 illustrates the effect and can be expressed by the relationship

$$P = P_B (1 - \phi_A)^2 \tag{73}$$

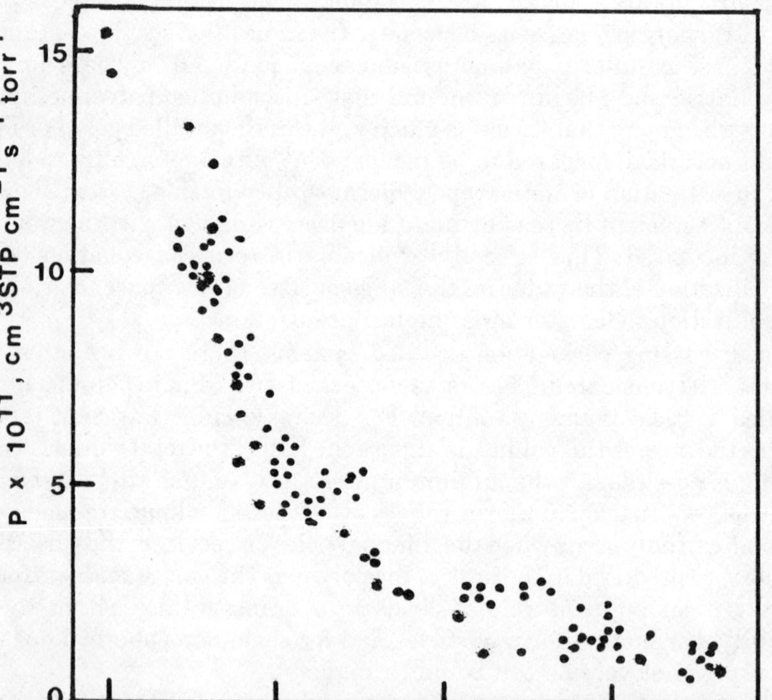

FIGURE 14.—Permeability coefficient of nitrogen in polyethylenes of various densities. Reproduced with permission from Alter, H.: 1962, "A critical investigation of polyethylene gas permeability." J. Polymer Sci. 57, p. 925. John Wiley and Sons Inc., Copyright owners.

It is well known that cis poly-1,4-dienes have much lower crystalline melting points than the corresponding trans compounds. This is clearly shown in the case of natural rubber and gutta percha but in the simpler polybutadiene systems it has proved possible to make polymers that have a range of cis/trans contents and it is found that up to about 70% trans, the polymers are amorphous at room temperature and thereafter they rapidly become crystalline. The increase in crystallinity causes a rapid fall in both vapour solubility and diffusion coefficients in these polymers (28,32). Table 7 shows the effects observed for benzene but Table 8 also indicates that the crystallite islands have a much greater blocking effect for large and inflexible molecules than they do for smaller or more flexible ones. This is clearly shown by the diffusion coefficient ratios and solubility ratios for the high percentage trans polymers compared with the 48% trans amorphous material. This effect has been observed in other systems and Michaels and Bixler (75), for instance, have proposed that as well as a tortuosity factor, θ, which may well be independent of penetrant size, there is a chain immobilization factor, β, which could depend on the type of diffusing molecule.

$$D = D_B/\theta\beta \tag{74}$$

The decrease in solubility of penetrant with increasing penetrant size means that the amorphous regions are less accessible to large penetrant molecules than they are to small ones. This could be interpreted in terms of differing accessibilities of the more or less ordered regions close to the crystallites, or at high crystallinities, the interstitial distances between the crystallites could be a critical factor.

TABLE 7. Interpolated equilibrium vapour sorption, c, at 30 torr vapour pressure and diffusion coefficient, D_0, at zero concentration of benzene at 25°C in poly(butadienes) of various trans contents (32)

Trans content, %	2	48	75	94	100
c, wt %	11.5	11.9	10.4	3.5	2.0
$10^{12}\ D_0$, m^2 s^{-1}	65	40	42	8.5	3.5

TABLE 8. Solubility coefficient ratios, s_{100}/s_{45}, and diffusion coefficient ratios, D_{100}/D_{45}, at zero concentration for various substances in 100% trans and 45% trans poly(butadienes) at 25°C (28,32)

Diffusant	CH_2Cl_2	C_6H_6	n-$C_{16}H_{34}$	DDT
s_{100}/s_{45}	0.19	0.15	—	—
D_{100}/D_{45}	0.117	0.088	0.045	0.016

Some investigations have shown that P, D, and s are very sensitive to annealing and to cold drawing. In many cases, these treatments do not alter the crystallinity very greatly but the morphology is considerably affected and it is clear that this has a very major effect at high crystallinities.

8.4. Permeability in Layered Structures

Perhaps the simplest form of heterogeneous membrane is one in which the system is made up of a series of layers having different properties. The permeability of these systems is of considerable interest.

Under steady state conditions the flux at each point in the membrane will be the same so that for any layer, i, thickness, h_i, and with permeability coefficient, P_i

$$J = P_i\,(\Delta a)_i/h_i \tag{75}$$

Here, $(\Delta a)_i$ is the drop in activity of diffusant across the layer. It is clear that the total drop of activity across the system is given by

$$a = \sum_i (\Delta a)_i = J \sum_i h_i/P_i \tag{76}$$

Since the overall permeability coefficient, P, is defined by

$$J = P\,a/h \tag{77}$$

where h is the overall thickness, it is clear that

$$h/P = \sum_i h_i/P_i \tag{78}$$

The permeation resistance, h/P, in a layered structure is thus additive and independent of the direction of permeation. This has been confirmed in many simple layered structures and was clearly shown for the diffusion of nitrogen in the binary polyethylene/glassine laminate over 25 years ago (77).

It is easy to see from equation 78 that even when the permeability coefficients of the components of a laminate satisfy an Arrhenius relationship

$$P_i = P_{0i} \exp\,(-\,\Delta\tilde{E}_i/RT) \tag{79}$$

this is not true for the overall laminate. Substitution of equation 79 into equation 78 gives

$$P = h/\sum\left[\,(h_i/P_{oi})\exp\,(\Delta\,\tilde{E}_i/RT)\right] \tag{80}$$

Bhargava et al. (78) have shown the considerable curvature of the Arrhenius plot that results when the $\Delta\tilde{E}_i$ terms have different signs in a polyethylene/glassine laminate.

Another interesting aspect of permeation in laminates is shown by the polyethylene/glassine structure. Even though in agreement with equation 78, the overall permeability coefficient for nitrogen is independent of the direction of flow through the laminate; it was found (77) that the permeability for water was greater when the flow occurred from the polyethylene side than from the glassine side. This arises because the permeability coefficient to water increases with concentration or activity of the water in the glassine but this is not true in the polyethylene. Consequently, the resistance to permeation of the polyethylene layer is independent of whether it is in contact with the zero vapour pressure side or the high vapour pressure side, while the glassine has a lower permeability when in contact with the low vapour pressure side and a higher one when it is in contact with the high vapour pressure side (78). In the ethyl cellulose/nylon 6 system, Rogers et al. (79) showed that the permeation of water vapour through an ethyl cellulose/nylon 6 composite could be as much as three and a half times as great from the ethyl cellulose to nylon direction, compared with the reverse direction. They also showed how a graphical technique could be used to calculate the overall flux rate from the individual diffusion coefficient and solubility data for each layer.

8.5. Diffusion Media with Continuously Variable Composition

Mathematical treatment of the simple system in which the diffusion coefficient is independent of concentration but depends on the distance through the membrane gives

$$c = J \int_0^x (1/D)\,\mathrm{d}x \tag{81}$$

for the concentration, c, against distance, x, relationship. Some concentration distance plots from this simple equation were evaluated by Barrer (80). Equation 81 shows that by putting $x = h$, permeability fluxes can be evaluated. A somewhat more complicated situation arises when the solubility of the permeant depends on distance. It is obvious that for sorption equilibrium with both faces of the membrane at the same activity, such a system will have concentration gradients without any flux. Such systems are best treated using the thermodynamic diffusion coefficient, D_T, of equations 14 and 55. It can be shown from these equations that

$$J = -D_T \left[\frac{\mathrm{d}c}{\mathrm{d}x} - c\,\frac{\mathrm{d}\ln \mathrm{s}}{\mathrm{d}x} \right] \tag{82}$$

where s is defined as c/a at the coordinate x. From this a positive flux can occur up a concentration gradient if the solubility coefficient changes rapidly enough with distance. Application of equations 81 and 82 leads to equal permeation rates for diffusants passing down the gradient of composition or up the gradient of composition in a membrane provided that the solubility coefficient and the diffusion coefficient are both independent of concentration, but as with the simple laminate, asymmetry of permeation rate occurs when these conditions are no longer met. A good example of this is given by the work of Rogers and co-workers who formed a graded composition membrane by γ-ray grafting of vinyl acetate monomer that was diffusing under steady state conditions through a polyethylene film (81,82). No asymmetry for the diffusion of simple gases was found in this system but on hydrolyzing the grafted vinyl acetate to vinyl alcohol, water vapour permeability coefficient up the gradient of poly (vinyl alcohol) concentration was only one-sixth of that in the reverse direction. Mathematical analysis of this situation is not easy but Sternberg and Rogers have made calculations for hypothetical membranes having solubility concentration relationships and diffusion coefficient concentration relationships that are given functions of distance (82,83). Application of Sternberg's and Rogers' relationships to, for instance, the diffusion of methanol through an asymmetric membrane of poly(vinyl alcohol) grafted onto polyethylene gave a good prediction of the magnitude of flux asymmetry in the system.

REFERENCES

1. Fick, A.: 1855, "Ueber Diffusion." Pogg. Ann. (Ann. der Phys. und Chem), 94, p. 59.
2. Crank, J.: 1975, "The Mathematics of Diffusion" (2nd ed.), Clarendon Press, Oxford.
3. Carslaw, H.S., and Jaeger, J.C.: 1959, "Conduction of Heat in Solids." Clarendon Press, Oxford.
4. Graham, T.: 1866, "On the absorption and dialytic separation of gases by colloidal septa." Phil. Mag. S. 4. 32, p. 401.
5. Daynes, H.A.: 1920, "The process of diffusion through a rubber membrane." Proc. Roy. Soc. A 97, p. 286.
6. Rogers, W.A., Buritz, R.S. and Alpert, J.: 1954, "Diffusion coefficient, solubility, and permeability for helium in glass." J. Appl. Phys. 25, p. 868.
7. Barrer, R.M. and Skirrow, G.: 1948, "Transport and equilibrium phenomena in gas–elastomer systems. I. Kinetic phenomena." J. Polymer Sci. 3, p. 549.
8. Yasuda, H. and Rosengren, Kj.: 1970, "Isobaric measurement of gas permeability of polymers." J. Appl. Polymer Sci. 14, p. 2839.
9. Ziegel, K.D., Frensdorf, H.K. and Blair, D.E.: 1969, "Measurement of hydrogen isotope transport in poly-(vinyl flouride) films by the permeation-rate method." J. Polymer Sci. A2, 7, p. 809.
10. Meares, P.: 1968, "Transport in ion exchange polymers," in "Diffusion in Polymers" (J. Crank and G.S. Park, Eds.) Academic Press, London/New York, Chapter 10.
11. Stannett, V. and Williams, J.L.: 1965, "The permeability of poly (ethyl methacrylate) to gases and water vapor." J. Polymer Sci. C 10, p. 45.
12. van Amerongen, G.J.: 1950, "Influence of structure of elastomers on their permeability to gases." J. Polymer Sci. 5, p. 307.
13. Barrer, R.M.: 1942, "Permeability in relation to viscosity and structure of rubber." Trans. Faraday Soc. 38, p. 322.
    1943, "The zone of activation in rate processes." Trans. Faraday Soc. 39, p. 237.
14. Wheeler, T.S.: 1938, "On the theory of liquids." Trans. Natl. Inst. Sci. India 1, p. 333.
15. Barrer, R.M. and Chio, H.T.: 1965, "Solution and diffusion of gases and vapors in silicone rubber membranes." J. Polymer Sci. C 10, p. 111.
16. Kumins, C.A. and Roteman, J.: 1961, "Diffusion of gases and vapors through polyvinyl chloride - polyvinyl acetate copolymer films. I. Glass transition effect." J. Polymer Sci. 55, p. 683.

17. Meares, P.: 1954, "The diffusion of gases through polyvinyl acetate." J. Am. Chem. Soc. 76, p. 3415;
1957, "The diffusion of gases in polyvinyl acetate in relation to the second-order transition." Trans. Faraday Soc. 53, p. 101.
18. van Amerongen, G.J.: 1964, "Diffusion in elastomers." Rubber Chem. Tech. 37, p. 1065.
19. Gee, G.: 1947, "Thermodynamic properties of high polymers. II. The solubility of gases in polymers." Quart. Rev. Chem. Soc. 1, p. 295.
20. Michaels, A.S. and Bixler, H.J.: 1961, "Solubility of gases in polyethylene." J. Polymer Sci. 50, p. 393.
21. Park, G.S. 1976, "Transport in polymer films," in "Treatise on Coatings, Vol. 2, Part II." (R.R. Myers and J.S. Long, eds.) Marcel Dekker, New York/Basel, Chapter 9, p. 506.
22. Garrett, T.A.: 1965, "The Diffusion of Benzene in Copolymers of Vinyl Acetate and Methyl Acrylate." M. Sc. Thesis, University of Wales.
23. Frensdorff, H.K.: 1964, "Diffusion and sorption of vapors in ethylene–propylene copolymers. II. Diffusion." J. Polymer Sci. A 1, p. 341.
24. Newns, A.C. and Park, G.S.: 1969, "The diffusion coefficient of benzene in a variety of elastomeric polymers." J. Polymer Sci. C 22, p. 927.
25. Hayes, M.J. and Park, G.S.: 1955, "The diffusion of benzene in rubber. Part I – Low concentrations of benzene." Trans. Faraday Soc. 51, p. 1134.
26. Fujita, H., Kishimoto, A. and Matsumoto, K.: 1960, "Concentration and temperature dependence of diffusion coefficients for systems polymethyl acrylate and n-alkyl acetates." Trans. Faraday Soc. 56, p. 424.
27. Prager, S., Bagley, E. and Long, F.A.: 1953, "Diffusion of hydrocarbon-vapors into polyisobutylene. II." J. Am. Chem. Soc. 75, p. 1255.
28. Jenkins, R.B. and Park, G.S.: 1983, "The effect of microstructure on the diffusion of n-hexadecane and DDT in poly (1,4-butadienes)." J. Membrane Science 15, p. 127.
29. Park, G.S. and Saleem, M.: 1984, "Diffusion of n-hexadecane and DDT in various poly(vinyl chloride)/(dialkylphthalate compositions." J. Membrane Science 18, pp. 177–185.
30. Kokes, R.J. and Long, F.A.: 1953, "Diffusion of organic vapors into polyvinyl acetate." J. Am. Chem. Soc. 75, p. 6142.
31. Brown, W.R.: 1970, "Diffusion in Stereoregular Polymers," Ph.D. Thesis, University of Wales.
32. Brown, W.R., Jenkins, R.B. and Park, G.S.: 1973, "The sorption and diffusion of small molecules in amorphous and crystalline polybutadienes." J. Polymer Sci. Symp. Edn., 41, p. 45.
33. Newns, A.C.: 1959, "The Sorption and Desorption Kinetics of Small Molecules in Polymers." Ph.D. Thesis, University of London.
34. Garrett, T.A. and Park, G.S.: 1967, "Solubility and diffusion behavior of benzene in a vinyl acetate-methyl acrylate copolymer." J. Polymer Sci. C 16, p. 601.
35. Brown, W.R. and Park, G.S.: 1970, "Diffusion of solvents and swellers in polymers," J. Paint Technol. 42, p. 16.
36. Saleem, M.: 1977, "Diffusion in Poly(Vinyl Chloride) Systems." PhD. Thesis, University of Wales.
37. Fujita, H.: 1961, "Diffusion in polymer-diluent systems." Fortschr. Hochpolym. Forsch. 3, p. 1.
38. Vrentas, J.S. and Duda, J.L.: 1977, a. "Diffusion in polymer–solvent systems. I. Reexamination of the free-volume theory"; b. "Diffusion in polymer–solvent systems. II. A predictive theory for the dependence of diffusion coefficients on temperature, concentration, and molecular weight." J. Polymer Sci. Physics Edn., 15, pp. 403 and 417.
39. Zimm, B.H. and Lundberg, J.L.: 1956, "Sorption of vapors by high polymers." J. Phys. Chem. 60, p. 425.
40. Lundberg, J.L.: 1972, "Molecular clustering and segregation in sorption systems." Pure and Applied Chem. 31, p. 261.
41. Crank, J. and Park, G.S.: 1951, "Diffusion in high polymers: some anomalies and their significance." Trans. Faraday Soc. 47, p. 1072.
42. Kokes, R.J., Long, F.A. and Hoard, L.J.: 1952, "Diffusion of acetone into polyvinyl acetate above and below the second order transition." J. Chem. Phys. 20, p. 1711.

43. Kovacs, A.J.: 1958, "La contraction isotherme du volume des polymères amorphes," J. Polymer Sci. 30, p. 131.
44. Crank, J.: 1953, "A theoretical investigation of the influence of molecular relaxation and internal stress on diffusion in polymer." J. Polymer Sci. 11, p. 151.
45. Petropoulos, J.H. and Roussis, P.P.: 1974, "A discussion of theoretical models of anomalous diffusion of vapors in polymers," in "Permeability of Plastic Films and Coatings" (H.B. Hopfenberg, Ed.) Plenum Press, New York and London, p. 219.
46. Park, G.S.: 1953, "An experimental study of the influence of various factors on the time dependent nature of diffusion in polymers." J. Polymer Sci. 11, p. 97.
47. Lewis, T.J.: 1978, "Diffusion of isopropyl nitrate, acetone, and water into nitrocellulose." Polymer 19, p. 285.
48. Cosgrove, J.D., Hurdley, T.G. and Lewis, T.J.: 1982, "The sorption of iso-propyl nitrate and acetone into nitrocellulose/nitroglycerine films." Polymer 23, p. 144.
49. Bagley, E. and Long, F.A.: 1955, "Two-stage sorption and desorption of organic vapors in cellulose acetate," J. Am. Chem. Soc. 77, p. 2172.
50. Fujita, H., Kishimoto, A. and Odani, H.: 1959, Progr. Theor. Phys. (Kyoto) Supp. No. 10, p. 210.
51. Park, G.S.: 1968, "The glassy state and slow process anomalies," in "Diffussion in Polymers" (J. Crank and G.S. Park, Eds.), Academic Press, London and New York, Chap. 5.
52. Alfrey, T., Gurnee, F.E. and Lloyd, W.G.: 1966, "Diffusion in glassy polymers." J. Polymer Sci. C 12, p. 249.
53. Frisch, H.L., Wang, T.T. and Kwei, T.K.: 1969, "Diffusion in glassy polymers. II." J. Polymer Sci. A2 7, p. 879.
54. Sarti, C.G.: 1979, "Solvent osmotic stresses and the prediction of Case II transport kinetics." Polymer 20, p. 827.
55. Peterlin, A.: 1977, "Diffusion with discontinous swelling. III. Type II diffusion as a particular solution of the conventional diffusion equation." J. Res. Nat. Bur. Std. 81A, p. 243. 1979, "Diffusion with discontinuous swelling. V. Type II diffusion into sheets and spheres." J. Polymer Sci. Physics Ed. 17, p. 1741.
56. Thomas, N.L. and Windle, A.H.: 1982, "A theory of Case II diffusion." Polymer 23, p. 529.
57. Barrie, J.A., Nunn, A. and Sheer, A.: 1974, "The sorption and diffusion of water in polyurethane elastomers," in "Permeability in Plastic Films and Coatings" (H.B. Hopfenberg, Ed.), Plenum Press, New York and London, p. 167.
58. King, G. and Cassie, A.B.D.: 1940, "Propagation of temperature changes through textiles in humid atmospheres. Part I. Rate of absorption of water vapor by wool fibres." Trans. Faraday Soc. 36, p. 445.
59. Taylor, R.L., Herrmann, D.B. and Kemp, A.R.: 1936, "Diffusion of water through insulating materials. Rubber, synthetic resins and other organics." Ind. Eng. Chem. 28, p. 1255.
60. Barrie, J.A. and Machin, D.: 1969, "The sorption and diffusion of water in silicone rubbers. Part I. Unfilled rubbers." J. Macromol. Sci. Phys. 3, p. 645.
61. Wellons, J.D. and Stannett, V.: 1966, "Permeation, sorption, and diffusion of water in ethyl cellulose." J. Polymer Sci. A1 4, p. 593.
62. Petropoulos, J.H. and Rousis, P.P.: 1969, "Anomalous diffusion of good and poor solvents or swelling agents in amorphous polymers." J. Polymer Sci. C 22, p. 917.
63. Vieth, W.R., Howell, J.M. and Hsieh, J.H.: 1976, "Dual sorption theory." J. Membrane Science 1, p. 177.
64. Stannett, V., Haider, M., Koros, W.J. and Hopfenberg, H.B.: 1980, "Sorption and transport of water vapor in glassy poly(acrylonitrile)." Pol. Eng. Sci. 20, p. 300.
65. Mauze, G.R. and Stern, S.A.: 1982, "The solution and transport of water vapor in poly(acrylonitrile). A re-examination." J. Membrane Science 12, p. 51.
66. Williams, J.L., Hopfenberg, H.B. and Stannett, V.T.: 1968, "Water transport and clustering in poly(vinylchloride), poly(oxymethylene), and other polymers." Polymer Prep. Am. Chem. Soc. Div. Polymer Chem. 9, No. 2, p. 1503.
67. Hermans, P.H.: 1948, "A Contribution to the Physics of Cellulose Fibres." Elsevier, Amsterdam, p. 23.

68. Maxwell, C.: 1873, "Treatise on Electricity and Magnetism." Vol. 1, Oxford Univ. Press, p. 365.
69. Fricke, H.: 1931, "The electronic conductivity and capacity of disperse systems." Physics 1, p. 106.
70. Hamilton, R.L. and Crosser, O.K.: 1962, "Thermal conductivity of heterogeneous two-component systems." Ind. Eng. Chem. Fundamentals 1, p. 187.
71. Barrer, R.M., Barrie, J.A. and Rogers, C.M.: 1963, "Heterogeneous membranes: Diffusion in filled rubber." J. Polym. Sci. A 1, p. 2565.
72. Barrie, J.A. and Machin, D.: 1969, "The sorption and diffusion of water in silicone rubbers. Part II. Filled rubbers." J. Macromol. Sci. Phys. 3, p. 673.
73. Kwei, T.K. and Kumins, C.A.: 1964, "Polymer–filler interaction: Vapour sorption studies." J. Appl. Polymer Sci. 8, p. 1483.
74. Alter, H.: 1962, "A critical investigation of polyethylene gas permeability." J. Polymer Sci. 57, p. 925.
75. Michaels, A.S. and Bixler, H.J.: 1961, "Flow of gases through polyethylene." J. Polymer Sci. 50, p. 413.
76. Gillespie, T. and Williams, B.M.: 1966, "Diffusion of water vapor through a hydrophilic polymer film," J. Polymer Sci. A1 4, p. 933.
77. Stannett, V. and Yasuda, H.: 1965, "Permeability," in "Crystalline and Olefin Polymers," High Polymers, Vol. 20, (R.A.V. Raff and K.W. Doak, eds.), Wiley, New York, p. 131.
78. Bhargava, R., Rogers, C.E., Stannett, V. and Swarc, M.: 1957, "Gas and vapor permeability of plastic films and coated papers. IV. Effect of a paper substrate." Tappi 40, p. 564.
79. Rogers, C.E., Stannett, V. and Szwarc, M.: 1967, "Permeability values." Ind. Eng. Chem. 49, p. 1933.
80. Barrer, R.M.: 1968, "Diffusion of permeation in heterogeneous media," in "Diffusion in Polymers" (J. Crank and G.S. Park, eds.), Academic Press, London & N.Y., Chap. 6, p. 199.
81. Rogers, C.E.: 1965, "Transport through polymer membranes with a gradient of inhomogeneity." J. Polymer Sci. C 10, p. 93.
82. Sternberg, S. and Rogers, C.E.: 1968, "Sorption and diffusion in asymmetric membranes." J. Appl. Polymer Sci. 12, p. 1017.
83. Rogers, C.E. and Sternberg, S.: 1971, "Transport through permselective membranes." J. Macromol. Sci. B. 5, p. 189.
84. Berens, A.R. and Hopfenberg, H.B.: 1982, "Diffusion of organic vapors at low concentrations in glassy PVC, polystyrene, and PMMA." J. Membrane Sci. 10, p. 283.

TRANSPORT PRINCIPLES - POROUS MEMBRANES

Peter M. Bungay

Biomedical Engineering and Instrumentation Branch
Division of Research Services
National Institutes of Health
Bethesda, Maryland 20892

This chapter begins with a summary of general expressions derived from linear, non-equilibrium thermodynamic theory for one-dimensional transport across membranes. These expressions provide a framework for many of the transport analyses presented throughout the book, for example, those of the solution–diffusion type. Transport in homogeneous membranes is discussed briefly as the limiting case for pure diffusion. The remainder of the chapter reviews the hydrodynamic approach to modeling combined diffusive and convective transport in porous membranes. The model for uniform diameter, cylindrical pores is examined in enough detail to indicate the origin of predictions for the global thermodynamic transport coefficients in the limit of very dilute solutions of neutral spherical solutes. Extension to less dilute solutions, non-spherical, flexible and charged solutes is summarized.

P. M. Bungay et al. (eds.), Synthetic Membranes: Science, Engineering and Applications, 109–153.

1. LINEAR, NON-EQUILIBRIUM THERMODYNAMIC CHARACTERIZATION

1.1. Global Transport Coefficients

It has been traditional to characterize the properties of a membrane with respect to transmembrane convective and diffusive transport of a neutral solute solution in terms of four parameters: hydraulic permeability, diffusive permeability and two reflection coefficients. These parameters are usually defined in terms of values of the intensive variables, concentration and pressure external to the membrane. In this context they are referred to as "global" transport coefficients. The hydraulic permeability is a measure of the ability to transport fluid volume under the action of a pressure difference across the membrane. This property is expressed as the proportionality coefficient between volumetric flux, J_V, and transmembrane pressure difference, Δp, when the solute is absent,

$$L_p = J_V/\Delta p \quad \text{for pure solvent,} \tag{1}$$

where

$$\Delta p = p' - p''. \tag{2}$$

The prime and double prime superscripts denote values within the fluid phases bathing the "feed" (high-pressure) and "permeate" (low-pressure) sides of the membrane, respectively. This definition applies as well in the presence of solute when the membrane is freely permeable to the solute. When differences in composition exist and the membrane is not freely permeable to the solute, the flux is affected by osmotic pressure differences across the membrane. The volumetric flux expression for a solution of the solute i generalizes from equation 1 to,

$$J_V = L_p\,(\Delta p - \sigma_d \Delta \pi_i), \tag{3}$$

in which $\Delta \pi_i = \pi_{\text{int}}' - \pi_{\text{int}}''$ is the difference between the osmotic pressures in the external solutions at the membrane/external-solution interfaces, and σ_d is the Staverman osmotic reflection coefficient for the solute i.

The diffusive permeability ω characterizes the transmissibility of the membrane for the solute i when the volumetric flux is constrained to be zero,

$$\omega = J_i/\Delta \pi_i \qquad \text{for } J_V = 0, \tag{4}$$

in which J_i is transmembrane molar flux of i. According to equation 3, this zero volumetric flux constraint implies that a pressure difference develops across the membrane unless it is freely permeable to the solute ($\sigma_d = 0$).

In general, when the constraints in equations 1 and 4 are removed, solute transport by both convection and diffusion occurs. Linear non-equilibrium thermodynamic treatments traditionally assume that these two contributions are additive. The analysis by Kedem and Katchalsky (1) for dilute solutions thus led to the generalization of equation 4 in the form,

$$J_i = (1 - \sigma_f) J_V\, c_{\text{avg}} + \omega R T \Delta c_i. \tag{5}$$

The van't Hoff law was invoked to express the flux relations in terms of concentrations,

$$\Delta \pi_i = R T \Delta c_i. \tag{6}$$

Expression 5 introduces the Staverman filtration reflection coefficient for the solute i, σ_f, as the fourth transport parameter.

The convective contribution represented by the first term on the right-hand side of equation 5 contains c_{avg}, which is some concentration of i intermediate between those of the two bathing solutions. Various averaging procedures have been proposed (2).

An expression of the form of equation 5 can be written for each constituent of the bathing media, including the solvent. For now we assume that the only components present in the system are the single solute i, the solvent w and the membrane. The flux equations for solute and solvent relative to the membrane are interrelated through

$$J_V = \tilde{V}_w J_w + \tilde{V}_i J_i, \tag{7}$$

in which $\tilde{V}_i$ and $\tilde{V}_w$ are the partial molar volumes of the solute and solvent. For dilute solutions, typically,

$$J_V \approx \tilde{V}_w J_w. \tag{8}$$

The solute and volumetric flux expressions 3 and 5 constitute the familiar Kedem–Katchalsky equations. Onsager reciprocity considerations have been invoked (1) to suggest the equality of the two Staverman reflection coefficients ($\sigma_f = \sigma_d$).

1.2. Differential Formulation, Intramembrane Coefficients and Variables

In the form shown, or in modified versions that account for effects such as mixtures of solute, the Kedem–Katchalsky equations have been used extensively in studies of both biological and synthetic membrane transport. These difference equations assume no variations in membrane properties in the directions parallel to the membrane. The structure of the membrane is unspecified and and spatial variations across the membrane are permissible. However, in general, c_{avg} in equation 5 is ambiguous. Patlak et al. (3) argued that it would be more appropriate to apply the thermodynamic approach to transport across an element of membrane of differential thickness. The resulting differential flux equations can be expressed using either pressure and solute concentration within the membrane phase, or in terms of external solution equivalent variables if the conditions at the membrane/solution interfaces are suitably specified. In terms of membrane phase variables, they may be formulated for a planar membrane as,

$$J_i = (1 - \sigma_f^{(m)})\, c_i\, J_V - (P^{(m)}/h)\, dc_i/d\zeta, \tag{9}$$

and,

$$J_V = -(k_V^{(m)}/\eta h)\,(dp/d\zeta - \sigma_d^{(m)}\, RT\, dc_i/d\zeta), \tag{10}$$

in which η is a viscosity, e.g., the external solution viscosity, h is the membrane thickness, $k_V^{(m)}$ is a hydraulic permeability substituting for L_p, and $P^{(m)}$ is a diffusive permeability, similarly replacing ω. With porous membranes a dimensionless tortuosity factor, θ, is often inserted as a constant multiplying the actual thickness h (see the chapter on "Selectivity" in the present volume).

The coordinate ζ is the dimensionless axis perpendicular to the faces of the membrane whose values range from zero to unity. The superscript (m) here signifies values based on the membrane phase variables.

As discussed by Lightfoot (4), in the absence of slow surface reactions resistance to transport across the membrane/solution interfaces is usually negligible and hence interfacial equilibria can be assumed. If, in addition, the solute is assumed to form an ideal dilute solution within the membrane, the equality in chemical potential for the solute and solvent across each interface can be expressed as

$$RT \ln (x_i'/x_i)_{\text{int}} + \tilde{V}_i(p' - p)_{\text{int}} = 0, \tag{11}$$

$$RT \ln (x_w'/x_w)_{\text{int}} + \tilde{V}(p' - p)_{\text{int}} = 0, \tag{12}$$

in which x represents mole fraction. For dilute solutions these equations, solved simultaneously, can be simplified to

$$c_i[0] = K_i\, c'_{\text{int}},\; p[0] = p'_{\text{int}} + (K_i - 1)\, c'_{\text{int}}\, RT, \tag{13a,b}$$

$$c_i[1] = K_i\, c''_{\text{int}},\; p[1] = p''_{\text{int}} + (K_i - 1)\, c''_{\text{int}}\, RT, \tag{14a,b}$$

in which K_i is the equilibrium solute distribution (partition) coefficient.

The values of the transport coefficient based on the external solution variables are related to the intramembrane values by

$$L_p = L_p^{(m)}, \tag{15}$$

$$\omega = K_i\, \omega^{(m)}, \tag{16}$$

$$1 - \sigma_f = K_i\, (1 - \sigma_f^{(m)}), \tag{17}$$

$$1 - \sigma_d = K_i\, (1 - \sigma_d^{(m)}). \tag{18}$$

Alternative definitions for the permeability coefficients have been utilized in equations 9 and 10 for the purpose of explicitly representing the dependence on thickness and viscosity. These are related to their counterparts in equations 3 and 5 by equations 15 and 16 and

$$L_p^{(m)} = k_V^{(m)}/\eta h, \tag{19}$$

$$\omega RT = P^{(m)}/h. \tag{20}$$

1.3. Solute Concentration Profile in a Uniform Membrane

The one-dimensional flux equations 9 and 10 implicitly assume that local components of flux in the plane of the membrane can be ignored relative to fluxes in the ζ direction. Under steady-state conditions the fluxes are independent of ζ. When the membrane properties, $k_V^{(m)}$, $P^{(m)}$ and $\sigma^{(m)}$, also do not vary spatially, these equations can be integrated directly. Solution of equation 9 yields a concentration profile for the solute within the membrane,

$$\frac{c_i[\zeta] - c_i[1]}{c_i[0] - c_i[1]} = \frac{1 - \exp[-Pe(1 - \zeta)]}{1 - \exp[-Pe]}, \tag{21}$$

in which the coefficients are combined in a dimensionless group, the axial Peclet number,

$$Pe \equiv (1 - \sigma_f^{(m)})\, J_V h / P^{(m)} = (1 - \sigma_f)\, J_V h / P. \tag{22}$$

The magnitude of the Peclet number indicates the importance of the convective relative to the diffusive process for solute transport. The volumetric flux, J_V, is positive for flow in the positive ζ direction, or negative for flow in the reverse sense (similarly for the Peclet number). The concentration profiles for representative values of Pe are illustrated in Figure 1. When diffusion is dominant (Pe close to zero) as in dialysis, the concentration varies nearly linearly in ζ between the surface values of $c_i[0]$ and $c_i[1]$ as would be expected. For large absolute values of the Peclet number, diffusive transport is significant only in a thin zone adjacent to the low pressure face of the membrane in which the concentration profile is very steep. Over most of the membrane volume, the solute concentration varies little from the value at the high pressure face.

1.4. Solute Rejection

Multiplying equation 9 by $d\zeta$, integrating from $\zeta = 0$ to $\zeta = 1$, and casting the result in the form of the difference equation 5 leads to the determination that,

$$c_{avg} = \frac{1}{K_i} \int_0^1 c_i[\zeta]\, d\zeta. \tag{23}$$

Substituting the concentration profile, equation 21, into the integrand of equation 23 provides an explicit expression for c_{avg}, which upon substitution in equation 5 yields, with the help of boundary conditions 13a and 14a,

$$\frac{J_i}{(1 - \sigma_f)\, J_V} = \frac{c'_{int} - c''_{int} \exp[-Pe]}{1 - \exp[-Pe]}. \tag{24}$$

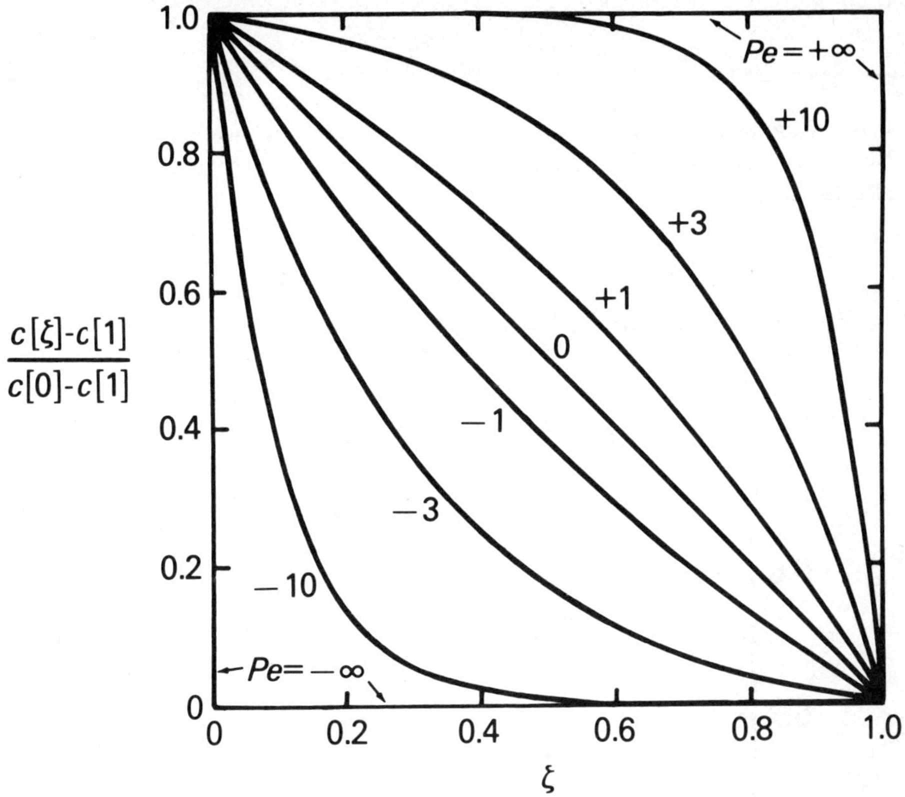

FIGURE 1.—Steady-state solute concentration profiles in simultaneous diffusion and convection across a planar membrane of uniform properties predicted from one-dimensional, linear, non-equilibrium thermodynamic theory. Numbers adjacent to profiles indicate values of the Peclet number whose sign depends upon the direction of the volumetric flux relative to the external solution concentration difference.

In ultrafiltration a useful measure of the separation achieved is the "observed solute rejection," which is defined in terms of the bulk concentration in the feed stream, c_i', and the permeate concentration c_i'', as,

$$\mathcal{R}_{obs} \equiv (c_i' - c_i'')/c_i'. \tag{25}$$

The true rejection by the membrane alone is correspondingly,

$$\mathcal{R}_i \equiv (c_{int}' - c_{int}'')/c_{int}'. \tag{26}$$

If we let

$$c''_{\text{int}} = J_i/J_V \tag{27}$$

in equation 24 and solve for the ratio $c''_{\text{int}}/c'_{\text{int}}$, we can express the contribution of the membrane, apart from boundary layer (concentration polarization) effects, as

$$\mathcal{R}_i = \sigma_f(1 - \exp[-Pe])/(1 - \sigma_f \exp[-Pe]). \tag{28}$$

It is usually desirable to have as high a rejection as practicable. As shown by the plot of equation 28 in Figure 2, the reflection coefficient is the upper

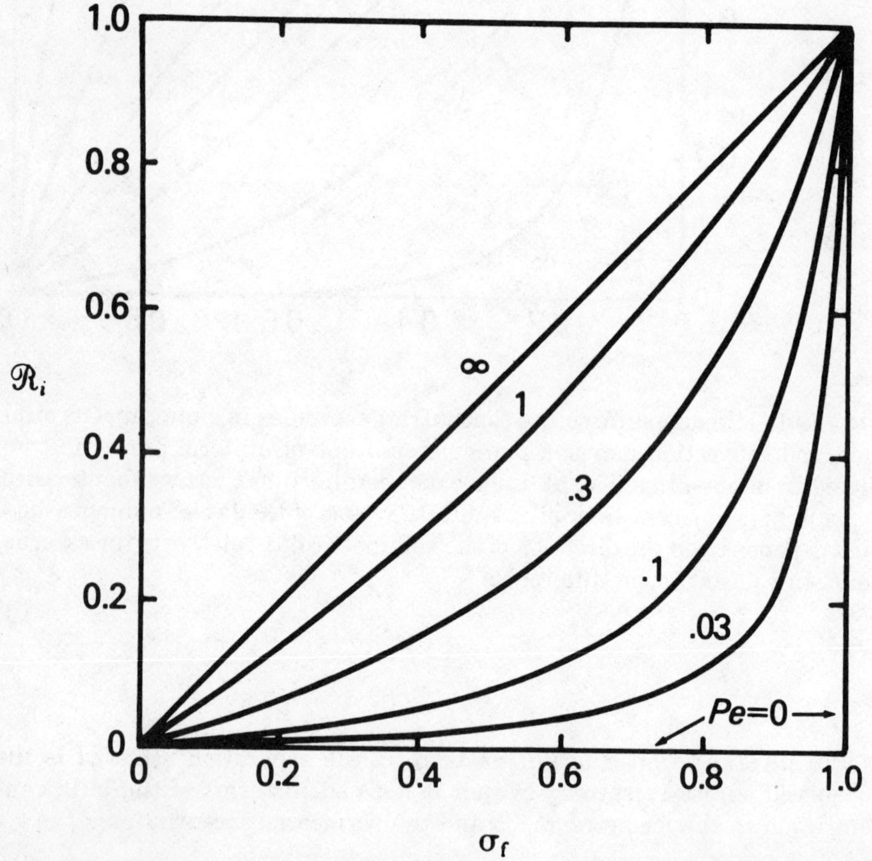

FIGURE 2.—Correspondence between the true or intrinsic rejection ($\mathcal{R}_i$) of solute and the filtration reflection coefficient (σ_f) for a planar membrane of uniform properties during steady-state diffusive and convective transport as predicted from one-dimensional, linear, non-equilibrium thermodynamic theory.

limit to rejection. But the rejection is equal to the reflection coefficient only for $Pe >> 1$. The rejection and hence the separation are reduced by the diffusional contribution to transport. This argument presupposes positive J_V and Pe, which is usually the case in ultrafiltration.

1.5. Membrane Configuration

The flux expressions in this section apply to flat membranes or configurations in which the membrane thickness is much less than the radius or curvature. This is usually, but not always the case. A uniform, i.e., non-asymmetric, thick-walled tubular membrane or hollow-fiber would be an exception. To express a difference across the membrane of an intensive variable, such as pressure, one needs to replace the Δp by $(r'' - r')\,\Delta p/r' \ln (r''/r')$ for fluxes based on the inner surface area of the tube. Here r' and r'' are the inner and outer tube radii, respectively.

1.6. Limitation

Within the constraints mentioned, the equations of this section place no requirements on the structure of the membrane. Since the performance of a membrane is related to its structure, the flux expressions by themselves have no predictive value. The values of the coefficients, when determined experimentally, provide only inferential evidence with regard to transport mechanisms. One of the ways in which these expressions are useful is as a framework into which structural information can be imbedded.

We turn now to two classes of membranes of specified structures for which it is possible to model the transport processes and predict the form of the transport coefficients.

2. HOMOGENEOUS MEMBRANE TRANSPORT

The class of membranes to be dealt with in this section are thought of as films of structureless continua. The most common approach to modeling transport in homogeneous membranes is the so-called "solution–diffusion" theory. Examples of this approach are given in the chapters on "Transport Principles," "Separation by Membranes," "Pervaporation," and "Membrane Gas Separations." Consequently, the theory will be given only a brief introduction here. Generally convective transport is not considered in solution –diffusion theory so that the differential form for one-dimensional flux of species i within the membrane is given by Fick's law as,

$$J_i = -\frac{D_i^{(m)}}{h}\frac{dc_i}{d\zeta} \tag{29}$$

from which, by comparison with equation 9 for $L_p = 0$, it can be seen that,

$$P^{(m)} = D_i^{(m)}. \tag{30}$$

For steady-state transport across uniform, homogeneous membranes, the integrated form of the above flux expression is,

$$J_i = P^{(m)}(c_i[0] - c_i[1])/h. \tag{31}$$

When it can be assumed that the partitioning between the membrane and external solutions is linear, the flux equation can be rewritten in terms of the external solution interface concentrations, using boundary conditions 13a and 14a:

$$J_i = P^{(m)}K_i\ (c'_{\text{int}} - c''_{\text{int}})/h, \tag{32}$$

with K_i being the constant solute equilibrium distribution or partition coefficient. Interfacial equilibrium is implicitly assumed as before. It is customary in solution–diffusion theory to denote the membrane permeability by

$$P = P^{(m)}K_i = D_i^{(m)}K_i, \tag{33}$$

which is consistent with the diffusive permeabilities of equations 16 and 20.

If the external media are mixtures of gases or vapors instead of liquid solutions, it is often more convenient to employ partial pressures in place of external concentrations. When the solubility of species i in the membrane is linear in external partial pressure, the steady-state flux expression analogous to equation 32 becomes,

$$J_i = P^{(m)}s_i\ (p'_{\text{int}} - p''_{\text{int}})/h, \tag{34}$$

in which p_{int} is the interfacial partial pressure of i and,

$$s_i = c_i[0]/p'_{\text{int}} = c_i[1]/p''_{\text{int}} \tag{35}$$

is a constant solubility coefficient. The definition of permeability complementary to equation 33 is then,

$$P = P^{(m)}\ s_i = D_i^{(m)}s_i. \tag{36}$$

Since K_i is dimensionless, whereas s_i is not, the units of P depends upon which of the two definitions, equation 33 or 36, is employed.

3. OPERATIONAL CRITERION FOR MEMBRANE POROSITY

In classifying membrane structure and the applicability of various theoretical approaches to predicting membrane transport, a common division is made between "porous" and "homogeneous (non-porous)" membranes. The latter class was treated in the previous section. The former category will be conceptualized in this section as impervious films pierced by solvent-filled channels that span the films. However, as the transverse dimensions of the channels decrease toward the size of individual molecules or molecular clusters and the impervious regions assume the transverse dimension of molecular chains or fibrils, the distinction loses its conceptual clarity. Consequently, in this structural range it is helpful to have an operational criterion for classifying the transport properties of a given membrane as being closer to one or the other of these two models.

It has been suggested by Bean (5) that an appropriate criterion can be developed by comparing the membrane permeability to pure solvent for transport driven by a pressure difference,

$$L_p = J_V/\Delta p \qquad \text{for} \qquad \Delta c_w = 0, \tag{37}$$

to the solvent permeability for transport driven by a concentration difference

$$\tilde{V}_w\, \omega_w = J_V^{\dagger}/RT\, \Delta c_w^{\dagger} \qquad \text{for} \qquad \Delta p = 0. \tag{38}$$

In the last equation the superscript, †, refers to labeled solvent (e.g., using radioisotopes) added in tracer amounts to otherwise pure solvent. Using thermodynamic arguments, Bean concluded that

$$L_p/\tilde{V}_w\, \omega_w \begin{cases} = 1 \text{ for non-porous membranes} \\ > 1 \text{ for porous membranes} \end{cases} \tag{39}$$

It is illuminating to apply this general thermodynamic criterion to a specific simple porous membrane model (Figure 3). If the channels in the porous membrane constitute a fraction, ϵ, of the membrane volume and each channel is modeled as a long, membrane-spanning, straight cylindrical pore of the same length, h, and radius, r_t, filled with a Newtonian fluid solvent, the hydraulic permeability can be approximated using the Hagen–Poiseuille law,

$$L_p \approx \epsilon\, r_t^2/8\eta_w h \qquad \text{for} \qquad h >> r_t. \tag{40}$$

Assuming that the solvent self-diffusion coefficient within the pores is the same as for the solvent in bulk,* the diffusive permeability can be estimated, again for long pores, as,

$$\omega_w = \epsilon\, D_w/RTh. \tag{41}$$

---

* This assumption may not be generally valid; it is invoked here without justification to permit the succeeding illustrative calculations.

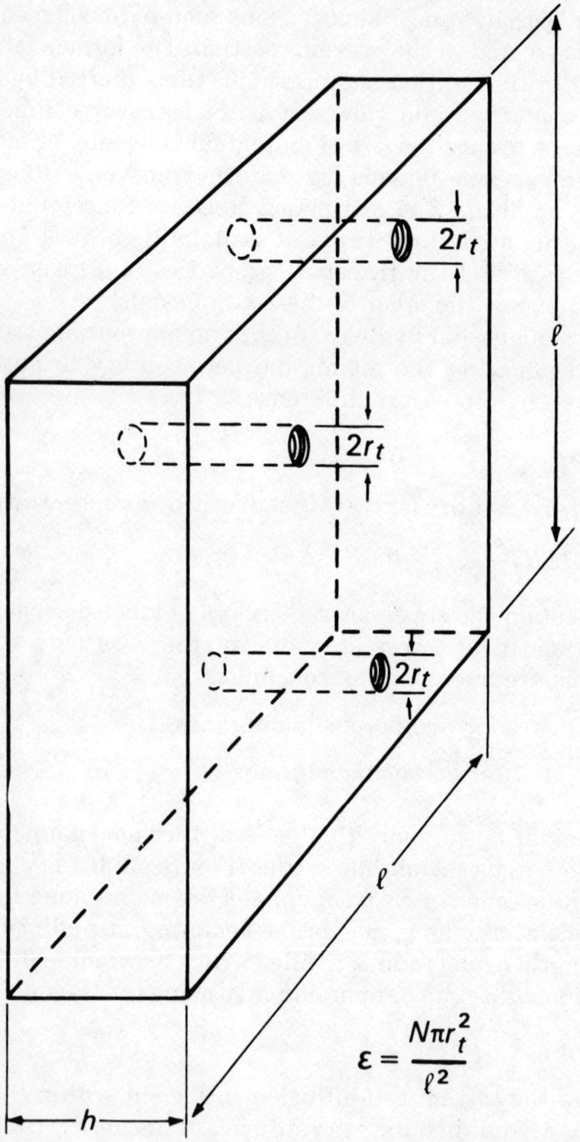

FIGURE 3.—Model homoporous membrane of thickness h, porosity ϵ, and pore radius r_t.

Combining equations 27–29 gives

$$\frac{L_p}{\bar{V}_w \omega_w} = \frac{r_t^2 RT}{8\bar{V}_w \eta_w D_w} = \frac{\pi r_t^2}{4}\left(\frac{\tilde{N}}{\bar{V}_w}\right)^{2/3} \tag{42}$$

The second quality above is derived by substituting a correlation for self-diffusion coefficients of liquids (6),

$$\frac{D_w \eta_w}{\kappa T} = \frac{1}{2\pi}\left(\frac{\tilde{N}}{\bar{V}_w}\right)^{1/3} \tag{43}$$

in which $\tilde{N}$ is Avogadro's number and

$$\kappa = R/\tilde{N} \tag{44}$$

is the Boltzmann constant. As a specific numerical example, let the solvent be water with $\bar{V}_w = 18\ \text{cm}^3/\text{mol}$. According to the operational criterion, a value of $L_p/\bar{V}_w\omega_w >> 1$ would be strong evidence that the membrane was porous. The value, $L_p/\bar{V}_w\omega_w = 10$, by equation 42, corresponds to an equiporous membrane of pore radius $r_t = 1.1$ nm. Note that the above porous membrane model yields a hypothetical equivalent pore radius, $r_t = 0.35$ nm, for the homogeneous membrane criterion of $L_p/\bar{V}_w\omega_w = 1$. For reference, this hypothetical radius is about twice the molecular radius of water molecules.

4. POROUS MEMBRANE TRANSPORT (Continuum Hydrodynamic Theory)

4.1. Bulk Solution

The theory for transport in porous membranes to be examined here could be said to have originated with Einstein's theory of Brownian motion in an unbounded solution (7). Einstein idealized solute molecules as rigid, neutrally buoyant spheres suspended in the solvent, which he treated as a continuous Newtonian fluid of infinite extent. This model led to what is now referred to as the Stokes–Einstein equation for the solute diffusion coefficient in dilute solutions,

$$D_i^\infty = \kappa T/\eta\, b_i f^\infty, \tag{45a}$$

$$f^\infty = 6\pi \tag{45b}$$

in which the characteristic molecular dimension was chosen as the molecular radius, b_i. The molecular radius is approximately proportional to the cube root of molecular weight. Diffusion coefficients for roughly spherical molecules in aqueous solution vary inversely with $\tilde{M}^{1/3}$, in accordance with equation 45, down to molecular weights in the range of 100–1000 daltons (8).

Equation 45 was applied by Einstein (7) to sucrose ($\tilde{M}$ = 342) with a molecular diameter of the order of 1 nm.

Generalization of equation 45 plays a fundamental role in the description of solute transport across porous membranes.

4.2. Linearity of Governing Equations

In the derivation of the Stokes–Einstein equation given in chapter 3, the denominator is termed a frictional resistance. This relates F, the viscous force (exerted by the surrounding fluid medium on the molecular particle), to the velocity of the particle relative to the fluid. The relationship defining f^{∞} is Stokes' law which, for parallel velocity vectors, can be written in a scalar form as,

$$F = -\eta b_i f^{\infty} (U_o - v_o). \tag{46}$$

Here, v_o is the velocity that the fluid would possess at the location of the particle center if the fluid were undisturbed by the presence of the particle. Because the particle length scale, b_i, is so small for molecules, the particle Reynolds number,

$$Re_i = |U_o - v_o| b_i/\eta, \tag{47}$$

is usually assumed to be extremely small, implying that the fluid behavior is dominated by viscous effects. Consequently, the equations governing the fluid motion simplify to the creeping motion equations (9) by neglect of the nonlinear fluid inertia effects. The creeping motion equations, together with "no-slip" boundary conditions on the surfaces of the solute particles and pore walls, are linear in the local continuum variables of velocity and pressure (and also in the fluid viscosity). This linearity of the governing equations facilitates the expression of generalized forms of equations 45 and 46.

4.3. Cylindrical Pore Model

We now combine the Einstein model for a dilute solution with the model membrane shown in Figure 3 with rigid cylindrical pores. The walls of the pores are taken to be impermeable, perfectly smooth, and motionless. This constrains the motion of the fluid inside the pores which, in turn, alters the hydrodynamic constraint on the particles. In general, the hydrodynamic resistance to particle motion increases and becomes dependent upon the direction of particle movement, and the size and the radial position of the particle relative to the pore. The orientation is important for non-spherical particles, but we will be concerned here with spherical solute particles whose orientation need not be explicitly considered. One such particle inside a right cylindrical pore is indicated in Figure 4. The coordinates (ϖ, ζ) specifying radial and axial location within the pore have been nondimensionalized using

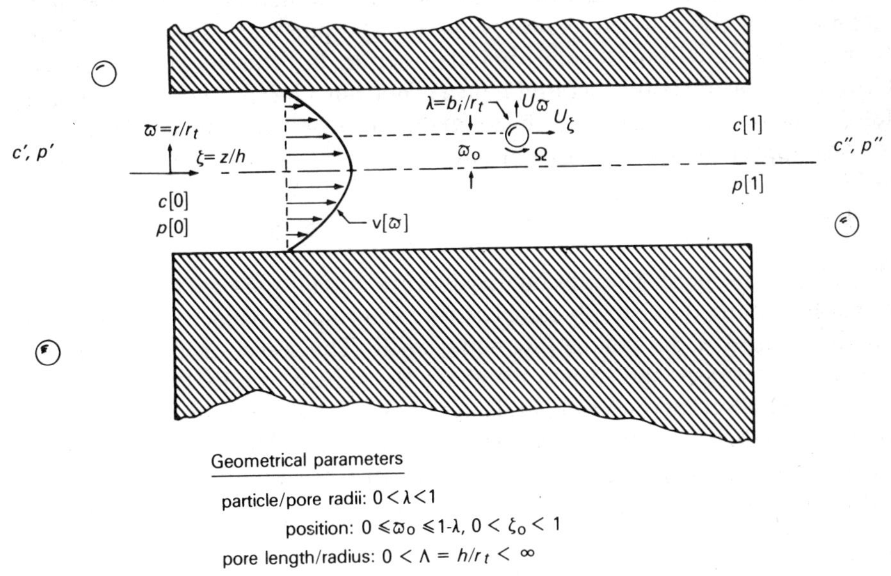

FIGURE 4.—Hydrodynamic continuum theory for convective and diffusive transport of a very dilute solution or suspension of neutral, rigid spheres within right circular cylindrical rigid pores of a homoporous membrane.

the pore radius, r_t, and length (membrane thickness), h, respectively. Dilute solutions bathe the faces of the membrane. Different pressures, p' and p'', in these external solutions result, according to equilibrium interfacial conditions analogous to equations 13b and 14b, in a pressure difference within the pore between the entrance and exit of $p^{(p)}[0] - p^{(p)}[1]$. Except near the entrances and exits, the solution within the pores responds to this pressure difference by flowing in Poiseuillean fashion, with mean velocity $\bar{v}$,

$$v = 2\,\bar{v}\,(1 - \varpi^2). \qquad (48)$$

The particle is at an arbitrary position within the pore, denoted by the coordinates (ϖ_o, ζ_o) of its center "o." The ratio of the particle radius, b, to the pore radius is given by λ. The impenetrability and rigidity of the particle and wall limit the range of particle radial position to between $\varpi_o = 0$ for the center of the pore axis and $\varpi_o = 1 - \lambda$ when the particle is in contact with the wall.

4.4. Resistance Coefficients for a Single Particle

The particle is in motion because of a combination of Brownian movement and convective displacement by the surrounding fluid. The particle motion relative to the membrane is indicated by radial and axial components, U_ϖ and U_ζ, of the particle center velocity. The particle is also undergoing Brownian movement in the meridional direction, but this does not contribute to transmembrane transport and hence is not considered. According to equation 48, the relative velocity in the axial direction between the particle center and the approach velocity of the fluid is

$$U_\zeta - v_o = U_\zeta - 2\bar{v}\,(1 - \varpi_o^2). \tag{49}$$

Within the pore, the viscous resistance to the particle's axial motion cannot be obtained by simply substituting the relative velocity, given above, into the infinite media Stokes' law, equation 46. The presence of the pore walls in general increases the resistance above that given by equation 46 and the degree to which it increases depends upon the relative position and size of the particle. Furthermore, the change in resistance caused by an increase in v_o is not the same as for an equivalent decrease in U_ζ. An analog of equation 46 within the pore can be written as,

$$F_\zeta = -\eta b_i f_\zeta\,[U_\zeta - 2(1 - \varpi_o^2)\,g_\zeta\,\bar{v}], \tag{50}$$

which retains the linearity of equation 46 in that the force resistance coefficient f_ζ and solute lag coefficient g_ζ are independent of the velocity components, U_ζ and $\bar{v}$. However, the resistance coefficients are no longer constants as is f^∞, but rather are dimensionless functions of the geometrical parameters,

$$f_\zeta = f_\zeta\,[\lambda, \Lambda, \varpi_o, \zeta_o], \quad g_\zeta = g_\zeta\,[\lambda, \Lambda, \varpi_o, \zeta_o]. \tag{51a,b}$$

In general, equation 50 should be altered to include a contribution to the viscous force linear in Ω which arises from particle rotation, and an expression similar in its linear form to equation 50 should be written for the viscous resistive torque that the fluid exerts on the particle (9). If no external torques act on the particle, the viscous torque vanishes and the torque equation can be combined algebraically with the force equation to eliminate Ω and reduce the force expression to the form of equation 50.

For particle translation in the radial direction, there is likewise a linear relation between force and velocity,

$$F_\varpi = -\eta b_i f_\varpi\,[\lambda, \Lambda, \varpi_o, \zeta_o]\,U_\varpi, \tag{52}$$

with the dimensionless radial resistance coefficient being a different function of the geometrical parameters than the axial coefficient, i.e., $f_\varpi \neq f_\zeta$.

In the vicinity of the particle and the pore openings the flow patterns deviate from equation 48 and alter the pressure distribution. It is convenient in the analysis to separate the contribution of disturbance flows from the Poiseuille component (10),

$$p^{(p)}[0] - p^{(p)}[1] = \Delta p_{\text{end}}^{+} + \Delta p_i^{+} - \frac{8\eta h\bar{v}}{r_t^2}, \tag{53}$$

in which $\Delta p_{\text{end}}^{+}$ and Δp_i^{+} are termed the "additional" pressure drops caused by the pore entrance effects and the presence of the solute particle, respectively.

The product of the solute additional pressure drop and the pore cross-sectional area, $A_p = \pi r_t^2$, is an axially directed force which, like the viscous force component, F_ζ, is linearly related to the axial particle velocity and the solution mean velocity,

$$\Delta p_i^{+} A_p = -\eta b_i \left[\mathcal{P}_U U_\zeta - 2(1 - \varpi_o^2)\, \mathcal{P}_v \bar{v}\right]. \tag{54}$$

The dimensionless coefficients, $\mathcal{P}_U$ and $\mathcal{P}_v$, like the force coefficients in equation 50, are independent of the velocities U_ζ and $\bar{v}$ and are functions of the geometrical parameters,

$$\mathcal{P}_U = \mathcal{P}_U[\lambda, \Lambda, \varpi_o, \zeta_o], \quad \mathcal{P}_v = \mathcal{P}_v[\lambda, \Lambda, \varpi_o, \zeta_o]. \tag{55}$$

4.5. Long Pore Simplifications

Because the pore Reynolds number, $2\rho\bar{v}r_t/\eta$, is typically very low, the entrance length, ℓ_e/r_t, for development of the parabolic velocity profile, is short (11) and probably a negligible fraction of the overall pore length $\Lambda = h/r_t$, at least for Λ significantly greater than unity. There are several important simplifications to equations 51–53 and 55 when the pore length is large in comparison with its radius ($\Lambda >> 1$). First, the resistance coefficients become independent of pore length, i.e., Λ, and axial position, ζ_o, of the solute particle (except for positions close to the pore ends):

$$f_\zeta = f_\zeta[\lambda, \varpi_o], \quad g_\zeta = g_\zeta[\lambda, \varpi_o],$$
$$\mathcal{P}_U = \mathcal{P}_U[\lambda, \varpi_o], \quad \mathcal{P}_v = \mathcal{P}_v[\lambda, \varpi_o] \text{ for } h >> r_t. \tag{56}$$

Second, the characteristic time for solute to diffuse a distance equal to the pore radius decreases relative to the mean solution transit time as $r_t/h \to 0$. The ratio of the former to the latter characteristic times, which is proportional to $(\bar{v}\, h/D^\infty)\,(r_t/h)^2$, should be much less than unity to justify employing one-dimensional membrane transport analyses of the type outlined below (11). A similar criterion holds when one applies the three-dimensional dispersion theory analysis of Brenner and Gaydos (12) to porous membranes.

4.6. Force Balance on Particle

When the solute is not influenced by external forces and torques—for example, an uncharged, non-polar, neutrally buoyant solute in an inert, uncharged pore—we may consider for heuristic purposes that the viscous force and torque on the particles are opposed only by the force and torque of Brownian origin. For the spherical solute, we need be concerned only with the components of the force balance,

$$F_{B\varpi} + F_{\varpi} = 0, \quad F_{B\zeta} + F_{\zeta} = 0. \tag{57a, b}$$

The Brownian force we take to be derived from the gradient of solute chemical potential,

$$F_{B\varpi} = -\frac{1}{\tilde{N} r_t}\frac{\partial \tilde{\mu}_i}{\partial \varpi}, \quad F_{B\zeta} = -\frac{1}{\tilde{N} h}\frac{\partial \tilde{\mu}_i}{\partial \zeta}, \tag{58a, b}$$

in which $\tilde{N}$ is Avogadro's number. For an ideal solution,

$$d\mu_i = RT \, d \ln c_i^{(p)} + \tilde{V}_i \, dp, \tag{59}$$

in which the local solute concentration in the pore, $c_i^{(p)}$, can be viewed here in a probabilistic sense (12) as being related to the probability of finding a particle center at the location (ϖ, ζ). Using equation 44 and the above, assuming the contribution from the pressure term will be small compared with the concentration term, gives,

$$F_{B\varpi} = -\frac{\kappa T}{r_t c_i^{(p)}}\frac{\partial c_i^{(p)}}{\partial \varpi}, \quad F_{B\zeta} = -\frac{\kappa T}{h c_i^{(p)}}\frac{\partial c_i^{(p)}}{\partial \zeta}. \tag{60a, b}$$

Substituting equations 50, 52, and 60 into 57 gives,

$$c_i^{(p)} U_{\varpi} = -\frac{\kappa T}{\eta b_i r_t f_{\varpi}}\frac{\partial c_i^{(p)}}{\partial \varpi}, \tag{61a}$$

$$c_i^{(p)} U_{\zeta} = 2(1 - \varpi_o^2)\, g_{\zeta}\, c_i^{(p)} \bar{v} - \frac{\kappa T}{\eta b_i h f_{\zeta}}\frac{\partial c_i^{(p)}}{\partial \zeta}. \tag{61b}$$

The left-hand side of equation 61b is the axial component of the net local solute flux, whereas the terms of the right-hand side represent convective and diffusive contributions to the flux. The factors multiplying the components of the concentration gradient function as diffusion coefficients. Let,

$$D_{\varpi} = \kappa T/\eta b_i f_{\varpi}, \tag{62a}$$

and,

$$D_\zeta = \kappa T / \eta b_i f_\zeta . \tag{62b}$$

These expressions are scalar forms of the generalized Stokes–Einstein equation (12). In contrast to the isotropic bulk solution diffusivity (equation 45), the appropriate diffusivity within the pore depends upon direction. According to equation 56, the local diffusion coefficients are not constants, but rather, vary with relative solute size and radial position,

$$D_\varpi = D_\varpi [\lambda, \varpi], \quad D_\zeta = D_\zeta [\lambda, \varpi]. \tag{63}$$

4.7. Pore Cross-Sectional Area Averaging

We will now derive a one-dimensional differential solute flux expression with the same form as equation 9, obtained from irreversible thermodynamic considerations. If we assume, as a first approximation, that $U_\varpi = 0$ throughout the pore, then we are led by integration of equation 61a to the local particle center distribution,

$$c_i^{(p)} [\varpi, \zeta] = \begin{cases} C_i^{(p)} [\zeta] , & 0 \leq \varpi \leq 1 - \lambda \\ 0 , & 1 - \lambda < \varpi \leq 1 \end{cases} \tag{64}$$

because the particle centers are excluded from an annular region adjacent to the pore wall. The pore cross-sectional area-averaged concentration* corresponding to the above profile is,

$$\bar{c}_i^{(p)} [\zeta] = \frac{1}{A_p} \int c_i^{(p)} dA_p = (A_i[\lambda]/A_p)\, C_i^{(p)} [\zeta], \tag{65a}$$

in which,

$$A_i[\lambda]/A_p = (1 - \lambda)^2 \tag{65b}$$

is the steric factor representing the fraction of the pore area accessible to the particle centers.

---

* Most previous treatments of transport in uniform pores utilize a pore concentration averaged over the cross-sectional area accessible to the sphere centers. Such approaches lead to definitions for the transport coefficients that differ from those given here by the multiplicative factor $(1 - \lambda)^2$. The appropriateness of averaging according to equation 65 is discussed by Brenner and Gaydos (12).

The pore cross-sectional area-averaged solute flux is defined as,

$$J_i^{(p)} = \frac{1}{A_p} \int c_i^{(p)} U_\zeta \, d A_p. \tag{66}$$

Substituting equations 61b, 62b, 64, and 65 into the above leads to the intrapore solute flux equation,

$$J_i^{(p)} = (1 - \sigma_f^{(p)}) \bar{c}_i^{(p)} J_V^{(p)} - (\overline{D}_i^{(p)}/h) \, d \bar{c}_i^{(p)}/d\zeta, \tag{67}$$

in which, $J_V^{(p)} = \bar{v}$,

$$1 - \sigma_f^{(p)}[\lambda] = \bar{g}_i^{(p)}[\lambda] = \frac{4}{(1 - \lambda)^2} \int_0^{1-\lambda} g_\zeta (1 - \varpi^2) \varpi \, d\varpi, \tag{68}$$

and,

$$\overline{D}_i^{(p)}[\lambda] = \frac{2}{(1 - \lambda)^2} \int_0^{1-\lambda} D_\zeta \varpi \, d\varpi. \tag{69}$$

4.8. Correspondence Between Hydrodynamic and Thermodynamic Models

Let the volume of pore per unit volume of membrane be ϵ. For the model membrane of identical, straight pores perpendicular to the membrane face, ϵ is also the pore area per unit surface area of membrane. The transmembrane solute molar flow rate per unit membrane area, J_i, is thus related to the solute flux within the pores by,

$$J_i = \epsilon J_i^{(p)}, \tag{70}$$

and similarly, the transmembrane volumetric flow rate per unit membrane area, J_V, is given by,

$$J_V = \epsilon J_V^{(p)}. \tag{71}$$

Furthermore, ϵ is the conversion factor between the pore area average concentration and the membrane (matrix + pore) area-averaged concentration, c_i,

$$c_i = \epsilon \bar{c}_i^{(p)}. \tag{72}$$

By means of the interrelationships 70–72, the thermodynamic solute flux equation 9 can be converted to the single pore form,

$$J_i^{(p)} = (1 - \sigma_f^{(m)}) \epsilon \bar{c}_i^{(p)} J_V^{(p)} - (P^{(m)}/h)\, d\bar{c}_i^{(p)}/d\zeta. \tag{73}$$

Comparison of the two forms of the pore solute flux equations (67 and 73) yields the correspondence between the thermodynamic coefficients based on the membrane area-averaged concentration and the hydrodynamically derived transport coefficients based on the pore area averaged concentration,

$$(1 - \sigma_f^{(m)}) \epsilon = 1 - \sigma_f^{(p)}. \tag{74}$$

$$P^{(m)} = \overline{D_i^{(p)}} \tag{75}$$

In terms of the more familiar global transport coefficients based upon the external solution variables and the membrane surface area (matrix + pore), the above correspondence can be expressed, using equations 16, 17 and 20, as,

$$1 - \sigma_f = (1 - \sigma_f^{(p)})\, K_i^{(p)}, \tag{76}$$

and

$$P = \omega RTh = \overline{D_i^{(p)}}\, K_i, \tag{77}$$

in which the pore distribution coefficient, $K_i^{(p)}$, is

$$K_i^{(p)} = \frac{K_i}{\epsilon} = \frac{\bar{c}_i^{(p)}[0]}{c'_{int}} = \frac{\bar{c}_i^{(p)}[1]}{c''_{int}}. \tag{78}$$

Combining equations 65 and 78 yields

$$C_i^{(p)}[0] = c'_{int},\quad C_i^{(p)}[1] = c''_{int}, \tag{79}$$

and

$$K_i^{(p)}[\lambda] = A_i[\lambda]/A_p = (1 - \lambda)^2. \tag{80}$$

Thus the pore distribution coefficient for the case of solutes that only interact hydrodynamically with the pore wall is equal to the steric factor.

From equations 76 and 77, the global diffusive permeability and reflection coefficients are each the product of an equilibrium thermodynamic term (the distribution coefficients) and a hydrodynamically derived term. For the reflection coefficient, the hydrodynamic term is the ratio of the averaged solute velocity to the averaged solution velocity in the pore for pure convection,

$$1 - \sigma_f^{(p)} = \overline{U}_\zeta/\bar{v} \quad \text{for} \quad \Delta c = 0. \tag{81}$$

For the diffusive permeability the product of the hydrodynamically derived averaged axial diffusion coefficient and the distribution coefficient is the porous membrane counterpart of homogeneous membrane permeability (equation 33).

4.9. Global Transport Coefficient Prediction

In principle, the hydrodynamic terms are calculable from the solutions of the creeping motion equations for a single sphere moving inside a fluid-filled tube. However, the complete solutions for the resistance coefficients f_ζ and g_ζ have not been evaluated for all ϖ_o and λ. Various asymptotic solutions for limiting cases of $\lambda << 1$ and $1 - \lambda << 1$ are known (9, 10, 13, 20, 21, 22) as well as accurate expressions (9) for the axisymmetric case, $\varpi_o = 0$. From these results, approximate calculations of the global parameter variations are available (5, 11, 12, 14–16) covering the full range of $0 < \lambda < 1$. The most detailed calculations, albeit only for the values $\lambda = 0.5$ and 0.9, were performed by Lewellyn (17), who also summarized and compared the prior work. Figure 5 illustrates the trends indicated by Lewellyn's analysis. In pure convection, the solute velocity is greater than the fluid velocity. However, the influence of the distribution coefficient is dominant for most λ leading to a monotonic increase in the reflection coefficient from $\sigma_f[0] = 0$ to $\sigma_f[1] = 1$. For small λ, the two factors have comparable, but opposite influences on σ_f, giving rise to the sigmoidal shape of the $1 - \sigma_f$ curve in Figure 5a. This trend is qualitatively similar to the broad variation in rejection with molecular weight observed for ultrafiltration membranes, as discussed in more detail in the chapter on selectivity.

The diffusive permeability in Figure 5b has been normalized to eliminate the influence of membrane thickness and porosity. Permeability decreases more rapidly with increasing solute size than predicted by the change in bulk solution diffusivity. This is referred to as "hindered" or "restricted" diffusion, with the ordinate in Figure 5b becoming the diffusive hindrance factor, H. By analogy, some would refer to the variation in σ_f with λ as hindered convection. From equations 45, 78, and 79, the diffusive hindrance factor is,

$$H[\lambda] = P[\lambda]/P[0] = \omega[\lambda]/\omega[0] = (\overline{D}_i^{(p)}/D_i^{\infty})K_i^{(p)}. \tag{82}$$

Both the axial diffusivity and the distribution coefficient decrease by comparable magnitudes for increasing λ, resulting in the correspondingly steeper decrease in the permeability. For $\lambda = 0.5$, Lewellyn predicted that the permeability would decrease to 4.7% of the $\lambda = 0$ value.

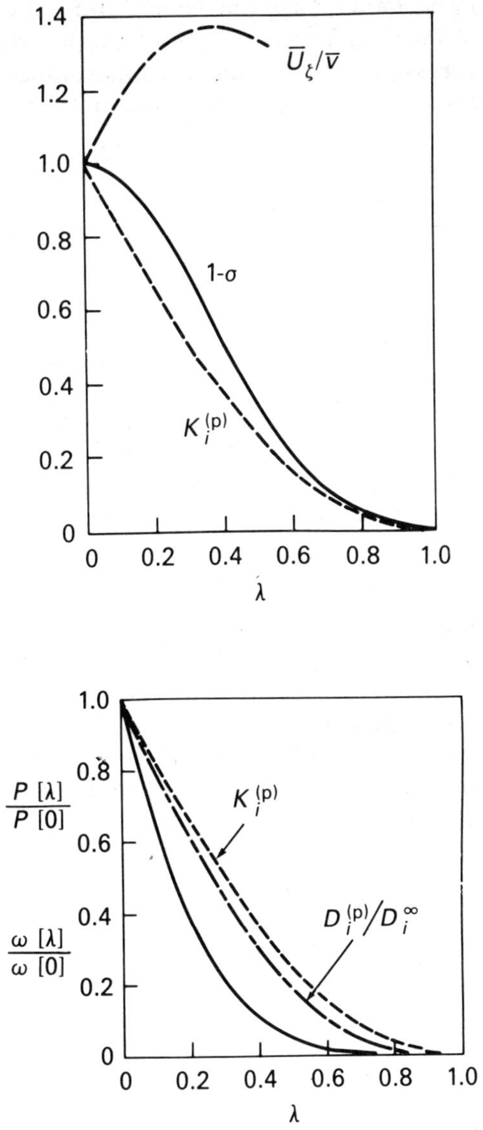

FIGURE 5.—Variation in transport coefficients with particle radius for a membrane having long right circular cylindrical pores as predicted by hydrodynamic theory for a very dilute solution or suspension of neutral, rigid spheres: (a) global reflection coefficient (solid curve) is equal to the product of the pore distribution coefficient (dashed curve) and the ratio of the pore cross-sectional area-averaged axial velocities (solute/solution); (b) normalized global diffusive permeability coefficient (solid curve) is equal to the product of the pore distribution coefficient (dashed curve) and the normalized pore cross-sectional area-averaged axial diffusion coefficient.

Lewellyn (17) also found a correspondence in form between the hydrodynamic equation 53 for the intrapore pressure drop and a non-equilibrium thermodynamic expression for the volumetric flux analogous to equation 3. In addition to confirming that in the dilute solution limit,

$$L_p = L_p^{(p)} \epsilon = r_t^2 \epsilon / 8 \eta h, \tag{83}$$

he formulated constraints under which Onsager reciprocity holds and

$$\sigma_d = \sigma_f. \tag{84}$$

An alternative proof of this equality for homoporous membranes had previously been offered by Levitt (18).

Lewellyn's approach to developing and refining the hydrodynamic transport theory, together with the antecedent treatment of Lightfoot *et al.* (16), starts with the Stefan–Maxwell equations as a rigorous alternative to the premises utilized in section 4.6. and its antecedents, e.g. (11).

4.10. Slit Pore Model

Hydrodynamic theory can be used in principle to predict the dilute solution transport coefficients in uniform pores of any shape. For example, Curry (15) and Ganatos *et al.* (19) evaluated the Kedem–Katchalsky coefficients for spherical solutes in slit-like pores representing the channels between adjacent endothelial cells lining blood capillaries. In place of equations 68 and 69, the pore averaging is accomplished according to

$$1 - \sigma_f^{(p)} = \frac{3}{2(1 - \lambda)} \int_0^{1-\lambda} g_\zeta \, (1 - \varpi^2) \, d\varpi, \tag{85a}$$

and,

$$\overline{D_i^{(p)}} = \frac{1}{1 - \lambda} \int_0^{1-\lambda} D_\zeta \, d\varpi, \tag{85b}$$

in which λ is the ratio of the solute radius to the pore half-width, and,

$$K_i^{(p)} = 1 - \lambda. \tag{86}$$

Figure 6 displays the more exact calculations of Ganatos *et al.* (19). Their tabulation of the results leaves some uncertainty as to the behavior of the coefficients for the limit $\lambda \to 1$, which is also the case for the cylindrical pore

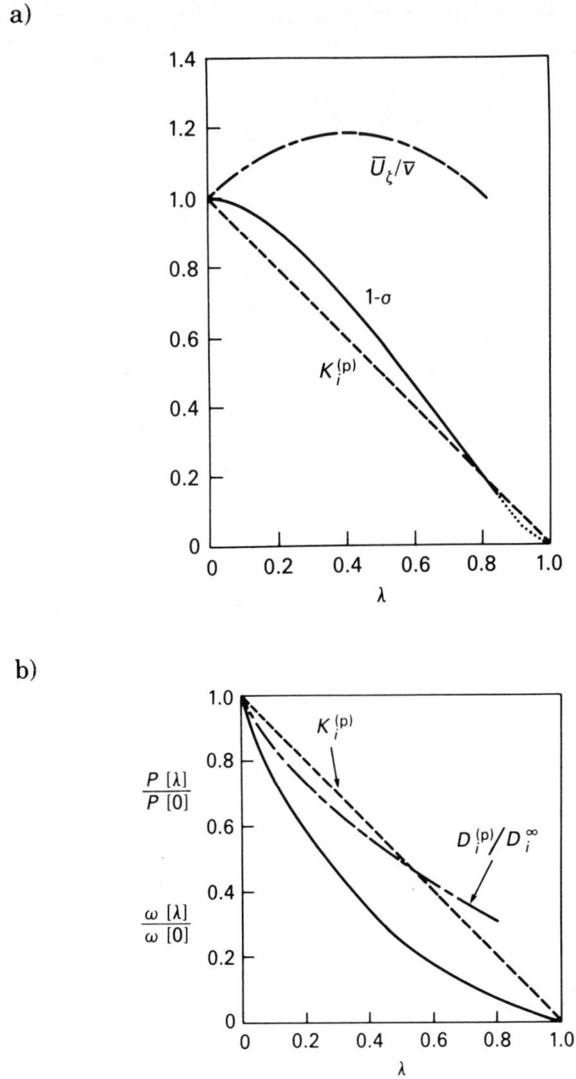

FIGURE 6.—Variation in transport coefficients with particle radius for a membrane having long, large aspect-ratio, slit-like pores as predicted by hydrodynamic theory for a very dilute solution or suspension of neutral, rigid spheres. (See Figure 5.)

predictions. The expression for the hydraulic permeability coefficient predicted from the plane Poiseuille flow equation is,

$$L_p = w^2\epsilon/12\ \eta h, \tag{87}$$

in which w is the pore width.

4.11. Incorporation of Non-Hydrodynamic Solute–Pore Wall Interactions

Other kinds of interactions, besides hydrodynamic, between a solute and the pore wall could alter the transport chracteristics of the pore. Intermolecular forces that give rise to attraction or repulsion between the solute and the membrane matrix are of particular importance. This might be most apparent for a charged solute interacting with a pore wall containing fixed charges, but the effect of van der Waals forces between neutral molecules could be appreciable. Such interactions might also produce torques on solutes with dipoles leading to preferred solute orientations with consequences for transport across the membrane.

In general, the force on a single, symmetric, spherical solute particle, $\boldsymbol{F}_{\mathrm{E}}$, can be expressed in terms of the gradient of the intermolecular particle–wall potential energy field, E,

$$\boldsymbol{F}_{\mathrm{E}} = - \nabla E. \tag{88}$$

For a long uniform pore the potential energy will only vary with radial position, except near the ends of the pore, and the force on the solute will be radially directed.,

$$F_{\mathrm{E}\varpi} = - (1/r_{\mathrm{t}})\, \partial E/\partial \varpi. \tag{89}$$

This force can then be included in the particle force balance, equation 57a. By an alternative procedure in which the method of moments was applied to the three-dimensional particle probability density conservation and continuity equations, Brenner and Gaydos (12) showed that, in the very dilute solution limit, equations 68 and 69 generalize to

$$1 - \sigma_{\mathrm{f}}^{(\mathrm{p})}[\lambda] = 2 \int_0^{1-\lambda} g_\zeta\, (1 - \varpi^2)\, \mathrm{e}^{-E[\varpi]/KT}\, \varpi\, \mathrm{d}\varpi \Big/ \int_0^{1-\lambda} \mathrm{e}^{-E[\varpi]/KT}\, \varpi\, \mathrm{d}\varpi, \tag{90}$$

and,

$$\overline{D_i^{(\mathrm{p})}}[\lambda] - \int_0^{1-\lambda} D_\zeta\, \mathrm{e}^{-E[\varpi]/KT}\, \varpi\, \mathrm{d}\varpi \Big/ \int_0^{1-\lambda} \mathrm{e}^{\,E[\omega]/KT}\, \varpi\, \mathrm{d}\varpi, \tag{91}$$

and the pore equilibrium distribution coefficient can be obtained from,

$$K_i^{(\mathrm{p})}[\lambda] = 2 \int_0^{1-\lambda} \mathrm{e}^{-E[\varpi]/KT}\, \varpi\, \mathrm{d}\varpi. \tag{92}$$

The corresponding global transport coefficients are given by equations 76 and 77 as in the neutral case, but with the pore coefficients replaced by equations 90–92. These results again apply only in the very dilute solution limit.

When the "external" force, $\boldsymbol{F}_E$, is negligible compared with the hydrodynamic and Brownian forces acting on the particles, the potential energy field takes on the values,

$$E[\varpi] = \begin{cases} 0, \text{ for } 0 \leq \varpi \leq 1 - \lambda \\ +\infty, \text{ for } 1 - \lambda < \varpi \leq 1, \end{cases} \tag{93}$$

with infinity corresponding to the impenetrability constraint. Positive potential values are associated with repulsion of the solute toward the centerline, negative potential with attraction toward the pore wall.

For equilibrium conditions, in very dilute solutions, the radial variation of solute probability density is proportional to the exponential term (the "Boltzmann factor") in equation 92. When the external forces are not negligible and the potential energy field differs from that in equation 93, the solute probability profile will be correspondingly altered. The equality, equation 80, between the pore distribution coefficient and the steric factor then no longer holds, in general.

4.12. Electrostatic Double-Layer Interaction Effects

According to equations 90–92, the hydrodynamic analyses utilized in predicting transport coefficient behavior for the very dilute neutral solute –neutral pore case can be similarly exploited to predict the coefficients when the solute and pore wall are charged, provided that the appropriate interaction potential field is known. Smith and Deen (23,24) followed this approach, together with derivation of interaction potential fields using the Gouy–Chapman theory for the diffuse double layers that form in solutions of electrolytes in proximity to charged surfaces.

The formation of a double-layer alters the potential energy field from that which would be predicted by the classical electrostatic theory for the interaction of charges embedded in a homogeneous dielectric medium. The spatial extent of the double layer is characterized by the Debye length. The ratio of the pore radius to the Debye length is a key dimensionless parameter for indicating the relative importance of electrostatic effects. In Smith and Deen's analysis this ratio is given by

$$\Xi = r_t \left[\frac{4\pi F^2}{\varepsilon_w RT} \sum_j (z_j^2 c_j) \right]^{1/2}, \tag{94}$$

in which F is the Faraday constant, ε_w is the solvent dielectric permittivity, and c_j and z_j are the molar concentration and valence, respectively, of the small electrolyte species j dissolved in the continuous liquid phase. The Debye length, and consequently the ratio Ξ and the electrostatic effects, are functions of the ionic strength of the solution as indicated by the summation in equation 94. A high ionic strength leads to a short Debye length and a weak effect of charge on transport behavior.

The linearized Poisson–Boltzmann equation was assumed to describe the variation of the electrical potential within the electrolyte solution filling the pore. A single rigid spherical particle, representing a large solute molecule or a colloidal particle, is located at an arbitrary position within the pore, as in Figure 4. Two different models for the particle were considered: one with uniform surface charge density (q_s) or uniform surface electrical potential (ψ_s), and the other a permeable particle with uniform volumetric charge density (q_V). The pore wall was taken to be impermeable, but characterized by either a uniform surface charge density (q_t) or uniform surface electrical potential (ψ_t). The dielectric properties of the sphere and the membrane matrix material were allowed to differ from each other and that of the electrolyte solution.

When the particle and pore surfaces have charges of similar sign, repulsive interaction reduces the pore distribution coefficient, as in Figure 7. At

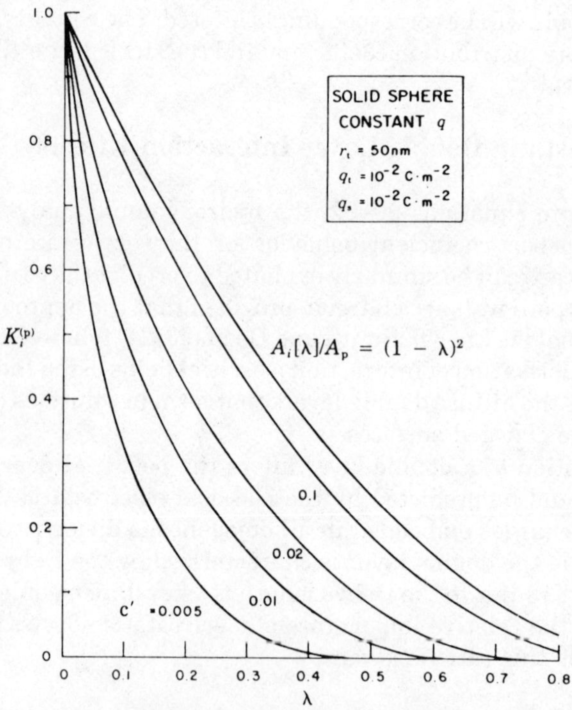

FIGURE 7.—Variation in the pore equilibrium distribution coefficient with particle radius for a charged solid sphere inside a charged pore. The surface charges are of *similar* sign and of constant density: q_t = cylindrical pore wall surface charge density, q_s = spherical particle surface charge density. The calculations assume the liquid phase to be an aqueous solution at 37°C of a 1:1 univalent electrolyte of molar bulk concentration c' outside the pore. Reproduced from Smith and Deen (22) with permission from Academic Press, Inc.

low ionic strengths (small Ξ) the interaction could be sufficiently strong to virtually exclude all particles from the pore. Electrostatic interaction effects vanish as $\Xi \to \infty$, although appreciable differences are predicted for values as large as $\Xi = 10$. An interaction is also predicted in situations in which one or the other surface has no fixed charge, because the charged surface induces a change in surface electrical potential in the other surface. According to the theory, only when the sphere is completely porous, as well as uncharged, will the interaction effect vanish if the pore wall is charged and the Ξ value is small or moderate.

When the sphere and the pore wall are oppositely charged, attractive interaction elevates the particle probability distribution near the pore centerline above that for the neutral case. The interaction becomes repulsive for radial positions near the wall and the probability distribution correspondingly diminishes below neutral case levels. The net effect for a pore cross-sectional area-averaged quantity like the distribution coefficient can either be an augmentation or a dimunition depending upon the various parameter values. The sample calculations of Figure 8, reproduced from Smith and Deen

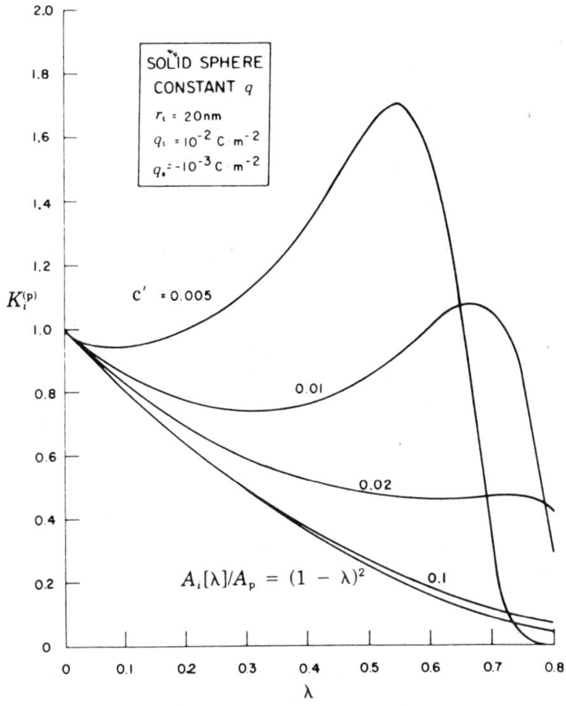

FIGURE 8.—Variation in the pore equilibrium distribution coefficient with particle radius for a charged solid sphere inside a charged pore. The surface charges are of *opposite* sign and of constant density. The calculations assume an aqueous solution at 37°C of a 1:1 univalent electrolyte at 37°C of molar bulk concentration c' outside the pore. Reproduced from Smith and Deen (22) with permission from Academic Press, Inc.

(24), exhibit a general enhancement of the distribution coefficient, but a complex dependence upon the ionic strength and particle-to-pore radius ratio.

An analysis of the electrostatic double-layer interaction effect on the diffusive hindrance factor, equation 82, was provided by Deen and Smith (25) for the situation in which the charge on the particle is of the same sign as that on the pore wall. These authors obtained good quantitative agreement between theory and experiments for negatively charged macromolecules in negatively charged track-etched capillary pore membranes. The diffusive permeabilitity decreases toward zero with decreasing ionic strength because of the increasing degree of exclusion of the macromolecules from the pores.

4.13. Concentration-Dependent Effects

The transport coefficient predictions thus far presented were based on the hydrodynamic interaction between an isolated sphere and a pore wall. As noted previously, they are applicable only to very dilute solutions in which the average particle-to-wall spacing is small compared with the average particle-to-particle spacing. Extensions of the theory are available which take both particle–wall and particle–particle interactions into account in a pair-wise additive manner. The particle–wall interaction is used to determine the potential energy, E. Simultaneous particle–particle and particle–wall interactions are expressed through a virial power series expansion in the external medium bulk concentration. For example, the pore equilibrium distribution coefficient expansion is,

$$K_i^{(p)} = K_0 + K_1\, c' + K_2\, (c')^2 + \ldots \tag{95}$$

The K_0 is the Henry's law constant (independent of concentration) whose variation with λ for the cylinder and slit pores was shown in Figures 5 and 6. The corresponding virial coefficients, K_1 and K_2 have been evaluated both by Anderson and Brannon (26) and by Glandt (27,28), showing that the pore distribution coefficient at finite concentrations is greater than the limiting value for $c' \to 0$.

As shown by Glandt (27), an interesting consequence of particle–particle interactions is that even in the absence of external forces acting on the solute, i.e. for an external potential field that is uniform across the pore interior, as in equation 93, the probability distribution function for particle centers varies in the radial direction. This is illustrated in Figure 9 by the family of dashed curves for the neutral case. The infinitely dilute concentration is uniform, but at finite external concentrations, the intrapore profile indicates increasing concentration with increasing radial position. The degree of increase at a given radius varies directly with the external concentration of the solute. Sample calculations are also displayed in Figure 9 for a charged solute in a pore of similar charge, showing again an augmentation of the ratio of intrapore to external phase solute concentration with increasing solute con-

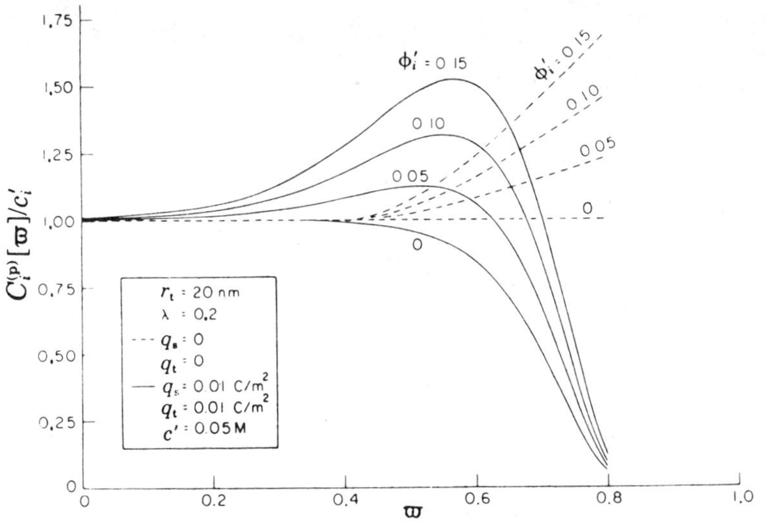

FIGURE 9.—Radial particle probability (concentration) profiles, $C_i^{(p)}[\varpi]/c_i'$, for a rigid spherical solute i of radius 4 nm inside a cylindrical pore of radius 20 nm. Solid curves are profiles for solute which has a uniform surface charge density, q_s, equal to that of the pore wall, $q_t = 0.01$ Coulomb/m^2. Dashed curves are solute profiles for uncharged solute and pore wall. The liquid phase is an aqueous solution at 25°C of a 1:1 univalent electrolyte with a bulk concentration outside the pore, $c' = 0.05$M. The various curves in each family correspond to different volume fractions, ϕ_i', of the solute i outside the pore in bulk solution. Reproduced from Mitchell and Deen (29) with permission from Elsevier Scientific Pub. Co.

centration. The elevation of the pore distribution coefficient by raising the external solute level tends to offset the depressing effect of the electrostatic particle–wall charge interaction.

As revealed by Figure 2, for $Pe >> 1$ there is little difference between the filtration reflection coefficient σ_f and the intrinsic solute rejection, $\mathcal{R}_i$, at least for dilute solutions. Equating the two coefficients is tantamount to neglecting the diffusive contribution to transmembrane transport. With this assumption, Mitchell and Deen (29) developed a virial expansion for the rejection coefficient in powers of the external solution solute concentration analogous to equation 95. Although their treatment assumes the lag coefficient g_ζ to be independent of radial position, the external solution concentration affects rejection through both the hydrodynamic and thermodynamic (partitioning) factors. Sample calculations displayed in Figure 10 suggest that both repulsive electrostatic and concentration effects tend to sharpen the range of solute size over which most of the variation in rejection occurs for the given

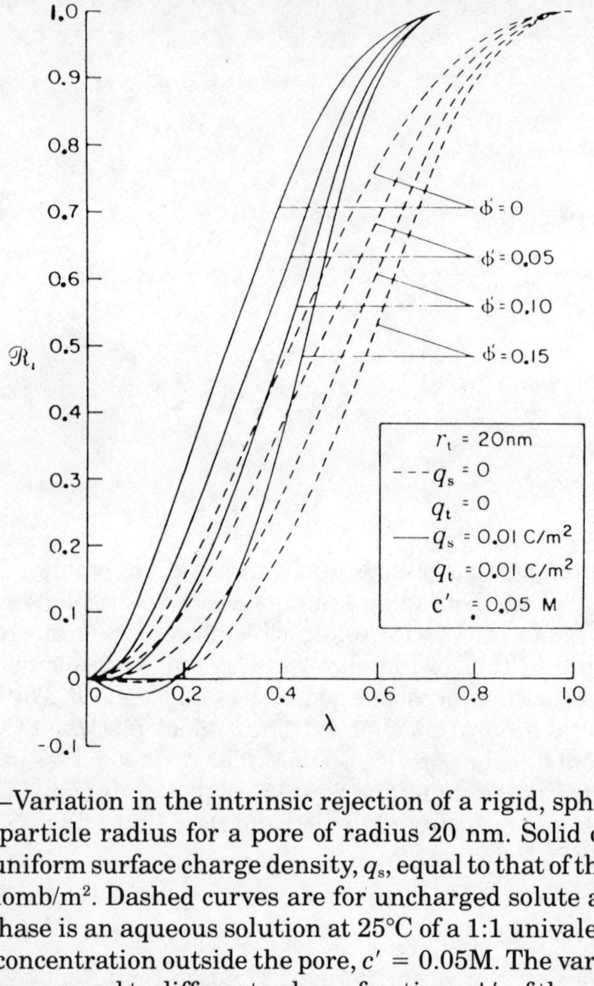

FIGURE 10.—Variation in the intrinsic rejection of a rigid, spherical solute i with solute particle radius for a pore of radius 20 nm. Solid curves are for solute with uniform surface charge density, q_s, equal to that of the pore wall, q_t = 0.01 Coulomb/m^2. Dashed curves are for uncharged solute and pore wall. The liquid phase is an aqueous solution at 25°C of a 1:1 univalent electrolyte with a bulk concentration outside the pore, c' = 0.05M. The various curves in each family correspond to different volume fractions, ϕ_i', of the solute i outside the pore in bulk solution. Reproduced from Mitchell and Deen (29) with permission from Elsevier Scientific Pub. Co.

pore size. This indicates one way in which the effects could be of interest in separation processes.

Other transport coefficients will exhibit solute concentration dependence. A virial expansion treatment for the osmotic reflection coefficient is provided by Adamski and Anderson (30), which supports the general decrease in the reflection coefficients with increasing solute concentration found by Mitchell and Deen (29). For transport involving concentration gradients in the axial direction within the pore, the analyses required appear too involved to consider in this brief survey (31).

The van't Hoff law, equation 3, is only valid for dilute solutions. Friedman and Meyer (32) have shown that the utilization of the van't Hoff law in the derivation of the Kedem–Katchalsky equations gives rise to a divergence in values between the osmotic (volume flux) and filtration (solute flux) reflection coefficients. These authors show that the two coefficients are simply related by,

$$\sigma_{\mathrm{d}} = \sigma_{\mathrm{f}}\,(1 + \Gamma)/(1 - \overline{\phi_i}), \tag{96}$$

in which

$$\overline{\phi_i} = \tilde{V}_i\,(c' + c'')/2 \tag{97}$$

is the mean volume fraction of solute in the external solutions, and

$$\Gamma = \overline{\mathrm{d}\ln\gamma_i/\mathrm{d}\ln c_i'} \tag{98}$$

is the corresponding mean measure of nonideality in the solution expressed by the concentration dependence of the solute activity coefficient, γ_i. If the Kedem–Katchalsky equations are expressed in terms of osmotic pressure differences, the osmotic reflection coefficient is redefined such that it retains its equality with σ_{f}. The latter exhibited little or no concentration dependence in measurements by the same authors (32) of sucrose transport across a Cuprophan membrane.

Dependence of transport coefficients of one solute upon the concentration of a second solute has been observed, for example, in the case of diffusive permeability (33). When a difference in concentration of the second solute exists between the two external media, the concentration variation of the transport coefficient may depend upon whether the net transport is in the same or opposite directions for the two solutes. The underlying effect, interaction between solutes moving at different velocities, has been termed "solute drag." It is an example of concentration-dependent interactions that occur in spatially inhomogeneous mixtures (34).

4.14. Heteroporosity and Non-Uniformity Effects

The availability of the track-etched capillary pore membranes described in Chapter 1 has been a major stimulus to research on transport in porous membranes. The regular geometry of the pores in these membranes is conducive to better controlled, more easily interpretable experiments. The membranes are nearly homoporous and their pore walls conform to a considerable degree to simple coordinate systems. Consequently, the single pore, uniform cross-section models surveyed here are more directly applicable to the phenomena observed with these membranes than most any other type of membranes.

Although the single pore transport equations can be used to construct models for heteroporous membranes, the common inverse procedure of fitting the homoporous Kedem–Katchalsky equations to experimental data obtained from a heteroporous membrane could be risky. Bresler *et al.* (35) pointed out that in such circumstances the range of validity of the solute flux equation may be small and only attainable under hidden constraints, such as a requirement for $\Delta p' >> \Delta\pi'$. Fortunately, the calculations of Wendt *et al.* (36) for various series-parallel arrays of pores with interconnections simulating both heteroporosity and non-uniformity in pore cross-sectional area indicate that, except in cases of large Peclet numbers or internal concentration polarization, the errors incurred could be confined within upper and lower bounds. These bounds are not very far apart, and Knierim *et al.* (37) showed that the bounds on cumulative pore-size distribution and on the solute flux –volume flow relationship could be considerably narrowed by judicious experimental measurements.

In addition, Friedman and Meyer (32) found that the relationship between the solute flow and volume flow reflection coefficients was independent of heteroporosity. Experiments with a variety of commercial, polymeric hemodialysis membranes support the utility of a homoporous, hydrodynamic model for making useful estimates of the dependence of transport coefficients on solute size (38,39).

4.15. External Boundary-Layer and Entrance Effects

In the external fluid medium in the vicinity of a pore entrance, the convergence or divergence of fluid streamlines alters the resistance to volumetric flux and the convergence of solute flux lines alters the diffusional resistance. These entrance effects are most pronounced relative to the membrane resistance for relatively short pores, i.e., $\Lambda = h/r_t = O(1)$ or less, and for relatively thick boundary layers, i.e., $\delta/r_t >> 1$ and $\delta_c/r_t >> 1$. Here, δ is the momentum boundary layer thickness and δ_c is the concentration boundary layer thickness. The ratio of the two boundary layer thicknesses generally varies according to (6),

$$\delta_c/\delta \propto (\eta/\rho\, D_i)^{-1/3}, \tag{99}$$

in which ρ and η are the density and viscosity of the external medium, respectively, and D_i is the solute binary diffusion coefficient. In liquid solutions generally $\delta_c < \delta$, hence $\delta_c/r_t >> 1$ is the more important constraint. The concentration boundary layer thickness for a constant diffusivity is given by,

$$\delta_c = D_i/k_i, \tag{100}$$

in which k_i is the solute mass transfer coefficient described in the chapter on Concentration Polarization.

In the limit as $\lambda \rightarrow 0$, the solute reflection coefficient goes to zero and the volumetric flux across the capillary pore membrane becomes uncoupled from effects of solute concentration, at least for the neutral solute–neutral pore case. On the assumption that the resistances are additive, the overall volumetric flux resistance, R_{VT}, can be decomposed into,

$$R_{VT} = (p' - p'')/J_V = R'_{V\delta} + 2\,R_{Ve} + R_{Vm} + R''_{V\delta}. \tag{101}$$

Resistances in the boundary layers, $R'_{V\delta}$ and $R''_{V\delta}$, do not seem to have been well described in the literature and will be neglected. The pore inlet and exit are taken to be symmetric with identical resistances R_{Ve}. Dagan *et al.* (40) showed theoretically that for widely spaced pores, the entrance resistance can be accurately approximated by,

$$R_{Ve} = 3\pi\eta/r_t\epsilon, \tag{102}$$

The membrane resistance as obtained from equations 3 and 83 is,

$$R_{Vm} = 8\eta h/r_t^2\epsilon, \tag{103}$$

and the hydraulic permeability is thus given by,

$$L_p = 1/(2R_{Ve} + R_{Vm}) = r_t^2\epsilon/(8\eta h + 3\pi\eta\, r_t). \tag{104}$$

The entrance resistance corresponds to a local pressure variation in a region extending approximately one pore diameter from the plane of the pore opening (40). The pressure differences from both the inlet and exit regions have been added to the pressure drop within the pore to arrive at this last expression.

The solute diffusional resistance in the neutral, small solute limit can similarly be represented in the absence of convection by

$$R_{iT} = (c' - c'')/J_i = R'_{i\delta} + 2R_{ie} + R_{im} + R''_{i\delta}, \tag{105}$$

in which the boundary layer resistances are

$$R'_{i\delta} = 1/k'_i = \delta'_c/D_i, \text{ and, } R''_{i\delta} = 1/k''_i = \delta''_c/D_i. \tag{106}$$

Malone and Anderson (41) found experimentally that for $8 < r_t < 500$ and $\epsilon < 0.5$, the entrance resistance is given to a good approximation by,

$$R_{ie} = \pi r_t/4\,D_i\,\epsilon, \tag{107}$$

and that it is additive with the membrane resistance given by

$$R_{im} = h/D_i\,\epsilon. \tag{108}$$

Consequently, the diffusive permeability in this limit is,

$$\omega RT = 1/(2R_{ie} + R_{im}) = D_i\epsilon/(h + \tfrac{1}{2}\pi r_t). \tag{109}$$

The entrance resistance given in equation 107 is the resistance for an isolated pore. Brunn (42) has presented an analysis for interaction between pores in pure diffusion which provides a correction with a dependence upon $\epsilon^{1/2}$ and the spatial distribution of the pores.

The entrance region for diffusion extends outward from the plane of the pore opening analogously to the situation for resistance to flow. This has implications for the interpretation of c_{int} and p_{int}.

The hydrodynamic analysis of situations in which $\lambda \neq 0$ is difficult because a particle must be introduced and the particle surface, together with the membrane face and the pore wall, constitutes a complex set of geometries. Nevertheless, significant progress has been made recently (43–45). These studies have yielded values for resistance coefficients, analogous to f_ζ and g_ζ, for a spherical particle outside a pore from which it is possible to predict, *inter alia,* the variation of the local solute diffusion coefficient in the ζ-direction in the entrance region. One of the most important findings is that in the entrance region particles migrate across steamlines. Even with a perfectly random distribution of particles in the external reservoir, the migration leads to a non-uniform probability distribution for particles in the radial direction across the plane of the pore opening. Interestingly, the analyses (43,45) suggest the distribution has a minimum at the pore axis and a maximum near the pore wall. However, for such a distribution to occur, convection would have to be dominant over diffusion. Munch *et al.* (46) note that an appropriate measure would be an entrance Peclet number much greater than unity, $r_t \overline{U}_\zeta/D_i \gg 1$. These authors argue that under such conditions one cannot assume equilibrium between the external medium and the solution just inside the pore, such as was invoked in arriving at equations 11 and 12. For long pores, the requirement that the entrance Peclet number be much less than unity is more stringent than the requirement for the one-dimensional transport assumption in section 4.5.

4.16. Solute and Pore Configuration Effects

The lack of spherical symmetry in solute particles can introduce a number of complications in analyzing transport across porous membranes. Elongated rods or circular disks are convenient shapes to consider in studying configuration effects since they retain line symmetry. Inside a pore the centers of such particles potentially have a larger accessible cross-sectional area than spheres of the same volume for small λ, i.e., a larger distribution coefficient. Because of rotational Brownian motion the predicted equilibrium distribution coefficient for a rod is considerably smaller than that for a volume equivalent sphere (47). However, when the volume flux is non-zero, the converging streamlines at the pore entrance can facilitate end-on particle orientation (48) and alter the sieving from the equilibrium level. This would be an example of a non-equilibrium entrance effect in addition to those mentioned in the previous section. For neutral, rigid rods in uniform pores of various shapes, Anderson (47) noted that under equilibrium entrance conditions the osmotic reflection coefficient could be correlated by

$$\sigma_d = (1 - K_i^{(p)})^2. \tag{110}$$

Long *et al.* (49) obtained qualitative agreement with equation 110 in rejection studies of the rod-like tobacco mosaic virus.

Inside the pore, shearing of the solvent (suspending medium) induces rotation of the particles. The rotational velocity of the particles is dependent upon orientation relative to the pore axis which leads rods to spend a larger fraction of time oriented parallel to the pore axis than perpendicular (50). Similarly, disks will spend more time moving edge-on, than face-on. The axial component of Brownian motion is more rapid for the edge-on and end-on orientations than for broad-side orientations. Orientation effects tend to be obscured by the randomizing rotational Brownian motions. Brownian motion is dominant when the local shear rate, G, is much less than the rotational diffusion coefficients, D_Ω (the ratio, G/D_Ω, is a rotational Peclet number). Rotational Peclet number criteria could also be relevant to orientation effects at the pore entrance such as was mentioned above.

Deformability can enhance orientation in pores and leads to radial migration of particles (51). Since orientation, deformability and radial migration are shear rate dependent (and G is proportional to $\bar{v}$), these effects can lead to transport coefficients that vary with the volumetric flux. Munch *et al.* (46) found that, through deformation, linear polyacrylamide macromolecules were able to pass through pores of radii smaller than the solute's Stokes–Einstein radius. Long and Anderson (52) found similar behavior for polystyrene solutions, although with larger pores the solutions behaved like rigid sphere suspensions as long as the solvent flow rate through the pore was sufficiently low. These authors introduced a criterion for deviation from the rigid sphere model based on the product of shear rate, G, and chain relaxation time, τ. Cannell and Rondelex (53) made measurements of polystyrene diffusion through pores of track-etched membranes and likewise were able to correlate the hindrance effect for $\lambda < 0.7$ using a rigid sphere model.

Studies of the transport of partially flexible, neutral macromolecules (dextrans) through pores corresponding to intermediate values of λ have yielded somewhat conflicting values for transport coefficients. Schultz *et al.* (54) conducted experiments in which the volumetric flux was in the direction opposite to the solute bulk concentration gradient and obtained osmotic reflection coefficient values lower than the rigid sphere hydrodynamic model predictions, which they suggested might be related to the lack of molecular rigidity. Long *et al.* (49) indicated agreement between the theory and their filtration reflection coefficient measurements with the same solutes and membranes. Based on pure diffusion measurements, Deen *et al.* (55) reported significantly less hindrance for dextran than predicted and more hindrance for cross-linked ficoll.

For flexible polyelectrolytes, pH and ionic strength can have dramatic effects on, at least, the size and probably the deformability of the molecules. The strong effect of ionic strength on the size and rejection of 5×10^6 dalton linear polyacrylamide polymer was shown by Munch *et al.* (46). Long *et al.* (49) found a noticeable ionic strength dependence on the reflection coefficient for a high molecular weight, charged dextran which they ascribed to molecular configuration changes.

4.17. Solute Adsorption Effects

Solute adsorption is a modifying influence in many membrane processes. As discussed in the Ultrafiltration chapter, macromolecular adsorption is a major component in fouling, which has important negative functional and economic consequences. In research on membrane phenomena, adsorption is a complicating factor, often compromising experiments or leading to equivocal interpretation of results. On the positive side, understanding of the mechanisms and dynamics of adsorption could have benefits in such diverse applications as biotechnological processes (enzyme immobilization), separation techniques (affinity chromatography), and the development of biocompatible synthetic materials. Capillary pore membranes, because of their controlled surface configurations, are useful tools in adsorption studies.

Surface physicochemical properties are important factors influencing adsorption. Most of the track-etched membranes used to date have been made from either of two materials. Muscovite mica has been used in custom-made membranes. Because of the crystal structure of this material, the pores are rhomboidal in shape in the cross-section perpendicular to the pore axis. Polycarbonate is the predominant material in the commercial track-etched membranes produced by Nuclepore Corporation (see the Microfiltration chapter). The pores in the polycarbonate films are circular, but taper slightly in progressing inward from the membrane face or faces exposed to the etching solution during pore formation.

Long *et al.* (49) reported measuring a net charge of about -0.05 μCoul./m^2 on the pore walls of clean Nuclepore membranes. Native mica pore walls are likewise negatively charged (46). Small neutral, inert, hydrophilic solutes in aqueous solution do not appear to bind to these membranes. For neutral macromolecules, Schultz *et al.* (54) noted a small apparent reduction in pore cross-sectional areas when neutral dextran T-500 was present as inferred from hydraulic permeability measurements. The hydraulic permeabilities were restorable to original values by flushing with water, suggesting reversible binding by the dextran. Other investigators (49, 53) have not detected consistent effects ascribable to neutral macromolecule adsorption in either aqueous or organic solvent solutions. A small reduction in apparent pore radius was noted (52) after ultrafiltration of polystyrene in mixed organic solvents using mica membranes, but this effect could have been due to plugging since no effect was detectable after soaking the membranes in polymer solution without transmembrane flow.

Strong adsorption has been reported for charged molecular solutes. Negatively charged, high molecular weight blue dextran exhibited multilayer adsorption to nominally negatively charged polycarbonate membranes (49). Greater reduction in apparent pore radius was observed during flow of the blue dextran solution through the pores over that observed under static conditions, although it is not clear whether that was due to increased adsorption, conformational change in the adsorbed layer or other factors. Protein adsorption appears to be much more complicated and similarly not well understood.

Generally, proteins in solution have a number of positively and negatively charged side groups. Since the pKs of the acidic and basic groups vary among themselves and are modified by the molecular environment, the numbers of negative and positive charges per molecule are a function of pH. The electrostatic effects as modified by molecular size and conformation changes and screening of the charges will vary as well with ionic strength. In anticipation of the important role of electrostatics in adsorption most studies at least control the pH and ionic strength.

A number of studies have utilized bovine serum albumin (BSA) as a common, readily available protein, even though its binding properties are complex. It serves, in part, as an important multifunctional carrier protein in blood plasma, and is suspected of being an important constituent of the blood vessel wall through binding to other constituents. BSA molecules are approximately prolate spheroidal in shape, with major and minor semi-axes of about 70Å and 40Å, and a Stokes–Einstein equivalent sphere radius of 37Å (5).

Wong and Quinn (56) used BSA reacted with L-cystine to block free sulfhydryl groups and stabilize against dimerization. Electrolyte diffusion or electrical conductivity measurements were used to determine pore radii. These indicated a reduction of 70Å in the apparent pore radii of mica membranes in the presence of BSA in solutions of various ionic strength at pH 8. The authors concluded that BSA adsorbs as a monolayer. Subsequent, more detailed investigations of BSA adsorption in mica membranes have been undertaken in the same laboratory (57).

Munch *et al.* (46) also studied BSA adsorption on mica membranes and used the flow of protein-free solution across the membrane and a modified Hagen–Poiseuille equation to determine pore radii. They found a mean equivalent pore radius decrease of 48Å after static exposure to BSA solutions at pH 6.8, which appeared to be independent of ionic strength. They suggested that this primary adsorption represented saturation of the mica-binding sites irrespective of the electrostatic interaction prior to adsorption between this net negatively charged solute and the negatively charged wall. When BSA solution was pumped through the pores, they found an additional time-dependent reduction in apparent pore radius which was dependent upon ionic strength. This secondary adsorption was at least partially reversible. They postulated that the secondary layer was "probably caused by flow-induced collisions between dissolved and adsorbed BSA molecules inside the pores."

Schultz *et al.* (54) used polycarbonate membranes and BSA solution, 8% by weight, in a buffer at pH 5.05. Their experiments indicated that albumin adsorption to the pore walls was probably irreversible with smaller reductions in apparent pore radius than those reported for mica membranes under the conditions of their osmotic flow experiments. However, they did not report measurements of pore reduction after soaking membranes in BSA solution to determine if uniform adsorption could have been obtained. For comparison they did determine greater apparent pore reductions for osmotic flow in the presence of gamma-globulin solutions than the reductions due to albumin.

In some of the earliest and most detailed tests of the hydrodynamic model for quantifying hindered diffusion in mica membranes, Beck and Schultz (58, 59) noted that a time dependence in the calculated diffusive permeability for the enzyme ribonuclease might be explained by adsorption or denaturation of the protein. Investigations of adsorption in capillary pore membranes is still largely at a qualitative observation stage. It is hoped that the coupling of refined experimental techniques, together with the introduction of double-layer models, other intermolecular interaction theories and quantitative transport phenomena concepts, will lead to improved understanding of the various processes involved.

5. SUMMARY

The Kedem–Katchalsky equations are a particular set of one-dimensional phenomenological equations for two components, a solute and a solvent, transported at constant temperature. Other, less familiar membrane transport equations have been formulated, some equivalent, some of greater generality. The interrelationships among various formulations have been concisely and rigorously shown by Mason and Viehland (60). The objective of most of these formulations has been to obtain generality independent of the structural details of the membranes. The utility of such approaches lies in providing a means to evaluate transport coefficients from experimental measurements. The coefficients indicate how well a given membrane system functions. However, to advance the science of membranes, one desires a better understanding of *how* membranes function, not just how well. The role of models, such as the continuum hydrodynamic model, is to provide a geometrical structure and mechanistic basis from which the membrane function can be predicted. The hydrodynamic model satisfies thermodynamic constraints as seen in section 4.8., and additional phenomena have been incorporated to enlarge its predictive capacity, e.g., the inclusion of electrostatic interactions discussed in section 4.11., and intermolecular interactions discussed in section 4.

While this chapter has focused on the hydrodynamic-based pore theory as a well-established model for transport in porous membranes, it is not the only one. Just as physiology has provided important applications for pore theory over the past thirty years (61, 62), so it is currently stimulating interest in fiber matrix theory (63) as an alternative model for describing transport across the endothelial lining of blood vessels and through the interstitial matrix surrounding cells in bulk tissue. The geometry in fiber matrix theory is similar to two-dimensional unconsolidated porous media. Concepts from studies of transport in three-dimensional porous media should find expanded utility in membrane science.

The focus of both pore theory and fiber matrix theory has been on transport between liquid phases bathing and imbibed in a porous membrane. Gas transport across porous membranes is also an old subject that is benefitting from recent refinements (64) and model development (65). The future should see continued advancement in these areas.

REFERENCES

1. Kedem, O., and Katchalsky, A.: 1958, "Thermodynamic analysis of the permeability of biological membranes to non-electrolytes." Biochim. Biophys. Acta 27, pp. 229–246.
2. Bresler, E.H., and Groome, L.J.: 1981, "On equations for combined convective and diffusive transport of neutral solute across porous membranes." Am. J. Physiol. 241 (Renal Fluid Electrolyte Physiol. 10), pp. F469–F476.
3. Patlak, C.S., Goldstein, D.A., and Hoffman, J.F.: 1963, "The flow of solute and solvent across a two-membrane system." J. Theor. Biol. 5, pp. 426–442.
4. Lightfoot, E.N.: 1974, "Transport Phenomena and Living Systems. Biomedical Aspects of Momentum and Mass Transport." New York, J. Wiley, p. 178.
5. Bean, C.P.: 1972, "The physics of porous membranes—neutral pores," in "Membranes, A Series of Advances, Vol. 1 Macroscopic Systems and Models." G. Eisenman, ed. New York, M. Dekker, p. 1–54.
6. Bird, R.B., Stewart, W.E., and Lightfoot, E.N.: 1960, "Transport Phenomena." New York, J. Wiley.
7. Fürth, R., and Cowper, A.D.: 1956, "Investigations on the Theory of the Brownian Movement by Albert Einstein." Edited translation of articles in Annalen der Physik (1906) 19, pp. 289–306; (1911) 34, pp. 591–592.
8. Stein, W.D.: 1967, "The Movement of Molecules Across Cell Membranes." New York, Academic Press.
9. Happel, J. and Brenner, H.: 1965, "Low Reynolds Number Hydrodynamics." Englewood Cliffs, N.J., Prentice-Hall.
10. Bungay, P.M. and Brenner, H.: 1973, "Pressure drop due to the motion of a sphere near the wall bounding a Poiseuille flow." J. Fluid Mech. 60, pp. 81–96.
11. Anderson, J.L., and Quinn, J.A.: 1974, "Restricted transport in small pores: A model for steric exclusion and hindered particle motion." Biophys. J. 14, pp. 130–150.
12. Brenner, H., and Gaydos, L.J.: 1977, "The constrained Brownian movement of spherical particles in cylindrical pores of comparable radius: Models of the diffusive and convective transport of solute molecules in membranes and porous media." J. Coll. Interface Sci. 58, pp. 312–356.
13. Bungay, P.M. and Brenner, H.: 1973, "The motion of a closely-fitting sphere in a fluid-filled tube." Int. J. Multiphase Flow 1, pp. 25–56.
14. Anderson, J.L., and Malone, D.M.: 1974, "Mechanism of osmotic flow in porous membranes." Biophys. J. 14, pp. 957–982.
15. Curry, F.E.: 1974, "A hydrodynamic description of the osmotic reflection coefficient with application to the pore theory of transcapillary exchange." Microvasc. Res. 8, pp. 236–252.
16. Lightfoot, E.N., Bassingthwaighte, J.B. and Grabowski, E.F.: 1976, "Hydrodynamic models for diffusion in microporous membranes." Ann. Biomed. Engng. 4, pp. 78–90.
17. Lewellen, P.C.: 1982, "Hydrodynamic Analysis of Microporous Mass Transport." Ph.D thesis, University of Wisconsin-Madison.

18. Levitt, D.G.: 1975, "General continuum analysis of transport through pores. I. Proof of Onsager's reciprocity postulate for uniform pore." Biophys. J. 15, pp. 533–551.
19. Ganatos, P., Weinbaum, S., Fischbarg, J., and Liebovitch, L.: 1980, "A hydrodynamic theory for determining the membrane coefficients for the passage of spherical molecules through an intercellular cleft." Adv. in Bioengng., Amer. Soc. Mech. Engrs., New York, pp. 193–196.
20. Hirschfeld, B.R., Brenner, H. and Falade, A.: 1984, "1st- and 2nd-order wall effects upon the slow viscous asymmetric motion of an arbitrarily-shaped, arbitrarily-positioned and arbitrarily-oriented particle within a circular-cylinder." PhysicoChem. Hyd. 5(2), pp. 99–133.
21. Wang, H. and Skalak, R.: 1969, "Viscous flow in a cylindrical tube containing a line of spherical particles." J. Fluid Mech. 38, pp. 75–96.
22. Paine, P.L. and Scherr, P.: 1975, "Drag coefficients for the movement of rigid spheres through liquid-filled cylindrical pores." Biophys. J. 15, pp. 1087–1091.
23. Smith III, F.G. and Deen, W.M.: 1980, "Electrostatic double-layer interactions for spherical colloids in cylindrical pores." J. Colloid Interf. Sci. 78, pp. 444–465.
24. Smith, F.G. and Deen, W.M.: 1983, "Electrostatic effects on the partitioning of spherical colloids between dilute bulk solution and cylindrical pores." J. Colloid Interf. Sci. 91, pp. 571–590.
25. Deen, W.M. and Smith III, F.G.: 1982, "Hindered diffusion of synthetic polyelectrolytes in charged microporous membranes." J. Membrane Sci. 12, pp. 217–237.
26. Anderson, J.L. and Brannon, J.H.: 1981, "Concentration dependence of the distribution coefficient for macromolecules in porous media." J. Polymer Sci: Polymer Phys. Ed. 19, pp. 405–421.
27. Glandt, E.D.: 1980, "Density distribution of hard-spherical molecules inside small pores of various shapes." J. Colloid Interf. Sci. 77, pp. 512–524.
28. Glandt, E.D.: 1981, "Distribution equilibrium between a bulk phase and small pores." AIChE J. 27, pp. 51–59.
29. Mitchell, B.D. and Deen, W.M.: 1984, "Theoretical effects of macromolecule concentration and charge on membrane rejection coefficients." J. Membr. Sci. 19, pp. 75–100.
30. Adamski, R.P. and Anderson, J.L.: 1983, "Solute concentration effect on osmotic reflection coefficient." Biophys. J. 44, pp. 79–90.
31. Anderson, J.L.: 1982, "Concentration effects on distribution of macromolecules in small pores." Adv. Colloid Interf. Sci. 16, pp. 391–401.
32. Friedman, M.H. and Meyer, R.A.: 1981, "Transport across homoporous and heteroporous membranes in nonideal, nondilute solutions-I. Inequality of reflection coefficients for volume flow and solute flow." Biophys. J. 34, pp. 535–544.
33. Van Bruggen, J.T., Boyett, J.D., van Bueren, A.L. and Galey, W.R.: 1974, "Solute flux coupling in a homopore membrane." J. Gen. Physiol. 63, pp. 639–656.
34. Cussler, E.L.: 1976, "Multicomponent Diffusion." Amsterdam, Elsevier Scientific Publ. Co.
35. Bresler, E.H., Mason, E.A. and Wendt, R.P.: 1976, "Appraisal of equations for neutral solute flux across porous sieving membranes." Biophys. Chem. 4, pp. 229–236.
36. Wendt, R.P., Mason, E.A. and Bresler, E.H.: 1976, "Effect of heteroporosity on flux equations for membranes." Biophys. Chem. 4, pp. 237–247.
37. Knierim, K.D., Waldman, M. and Mason, E.A.: 1984, "Bounds on solute flux and pore-size distributions for non-sieving membranes." J. Membr. Sci. 17, pp. 173–203.
38. Wendt, R.P., Klein, E., Bresler, E.H., Holland, F.F., Serina, R.M. and Villa, H.: 1979, "Sieving properties of hemodialysis membranes." J. Membr. Sci. 5, pp. 23–49.
39. Klein, E., Holland, F.F. and Eberle, K.: 1979, "Comparison of experimental and calculated permeability and rejection coefficients for hemodialysis membranes." J. Membr. Sci. 5, pp. 173–188.
40. Dagan, A., Weinbaum, S. and Pfeffer, R.: 1982, "An infinite series solution for the creeping motion through an orifice of finite length." J. Fluid Mech. 115, pp. 505–523.
41. Malone, D.M. and Anderson, J.L.: 1977, "Diffusional boundary-layer resistance for membranes with low porosity." AIChE J. 23, pp. 177–184.
42. Brunn, P.O.: 1984, "Interaction between pores in diffusion through membranes of arbitrary thickness." J. Membr. Sci. 19, pp. 117–136.

43. Davis, A.M.J., O'Neill, M.E. and Brenner, H.: 1981, "Axisymmetric Stokes flows due to a rotlet or stokeslet near a hole in a plane wall: filtration flow." J. Fluid Mech. 103, pp. 183–205.
44. Dagan, Z., Weinbaum, S. and Pfeffer, R.: 1982, "General theory for the creeping motion of a finite sphere along the axis of a circular orifice." J. Fluid Mech. 117, pp. 143–170.
45. Dagan, Z., Weinbaum, S. and Pfeffer, R.: 1982, "Theory and experiment on the three-dimensional motion of a freely suspended spherical particle at the entrance to a pore at low Reynolds number." Chem. Engng. Sci. 38, pp. 583–596.
46. Munch, W.D., Zestar, L.P. and Anderson, J.L.: 1979, "Rejection of polyelectrolytes from microporous membranes." J. Membr. Sci. 5, pp. 77–102.
47. Anderson, J.L.: 1981, "Configurational effects on the reflection coefficient for rigid solutes in capillary pores." J. Theor. Biol. 90, pp. 405–426.
48. Auvray, L.: 1981, "Solutions de macromolécules rigides: effets de paroi, de confinement et d'orientation par un écoulement," J. Physique 42, pp. 79–95.
49. Long, J.D., Jacobs, D.L. and Anderson, J.L.: 1981, "Configurational effects on membrane rejection." J. Membr. Sci. 9, pp. 13–27.
50. Goldsmith, H.L. and Mason, S.G.: 1967, "The microrheology of dispersions." In "Rheology: Theory and Applications," vol. 4, Eirich, F.R. (ed.), New York, Academic Press, pp. 85–250.
51. Brenner, H.: 1966, "Hydrodynamic resistance of particles at small Reynolds number." In "Advances in Chemical Engineering," vol. 6, New York, Academic Press, pp. 287–438.
52. Long, T.D. and Anderson, J.L.: 1984, "Flow-dependent rejection of polystyrene from microporous membranes." J. Polymer Sci.: Polymer Phys. Ed. 22, pp. 1261–1281.
53. Cannell, D.S. and Rondelez, F.: 1980, "Diffusion of polystyrenes through microporous membranes." Macromolecules 13, pp. 1599–1602.
54. Schultz, J.S., Valentine, R. and Choi, C.Y.: 1979, "Reflection coefficients of homopore membranes: Effect of molecular size and configuration." J. Gen. Physiol. 73, pp. 49–60.
55. Deen, W.M., Bohrer, M.P. and Epstein, N.B.: 1981, "Effects of molecular size and configuration on diffusion in microporous membranes." A.I.Ch.E.J. 27, pp. 952–959.
56. Wong, J. and Quinn, J.A.: 1976, "Hindered diffusion of macromolecules in track-etched membranes." In "Colloid and Interface Science," vol. 5: "Biocolloids, Polymers, Monolayers, Membranes and General Papers," Kerker, M. (ed.). New York, Academic Press, pp. 169–180.
57. Ph.D. thesis research under the supervision of Prof. J.A. Quinn, University of Pennsylvania, Yavorsky, D.Y.: 1981, "Static and hydrodynamic studies of the conformation of adsorbed macromolecules at the solid/liquid interface." Rodilosso, P.D.: 1984, "Determination of the partition coefficient for macromolecules in porous media; Potential flows of mass and charge about solute obstacles in model membranes."
58. Beck, R.E. and Schultz, J.S.: 1970, "Hindered diffusion in microporous membranes with known pore geometry." Science 170, pp. 1302–1305.
59. Beck, R.E. and Schultz, J.S.: 1972, "Hindrance of solute diffusion within membranes as measured with microporous membranes of known pore geometry." Biochim. Biophys. Acta 255, pp. 273–303.
60. Mason, E.A. and Viehland, L.A.: 1978, "Statistical-mechanical theory of membrane transport for multicomponent systems: Passive transport through open membranes." J. Chem. Physics 68(8): pp. 3562–3573.
61. Renkin, E.M. and Curry, F.E.: 1979, "Transport of water and solutes across capillary endothelium." In "Membrane Transport in Biology," Vol. IVA, G. Giebisch et al. (eds.), Heidelberg, Springer-Verlag, pp. 1–45.
62. Crone, C. and Christensen, O.: 1979, "Transcapillary transport of small solutes and water." In "International Review of Physiology, Cardiovascular Physiology III, Vol. 18, Guyton, A.C. and Young, D.B. (eds.), Baltimore, University Park Press, pp. 149–213.
63. Curry, F.E.: 1984, "Mechanics and thermodynamics of transcapillary exchange." In "Handbook of Physiology," Sect. 2, "The Cardiovascular System," Vol. 4, "Microcirculation," Ch. 8, pp. 309–374.
64. Matson, S.L. and Quinn, J.A.: 1977, "Knudsen diffusion through non-circular pores: Textbook errors." A.I.Ch.E.J. 23, pp. 768–770.
65. Mason, E.A. and Malinauskas, A.P.: 1983, "Gas Transport in Porous Media: The Dusty-Gas Model." Amsterdam, Elsevier Scientific Publ. Co.

SYMBOLS

A_p pore cross-sectional area
c_{avg} concentration integrated across the membrane, equations 5 and 23
E external potential energy
f force resistance coefficient, equation 50
F hydrodynamic force exerted by solvent (suspending medium) on solute (suspended particle)
$\boldsymbol{F}_E$ vector force exerted on the solute (suspended particle) due to external potential
g solute (particle) lag coefficient
G local shear rate (in a uniform pore, dv/dr)
H diffusive hindrance factor, equation 82
k_V hydraulic permeability coefficient, equations 10 and 19
K_n virial coefficients for the pore distribution coefficient, equation 95 $n = 0,1,2,\ldots$
ℓ_e entrance length
$\mathcal{P}$ pressure drop resistance coefficient
Pe axial Peclet number, equation 20
q_s spherical particle surface charge density
q_t pore wall surface charge density
q_V volumetric charge density
R_i solute transport resistance
R_V hydraulic resistance
δ momentum boundary layer thickness
δ_c concentration boundary layer thickness
ζ dimensionless coordinate axis perpendicular to membrane, x/h
λ ratio of particle radius to pore radius
Λ pore length-to-radius ratio, h/r_t
Ξ dimensionless ratio of pore radius to Debye length, equation 94
σ_d osmotic reflection coefficient, dimensionless
σ_f filtration reflection coefficient, dimensionless
τ polymeric chain relaxation time
ψ_s spherical particle surface electrical potential
ψ_t pore wall surface electrical potential
ϖ dimensionless pore radial coordinate, r/r_t
ε_w solvent dielectric permittivity

Diacritical Marks

$\bar{\ }$ pore cross-sectional area average value

Superscripts

(m) intramembrane value based on combined matrix and pore cross-sectional area
(p) intrapore value based on pore cross-sectional area

†	labelled solvent
∞	unbounded medium (very dilute bulk solution)
+	increment above Poiseuillean contribution due to presence of particle

Subscripts

B	Brownian motion component
e	entrance
E	external potential energy
o	spherical particle center
t	tube (pore)
δ	boundary layer
ζ	scalar coefficient of axially directed vector component
ϖ	scalar coefficient of radially directed vector component

SEPARATION BY MEMBRANES

P. Meares

Head, Department of Chemistry
University of Aberdeen, Scotland

Separation does not take place spontaneously and to drive any separation process, work must be done on the overall system to supply the necessary increase in free energy. The forces—pressure, electric potential and concentration gradients—act differently on the components to be separated and constitute an important variable in membrane and process selection. When several transportable substances are present together, they interact to influence one another's permeation properties in many ways. They may affect the solubility and diffusion coefficients of all the substances in the membrane and their fluxes may interact by direct exchange of molecular momentum.

Because of the above complexities a too general treatment of the theory of membrane separation processes is not useful. Instead a relatively simplified steady state approach is adopted here based on Fick's first law. It is used first to analyze gas separation.

The possibility of improving gas separation by the incorporation of specific molecular carriers in the membrane is considered, with attention confined to principles and not to specific examples.

Permeation between liquid phases takes place under pressure and osmotic gradients. The subject is introduced by treating the permeation of a single component under pressure and the comparison of this process with vapour permeation and pervaporation. The importance of an apparent convective contribution to the flux is indicated.

P. M. Bungay et al. (eds.), Synthetic Membranes: Science, Engineering and Applications, 155–168.

Normal osmotic flow of solvent and solute through imperfectly semipermeable membranes is considered next as a prelude to a discussion of solute, particularly salt, removal by reserve osmosis in which the importance of ion exclusion by the membrane is indicated.

1. INTRODUCTION

The principal use of synthetic membranes is to achieve the separation of mixtures into their components. Separation is inherently an entropy-decreasing, i.e., free energy-increasing, process and will not take place spontaneously. Therefore, to run a continuous separation process, one needs a source of free energy that can interact with the membrane system—a matter that will receive considerable attention. When the upstream and, perhaps, the downstream phases in contact with the membrane are mixtures, it is necessary to understand their thermodynamic properties if one is to make an accurate assessment of the differences in the chemical potentials of the components on opposite sides of the membrane. When working with mixtures, two especially complex problems have to be taken into account. Except in the case of an almost perfect molecular sieve, such as the separation of hydrogen from other gases by passage through heated palladium, more than one substance will be taken up by and transported in the membrane. The presence of any substance in the membrane may affect its power to take up another; in other words, the meaning of the solubility coefficient s becomes complex. Just as the diffusion coefficient of any substance in the membrane may be a function of its own concentration in the membrane, so the diffusion coefficient of any component in a mixture may be a function of the concentrations of all the substances present in the membrane.

The situation is further complicated by the fact that the transport of any substance in the membrane may be influenced not only by the presence of other substances in the membrane but also by their motions. Thus the fluxes of different substances may interact so as to affect the extent of separation achieved.

Faced with so great a range of interdependent factors it is clear that a completely general treatment of separation by membranes is neither possible at the present time nor would it prove particularly useful to membrane scientists and technologists because a vast number of empirical parameters would inevitably appear in the final equations; moreover, most of these parameters would be far more difficult to measure than the fluxes one would be seeking to predict and interpret. Instead one is forced to consider particular types of systems where some simplifications are permissible.

In the preceding chapter, Dr. Park has used the solution–diffusion model, with constant solubility and diffusion coefficients, to express the flux of the permeant from a region of high concentration to one of low concentration and has compared the predictions of this ideal model with observed behaviour in order to elucidate how and why s and D vary with permeant and polymer. He has shown also how the solution–diffusion model has to be modified when one progresses from a homogeneous and isotropic membrane to one having a heterogeneous or anisotropic structure. Although we have to adopt a less detailed approach when dealing with mixtures of permeants, all of these molecular influences on the mechanism of transport must be borne in mind because they give us qualitative guidance as to how to approach the separation of mixtures.

2. FLUXES AND FORCES

It is often useful to start the consideration of membrane transport by writing the following simple expression for each component

$$\text{flux} = \text{force} \times \text{concentration} \times \text{mobility} \qquad (1)$$

and then seeking to identify the three factors on the right side. Although one might think of the *difference* in the chemical potential across the membrane as driving force, in most cases the concentration will vary with distance through the membrane and it is appropriate therefore to treat equation 1 as a local equation and to integrate it across the membrane under the condition that in steady state the flux is constant. In this case the local forces are the *gradients* of chemical potential $d\mu_i/dx$ of every component that can be transported. In the terms of the solution–diffusion model the membrane material is an active thermodynamic component. When writing the Gibbs-Duhem equation for the membrane phase, the membrane material must be included among the components. Thus the chemical potentials of the transported components can be regarded as a set of independent variables.

The variation of the generalized chemical potential of component i can be expressed as a sum of terms; thus

$$d\mu_i = RT\ d\ln c_i + RT\ d\ln\gamma_i + \tilde{V}_i\ dp + z_i F\ d\psi. \qquad (2)$$

Whereas the concentration c_i and activity coefficient γ_i are not under the arbitrary control of the experimenter, the pressure p and electric potential ψ

can be varied at will by external means. Hence, by choosing the pressure and potential differences across the membrane not only can free energy be supplied to the membrane system in the form of mechanical or electrical work on the transported components, but also the forces on these mobile components can be varied relative to one another in order to improve the separation achieved. There are, however, severe limitations in that an applied pressure acts on every component in proportion to its molar volume while an electric field acts on every ionic species according to its charge sign and number and does not affect non-ionic species.

These ideas can be explored further most conveniently by examining some of the wide variety of systems that are of practical interest (1).

3. THE SEPARATION OF GASES

Gases, in contrast with easily condensible vapours, have low solubilities in polymers. Their solubility usually obeys Henry's law relatively closely and the dissolved gas is insufficient to plasticize the polymer significantly, so that the diffusion coefficient is nearly constant. When two gases are present they behave independently to a good approximation and in most cases the non-ideality of the mixed gas phases is either negligible or easily corrected for. Thus for a pair of gases i and j with partial pressures p_i' and p_j' on the feed side and p_i'' and p_j'' on the product side of the membrane, one may write for the transmembrane molar flux density

$$J_i = -D_i s_i (p_i'' - p_i')/h \tag{3}$$

$$J_j = -D_j s_j (p_j'' - p_j')/h. \tag{4}$$

Generally the composition of the product phase is determined by the fluxes, i.e.,

$$J_i/J_j = p_i''/p_j''. \tag{5}$$

In any experiment the separation factor α_{ij} is defined by (2)

$$\alpha_{ij} = \frac{p_i''/p_j''}{p_i'/p_j'} = \frac{J_i/p_i'}{J_j/p_j'} \tag{6}$$

while the membrane selectivity coefficient Γ_{ij} is defined by

$$\Gamma_{ij} = D_i s_i / D_j s_j. \tag{7}$$

It is a function only of the membrane material, permeants and temperature when the diffusion and solubility coefficients are constants.

Combining these equations shows that

$$\alpha_{ij} = \Gamma_{ij} \left[\frac{1 - (p_i''/p_i')}{1 - (p_j''/p_j')} \right] . \tag{8}$$

In a typical process the partial pressures of both gases in the product will be small compared with those in the feed and $\alpha_{ij} \rightarrow \Gamma_{ij}$. If i and j are assigned to ensure $\Gamma_{ij} > 1$, then $(p_i''/p_i') > (p_j''/p_j')$ and it can be seen that the equality $\alpha_{ij} = \Gamma_{ij}$ represents the upper limit of the separation achievable in a single-pass gas separation in which Fick's and Henry's laws are obeyed with constant D and s. It will also be noted that neither area nor thickness appears in equation 6; thus an increase in flow can be achieved by increasing membrane area and decreasing thickness without loss of separation until mechanical weakness or imperfections set a limit or until concentration polarization in the gas phases becomes controlling.

The only way of varying the concentration of a gas in the feed is by varying its partial pressure and hence, in a gas phase membrane process, concentration and pressure cannot be regarded as independent variables. The sources of free energy required to drive such a gas separation are the work required to maintain the pressure on the feed side and the entropy increase due to the expansion of the gases during permeation. Further work must then be done when the product gas, at low pressure, is recompressed.

3.1. Optimization of Gas Separation

Because there is but little flexibility in the fundamental control of gas separation, success depends heavily on clever chemical engineering and on the choice of membrane for the particular gases to be separated. Γ_{ij} can be thought of as the product of a thermodynamic factor s_i/s_j and a kinetic factor D_i/D_j. If s_i and D_i were each larger than s_j and D_j, respectively, a high selectivity for transport of i would result. Unfortunately when dealing with simple gases it is usual to find s_i an increasing, and D_i a decreasing, function of molecular weight. The range of solubility coefficients is smaller than the range of diffusion coefficients, as the data in Table 1 show. Selectivity is thus controlled mainly through the kinetics of transport and is opposed by thermodynamic factors that influence the uptake of gases by the membrane.

For a given pair of gases the ratio of permeability coefficients is greater the harder and more polar the polymer (4) and the lower the temperature, which in practice has to be above T_g if the membrane toughness is to be adequate. Unfortunately these are just the criteria that lead to a very low permeation flux and one is forced to compromise between production rate and separation.

Because there seems to be an almost unbreakable correlation between flux rate and separation when homogeneous one-component polymer mem-

TABLE 1. Solubility and diffusion coefficients of neon and nitrogen in some polymers at 25°C (3)

Polymer	Ne $10^2\ s$	Ne $10^6\ D$	N_2 $10^2\ s$	N_2 $10^6\ D$
	cm^3 (STP) cm^{-3} atm^{-1}	$cm^2\ s^{-1}$	cm^3(STP)cm^{-3} atm^{-1}	$cm^2\ s^{-1}$
poly(dimethyl siloxane) (20°C)	9.0	16.1	20.2	8.5
cis poly(butadiene)	2.2	6.6	4.9	3.0
poly(ethylene-co-propylene) (40/60)	2.0	4.5	5.5	0.67
poly(ethyl methacrylate)	1.5	1.5	7.5	0.022
poly(vinyl chloride)	1.2	0.25	2.4	0.0038

branes are being used, it is appropriate to consider modifications that might prove more satisfactory. The most promising route is to seek a more favourable solubility ratio by modifying the membrane to absorb one of the gases in a highly preferential way. It is useless, for example, to incorporate in the membrane a powdered zeolite which absorbs one gas and rejects the other, thus increasing the apparent solubility in the membrane of one gas relative to the other, because the extra gas thus taken up is immobilized. One is seeking, rather, a mobile carrier molecule that can interact reversibly with one component gas to form a complex that can diffuse within the membrane but not escape from it.

3.2. Facilitated Gas Transport

Such carrier molecules exist—among the best known is haemoglobin as an oxygen carrier—but they are relatively large and diffuse only very slowly in solid polymers. In order to exploit them one has to make use of the fact that diffusion coefficients in liquids are larger by several orders of magnitude than in polymers. The challenge is therefore to produce a membrane with the lifetime and mechanical properties of a polymer and the internal viscosity of a liquid in which a suitable involatile carrier for one of the gases to be separated can be dissolved. Promising candidates include rubbers, particularly polydimethyl siloxane, plasticized polymers, especially using polymeric plasticizers to ensure involatility, and carrier moities attached as terminal groups to flexible graft chains to ensure permanence in the membrane.

Little fundamental work has been done along these lines and it is not the purpose of these introductory chapters to become involved with the specific systems that will be discussed later. A brief theoretical analysis of carrier transport is appropriate because the same principles are applicable to facilitated transport between liquid phases.

Imagine a mixture of gases i and j of which i only can form a complex, ik, with a carrier species k present in the membrane. If k is to be an efficient carrier, the equilibrium

$$i + k \leftrightarrows ik$$

must be established rapidly compared with the time of transport of k or ik across the membrane. The association constant K_{eq} of the reaction must be large enough for a substantial amount of association to occur at the partial pressure of i expected in the feed gas, but not so large that the position of equilibrium lies far towards the right at the partial pressure of i on the product side.

Figure 1 shows the concentration profiles existing in the system. The concentration ik is higher at the feed than at the product side and a flux of ik flows down this gradient. Similarly, the concentration of unloaded carrier k is higher at the product side and a flux of k, which in the steady state must be equal to that of ik, travels back towards the feed side. The forward flux of i carried in this way is additional to the normal diffusive flux of uncomplexed molecules of i dissolved in the membrane. Similarly, there is an ordinary diffusive flux of j which does not complex with k.

If c_k^T *and* D_k are the total concentration and diffusion coefficients of carrier in the membrane and the diffusion coefficients of k and ik are assumed equal it is found that (5)

$$J_i = \frac{1}{h}\left[D_i s_i + \frac{K_{eq} D_k c_k^{\mathrm{T}}}{(1+K_{eq}p_i')\,(1+K_{eq}p_i'')}\right](p_i' - p_i''). \tag{9}$$

It can be seen that if $K_{eq}p_i' >> 1$ and p_i'' is very small, the carrier flux approaches the constant value $D_k c_k^T/h$, i.e., the carrier is fully saturated at the feed side. If the feed pressure were increased above this level, the flux of j relative to i would be increased and separation reduced.

When the membrane is made so thin, in order to increase the flux, that membrane transport becomes rapid compared with the establishment of the association–dissociation equilibrium at the membrane faces, the theoretical analysis becomes more complex (q.v. the chapter on carrier-mediated transport). The rates of the separate reactions then enter into the flux equations together with the diffusion parameters. The net result is, however, to reduce the separation factor achieved while increasing the flux. Once again one is faced with selecting a compromise condition based on optimizing the economic factors involved in the particular process. Some particular examples of gas separations enhanced by carrier-mediated fluxes are referred to in the chapter by Dr. Ward.

i,j: feed gas components, k: membrane carrier component

Complex formation:

$$i + k \rightleftharpoons ik$$

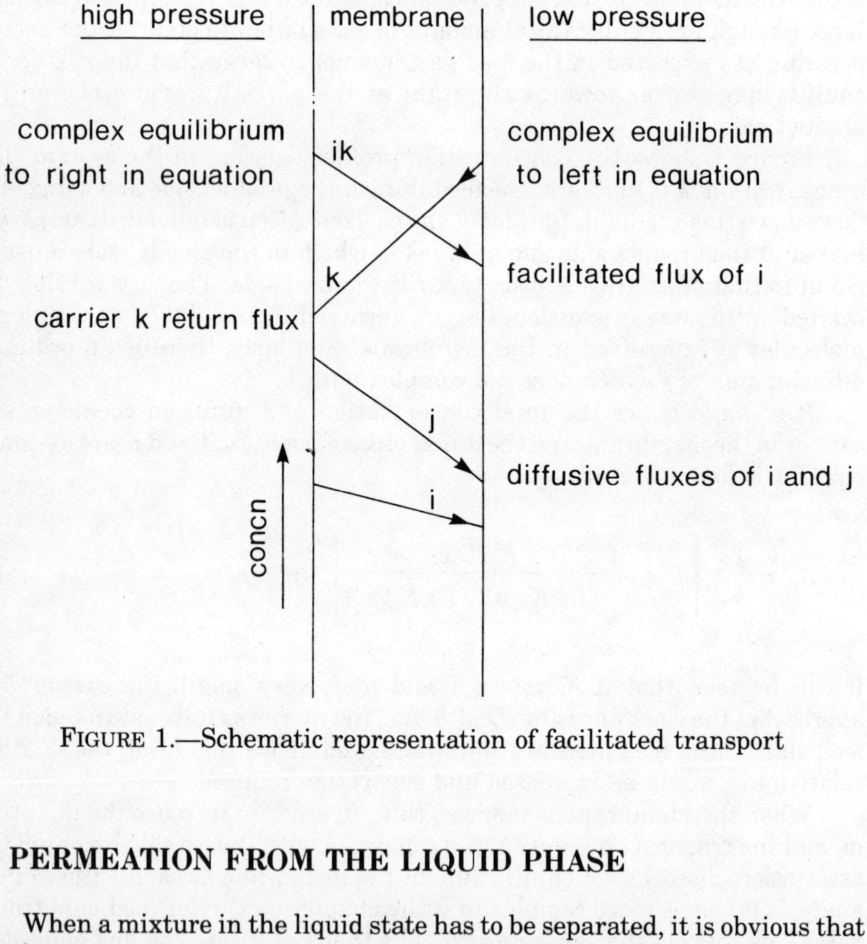

FIGURE 1.—Schematic representation of facilitated transport

4. PERMEATION FROM THE LIQUID PHASE

When a mixture in the liquid state has to be separated, it is obvious that the free energy needed to drive the process could be supplied by applying a pressure to the feed side and doing mechanical work on the system through the diminishing volume of the feed phase. This description of the process energetics applies whenever feed and product phases are both liquids, i.e., in normal filtration, ultrafiltration and hyperfiltration or reverse osmosis. As the last-named can separate molecules, it is especially interesting and valuable.

The liquid feed might be a solution of an ionic or non-ionic solid solute, in which case the mole fraction of one component is likely to be relatively small. Alternatively, a liquid mixture in which both components have comparable mole fractions can be considered. Then the distinction between solute and solvent loses its value.

The state of a permeant molecule in a polymer is determined by the local composition and the local values of the temperature and pressure. Because the nature of the phase from which the molecule originated plays no part in determining its state, it might appear that nothing new can be learned by studying permeation from a liquid phase rather than from a vapour, at least for reasonably volatile substances. However, transport between liquid phases is qualitatively distinguishable from that between vapour phases in that the fluxes are carried by large concentrations of the permeants moving under small thermodynamic forces (6).

An isothermal gradient of chemical potential between two phases of a single liquid can be created only by a pressure difference. With a molar volume of 200 ml, an applied pressure of almost 300 atm is required to produce the same thermodynamic force as a gas phase pressure ratio of 10 at ordinary temperatures. In the case of water, 100 atm produces the same driving force as an 8% difference in vapour pressure.

Instead of increasing the chemical potential at one side by applying a pressure, one may lower it at the other by adding a solute, and thereby lowering the mole fraction of the substance of interest. If the other substance cannot permeate the membrane at all, an ideally semi-permeable osmotic system is set up and the chemical potential difference is created by a modest concentration difference and without the mechanical consequences of a large hydrostatic pressure.

In liquid permeation the membrane will be saturated at one face by a liquid at unit activity. This may involve a considerable degree of swelling of the polymer. It will be only slightly less swollen at the downstream face. If a significant flux is to be achieved with the pressure forces normally at one's disposal, the amount of liquid taken up throughout the membrane must amount to at least a few per cent by weight. The plasticizing effect of this absorbed liquid on the diffusion coefficients in the polymer must be taken into account. The composition difference across the membrane may be sufficiently small that, especially with polymers that are rubbery in the swollen state, the decrease in diffusion coefficient across the membrane from the upstream to the downstream side may not be very great. For example, this appears to hold satisfactorily for water in cellulose acetate under typical reverse osmosis conditions. Although that is not a rubbery system, clustering of the water molecules at high humidity may help to hold the diffusion coefficient constant.

Where transport occurs from the liquid into a vapour phase, the transport configuration that has come to be known as pervaporation, the chemical potential in the vapour can easily be made very low and a large force obtained without the application of a pressure to the liquid. In these circumstances the concentration of permeant in the polymer at the vapour side is very low and a huge variation in diffusion coefficient across the membrane may be found. The precise quantitative analysis of the flux–force relationship in these circumstances has proved to be very difficult and we shall not consider it further.

4.1. Transport of a Single Pure Liquid Under Pressure

The flux of water through dense membranes under pressure has been studied widely (7). The permeation of some organic liquids has also been measured recently (8). The linear relation between volume flux and pressure

$$J_V = L_p \, \Delta p, \tag{10}$$

Where L_p is called the filtration coefficient, appears to hold for water in cellulose acetate and other hard polymers up to at least 100 atm (9). In softer polymers, the curvature of J_V versus Δp is noticeable even at quite low pressures. A typical example is n-hexane in natural rubber (10).

In such experiments the membrane has to be supported on a porous plate and is subjected to compression by the applied pressure. Although at 100 atm the compression will be less than 1% with soft polymers and probably less than 0.1% with hard ones, this represents a substantial reduction of the free volume in the material (which at the glass temperature amounts to about 2.5%). As a result a marked reduction in diffusion and filtration coefficients may be expected.

The application of pressure will affect the swelling of the polymer if the molar volume of the penetrant is different in the polymer $\tilde{V}_i$ and in the liquid $\tilde{V}_i'$. Thus if $w_i(p)$ is the weight fraction sorbed at pressure p and $w_i(0)$ that at the reference pressure, usually 1 atm,

$$w_i(p) = w_i(0) \exp \left[(\tilde{V}_i' - \tilde{V}_i) \, \Delta p / RT \right]. \tag{11}$$

This effect will influence the flux through both concentration and concentration-dependent mobility factors.

Potentially more important is the deswelling that Paul (11) has demonstrated at the low pressure face of a rubber membrane, because this creates a concentrated gradient inside the membrane. Paul argues that the pressure in a rubber membrane is uniformly that applied to the high pressure face, whereas the liquid just outside the downstream face is at the downstream pressure. If the activity of the permeant in the membrane at the upstream face $a_i(0)$ is unity, at the downstream face it is given by

$$a_i(h) = \exp \left[-\tilde{V}_i'' \, \Delta p / RT \right]. \tag{12}$$

Diffusion down the concentration gradient thus created will contribute to the observed flux. By estimating this normal diffusion flux from the known diffusion coefficient of hexane in rubber, it can be shown that about half the observed flux can be accounted for in this way (6). A similar study of hexane in polyethylene showed that about one-third of the flux was due to diffusion (12).

It may be questioned whether the same pressure distribution applies in harder membranes. If the flow of water in homogeneous cellulose acetate membranes is correlated with the diffusion coefficient from vapour permeation studies, it is found that the observed flux under pressure exceeds that calculated from the induced concentration gradient by more than an order of magnitude (6).

Such observations on relative flows down concentration and pressure gradients have encouraged the adoption of a simple non-rigorous mental picture that the flow can be thought of as made up of two contributions: a diffusive flow and a convective flow. No formal separation can be made unless there is a rigid and known pore structure in the membrane, but a useful procedure has been developed by combining diffusion flows measured with radio-tracers and pressure-driven flows to evaluate membrane–permeant and permeant–permeant friction coefficients in the membrane (13). The results of this method have shown the importance of direct transfer of momentum between permeant molecules in flow at high membrane concentrations (14).

4.2. Osmotic Phenomena

It has been demonstrated many times that, provided the membrane is completely impermeable to the solute used, an osmotic pressure difference $\Delta\pi$ between pure solvent and solution produces exactly the same flow as a hydrostatic pressure equal to $\Delta\pi$. This is a further proof that the effect of a pressure in generating a flow arises purely from its effect on the chemical potential of the liquid.

These experiments have all been conducted with hard polymeric membranes and there does not appear to be a demonstration of the exact equivalence of osmotic and hydrostatic pressures in producing flow through swollen rubbers. The choice of a non-permeating solute might not be easy in such cases.

Usually the solute is not completely excluded by the membrane, and oppositely directed fluxes of solvent and solute occur. These fluxes may interact mutually unless they occur by separate pathways in the membrane.

The formulation of these phenomena through linear non-equilibrium thermodynamics is well known. Three coefficients are needed: the filtration coefficient L_p, solute permeability coefficient ω and reflection coefficient σ. Assigning subscripts w for solvent and i for solute, the flux equations are

$$J_v = J_w\tilde{V}_w + J_i\tilde{V}_i \tag{13}$$

$$J_V = L_p[(\Delta p - \Delta\pi) + (1-\sigma)\Delta\pi] \tag{14}$$

$$J_i = \bar{c}_i(1-\sigma)J_V + \omega\Delta\pi \tag{15}$$

where $\bar{c}_i$ is the mean concentration of solute in the feed and product.

For ideal semi-permeability σ is unity. In equation 15, $\omega\Delta\pi$ measures the diffusional flux of solute, and $\bar{c}_i(1-\sigma)J_V$ measures the flux of solute contributed by coupling with solvent.

Despite their comparatively great age, these equations have been used only rarely in practice and then mainly in the study of electrolytes as solutes in asymmetric membranes.

There is one set of data on organic solutes in rather porous cellulose acetate membranes (15). It shows values of σ ranging from >0.9 for raffinose to >0.3 for urea. This indicates that the coupling of solute and solvent flows can be an important phenomenon and requires more study.

Much practical evaluation has been done on the removal of trace organic contaminants from water by reverse osmosis through cellulose ester and polyamide membranes (16). The results have shown that the more strongly the solutes can interact with the hydrogen bonding sites in the membrane, the more they permeate, i.e., the lower their rejection. The quantity measured in such experiments is a function of both ω and σ and they cannot be separately evaluated.

The relative permeabilities of a wide range of organic solutes permeating polyethylene membranes from one aqueous phase to another have been found to correlate closely with their solubilities in hexane (17). This suggests that the permeation of relatively large solute molecules in polymers is dominated by their concentrations or solubilities in the membrane rather than by their mobilities in the membrane. These observations seem to confirm the segmental mechanism, already widely accepted, for the movement of molecules comparable in size with the kinetic segments of polymer chains.

4.3. Separation by "Reverse Osmosis"

Although hyperfiltration can in principle be applied to non-aqueous as well as to aqueous solutions and to the separation of non-ionic and ionic solutes, its outstanding success has been in the separation of salts from water. The first successful membranes were made of cellulose acetate. Other polymers have since been discovered which have equally good, though rarely better, separating capabilities combined with improved mechanical and chemical stabilities.

By considering the separation of aqueous sodium chloride solution through cellulose acetate, the most intensively studied system, it is possible to demonstrate the interplay of the thermodynamic and kinetic factors involved. If the mole fraction of sodium chloride in the feed were 0.01, comparable with that in sea water, and 99% were removed, a quite modest achievement judged by the best standards, the mole fraction of salt in the product would be 0.0001, i.e. about 300 ppm. A typical operating pressure might be 10 MPa (about 100 atm) at 25°C.

Thus the thermodynamic forces per mole on solvent and solute are

$$\Delta\mu_w = 10^7\tilde{V}_w - 2\,RT \ln (0.9999/0.99) \quad (16)$$

$$\Delta\mu_i = 10^7\tilde{V}_i - RT \ln (0.0001/0.01) \quad (17)$$

expressed in the direction from feed to product. Letting $\tilde{V}_w$ be 18 $cm^3\ mol^{-1}$ and $\tilde{V}_i$ be 30 $cm^3\ mol^{-1}$ gives at 25°C

$$\Delta\mu_w = 130\ J\ mol^{-1};\ \Delta\mu_i = 11710\ J\ mol^{-1} \quad (18)$$

Thus the force driving solute across the membrane is 90 times that driving solvent. Because $\tilde{V}_i > \tilde{V}_w$ and while the water is flowing against its concentration gradient the salt is diffusing across the membrane downhill, no matter how great the pressure, the force on the salt will always be the greater. Consequently it is seen that separation must be due to the differences in the concentrations and mobilities of salt and water in the membrane.

Cellulose acetate absorbs water more readily than it absorbs ions, the latter being excluded on account of their Coulombic and hydration free energies in water, which has a far higher dielectric constant than has the polymer. The molar ratio of salt to water in the membrane is about one-twentieth of that in solution. This exclusion of salt by the membrane is insufficient to account for the separation observed in the face of the unfavourable ratio of forces. The system under study is a rare example of those in which the thermodynamic and kinetic factors enhance rather than cancel one another. The mobilities of water, Na^+ and Cl^- in cellulose acetate in contact with aqueous solutions can be measured by using radiotracers. The experiments show that the mobility of water exceeds that of sodium chloride by a factor of about 600 (18). Thus the ratio of their permeabilities is about $20 \times 600 = 12{,}000{:}1$, sufficient to account for the observed separation factor.

It is interesting to note that the fluxes of salt and water calculated from these solubility and diffusion coefficients are only about 40% of those observed in the hyperfiltration experiments, thus demonstrating that their flows are coupled to a significant extent and that there is a convective contribution to the total transport.

In fact this simplified account of desalination omits an effect that is particularly significant in the polymers developed more recently for this process. Where the membrane polymer has chemically bound to it ionic groups, these can play an important role in excluding salt from the membrane. The consideration of transport in membranes of ionizable polymers is rather complex and forms the subject of another chapter.

REFERENCES

1. Hwang S.-T. and Kammermeyer K.: 1975, "Membranes in separations," in Techniques of Chemistry, Vol. 7. Wiley-Interscience, New York; Meares P. (ed.): 1976, Membrane Separation Processes. Elsevier, Amsterdam.
2. Meares P.: 1975, "Selective transport processes in polymers." In Proc. Internat. Symp. Macromol. Mano, E.B. (ed.) Elsevier, Amsterdam, pp. 131-152.
3. Yasuda H. and Stannet V.: 1975, "Permeability coefficients." In Polymer Handbook, 2nd Ed. Brandrup, J. and Immergut, E.H. (eds.) Wiley-Interscience, New York, pp. III 229-240.
4. Stern S.A.: 1972, "Gas permeation processes," in Industrial Processing with Membranes. Lacey, R.E. and Loeb, S. (eds.) Wiley-Interscience, New York, Ch. 13.
5. Ward W.J.: 1970, "Analytical and experimental studies of facilitated transport." A.I.Ch.E.J. 16, pp. 405-410.
6. Meares P.: 1979, "Transport through polymer membranes from the liquid phase." Ber. Bunsenges. Phys. Chem. 83, pp. 342-351.
7. Reid C.E. and Kuppers J.R.: 1959, "Physical characteristics of osmotic membranes of organic polymers." J. Appl. Polymer Sci. 2, pp. 264-272.
8. Adam W.J., Lüke B. and Meares P.: 1983, "The separation of mixtures of organic liquids by hyperfiltration." J. Membrane Sci. 13, pp. 127-149.
9. Meares P., Craig J.B. and Webster J.: 1970, "Diffusion and flow of water in homogeneous cellulose acetate membranes," in Diffusion Processes. Vol. 1. J. Sherwood et al. (eds.), Gordon and Breach, London, pp. 609-627.
10. Paul D.R. and Paciotti J.D.: 1975, "Driving force for hydraulic and pervaporative transport in homogeneous membranes." J. Polymer Sci. Polym. Phys. Edn. 13, pp. 1201-1214.
11. Paul D.R.: 1972, "The role of membrane pressure in reverse osmosis." J. Appl. Polymer Sci. 16, pp. 771-782.
12. Greenlaw F.W., Prince W.D., Shelden R.A. and Thompson E.V.: 1977, "Dependence of diffusive permeation rates on upstream and downstream pressures. I." J. Membrane Sci. 2, pp. 141-151.
13. Thau G., Bloch R. and Kedem O.: 1966, "Water transport in porous and non-porous membranes." Desalination 1, pp. 129-138.
14. Meares P.: 1977, "The mechanism of water transport in membranes." Phil. Trans. Roy. Soc. (London) B278, pp. 113-154.
15. Jonsson G. and Boesen C.E.: 1975, "Water and solute transport through cellulose acetate reverse osmosis membranes." Desalination 17, pp. 145-165.
16. Sourirajan S.: 1977, Reverse Osmosis and Synthetic Membranes. N.R.C. Ottawa Ch. 2,3.
17. Nasim K., Meyer M.C. and Autian J.: 1972, "Permeation of aromatic organic compounds from aqueous solutions through polyethylene." J. Pharm. Sci. 61, pp. 1775-1780.
18. Meares P.: 1977, Membrane Separation Processes: Principles, Practice and Prospects. Interdiscip. Sci. Rev. 2, pp. 327-336.

SYMBOLS

Diacritical Mark

— mean between feed and product stream values

Subscripts

i, j	permeant gas or solute species
k	carrier species
w	solvent

TRANSPORT IN ION EXCHANGE MEMBRANES

P. Meares

Head, Department of Chemistry
University of Aberdeen, Scotland

A polymeric matrix to which ionizable groups are covalently bound swells and imbibes water when immersed in an aqueous solution. Dissociation of the ionizable groups leaves charges bound to the matrix and counterions free in the absorbed water. When prepared in membrane form such polymeric matrixes are called ion exchange membranes. They can convey high flux densities of the counterions, either under an applied electric field or in exchange for ions of another type, with the same charge sign, from an external source. Co-ions, of charge opposite to the counterions, are excluded from the membrane, thus giving it selective transport properties.

Ion exclusion can be accounted for by the Gibbs-Donnan equilibrium principle and this is set down in thermodynamic terms. There are significant departures from the quantitative predictions of the Gibbs-Donnan principle, especially at low concentrations. The reasons for these are discussed and some semi-empirical improvements are made to the theory which bring it into closer agreement with experience.

The description of ion transport in such membranes frequently starts from the Nernst–Planck equation. This is a linear superposition of a Fickian diffusion flux and a Faraday ion conduction flux. After presenting and discussing this equation, its deficiencies and the reasons for them are explained. In particular, the frictional effect of the flux of osmotically and electro-osmotically driven water on the fluxes of counterions and co-ions is explained and a correction term inserted in the Nernst–Planck equation.

P. M. Bungay et al. (eds.), Synthetic Membranes: Science, Engineering and Applications, 169–179.

Finally, the role of a completely general treatment, including ion–ion as well as ion–water flux coupling, via non-equilibrium thermodynamics, is introduced but not pursued because the complexities involved in its proper application to practical separation processes appear to be prohibitively great.

1. INTRODUCTION

This chapter will be confined to transport, principally ion transport, through what are usually called ion-exchange membranes. These are membranes that can transmit fluxes large enough to be of practical importance in technical and industrial processes such as desalting and alkali manufacture. The membranes must have low electrical resistances and good, though not necessarily perfect, ion selectivities. They differ from the membranes used in ion-selective electrodes, which must have extremely high selectivity and which, frequently, have high electrical resistances and transmit only minute fluxes.

The ion-exchange membranes in large-scale use consist of an organic polymeric matrix to which ionizable groups are covalently bound. These groups must be sufficiently dissociated to create on the matrix a net charge which is just balanced by the charge on the counterions released. These counterions carry the principal ion fluxes in the membrane. It is necessary therefore that they have a reasonably large mobility within the matrix.

A variety of polymers has been used as the matrix (1). The earliest were polycondensates of phenol and formaldehyde. This material is polar and hydrophilic. Later membranes have made use of more hydrophobic matrixes such as polystyrene and polyethylene. Some matrix polymers have very flexible chains while others, e.g., polyphenylene sulphone, have stiff chains. Most recently, membranes made from copolymers of tetrafluoroethylene and substituted perfluorinated vinyl ethers have been introduced. They are extremely hydrophobic and have a very low dielectric constant.

It should not be expected that such diverse polymers would provide entirely similar environments for the counterions and fixed charges. A comprehensive theory of ion-exchange membranes would be able to take the properties of the matrix into account but this has not yet been achieved quantitatively.

Four types of bound ionogenic groups have outweighed all others in importance. They are carboxylic and sulphonic acids and their salts, quaternary derivatives of amino-nitrogen atoms and pyridinium salts.

Ion-exchange membranes are typically from 100–500 μm thick. When such a polymeric membrane is immersed in water or an aqueous salt solution, it swells and imbibes water to an extent that depends on the concentration of ionized groups on the matrix and on the hydrophobic–hydrophilic balance of the polymeric material. The first water molecules enter primarily due to enthalpic influences and they hydrate the ions present in the membrane. Further water molecules enter for primarily entropic reasons, since they increase the configurational entropy of the polymer and of the dissociated counterions by increasing the volume available to them. Swelling is limited ultimately because the increase in entropy, when more water enters the polymer, is insufficient to offset the interference with the hydrogen bonding possibilities of the water molecules by the hydrophobic segments of the polymeric matrix.

If the density of ionic groups on the polymer were large enough and the matrix were reasonably hydrophilic, complete dissolution of the membrane might occur. This is prevented by the controlled introduction of chemical crosslinks between the polymer chains during manufacture. Where the matrix is very hydrophobic, such as the fluorocarbon-based membranes, or readily forms crystallites, as in polyethylene-based membranes, no crosslinks are required and the degree of swelling is controlled by the concentration of the ionic groups, i.e., by the mass of polymer per equivalent of ions.

The absorbed water performs two essential functions in the membrane. It lowers the electrostatic free energy in the region of the ionogenic groups and so permits the ion pairs to dissociate, thus liberating the counterions. By plasticizing the polymeric matrix, the water increases the mobilities or diffusion coefficients of the counterions by several orders of magnitude and permits the polymeric segments to take some part in the micro-Brownian motion of the whole system. When the volume fraction of water exceeds 20–25% there are probably some continuous, though tortuous, aqueous-filled pathways within the matrix along which ion transport is particularly favourable. In some membranes, where the water content may reach 60–70% of the total volume, one can think of the swollen membrane as being a concentrated aqueous solution interpenetrated by an open network of polymeric chains. It is clear that in all cases the precise properties of the swollen matrix will be strongly dependent on how far the distribution of water is uniform.

2. THE FIXED CHARGE THEORY

About ten years before the first ion-exchange membranes became available, the so-called fixed-charge theories of Teorell (2) and Meyer and Sievers

(3) were developed to explain ion transport in biological membranes. They have proved to be of great value in understanding the behaviour of ion-exchange membranes. Biologists had recognized for many years that the control certain membranes exercised over the transport of ions of differing charges was greatly influenced by the presence of ionic groups attached to the membrane components either by physical adsorption or by chemical bonds. The fixed-charge theory put these ideas on a quantitative basis by combining two well-established physico-chemical principles: the Donnan ion distribution and the Nernst–Planck flux equation.

An important feature of this theory is the assumption that only processes within the membrane control the rates of transport between the phases on opposite sides of it. Thermodynamic equilibrium is assumed to exist across each interface between membrane and solution. In practice, at high flux densities, this assumption may fail.

Each interface can be thought of as a membrane separating the solution from the polymeric phase. Such an interface is permeable to solvent, to counterions and to small ions of opposite charge, i.e., co-ions. It is impermeable to the ionic groups because they are bound to the polymeric matrix. Thus there is a close analogy with the Donnan membrane equilibrium of classical colloid chemistry, wherein large impermeant ions are present in a solution phase on one side of a membrane. Ions of each species distribute themselves between the two phases so that the chemical potential of each "neutral electrolyte" has the same value on both sides of the membrane. In the case we are considering, there are more counterions on the matrix side of the interface, to balance the fixed charges, than on the solution side. Consequently, there must be fewer co-ions in the polymeric phase relative to their concentration in the solution. This is the so-called Donnan exclusion effect.

Both macroscopic phases must remain electrically neutral overall but charged ionic double-layers develop on each side of the interface, leading to the Donnan potential difference between the two phases. The separate interfacial Donnan potentials cannot be directly measured by inserting an electrode into the polymeric matrix; nevertheless, the vector sum of the two interfacial Donnan potentials forms a part of the observable potential difference between the solutions on opposite sides of an ion exchange membrane. A means is required therefore for estimating these potentials. The concentrations of ions on the matrix side of the interfaces are also required because they determine the boundary conditions of the flux-determining processes taking place within the membrane.

3. THE DONNAN DISTRIBUTION

It is worthwhile looking in some detail at the ion distributions because they are always important, regardless of the means used to formulate the transport processes within the membrane. In order to treat the distributions of the individual ionic species, one must write and equate expressions for their electrochemical potentials in the solution and in the membrane (4).

Quantities referring to the external solution will be indicated by a prime superscript and the ° superscript will indicate standard state values. Thus for ionic species i

$$\mu_i = \mu_i^{\circ} + RT \ln m_i + RT \ln \gamma_i + z_i F \psi \qquad (1)$$

in which activity has been replaced by the product of molal concentration m_i and activity coefficient γ_i. When equating the expressions for μ_i' and μ_i it is desirable to be able to choose the same reference state for both phases. Then $\mu_i'^{\circ} = \mu_i^{\circ}$ and the distribution of concentrations across the interface is given by

$$m_i/m_i' = (\gamma_i'/\gamma_i) \exp(-z_i F\psi_D/RT) \qquad (2)$$

where $(\psi - \psi')$ is the Donnan potential ψ_D. This equation can be used only when γ_i and ψ_D are both known.

Expressions like equation 2 can be written for the counterions, subscript g, and co-ions, subscript n. The two expressions can be combined to eliminate ψ_D which gives

$$m_n^{\nu_n} m_g^{\nu_g} = m^{\nu_n} (m + M/\nu_n\nu_g)^{\nu_g} = (\gamma_{\pm}'/\gamma_{\pm})^{\nu}(m')^{\nu}. \qquad (3)$$

Here ν_n and ν_g are the numbers of the ions indicated in the stoichiometric neutral electrolyte molecule and $\nu = \nu_n + \nu_g$. M is the molal concentration of fixed charges, assumed univalent, in the swollen membrane; m' is the molality of electrolyte in the external solution and m, defined by m_n/ν_n, that in the membrane. $\gamma_{\pm}'$ and $\gamma_{\pm}$ are the mean molal activity coefficients in the external and membrane phases, respectively. Qualitatively, it is readily seen that: the higher the valence of the co-ions, the more effectively they are excluded; the higher the valence of the counterions, the less well are the co-ions excluded; and the greater the ratio of the molal concentrations of fixed ions and bathing solution, the greater is the Donnan exclusion.

Quantitatively, nothing can be deduced without a knowledge of the ionic activity coefficients in the membrane. Three factors must be borne in mind in discussing these: the high ionic strength in the membrane, the polyelectrolyte nature of the polymer chains and the effect of the choice of reference states to make $\mu_i'^{\circ} = \mu_i^{\circ}$.

The ionic strengths are typically 2–10 molal and the interionic attraction theory can give little help regarding activity coefficients. Comparison with free aqueous solutions at similar concentrations shows that the activity coefficients are highly specific to individual electrolytes. It has long been known that in polyelectrolyte solutions the activity coefficients of the co-ions do not appear particularly abnormal, whereas the activity coefficients of the counterions are greatly depressed due to the cooperative effect on them of the charges distributed along the polyions. Indeed, if the charge density on the polyions is large enough, counterions will condense onto the polymer chains, effectively resulting in incomplete dissociation or non-specific ion pairing. We expect to find therefore $\gamma_g < \gamma_n$.

The effect of choosing a reference state that makes $\gamma_{\pm}$ approach unity at infinite dilution in pure water is hard to express quantitatively. The electrostatic free energies of the ions are strongly dependent on local dielectric constant. Certainly this is much lower in the organic matrix, even when considerably swollen by water, than in pure water. The ions are, however, sensitive to their local environment and do not see the membrane as a dielectric continuum. An attempt by Glueckauf (5) to make use of this effect to explain exclusion of ions from uncharged reverse osmosis membranes has shown that the activity coefficients of individual ions in the membrane will be greater than in free solution but that ion pairing will also be greater. The consequences of ion pairing and of counterion condensation will be to reduce the effective value of the counterion activity coefficient in the membrane.

In such complex circumstances it is not possible to predict the overall values of the activity coefficients, and it is necessary to measure the co-ion uptakes m_n as a function of m' to assess the relative importances of the various influences.

It might be expected that the ratio $\gamma_{\pm}'/\gamma_{\pm}$ would be relatively insensitive to m' and be determined mainly by the characteristics of the polymeric matrix. Thus the ratio of activity coefficients could be replaced by some constant α for any given counterions. According to equation 3, a plot of $(1/\nu)\log [m^{\nu_n}(m + M/\nu_n\nu_g)^{\nu_g}]$ versus $\log m'$ should give a straight line of slope unity and intercept $\log \alpha$ at $\log m' = 0$.

The accurate determination of the co-ion uptakes by membranes, especially at small values of m', is not an easy matter and there is relatively little data in the literature. The results (4) on sodium and magnesium chlorides in the phenol-formaldehyde-based cation exchanger Zeo-Karb 315 show straight line relationships, as expected, but the slope is only 0.83. The intercepts at $\log m' = 0$ show that α is close to unity for sodium chloride but a little less for magnesium chloride.

The low slope means that the co-ion uptake does not fall off as sharply as the Donnan expression predicts at low values of m'. Similar observations have been made by all workers who have measured co-ion uptakes. One reason for this discrepancy is not hard to find. The only membrane parameter that appears in the Donnan distribution expression is the fixed charge concentration M. These fixed charges are discrete and are separated by distances that are large compared with the thickness of the ionic atmospheres around them. Thus, at the molecular level in the membrane, there are local fluctuations in the electric potential. Because of the exponential influence of the potential in the ion distribution equation 2, a region of lower than average potential enhances the co-ion uptake to a greater extent than a region of greater than average potential reduces it. Thus any effect that causes fluctuations in the local potential about the average will lead to m_n being greater than the predicted ideal value.

In addition to the fixed charges being spaced along the chains, the chains themselves will be separated by water and ions in the swollen membrane and their distribution in space will not be uniform if the crosslinks, entanglements, crystallites, etc., that maintain the integrity of the membrane are not uniformly distributed. Efforts have been made to use uptake data to analyse the micro-heterogeneity of the matrix (6) but the distribution function needed to model a range of membranes is too complex to be useful. Instead one may recast equation 3 in the empirical form suggested by the data

$$m^{\nu_n} (m + M/\nu_n \nu_g)^{\nu_g} = (\alpha m')^{\beta\nu} \quad (4)$$

where β, which is given by the slope of the linear log–log plot, is an empirical measure of the heterogeneity of any particular membrane. The greater (1−β), the greater is the heterogeneity.

The two parameters α and β can be evaluated from only a limited amount of co-ion sorption data. These parameters enable not only the co-ion concentrations in the membrane to be related to the external solution concentrations and the analytically determinable fixed charge concentration; they also enable the Donnan potentials to be estimated from equation 2 when recast in the form:

$$\psi_D = (RT/z_i F) \ln (\alpha m'/m). \quad (5)$$

It may be noted that, although α is not strictly equal to γ_n'/γ_n, when the Donnan potentials at the opposite faces of the membrane are summed in setting down the overall membrane potential, the terms in α cancel out.

There is no explicit general solution to equation 4 and a complete theory of transport in a membrane system will have to consist of two parts, each requiring separate handling. One will be concerned with the interfacial equilibria and the other with the rate-limiting transport processes.

4. COUNTERION SELECTIVITY

Before leaving the subject of the interfacial equilibria, the important case of a mixture of two counterions and a single co-ion should be mentioned because it forms the basis of many interesting separation possibilities. In principle, the Donnan equation and the concept of electrically neutral phases can be used to describe the distribution of the counterions between the membrane and solution. This approach is, however, not instructive if all components are assumed to behave ideally and the two counterions have equal valencies. It predicts only that the mole fractions of the ions in the solution and membrane in contact are equal. This is certainly not the case and a satisfactory treatment of ion-exchange equilibria demands the inclusion of the activity coefficients of the counterions. Nor can ion-exchange equilibria be represented by mole-fraction independent equilibrium or selectivity constants, because the activity coefficients in the membrane phase of the preferred and non-preferred counterions are strongly composition dependent (7).

The Donnan treatment is more informative in the case of a pair of counterions of different valencies. It shows that the counterions of higher valence will be relatively more concentrated in the membrane than in the solution and that the effect is greater in dilute solutions than in more concentrated ones.

The complex nature of these counterion exchange distribution functions has resulted in the quantitative formulation of membrane transport in bi-ionic systems lagging far behind that in systems of a single electrolyte at different concentrations on opposite sides of the membrane.

5. NERNST–PLANCK ION FLUX EQUATION

The fixed-charge theory of membrane phenomena expressed the flux J_i of ions i in the membrane in terms of the Nernst–Planck equation 6. This is, in effect, a superposition of Fickian diffusion and Faraday ion conductance.

$$J_i = -D_i \frac{\mathrm{d}c_i}{\mathrm{d}x} - z_i F\, u_i c_i \frac{\mathrm{d}\psi}{\mathrm{d}x}. \tag{6}$$

Descriptions of overall transport rates involve membrane thickness and area. Concentrations must therefore be expressed on the molar scale. Conversion from molal m_i in the ion uptake equations to molar c_i in the transport equations must be done with care.

In a rapidly relaxing medium such as a swollen polymeric membrane, the Nernst–Einstein relation holds and $D_i = u_i RT$. Equation 6 can thus be recast as

$$J_i = -u_i c_i\,(RT\,\mathrm{d}\ln c_i + z_i F\,\mathrm{d}\psi)/\mathrm{d}x \tag{7}$$

which has the form

flux = mobility × concentration × force.

A similar equation can be written for each ionic species in the membrane. These ion flux equations can be combined with an expression for the electric current

$$\Sigma J_i z_i F = I \tag{8}$$

and another for local electrical neutrality in the membrane

$$\Sigma z_i c_i\,\mathrm{d}\ln c_i/\mathrm{d}x = \mathrm{d}(\rho_w M)/\mathrm{d}x. \tag{9}$$

Here ρ_w is the mass of water per unit volume of swollen membrane. When there is no electric current $I = 0$, and when the membrane is uniform $\mathrm{d}(\rho_w M)/\mathrm{d}x = 0$.

The Nernst–Planck equation is a differential equation and to evaluate fluxes and potentials the set of equations 6 must be integrated across the thickness of the membrane. For this purpose the boundary values of the concentrations must be known. Two or more ion flux equations may be combined to eliminate either the potentials or the fluxes.

When carrying out the integration it is usually assumed that the ionic mobilities are constants everywhere in the membrane, despite a good deal of evidence showing that for the counterions the mobility increases considerably with increasing concentration (8). This appears to be caused by the sorbed co-ions smoothing out local fluctuations in the electric potential. Similar effects have been noted in studies of the tracer diffusion coefficients and electrical conductances of counterions in polyelectrolyte solutions.

Equation 7 may be seen to be inadequate on two grounds. The contribution $RT\,\mathrm{d}\ln c_i/\mathrm{d}x$ to the driving force should incorporate also the effects of non-ideality and so be replaced by $RT\,\mathrm{d}\ln a_i/\mathrm{d}x$. In most cases the omission of this non-ideality correction is probably of minor importance where a single electrolyte is concerned, but may be more significant in bi-ionic systems in which the activity coefficient of a low concentration of the preferred ions may become very low indeed, while the converse is true for the non-preferred ions.

Usually there is a substantial electro-osmotic transfer of solvent when an electric current is passed through an ion exchange membrane. The osmotic flows through such membranes are also very anomalous. These observations indicate strong coupling between ion and water flows in membranes. This coupling occurs because the ions and water tend to be concentrated into the same volume elements of the membrane. Thus a high proportion of the frictional retarding forces experienced by the ions results from their interactions with the surrounding water molecules. Accordingly, any net motion of the water due to osmosis or electro-osmosis influences these frictional interactions and should appear in the equations of motion of the ions.

To take approximate account of this effect, the Nernst–Planck flux equation can be appropriately modified to read

$$J_i = -u_i c_i (RT\,\mathrm{d}\ln c_i + z_i F\,\mathrm{d}\psi)/\mathrm{d}x + \theta_i J_V c_i \tag{10}$$

where J_V is the volume (mainly solvent) flux density expressed as volume transported/area × time and θ_i represents the tightness of coupling of the flows of water and ions i.

Equation 10 introduces two new difficulties in addition to the need either to measure J_V or to produce a theory relating it to the concentrations of the solutions bathing the membrane. The variables in equation 10 cannot be separated for integration and the concentration profile c_i versus x must first be determined. This creates a mathematical inconvenience. The other problem is that the coefficients θ_i are not known and will take different values for different ions. Because equation 10 is concerned with quantities averaged over the plane x and the microscopic values of θ_i will vary with the location of an ion relative to the matrix, it is probable that θ_i will also vary with c_i.

Whenever equation 10 has been used, the factors θ_i have usually been omitted, i.e., they have been set equal to unity. This amounts to treating all the ions in the membrane as being fully immersed in a water stream moving with velocity J_V relative to the fixed matrix. This gross approximation is a considerable improvement on the simple Nernst–Planck equation, although quantitative fits between calculated and observed fluxes are better still if θ_i is set arbitrarily at 0.5 (9).

6. NON-EQUILIBRIUM THERMODYNAMICS

The function of the volume flow term is to take account of one of the main flow couplings that is omitted from the original fixed-charge theory and the Nernst–Planck equation. Flows of like-charged and oppositely charged ions may interact also through the Coulombic forces between these particles without requiring direct frictional contact between them. It is not practicable to include all such pair-wise couplings by extending the Nernst–Planck equation.

The formalism of non-equilibrium thermodynamics provides an alternative set of flux equations in which all these vectorial coupling processes are included. The use of this formalism in the quantitative interpretation of membrane processes has been limited because instead of requiring a single mobility coefficient to characterize each mobile component, i.e., n components requiring n coefficients, $\frac{1}{2}(n^2 + n)$ independent coefficients are needed, i.e., six instead of three for a system of one counterion, co-ion and water; and ten instead of four when there are two counterions, co-ion and water. Of these coefficients, n correspond approximately with the normal mobilities; the remainder deal with couplings between pairs of components. The experimental problem of determining so many independent coefficients is made worse by the fact that, although they are independent of the local gradients of the intensive variables, they may depend on their local values and certainly depend on concentration (10). Thus the coefficients vary along the concentration profile in the membrane.

We cannot pursue the non-equilibrium thermodynamic theory here. A full range of the experimental data needed to apply it has never been acquired for a case more complex than two mobile ions and water (11). Even in that restricted case a long and complex series of measurements of different types of fluxes is needed. At the present time the rate of appearance of new membranes from the manufacturers is very high, and this acts as a disincentive to any user to invest a large experimental programme in the study of a membrane that may quickly pass out-of-date. Instead it appears that in the foreseeable future we shall have to remain content with a partial characterization in which analytical and electrochemical determinations of the effective exchange capacity are compared to gain insights into the counterion mobility and distribution. It may prove possible to model some unexpected features of membrane behaviour as being due to partial dissociation of the fixed charges and counterions (12) rather than by attempting a more realistic interpretation of complete dissociation combined with subtle variations in pair-wise ionic interactions.

REFERENCES

1. Meares P.: 1983, "Ion-exchange membranes," in Mass Transfer and Kinetics of Ion Exchange. Liberti, L. and Helfferich, F.G. (eds.), Martinus Nijhoff, The Hague, pp. 329–366.
2. Teorell T.: 1935, "An attempt to formulate a quantitative theory of membrane permeability." Proc. Soc. Exptl. Biol. Med. 33, pp. 282–285.
3. Meyer K.H. and Sievers J.-F.: 1936, "La perméabilité des membranes. I. Théorie de la perméabilité ionique." Helv. Chim. Acta 19, pp. 649–664.
4. Meares P.: 1973, "The permeability of charged membranes," in Transport Mechanisms in Epithelia. Ussing, H.H. and Thorn, N.A. (eds.) Munksgaard, Copenhagen, pp. 51–67.
5. Glueckauf E. and Watts R.E.: 1962, "The Donnan law and its application to ion exchange polymers." Proc. Royal Soc. (London) A268, pp. 339–349.
6. Glueckauf E.: 1962, "A new approach to ion exchange polymers." Proc. Royal Soc. (London) A268, pp. 350–370.
7. Meares P. and Thain J.F.: 1968, "The thermodynamics of cation exchange. VI. Selectivity and activity coefficients in moderately concentrated solutions." J. Phys. Chem. 72, pp. 2789–2797.
8. Meares P.: 1968, "Transport in ion-exchange polymers," in Diffusion in Polymers. Crank, J. and Park, G.S. (eds.) Academic Press, London, Ch. 10.
9. Meares P.: 1976, "Some uses for membrane transport coefficients," in Charged and Reactive Polymers, Vol. 3, Part I. Selegny, E. (ed.) Reidel, Dordrecht, pp. 123–146.
10. Krämer H. and Meares P.: 1969, "Correlation of electrical and permeability properties of ion-selective membranes." Biophysical J. 9, pp. 1006–1028.
11. Foley T., Klinowski J. and Meares P.: 1974, "Differential conductance coefficients in a cation-exchange membrane." Proc. Royal Soc. (London) A336, pp. 327–354.
12. Toyashima Y., Yuasa M., Kobatake Y. and Fujita H.: 1967, "Studies of membrane phenomena." Parts 3, 4 and 5. Trans. Faraday Soc. 63, pp. 2803–2813, 2814–2827, 2828–2838.

SYMBOLS

M	molal concentration of fixed charges in the membrane
α	constant $= \gamma_\pm'/\gamma_\pm$
β	membrane heterogeneity parameter, equation 4
$\gamma_\pm$	mean activity coefficient, $\sqrt{\gamma_g \gamma_n}$
θ_i	flux coupling coefficient, equation 10
ν	number of ions formed by dissociation per molecule of neutral species i
ν_g	number of counterions per molecule of species i
ν_n	number of co-ions per molecule of species i
ρ_w	mass of water per unit volume of swollen membrane
ψ_D	Donnan electric potential

Subscripts

i	species
g	counterion
n	co-ion
D	Donnan

MEMBRANE ELECTRODES

P. Meares

Head, Department of Chemistry
University of Aberdeen, Scotland

When the substances to which a membrane is selectively permeable are ionic, their virtual fluxes can be easily and quantitatively detected by the electrical potential differences to which they give rise. Membranes when used in analysis in this way constitute membrane electrodes. Membrane electrodes have been widely used throughout this century, the pH-sensitive glass electrode being the most familiar. Many types of membrane electrodes exist; this chapter is restricted to those which incorporate a solid or semi-solid synthetic membrane as active element.

After a brief account of the origin of the membrane potential, the class of electrodes in which the synthetic material has the least active role is dealt with. The ion-selective material is an insoluble crystalline inorganic salt. It is suspended in a highly divided state and at high loading in an inert and hydrophobic polymeric matrix. The theory of the behaviour of such electrodes follows the well-known example of the silver halides.

When ionizable groups are chemically bound to the polymeric matrix, as in ion exchange membranes, selective ion transport occurs. The chemical requirements of the membrane required to maximize this selectivity are quite unlike those of high flux ion exchange membranes. A hydrophobic material with a high dielectric constant and sufficiently soft to impart a significant mobility to ions in the matrix is required. Polyvinyl chloride plasticized by various esters or ethers has been the material most frequently used to date. The fixed ions may either be chemically linked to the polymer or

P. M. Bungay et al. (eds.), Synthetic Membranes: Science, Engineering and Applications, 181–196.

incorporated as large organic ions which are practically unable to move in the membrane and which are not extracted by contact with surrounding solutions. By a suitable choice of plasticizer and ionic sensor to complex specific counterions, highly selective electrodes have been produced and the origin and formulation of this selectivity are discussed.

If the electrode matrix material were sufficiently swollen with plasticizer and the ion sensor were dissolved in it, the membrane would be essentially a liquid film in which the sensor ions as well as the counterions would be mobile. The selectivities of such electrodes follow different rules from those with fixed-site membranes. These differences are described and explained. It appears that in plasticized polyvinyl chloride membranes the mobility of the sensor ions is not a major factor.

Finally the case of membranes containing electrically neutral organic substances, which can bind powerfully and in a very specific way to certain ions, is introduced. A problem of particular interest in these membranes is the mechanism of ion transport, which must be reconciled with the thermodynamic requirements of local electrical neutrality. Probably available matrix polymers always carry a few ionic impurities and the problem may therefore be more apparent than real.

1. INTRODUCTION

Any typically thin material that can regulate the transfer of a range of substances can be regarded as a membrane with selective properties. When the substances it can transport include ions, its selective transport properties can give rise to electrical effects. These can be measured with relative ease and used to give information on the activities and concentrations of the substances, provided the behaviour of the membrane is properly understood. Probably the easiest electrical effect to measure accurately is potential difference at zero current. Thus the absolute permeability of the membrane can be very low without significant disadvantage. This freedom permits electrode designers to concentrate on the search for maximum selectivity.

Although electrical effects associated with biological membranes had been noted 100 years earlier by Galvani (1), the electrochemical study of membranes was started by Ostwald around 1890 (2). In 1906 Cremer (3) noted the electrical response of a glass membrane to changes of acidity. Since that time the glass pH electrode has become a familiar and precise analytical tool, joined in more recent times by glass electrodes responsive also to some alkali and alkaline earth cations (4).

Following perhaps the biological lipid membrane concept, Sollner studied the behaviour of organic electrolytes dissolved in organic solvents and found that these could respond to anions as well as to cations (5). In this way electrodes based on liquid ion exchangers were born.

The attempt to turn these discoveries into convenient analytical devices applicable to a wide range of substances has led to the development of a major industry and very active area of scientific research in the space of 15 years. Here we will consider electrodes that use as their active element a solid or semi-solid synthetic membrane. Glass electrodes will be excluded. This leaves membranes based on synthetic polymers incorporating chemically or physically fixed ion sites or swollen by a liquid ion exchanger or by a neutral ion carrier.

In the practical use of ion-selective electrodes as analytical tools a potential difference has to be measured between a pair of electrodes immersed in the test solution. One of these is a reference electrode whose potential relative to that of the test solution is intended to remain constant. The saturated calomel electrode, in which contact with the test solution is made via a saturated potassium chloride solution in some form of liquid junction, is probably the most familiar example. In fact the greatest source of uncertainty in the measurement and interpretation of the potential differences arises probably from this liquid junction. This is not directly relevant to the membrane phenomena in the sensor electrode, however, and it will not be covered here.

2. THE ORIGIN OF THE MEMBRANE POTENTIAL

In the most common and best understood membrane electrodes the membrane stands between the test solution and a solution of known constant composition contained in the electrode body. Dipped into this filling solution is a metal electrode that takes up a constant reversible potential relative to the solution. Characteristically, the metal electrode can be a chloridized silver wire with the filling solution containing a set concentration of chloride; or a redox system can be used, which has the advantage that it can be made insensitive to changes in temperature. Thus we are concerned only with the potential difference established between the filling and the test solutions by the selective permeability properties of the membrane and the way in which this potential difference depends upon the composition of the solutions.

In some electrodes an attempt has been made to dispense with the filling solution and to coat the membrane directly on to the metal electrode (6). As yet there is no generally successful method of ensuring equilibrium between the metal and the membrane, which would yield a constant reproducible potential difference. This is not considered further here.

All true membrane electrodes consist of a sheet of material of which both surfaces are in contact with solutions. Ions may be transported across these interfaces and also in the interior of the membrane, but not necessarily the same ions. Potential differences will be generated across each interface and a diffusion potential may arise in the bulk of the membrane. Figure 1 shows how the total membrane potential is the sum of these three effects.

The interfacial potentials can be derived by assuming thermodynamic equilibrium between the membrane and solution because the non-equilibrium transport processes are very slow compared with the rate of attainment of equilibrium. Complications arise only when several kinds of ions may cross the interfaces or complex ionic equilibria exist in the solution. Where internal diffusion potentials exist, precise descriptions are difficult to give and some mobility parameters may be almost impossible to measure. The difficulties are similar in nature to those met in formulating liquid junction potentials. Instead of attempting a comprehensive theory we shall deal with the potential determining mechanisms for each of the electrode types considered.

3. HETEROGENEOUS PRECIPITATE-BASED MEMBRANE ELECTRODES

Attempts to prepare homogeneous pressed films of the silver halides, which would function as anion-responsive membranes, were not successful. More recently (7), it was discovered that by mixing the halide with the much softer and less soluble salt, silver sulphide, strong and coherent films could be made. These form the basis of several commercially available electrodes. Concurrently Pungor and co-workers (8) used hydrophobic polymers such as polyethylene, polyvinyl chloride and, especially, silicone rubber as binders for powdered silver halides and found that, provided certain precautions were observed, satisfactory and responsive membranes could be produced.

In the 20 years since that work, membranes sensitive to a range of cations and anions have been prepared by incorporating insoluble salts into inactive polymeric binders (9). Relatively few have, however, provided sufficiently reliable, reproducible and selective for practical use in potentiometry.

The salt employed must have a very low solubility in the contacting solutions. It must possess electrical conductivity, usually by ion transport facilitated by Frenkel defects. It must reach ion exchange equilibrium with the solutions at its surfaces very rapidly.

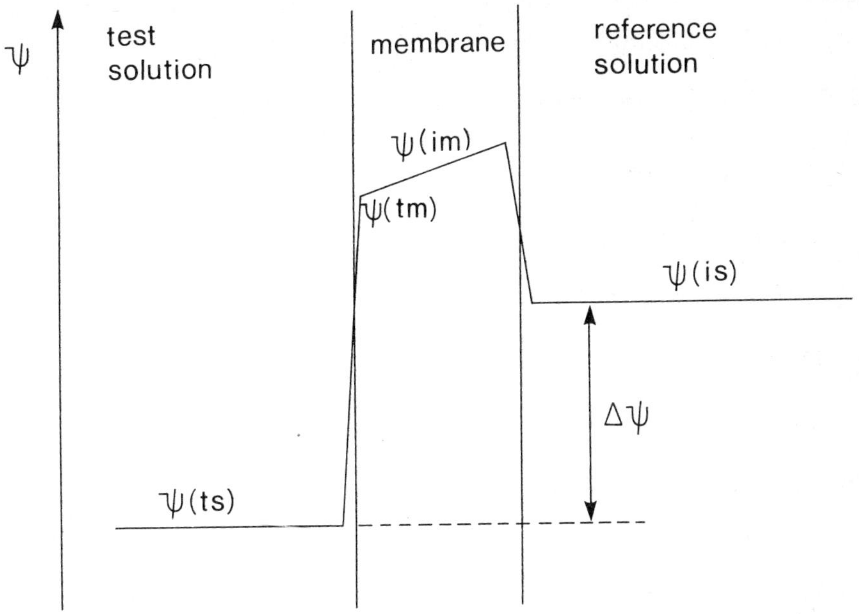

FIGURE 1.—Diagrammatic representation of potential profile

The polymeric binder must be tough enough to withstand clamping as a thin film. It must not absorb salts or water which would provide an alternative conductance chain. It should "wet" the crystalline particles so as to eliminate the tendency of the aqueous solutions to creep along the polymer/particle interfaces and cause electrical leaks. Such a polymer must be an insulator and the most favoured examples have been polyethylene and silicone rubber, but there is probably room still for development in this area.

To provide satisfactory trans-membrane conductance the crystalline particles should be uniform in size and shape. The size range 1–10 nm is satisfactory. The loading of the polymer must be high to ensure conducting chains of particles directly in contact from one face of the membrane to the other. A volume fraction of solid not less than 0.5 is desirable to achieve this.

The mechanism of the response of the potentials and of the selectivity of such membranes is relatively straightforward to explain. We may take the silver halides as an example. They are used to determine the activities of the halide anions but the internal conductance is due to silver ions; hence there is no diffusion current and potential. Only the surface potentials have to be considered.

The chemical potential of chloride ions in the internal filling solution (is) can be written

$$\mu_{Cl}(is) = \mu^{o}_{Cl}(s) + RT \ln a_{Cl}(is) - F\psi(is). \qquad (1)$$

In the membrane (im) at the interior side the chloride chemical potential is

$$\mu_{Cl}(im) = \mu^{o}_{Cl}(m) - F\psi(im), \tag{2}$$

where $\mu^{o}_{Cl}(s)$ and $\mu^{o}_{Cl}(m)$ are the standard state chemical potentials. At the other side of the membrane, in contact with the test solution (ts), the corresponding equations are

$$\mu_{Cl}(ts) = \mu^{o}_{Cl}(s) + RT\ln a_{Cl}(ts) - F\psi(ts), \tag{3}$$

$$\mu_{Cl}(tm) = \mu^{o}_{Cl}(m) - F\psi(tm). \tag{4}$$

In the absence of a diffusion potential

$$\psi(im) = \psi(tm) \tag{5}$$

and at both faces the chloride chemical potentials are equal; therefore

$$\mu_{Cl}(is) = \mu_{Cl}(im), \tag{6}$$

$$\mu_{Cl}(ts) = \mu_{Cl}(tm). \tag{7}$$

Hence

$$\Delta\psi = \psi(is) - \psi(ts) = \frac{RT}{F}\ln\frac{a_{Cl}(is)}{a_{Cl}(ts)}. \tag{8}$$

If the electrode is transferred from a test solution in which the activity is a'_{Cl} to another of activity a''_{Cl} the increase in potential difference, measured with respect to a reference electrode, ΔE is given by

$$\Delta E = \Delta\psi'' - \Delta\psi' = \frac{RT}{F}\ln\frac{a''_{Cl}}{a'_{Cl}}. \tag{9}$$

Thus the electrode response is Nernstian. In practice it is important to know how the potential may be influenced by the presence in the test solution of other anions that can form insoluble salts with silver as, for example, bromide.

At equilibrium, at the interface between membrane and solution

$$a_{Ag}(ts)\,a_{Cl}(ts) = K_s(AgCl) \tag{10}$$

holds where K_s(AgCl) is the solubility product of silver chloride. Hence

$$a_{Ag}(ts) = K_s(AgCl)/a_{Cl}(ts). \tag{11}$$

When the solubility product K_s(AgBr) is not exceeded, i.e., when

$$a_{Br}(ts) < a_{Cl}(ts)K_s(AgBr)/K_s(AgCl), \tag{12}$$

no bromide will enter the solid and the potential responds only to a_{Cl}. When a_{Br} exceeds that critical value, the potential will be independent of a_{Cl} and will depend on the bromide activity. In practice, the transition between chloride and bromide responsiveness is not sharp and the potential difference E with respect to a reference electrode can be expressed approximately by

$$E = E_o + \frac{RT}{F}\ln\left[a_{Cl} + \frac{K_s(AgCl)}{K_s(AgBr)}a_{Br}\right], \tag{13}$$

where E_o is determined by constant factors involving the membrane composition, the filling solution and metal wire of the ion selective electrode, and the nature of the reference electrode. Figure 2 illustrates diagrammatically the effect of an e-fold increase in a_{Br} on the behaviour of a Cl^- electrode.

4. MEMBRANE ELECTRODES WITH FIXED IONIC SITES

Probably the most immediately obvious electrically responsive and ion-selective membranes, as seen by synthetic membrane chemists, would be ion exchange membranes used, for example, in electrodialysis. The generation of potential differences across membranes was studied for many years by biologists before the Teorell-Meyer and Sievers fixed-charge theory of the 1930's provided a quantitative theory of concentration and bi-ionic potentials related to structural parameters of the membranes (10,11). This theory was tested with oxidized collodion membranes, which are selective to cations, and protamine-doped collodion, which is selective to anions (12). Later, ion exchange resins and membranes formed from chemically substituted polymers such as polystyrene became available and more rigorous testing of the fixed-charge membrane theory became possible. As a result it was found that the behaviour of the membrane potentials, which is extremely complex, could not be described precisely by equations as simple as the Nernst equation, even in an extended form such as equation 13. As a result, membranes composed of polymeric organic ion exchange resins have not found much application in electrodes for use in analysis.

It is worthwhile to record the reasons for this complex behaviour. The origin of selectivity is the Donnan exclusion of the co-ions due to the ther-

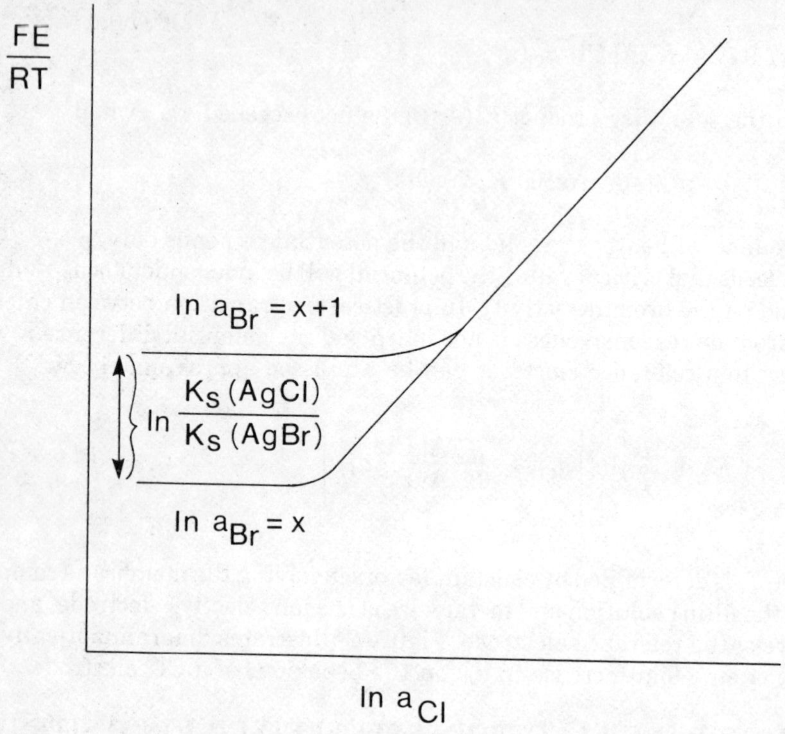

FIGURE 2.—Response of a Ag/AgCl electrode to a_{Cl} at constant a_{Br}

modynamic activity of the counterions. Denoting these by subscripts n and g, respectively, equilibrium at the membrane/solution interface occurs when

$$a'_n a'_g = a_n a_g \tag{14}$$

where activities in the solution are denoted by a'. Equation 14 is written for univalent ions. If the activity coefficients were all equal we could write

$$c'^2 = c_n c_g \tag{15}$$

where c' is the concentration of salt in the solution. Letting the concentration of fixed charges in the membrane be c_x then, for electroneutrality to hold

$$c_n = c_g - c_x. \tag{16}$$

Hence combining equations 15 and 16 shows that, while $c_n < c' < c_g$, c_n is not negligible unless $c' << c_x$. This is a seriously restrictive condition because c_x

is usually about 1M. Furthermore, the activity coefficients are not equal and vary in such a way that, at low values of c', c_n is much larger than these simple equations predict (13).

The presence of the co-ions allows the system of a membrane separating two solutions at different concentrations to behave like a concentration cell with transport. The membrane potential falls below the Nernst value by an amount determined by the transport numbers of co-ions and counterions in the membrane. The problem is made worse by the relatively greater retardation of the counterions than the co-ions by interaction with the fixed charges and, even when $c_n << c_g$, the transport number t_n is appreciable and is a strong and complex function of concentration.

When there is more than one type of counterion present in the system, a further complication arises because the exchange sites show relatively weak specificity for association with different counterions of the same valency. Such specificity is normally expressed through an ion exchange selectivity coefficient, which varies with the mole fractions of the counterions according to thermodynamic factors that are unique to each exchanger. Even where the different counterions have different valencies the electro-selectivity towards those of higher valence is insufficient to allow those of lower valence to be ignored. Thus interferences are large and unpredictable and such membranes cannot be used to make analytical electrodes except for use in pure and very dilute solutions.

Membranes of the type we are discussing swell and imbibe considerable amounts of water from the solutions; this enables the counterions to be dissociated from the fixed charges and so propagate the large ion fluxes needed in electrodialysis. As a matter of fact, there is a coupling between the ion and water flows, as in electro-osmosis, which adds a further complication to the membrane potentials.

If the membrane polymer were swollen instead with a suitable hydrophobic organic liquid, the low dielectric constant of the matrix might be just sufficient to dissociate a few counterion/fixed-charge ion pairs. The minute conductance needed to convey the electrical information from one membrane surface to the other would thus exist but the factors controlling selectivity and specificity would be greatly altered. The remaining three types of membrane electrode we shall consider all make use of matrices of this type. Although a variety of polymers and swelling liquids has been examined, one particular system has dominated the field and we shall concentrate attention on it.

Polyvinyl chloride is a commonly available hydrophobic polymer with a fairly large dielectric constant. Being a hard polymer at ordinary temperatures, ionic mobilities in pure polyvinyl chloride are too low to be useful. This polymer is compatible with a wide range of water-insoluble plasticizers, particularly polyfunctional esters, which soften it sufficiently to provide adequate ionic mobilities and which themselves have dielectric constants and ion solvating capabilities sufficient to promote some dissociation of ion pairs (14).

The three types of electrode differ in the means adopted to bring into the membrane and transmit through it the desired ions while meeting the requirements of overall and local electroneutrality. Although these requirements are relaxed in certain respects in the ultrathin lipid bilayers and biological membranes, they must be met in the macroscopically thick membranes used in practical electrodes.

We shall consider first electrodes in which there are charged sites chemically bound to the polyvinyl chloride. There are relatively few examples. The problem is to incorporate a small concentration of uniformly distributed charged sites into the polyvinyl chloride. Only a small number of sites is required to produce a sufficiently high conductance for a membrane electrode. If their concentration is too high they may form clusters in the way typical of ionomers and they also encourage water to enter the membrane along with other ions. This degrades its selectivity.

The desired result can be achieved by using the radical ion SO_3^- as polymerization initiator or a tertiary amine as chain transfer agent during the polymerization of vinyl chloride (15). In the first case the chains bear a terminal $-SO_3H$ group, and in the second case a tertiary amino-group, which can be quaternized to yield the $-\overset{+}{N}R_3$ cation. By controlling the conditions of polymerization, the polymer molecular weight and hence the concentration of ionic end groups can be controlled.

In another method a copolymer of vinyl chloride and vinyl acetate, containing little of the latter, is hydrolyzed to give free hydroxyl groups to which a mono-alkyl ester of phosphoric acid is bound by esterification (16). A small amount of this copolymer is mixed with normal polyvinyl chloride in order to keep the concentration of charged groups low. To impart sufficient mobility to the ions in the membrane the polyvinyl chloride is swollen with a plasticizer, commonly to the extent of 50–60% by weight. Many plasticizers have been studied but the alkyl, aryl and mixed phosphates and phosphonates have been found widely useful. Other ester plasticizers are also effective, including polyesters that function as polymeric plasticizers.

The relatively simple charged sites do not have much ion exchange selectivity but this is unimportant in determining the electrode specificity. Writing the response to an ion i and an interfering ion j in the form of the Nicolsky equation,

$$E = E_o \pm \frac{RT}{F} \ln (a_i + \Gamma_{ij} a_j), \tag{17}$$

the selectivity coefficient can be expressed by

$$\Gamma_{ij} = u_i K_i / u_j K_j \tag{18}$$

where u_i and u_j are the mobilities of ions i and j in the membrane and K_i and K_j are the partition coefficients of i and j between the solution and membrane phases.

The ratio u_i/u_j does not vary greatly with the choice of plasticizer or, over the experimentally useful range, with the amount of plasticizer. Inasmuch as K_i/K_j can be very sensitive to the choice of plasticizer, it is this factor that is exploited to develop good specificity for particular ions in electrodes containing a liquid ion-exchanger as ion carrier instead of fixed ion sites.

The great advantage of the fixed site membranes is that the sensor material cannot be extracted into the solution phases. They are particularly suited for use in studies of ionic surfactants above the critical micelle point and for use in mixed or non-aqueous solvents (17). They possess excellent selectivities for organic ions, which are highly soluble in the plasticized polymeric matrix, relative to small inorganic ions. Their selectivities between closely similar organic ions are not outstanding, e.g., between a pair of surfactants with identical head groups and differing only in the length of the hydrophobic chain. At high plasticizer content selectivity is in favour of the larger ion (i.e., longer chain) because its more hydrophobic character increases its solubility in the membrane. At low plasticizer content the difference between the mobilities, the larger ion being the slower, assumes increasing importance but it has not proved possible to reverse the order of selectivity without increasing the electrical impedance of the membrane to unacceptably high values; hence this trend in selectivity with plasticizer content is not useful in practice.

5. MEMBRANE ELECTRODES WITH MOBILE IONIC SITES

Following the early work of Sollner on the potentials developed across organic liquids of high dielectric constant containing dissolved organic electrolytes contained in U-tubes (5), liquid membrane electrodes were developed and marketed in which the liquid was supported in the pores of an ultrafilter disc (7). Such liquid membranes, which are not robust physically, are maintained by allowing the liquid to "wick" continuously through the filter from a peripheral reservoir. Shortly after the appearance of these membrane electrodes, it was found that the same liquid membrane components could be homogeneously dispersed in a polyvinyl chloride matrix to give rather similar electrical responses and selectivities to those of the liquid membranes but membrane lifetimes were greatly improved (18).

The method of making these electrodes is very simple. About 40% w/w high molecular weight linear, gel-free, polyvinyl chloride is mixed with a plasticizer, commonly alkyl and aryl phosphonates, phosphates, phthalates or adipates, and the liquid ion exchanger and, if required, a mediator to complex with and increase the solubility of the ions to be assayed in the membrane. The proportions of these liquid components are adjusted to maximize selectivity and compatibility with the polymer. Very many recipes for particular cations and anions have now been published in the chemical literature (19). The components are all dissolved in tetrahydrofuran to give a solution which is cast on a clean glass plate. The solvent is allowed to evaporate under dry, dust-free conditions and the resulting membrane, desirably about 0.1 mm thick, is peeled from the plate and mounted in the electrode body.

The behaviour of membranes with mobile sites poses a complex problem, which affects particularly their selectivity for competing ions (20). The problem can be formulated without difficulty. Consider first a single cationic counter-ion i^+ at activity a_i' in the solution. Its concentration c_i in the membrane at the interface will be given by

$$c_i = K_i a_i'. \tag{19}$$

In the membrane the dielectric constant will be relatively low and ion association will occur between i^+ and the mobile exchanger ions m^-

$$i^+ + m^- \rightleftharpoons im \tag{20}$$

so that an association constant K_{im} is given by

$$K_{im} = c_{im}/c_i c_m \tag{21}$$

and for electroneutrality to hold

$$c_i = c_m. \tag{22}$$

On combining these equations one finds

$$c_{im} = K_{im} K_i^2 a_i'^2. \tag{23}$$

Thus, with an activity difference from a_i' to a_i'' between the solutions on opposite sides of the membrane, a concentration difference of the neutral species im is set up inside the membrane. Because im is mobile this gives rise to a flux J_{im} from the concentrated to the dilute side. Since the liquid ion exchanger is confined to the membrane, an equal and opposite flux of the ions m^- must occur in the opposite direction. Thus

$$J_{im} = -D_{im}\,\Delta c_{im}/h = -J_m \tag{24}$$

where D_{im} is the diffusion coefficient of the associated species and h the membrane thickness. The flux of m^- is influenced by the potential gradient and by its concentration gradient and, from the Nernst–Planck equation,

$$J_m = -D_m\left(\frac{\mathrm{d}c_m}{dx} - c_m \frac{F}{RT}\frac{d\psi}{\mathrm{d}x}\right). \tag{25}$$

These equations concerning the motions of the ionic sites have to be combined with the usual Nernst–Planck equation for the flux of the counterions in order to find the internal diffusion potential, which when added to the difference in the Donnan potentials, gives the determinable membrane potential.

In fact, when only a single ionic species i^+ which can permeate through the membrane is present, the potential is not affected by the molecular details of the processes in the membrane and the response is given by the Nernst equation 9.

When a second permeating species j^+ is present, which competes with i^+, the parameters determining the membrane response include u_i, u_j, K_i and K_j, as for a fixed site membrane, and also the two association constants K_{im} and K_{jm}, the diffusion coefficients D_{im} and D_{jm} and the site mobility u_m ($=D_m/RT$). The algebraic solution in the general case is very complex and, as most of the parameters cannot be determined separately, the equation cannot readily be checked. The response, in general, is not predicted to follow the Nernstian equation in the presence of an interfering ion and the response cannot be expressed in terms of a Nicolsky equation 17 with a constant selectivity coefficient Γ_{ij}.

The matter of main importance here is whether the mobility of the sites in a plasticized polyvinyl chloride membrane is comparable with that in a normal liquid membrane. If the association constants K_{im} and K_{jm} are both large, then their ratio dominates the selectivity of a liquid membrane. An approximate selectivity parameter given by $K_{im}D_{im}/K_{jm}D_{jm}$ is useful and $D_{im} \simeq D_{jm}$ for small ions i and j combined with a large site ion m. But if the site mobility is low then the fluxes of the complexes are small compared with those of the dissociated ions and hence the selectivity is determined by Γ_{ij} as defined in equation 18.

Comparison of these two cases shows that if the site mobility is significant, the selectivity is heavily dependent on the choice of liquid ion exchanger because this controls K_{im}/K_{jm}, but with poorly mobile sites the ratio of the distribution coefficients K_i/K_j is important, i.e., the nature of the plasticizer will influence selectivity but the choice of ion exchanger will not.

The level of understanding in this area is not yet satisfactory because there has been no thorough study of the progressive transition from a truly liquid membrane with mobile sites to a concentrated polyvinyl chloride membrane with low site mobility. The experimental evidence is that in polyvinyl chloride membranes of the normal composition, the site mobility is low and their behaviour is close to that of fixed site membranes. Thus the choice of swelling agent (i.e., plasticizer plus any other liquid component that is added to solvate the ions and encourage ionization in the membrane) is important but the choice of the liquid ion exchanger itself is not.

6. MEMBRANE ELECTRODES WITH NEUTRAL ION CARRIERS

It has been found that a variety of naturally occurring macrocyclic compounds can act as ligands, which powerfully bind cations that can fit well into the polar internal cavity of the macrocycle. When these molecules have a non-polar exterior structure they have the effect of rendering the cations soluble in a non-polar organic phase. Such compounds include depsipeptides, macrotetralides, peptides and cyclic polyethers, among others. In addition to the naturally occurring ligands many synthetic examples have now been discovered, including some in which the cation is sandwiched between two rings. The size, polarizability and charge of the cations determine their free energy of transfer from an aqueous solution into the ligand cavity. Consequently, very high ionic selectivities can be achieved in many cases. A frequently quoted example is the greater than 10,000:1 selectivity shown by the naturally occurring depsipeptide, valinomycin, for potassium relative to sodium.

There has been a great deal of research in this area (21). Much has been concerned with the ligands dispersed in lipid bilayers which can transmit ions of one charge sign without having to comply with the constraint of bulk electroneutrality.

Thick layers of lipid containing a small concentration of the ligand also develop potentials and function as selective electrodes when separating solutions at different concentrations. There has been considerable debate as to how their selectivity can be reconciled with the requirement of electroneutrality in thick films.

The neutral carriers can also be dissolved in plasticized polyvinyl chloride membranes. The methods of manufacture of the polymer frequently result in the incorporation of anionic end groups, or an oil-soluble anion such as tetraphenyl borate, which has a very low mobility in plasticized polyvinyl chloride, can be included. Thus there is no mystery about the cation conductance of such membranes; they behave like ordinary ion exchange membranes of low exchange capacity in which the counterions, instead of being solvated by water, are "solvated" by the dipoles of the macrocyclic ligand. Because the concentration of the ligand is generally low it may become fully complexed with ions at relatively low external concentration. This sets an upper limit on the working range of concentrations for a Nernstian response.

When two cations compete for a ligand, the sizes and hence mobilities of the competing complexes are identical and the electrode selectivity is determined only by the ratio of the stability constants of the competing complexes. Over their useful concentration ranges these electrodes respond according to the Nernst equation and interference can be handled satisfactorily with the Nicolsky equation and a selectivity coefficient Γ_{ij}, which is effectively a constant.

7. CONCLUSION

Four classes of polymer-based ion-selective membranes have been described here, all of which can be used to build conventional ion selective electrodes containing a filling solution and reversible, metal $\rightleftharpoons$ e + ion, contact with an external circuit. These electrodes are easy to use, relatively inexpensive and rugged. Response times, which have not been discussed here, are satisfactory in well-designed electrodes.

The ease of depositing the plasticized polymer by solvent casting has also led to the development of electrodes without an internal reference solution in which the polymeric system is coated directly on to a suitably treated electron conducting support (6). The use of such electrodes in thermodynamic measurements requires closer attention but they have been found simple and convenient as indicator electrodes in potentiometric titrations.

REFERENCES

1. Galvani L: 1791, De Viribus Electricitatis in Motu Musculari Commentarius. Bologna.
2. Ostwald W: 1890, "Elektrische Eigenschaften Halbdurchlässiger Scheidewände." Z. Physik. Chem. 6, pp. 71–82.
3. Cremer M: 1906, "Uber die Ursache der elektromotorischen Eigenschaften der Gewebe, zugleich ein Beitrag zur Lehre von den polyphasischen Elektrolytketten." Z. Biol. 47, pp. 562–608.
4. Eiseman G: 1969, Glass Electrodes for Hydrogen and Other Cations. Dekker, New York.
5. Sollner K: 1971, "The basic electrochemistry of liquid membranes," in Diffusion Processes, Vol. 2. Sherwood, J.N. et al. (eds.) Gordon & Breach, London, pp. 655–730.
6. Freiser H: 1980, "Coated wire ion-selective electrodes," in Ion-Selective Electrodes in Analytical Chemistry, Vol. 2. Freiser, H. (ed.) Plenum, New York, Ch. 2.
7. Ross J.W: 1969, "Solid state and liquid membrane ion selective electrodes," in Ion Selective Electrodes. Durst, R.A. (ed.) NBS Special Publ. No. 317, Washington D.C., Ch. 2.
8. Pungor E. and Toth K: 1970, "Ion-selective membrane electrodes." Analyst 95, pp. 625–648.
9. Pungor E: 1978, "Precipitate-based ion-selective electrodes," in Ion-selective Electrodes in Analytical Chemistry, Vol. 1 Freiser, H. (ed.) Plenum, New York, Ch. 2.
10. Teorell T: 1935, "An attempt to formulate a quantitative theory of membrane permeability." Proc. Soc. Exptl. Biol. Med. 33, pp. 282–285.
11. Meyer K.J. and Sievers J.-F: 1936, "La perméabilité des membranes. I. Théorie de la perméabilité ionique." Helv. Chim. Acta 19, pp. 649–664.
12. Sollner K: 1945, "The physical chemistry of membranes with particular reference to the electrical behaviour of membranes of porous character. II." J. Phys. Chem. 49, pp. 171–191.
13. Helfferich F.G.: 1962, Ion Exchange. McGraw Hill, New York, pp. 133–147.
14. Moody G.J. and Thomas J.D.R: 1978, "Poly(vinyl chloride) matrix membrane ion-selective electrodes," in Ion-selective Electrodes in Analytical Chemistry, Vol. 1, Freiser, H. (ed.) Plenum, New York, Ch. 4.
15. Cutler S.G., Hall D.G. and Meares P: 1977, "Surfactant sensitive polymeric membrane electrodes." J. Electroanal. Chem. 85, pp. 145–161.
16. Keil L., Moody G.J. and Thomas J.D.R: 1977, "Ion-selective electrode membrane system based on a vinyl chloride-vinyl alcohol copolymer matrix and its role for accommodating grafted alkyl phosphate sensors." Analyst 102, pp. 274–280.
17. Cutler S.G., Meares P. and Hall D.G: 1978, "Ionic activities in sodium dodecyl sulphate solutions from electromotive force measurements." J. Chem. Soc. Faraday Trans. I 74, pp. 1758–1767.
18. Moody G.J. and Thomas J.D.R: 1976, Selective Ion Sensitive Electrodes. Merrow, Watford.
19. Buck R.P., Thompson J.P. and Melroy O.R: 1980, "Compilation of ion-selective membrane electrode literature," in Ion-selective Electrodes in Analytical Chemistry, Vol. 2, Freiser, H. (ed.) Plenum, New York, Ch. 4, Tables 11 and 11a.
20. Bloch R. and Lobel E: 1976, "Recent developments in ion-selective membrane electrodes," in Membrane Separation Processes, Meares, P. (ed.), Elsevier, Amsterdam, Ch. 12.
21. Morf W.E. and Simon W: 1977, "Transport properties of neutral carrier membranes" in Ion Selective Electrodes Symp. 2nd, Pungor, E. and Buzas, I. (eds.), Akademiai Kiado, Budapest, pp. 25–40.

SYMBOLS

E	observed electrical potential with respect to a reference electrode
ΔE	observed electrical potential difference
E_o	electrode constant; equations 13 and 17
im	membrane surface in contact with interior filling solution of electrode
i^+	mobile cationic species
is	internal filling solution
K_{im}	association constant for species i and m, equation 21
K_s	solubility product
m^-	mobile anion exchanger confined in membrane
s	solution of specific ion
tm	membrane surface in contact with test solution
t_i	transference number with subscript, i = g, n
ts	test solution

Superscripts

prime and double prime distinguish different test solutions

Subscripts

cl	chloride
i	mobile ionic species
im	neutral species obtained by association of i and m
g	counterion
j	interfering ionic species
m	mobile ion exchanger species confined in membrane
n	co-ion
x	fixed charge

ELECTRODIALYSIS

H. Strathmann

Fraunhofer-Institut für Grenzflächen-
und Bioverfahrenstechnik
Nobelstrasse 12
7000 Stuttgart 80
Federal Republic of Germany

Electrodialysis is a process by which electrically charged membranes are used to separate ions from an aqueous solution by an electrical potential driving force. Given its separation characteristics, electrodialysis is used today for the production of potable water by desalination, the recovery of water and valuable metal ions from industrial effluents, the removal of salts and acids from pharmaceutical solutions and in food processing, and for the production of salts from sea water. The principle of electrodialysis is described and the various parameters determining the technical feasibility and the economics of the process as well as the design of large-scale systems and operational problems are discussed. Examples for the use of electrodialysis in the food and drug industry and in advanced waste water treatment are given. Finally, recent developments and future prospects of electrodialysis and related processes are discussed.

1. THE PRINCIPLE OF ELECTRODIALYSIS

2. ION-EXCHANGE MEMBRANES AND THEIR PROPERTIES

P. M. Bungay et al. (eds.), Synthetic Membranes: Science, Engineering and Applications, 197–223.

Electrodialysis is a process by which an electrical potential difference and ion-exchange membranes are used to separate charged components from uncharged molecules. The largest application of electrodialysis is the desalination of brackish water, as an efficient and reliable method for the production of potable water. Such plants are in operation around the world and in many places electrodialysis is the sole source of potable water. But there are many more applications in the food and drug industry, and for the economic and efficient treatment of certain industrial waste waters, which have only recently been explored. The principle of the process has been known for more than 50 years. Large-scale industrial utilisation, however, did not become possible until the development of the multi-cell stack design and efficient ion-exchange membranes with high selectivity, low electric resistance, and good chemical and mechanical stability.

The basic principles of electrodialysis will be reviewed briefly. The main process parameters, such as membrane properties and energy requirements, will be discussed; a description of the main components of a plant and a cost analysis of the process will also be provided. Finally, examples of today's application of electrodialysis in water desalination and in the food and drug industry will be described along with developments of electrodialysis and electrodialysis-related processes that can be anticipated in the future.

For a better understanding of the electrodialysis process and its industrial potential a brief review of the basic physicochemical relations of the process is necessary.

1. THE PRINCIPLE OF ELECTRODIALYSIS

Electrodialysis is a process by which electrically charged membranes are used to separate ions from an aqueous solution under the driving force of an electrical potential difference. The principle of the process is illustrated in Figure 1, which shows a schematic diagram of a typical electrodialysis cell arrangement consisting of a series of anion- and cation-exchange membranes arranged in an alternating pattern between an anode and a cathode to form individual cells. An ionic solution such as an aqueous salt solution is pumped through these cells. When a direct current potential is applied between an anode and a cathode, the positively charged cations in the solution migrate toward the cathode. These ions pass easily through the negatively charged cation-exchange membrane but are retained by the positively charged anion-

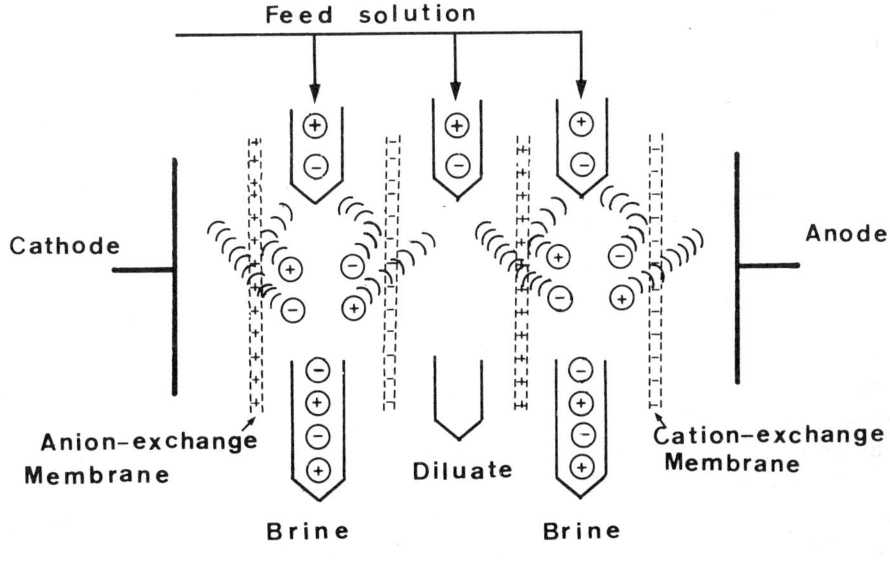

FIGURE 1.—Schematic diagram of the electrodialysis process

exchange membrane. Likewise, the negatively charged anions in the solution migrate toward the anode, pass through the anion-exchange membrane, and are retained by the cation-exchange membrane. The overall result is an ion concentration increase in alternate compartments, with a simultaneous depletion of ions in the other compartments. The depleted solution is generally referred to as the diluate and the concentrated solution as brine.

In a practical electrodialysis system, 200 to 400 cation- and anion-exchange membranes are installed in parallel to form an electrodialysis stack with 100 to 200 cell pairs. With multicompartment electrodialysis, the irreversible energy-consuming processes represented by the formation of hydrogen and oxygen or chlorine at the electrodes can be distributed over many demineralizing compartments and thus be minimized.

2. ION-EXCHANGE MEMBRANES AND THEIR PROPERTIES

The technical feasibility of electrodialysis as a mass separation process, i.e., its capability of separating certain ions from a given mixture with other molecules, is mainly determined by the ion-exchange membranes used in the system. Consequently, the membranes are the most important components in an electrodialysis unit. They should have a high selectivity for oppositely charged ions and a high ion permeability, i.e., a low electric resistance. Furthermore, they should have high form-stability, i.e., low degree of swelling and good mechanical strength.

Ion-exchange membranes are ion-exchangers in film form. There are two types: anion-exchange and cation-exchange membranes. Anion-exchange membranes contain cationic groups fixed to the resin matrix. The fixed cations are in electrical equilibrium with mobile anions in the interstices of the resin. When such a membrane is immersed in an electrolyte solution, the anions in the solution can intrude into the resin matrix and replace the anions initially present, whereas the cations are prevented from entering the matrix by the repulsion of the cations affixed to the resin.

Cation-exchange membranes are similar. They contain fixed anionic groups that permit intrusion and exchange of cations from an external source, but exclude anions. This type of exclusion is usually called Donnan exclusion in honor of the pioneering work of F. G. Donnan.

Methods of making ion-exchange membranes are described in detail in the literature. There are basically homogeneous and heterogeneous types of membranes. Heterogeneous membranes have been made by incorporating ion-exchange particles into film-forming resins by: (a) dry molding or by calendering mixtures of the ion-exchange and film-forming materials; (b) dispersing the ion-exchange material in a solution of the film-forming polymer, then casting films from the solution and evaporating the solvent; and (c) dispersing the ion-exchange material in a partially polymerized film-forming polymer, casting films, and completing the polymerization. Heterogeneous ion-exchange membranes have several great disadvantages, the most important of which are relatively high electrical resistance and poor mechanical strength when highly swollen in diluted salt solutions.

Homogeneous ion-exchange membranes have significantly better properties in this respect, since the fixed ion charges are distributed homogeneously over the entire polymer matrix. The general methods of preparing homogeneous membranes are as follows:

a) Polymerization of mixtures of reactants (e.g., phenol, phenolsulfonic acid, and formaldehyde) that can undergo condensation polymerization. At least one of the reactants must contain a moiety that either is, or can be made, anionic or cationic.
b) Polymerization of mixtures of reactants (e.g., styrene, vinylpyridine, and divinylbenzene) that can polymerize. At least one of the reactants must contain an anionic or cationic moiety, or one that can be made to do so.
c) Introduction of anionic or cationic moieties into a polymer or preformed films by techniques such as imbibing styrene into polyethylene films, polymerizing the imbibed monomer, and then sulfonating the styrene. Other similar techniques, such as graft polymerization, have been used to attach ionized groups onto the molecular chains of preformed films.
d) Introduction of anionic or cationic moieties into a polymer chain such as polysulfone, followed by the dissolving of the polymer and casting into a film. Membranes made by any of the above methods may be cast or formed around screens or other reinforcing materials to improve their strength and dimensional stability. Anionic or cationic moieties most commonly found in commercial ion-exchange membranes are $-SO_3^-$ or $-NR_3^+$. However, other charged groups such as $-COO^-$, $-PO_3^{2-}$, and $-HPO_2^-$ as well as various tertiary and quaternary amines are used in ion-exchange membranes. The resistance of ion-exchange membranes used today is in the range of 2–10 Ω cm^2 and the fixed charge density is about 1–2 m equiv/g.

3. ENERGY REQUIREMENTS

The energy required is an additive of two terms: One is the electrical energy to transfer the ionic components from one solution through the membranes into another solution. The second term is the energy required to pump the solutions through the electrodialysis unit. Depending on various process parameters, particularly the feed solution concentration, either one of the terms may be dominant, thus determining the overall energy costs.

3.1. Energy Requirements for Transfer of Ions

The energy necessary to remove salts from a solution is directly proportional to the total current flowing through the stack, and the voltage drop between the two electrodes in a stack. The energy consumption in a practical electrodialysis separation procedure can be expressed as

$$\dot{E} = I^2 n R_e t. \tag{1}$$

Here $\dot{E}$ is the energy consumption, I the electric current through the stack, R_e the resistance of the cell, n the number of cells in a stack, and t the time. The electric current needed to desalt a solution is directly proportional to the number of ions transferred through the ion exchange membranes from the feed stream to the concentrated brine. It is expressed as

$$I = \frac{zFQ\Delta c'}{\xi}. \tag{2}$$

Here F is the Faraday constant, z the electrochemical valence, Q the total feed solution flow rate, $\Delta c'$, the concentration difference between the feed solution and the diluate, and ξ the current utilization. Current utilization is directly proportional to the number of cells in a stack and is governed by the current efficiency. In practical electrodialysis, the efficiency with which ions can be separated from the mixture by the electrical current is usually less than 100%.

A combination of equations 1 and 2 gives the energy consumption in electrodialysis as a function of the current applied in the process, the electrical resistance of the stack (i.e., the resistance of the membrane and the electrolyte solution in the cells), current utilization, and the amount of salt removed from the feed solution:

$$\dot{E} = \frac{InR_e tz\ F\ Q\ \Delta c'}{\xi}. \tag{3}$$

Equation 3 indicates that the electrical energy required in electrodialysis is, therefore, directly proportional to the amount of salts that has to be removed from a certain feed volume to achieve the desired product concentration. Energy consumption in electrodialysis is directly proportional to the feed solution concentration. Energy consumption is also a function of the number of cells in a stack and the electrical resistance in a cell. Electrical resistance, again, is a function of individual resistances of the membranes and of the solutions in the cells. Furthermore, because the resistance of the solution is inversely proportional to its ion concentration, the overall resistance of a cell will in most cases be determined by the resistance of the diluate solution. This has some consequences for the design of an electrodialysis stack, as will be shown later. A typical value for the resistance of an electrodialysis cell pair, i.e., the cation- and anion-exchange membrane plus the diluate and concentrated solution, e.g., in the desalination of brackish water, lies within the range of 10 to 50 Ω cm^2.

3.2. Process Efficiency and Current Utilization

A rather important parameter in equation 3 is the current utilization. Decreasing current utilization means higher energy costs. The current utilization, which is always less than 100%, is affected by three factors:

1) In general, the membranes are not strictly semipermeable, that is, the co-ions that carry the same charge as the membrane are not completely excluded, especially at high feed solution concentrations.
2) Some water is generally transferred through the membrane by osmosis and with the solvated ions.
3) There may be partial electrical current flow through the stack manifold.

The total current utilization can, therefore, be expressed by the following relation:

$$\xi = n\, \eta_s\, \eta_w\, \eta_m. \tag{4}$$

Here ξ is the current utilization, n the number of cells, and η_s, η_w and η_m are always smaller than one; η_m is determined by the cell design and in modern electrodialysis systems it can be kept close to one. η_w is determined by the water transferred with the hydration shell of the ions. For feed solutions with low ion concentrations it is also close to one. For feed solutions with very high salt concentrations it might well be significantly smaller than one. Finally, η_s is a membrane constant which, however, depends strongly on the feed solution salt concentration. This is due to a phenomenon referred to as the Donnan equilibrium relationship, which can be expressed for a diluted solution of a univalent electrolyte (e.g., NaCl) by the following equation,

$$c_{\text{co-ion}} = \frac{(c'_{\text{co-ion}})^2}{M_R}\left(\frac{\gamma'_\pm}{\gamma_\pm}\right)^2, \tag{5}$$

where primed quantities relate to the external solution and unprimed to the membrane phase, M_R is the concentration of fixed charges per unit volume of solvent in the membrane, and $\gamma_\pm$ the mean activity coefficient of the salt in the designated phase.

Equation 5 indicates that the concentration of the co-ion within the membrane, that is, the ion that carries the same charge as the membrane, is a function of the concentration of the fixed charges in the membrane and the concentration of the co-ions in the feed solution.

Assuming a typical situation in which a membrane with a fixed ion concentration of 5×10^{-3} equiv/cm^3 of internal solvent ($M_R = 5 \times 10^{-3}$ equiv/cm^3) is immersed in a solution of 0.01 N NaCl ($c = 1 \times 10^{-5}$ equiv/cm^3), the co-ion concentration in the membrane is only 2×10^{-8} equiv/cm^3 of solvent.

Equation 5 indicates that for a cation-exchange membrane with a diluted external solution, the anions are excluded by the fixed charges of the membrane to such an extent that the concentration of the anions in the pore liquid is proportional to the square of the concentration in the external solution. The cations, however, may move freely into the pore liquid. Since the transfer of ions through the membrane is proportional to their concentration in the membrane phase, the anions will be rejected by the cation-exchange membrane. Thus for a diluted feed solution, i.e., the feed solution concentration is much lower than the concentration of the fixed charges in the membrane, a cation-exchange membrane is strictly semipermeable, being permeable to cations only. For anion-exchange membranes the reverse holds true. These membranes are permeable to anions only. When, however, the ion concentration in the feed solution is of the same order as that of the fixed charges in the membrane, co-ions may also enter the membrane which will then lose its selectivity; consequently, the current efficiency as expressed in equation 4 becomes very small and the energy cost of electrodialysis as expressed in equation 3 prohibitively high.

In the above calculation an activity coefficient ratio of 1 was assumed. Although activity coefficients generally approach unity in diluted solutions, there is considerable evidence that this is not necessarily the case within the membrane. Osmotic swelling, specific interactions of the ions with the fixed charges or other groups on the resin matrix, and effects of size and charge of the ions all combine to make resins of a given composition more selective to some ionic species than to others. For example, there are commercially available membranes that will selectively transport univalent ions and reject (or partially reject) multivalent ions in electrodialysis. With these membranes, not only the concentrations but also the compositions of electrolyte solutions can be altered by electrodialysis.

3.3. Pump Energy Requirements

In an electrodialysis system generally three pumps are necessary to circulate the diluate (the solution depleted of ions), the brine (the solution into which the ions are transferred), and the electrode rinse solutions. The energy required for pumping these solutions is determined by the volumes to be circulated and the pressure drop in the electrodialysis unit. It can be expressed by the following relation:

$$\dot{E}_{\mathrm{P}} = \eta_{\mathrm{D}}\, Q_{\mathrm{D}}\, \Delta p_{\mathrm{D}} + \eta_{\mathrm{B}}\, Q_{\mathrm{B}}\, \Delta p_{\mathrm{B}} + \eta_{\mathrm{E}}\, Q_{\mathrm{E}}\, \Delta p_{\mathrm{E}} \qquad (6)$$

Here $\dot{E}_{\mathrm{P}}$ is the pump energy consumption rate, η_{B}, η_{D}, and η_{E} are constants referring to the efficiency of the pumps, Q_{D}, Q_{B}, and Q_{E} are volume flows of the diluate, brine and electrode rinse solutions, and Δp_{D}, Δp_{B}, and Δp_{E} are the pressure losses in the diluate, in brine and in electrode cells. The pressure losses in the various cells are determined by solution flow velocities and the cell design. The energy requirements for circulating the solution through the system may become significant or even dominant when solutions with rather low salt concentration (i.e., less than 500 ppm) are processed.

4. CONCENTRATION POLARIZATION AND LIMITING CURRENT DENSITY

Concentration polarization is a phenomenon that affects all membrane separation processes. In electrodialysis, concentration polarization limits the current density that can be applied in a certain separation problem and thus affects the membrane areas required in this process. Concentration polarization in electrodialysis leads to an increase or a decrease of the ion concentrations in the laminar boundary layer on the two different sides of the membrane, as indicated in Figure 2. This figure shows the concentration of a cation in the boundary layer at the surface of a cation-exchange membrane during an electrodialysis desalting process.

The transport of charged particles to the anode or cathode through a set of ion-exchange membranes leads to a decreased concentration of counter-

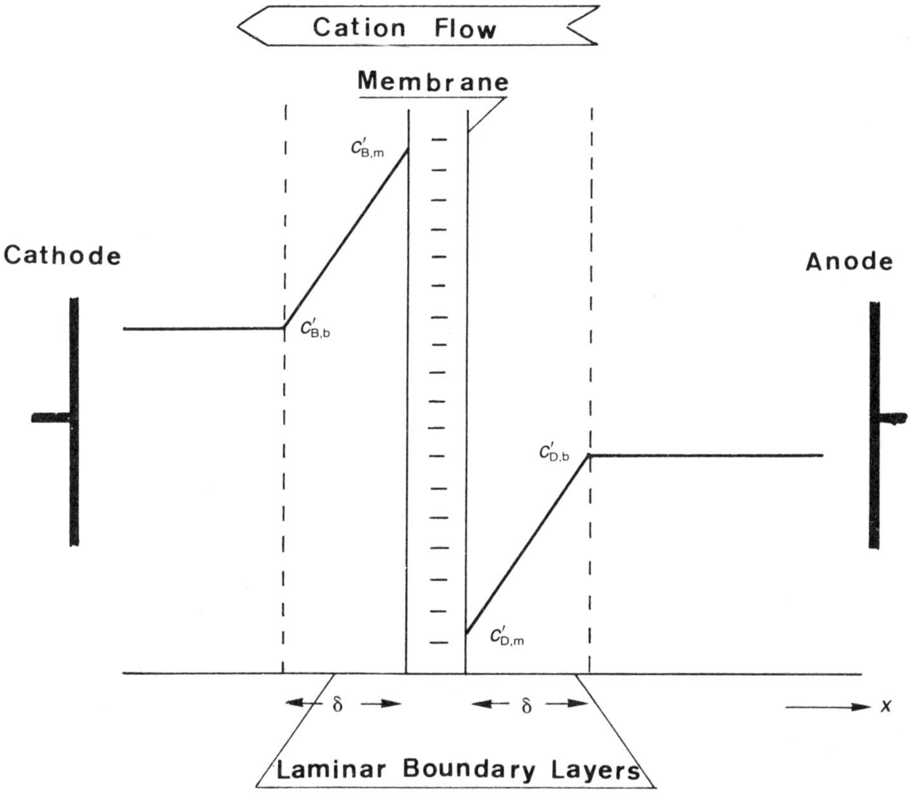

FIGURE 2.—Schematic diagram of the concentration profiles of the cations in the boundary layer at both surfaces of a cation-exchange membrane during electrodialysis

ions, i.e., the ions that will pass the membrane, in the laminar boundary layer at the membrane surface facing the diluate cell and an increase at the surface facing the brine cell. The effect of concentration polarization on the brine side is generally not severe, although it affects the current utilization to some extent. A decrease in the concentration of counterions on the diluate side directly affects the limiting current density and increases the electrical resistance of the solution in the boundary layer.

The total steady-state flux of a cation through the membrane and through its adjacent laminar boundary layers is given by

$$J = T_+ \frac{i}{z_+ F} = T_+' \frac{i}{z_+ F} - D' \frac{\mathrm{d}\, c'}{\mathrm{d}\, x} . \tag{7}$$

Here T_+ and T_+' are the transport numbers of the cation in the membrane and in the solution, respectively; i is the current density, F the Faraday constant, D' the diffusion coefficient of the cation in the solution, z_+ the electrochemical valence of the cation, and $\mathrm{d}c'/\mathrm{d}x$ the concentration gradient of the cation in the laminar boundary layer. With most ion-exchange membranes, the ion transport number in the membrane is much larger than that in the solution, i.e., $T_+ >> T_+'$. The transport under the driving force of an electrical potential gradient in the membrane is, therefore, fast compared with that in the solution. This leads to a depletion of ions on one side of the membrane and to an accumulation of ions on the other side, until concentration gradients are established in the boundary layers and diffusion provides the additional flux needed to maintain steady-state conditions in the boundary layers.

The integration of equation 7 yields a simple relationship between boundary layer thickness, the current density, and the difference in concentrations between the membrane surface and the bulk solution. Employing the following boundary conditions,

$$c' = c'_m \text{ at } x = 0,$$

$$c' = c'_b \text{ at } x = \delta,$$

leads to

$$c'_{B,m} = c'_{B,b} + (T_+ - T_+') \frac{i\,\delta}{z_+ F D'}, \tag{8}$$

$$c'_{D,m} = c'_{D,b} - (T_+ - T_+') \frac{i\,\delta}{z_+ F D'}. \tag{9}$$

Here $c'_{B,m}$, $c'_{D,m}$, $c'_{B,b}$ and $c'_{D,b}$ are the cation concentrations in the cells of the concentrated and depleted solutions at the membrane surfaces and in the bulk solutions, and δ is the thickness of the boundary layer.

For a given system geometry and feed flow velocity, there is a limiting current density at which the ion concentration at the membrane surface in the cell with the depleted solution will approach zero. This limiting current density, i_{lim}, in the case of a cation-permeable membrane is obtained by rearranging equation 9:

$$i_{\mathrm{lim}} = \frac{c'_{\mathrm{D,b}}\, D'\, z_+\, F}{\delta\,(T_+ - T_+')}. \tag{10}$$

If in electrodialysis the limiting current density is exceeded, the process efficiency will be drastically diminished because of the increasing electrical resistance of the solution and because of water splitting, which leads to both pH changes and additional operational problems. The limiting current density determines the membrane area necessary to achieve a certain desalting effect, and therefore to a large extent determines the investment costs for an electrodialysis plant.

5. SYSTEM DESIGN

In addition to the stack of cells, an electrodialysis plant consists of several components essential for proper operation and of certain feed water pretreatment procedures, which may or may not be necessary depending on the feed solution composition and the membrane stack design.

A flow diagram of a typical electrodialysis plant as used for desalination of brackish water is shown in Figure 3. After proper pretreatment, which may consist of flocculation, decarbonization, pH control and prefiltration, the feed solution is pumped through the actual electrodialysis unit, which generally consists of one or more stacks in series or parallel. A de-ionized solution and a concentrated brine are obtained. The concentrated and depleted process streams leaving the last stack are collected in storage tanks when the desired degree of concentration or depletion is achieved, or they are recycled if further concentration or depletion is desired. Often acid is added to the concentrated stream to prevent scaling of carbonates and hydroxides. To prevent the formation of free chlorine by anodic oxidation, the electrode cells are generally rinsed with a separate solution that does not contain any chloride ions.

In addition to the conventional electrodialysis process shown in Figure 3, there are several variations to the basic scheme. One of the more important ones is a process design using so-called reverse polarity. In this scheme the polarity of the current is changed at specific time intervals ranging from a few minutes to several hours. The role of each cell is correspondingly reversed, i.e., the dilute cell will become the brine cell and vice versa. The advantage of the reverse polarity operating mode is that precipitations in the brine cells are to a large extent prevented, or if there is some precipitation, it will be redissolved when the brine cell becomes the diluate cell in the reverse operating mode.

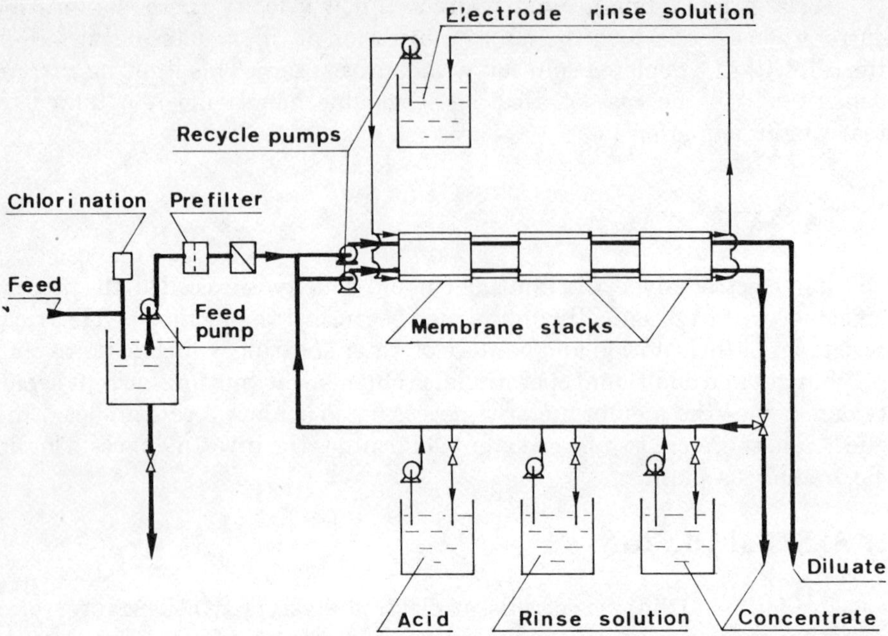

FIGURE 3.—Flow diagram of a typical electrodialysis desalination plant

An electrodialysis plant consists of four basic components: the membrane stack, the electric power supply, the hydraulic flow system, and process control devices.

5.1. The Electrodialysis Stack

The unique element in an electrodialysis plant is the membrane stack; the other components are common in the process industry. An electrodialysis stack is essentially a device to hold an array of membranes between electrodes in such a way that the streams being processed are kept separated. Figure 4 is an exploded view of part of an electromembrane stack that shows the main constituents.

A typical electrodialysis stack consists of 200 to 400 cation- and anion-exchange membranes arranged in an alternating pattern between two electrodes which are generally assembled in a separate cell. The membranes are separated by suitable gaskets, with two membranes forming a cell pair. The gaskets not only separate the membranes but also contain manifolds to distribute the process fluids in the different compartments. The supply ducts for the diluate and the brine are formed by matching holes in the gaskets, the membranes and the electrode cells. The distance between the membrane sheets, i.e., the cell thickness, should be as small as possible, because water,

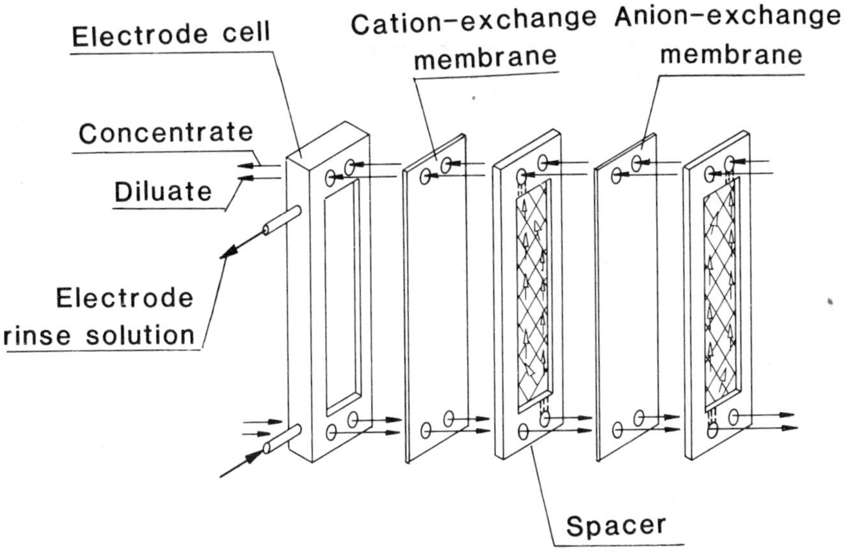

FIGURE 4.—Exploded view of components in an electromembrane stack

even with salt in it, has a relatively high electrical resistance. In most stack designs, some type of spacer is introduced between the individual membrane sheets both to support the membrane and to help control the feed solution flow distribution. The most serious design problems for an electrodialysis stack are those of assuring uniform flow distribution in the various compartments and of preventing any internal leakage, especially leakage of liquid from the compartment with the concentrated solution into the compartment with the diluate. Most stack designs used in today's large-scale electrodialysis plants are one of two basic types: tortuous-path or sheet-flow. These designations refer to the type of solution flow in the compartments of the stack.

In the tortuous-path stack, the membrane spacer and gasket have a long serpentine cut-out which defines a long narrow channel for the fluid path. The objective is to provide a long residence time for the solution in each cell in spite of the high linear velocity that is required to limit polarization effects. A tortuous-path spacer gasket is shown schematically in Figure 5. The actual flow channel makes several 180° bends between the entrance and exit ports which are positioned in the middle of the spacer. The channels contain cross-straps to promote turbulence of the feed solution.

In stack designs employing the sheet-flow principle, a peripheral gasket provides the outer seal and a plastic net or screen is used as a turbulence promoter and as a spacer which prevents the membranes from touching each other. The solution flow in a sheet-flow type stack is approximately in a

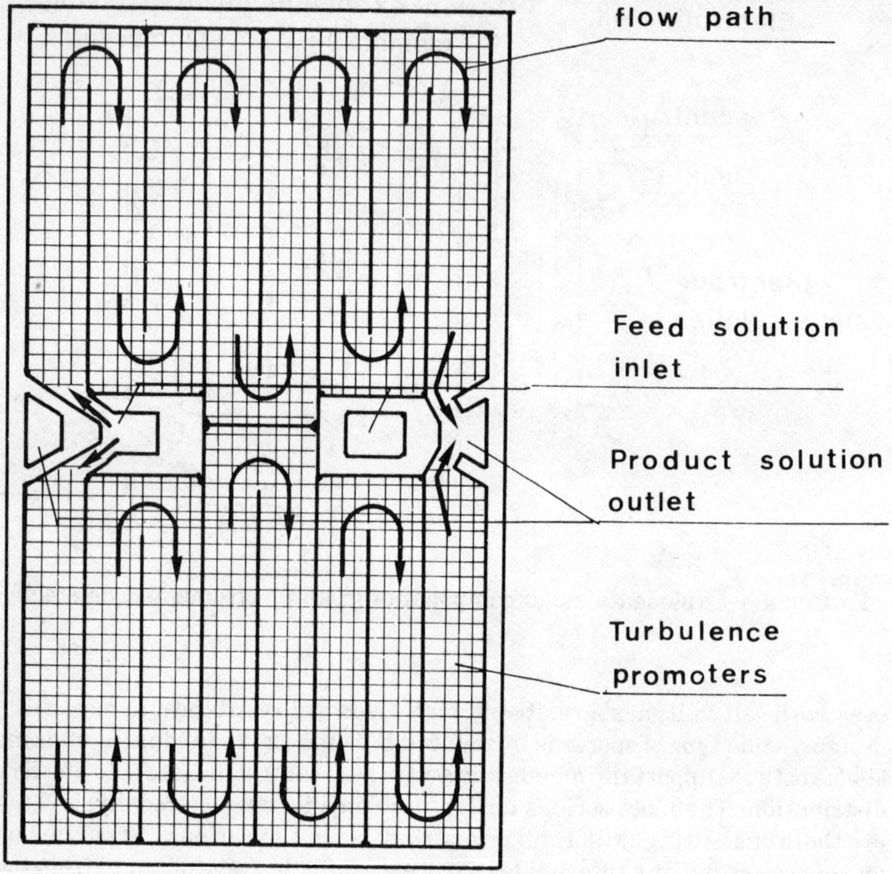

FIGURE 5.—Schematic diagram of a tortuous-path electrodialysis spacer gasket

straight path from the entrance to the exit ports which are located on opposite sides in the gasket. This is illustrated in Figure 6, which shows the schematic diagram of a sheet-flow spacer of an electrodialysis stack.

Solution flow velocities in sheet-flow stacks are typically between 5 and 10 cm/sec whereas in tortuous-path stacks, solution flow velocities of 30 to 50 cm/sec are obtained. Because of the higher flow velocities and longer flow paths, higher pressure drops in the order of 5 to 6 bars are obtained in tortuous-path stacks than in sheet-flow systems, where pressure drops of 1 to 2 bars occur. The membranes used in tortuous-path stacks are generally much thicker and more rigid than those used in sheet-flow systems.

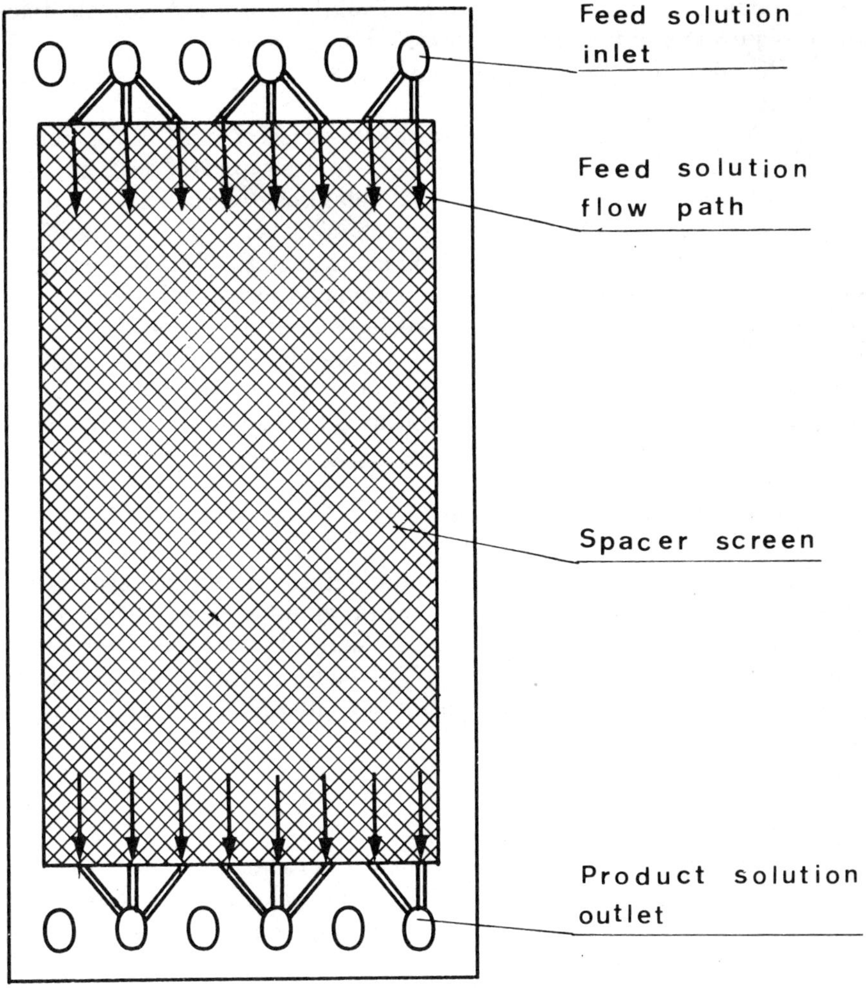

FIGURE 6.—Schematic diagram of a sheet-flow electrodialysis spacer gasket

The final design of a stack is almost always a compromise between a number of conflicting criteria and considerations, e.g., short distances between membranes for low electrical resistance, high feed flow velocities, and high turbulence for control of concentration polarization effects, and low pressure losses in the solution pumped through the stack.

5.2. The Electric Power Supply

The energy requirements for electrodialysis are entirely electrical. These requirements may be categorized in two groups:

1) DC power for the electrodialysis stack
2) AC power for pumping

The DC power is usually supplied on-site by use of an AC-to-DC converter. Conversion efficiencies of about 90 percent are typical. Constant voltage regulators are utilized to maintain stable plant operation. They are used to prevent stack damage. Stack resistance changes occur as a result of scale formation, membrane deterioration and changes in the fluid concentrations within the stack. The voltage regulator is adjusted periodically to compensate for these changes. Since in conventional electrodialysis stacks, 200 to 300 cell pairs may be arranged in parallel, with a voltage drop of 1–2 volts per cell pair, the DC voltage is on the order of 200 to 600 volts and the current up to 200 amperes for a stack where the membrane area of a cell is about 1 m^2, allowing for a current density up to 20 mA/cm^2.

5.3. The Hydraulic Flow System

The primary considerations in designing the hydraulic flow system are to obtain low hydraulic pressure drops while simultaneously achieving the required high volume flow rates. The pressure drop in the electrodialysis system is on the order of 2 to 8 bars. Therefore, simple plastic centrifugal pumps are generally used for circulating the different solutions through the stack.

5.4. Process Control Devices

The following variables are usually measured or controlled, or both:

1) DC voltage and current supplied to each electrodialysis unit
2) Flow rates and pressures of the depleting and concentrating streams, and of the electrode rinse streams
3) Electrolyte concentrations of the depleting and concentrating streams at the inlets and outlets to the electrodialysis stacks
4) pH of the depleting stream and the electrode rinse streams
5) Temperature in the feed stream

All the above variables are interrelated. Automatic control of the flows of the depleting and concentrating streams can be achieved by the use of flow-type conductivity cells in the effluent depleting and concentrating streams, along with a controller that compares the conductivites of the depleting and concentrating streams with that of a preset resistance and actuates flow-control valves in the liquid supply lines.

To prevent damage to the membranes or other components of an electromembrane stack in the event of stoppage of liquid flow in the stacks, the equipment should be provided with fail-safe devices that will turn off the power to the stacks and pumps.

6. PROCESS ECONOMICS

The total cost of the electrodialysis process is the sum of fixed charges associated with amortization of the plant capital cost and operating costs, such as energy requirements, spare parts costs, chemical treatment and operation labor costs. Membrane replacement cost will be regarded as a separate item because of the relatively short life of the membrane.

6.1. Investment Costs

Capital cost items may be divided into two categories: depreciable items and nondepreciable items. Depreciable items include the electrodialysis stacks, pumps, electrical equipment, and membranes, etc. Nondepreciable items include land and working capital. The nondepreciable items vary significantly with the construction site and the local policy and are therefore not discussed in this outline.

Capital costs of depreciable items The capital costs of an electrodialysis plant will strongly depend on the number of ionic species to be removed from a feed solution. This can easily be demonstrated for an electrodialysis plant producing potable water from saline water sources. The total membrane area required by a certain plant capacity is given by:

$$A = \frac{z\, F\, Q\, \Delta c'\, n}{i\, \xi} \tag{11}$$

Here A is the membrane area, z the chemical valence, Q the volume of the produced potable water, $\Delta c'$ the difference in the salinity of feed and product water, n the number of cells in a stack, i the current density, and ξ the current utilization.

For a certain plant capacity, i.e., Q = constant, and a given current density i, which is determined mainly by the product water concentration, the required membrane area is directly proportional to the feed water concentration. This is demonstrated in Figure 7.

For typical brackish water of ca. 5000 ppm (total dissolved solids (TDS) and an average current density of 6 mA/cm^2, the required membrane area for a plant capacity of 1 m^3 product per day is ca. 1.2 m^2 of each cation- and anion-exchange membrane. Since other items, such as pumps and electric power supply, do not depend at all or only very slightly on feed water salinity and since, with increasing feed water salt content, the average limiting current density is increased, the dependence of the total capital costs on the feed water salinity is less than linear; of course, the total capital cost will also depend on

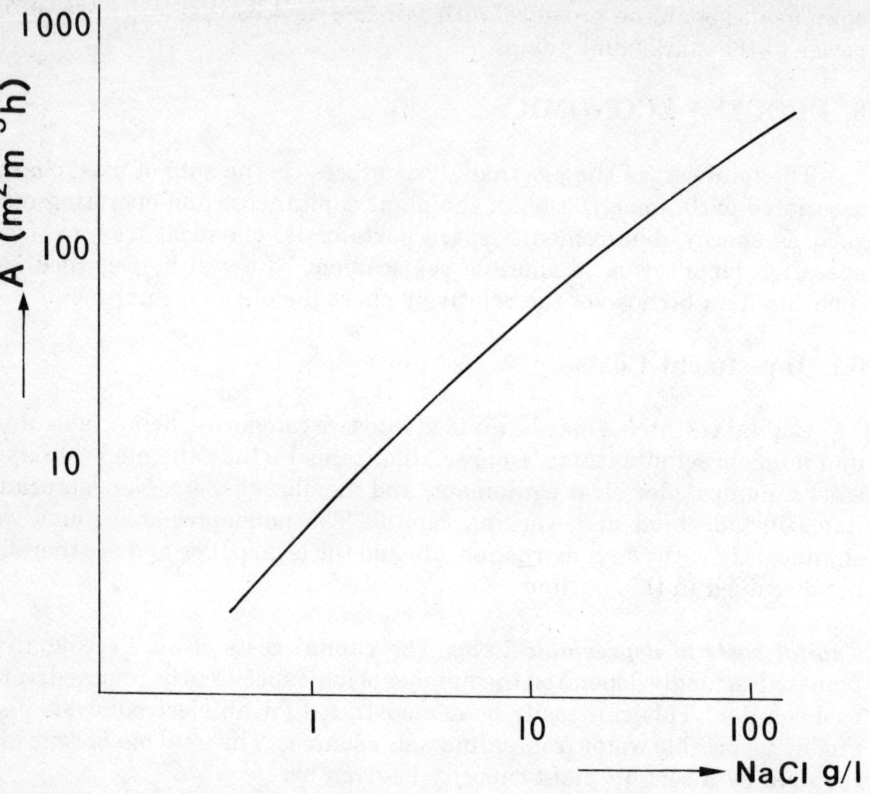

FIGURE 7.—Schematic diagram of the required membrane areas in electrodialysis desalination as a function of the feed water concentration at constant current density and plant capacity

the plant capacity. For desalination of brackish water with a salinity of ca. 5000 ppm, the total capital costs for a plant with a capacity of 1000 m^3/day will be in the range of \$600 to \$700 per m^3/d capacity. The cost of the actual membrane is less than 30% of the total capital costs. Assuming a useful life of the membranes of five years and of the rest of the equipment of 10 years and an availability of the plant of 95% and a 24-hour operating day, the amortization of the investment per m^3 potable water obtained from 5000 ppm brackish water is in the range of \$0.23 to \$0.27 per m^3 potable water. These costs do not include any interest, taxes, housing or extensive pretreatment.

Capital costs of non-depreciable items These items will include the land, the building, interest on capital investment, taxes, etc., which may differ significantly from country to country. In general they are in the same order of magnitude as the depreciable capital costs.

6.2. Operating Costs

The largest single item of the operating costs is the required energy. All other operating costs are minor, at least for large-scale plants.

Energy requirements The energy costs in electrodialysis are determined by the electrical energy required for the actual desalting process and the energy necessary for pumping the solution through the stack. The energy can be determined from equations 3 and 6. According to these equations the energy for the actual desalting process for a given plant capacity is directly proportional to the number of ionic species removed. The energy requirements for the production of potable water with a TDS of <500 ppm of saline feed water are shown schematically as a function of the feed water concentration of Figure 8.

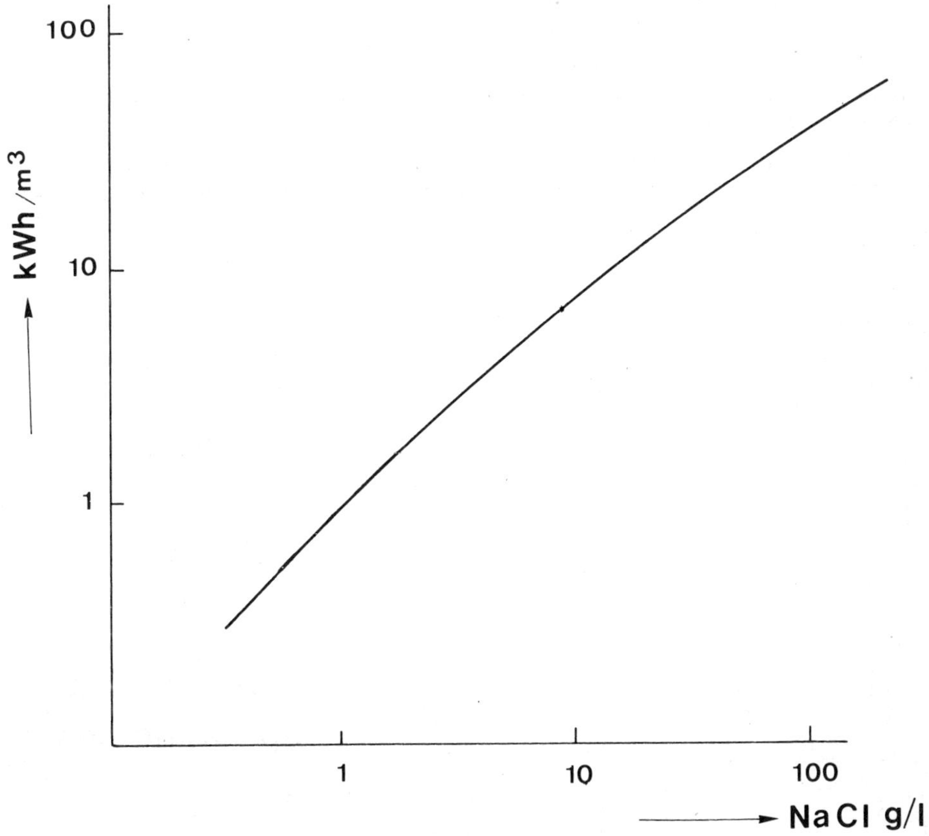

FIGURE 8.—Energy requirements for the production of potable water with a solid content of < 500 ppm as a function of the feed solution concentration ($\Delta\psi$ per cell pair = 0.8 V)

The pumping energy is independent of the feed solution salinity. Assuming a pressure drop in the unit of ca. 400 KPa (4 bar) and a pump efficiency of 70%, and 50% recovery, the total pumping energy will be about 0.32 kWh/m^3. This indicates that at low feed water salt concentration, the cost for pumping the solution through the unit might become quite a significant contribution to the total operating costs.

Chemical treatment and operating labor The cost for a chemical pretreatment will depend very much on the feed water composition. Some feed solutions may require a significant amount of pretreatment while other feed solutions can be processed without any chemical treatment.

Again for feed solutions with low salinity, chemical pretreatment and general maintenance and operating costs may become a significant portion of the total operating costs.

6.3. Total Electrodialysis Process Costs

The total costs of electrodialysis are shown in Figure 9 as a function of the applied current density for a given feed solution calculated according to equations 3 and 11.

This graph indicates that the capital costs are decreasing with increasing current density while the energy costs are increasing with increasing current

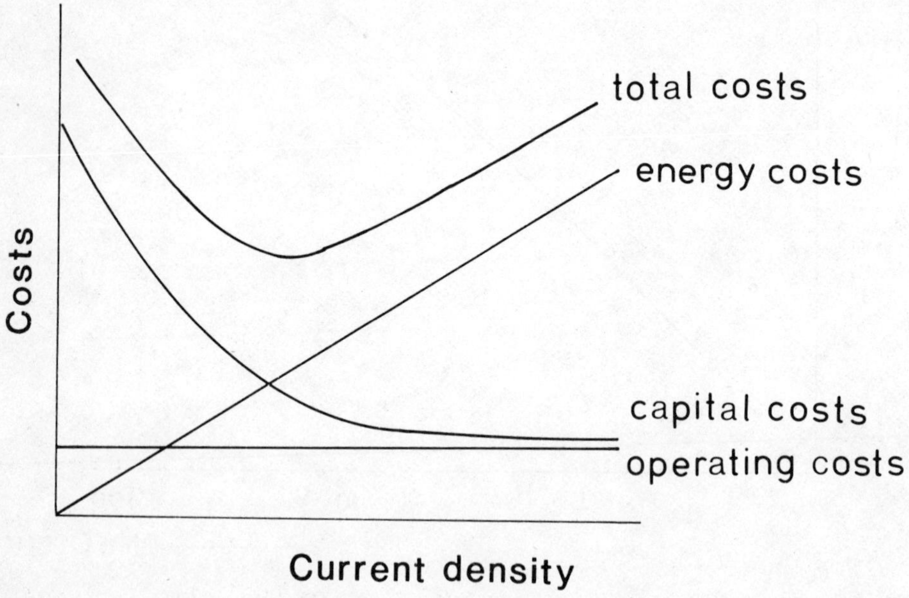

FIGURE 9.—Schematic diagram of the electrodialysis process costs as a function of the applied current density

density, and other cost items are more or less independent of the current density. In general, for a given feed solution there is a current density where the total electrodialysis process costs will reach a minimum.

Quite interesting, furthermore, is a comparison of the cost of desalination by various processes as a function of the feed water salinity, as shown in Figure 10.

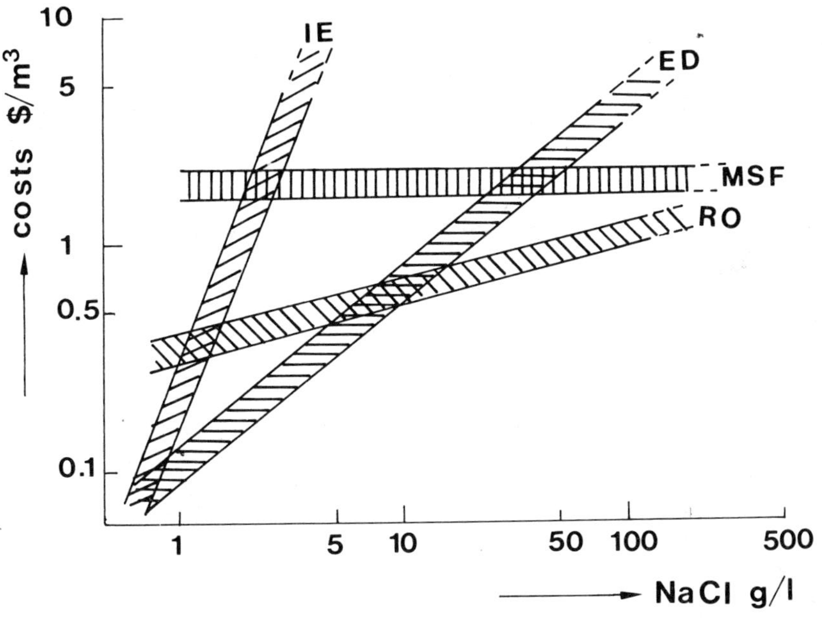

FIGURE 10.—Costs of desalination of saline water as a function of the feed solution concentration for ion-exchange (IE), electrodialysis (ED), reverse osmosis (RO), and multistage flash evaporation (MSF)

These graphs indicate that at very low feed solution salt concentration, ion-exchange is the most economical process. But its costs increase sharply with the feed solution salinity and at about 500 ppm TDS, electrodialysis becomes the more economical process, while at ca. 5000 ppm reverse osmosis is the least costly process. At very high feed solution salt concentrations in excess of 100,000 ppm, multistage flash evaporation becomes the most economical process. The costs of potable water produced from brackish water sources are in the range of \$0.2 to \$0.5 per m^3.

7. LARGE-SCALE APPLICATIONS

Electrodialysis has found a significantly smaller number of applications than reverse osmosis and ultrafiltration since it can only remove low molecular weight ionic components. Electrodialysis is mainly used today to produce potable water from brackish water sources, to treat certain industrial effluents, and to demineralize certain products in the chemical, food, and drug industries. Most modern electrodialysis units operate in the reverse polarity mode described in section 3.1 to prevent scaling due to concentration polarization effects.

7.1 Desalination of Brackish Water by Electrodialysis

The most important large-scale application of electrodialysis is the production of potable water from brackish water. Here, electrodialysis is competing directly with reverse osmosis and multistage flash evaporation. For water with relatively low salt concentration (less than 5000 ppm), electrodialysis is generally the most economic process, as indicated earlier. One significant feature of electrodialysis is that the salts can be concentrated to comparatively high values (in excess of 18 to 20 wt.%) without affecting the economics of the process severely. Because of these process characteristics, electrodialysis is also used (mainly in Japan) for the production of salt from sea water.

7.2 Electrodialysis in Waste Water Treatment

The main application of electrodialysis in waste water treatment systems is in processing rinse waters from the electroplating industry. Here, complete recycling of the water and the metal ions is achieved by electrodialysis. Compared with reverse osmosis, electrodialysis has the advantage of being able to utilize thermally and chemically more stable membranes, so that processes can be run at elevated temperatures and in solutions of very low or high pH values. Furthermore, the concentrations achievable in the brine product can be significantly higher. The disadvantage of electrodialysis is that only ionic components can be removed and additives usually present in a galvanic bath cannot be recovered.

An application that has been studied in a pilot plant stage is the regeneration of chemical copper plating baths. In the production of printed circuits, a chemical process is often used for copper plating. The components to be plated are immersed in a bath containing, besides the copper ions, a strong complexing agent, for example, ethylenediaminetetraacetic acid (EDTA), and a reducing agent such as formaldehyde. Because all constituents are used in relatively low concentrations, the copper content of the bath is soon exhausted and $CuSO_4$ has to be added. During the plating process, formaldehyde is oxidized to formate. After prolonged use, the bath becomes enriched with Na_2SO_4 and formate and consequently loses useful properties. By applying electrodialysis in a continuous mode, the Na_2SO_4 and formate can be selectively removed from the solution, without affecting the concentrations of formaldehyde and the EDTA complex, thereby significantly extending the useful life of the plating solution. Several other successful applications of electrodialysis in waste water treatment systems which have been studied on a laboratory scale are reported in the literature. However, large commercially operated plants are, at present, rare.

7.3 Electrodialysis in the Chemical, Food, and Drug Industries

The use of electrodialysis in the food, drug, and chemical industries has been studied quite extensively in recent years. Several applications seem to have considerable economic significance. One is the demineralization of cheese whey. Normal cheese whey contains between 5.5 and 6.5% dissolved solids in water. The primary constituents in whey are lactose, protein, minerals, fat and lactic acid. Whey provides an excellent source of protein, lactose, vitamins, and minerals, but in its normal form it is not considered a proper food material because of its high salt content. With the ionized salts substantially removed, whey approaches the composition of human milk and therefore provides an excellent source for the production of baby food. The partial demineralization of whey can be carried out quite efficiently by electrodialysis. The process has been studied extensively and is described in detail in the literature.

The removal of tartaric acid from wine is another possible application. In the production of bottled champagne, especially, it is necessary to avoid the formation of crystalline tartar in the wine and tartaric acid must therefore be reduced to a level that does not exceed the solubility limit. This again can be done efficiently by electrodialysis.

Desalting of dextran solutions, another application for electrodialysis, is of technical significance as a large-scale industrial process.

Several other applications of electrodialysis in the pharmaceutical industry have been studied on a laboratory scale. Most of these applications are concerned with desalting of solutions containing active agents which have to be separated, purified or isolated from certain substrates. Here, electrodialysis is often in competition with other separation procedures such as dialysis and solvent extraction. In many cases, electrodialysis is the superior process as far as the economics and the quality of the product are concerned.

7.4. Other Electrodialysis-related Processes

In addition to conventional electrodialysis, several processes closely related to electrodialysis have been discussed in the literature. Most of these processes are today still in a laboratory stage, but several show a significant industrial potential and are briefly reviewed.

Donnan-dialysis with ion-selective membranes In Donnan-dialysis the ion-concentration difference in two phases separated by an ion-exchange membrane is used as the driving force for the transport of ions with the same electrical charges in opposite direction. The principle of the process is shown schematically in Figure 11.

This figure shows as an example two solutions of $CuSO_4$ and H_2SO_4 separated by a cation-exchange membrane. Since the H^+-ion-concentration in solution ′ is significantly higher (pH = 1) than the H^+-ion-concentration in solution ″ ($p_H = 3$), there will be a constant driving force for the flow of H^+-ions from solution ′ into solution ″. Since the membrane is permeable for cations only, there will be a build-up of an electrical potential (diffusion potential) that will counterbalance the concentration difference driving force of the H^+-ions, and ions of the same charge will be transported in the opposite direction as indicated by the flow of Cu^{++}-ions from solution ″ into solution ′ in Figure 11. As long as the H^+-ion-concentration difference between the two phases separated by the cation-exchange membrane is kept constant, there will be a constant transport of Cu^{++}-ions from solution ″ into solution ′ until the Cu^{++}-ion-concentration difference reaches the same order of magnitude as the H^+-ion-concentration difference, i.e., Cu^{++}-ions can be transported against their concentration gradient driving force by Donnan-dialysis.

The same process can be carried out accordingly with anions through anion-exchange membranes. An example of anion Donnan-dialysis is the

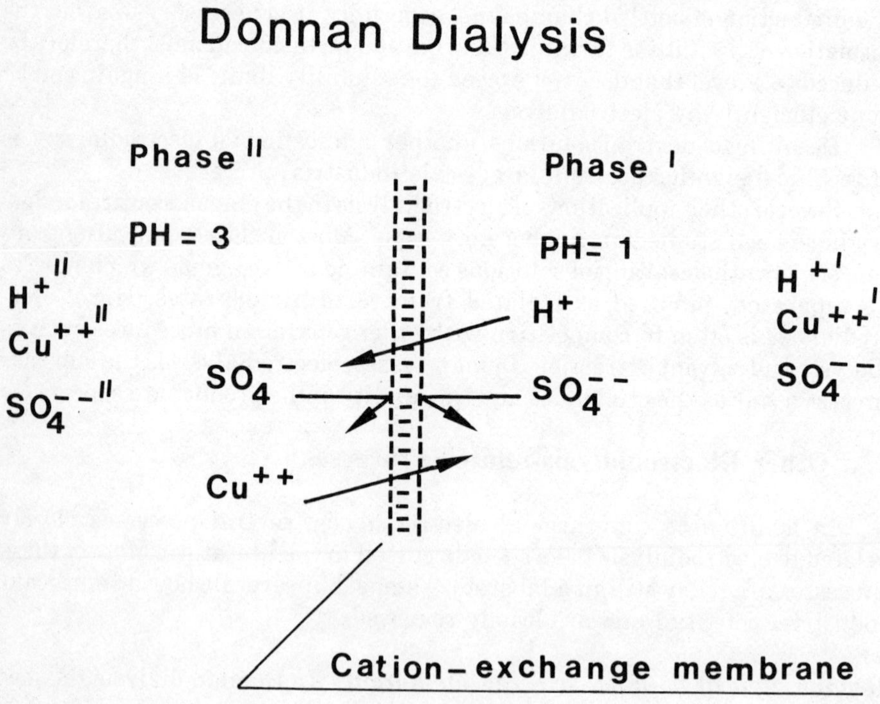

FIGURE 11.—The principle of Donnan-dialysis

sweetening of citrus juices. In this process, hydroxyl ions furnished by a caustic solution replace the citrate ions in the juice.

Effluent-free regeneration of ion-exchange resins A virtually effluent-free regeneration of a charged ion-exchange resin by recovering the extracted metal in solid form can in some cases be achieved with an electrodialytic procedure, the principle of which is shown in Figure 12.

A cation-exchange resin that is charged with metal-ions is placed between two cation-exchange membranes and two electrodes. The electrode compartments contain electrolyte solutions, e.g., sulfuric acid, to provide the necessary conductivity. By applying a direct current, H^+-ions generated at the anode migrate through the cation-exchange membrane into the resin and replace metal-ions, which migrate through the opposite membrane toward the cathode where they are electrochemically reduced and precipitated as solid metal. The process can be continued until all metal-ions in the resin are replaced by H^+-ions and precipitated at the cathode.

Water splitting with bipolar membranes A process referred to as "electrodialytic water splitting", which employs bipolar membranes to produce acids and bases from salts, has been suggested for a number of years. So far, however, the process has only been evaluated on a laboratory scale and an industrial application has not yet been realized, although the process shows significant advantages in terms of energy requirements and capital costs over

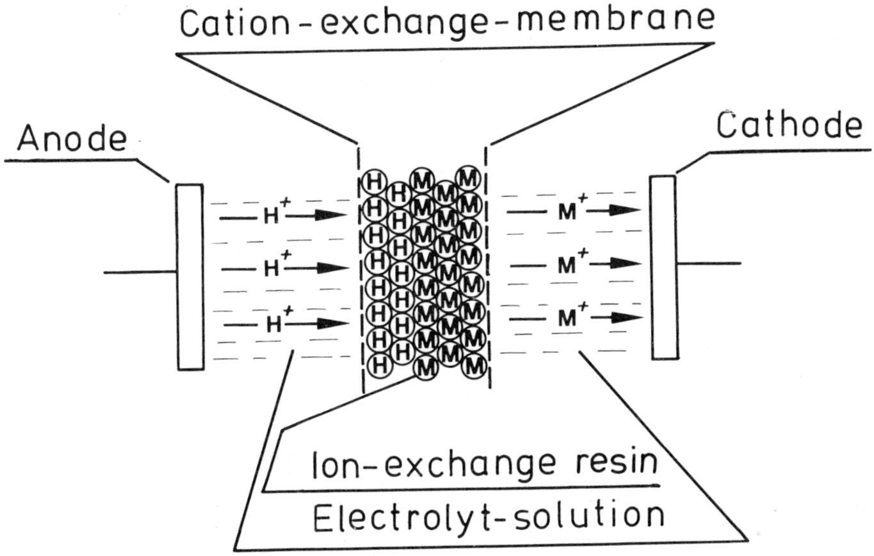

FIGURE 12.—Schematic diagram of the electrodialytic regeneration of a metal-ion charged cation-exchange resin

conventional acid and base production procedures. The process is conceptually rather simple as indicated in the schematic diagram in Figure 13.

A cell system consisting of an anion-, a bipolar- and a cation-exchange membrane as a repeating unit is placed between two electrodes. The sodium sulfate solution is placed in the outside phase between the cation- and anion-exchange membranes. When a direct current is applied, water will dissociate in the bipolar membrane to form an equivalent amount of hydrogen and hydroxyl ions. The hydrogen ions will permeate the cation-exchange side of the bipolar membrane and form sulfuric acid with the sulfate ions provided by the sodium sulfate from the adjacent cell. The hydroxyl ions will permeate the anion-exchange side of the bipolar membrane and form sodium hydroxide with the sodium ions permeating into the cell from the sodium sulfate solution through the adjacent cation-exchange membrane. The most important part in this cell arrangement is the bipolar membrane, which generally consists of a cation- and an anion-exchange membrane laminated together. The membrane should have good chemical stability in acid and base solutions and low electrical resistance. Laboratory tests have demonstrated that production costs for caustic soda by utilizing bipolar membranes are only one third to half of the costs of the conventional electrolysis process.

This brief outline has given a few typical and well-known applications of electrodialysis. Many more are possible and have already been evaluated on a laboratory scale. With the development of more efficient membranes the future will certainly see an increasing use of electrodialysis, which will go far beyond its present application in the desalination of brackish water.

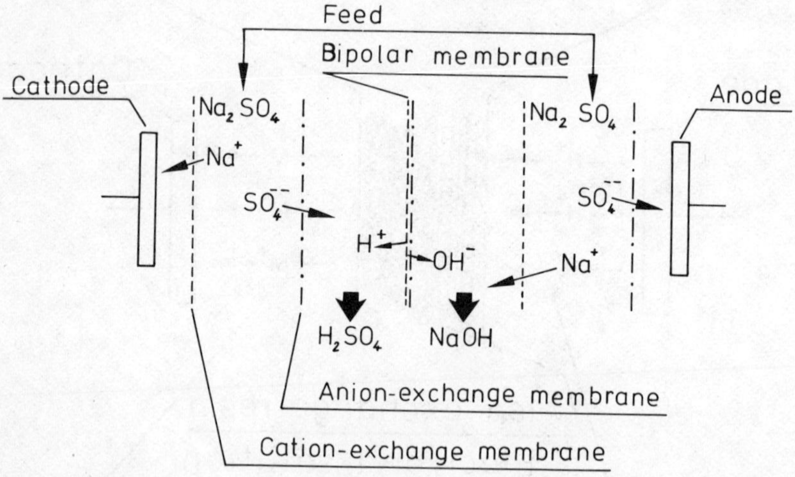

FIGURE 13.—Schematic diagram showing the electrodialytic regeneration of sulfuric acid and sodium hydroxide from sodium sulfate employing bipolar membranes.

SELECTED BIBLIOGRAPHY

1. Helfferich, F.: 1962, "Ion-Exchange," McGraw-Hill, New York.
2. Spiegler, K. S.; 1966, "Principles of Desalination," Academic Press, New York.
3. Wilson, J. R.: 1960, "Demineralization by Electrodialysis," Butterworths, London.
4. Tuwiner, S. B.: 1962, "Diffusion and Membrane Technology," Reinhold, New York.
5. Lacey, R. E., Loeb, S. Eds.: 1972, "Industrial Processing with Membranes," Wiley-Interscience, New York.

SYMBOLS

A	membrane area
M_R	fixed charges
T	ion transport number
n	number of cells in a stack

Subscripts

B	brine
C	concentrate
D	diluate
E	electrode rinse solution
M	membrane phase
b	bulk solution
n	current flow through stack manifold
+, −	anion, cation
m	membrane surface
s	membrane selectivity
w	water transfer

MICROFILTRATION

Mark C. Porter

Consultant
3449 Byron Court
Pleasanton, California 94566

This survey of the most common pressure-driven membrane process, microfiltration, begins with a historical review. Membrane structure, materials and fabrication are discussed along with methods of pore size determination. Factors that influence the retention of particulates and microorganisms are balanced against those affecting membrane plugging and throughput. The survey ends with a discussion of equipment design and industrial applications.

P. M. Bungay et al. (eds.), Synthetic Membranes: Science, Engineering and Applications, 225–247.

1. INTRODUCTION

Pressure-driven membrane processes have experienced dramatic growth in the last 30 years. The current worldwide market for the three most common processes is estimated to be in excess of $600 millions annually (see Table 1), whereas in 1952, the total market was probably less than $60,000 annually. This represents an averge annual growth rate of 36%.

Microfiltration (MF) has the largest share of this market, being in excess of $300 millions. This is not surprising since the commercialization of MF began in 1927 by Sartorius-Werke in Germany, whereas RO and UF were commercialized some 40 years later.

TABLE 1. Annual worldwide market for pressure-driven membrane filtration in 1982

Membrane Process		Market Size ($ Million/Year)
Microfiltration (MF)		300
Ultrafiltration (UF)		60
Reverse Osmosis (RO)		240
	Total	$600 Million/Year

However, the RO market is growing at a faster rate due to desalting demands and the larger size of RO plants.

Microfiltration (MF) may be distinguished from reverse osmosis (RO) and ultrafiltration (UF) by the size of the particle it is capable of retaining—as in Figure 1.

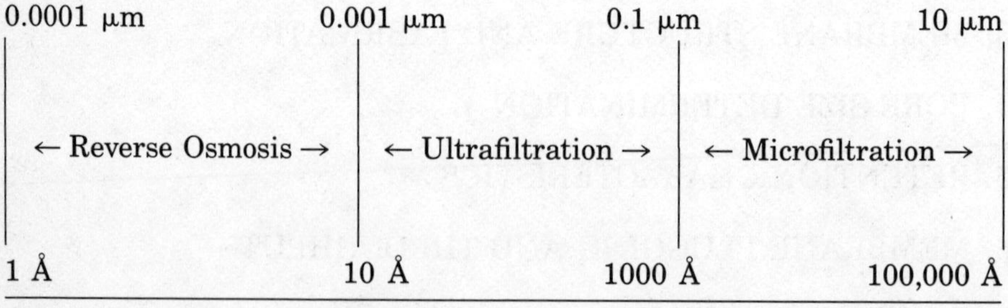

FIGURE 1.—Definition of pressure-driven membrane processes (based on size of particle or solute retained).

Further, since microfiltration membranes have no retention for salt, there is no osmotic pressure to be overcome, and the filtration can be carried out at relatively low pressures—below 50 psi (or 3 bar).

2. HISTORY

Fick (1) was probably the first to make a synthetic MF membrane (out of nitrocellulose) in 1855. But it was not until the early 1900s that Bechhold (2) began to sort out the variables affecting membrane characteristics; he was the first to produce a graded series of MF membranes with varying permeability. To estimate pore diameter, Bechhold measured the air pressure required to overcome capillary forces and expel water from water-wet pores (referred to as the "bubble-point" measurement today) and the water flow rate as a function of pressure. Using the capillary equation and Poiseuille's law, he was able to obtain the pore size.

In 1918, Zsigmondy and Bachmann (3) began to use the findings of Bechhold and others (notably Bigelow, Gemberling, Schoep, and Brown) to develop the manufacturing technology for production of nitrocellulose and cellulose-ester membranes on a semi-commercial scale. Dr. Zsigmondy, director of the Institute of Colloid Chemistry at the University of Göttingen, obtained a U.S. patent on the process in 1921.

In 1927, the Sartorius-Werke Aktiengesellschaft and Company refined the Zsigmondy process and began commercial production of membrane filters on a small scale. The product was only of interest to the research community. The membranes were used to remove particles, microorganisms and viruses from liquids; for diffusion studies; and for sizing of proteins.

During World War II, many of the German water supply systems were destroyed in air-raids. The authorities were forced to check drinking water for contamination at frequent intervals. The standard methods of bacteriological analysis were time-consuming and cumbersome—taking up to 96 hours.

Dr. Gertrud Müller and associates at the Hygiene Institute of the University of Hamburg developed membrane filter techniques (4) suitable for the bacteriological analysis of water. A 500-ml sample of water could be filtered through the membrane trapping all of the bacterial flora on the surface. The membrane was subsequently placed in contact with a nutrient medium so the organisms could grow into visible colonies within 12 to 24 hours.

After the war (in 1946), the Joint Intelligence Objectives Agency of the U.S. Armed Services commissioned Dr. Alexander Goetz to visit Germany and obtain information on its membrane technology. Goetz visited Membranfiltergesellschaft Sartorius-Werke (Göttingen), the Hygienisches Institut der Universität (Göttingen), and the Hygiene Institut der Universität (Hamburg). Goetz made a full report (5) on Germany's membrane technology and generated considerable interest in the Biological Department of the U.S. Army Chemical Corps, which subsequently awarded research contracts to Goetz at the California Institute of Technology to improve the technology.

Within three years (1950), Goetz had developed new production methods capable of producing membranes with improved flow rates and uniformity of pores.

Following the Goetz contract, the Chemical Corps funded the development of manufacturing technology for the production of these membranes at the Lovell Chemical Co., Watertown, Massachusetts. In 1954, the manufacturing equipment developed during this program was sold to the newly formed Millipore Corporation. Within months, Millipore had set up to produce eight different pore sizes, from below 0.1 microns to 10 microns. However, the first major market did not emerge until 1957 when the U.S. Public Health Service and the American Water Works Association officially accepted the membrane filter procedure for recovery of coliform bacteria.

Until 1963, microfilters were predominately nitrocellulose or mixtures of cellulose esters. As new applications began to emerge, the need for membranes with improved chemical resistance and heat stability prompted investigations of other materials and methods of fabrication. Today there is a wide variety of media available.

3. MEMBRANE STRUCTURE AND FABRICATION

Microfiltration membranes can be classified as to pore configuration into two generic types: "tortuous-pore" and "capillary-pore" (see Figure 2). "Tortuous-pore" membranes are the most common. Their structure resembles that of a sponge with a tortuous labyrinth of pores. Such membranes can be solution-cast from a variety of polymers including cellulose esters, polyvinylchloride (PVC), PVC-acrylonitrile copolymers, polyamides, polyvinylidene fluoride, etc. Controlled evaporation of solvents in a humid atmosphere precipitates the polymer around residual solvents creating the open-celled porous structure. Solution-cast membranes of this type can be also cast on fabrics which provide reinforcement and permit pleating and other manipulations not otherwise possible. However, in the last five years non-reinforced membranes have been produced which are pleatable and much more rugged than the common cellulose ester membranes.

In addition to solution-cast tortuous-pore membranes, polytetrafluoroethylene (PTFE) and polypropylene tortuous-pore membranes have been manufactured using a stretching process (Figure 3) (6).

In the mid-1970's, Hydronautics developed a unique porous polypropylene tube formed by dissolving polypropylene in an organic solvent at elevated temperatures and then cooling the polymeric solution until the polypropylene precipitates out around the remaining solvent in two continuous phases. This "thermal-inversion" process has been further developed by Membrana (Enka).

The "capillary-pore" structure is currently available only from Nuclepore Corporation. (Other firms are now free to pursue this technology since the basic patent has expired). Nuclepore makes polycarbonate and polyester

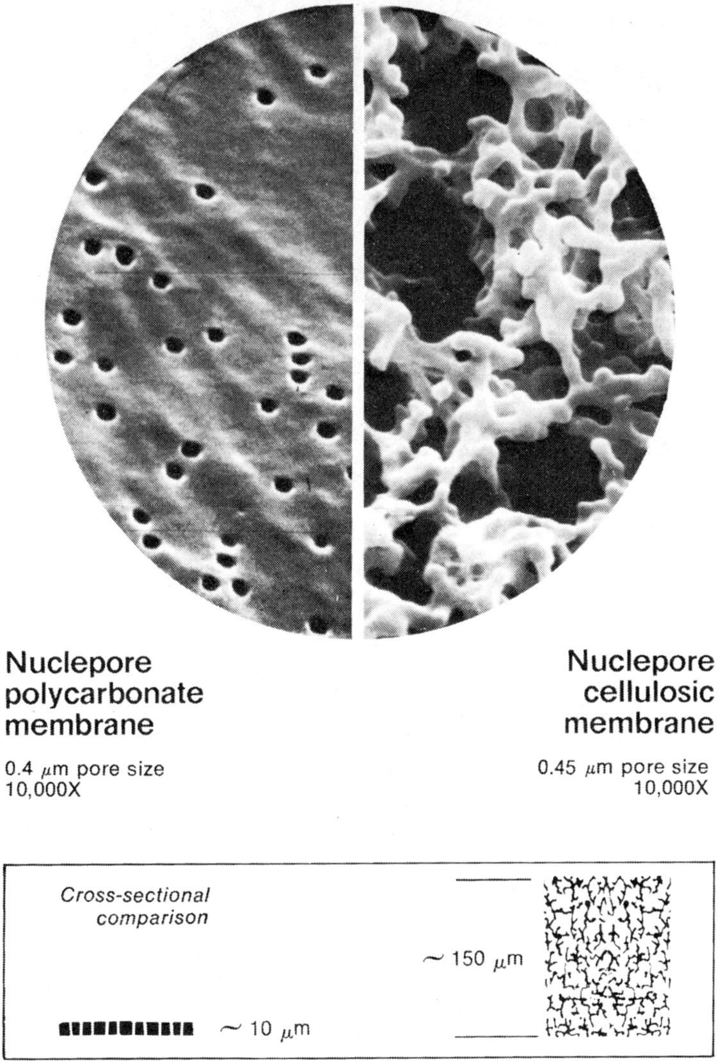

FIGURE 2.—Pore configuration of two generic types of microfiltration membranes: "capillary-pore" (left) and "tortuous-pore" (right).

"capillary-pore" membranes (Figure 4) using a "track-etch" process. The pores are not tortuous but straight through cylindrical pores. (More often than not they are of a "barrel" or "hour-glass" shape).

The "track-etch process" is a two-stage process (Figure 5). In the first stage, polycarbonate or polyester film is irradiated with a collimated beam of fission fragments, which leave "damage-tracks" in their path as they pene-

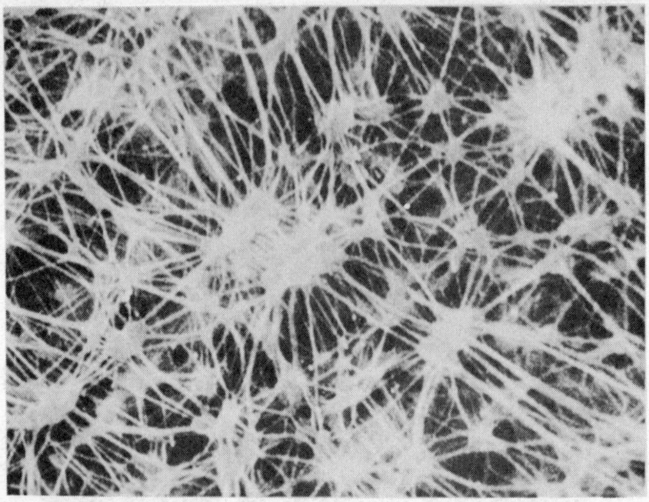

FIGURE 3.—Illustration of tortuous-pore membranes manufactured by a stretching process.

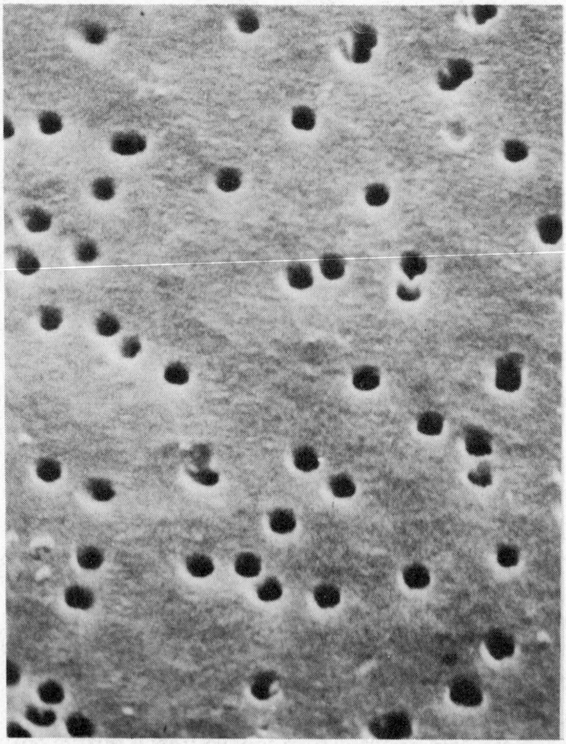

FIGURE 4.—Nuclepore "capillary-pore" membrane, made by a "track-etch" process.

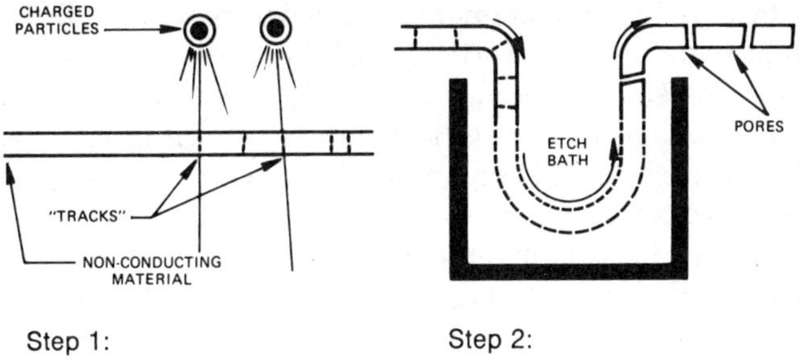

Step 1: *In the Nuclepore two-step process, polycarbonate or polyester film is first exposed to collimated, charged particles in a nuclear reactor. As particles pass through the material, they leave sensitized tracks. Pore density (pores/cm^2) is controlled by the residence time in the irradiator.*

Step 2: *In the second step, the tracks left by the particles are preferentially etched into uniform, cylindrical pores. By controlling the length of the etching process, a specified pore size is produced.*

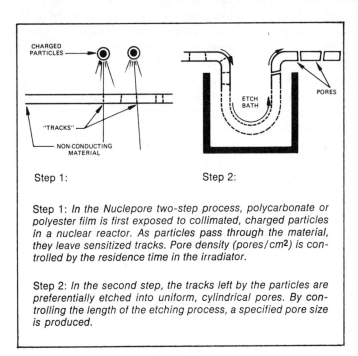

FIGURE 5.—Schematic of the two-stage track-etch process used by Nuclepore.

trate the film. (If the film thickness exceeds 15 microns, sizeable numbers of the fission fragments will not pass through the total thickness). In the second stage, those ionized tracks are preferentially etched in hot caustic to form the pores. Obviously, the pore density (number per sq cm) is determined by the residence time in the irradiator, whereas the pore size is independently controlled by the residence time in the etch bath.

To date, the capillary-pore membranes have found niches in specialized laboratory applications, such as particle fractionation, gravimetric and trace element analysis, microscopic analysis, and blood rheology studies. As of this date, most of the large-scale industrial applications are better served by tortuous-pore membranes, as will become obvious.

4. PORE SIZE DETERMINATION

Classically, the limiting pore size of microfiltration membranes has been determined by challenging the membrane with various organisms. For example, the smallest *Serratia marcescens* organism is about 0.45 microns in diameter whereas the smallest bacteria, *Pseudomonas Diminuta,* is about 0.22 microns. Although membrane lots may be characterized by the manufacturer using this approach, it is laborious and tedious.

A non-destructive test commonly applied by both manufacturer and user is the bubble point test. A wetting fluid is used to wet-out all the pores. Subsequently, one side of the membrane is pressurized with a gas until the gas-pressure overcomes the capillary forces retaining liquid in the pores (Figure 6). The pressure at which the first bubble appears is termed the "bubble point pressure". Of course this represents the maximum pore size since the capillary force is inversely proportional to the pore size. The theoretical bubble point may be calculated from the equation

$$p_{bp} = \frac{4\ \gamma \cos \theta}{d_p} \tag{1}$$

where p_{bp} = bubble point gauge pressure
γ = liquid surface tension
θ = liquid–solid contact angle
d_p = pore diameter

In practice, no membrane exhibits a bubble point equivalent to that of the rated pore size. This is because tortuous-pore membranes always have a few pores larger than the rated maximum and capillary-pore membranes have a few doublets and triplets. This is confirmed by so-called "grow-through" experiments where, given enough time, bacteria do pass membranes that are supposed to retain them.

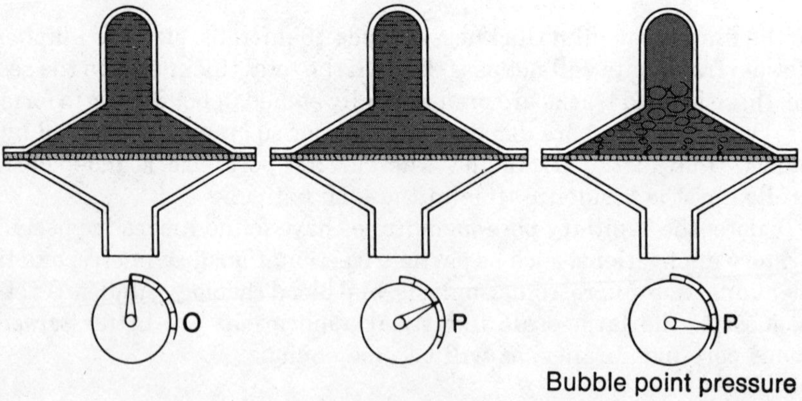

FIGURE 6.—Schematic of the "bubble point" test. When the air pressure against a wet membrane filter is increased, the pressure at which the largest pore begins to pass air is the bubble point.

When bubble point testing large areas of membrane, one must take into account gas passage through liquid-filled pores due to diffusion. The gas dissolves in the liquid on the high-pressure side of the membrane, and after diffusing across the liquid-filled pores, comes out of solution on the low-pressure side of the membrane. Gas diffusion rates are typically one ml/min/ft^2 at a gas pressure of 10 psi and are orders of magnitude below the gas permeation rate at the bubble point.

Mercury intrusion porosimeters are also used to determine pore size distributions but measure pits as well as pores.

Scanning electron micrographs are virtually useless in determining pore size except with the capillary-pore membranes.

5. RETENTION CHARACTERISTICS

The retention of organisms or particles is dependent on other factors in addition to the pore size. For example, the deformability of red cells markedly affects their transmission through pores [Figure 7 (7)]. Further, some particles much smaller than the rated pore size may be adsorbed within the medium.

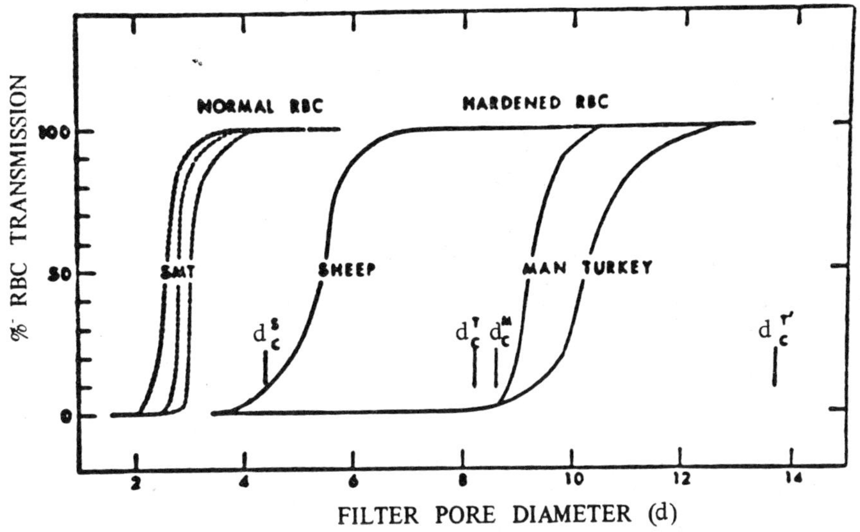

FIGURE 7.—Transmission of normal and hardened red blood cells from sheep, man, and turkey through polycarbonate sieves. Note the transmission of normal red blood cells through pores with diameters smaller than cell diameter (d_c). Hardened cells generally require pore diameter larger than cell diameter for their passage. For the turkey, the major ($d_c^{T\prime}$) and minor (d_c^{T}) diameters of the oval are given, and it seems that the minor diameter is the major determinant for cell transmission (7).

Table 2 (8) shows retention data for gold colloids filtered through tortuous-pore and capillary-pore membranes. The much higher retentions in tortuous-pore membranes reflect the fact that they are 10 to 15 times thicker and have 25 to 50 times more internal area for adsorption than do capillary-pore membranes. It is generally true that if the objective of filtration is fractionation of particles, capillary-pore membranes are best. On the other hand, most industrial-scale filtration has as its objective the removal of *all* particles. This is one reason why tortuous-pore membranes are more often chosen for large-scale filtration.

TABLE 2. Percent retention of Au-colloids filtered through cellulosic and Nuclepore membranes (8)

	Colloid Size (μm)	
	0.05	0.005
0.1μm Nuclepore	1.2	0.2
0.1μm Cellulosic	92.0	8.2
0.4μm Nuclepore	1.3	0.2
0.45μm Cellulosic	46.9	12.7
1.0μm Nuclepore	0.7	0.3
1.2μm Cellulosic	46.5	26.7
3.0μm Nuclepore	0.4	0.2
5.0μm Cellulosic	59.3	17.9

Other factors, such as the "zeta-potential" of the membrane medium, can also have a profound effect on retention of oppositely charged particles. Cuno Corp. advertises substantial retention of pyrogens with its Zetapore 0.2 micron medium, while cellulose acetate membranes show no retention for pyrogens.

In the filtration of aerosols, there are three mechanisms of retention illustrated by the efficiency of particle capture data in Figure 8. Particles of radius (b) larger than the rated pore radius (r_p = 4 microns) are "intercepted." Particles smaller than the rated pore size may be captured by "inertial-impaction" or by "diffusion" (9).

In "inertial-impaction" the particle is smaller than the pore size but too large to follow the stream-lines of the gas. Consequently, its inertia results in impaction on the pore walls where it is captured. The higher the gas velocity, the greater the number of particles captured in this way.

Extremely small particles may also be captured by "diffusion" since the diffusivity is inversely proportional to the particle size. The lower gas velocities enhance capture by diffusion since the residence time of the particle in the pore is longer.

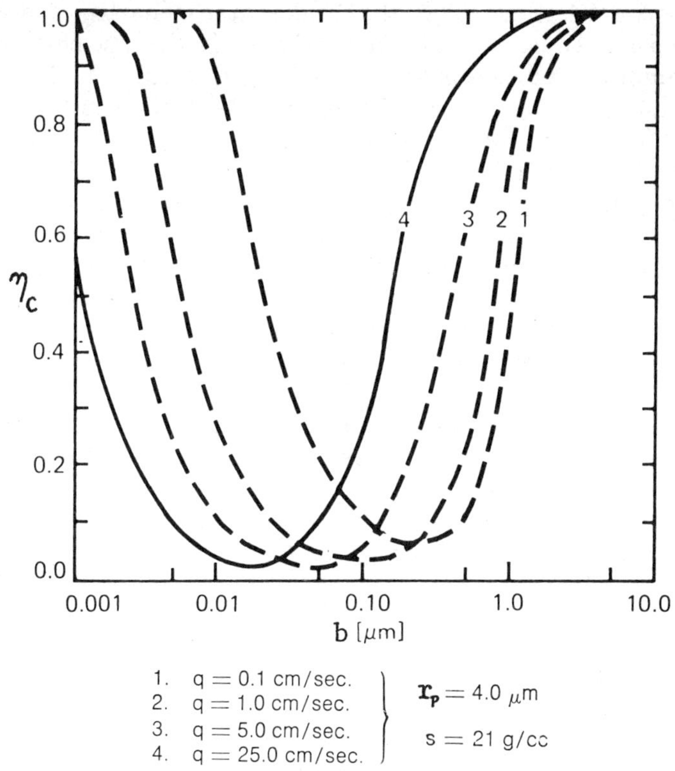

FIGURE 8.—Spurny et al. (8) have carefully characterized the retention efficiency of the Nuclepore screen membrane when filtering a gas, as illustrated. Particle radius = b, pore radius = r_p, particle density = s, superficial face velocity of gas = q, particle collection efficiency = η_c.

Aerosol particle capture is greater with tortuous-pore membranes than with capillary-pore membranes because of the more tortuous path (favoring inertial impaction) and longer pore lengths (favoring capture by diffusion). Indeed, minima like those observed with capillary pore membranes in Figure 8 are unheard of with tortuous-pore membranes. Again, fractionation of aerosol particles is generally possible only with capillary pore membranes; tortuous-pore membranes collect most of the particles, regardless of size.

6. MEMBRANE PLUGGING AND THROUGHPUT

Although microfiltration membranes offer more "absolute" filtration than conventional "depth" filters, they plug much more rapidly. The depth media are generally fibrous with an enormous "dirt-loading" capacity. Membranes collect much of the dirt on the surface of the membrane and very little within the pores. Thus, a combination of the two is often the best choice. The depth medium is used as a prefilter while the membrane is the final filter capturing any particles that have slipped through the depth filter or any fibers shed from it via media migration.

The matching of the right prefilter with the final membrane can generally only be determined from actual experimental data. The prefilter should remove the bulk of the particulate load while reserving the membrane for the final polishing. However, if the prefilter is too tight, it can plug before the membrane. If it is too loose, the membrane will plug first. Usually, the most cost-effective combination is when both plug at the same time. The nominal pore size of the prefilter and its area can be tailored to match the final membrane of fixed area.

Often, in the pharmaceutical industry, a series of prefilters or of membranes provides the maximum life and minimum filtration cost.

Obviously, the pore configuration of the membrane has a dramatic effect on membrane throughput (the volume that can be processed before the membrane plugs). Figure 9 shows typical data taken at a constant filtration

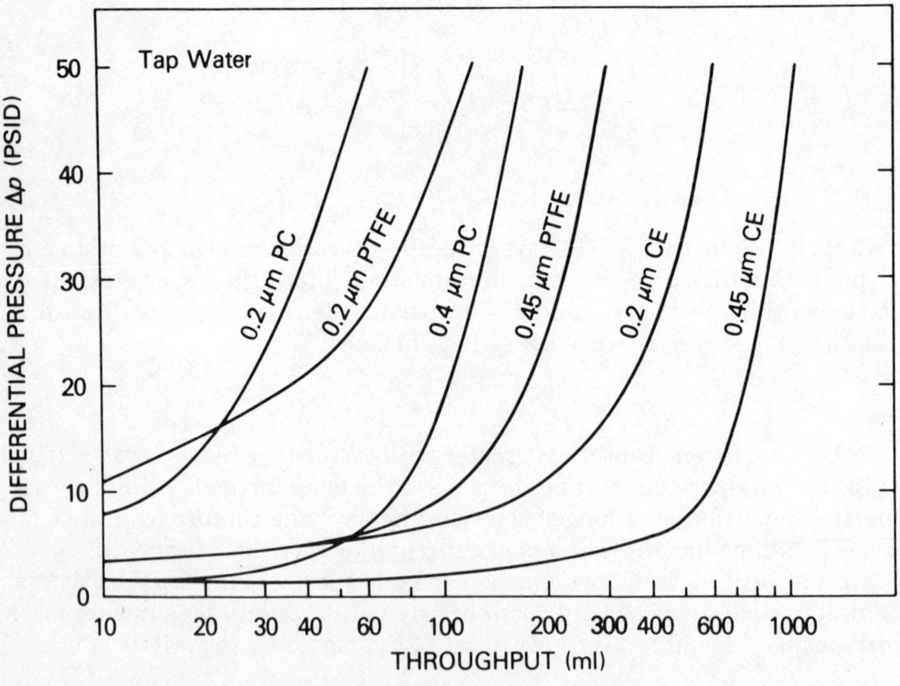

FIGURE 9.—Increase in transmembrane differential pressure as a function of cumulative volume of tap water filtered through various microfiltration membranes at the same constant volumetric flow rate.

rate. As the membrane plugs, the differential pressure across the membrane must increase to maintain the same flow rate. Eventually, a maximum pressure is reached where the pump can no longer supply the specified flow-rate. At this point, the filter must be replaced; it is plugged (typically when the Δp rises to 30 psi).

Figure 9 shows that polycarbonate (PC) capillary-pore membranes have considerably less throughput per unit area than do the tortuous-pore membranes (cellulose ester, CE, and teflon, PTFE). This is because of the lower dirt-holding capacity of these membranes (only 3% of the internal area of the cellulose ester membrane).

In addition to differences in dirt-holding capacity for the various media, the flux (filtration-rate) has a profound effect on throughput. Figure 10 indicates that higher throughputs may be obtained with lower flux values.

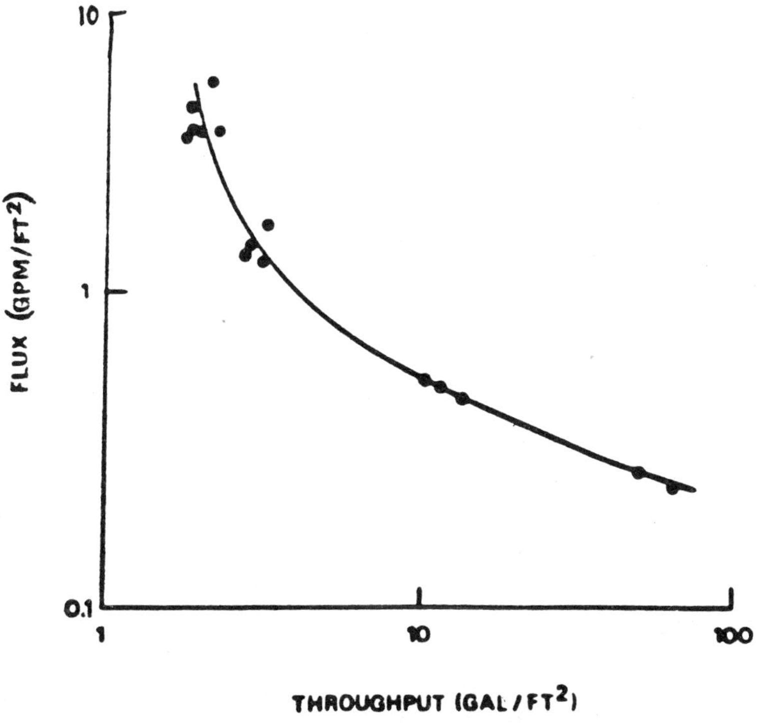

FIGURE 10.—Interrelationship between total cumulative filtered volume (throughput) per unit of membrane area and transmembrane flux.

This is a direct consequence of the higher pressure drops (with attendant impaction of particles) resulting from the higher flow rates through the media. This may be demonstrated mathematically as follows:

$$J_V = \frac{\Delta p}{R_c + R_m} \tag{2}$$

where J_V = volumetric flux or filtration rate per unit area
Δp = differential pressure drop across the media
R_m = resistance due to the membrane alone
R_c = resistance due to the cake build-up on and in the membrane

$$R_c = \frac{\alpha\, w\, V_t\, (\Delta p)^{\beta}\, \eta}{A} \tag{3}$$

where β = compressibility of the cake (deformation of deposited particles)
η = viscosity of the liquid
w = concentration of particles in the fluid stream per unit volume of fluid
V_t = volumetric throughput up to any given time
α = constant dependent on the packing density of the particles deposited in the cake
A = membrane area over which the particles are deposited

Combining equations 2 and 3

$$J_V = \frac{\Delta p}{\dfrac{\alpha w V_t (\Delta p)^{\beta} \eta}{A} + R_m} \tag{4}$$

As the membrane begins to plug $R_m << R_c$

$$J_V = \frac{A\,(\Delta p)^{1-\beta}}{\alpha w V_t \eta} \tag{5}$$

or

$$V_t = \frac{A\,(\Delta p)^{1-\beta}}{\alpha w\, J_V \eta} \tag{6}$$

From equation 6 one can infer that the throughput is directly proportional to the membrane area squared by replacing J_V by transmembrane flow rate (held constant) divided by membrane area. Thus, for a given flow rate to be processed, one can quadruple the throughput (membrane life) by doubling the membrane area. There is then an optimum flow rate where the total cost will be minimized. As we decrease the flux, the membrane replacement costs are reduced (longer life), but the first costs (capital for more housings and membranes) are increased.

Other techniques have also been developed to increase membrane life. Backwashing has been successful in some applications with capillary-pore membranes—notably in the filtration of liquid sugar and radioactive wastes. Of more universal application is the technique of cross-flow filtration.

Recent publications leave the impression that cross-flow filtration is a revolutionary fluid management technique that is of recent origin. Actually, the technique was first developed in the early sixties for reverse osmosis and ultrafiltration, where it is essential to control concentration polarization. More recently the technique has been applied to microfiltration with considerable success.

Cross-flow filtration extends membrane life by flowing the process stream tangential to and across the membrane surface to sweep away the accumulating particulates retained by the membrane. Thus, unlike through-flow filtration, there are two streams out of the filter. One exit stream is the filtrate while the other is a recirculating stream utilized to provide the sweeping action across the membrane surface (Figure 11). In this system, mechanical pumping energy is expended to prolong membrane life.

For homogeneous membrane media there is still some flux decay with time due to internal pore fouling (Figure 12). Nevertheless, a steady state flux is often obtained, which prolongs membrane life significantly. In some industrial applications, a frequent back-wash cycle coupled with cross-flow restores the flux to near its original value (Figure 13).

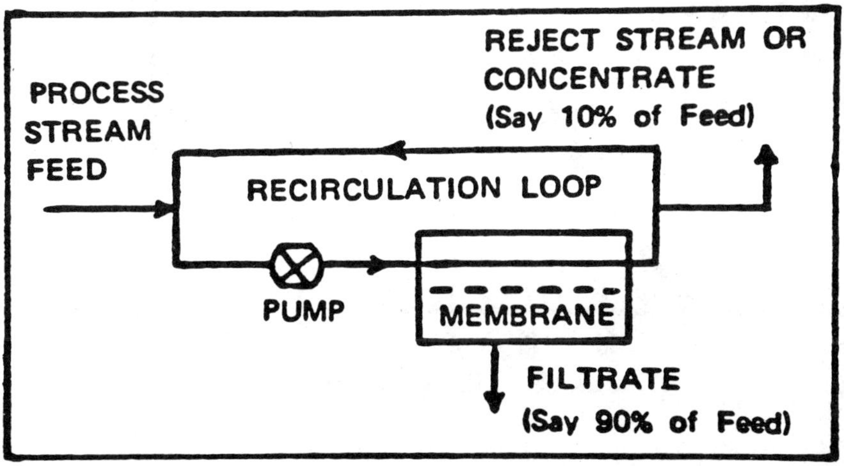

Fig. 1: Schematic of cross-flow microfiltration

FIGURE 11.—Schematic of cross-flow microfiltration.

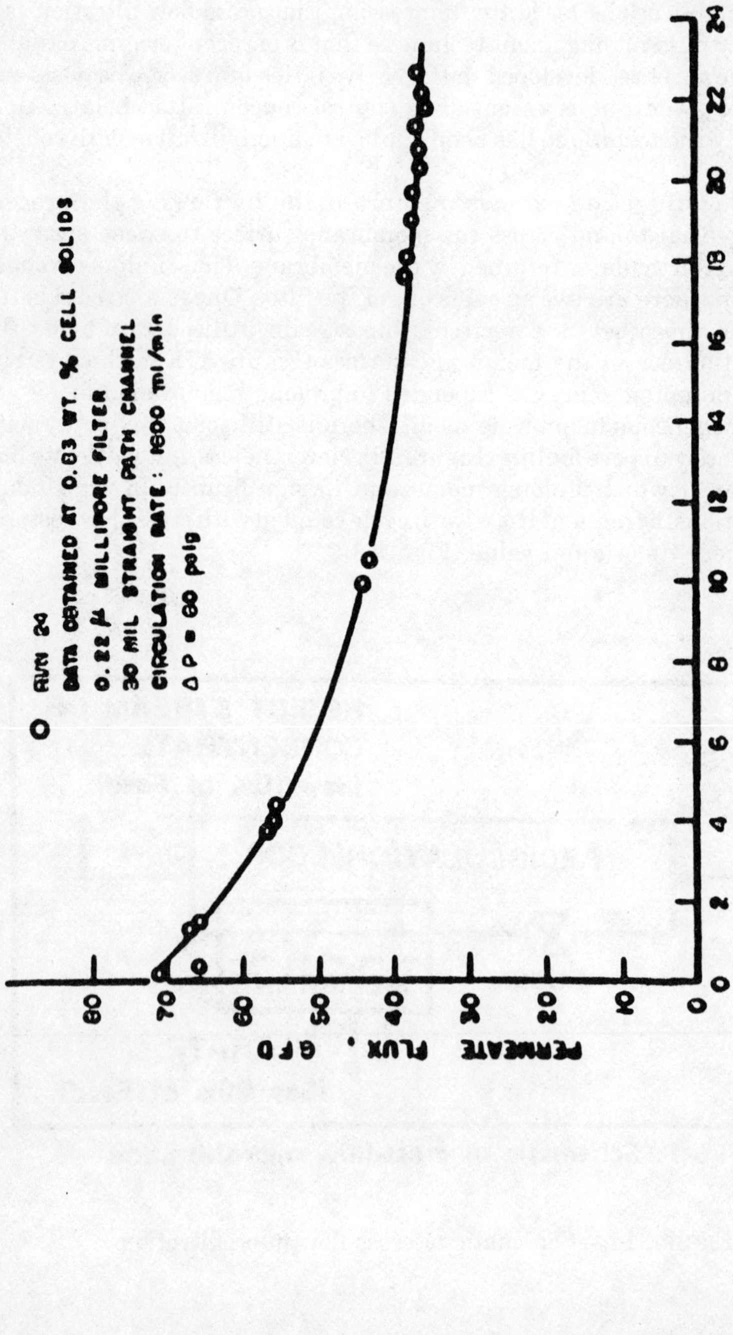

FIGURE 12.—Cross-flow filtration at constant transmembrane pressure: dimunition in permeate flux over time.

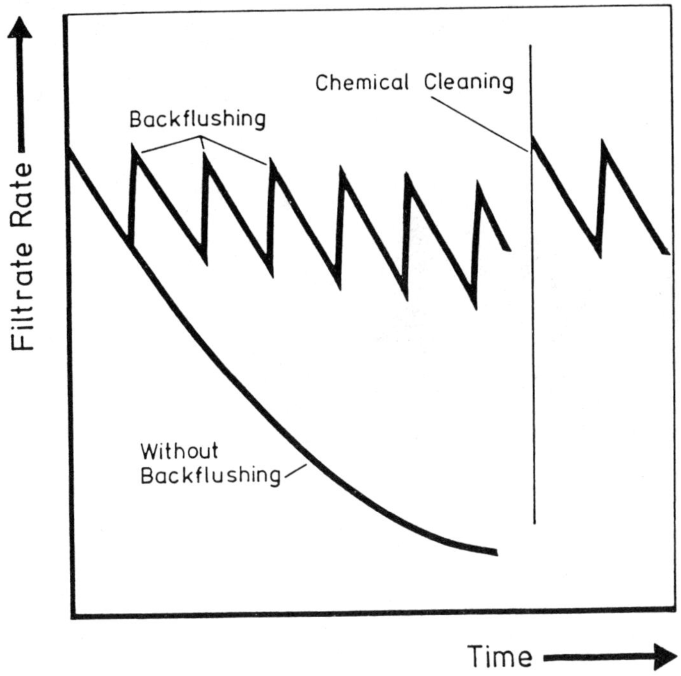

FIGURE 13.—Effect of membrane cleaning procedures on microfiltration system performance.

In the last two years, considerable progress has been made in developing anisotropic microfiltration media (both polymeric and inorganic), which essentially eliminates internal pore fouling and thereby prevents flux decay. The inorganic medium is Al_2O_3 and can be cleaned with caustic and acids. The fine pore structure on the active surface is supported by a much more open structure underneath. Thus, any particle penetrating the active surface will likely pass through into the filtrate rather than lodging within the pore matrix.

7. EQUIPMENT DESIGN

For through-flow microfiltration, the most common configurations are stacked plate systems (designed for 293-mm diameter membrane discs) and pleated cartridges.

Figure 14 shows a simple seven-plate system with its flow path. Incoming flow is diverted radially outward, then enters all the support plates simultaneously. The filtrate collects in the center cylinder and exits through the center post. Each 293-mm membrane has an area of about 0.6 sq ft.

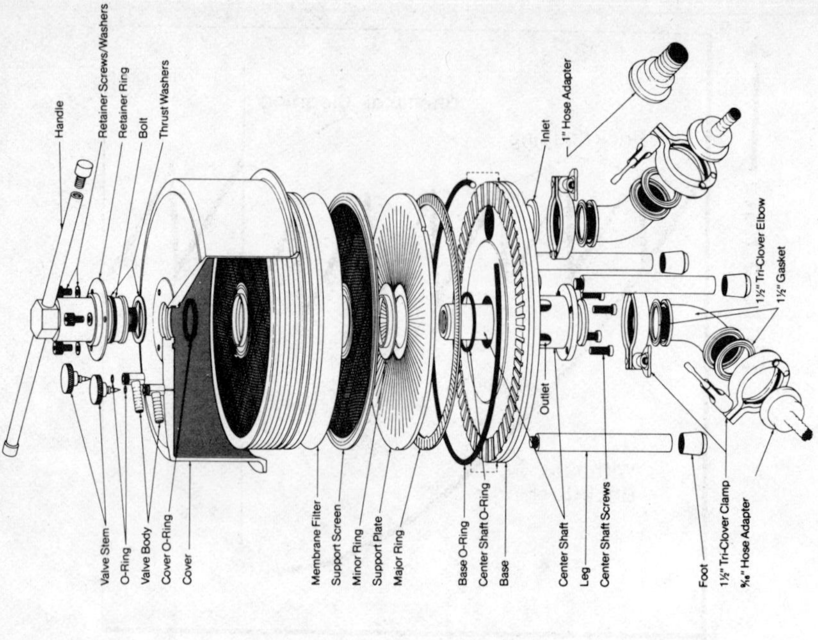

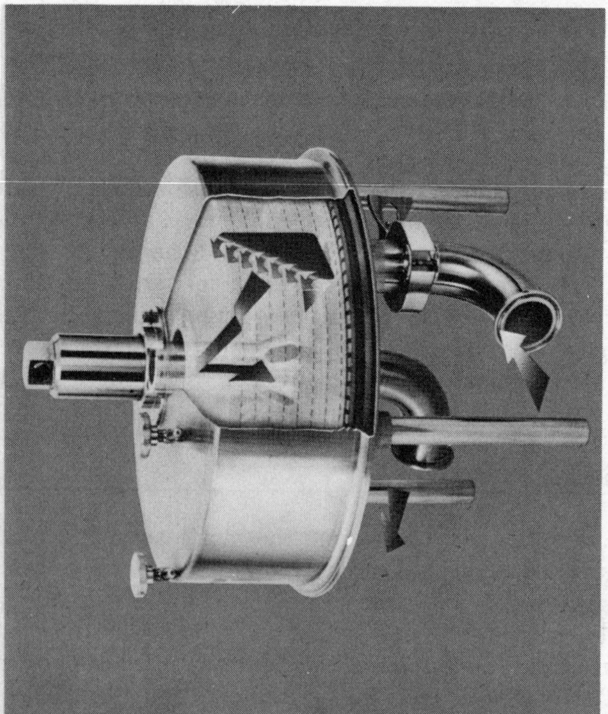

FIGURE 14.—Stacked plate module.

Figure 15 shows a typical pleated cartridge. A ten-inch high pleated cartridge has anywhere from 5 to 20 sq ft of membrane area, depending on the media. With tortuous pore media, the area is typically less than 10 sq ft. With the thinner capillary pore media, the area can approach 20 sq ft. This explains why pleated capillary-pore membrane cartridges can be competitive with the higher throughput tortuous-pore cartridges; they have more area.

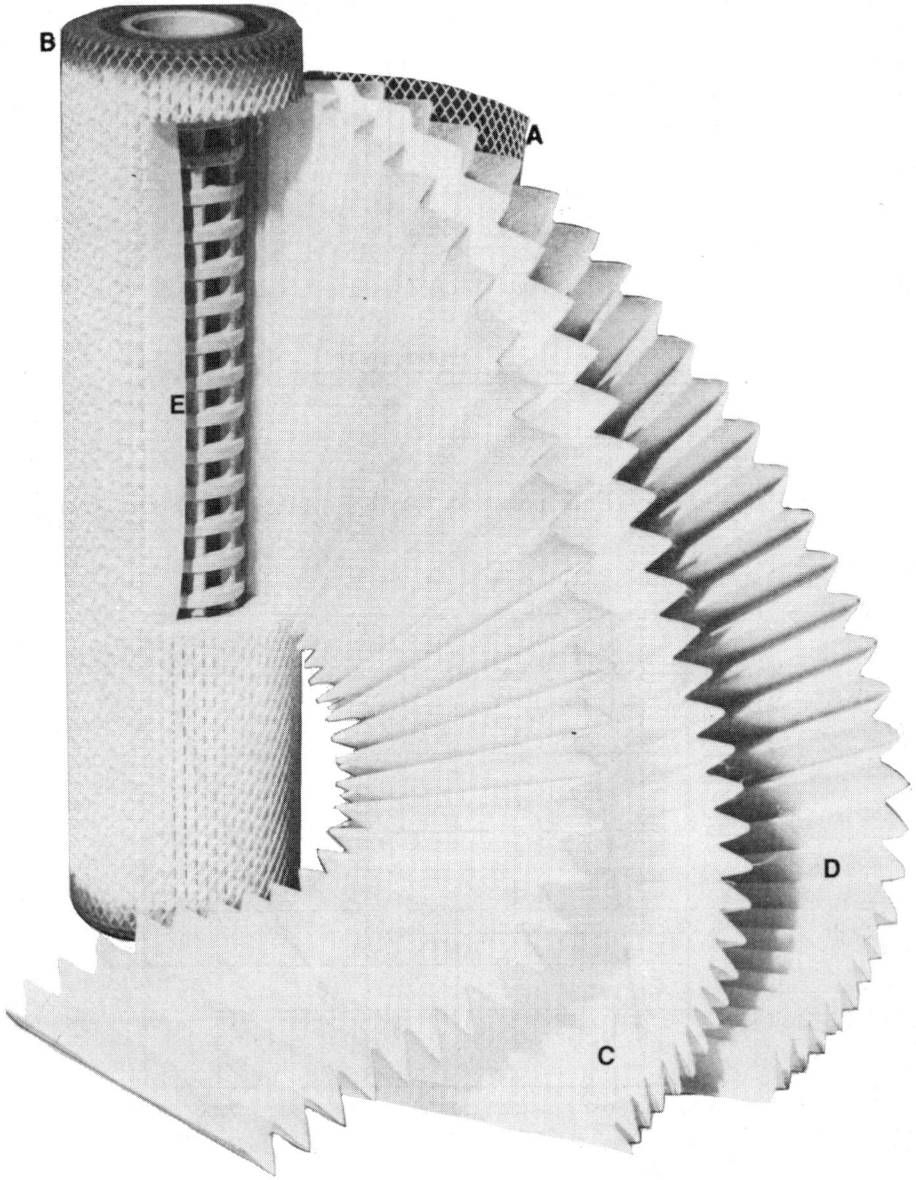

FIGURE 15.—Filter with pleated membrane sandwich and core wrap.

Pleated cartridges have also been modified for use in cross-flow operation (Figure 16) (10). By this simple change in fluid management, the throughput (Figure 17) can be improved by a factor of 50. However, more efficient cross-flow systems utilize the tubular configuration.

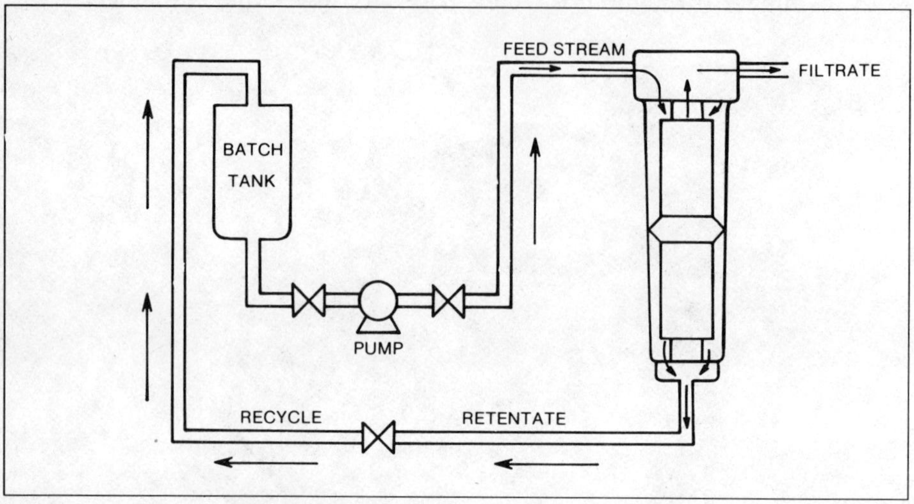

FIGURE 16.—Flow paths in crossflow cartridge system.

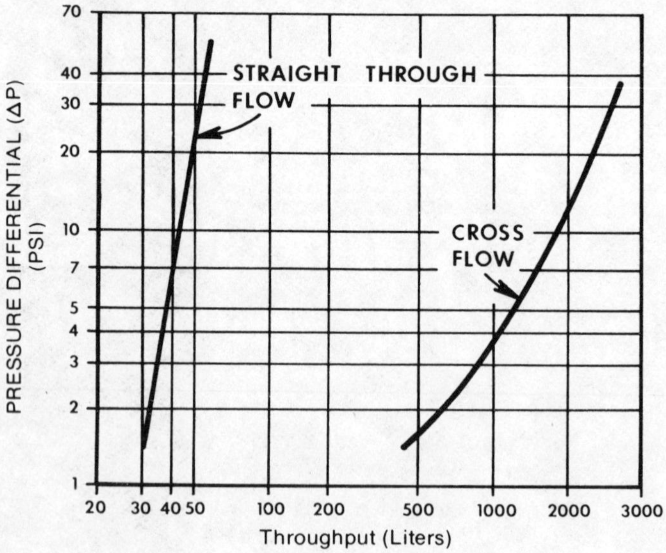

FIGURE 17.—Throughput of 0.5% yeast in de-ionized water of Nuclepore QR 0.2 μm cartridge in straight-through and cross-flow configurations.

8. INDUSTRIAL APPLICATIONS

Space will not allow even a cursory survey of the hundreds of laboratory applications for microfiltration. Suffice it to say that water microbiology and related analytical applications will represent the largest segment of the laboratory market.

Perhaps the largest single industrial market is found in the sterilization of beverages and pharmaceuticals. Gradually these industries seem to be shifting from stack-plate systems to cartridges because of the labor saving in membrane replacement. However, the 293-mm diameter discs are considerably less expensive and some wineries, e.g., Gallo, claim that stack-plate systems are more cost-effective. In addition, they can be more reliably bubble point tested. It is only recently that Pall Corporation and Millipore Corporation have introduced pleated cartridges that can be truly bubble point tested.

Another large application for microfiltration membranes is found in the processing of ultrapure water—primarily in the semi-conductor industry but also in pharmaceutical plants. The individual microelectronic components of semiconductor chips are so infinitessimally small that even submicron particulates can result in a malfunction. The water used to rinse these chips must have a resistivity near the theoretical limit (18 meg. ohm) and be particle-free. Yields in semi-conductor production are a strong function of the purity of the rinse water.

Figure 18 is a typical process diagram for an ultrapure water system. The heart of the system is the ion exchange resin, which makes possible the demineralization of the water to 18 meg ohm. Reverse osmosis is used prior to the cation and anion exchange units to reduce the load. The combination of RO and ion exchange is more cost effective. Unfortunately, the resin beds are notorious breeding grounds for bacteria. Therefore, the product water has

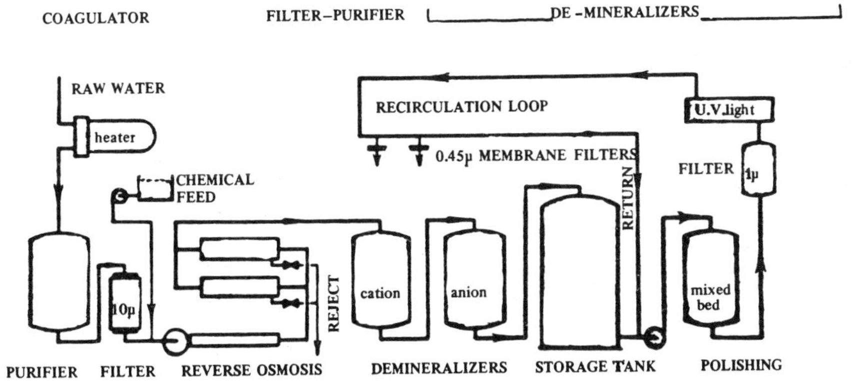

FIGURE 18.—Schematic diagram of process for production of ultrapure water.

resin fines and bacteria. A U.V. light in the recirculating loop kills some of the bacteria and keeps the population under control. The final microfiltration membrane cartridges (usually 0.2-micron pore size) remove all particulates and microorganisms.

Cross-flow applications are beginning to emerge. In the medical area, plasmapheresis, the separation of blood cells from plasma, can be accomplished by carefully controlled microfiltration (0.6-micron pore size). Return of the blood cells to the donor during the plasma collection process makes it possible for each donor to donate plasma more frequently. Therapeutic plasmapheresis appears to have enormous potential. Large-scale industrial plasmapheresis is also in the planning stage.

Other cross-flow applications include metals recovery (as the precipitated oxide), waste water clean up, and the harvesting and washing of cells. In the future, I expect to see a rapidly increasing demand for cross-flow microfiltration in the emerging biotechnology companies. For example, microfiltration membranes are expected to be an integral part of the new continuous membrane fermentors (Figure 19).

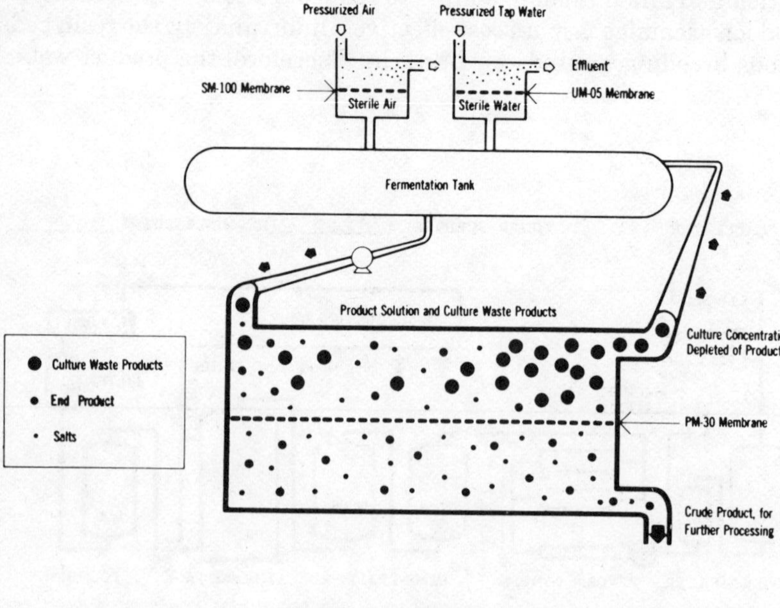

FIGURE 19.—Continuous fermentation process.

REFERENCES

1. Fick, A.: 1855, "Uber Diffusion." Pogg. Ann. 94, pp. 59–86.
2. Bechhold, H.: 1907, "Kolloidstudien mit der Filterationes-methode," Z. Phys. Chem. 60, pp. 257–318.
3. Zsigmondy, R. and Bachmann, W.: 1918, "Ueber neue Filter," Z. Anorg. Allgem. Chem. 103, pp. 119–128.
4. Müller, G.: 1947, "Lactose–fuchsin plate for detection of Coli in drinking water by means of membrane filters." Z. Hyg. Infektionskr 127, pp. 187–190.
5. Goetz, A.: 1947, "Materials, Techniques, and Testing Methods for the Sanitation (Bacterial Decontamination) of Small-Scale Water Supplies in the Field Used in Germany During and After the War", Final Report 1312, Joint Intelligence Objectives Agency, Washington, D.C.
6. Gore, R. W.: "Process for Producing Porous Products", U.S. Pat. 3,953,566, April 27, 1976.
7. Chien, S., Luse, S. A., and Bryant, C. A.: 1971, "Hemolysis during filtration through micropores." Microvasc. Res. 3, pp. 183–203.
8. Davis, M. A., Jones, A. G., and Trindade, H.: 1974, "A rapid and accurate method for sizing radiocolloids." J. Nucl. Med. 15, pp. 923–928.
9. Spurny, K. R., Lodge, J. P., Jr., Frank, E. R., and Sheesley, D. C.: 1969, "Aerosol filtration by means of Nuclepore filters: Structural and filtration properties." Environ. Sci. and Tech. 3, pp. 453–464.
10. Porter, M. C. and Olson, W. P.: "Non-Clogging Microporous Filter With Cross-Flow Operation", U.S. Patent 4,178,248, Dec. 11, 1979.

SYMBOLS

d_c	red blood cell diameter
p_{bp}	bubble point gauge pressure
q	face velocity through membrane
s	particle density
V_t	volumetric throughput up to time t
w	suspension particle concentration
α	packing density coefficient
β	cake compressibility
η_c	particle collection efficiency
θ	liquid–solid contact angle

ULTRAFILTRATION

Philippe Aptel and Michael Clifton

Laboratoire de Génie Chimique (CNRS L.A. 192)
Université Paul Sabatier
118, route de Narbonne
31062 Toulouse CEDEX, France

The ultrafiltration process, its basic principles, its capabilities and limitations are presented in an overall view intended as a general introduction to this versatile unit operation process. A description of available ultrafiltration membranes and modules is given together with a discussion of techniques for membrane characterization. The various operating modes and system designs at present in use are considered and both developed and developing applications of the process are explained.

P. M. Bungay et al. (eds.), Synthetic Membranes; Science, Engineering and Applications, 249–305.

1. INTRODUCTION

1.1. Principle

Ultrafiltration is a membrane separation process that can be grouped together with reverse osmosis and microfiltration. In these three processes, the liquid being treated is circulated under pressure in contact with a membrane through which the solvent and certain dissolved species are able to pass while the remaining components are held back. These processes are distinguished from one another essentially according to the size of the particles or molecules that are retained by the membrane. In this way, ultrafiltration covers the zone between the other two processes; the particles retained range from small macromolecules (molecular weight 500) up to colloidal particles with a diameter of 0.2 μm. The actual separation on a microscopic level can be thought of as a sieving mechanism in which molecules are distinguished according to their size and shape; larger, wider molecules are unable to pass through the pores in the membrane while smaller, narrower molecules pass more easily. This is not quite the same principle as in reverse osmosis, where the chemical affinities of the solute molecules play an important role in determining their degree of retention. Other points of difference between reverse osmosis and ultrafiltration are that in the latter process the operating pressures required are much lower and the membranes used are highly permeable.

1.2. Historical Development

Ultrafiltration first appeared at the end of the last century and was for a long time used as a concentration and purification technique in laboratory preparations. However, the membranes used were incapable of offering the chemical and mechanical solidity and high permeability required for applications on an industrial scale.

This situation was suddenly altered in the early 1960's when Loeb and Sourirajan developed the first asymmetric reverse osmosis membranes. These membranes were much more permeable than the previously used homogeneous membranes but were still capable of assuring a good selectivity. The technique used in their preparation made it possible to prepare membranes with widely varying properties and in particular, membranes suitable for use in ultrafiltration. These early membranes made of cellulose acetate still had a fairly low resistance to chemical attack but their hydraulic permeability was sufficiently high to make industrial application of ultrafiltration worthwhile. By the end of the decade, the first industrial ultrafiltration plants had been installed.

The early cellulose acetate membranes were very limited in the chemical conditions under which they could operate and also, with use, were liable to undergo compaction. It was soon found that other, more inert polymers could be used in the place of cellulose acetate, since the asymmetric structure is determined more by the preparation technique than by the polymer involved. Commercial ultrafiltration membranes are now available which offer good selectivity, a high permeability and considerable chemical stability. One quite interesting recent development has been the appearance on the market of inorganic membranes capable of withstanding high temperatures and pressures.

1.3. Applications

The first industrial applications of ultrafiltration were: (1) enzyme recovery, (2) protein recovery from cheese whey, (3) electropaint recovery. Other more recent applications include: various product recoveries, effluent treatment, milk concentration for cheese making, PVA recovery from textile desize liquors, and soluble-oil emulsions.

So far, in all applications the major component has been water. In principle, non-aqueous solvents can also be used if appropriate membranes are developed, for example, in purification of lubricating oils dissolved in toluene (polyacrylonitrile membranes), or in high temperature oil purification with inorganic membranes.

1.4. Concentration Polarization

While in the case of reverse osmosis, plant productivity continues to be controlled by membrane permeability, a completely different situation has arisen with ultrafiltration. Ultrafiltration membranes have now improved in permeability to the point where the process is now limited rather by the rate of mass transfer at the membrane–solution interface. This problem is not entirely new; a similar limitation had already been encountered many years before in the case of electrodialysis.

The solute retained by the membrane tends to accumulate at the surface of the membrane and constitutes a supplementary barrier opposing resistance to the passage of the solvent. The most important consequence of this "concentration polarization" effect is the limit that it places on the increase in permeate flux in response to an increase in pressure. At low operating pressures, the permeate flux increases almost linearly with the applied pressure. But as the pressure rises, the permeate flux increases more and more slowly and finally a pressure is reached beyond which no further increase in flux is observed. At this point the limiting flux has been reached. This behavior is illustrated in Figure 1 (1).

1.5. Membrane Fouling

It is found, particularly when feed solutions of biological origin are being treated, that the permeation rate is not limited in a constant manner. When an ultrafiltration module is operated under limiting conditions over long periods (the usual industrial situation), the flux decreases steadily with time. This observation is not simply explicable in terms of concentration polarization but rather is due to chemical modifications occurring in the layer of concentrated solution or gel at the membrane surface. In solutions of biological origin, the effect of the conditions found in the polarization layer (high concentration, shear stress and possible variation in ionic strength) is to

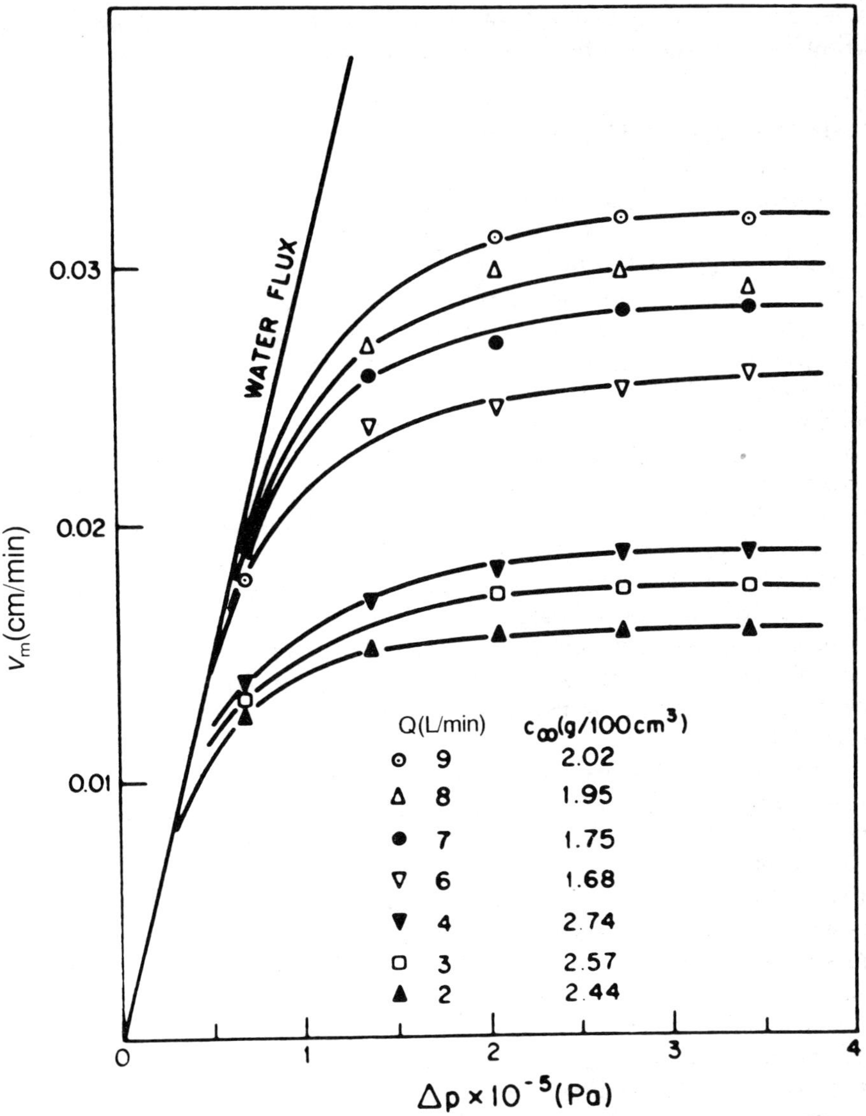

FIGURE 1.—Variation of permeation rate with applied pressure: data for ultrafiltration of solutions of bovine serum albumin in plane laminar flow (1). Reprinted by courtesy of the American Chemical Society.

cause proteins to be denatured so that they precipitate on the membrane surface. The hydraulic resistance of this layer increases with time and reduces the effective permeability of the membrane, so that the permeation rate through the membrane declines, as shown in Figure 2 (2).

2. MEMBRANES AND MODULES

2.1. Commercial Membranes

In the course of the last ten years, the cellulose derivatives originally used in preparing ultrafiltration membranes have gradually been replaced by other synthetic polymers. Following on from the work of Michaels, in the late 1960's other investigators produced the first membranes formed by combining two polyelectrolytes carrying opposite charges. It had long been known that if a solution of a cationic polymer was mixed with a solution of an anionic polymer, then a precipitate was formed. This type of polymer, known as a polysalt or a polyelectrolyte complex, is insoluble in all common solvents. However, if the initial solution contains a highly dissociated electrolyte and a suitable solvent, then the various ionic and organic groups of the two chains are strongly solvated. Under these conditions, the polysalt stays in solution and the collodion obtained can be used for preparing an asymmetric membrane. Among the membranes of this sort are the Amicon UM series and the Rhone-Poulenc IRIS 3042 and 3038 series; they can be used in continuous operation in the pH range from 1 to 10 but cannot withstand temperatures above 50°C.

Other membranes with very satisfactory characteristics have been obtained using aromatic polycondensates. They can be of polyamide (Berghof series BM and BR, the Kalle PA series. . .) of fluoropolymers (Abcor HFM series, Nuclepore F series) or more often of polysulfone (Osmonics PS series, the GR series from DDS, the Amicon PM series, the Wafilin WFS series. . .). These membranes can be used over a wide range of pH from 1 to 13 and at a temperature of 80°C for the polysulfones and higher than 100°C for the fluoropolymers.

A new range of possibilities is opened up by the inorganic Carbosep membranes manufactured by SFEC, which consist of a tubular microporous carbon support, with an internal diameter of 6 mm, on the inside of which is deposited a 20 μm layer of zirconium oxide. These membranes are capable of retaining particles ranging from molecules of molecular weight 20,000 up to 0.1 μm particles. Their great advantage is their very wide zone of operating conditions: pH 0 to 14, pressures up to 20 bars (which makes possible the treatment of viscous feed solutions) and temperatures as high as 120°C (so the membranes can be steam sterilized). In fact the membranes themselves can stand much higher temperatures and the company is at present developing modules for operation at 300°C, which could be used for de-asphalting heavy oils (3).

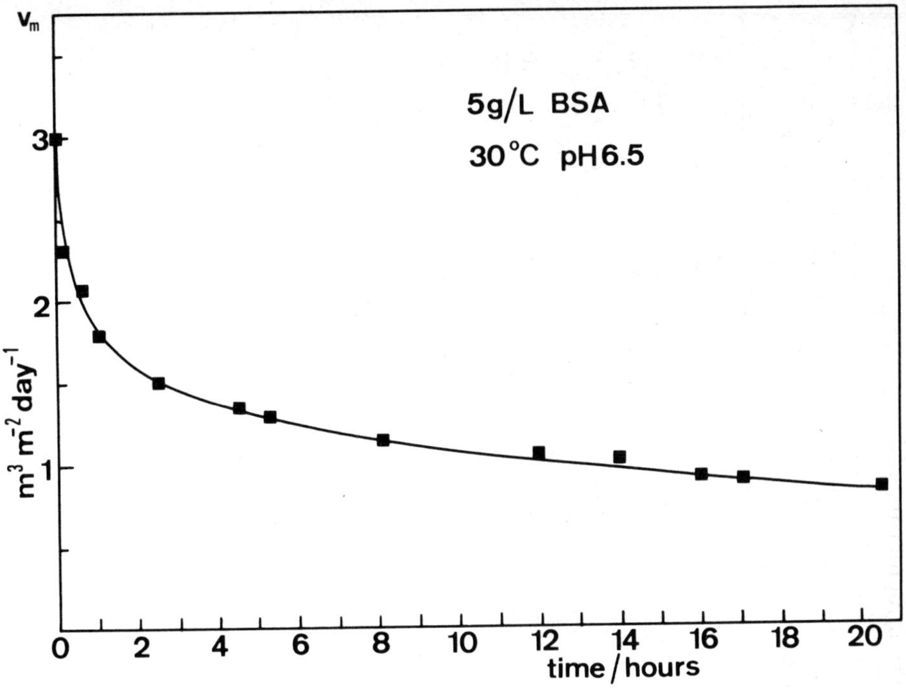

FIGURE 2.—Flux decline due to membrane fouling (2).

The characteristics of commercial membranes on the market at present are given in Table 1.

2.2. Membrane Characterization

The performance of ultrafiltration membranes is commonly quoted by manufacturers in terms of the pure water flux and the so-called "cut-off". On the basis of such data it is in fact quite impossible to predict the ability of a membrane to filter out a given solute from an aqueous or nonaqueous solution. Our aim in this section is to define parameters easily obtained experimentally, which provide better characterization of membranes and so allow comparison of membranes among themselves and prediction of the rejection behavior of a membrane towards a given solute.

Permeability As has already been mentioned in preceding chapters, the pure water flux through ultrafiltration membranes is directly proportional to the driving force (Δp); the proportionality coefficient (L_p) is called the hydraulic permeability coefficient:

$$(v_m)_w = L_p \, \Delta p \tag{1}$$

TABLE 1. Characteristics of commercially available ultrafiltration membranes

Manufacturer	Type	Chemical Composition[1]	Geometry[2]	Water Permeability (m^3/day m^2 atm)	Cut-off. (dalton $\times 10^{-3}$)	Resistance			
						pH	Temp. (°C)	Pressure (atm)	Cl (ppm)
ABCOR	HFA	CA	P	1.4	15	2-8	50	5	2
			T	1.4	15	2-8	50	4	2
	HFD	PA (?)	P	1.4	8	2-12	80	5	0
			T	3.5	8	2-12	80	4	0
	HFM-100	PVDF	P	1.8	10	0.5-13	90	6	10
			T	5.9	10	0.5-13	90	4	10
ABCOR	HFM-180	PVDF	P	1.8	18	0.5-13	90	6	10
	HFM-251		T	7.0	18	0.5-13	90	4	10
	HFK	?	P	1.8	5-10	2-14	90	6	
			T	6.0	5-10	2-14	90	4	
AMICON	YCO 5	?	P	0.2	0.5				
	YM 2	regenerated cellulose (low	P	0.1	1				
	5	protein binding		0.4	5				
	10	properties)		0.6	10	2-11	75	5	
	30			5	30				
AMICON	PM 10	PS	P	10	10	1.5-13	75		
	30			30	30				
	XM 50			6	50			4	
	XM 100 A	Dynel	P	25	100	1.5-13	50	1	
	X 300			25	300			1	

AMICON	P 1		H.F. (1.1)	1	1				
	2		(1.1)	1	2				
ROMICON	5		(1.1/0.5)	2	5				
	10	PS	(1.1/0.5/0.2)	3	10	1.5-13	75	1.8	
	30		(1.1)	3	30				
	50		(1.1)	5	50				
	100		(0.5/1.1)	6	100				
	X 50		(0.5/0.2)	3-6	50	1.5-13	45	1.8	
	100	Dynel	(0.5/0.2)	3-6	100				
ROMICON	CX	Acrylic Copo. (cationic)	(1.1)						
ASAHI	HL-HC-HH	PAN	HF(0.75)	1-7	6-50	2-10	50	2.9	
		PS			3-6				
	0.5			0.08	0.5				
BERGHOF/	BM 1		P	0.15	1				
NUCLEPORE	or A 5			0.7	5				
	10	PA	P	1.0	10	2-12	80		0
	50			3.2	50				
	100			5.5	100				
	BPR 2	PA	HF	0.25	2				
	or A 10		(0.6/1.1)	3.0	10	2-12	80	2	0
	50			6.0	50				

(1) PS : Polysulfone; PI : Polyimide; PAN : Polyacrylonitrile; PA : Polyamide; PO : Polyolefin; CA : Cellulose acetate; CTA : Cellulose triacetate; PVDF : Polyvinylidene fluoride.
(2) P : Flat; T : Tubular (inside diam. in mm); HF : Hollow-fiber (inside diam. in mm).
Water permeability—Conversion factor : 1 m^3/m^2 day atm = 1.15 10^{-8} cm/s.Pa.
Resistance : Except for free chlorine, data are relative to maximum for continuous operation.

TABLE 1. Characteristics of commercially available ultrafiltration membranes (continued)

Manufacturer	Type	Chemical Composition[1]	Geometry[2]	Water Permeability (m^3/day m^2 atm)	Cut-off. (dalton $\times 10^{-3}$)	Resistance			
						pH	Temp. (°C)	Pressure (atm)	Cl (ppm)
BERGHOF/ NUCLEPORE	F 5		P	0.7	5				
	10			1.0	10				
	50			6.0	50				
	100	Fluoro. Polymer	P	9.0	100	1-14	130		high
	300			20.0	300				
	500			25.0	500				
	1000			40.0	1000				
NUCLEPORE	C 0.5			0.08	0.5				
	1			0.15	1				
	5	Modified cellulosic		0.8	5				
	10		P	1.6	10	2-10	90		0
	20	(low adsorption-		2-5	20				
	50	hydrophilic)		8.0	50				
	100			16.0	100				
DAICEL	DUY-NH			0.5	5			30	
	DUY-H	PAN	T	1	10		45	10	
	DUY-M			1.6	20			10	
	DUY-L		(14.5)	2.6	40			10	
	DUS-40	PS (polyethersulfone)	T	4	40		90	10	
	PLS-HHA			0.3	5				
	PLS-HA	PAN	P	0.6	10		45	10	
	PLS-MA			1.2	20				
	PLS-LA			2	40				

D.D.S.	CA 600 PP	CA	P	1	20	2-8	50	10	20
	800 PP			0.5	8	2-8	50	20	20
	GR 51 PP		P	2.4	50			10	200
	60 PP	PS		1.8	25	1-13	80	10	50
	61 PP			2.4	20			15	200
	81 PP			1.2	6			15	200
	FS 50 PP			2.0	30			10	
	61 PP	? (hydrophilic)	P	2.0	20	1-12	80	10	high
	81 PP			1.2	6			15	
	GS 61 PP			1.8	20				
	81 PP	? (modified PS)	P	1.8	6	1-13	80	15	
	90 PP			0.7	1.5	2-12	60	20	
DESALINA-TION SYSTEM	A	CA	P			3-9	40		
	E	modified PS				1-13	100		
DORR-OLIVER	C 2				1				
	5	CA	P		5	3.5-10.5	75		
	10				10				
	30				30				
	D 50				50				
	100	DYNEL	P		100	2-12	60		
	300				300				
	S 10	PS	P		10	1-13	75		
	30				30				
KALLE	TU AF 20...	CA	T		2 to	2-8	40	20	
	AG 20...		P		100				
	TU AK 30...	PA	P		20 to	2-12	60	10	
	AL 30...		T		100				

TABLE 1. Characteristics of commercially available ultrafiltration membranes (continued)

Manufacturer	Type	Chemical Composition(1)	Geometry(2)	Water Permeability (m^3/day m^2 atm)	Cut-off. (dalton $\times 10^{-3}$)	Resistance pH	Resistance Temp. (°C)	Resistance Pressure (atm)	Resistance Cl (ppm)
KALLE	TU AN 4...	PS	P		8 to	1-14	90	10	
	AO 4...		T		25				
MILLIPORE	PT GC		P	4.8	10				
	TK	PS			30	1-4		7	50
	HK			12.0	100				
	PS AC			0.8	1				
	PS ED	?	P	1.6	25	2-10			
	PS VP			2.4	100				
NITTO	NTU 2006	PO (hydrophilic)	T (11.5)	0.2	6				
	2020			1	20	1-13	50	10	
	20100			2	100				
	NTU 3020	PS	T (11.5)	4.5	20	1-13	60	10	high
	NTU 4206			0.15	6				
		PI	T (11.5)			2-8	40	20	
	4220			0.3	20				
	NTU 8010	PA	HF	5	10		60	3	
	8050			12	50				
OSMONICS	SEPA CA 0		P	0.6	1			14	
	20 K	CA		2.3	20	2-8		7	5
	50			0.1	0.6			50	
	PS 0	PS	P	1.2	1	0.5-13			50
	20 K			4.6	20				
	50 K				50				
	PI 50	PI	P	0.05	0.6	4.5-11			0

PATTERSON	T 2 A		T (12.5)	0.1	1	2-8	30	25	1
CANDY	T 4 A	CA		0.6	10			10	1
	T 6 B	?	T (12.5)	0.3	20	2-11	60	10	10
RHÔNE	IRIS 3038	PAN Copolymer	P	6	15	3-10	40	4	high
POULENC	3042	(polyelectrolyte)		11	20		50	4	
	3050	(complex)		10	20	3-10	40	4	high
		(cationic)							
SARTORIUS	SM 145 39	CTA	P	6	10	4-8	50	4	
	49				20				
SFEC	Carbosep	non-organic	T	1	10	1-13	95-120	20	high
		(ZrO_2/Carbon)	(6)		20		or		
		(ZrO_2/Alumina)	(15)		100		300		
TEIJIN	TU 10T			3.6					
	10L	PS	T	9			90	10	
	10S			9					
WAFILIN	WFS 8010		T (14-4)	1.2	20			10	
	6010	? (non-cellulosic)		2	35	2-12	85	7	100
	5010			4	100			3	
	WFA 7010		T (14-4)	0.6				—	
	5010	? (non-cellulosic)		1.2		2-10	55	—	30
	4010			2				9	
	3010			4				5	

(1) PS : Polysulfone; PI : Polyimide; PAN : Polyacrylonitrile; PA : Polyamide; PO : Polyolefin; CA : Cellulose acetate; CTA : Cellulose triacetate; PVDF : Polyvinylidene fluoride.
(2) P : Flat; T : Tubular (inside diam. in mm); HF : Hollow-fiber (inside diam. in mm).
Water permeability—Conversion factor : 1 m^3/m^2 day atm = 1.15 10^{-8} cm/s.Pa.
Resistance : Except for free chlorine, data are relative to maximum for continuous operation.

Since the water flux measured at different temperatures is inversely proportional to the viscosity (4), it is useful to define a "standard" permeability coefficient L_p^* by the following relationship:

$$(v_m)_w = L_p^* \, \Delta p/\eta_w \tag{2}$$

The standard permeability coefficient is characteristic of the system membrane/water and is independent of the pressure and temperature.

The availability of membranes resistant to organic solvents has increased the interest in applying ultrafiltration to non-aqueous media. This raises the problem of defining a parameter to characterize the behavior of membranes towards organic solvents (4, 5). One possibility is to introduce a dimensionless parameter α into equation 2:

$$(v_m)_s = \alpha \, L_p^* \, \Delta p/\eta_s \tag{3}$$

α is called the "permeability ratio" and is defined by the expression:

$$\alpha = (L_p^*)_s/L_p^* \tag{4}$$

where $(L_p^*)_s$ is the standard permeability coefficient of the binary system membrane/solvent s.

It must be pointed out that α can only be defined in the case where a stable solvent flux is obtained after the membrane has undergone suitable conditioning operations. In the case of these solvents, the α ratio can be seen as a measure of the effect of the organic solvent on the structure of the membrane. For example, $\alpha > 1$ will mean that the liquid "sees" a looser membrane than water does, while $\alpha = 1$ indicates that the permeability is not modified when the membrane is used first in water and then in the solvent. Table 2 gives values for the two intrinsic coefficients L_p^* and α for some commercial membranes.

Pore-size distribution The evaluation of the distribution of pore sizes in the skin of an ultrafiltration membrane has given rise to a great number of important investigations (see the article by Dr. G. Jonsson in the present volume). We shall only examine here two promising experimental methods.

The first technique is high-resolution electron microscopy (by scanning or transmission), by which it is possible to obtain a direct measurement of the surface porosity of membranes having pores larger than 10 nm in diameter. Figure 3 shows pore-size distributions determined in this way for a commercial membrane (6).

The second method, thermoporometry, is a calorimetric technique developed in recent years by Brun and Eyraud (7–9). The method is based on the fact that the width of the solidification thermogram of a pure substance held

TABLE 2. Values of L_p^* and α for some commercial membranes (4)

	Amicon	Rhône-Poulenc		
10^{13} L_p^* (m)	UM 10 3.0	IRIS 3038 10.0	IRIS 3042 4.7	IRIS 3069 0.5
Solvent	α			
methanol	1.7	1.0	1.0	1.0
ethanol	1.7	1.0	1.0	1.0
n-butanol		1.0	1.25	1.0
acetone		1.0	1.0	
n-heptane		0.7	1.4	
n-decane		0.8	1.4	
benzene		0.9	1.4	
toluene		0.9	1.4	
chloroform		0.45	0.9	
dioxane		0.05	0.2	

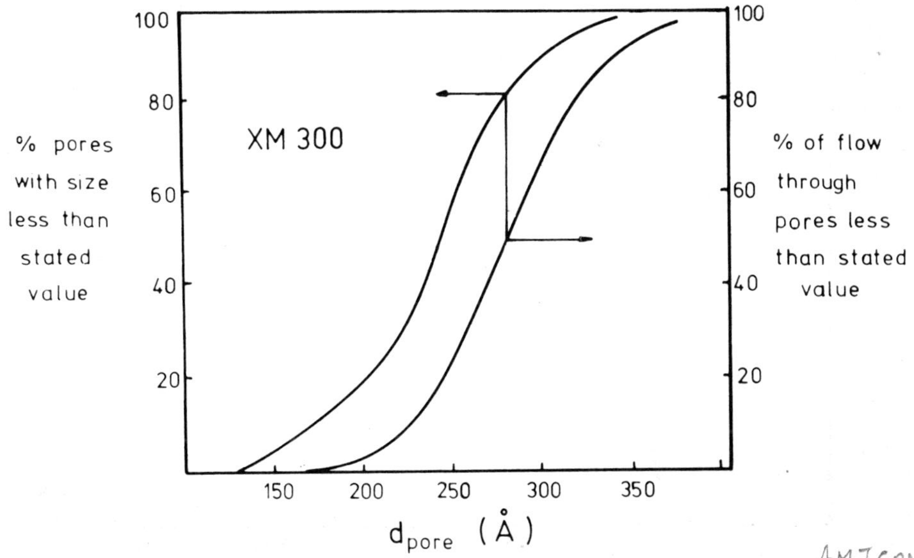

AMICON

FIGURE 3.—Distribution of pore diameters and calculated flow for an XM300 membrane (6). Reprinted by courtesy of Elsevier Science Publishers Company, Inc.

in a porous material depends on the pore-size distribution. Therefore, from calorimetric measurements of transition energy and temperature, it is possible to determine the pore-size distribution of a porous sample. As distinct from mercury porosimetry, physical adsorption porosimetry or gas permeability methods, thermoporometry requires no assumptions concerning pore shapes. It can be considered that at every moment solidification occurs wherever pores are large enough to enable the formation of a solidification germ. Examples of pore-size distributions determined by this technique are shown in Figures 4 and 5 for a homogeneous hemodialysis membrane and for

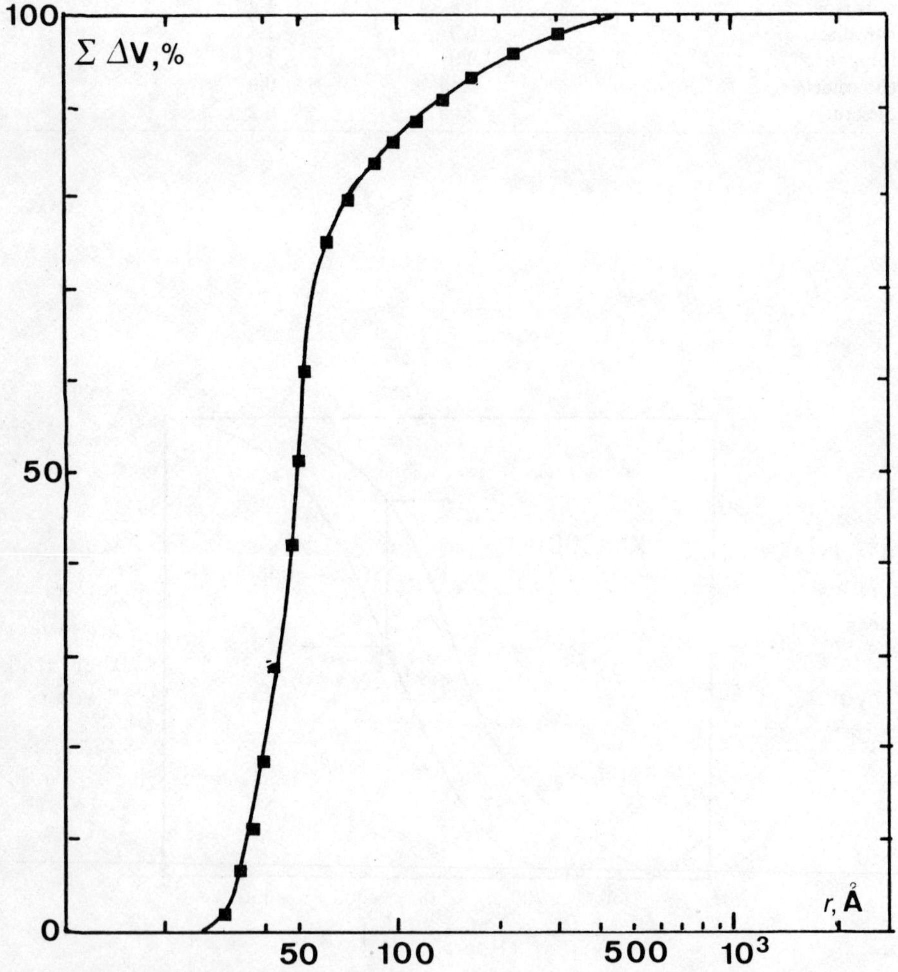

FIGURE 4.—Distribution of pore radii obtained by thermoporometry (8). ΔV is the volume percentage of pores with size less than the stated value: hemodialysis membrane, Rhône-Poulenc IRIS 3069.

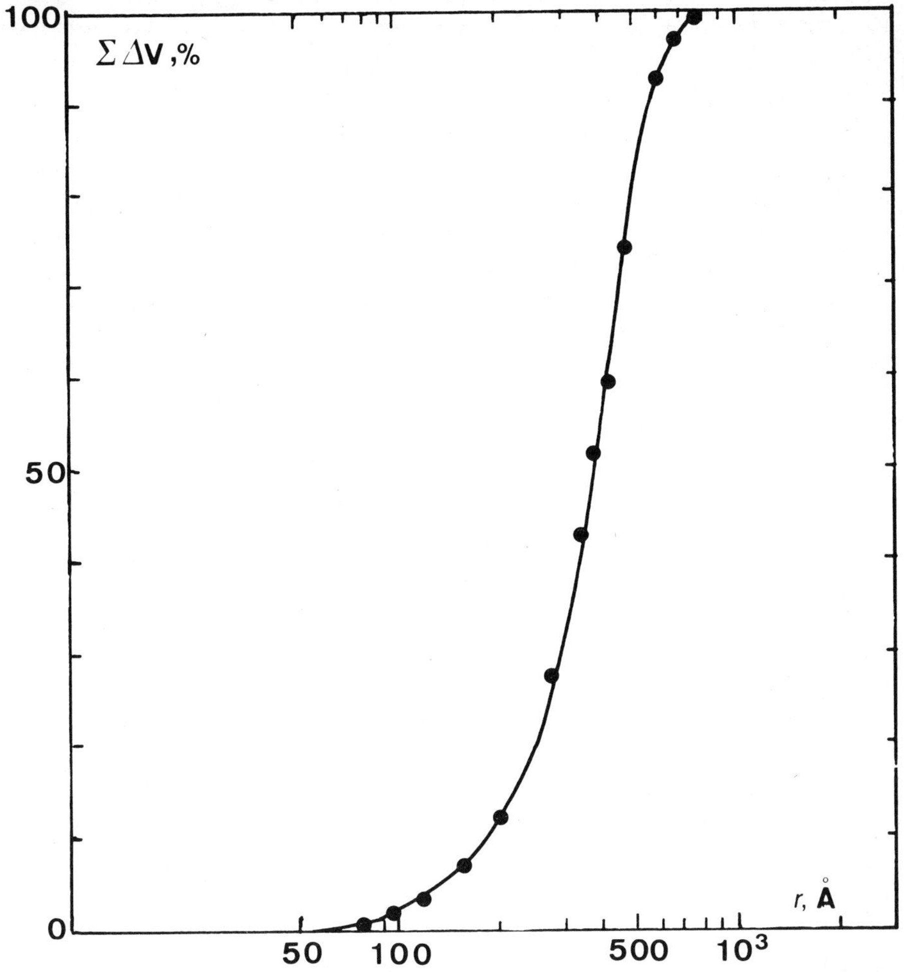

FIGURE 5.—Pore radius distribution obtained by thermoporometry: ultrafiltration membrane, Rhône-Poulenc IRIS 3042 (from Desbrières, 1980, thesis, Université de Grenoble).

an asymmetric ultrafiltration membrane, respectively. The main drawback of this technique is the need for a sample of at least 50 mg to obtain a quantitative measurement of the pore-size distribution. Though this quantity may seem small, it means that for an asymmetric membrane having a skin 0.1 μm thick, a membrane surface area of more than 0.5 m^2 is required.

Rejection coefficient and cut-off In pressure-driven membrane processes, the observed rejection coefficient $\mathcal{R}_{obs}$ of a solute is defined by the equation:

$$\mathcal{R}_{obs} = 1 - c''/c' \tag{5}$$

The ratio of the ultrafiltrate to retentate concentrations, c''/c', is the "sieving coefficient" Θ:

$$\Theta = c''/c' \tag{6}$$

Normally, membrane manufacturers designate an upper molecular-weight limit for solute passage, above which less than 5 or 10% solute passage occurs. This so-called "cut-off" is given in daltons and is reported for purified proteins because these molecular-weight probes are monodisperse (Figure 6).

Note that the cut-off is not an intrinsic property of the membrane, because the rejection coefficient depends on concentration polarization, the chemical nature of the macromolecule, the solvent used, the conformational changes in the macrosolute, etc.

To take into account the effect of concentration polarization, a true rejection coefficient $\mathcal{R}$ is defined by the relationship:

$$\mathcal{R} = 1 - c''/c'_{\text{int}} \tag{7}$$

in which c'_{int} is the concentration at the membrane–solution interface on the feed side.

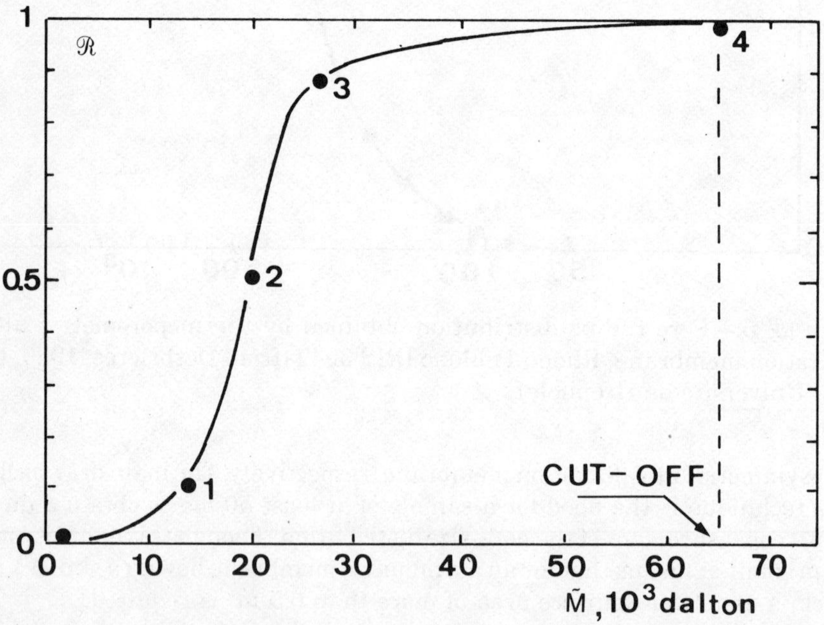

FIGURE 6.—Variation of rejection with solute molecular weight: 1. cytochrome C ($\tilde{M}=14400$), 2. trypsin ($\tilde{M}=20000$), 3. α-chymotrypsin ($\tilde{M}=24500$), 4. bovine serum albumin ($\tilde{M}=67000$). Membrane: Millipore PSED 25000.

For a stirred batch ultrafiltration cell, $\mathcal{R}$ can be calculated from the apparent rejection coefficient $\mathcal{R}_{obs}$ by assuming a film model for mass transfer in the polarization layer (10, 11):

$$\mathcal{R} = \frac{\mathcal{R}_{obs} \exp(v_m/k)}{1 - \mathcal{R}_{obs}[1 - \exp(v_m/k)]} \tag{8}$$

where the mass-transfer coefficient, k, can be estimated from the relationship:

$$k = 0.0443 \, [D/r_m] \, [\nu/D]^{0.333} \, [\omega r_m^2/\nu]^{0.746} \tag{9}$$

where r_m is the radius of the circular segment of exposed membrane in the cell and ω is the stirrer speed. Even though the model used to account for concentration polarization is only an approximate one, it still shows clearly that to obtain an experimental value for $\mathcal{R}_{obs}$ close to $\mathcal{R}$, the ratio v_m/k must be minimized by operating at low pressure (low flux), at high tangential velocity and at low concentration (low viscosity).

It is well known that linear, flexible macromolecules tend to be retained to a lesser degree than the more highly structured proteins of the same molecular weight (12, 13). Also, the rejection of a polymer may be very sensitive to the nature of the solvent (14). Two examples illustrate the importance of the flexibility parameter.

The first of these concerns solutions of poly(L-glutamic acid) (14). In acid medium, this homopolypeptide presents a rigid α-helical conformation due to the intramolecular hydrogen bonds that are formed between the carboxylic and the amide groups. In basic solutions the $-COOH$ groups are ionized, and as a result, the repulsive forces between the lateral groups disorganize the helical structure to give a flexible random coil. This conformational change explains why the rejection decreases from 0.7 to 0.25 when the pH of the solution is changed from 4.9 to 8.4.

The second example is the ultrafiltration of amylose–iodine solutions (15). Amylose is a poly(anhydro-1,4-D-glucopyranose), which has a loose, rather flexible conformation when alone in solution, but becomes denser and stiffer in the presence of iodine–iodide ions because of the formation of a complex in which iodine is trapped in the central cavity of the amylose molecule, forming a tight helix. About half of the amylose molecules were retained by an IRIS 3038 membrane in the absence of iodine, whereas the complexed form of amylose was perfectly retained by the same membrane (Figure 7).

The above brief discussion has shown some of the difficulties in predicting the retention capability of an ultrafiltration membrane. A method has recently been published (16, 17), still based on rejection measurements, but which in principle allows the determination of the complete sieving curve from only two experimental measurements. The method consists of plotting

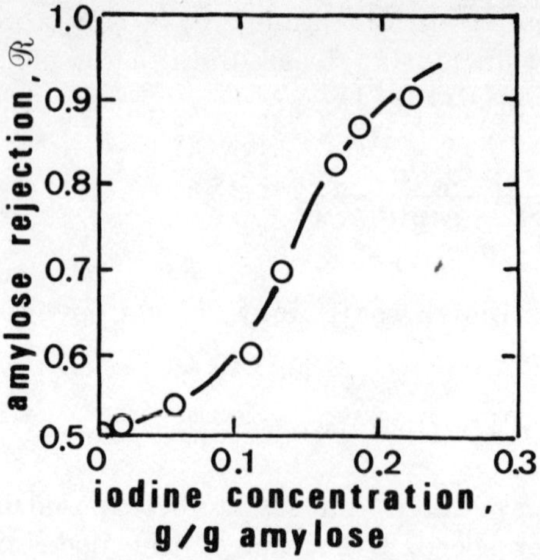

FIGURE 7.—Variation of amylose rejection with iodine concentration (15).

the rejection coefficient against the Einstein–Stokes radius (ESR) of the macrosolute on a log-normal scale. The ESR is the "apparent equivalent spherical radius" of the macrosolute, b, as computed from the measured diffusivity, D, of the molecule in free solution, using the Einstein–Stokes equation:

$$D = \frac{\kappa T}{6\pi\eta b} \tag{10}$$

The ESR has an unequivocal physical meaning only for a rigid, spherical particle in a fluid continuum and its significance for asymmetric, solvated, free-draining or compliant-chain macromolecules is ambiguous.

For aqueous-phase gel-permeation chromatography it is customary to calibrate the chromatographic column with a series of monodisperse proteins of known ESR and to ascribe to the chromatographic fractions eluted from the column the same ESR as that of a protein displaced at the corresponding elution volume. The ESR values determined by GPC are, therefore, empirical characterization parameters related in some complex way to the "effective molecular size".

By plotting the sieving coefficient Θ as normal probability against log b, the logarithm of the ESR of the permeating macromolecule, Michaels (17) has shown how straight lines were obtained for a number of synthetic ultrafiltration membranes and for a variety of biological membranes. These sieving curves (Figure 8) conform closely to a log-normal-probability relationship between Θ and b. This seems to show that the selectivity of a membrane could

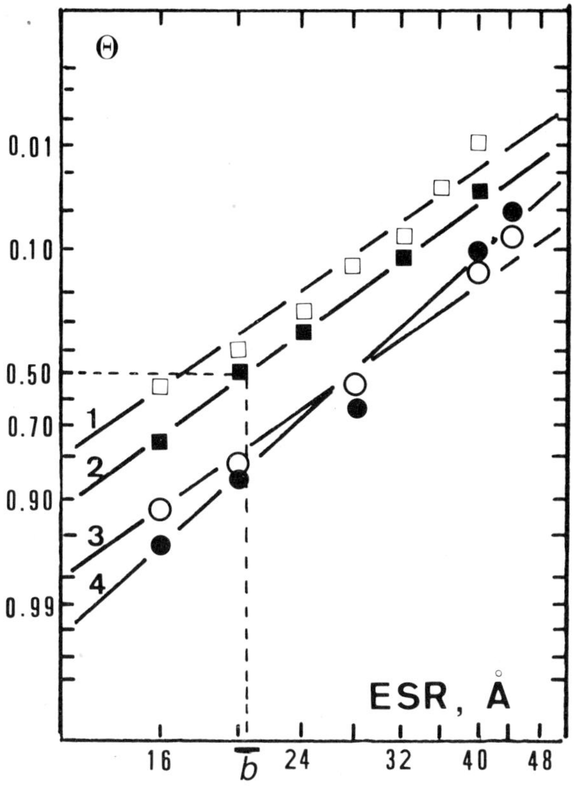

FIGURE 8.—Sieving curves for synthetic ultrafiltration membranes (17): 1. Cuprophane, 2. and 4. Amicon X-50, 3. Rhône-Poulenc IRIS 3069.

easily be characterized by two parameters: $\bar{b}$, which is the ESR of the "mean" molecule, for which $\Theta = 0.5$, and σ_b, the geometric standard deviation. The σ_b value is a measure of the sharpness of the rejection spectrum of the membrane, whereas $\bar{b}$ is related to the pore size.

Thus for practical purposes, ultrafiltration membranes could be characterized and compared with each other by the use of three parameters, which can be obtained from only three measurements: L_p (one measurement of water permeability), $\bar{b}$ and σ_b (two measurements of rejection at low concentration, low pressure and high stirring speed, for two different solutes of known ESR). When a solvent other than water is used, a fourth parameter, the permeability ratio α, can be specified (one measurement of permeability for each solvent).

2.3. Module Configurations

The basic sub-units of a battery of ultrafiltration membranes are arranged in a modular form (3, 18-20). Each module is designed as a more or less self-contained pressure vessel containing the array of membranes and their supports, together with a minimum of conduits for liquid distribution and collection. There are two essential requirements for every module design: (1) to ensure a sufficient flow rate past the membrane surface so as to limit as much as possible the formation of concentration polarization layers and to avoid forming dead spots, (2) to be as compact as possible in form, i.e., containing a maximum of membrane surface in a given volume.

Besides these two essential qualities, other desirable characteristics are:

—ease of disassembly for membrane replacement,

—ease of cleaning and sterilization, which is of great importance for the pharmaceutical and food industries,

—good chemical compatibility and pressure resistance.

As in reverse osmosis, there are four designs in use (Table 3): spiral-wound, hollow-fiber, tubular and plate-and-frame modules (schematic diagrams are presented in the chapters by Dr. H. Lonsdale and Prof. R. Rautenbach).

One might be tempted to ask how so many different configurations could have been designed for ultrafiltration modules that are all intended for the same applications. The first reason seems to be a commercial one. Because of

TABLE 3. Main advantages and disadvantages of different classes of modules (18)

	Advantages	Disadvantages
TUBULAR	–Least susceptible to plugging –Prefiltration usually not needed –Small replacement area –Easy cleaning (including mechanical: sponge ball)	–Capital cost per unit area –Pumping energy requirement
FLAT PLATE	–Small replacement area –Low cost for membrane replacement –Reasonably compact	–Prefiltration often needed –Membrane replacement difficult
SPIRAL WOUND	–Capital cost per unit area –Low energy cost –Compact	–Prefiltration necessary –Leak detection difficult –High area per module
HOLLOW FIBER	–Capital cost per unit area –Easy replacement –Compact –Back flushing possible	–Prefiltration necessary –Strength of the fibers limits operating pressure

the size of the potential market for ultrafiltration equipment, there is strong competition among manufacturers, so they each have their own particular membrane and module fabrication processes, which are different from one another for reasons of patent protection. The second reason is a scientific one. Up to the present time there exists no strong scientific basis for optimizing mass transfer in the ultrafiltration process: for the same application, some modules are operated in laminar flow conditions, and others (tubular modules) in turbulent flow!

Tubular modules (Table 4) This was the most popular configuration in the 1970's. The membrane is inserted into a tubular support or cast directly onto its internal surface. The pressure support is a perforated stainless steel or fiber-glass tube. The cost of circulating the process solution through the tubular system has led manufacturers to change to smaller diameters (6 or 12 mm rather than 25 mm) and greater lengths (or more tubes in series in the pressure vessel) so as to increase the recovery for a single pass. For example, one supplier (Paterson Candy International) has recently changed its configuration to one with more tubes in series within each module, claiming that this reduces the energy consumption by 33%.

Plate-and-frame modules (Table 5) This is probably the most widely used ultrafiltration configuration at the present time. Two suppliers (Dorr Oliver and Millipore) have designed a system in which a stack of flat leaves is assembled into a cartridge; the feed solution flows in parallel along the plates. DDS and Rhône-Poulenc modules consist of a number of plates covered on both sides with membranes. Each membrane support plate is equipped with separate permeate outlets allowing local inspection of the permeate; the plates are mounted in parallel and in series. Rhône-Poulenc and Sartorius have designed special support plates with a turbulence-promoting rippled surface.

TABLE 4. Commercially available tubular modules

Supplier	Internal Diameter	Length (m)	Arrangement No. of Tubes
SFEC	6 mm (15 mm)	1.5 (4.5)	100
TEIJIN		3	16
NITTO	12 mm	3	4-18
P C I		3.6	18
WAFILIN	17 mm	6	7-18
ABCOR	25 mm	3	8-16
KALLE		membrane only	

TABLE 5. Commercially available plate-and-frame modules

Supplier	Basic Element	Module (max. area)	
DORR-OLIVER	–"cartridge"	–"IOPLATE"	
	–2-3 mm spacing	–12 cartridges in series	
	–Rectangular plates		
	–Parallel flow		
	–Horizontal		
	–1.3 m^2	(16 m^2)	
MILLIPORE	–"cassette"	–"PELLICON"	
	–Rectangular plates	–Cassettes in series or parallel	
	–Parallel flow		
	–Horizontal		
	–0.5 m^2		
D.D.S.	–0.7 mm spacing	–Module	Area (m^2)
	–Ellipsoidal plates	35	42
	–Rectangular flow channel	37	?
	(6-60 cm long)	40	28
	–Vertical	–Plates in parallel or in series	
RHÔNE-POULENC	–1.5 mm spacing	–Module	Area (m^2)
	–Rectangular flow channel	UFP 70	50
	(90 × 10 cm)	UFP 71	21
	–Turbulence-forming shoulders	–Plates in parallel or in series	
	–Vertical		
	–0.4 m^2		
SARTORIUS	–0.1 mm spacing	–"SARTOCON" (2.5 m^2)	
	–Rectangular flow		
	–Turbulence-forming shoulders		
	–0.5 m^2		

Spiral-wound modules (Table 6) The spiral wound elements used in ultrafiltration are similar to those used in reverse osmosis. A pressure vessel can contain up to six elements in series, giving a total area per module of up to 30 m^2.

Hollow-fiber modules (Table 7) This configuration has gained a wide acceptance in the past few years as membrane materials and systems design technologies have improved. There are now four manufacturers; the differences between them are not nearly as marked as in the case of the tubular and plate-and-frame configurations. The advantage of this configuration is a large membrane area per unit volume and the possibility of operating with back pressure for flushing the module.

TABLE 6. Commercially available spiral-wound elements

Supplier	Element						Module
ABCOR	Length: 94 cm Diam.: 11 cm Area: 5.6 m^2						2-3 elements in series in a pressure vessel
MILLIPORE	Length: 30-55 cm Diam.: 7.4-8.7 cm Area: $1.4m^2$-$5.6m^2$						
OSMONICS	"Osmo":	52	112	192	554	3258	Up to 6 elements in series in a pressure vessel
	Length: (cm)	33	66	99	99	102.6	
	Diam.: (cm)	5	5	5	10.2	12.4	
DESALINATION SYSTEM	"U"	6	60	90	225	350	—
	Length: (cm)	30	63	101	63	101	
	Diam: (cm)	5	10.2	10.2	20.3	20.3	
	Area: (m^2)	0.4	5.1	8.3	20	31	

TABLE 7. Commercially available hollow-fiber modules

Supplier	Module	Membrane Area (m^2)	Cartridges Diam. (cm)	Cartridges Length (cm)	Fibers No.	Fibers Internal Diam. (cm)
ROMICON	HF 2.5-20	0.23	2.5	63.5	250	0.05
	HF 30-20	2.8	7.6	"	2930	"
	HF 1.1-45	0.1	2.5	"	50	0.11
	HF 15-45	1.4	7.6	"	660	"
	HF 26.5-45	2.5	7.6	109	660	"
	HF 39-20	3.6	7.6	"	2147	0.05
	HF 55-20	4.9	7.6	"	2940	"
BERGHOF	BST	0.5-1				0.6
	(PA)					1.6
ASAHI	H1-1					
	HC-2					
	HC-5					
	HH-1					.75
NITTO Elect.	NTU-8010					
	NTU-8050					

3. PLANT OPERATION

3.1. Defining the Purpose of the Operation

The components of an ultrafiltration feed solution can be divided into three categories:

(M) macromolecular solutes or colloidal particles
(A) small solutes
(W) solvent (water)

A distinction is often made between three different types of operation that are possible with ultrafiltration:

(1) Enrichment of a solution in macromolecular solutes: the aim is to obtain a retentate whose concentration c_M is higher than in the feed; the value of c_A is unimportant,

(2) Purification of a solvent: the product required is a permeate for which c_M is as low as possible; the value of c_A is hardly changed,

(3) Fractionation of a mixture of solutes: the retentate should have a high concentration ratio c_M/c_A, whereas for the permeate this ratio should be small.

Although arbitrary, this division has some value in indicating the versatility of the ultrafiltration process and it can also help clarify the priorities to be respected. The choice of the type of operation depends on the possible economic value of the three classes of components and on the use to which the two product streams are to be put. For example, in some cases it can be important to obtain the macromolecular solute in a solution free of microsolutes, whereas in other cases it is more useful to produce a permeate without any colloidal particles in it.

Another operation that should, in principle, be possible with ultrafiltration is the fractionation of mixtures of macromolecules, e.g., the separation of albumin from globulins in blood plasma. So far this sort of operation has not been possible because of the adsorption of macromolecules onto the membrane, which alters the distribution of pore sizes, and because macromolecular species often interact with each other.

Permeation rate Having chosen the membrane that is capable of giving the required separation, one's next step in designing an ultrafiltration plant is to obtain a permeation rate that will stay at high enough values over a sufficiently long period of time.

The permeation rate obtained in ultrafiltration with pressures in the range from 100 to 600 kPa and at temperatures up to 60° C, when concentration polarization is present, is normally between 1 and 100 L m^{-2} hr^{-1} (0.3 to 30 $\mu m\ s^{-1}$). These values should be compared with the maximum flux, which is given by the intrinsic membrane permeability, and which lies in the range from 100 to 500 L m^{-2} hr^{-1}. The great difference between the maximum flux and the usually observed flux is mainly due to concentration polarization. So it is easy to see how important this phenomenon is in ultrafiltration.

So far no definitive theory exists to explain concentration polarization in ultrafiltration and allow an exact calculation of its effects. The two explanations generally given can be called the "gel layer" theory and the "osmotic pressure" theory. The gel theory, treated in some detail in the chapter by Dr. M. C. Porter, is the most generally used explanation. It has the advantage of providing a fairly simple means for calculating the effects of concentration polarization, but these calculations are often inaccurate. It also tends to be misleading in its implications concerning the physicochemical properties of macromolecular solutions. For example, the "gel concentrations" that it gives for these solutions are often unrealistic and generally vary according to the ultrafiltration apparatus with which the measurements are made. It is also based on the assumption that concentrated solutions of macromolecules have negligible osmotic pressures, whereas in fact these are generally quite comparable with the operating pressures used in ultrafiltration.

The osmotic pressure model, which cannot be formulated in such a succinct form as the gel model, is less useful for rapid calculations. It also requires a knowledge of the variation of solution properties with concentration; the diffusion coefficient, the osmotic pressure (and sometimes the viscosity) of the macromolecular solutions must be known over a wide range of concentrations. But in many cases, the osmotic pressure model is of considerable aid in studying the basic nature of concentration polarization.

When a semi-permeable membrane separates two solutions with different osmotic pressures, the permeation rate through the membrane is given by the following expression:

$$v_{\mathrm{m}} = L_p\,(\Delta p - \Delta\pi) \qquad (11)$$

Figure 9 shows how this relationship can be used to explain the limiting flux in ultrafiltration. As the applied pressure across the membrane is increased, the concentration of macromolecules at the membrane surface rises. Now the effective pressure difference across the membrane, which is responsible for the permeation, is no longer Δp but rather $\Delta p - \Delta\pi$. As can be seen in Figure 10 for a particular macromolecule, the osmotic pressure of a macromolecular solution rises more and more steeply as the concentration increases. Beyond a certain value of Δp, any further increase in pressure will cause a rise in concentration at the membrane sufficiently great for the increase in osmotic pressure to counteract completely the increase in Δp.

A particularly interesting formulation of this model, for a thin-channel cell, has been given by Leung and Probstein (21). Using an "integral" approach very similar to that previously suggested by Doshi et al. (22) in a treatment of concentration polarization in reverse osmosis, Leung and Probstein found a very good agreement between the osmotic-pressure model and their experimental data. These results, together with observations by other authors (23), suggest that the osmotic-pressure model should in many

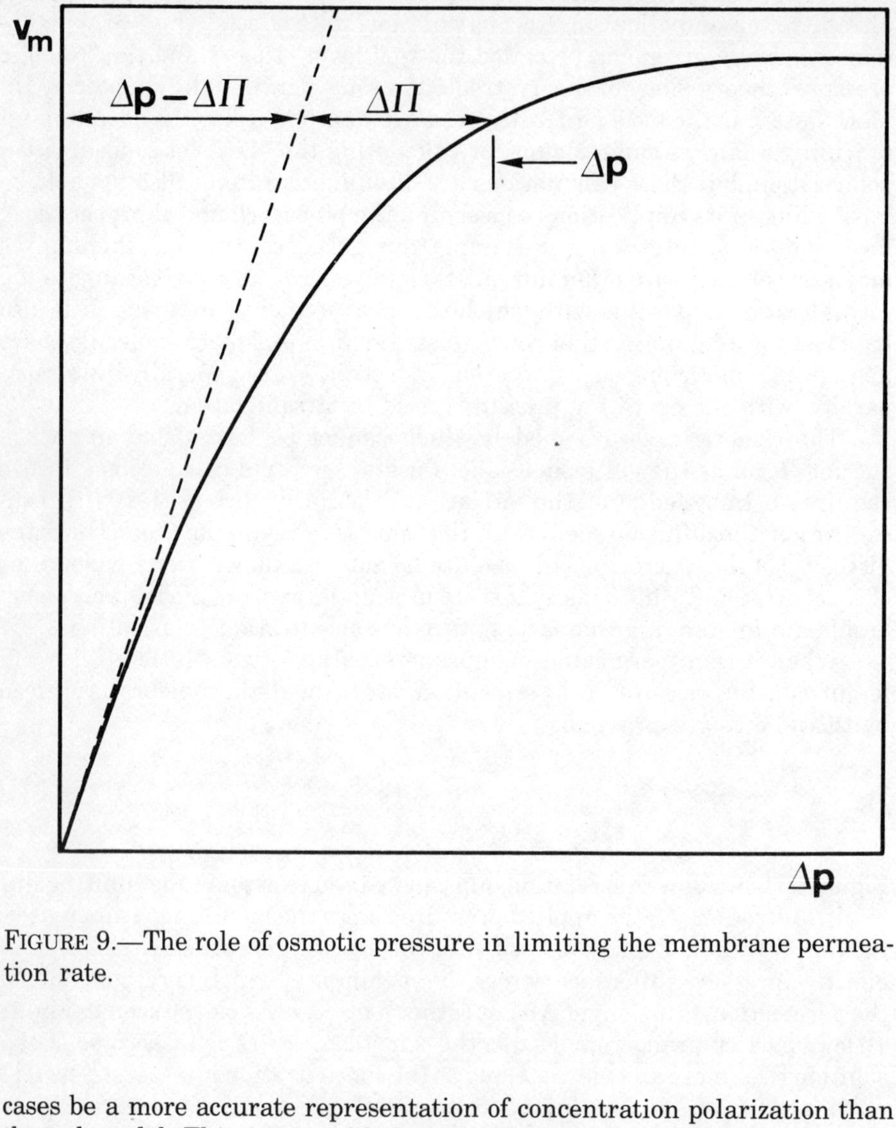

FIGURE 9.—The role of osmotic pressure in limiting the membrane permeation rate.

cases be a more accurate representation of concentration polarization than the gel model. This is most likely to be the case in systems operating at relatively low applied pressures (e.g., in hollow-fiber modules).

It is important to consider the effects of various parameters on the permeation rate that can be obtained.

Membrane permeability It has already been mentioned that the use of high-permeability membranes has meant that, paradoxically, the intrinsic membrane permeability no longer plays an important role in determining the

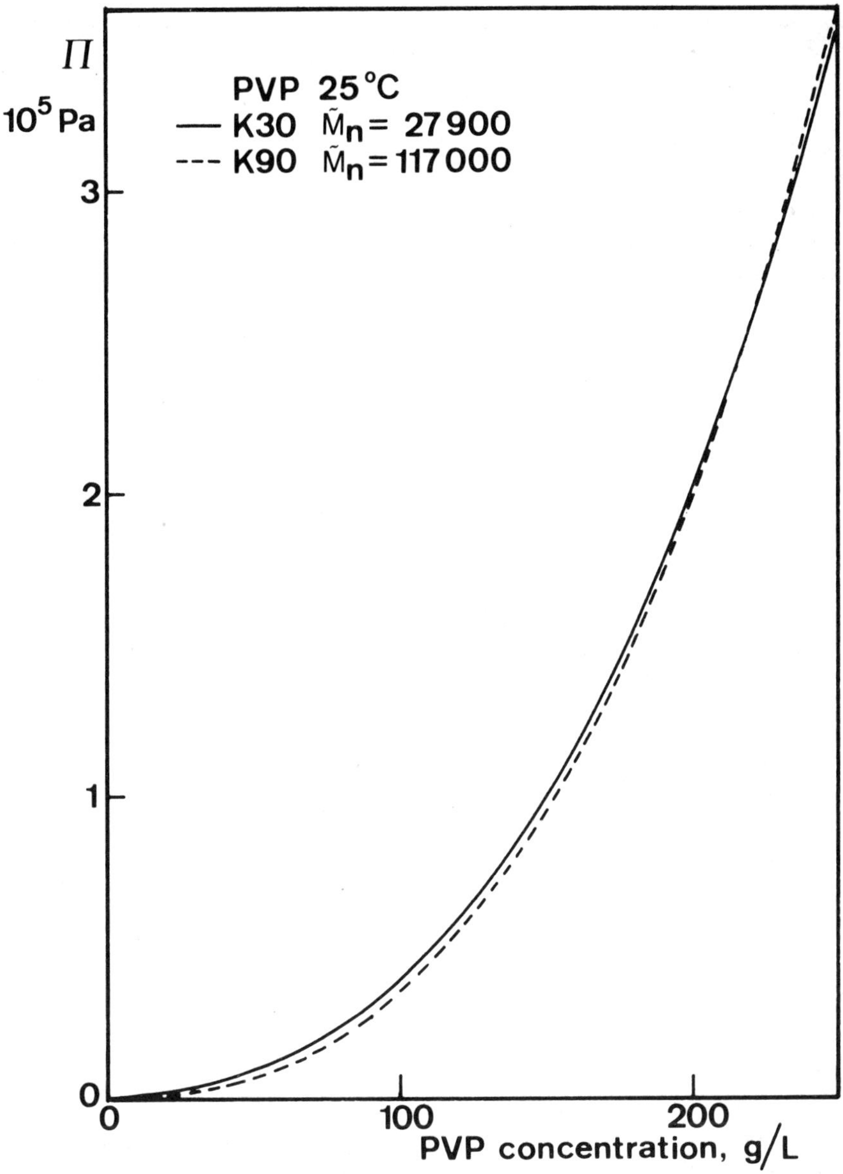

FIGURE 10.—Variation of osmotic pressure with concentration of macromolecules for two fractions of polyvinylpyrrolidone, calculated from virial coefficients determined by Vink, H.: 1971, Eur. Polym. J. 7, pp. 1411–1419.

permeation rate that can be obtained in an ultrafiltration plant. This can be seen in the results shown in Figure 11, where the limiting flux for two membranes of different permeabilities is the same (6). One possible advantage to be obtained from increasing membrane permeability is that the limiting flux can be attained at lower operating pressures; this could mean a saving in pumping energy. In industrial systems, however, membrane fouling generally modifies the effective membrane permeability to such an extent that this distinction is rarely observable in practice.

Flow velocity A high tangential velocity parallel to the membrane surface is still, at present, the surest way of controlling concentration polarization. The higher permeation rate obtained is paid for by an increase in energy expenditure in the pumps. Manufacturers of ultrafiltration equipment usually specify the optimum range of flow rate for each particular module. In laminar flow, in thin-channel modules, velocities of 1 or 2 m/s are used. In tubular modules, turbulent flow may be generated with velocities up to 5 m/s.

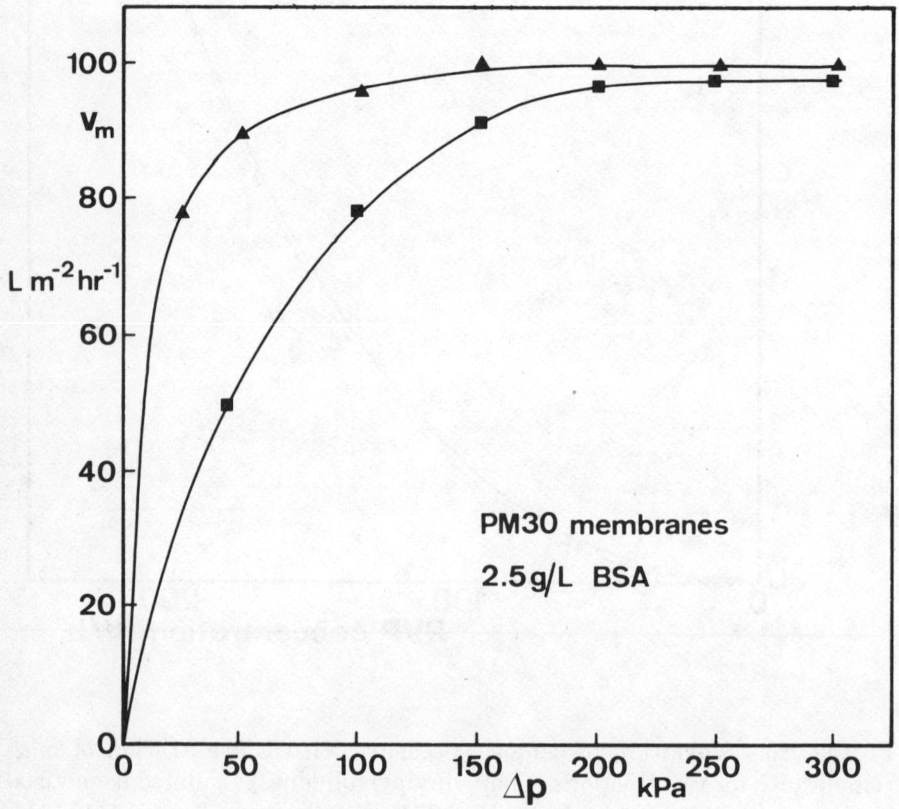

FIGURE 11.—Variation of permeation rate with applied pressure for two membranes of different permeabilities (6).

Transmembrane pressure To maximize plant output, equipment is normally operated at pressures giving a permeation rate close to the limiting flux. Once this limit has been attained, any increase in pressure is a source of inefficiency, because the energy input increases for no increase in production rate. It is also dangerous to increase pressure beyond this point because membrane fouling becomes increasingly important and the flux decline is accelerated. The ideal situation would be to have the whole membrane surface within each module operating just below the limiting flux. In practice, the pressure loss within the module, which is related to the need to control concentration polarization, implies that the inlet pressure to the module must be well above the pressure required for limiting flux, so that the pressure at the module outlet can still be high enough to ensure a permeation rate close to the limiting value. This is shown in Figure 12. The unbroken line represents the

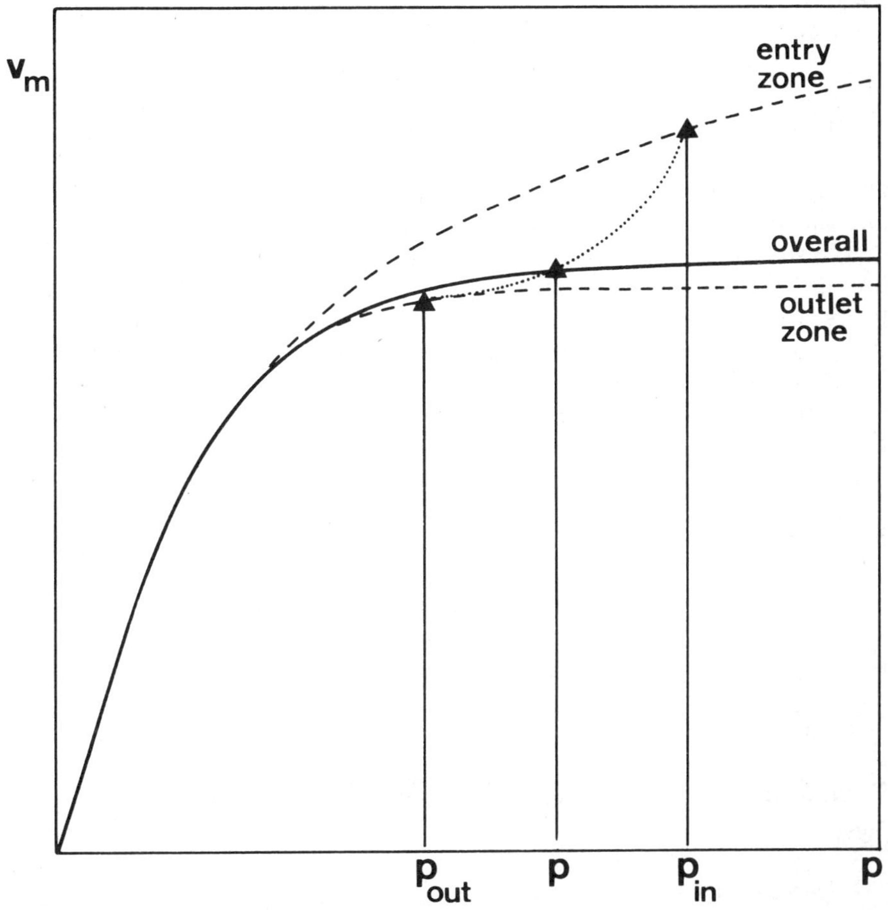

FIGURE 12.—Variation of permeation with applied pressure: in a zone near the module inlet, in a zone near the module outlet and the overall result for the whole module. P_{In} is the inlet pressure; P_{Out} is the outlet pressure.

overall behaviour of the system, while the dashed lines show local permeation rates at the inlet and outlet of the module. Near the inlet the permeation rate is above the average value because the polarization layer is thinner at that point, whereas in the outlet zone the permeation rate is somewhat lower than the average because the polarization layer is at its maximum thickness. It can be seen that the maximum overall flux is only obtained when the outlet zone of the module is operating in the fully polarized region.

Temperature When the operating temperature is raised, the retentate viscosity is reduced and diffusion coefficients of macromolecules increase. The effect of these two factors is to enhance mass transfer and so increase the permeation rate. However, certain limitations are imposed by the stability of colloids and macromolecules, e.g., enzymes (<25° C), electropaints (<30° C), proteins (<55° C), etc. In the dairy industry, the normal operating temperature is 50 to 55° C. Maubois (24) explains this choice as a compromise between the following three considerations:

(a) the permeation rate rises by 3% per degree C,

(b) at these temperatures, no bacterial growth is observed in a period of 3 to 6 hours, depending on the nature of the initial bacterial population,

(c) the combined effects of temperature and residence time must be below 65° C and 30 minutes, because beyond these limits the proteins are denatured and the viscosity of the retentate rises abruptly.

In some cases the possibility of treating a high-temperature feed can be useful for reasons other than the higher permeation rate. This is the case in the recovery of PVA from textile desize liquors by ultrafiltering at 85° C. Here the permeate that leaves the ultrafiltration plant is quite hot and can be recycled without reheating.

A more extreme case of high-temperature ultrafiltration is the regeneration of used lubricating oils by ultrafiltration with inorganic membranes at 330° C. This process is at present operating at the pilot-plant stage.

Retentate concentration When the concentration of the retentate is raised, its viscosity also increases and the polarization layer thickens. This means that mass transfer rates decrease and so the permeation rate is also lowered. The gel theory of polarization predicts that the limiting flux should vary with the inverse logarithm of concentration and this relationship is often reasonably accurate. Some departures from this law are observed at very high retentate concentrations.

Fouling and flux decline It is often considered that membrane fouling takes place in three stages:

(a) concentration polarization

(b) adsorption of macrosolutes

(c) polymerization of the adsorbed layer.

Even though there is some disagreement as to the relative importance of the first two steps, this mechanism does explain the observed sudden decrease in flux in the first minutes of operation, followed by a slow decline, which continues over many hours.

Howell et al. (2) showed that when a protein solution (5g/L albumin) is ultrafiltered with a membrane whose average pore size is slightly smaller than the solute diameter, the membrane first allows a certain amount of protein to pass because the largest pores in the membrane are larger than the protein molecules. As the protein is adsorbed onto the membrane, the larger pores are gradually closed and within a few minutes the albumin is completely retained by the membrane.

Fouling is particularly a problem in the case where the feed solution is of biological origin. In such solutions there is generally a complex equilibrium set up; any changes in protein concentration or ionic strength can destroy this equilibrium and cause the protein to be denatured and precipitated. The concentration changes in the polarization layer are generally quite considerable, so that the precipitation of protein on the membrane surface is not at all surprising.

Colloidal systems such as electropaint baths are often simpler systems and are less susceptible to fouling. This is particularly so if the ultrafiltration membrane carries fixed charges of the same sign as the charges on the suspended particles. Ultrafiltration plants used to reconcentrate electropaint baths can often operate for several months without cleaning and show no flux decline at all.

The importance of fouling in the food and pharmaceutical industries (where it is severest) is limited by the fact that ultrafiltration plants working in these industries have to be cleaned every 20 hours (or more often) in order to control bacterial growth. This cleaning is usually quite effective in restoring ultrafiltration fluxes.

Pretreatment It can sometimes be useful to pretreat the feed solution before ultrafiltration to minimize the effects of membrane fouling. Adjustment of pH can be used so that protein solutes will be far from their isoelectric point and so increase their solubility. Heat pretreatment can be used to accelerate the precipitation of potentially fouling substances. For example, if cheese whey is kept for one hour at 65° C, the immunoglobulins and the fat, the two least soluble components of whey, are precipitated. Prefiltration or centrifugation can be used to eliminate the largest suspended particles. Dairy products often give better results after being demineralized, though the type of demineralization procedure can be important. It is most important to eliminate the Ca^{2+} and Mg^{2+} ions and this is best done by ion exchange; electrodialysis, on the other hand, tends to remove Na^{+} and K^{+} ions preferentially and can often increase the danger of fouling. A complexing agent can also be used to prevent the precipitation of certain ionic species. Pretreatment should, however, be applied with caution, because it often adds considerably to the overall cost of the treatment. Its use is usually not justified if its only function is to improve ultrafiltration performance. This is particularly true of the more elaborate treatments (e.g. centrifugation, ion exchange).

Turbulence promoters and other means of reducing polarization It is often suggested that ultrafiltration performance could be improved by including small obstacles in the retentate flow channel so as to disrupt the stable laminar flow and enhance mixing. Hiddink et al. (25) have published an interesting study in which Kenics static mixers were used in a tubular ultrafiltration module. The feed solutions were skim milk and whey. It was found that for a given tangential velocity, the turbulence promoters raised the permeation rate by a factor of 1.5 to 3, but they also increased the pressure losses. An equivalent increase in permeation rate could also be obtained with empty tubes by increasing the pumping rate. The authors concluded that the performance obtained for a given energy input was the same, whether turbulence promoters were used or not.

Shen and Probstein (26) studied the use of turbulence promoters in a thin-channel cell. Their economic analysis takes into account both the capital cost and the energy cost and they conclude that turbulence promoters can give worthwhile improvements in performance.

The only industrial modules in which turbulence promoters are always included are the spiral-wound modules where the mesh used as a membrane spacer should act to enhance mixing. A similar result is obtained with tubular modules, which are normally operated in the turbulent flow regime. The problem with all forms of turbulence is that energy is dissipated mainly in mixing up the bulk solution, which does not need mixing, whereas the turbulence hardly penetrates into the polarization layer, which in ultrafiltration is extremely thin (only a few micrometers).

A number of authors have proposed other techniques for enhancing the mixing of the polarization layer. Lopez-Leiva (27) used a rotating cylinder as a membrane support with the membrane mounted on the outside. This creates centrifugal forces that act on the difference in density between the polarization layer and the bulk solution. The effects of concentration polarization can be almost completely eliminated by this technique but it introduces an undesirable mechanical complexity into the ultrafiltration module. A similar technique was more recently proposed by Robertson et al. (28).

Since most of the particles to be retained by ultrafiltration membranes are electrically charged (colloids, proteins), another way of reducing polarization would be to apply an electric field so that electrophoretic migration would act against the formation of polarization layers. Using this technique, Radovich and Sparks (29) were able to obtain considerable improvements in performance when ultrafiltering protein solutions. They suggest that the use of this technique could allow ultrafiltration to be used for protein fractionation, because the differences in electrophoretic mobility of the various proteins could be used to enhance separation and eliminate the adsorption of the retained species.

3.2. System Design

All ultrafiltration plants have a modular structure, which offers two advantages. Firstly, the individual modules can be arranged in many different ways: in series, in parallel and in various combinations of these. Secondly, each module has a relatively small membrane area, so it is easy to build a pilot plant with an industrial type of module and operate it under industrial conditions. This considerably simplifies the process of scaling-up to the full-size plant.

Ultrafiltration plants can be operated in two different modes: batch or continuous. Batch operation is suitable for feed material originating from small-scale batch production, such as fermentation. To control polarization, the tangential flow inside the module must be kept high. This is possible in an open loop, such as the one shown in Figure 13a, but it is usually preferable to use a "feed and bleed" arrangement (Figure 13b), which has the advantage of reducing pumping energy and if prefiltration is required, only the feed flow need be treated. To ensure a fairly uniform concentration at all points in the loop, a recirculation flow rate about ten times greater than the permeate flow rate should be used.

Figure 14 shows an example of a batch concentration by ultrafiltration. It is important to note how the permeation rate decreases as the retentate concentration rises. If we compare this with a loop operating in the continuous mode over the same concentration difference, we can see that the batch

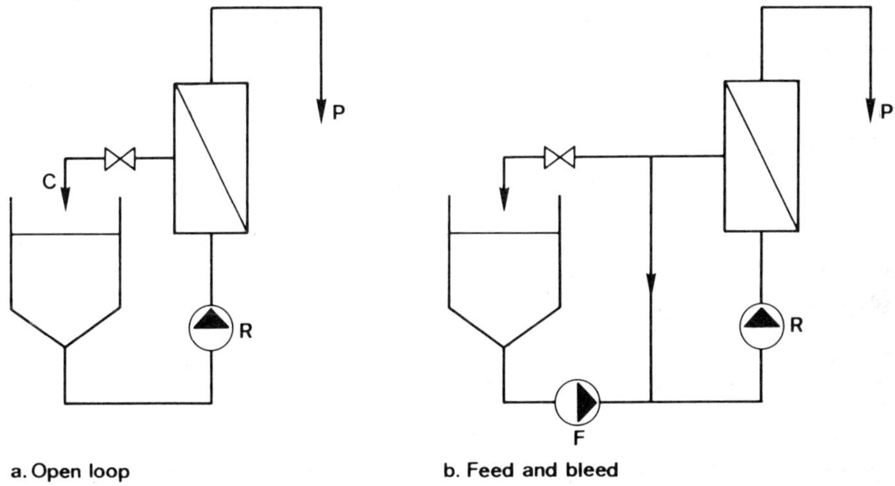

FIGURE 13.—Batch operation of an ultrafiltration unit. C: concentrate, P: permeate, F: feed pump, R: recirculation pump.

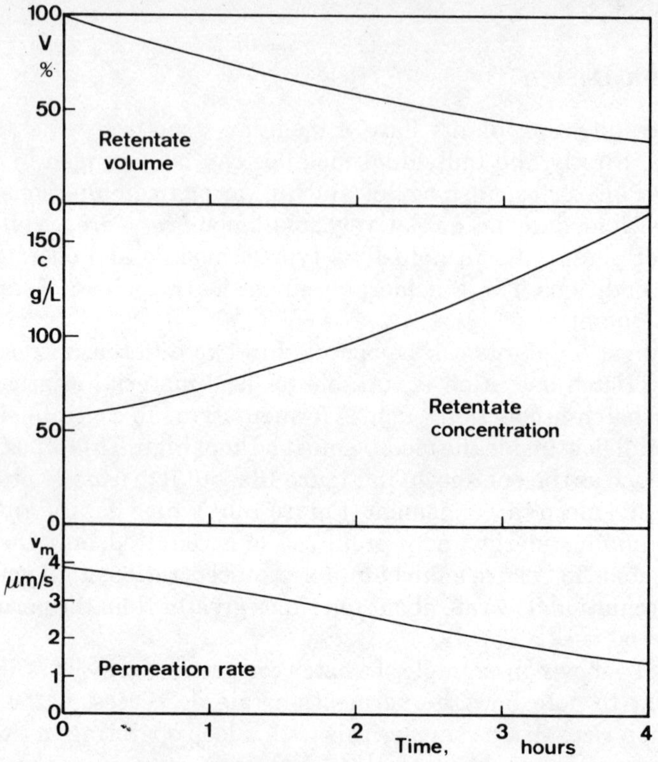

FIGURE 14.—An example of batch concentration, showing the variation in time of (1) retentate volume, (2) retentate concentration, (3) permeation rate, for laboratory-scale ultrafiltration of bovine plasma.

operation gives a higher average permeation rate, because in the continuous mode the retentate concentration is always at the final value, which is the maximum, so the permeation rate is always at the lowest point.

To calculate operation times for batch concentration it is necessary to know how the retentate concentration will vary in time. The basic equation describing this process is the following:

$$\frac{\mathrm{d}c}{\mathrm{d}t} = \mathcal{R}A\,\frac{c\,v_{\mathrm{m}}}{V} \qquad (12)$$

where c and V are functions of time and v_{m} is a function of c. In the case of a perfectly retained solute ($\mathcal{R} = 1$) a simpler expression is obtained:

$$\frac{\mathrm{d}c}{\mathrm{d}t} = \frac{A}{m}\,c^2\,v_{\mathrm{m}} \qquad (13)$$

Even if the variation of v_m with c is assumed to follow the simple relationship given by the gel polarization theory, an analytical solution to these equations can only be obtained by making further approximations.

Residence times are generally long in batch operation but much shorter in continuous operation. This could be important in cases where microbial growth and product denaturing could be a problem. Continuous operation is usually necessary for large-scale plants so as to avoid the necessity of having holding tanks of a prohibitive size. This mode requires less pumping energy than batch operation does, but the average permeation rates are also lower. Recirculation is usually necessary to maintain a sufficiently high tangential flow.

Figure 15 represents a single loop operating in the continuous mode. The ratio of outlet to inlet concentrations is given by the following expression:

$$c_1/c_0 = Q_0/[Q_1 + Q_p(1 - \mathcal{R})] \quad (14)$$

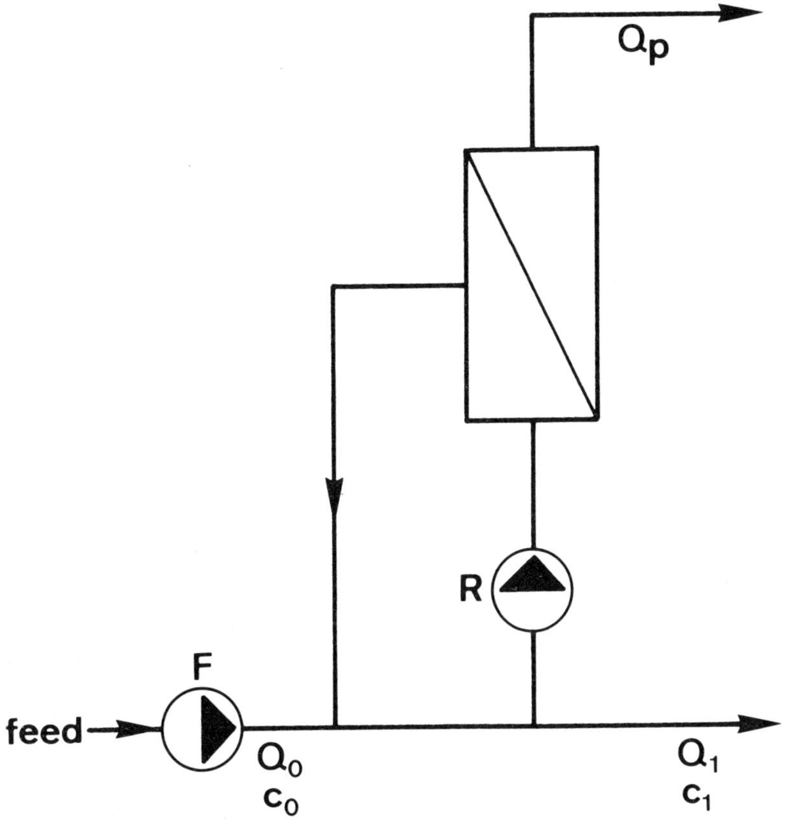

FIGURE 15.—Continuous operation of an ultrafiltration unit: single stage. F: feed pump, R: recirculation pump.

Continuous operation in a single stage is inefficient because the membrane always operates at the final concentration, for which the permeation rate is lowest. It is preferable to connect a number of recirculation loops in series, with each stage operating at successively higher concentrations (Figure 16). Three stages operating in series have a membrane productivity (mean permeation rate), which is 80% of the value for batch operation. This value approaches 100% as the number of stages is increased, with three stages usually being the minimum.

Some of the implications of ultrafiltration treatment can be seen very easily from the hypothetical example shown in Table 8, which was taken from

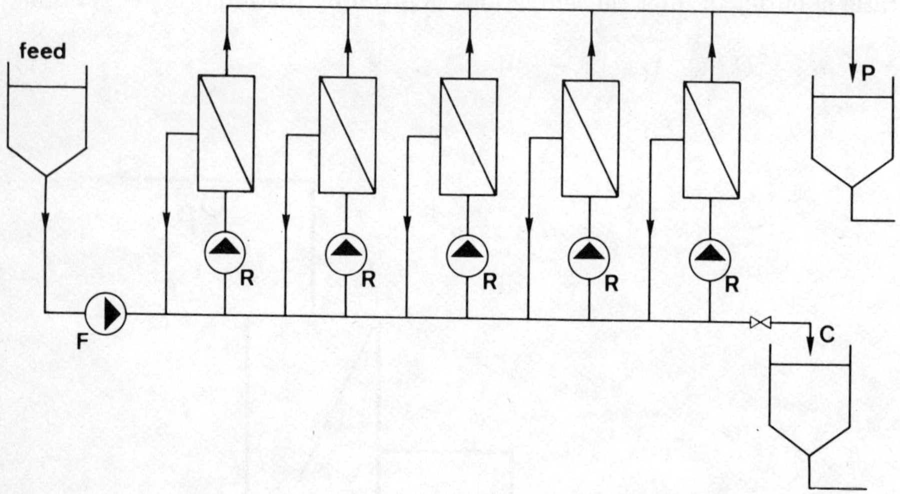

FIGURE 16.—Ultrafiltration plant in continuous operation: several stages in series. C: concentrate, P: permeate, F: feed pump, R: recirculation pump.

TABLE 8. Concentration and purification by ultrafiltration (30)

Feed: $c_M = 1.0\%$ $c_A = 4.0\%$
$\mathcal{R}_M = 1$ $\mathcal{R}_A = 0$

	CF	PR %	Permeate c_M	Permeate c_A	Retentate c_M	Retentate c_A	c_A/c_M	Total solids %	c_M/solids %
Feed	1				1.0	4.0	4.00	5.0	20.0
	5	80.0	0	4.0	5.0	4.0	0.80	9.0	55.6
	10	90.0	0	4.0	10.0	4.0	0.40	14.0	71.4
	15	93.3	0	4.0	15.0	4.0	0.27	19.0	79.0

CF: concentration factor
PR: permeate recovery

an article by Breslau (30). The feed solution contains a macrosolute M at a concentration of 1% and a microsolute A at a concentration of 4%. While M is completely retained by the membrane, the solute A is not retained at all. As the solution is concentrated 5,10,15 times, not only does the total solids content of the product increase, but the relative importance of the two solutes is also considerably modified. The macrosolute is not only being concentrated but is also being purified, because the microsolute A is not undergoing any concentration. If the aim of the ultrafiltration operation is to purify the macrosolute, then a straight concentration operation is usually not sufficient. The solution finally obtained at the end of Table 8 is unsuitable for further concentration because the total solids content will continue to rise, thus giving unsatisfactory permeation rates.

If the macrosolute is to be further purified, another ultrafiltration operation, "diafiltration," should be used. Figure 17 represents a semi-continuous batch diafiltration operation. An ultrafiltration loop is operated in the batch mode and water is added to the retentate tank at a rate equal to the permea-

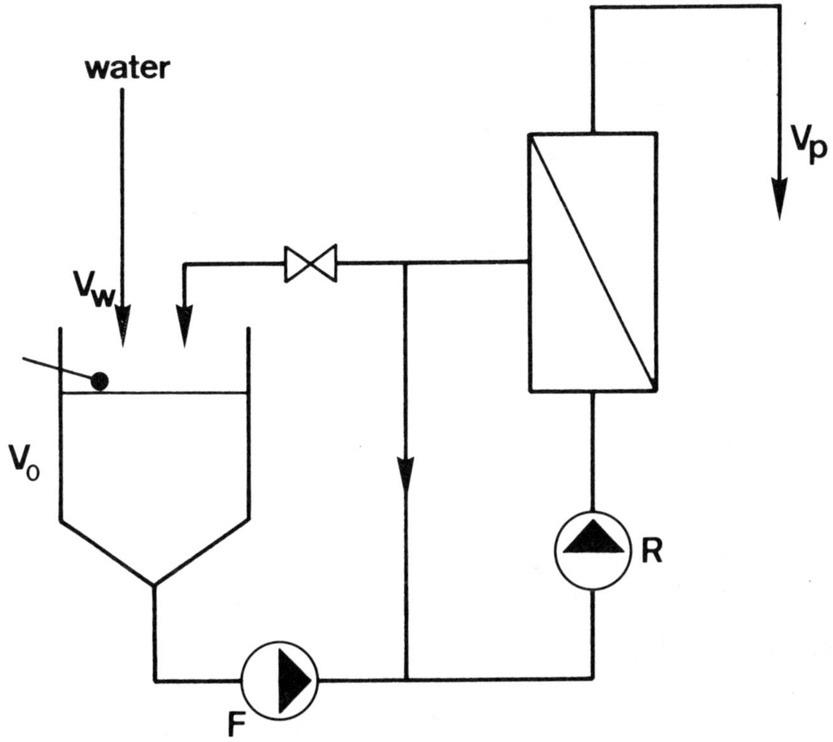

FIGURE 17.—Semi-continuous batch diafiltration. V_0: volume of retentate, V_w: volume of water added, V_p: volume of permeate removed, F: feed pump, R: recirculation pump.

tion rate, thus maintaining a constant total retentate volume. The ratio of the microsolute concentration c_A to the macrosolute concentration c_M varies with the volume of water added V_W according to the following relationship:

$$\left(\frac{c_A}{c_M}\right)_f = \left(\frac{c_A}{c_M}\right)_0 \exp\left[-\frac{V_W}{V_0}(\mathcal{R}_M - \mathcal{R}_A)\right] \quad (15)$$

This shows that even if the macrosolute is not entirely retained ($\mathcal{R}_M \neq 1$), it is still possible to improve its purity by increasing the amount of water added; however, care must be taken in this case because macrosolute is being lost through the membrane. If the macrosolute is completely retained by the membrane ($\mathcal{R}_M = 1$), then a simpler formula may be used:

$$(c_A)_f = (c_A)_0 \exp\left[-\frac{V_W}{V_0}(1 - \mathcal{R}_A)\right] \quad (16)$$

Table 9 shows what could be obtained if the final product from Table 8 were treated by diafiltration. It can be seen that the purity of the macrosolute very rapidly approaches 100%.

TABLE 9. Purification of a macromolecule by diafiltration (30)

$\mathcal{R}_M = 1$ $\mathcal{R}_A = 0$

V_W/V_0	Retentate c_M %	Retentate c_A %	c_A/c_M	Total solids %	c_M / solids %
0	15.00	4.00	0.27	19.0	79.0
1	15.00	1.50	0.10	16.5	90.9
2	15.00	0.54	0.04	15.5	96.8
3	15.00	0.20	0.01	15.2	98.7

There are three possible modes of diafiltration:

(a) predilution of the feed, followed by a batch reconcentration

(b) semi-continuous batch operation (Figure 17)

(c) a continuous cascade operation with addition of water between the stages. The latter process is shown in Figure 18.

Diafiltration can also be used for microsolute exchange. In that case, instead of adding water, a solution of microsolutes is added.

3.3. Cleaning and Disinfecting

The coating of the membrane surface and the plugging of pores due to fouling implies the necessity of regular cleaning. The cleaning procedure should be adapted to the nature of the product being treated. The most common procedure is to use chemical cleaning agents. For example, electropaint deposits can be removed using detergent solutions with or without a

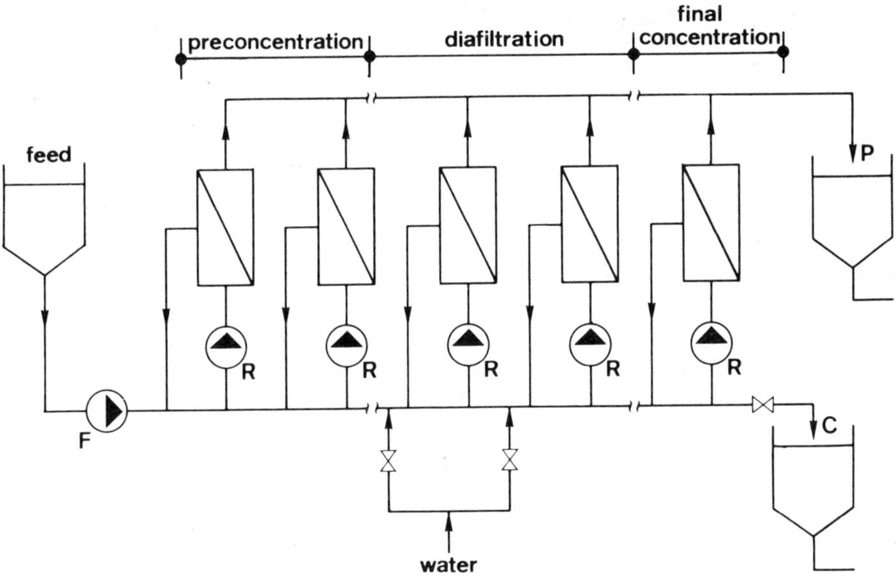

FIGURE 18.—Continuous diafiltration with water addition between the stages. C: concentrate, P: permeate, F: feed pump, R: recirculation pump.

"bridging" solvent. In the food and pharmaceutical industries the choice is much wider: enzymes, detergents, alternated acid/base washing. A cleaning phase is normally included in the operating cycle of the plant. The duration and frequency of the cleaning depend on the severity of the fouling.

Among the disinfecting agents used, the most common is chlorine, but peroxides and iodophors are also used. These agents pass readily through the membrane and so disinfect both the upstream and downstream sides of the system.

A typical cleaning phase for an ultrafiltration unit used for concentrating cheese whey might consist of the following washing steps: water, 0.3% nitric acid, water, 0.5% sodium hydroxide with 0.5% EDTA, water, 0.1% hydrogen peroxide, water. These cleaning solutions are normally used at high temperatures: 50 to 75° C. Products less strongly fouling than cheese whey require simpler procedures.

In some plants, cleaning techniques other than chemical cleaning can be used. Tubular modules can be cleaned mechanically, whereas in hollow-fiber systems backpressure flushing is possible since the membrane is self-supported. Steam sterilization can be used with inorganic membranes and with polysulfone membranes.

3.4. Membrane Durability

Each manufacturer guarantees membrane durability over a specified period: usually 12 to 18 months. For the guarantee to remain valid, certain restrictions on flow conditions, pressure, pH, temperature and the presence of chlorine or organic solvents have to be observed. It is also important to avoid errors in operation and in cleaning. The guaranteed lifetime should normally be considered as a probable minimum durability; the actual life expectancy of the membranes is generally longer.

3.5. Operating Costs

Figure 19 shows the relative importance of various contributions to the operating cost of an ultrafiltration unit. It is based on figures given by Beaton and Steadly (31) for units with an installed membrane area of more than 100 m^2. Variations in these figures can be expected according to the particular application, the capacity of the unit and the economic environment. It is particularly interesting to note that in the operating cost the dominant element is the capital charge, whereas the energy cost is relatively small.

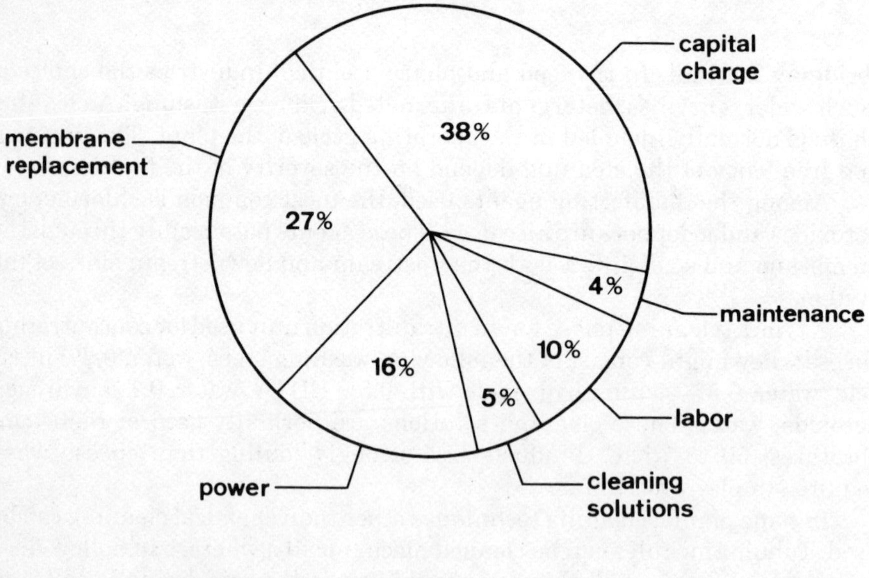

FIGURE 19.—The relative importance of the various components in the operating cost of an ultrafiltration plant (31).

4. APPLICATIONS

Ultrafiltration is a relatively recent addition to the array of industrial processes at the chemical engineer's disposal. As such it has had to enter into competition with the already established processes: this means that even if a particular application is found technically feasible, ultrafiltration will not necessarily be economically competitive when compared with alternative processes. The history of the spread of ultrafiltration in industry, like that of other membrane processes, is typical in this respect. The pay-back on the investment is an important aspect to remember when considering the growth of ultrafiltration in industrial applications; the faster the pay-back, the more willing a potential user will be to risk investing in a new process. In fact, the first applications for ultrafiltration were all ones for which the economic stimulus was particularly strong.

Ultrafiltration is an attractive process because it makes possible the following operations:

— recovery of high-value products,
— recycling of the permeate,
— pollution control,
— energy savings when the permeate that can be recycled is hot.

In biotechnology two other prime advantages of ultrafiltration are its ability to concentrate at ambient temperature and its simultaneous purification capability (salts or other undesirable low-molecular-weight solutes are eliminated with the solvent). In Table 10, the main applications of ultrafiltration are listed, both those already established and those that have only recently appeared (3, 18, 19, 30, 31).

4.1. Electropaint Recovery

Ultrafiltration has been employed in electro-deposition painting for over ten years. During this time, ultrafiltration has evolved from a simple means of controlling excess bath conductivity to a highly sophisticated technology which also controls the paint concentration, produces filtrate for countercurrent rinsing of painted components and therefore plays an equally important role in water recycling, paint recovery and pollution abatement.

The electrocoat market constitutes the largest single application of ultrafiltration technology in the world. More than 1000 ultrafiltration units are currently operating on electrodeposition lines today and in Europe alone 20,000 m^2 of membranes have been installed.

In this painting process, the piece to be coated functions as an electrode and the paint tank as a counter-electrode. In anodic paint, the work piece is the anode, while in cathodic paint, it becomes the cathode. Figure 20 is a schematic view of the process: by means of ultrafiltration, water, solvents and other low-molecular-weight species are separated from the paint feed as filtrate and are used for rinsing the coated pieces before being returned to the

TABLE 10. Applications of ultrafiltration (18)

Industry	Existing	Emerging
Metal finishing	Electropaint Oil/water emulsions	Spray paint
Metal working	Oil/water emulsions	Spent lubricating oils
Dairy	Whey proteins Milk	Protein hydrolysis
Pharmaceutical	Enzymes Vaccines Plasma proteins Antibiotics Pyrogens	Membrane reactors
Food	Potato starch Egg white Gelatin Juice clarification	Blood (abattoirs) Vegetable oils
Textile	Sizing chemicals Indigo Wool scouring	
Pulp/Paper	Lignin compounds	
Chemicals	Waste latex In-process latex	In-process and waste polymers
Leather Working		Tannery wastes
Sewage	Sewage treatments for buildings	Municipalities
Water	Water purification —for various industries —for RO pretreatment	

paint tank. The ultrafiltration concentrate is sent back to the tank. Table 11 gives some typical formulations of paint baths.

The cathodic process has proven its superiority over the anodic one; the cathode is passive (the coating is deposited by reduction), thereby avoiding

TABLE 11. Typical composition of electro-paint baths (32)

Anodic Formation		Cathodic Formation	
Component	Concentration Volume %	Component	Concentration Volume %
butyl cellosolve	5	butyl cellosolve	2
n-butyl alcohol	5	lactic acid	4
alkanolamine	2.5	cellosolve acetate	2
anionic surfactant	2.5	cationic surfactant	1
water	85	water	91

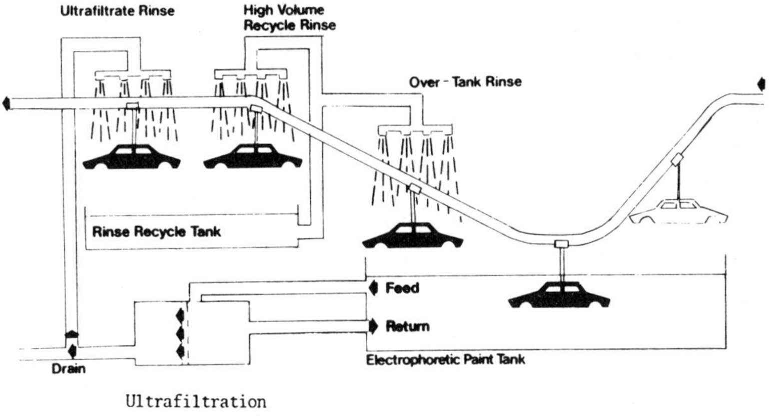

FIGURE 20.—Flow scheme for incorporating ultrafiltration into electropainting operations (31). Reprinted by courtesy of CRC Press.

the problem posed by the oxidation reaction that occurs at the anode. Because of the improved corrosion resistance obtained, large-scale users, particularly automotive manufacturers, are rapidly converting their anodic systems to the cathodic one. This conversion has created a severe problem of fouling with ultrafiltration membranes that were either neutral or negatively charged. As a result, strong interactions occurred with the positively charged paint particles and pretreatment contaminants. New membranes with positive fixed charges were designed and it has been found that they could actually repel these particles, thus reducing or even eliminating the precipitation of insoluble complexes on the membrane skin.

4.2. Waste Oil/Water Emulsions

In the metal-finishing industry, ultrafiltration is used for the treatment of oil/water emulsions. Prior to painting, the metal pieces are degreased before undergoing a chemical treatment, such as phosphatizing, which prepares the metal surface for bonding with the paint or any other primary coating. Ultrafiltration serves to prolong the life of the detergent bath by removing oil; it also solves a pollution problem when the bath is dumped.

Another well established application concerns machining coolants used in the metal-working industry. These emulsions have to be changed regularly as they constitute a medium in which bacteria multiply readily. They then become a pollution problem because control of oil in waste discharges is often particularly strict. Two kinds of emulsion are currently used: mineral oil/water and synthetic emulsions. In the first case, the use of ultrafiltration simply serves to avoid a pollution problem, while with synthetic emulsions, ultrafiltration can make it possible to recycle the permeate and the concentrated oil. It has been estimated that 100 to 200 plants (generally small in size: around 10 m^2) are operating in this application throughout the world. The oil can be concentrated up to levels of about 30 to 60%.

4.3. Cheese Whey

Cheese whey is the supernatant liquid produced in cheese-making or in the casein process. The world production per year is around 3 million tons of dry matter (33). In the late sixties, a large part of the whey produced was discharged into rivers, causing a loss of valuable nutrients and a severe pollution problem. The first solution found for this was to spray-dry the whey and sell the whey powder as a source of protein and sugar for livestock feed. The value of the dried form can be substantially improved by ultrafiltration which can increase the concentration of proteins selectively prior to drying. In fact, a new industry for treating whey protein concentrates, with a protein content in the range 35 to 80% of dry matter, is coming into existence.

Figure 21 (24) shows the development of the surface area of membranes installed in the world dairy industry. About 90% of this surface area is in ultrafiltration and the rest is for reverse osmosis. On the basis of these figures and assuming average operating productivities, the overall amount of whey treated daily by ultrafiltration is an estimated 50,000 tons of dry matter, roughly 15% of the total world production.

Typical compositions of sweet and acid wheys are shown in Table 12. it can be seen that by direct spray-drying, the powder produced contains only 10% protein. Using ultrafiltration to eliminate 75% of the volume, one can produce a retentate with a protein content similar to that of milk, which can be used as a milk substitute in cattle feed. Diafiltration can be used to increase the protein content in the final product, and the whey protein concentrates (50 to 80% protein) are then used for their functional properties (high water solubility, foamability, emulsification, gelation) and become widely used in the food industry for human nutrition (24): beverages, cakes, ice cream, sausages, milk desserts, baby and dietary foods.

TABLE 12. Typical whey compositions from cows' milk (34)

Component	Sweet Whey Concentration (g/L)	Acid Whey Concentration (g/L)
Lactose	45–48	45–48
Protein	5.5–5.8	5.8–6.2
Non-protein nitrogen	0.3	0.3
Ash	4.5–5.0	7
Fat	0.5	0.5
Dry matter (total)	55–60	60
pH	6.6	4.4–4.6

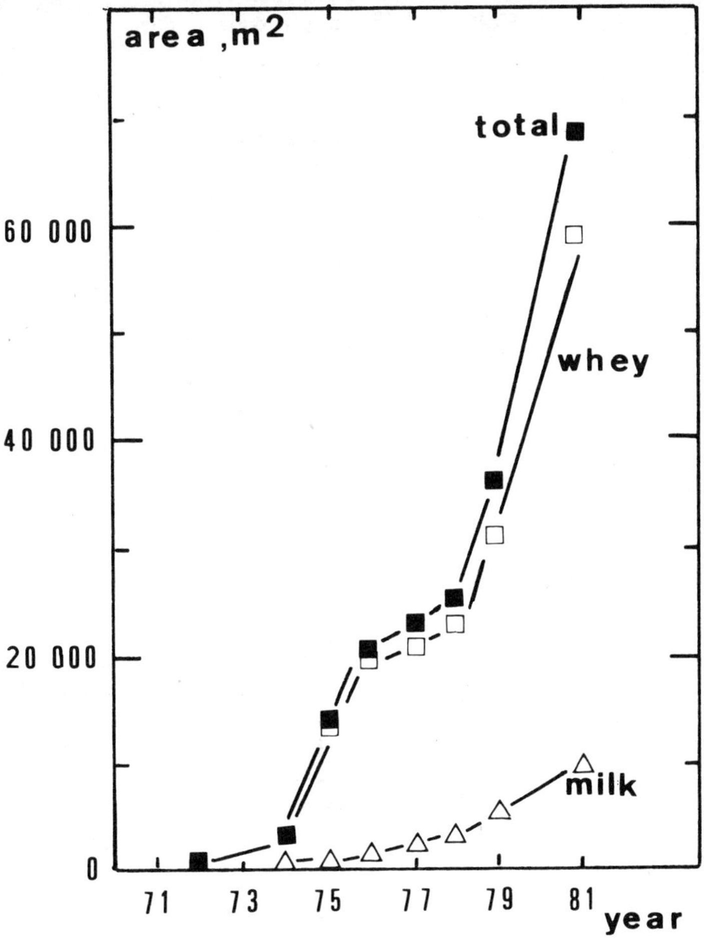

FIGURE 21.—Development of installed membrane area in the dairy industry (24).

4.4. Milk Processing

Ultrafiltration of milk is now used to produce a protein-enriched milk stream for the manufacture of cheese or other fermented dairy products. This technique has taken two principal forms: 1) in the production of a 1.3- to 2-fold protein-enriched milk stream which can be used in conventional cheese-making facilities; and 2) in the production of a high-protein liquid "pre-cheese," which is used in the MMV process (Figure 22) (35).

Ultrafiltration of milk offers the following advantages:

— the protein yield is higher;
— a smaller plant is required for the manufacture of products;
— less rennet, starter . . . etc. is required;
— the continuous process is made easier.

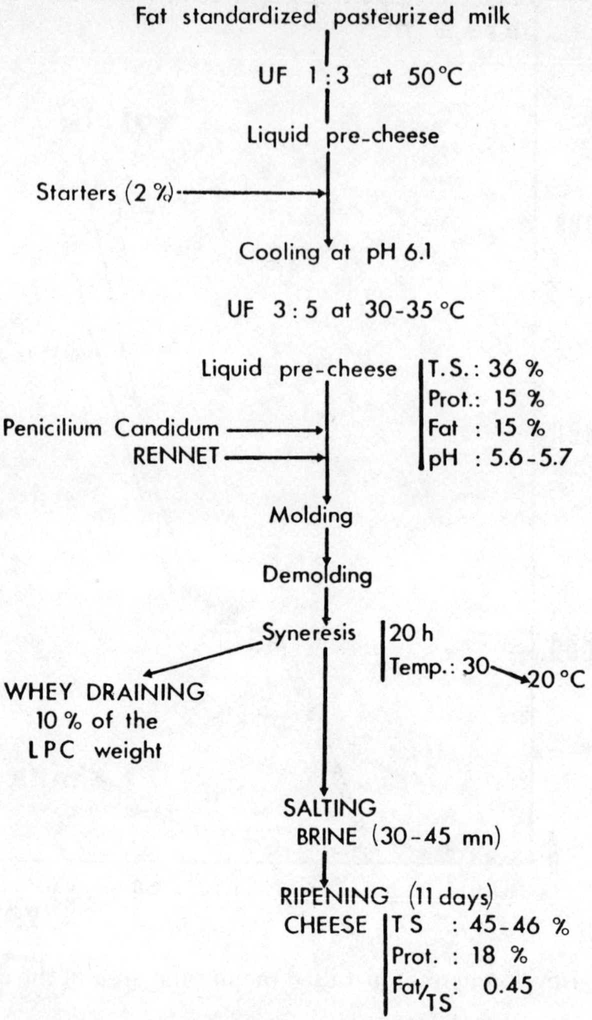

FIGURE 22.—Ultrafiltration process for Camembert cheese production (35). Reprinted by courtesy of Plenum Press.

Milk ultrafiltration for cheese making is now widely implanted (120,000 tons produced in 1981). The most important item in this production is in the Feta cheese produced in Denmark: 92,000 tons in 1981, which represents 32% of the cheese production in that country (24). The generalization of such an operation to other cheeses (such as Camembert) will require important advances in cheese technology. In fact it appears that the real problems to be solved in this case are not in the ultrafiltration stage but rather in the other steps: renneting formulations, mould implantation, etc.

4.5. Pharmaceutical Industry

The pharmaceutical industry offers enormous possibilities for the application of ultrafiltration. Since the early seventies, ultrafiltration has been routinely applied to the purification and concentration on an industrial scale of enzymes produced by fermentation. Other common products to which it has been applied include vaccines, antibiotics, human plasma, protein solutions and pyrogen-free solutions.

Figure 23 shows the increasing impact of ultrafiltration in this field. The figures on the installed square meters of membrane area were obtained from

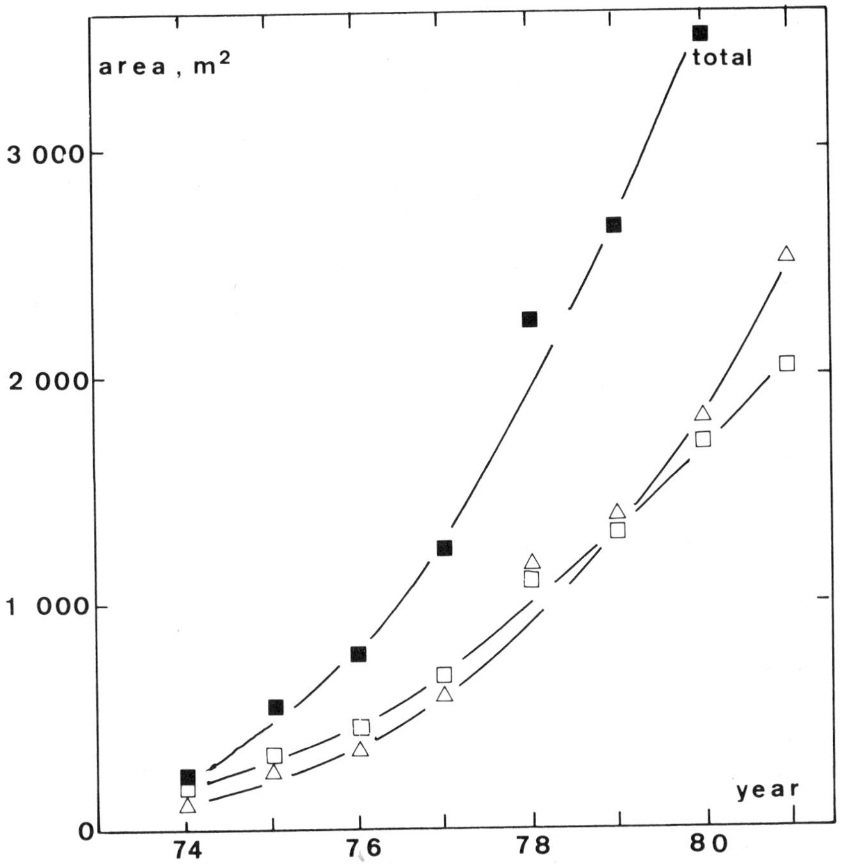

FIGURE 23.—Development of installed membrane area in the pharmaceutical industry (data supplied by two manufacturers: DDS and Romicon).

only two membrane manufacturers. A large part of the market is still in the domaine of enzyme processing. Most industrially produced microbial enzymes are prepared in relatively small batch-fermentation operations. Enzyme fermentation broths are normally clarified by conventional filtration, to remove cells and cell debris, before undergoing ultrafiltration. The concentration can be conveniently performed in the batch operation mode in plants containing up to 300 m^2 of membrane area.

One advantage of ultrafiltration compared with conventional vacuum evaporation systems is an improved activity recovery (95% and 60 to 90%, respectively). The degree of enzyme purification can be increased by diafiltration. The most common enzymes that are produced using ultrafiltration include: proteases, pectinases and amylases.

Human plasma is a very complex mixture, containing over 100 different proteins. Ultrafiltration has been introduced into the Cohn fractionation process (36), both in the form of diafiltration for the removal and exchange of low-molecular-weight solutes (salts, ethanol) and as a concentration step (Figure 24). Among the disadvantages of the conventional Cohn process are: the high cost of lyophilization, the pollution caused by the disposal of concentrated salt solutions and the increased risk of denaturation caused by the repeated steps of precipitation and dissolution. However, the introduction of ultrafiltration into the process does create some severe problems. The material treated is of high value, so the protein loss should be kept to a minimum

<table>
<tr><th>a)</th><th>b)</th><th>c)</th><th>d)</th></tr>
<tr><td>DISSOLUTION</td><td>DISSOLUTION</td><td>DISSOLUTION</td><td>DISSOLUTION</td></tr>
<tr><td>FILTRATION</td><td>FILTRATION</td><td>FILTRATION</td><td>FILTRATION</td></tr>
<tr><td>PRECIPITATION</td><td>PRECIPITATION</td><td rowspan="3">GELFILTRATION</td><td rowspan="6">DIAFILTRATION
+
ULTRA-FILTRATION</td></tr>
<tr><td>CENTRIFUGATION</td><td>CENTRIFUGATION</td></tr>
<tr><td>DISSOLUTION</td><td>DISSOLUTION</td></tr>
<tr><td>FILTRATION</td><td>FILTRATION</td><td rowspan="3">ULTRA-FILTRATION</td></tr>
<tr><td>LYOPHILIZATION</td><td rowspan="2">THIN LAYER EVAPORATION</td></tr>
<tr><td>DISSOLUTION</td></tr>
<tr><td>FILTRATION</td><td>FILTRATION</td><td>FILTRATION</td><td>FILTRATION</td></tr>
</table>

FIGURE 24.—Processing of ethanolic albumin precipitate (Cohn fraction IV, precipitate c) to clinical 25% solution by: (a) conventional method, (b) thin-layer evaporation, (c) gel- and ultra-filtration, (d) dia- and ultra-filtration (36). Reprinted by courtesy of Plenum Press.

and the hold-up volume of the equipment should be as small as possible. For example, if the retention is only 99%, the equivalent of 50 donations is lost from a batch of 200 liters of 25% albumin solution, and one liter of hold-up volume corresponds to a loss of approximately 25 donations.

Water quality is of great importance in the pharmaceutical industry. Here ultrafiltration can be used for processing tap water or deionized water to remove all suspended particles and colloids (including colloidal silica or iron) as well as micro-organisms and spores. Pyrogen removal from water is an application of particular importance. Pyrogens are debris from the breakdown of bacteria, which cause fever when solutions containing them are injected. Here ultrafiltration offers an economical alternative to distillation, reverse osmosis or adsorption by asbestos or other media. The applications of pyrogen-free water are in the preparation of parenteral solutions of low-molecular-weight species, clinical solutions such as electrolytes, radio-opaque markers, sugars and antibiotics.

4.6. Textile Industry

The textile industry offers some interesting opportunities for ultrafiltration and some applications have now reached the commercial scale.

Polyvinyl alcohol and carboxymethylcellulose solutions are used for sizing (lubricating) cotton or cotton/synthetic blends prior to weaving. A washing operation is carried out before dyeing or printing and consumption of 200 m^3 of water at 80°C per cycle is typical. Figure 25 shows how ultrafiltration can be incorporated into this process so as to recover the sizing chemicals from the wash water. This application has the same attractive aspects as does the electro-paint recovery application: product recovery, pollution control, and water recycle (but allows energy savings as well, because the recyclable rinse water is hot).

A number of dyes might be recovered by ultrafiltration, but the only existing industrial application seems to be indigo recovery. This is explained by the high price of this dye due to the increasing demand for it in the manufacture of blue jeans. Here ultrafiltration offers the same advantages as in the treatment of desizing liquors.

4.7. Sewage Treatment

The use of ultrafiltration to handle toilet flushings and sink water directly in apartment buildings is becoming more and more common in Japanese cities where there are restrictions on the volume of waste water that can be discharged into the sewer. Ultrafiltration is coupled with an activated sludge reactor in the basement of the building. The permeate is re-used as toilet flush water in a separate piping network.

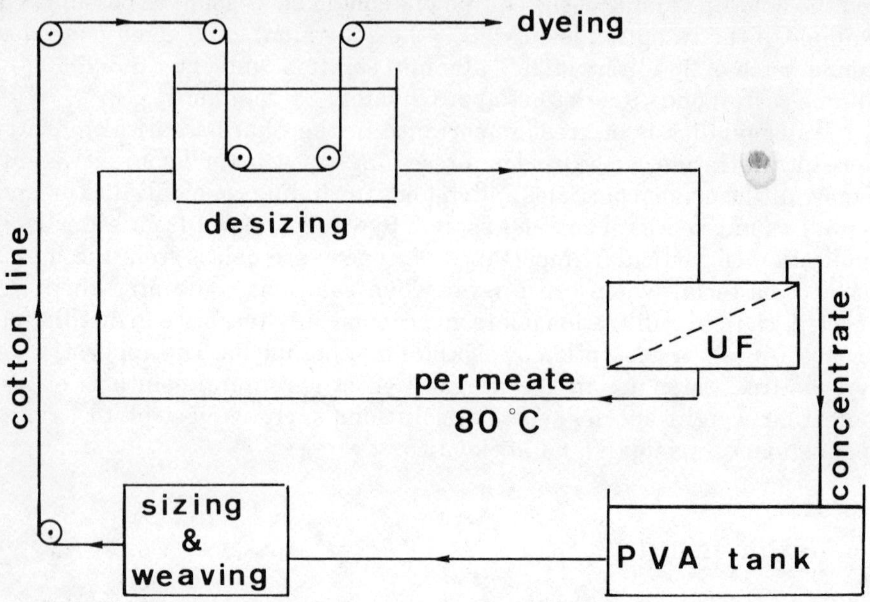

FIGURE 25.—Schematic representation of the use of ultrafiltration to recover sizing chemicals.

4.8. Pure Water Production

High-purity water is required for various industrial uses: rinsing of electronic components, boiler feed water, beverage formulation, etc. Ultrafiltration offers an economic means of removing colloidal materials (iron, silica, manganese), suspended solids or micro-organisms. It is often used as a pretreatment process for reverse osmosis or ion exchange to protect these systems from fouling. Ultrafiltration is also increasingly being applied as pretreatment for desalination.

4.9 Pulp and Paper Industry

The development work on the application of ultrafiltration to the treatment of effluents from the pulp and paper industry began in the early seventies. These industries need large amounts of water, especially for bleaching processes, and this poses some serious problems (BOD, COD, colored effluents). Moreover, some valuable products can be recovered from these wastes: lignin from kraft black liquor, lignosulfonate from spent sulfite liquors. Diafiltration could also be used to purify these raw materials to the extent necessary for the intended application (e.g., adhesives for plywood to replace the more expensive phenol-formaldehyde resins). Figure 26 shows the

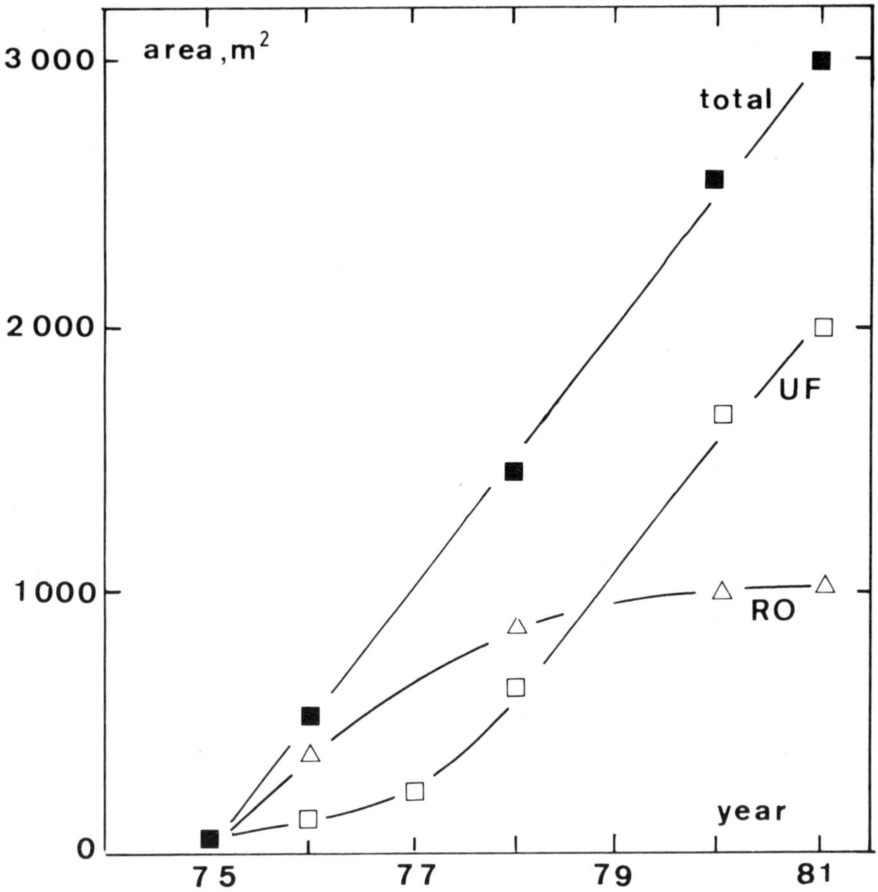

FIGURE 26.—Development of installed membrane area in the pulp and paper industry (data supplied by one manufacturer).

growth of ultrafiltration membrane areas installed by one manufacturer operating in this field.

5. CONCLUSION

Though industrial ultrafiltration is already widely accepted as a full-fledged unit operation, it can be expected that better control of polarization and fouling, together with a reduction of membrane replacement costs, will widen its field of application.

Among its future applications, we have already mentioned the treatment of organic-solvent solutions. In other potential applications, ultrafiltration will be combined with other processes; complexation-ultrafiltration (37, 38) and biospecific ultrafiltration (39), for example, have recently been proposed. This will convert ultrafiltration from being a simple concentration to a process capable of fractionation.

REFERENCES

1. Probstein, R.F., Leung, W.F., Alliance, Y.: 1979, "Determination of diffusivity and gel concentration in macromolecular solutions by ultrafiltration." J. Phys. Chem. 83, pp. 1228-1232.
2. Howell, J.A., Velicangil, O., Le, M.S., Herrera Zeppelin, A.L.: 1981, "Ultrafiltration of protein solutions." Ann. N.Y. Acad. Sci. 369, pp. 355-366.
3. Noelle, N.: 1983, "Comment choisir un module d'ultrafiltration." Usine Nouvelle No. 6, p. 106.
4. Nguyen, Q.T., Aptel, P., Néel, J.: 1979, "Characterization of ultrafiltration membranes. Part I. Water and organic-solvent permeabilities." J. Membrane Sci. 5, pp. 235-251.
5. Nguyen, Q.T., Aptel, P., Néel, J.: 1980, "Characterization of ultrafiltration membranes. Part II. Mass transport measurements for low and high molecular weight synthetic polymers in water solutions." J. Membrane Sci. 7, pp. 141-155.
6. Fane, A.G., Fell, C.J.D., Waters, A.G.: 1981, "The relationship between membrane surface pore characteristics and flux for ultrafiltration membranes." J. Membrane Sci. 9, pp. 245-262.
7. Brun, M., Lallemand, A., Quinson, J.F., Eyraud, C.: 1977, "A new method for the simultaneous determination of the size and shape of pores: thermoporometry." Thermochim. Acta 21, pp. 59-88.
8. Desbrières, J., Rinaudo, M., Brun, M., Quinson, J.F.: 1981, "Relation entre le taux de rétention et la distribution de pores dans une membrane d'ultrafiltration." J. Chim. Phys, 78, pp. 187–191.
9. Brun, M., Quinson, J.F., Eyraud, C.: 1979, "Détermination des rayons de pore par thermoporométrie en milieu liquide." Actualité Chimique 8, pp. 21-26.
10. Colton, C.K., Friedman, S., Wilson, D.E., Lee, R.S.: 1972, "Ultrafiltration of lipoproteins through a synthetic membrane. Implications for the filtration theory of atherogenesis." J. Clin. Invest. 51, pp. 2472-2481.
11. Klein, E., Holland, F.F., Eberle, K.: 1978, "Rejection of solutes by hemofiltration membranes." Trans. Amer. Soc. Artif. Int. Organs 24, p. 662.
12. Baker, R.W., Strathmann, H.: 1970, "Ultrafiltration of macromolecular solutions with high-flux membranes." J. Appl. Polym. Sci. 14, p. 1197.
13. Porter, M.C.: 1977, "What, when, and why of membranes MF, UF, and RO." AIChE Symp. Ser. 73, No. 171, pp. 83-103.
14. Nguyen, Q.T., Néel, J.: 1983, "Characterization of ultrafiltration membranes. Part III. Role of solvent media and conformational changes in ultrafiltration of synthetic polymers." J. Membrane Sci. 14, pp. 97-128.
15. Nguyen, Q.T., Aptel, P., Néel, J.: 1976, "Investigation of the amylose-iodine complexation in aqueous solution by ultrafiltration." Biopolymers 15, pp. 2097-2100.
16. Cooper, A.R., Van Derveer, D.S.: 1979, "Characterization of ultrafiltration membranes by polymer transport measurements." Sep. Sci. Tech. 14, pp. 551-556.
17. Michaels, A.S.: 1980, "Analysis and prediction of sieving curves for ultrafiltration membranes: A universal correlation?" Sep. Sci. Tech. 15, pp. 1305-1322.
18. Horton, B.S.: 1980, "Improvements needed in UF to increase the recycle of water." NTIS PB 81-212342.
19. Dabaji, M.: 1983, "Ultrafiltration: un pas de plus grâce aux membranes minérales." Usine Nouvelle No. 5, pp. 65-67.
20. Spatz, D.D.: 1981, "Multiyear experience with oily and organic chemical waste treatment using reverse osmosis." in Synthetic Membranes Vol. II, A.F. Turbak ed., ACS Symposium Series No. 154, p. 221.

21. Leung, W.F., Probstein, R.F.: 1979, "Low polarization in laminar ultrafiltration of macromolecular solutions." Ind. Eng. Chem. Fundam. 18, pp. 274-278.
22. Doshi, M.R., Dewan, A.K., Gill, W.N.: 1971, "The effect of concentration-dependent viscosity and diffusivity on concentration polarization in reverse osmosis flow systems." AIChE Symp. Ser. 68, No. 124, pp. 323-339.
23. Vilker, V.L., Colton, C.K., Smith, K.A.: 1981, "Concentration polarization in protein ultrafiltration." AIChE J. 27, pp. 632-645.
24. Maubois, J.L.: 1982, in "Protéines Animales." ed. C.M. Bourgeois and P. Leroux, Lavoisier, Paris. pp. 172-190.
25. Hiddink, J., Kloosterboer, D., Bruin, S.: 1980, "Evaluation of static mixers as convection promoters in the ultrafiltration of dairy liquids." Desalination 35, pp. 149-167.
26. Shen, J.J.S., Probstein, R.F.: 1979, "Turbulence promotion and hydrodynamic optimization in an ultrafiltration process." Ind. Eng. Chem. Process Des. Dev. 18, pp. 547-554.
27. Lopez-Leiva, M.: 1980, "Ultrafiltration at low degrees of concentration polarization: technical possibilities." Desalination 35, pp. 115-128.
28. Robertson, G.H., Olieman, J.J., Farkas, D.F.: 1982, "Concentration-polarization reduction in a centrifugally driven membrane separator." AIChE Symp. Ser. 78, No. 218, pp. 129-137.
29. Radovich, J.M., Sparks, R.E.: 1980, in "Ultrafiltration Membranes and Applications." ed. A.R. Cooper, Plenum Press, Palo Alto, pp. 249-268.
30. Breslau, B.R.: 1982, "Ultrafiltration theory and practice." Paper distributed by Romicon, presented at the Corn Refiners' Association Scientific Conference, Lincolnshire, Illinois, 16-18 June 1982.
31. Beaton, N.C., Steadly, H.: 1982, "Industrial ultrafiltration." in Recent Developments in Separation Science, Vol. 7, ed. N.N. Li, CRC Press, Boca Raton, Florida, pp. 1-29.
32. Milnes, B.A.: 1982, "Practical considerations of electrocoat paint ultrafiltration." Paper distributed by Romicon.
33. Gosset, C.: 1981, "Aspects statistiques et économiques de l'industrie du lactosérum." Technique Laitière 952, pp. 13-15.
34. Malige, B.: 1982, in "Protéines Animales." ed. C.M. Bourgeois and P. Leroux, Lavoisier, Paris, pp. 191-201.
35. Maubois, J.L.: 1980, "Ultrafiltration Membranes and Applications." ed. A.R. Cooper, Plenum Press, Palo Alto, pp. 305-318.
36. Friedli, H., Kistler, P.: 1980, ibid., pp. 565-573.
37. Nguyen, Q.T., Aptel, P., Néel, J.: 1980, "Application of ultrafiltration to the concentration and separation of solutes of low molecular weight." J. Membrane Sci. 6, pp. 71-82.
38. Aulas, F., Rumeau, M., Renaud, M., Tyburce, B.: 1980, "Application de l'ultrafiltration à la récupération de cations métalliques en solution." Informations Chimie 204, pp. 145-151.
39. Adamski-Meda, D., Nguyen, Q.T., Dellacherie, E.: 1981, "Bio-specific ultrafiltration: a promising purification technique for proteins?" J. Membrane Sci. 9, pp. 337-342.

SYMBOLS

Values are given in SI units unless otherwise indicated: length (m), mass (kg), time (s), temperature (K), energy (J), pressure (Pa)

b	Einstein–Stokes radius of macromolecule (nm), equation 10
c	concentration (g/L, kg m^{-3})
L_p^*	standard hydraulic permeability (m), equation 2
m	mass of macrosolute (kg)
α	permeability ratio, equation 4
Θ	sieving coefficient
σ_b	standard deviation in distribution of pore radii (nm)
ω	angular velocity, stirrer speed (rad s^{-1})

Subscripts

A	microsolute
f	final product
M	macrosolute
s	organic solvent
w,W	water

APPENDIX: MANUFACTURERS OF ULTRAFILTRATION MEMBRANES AND MODULES

ABCOR, INC.
850 Main Street
Wilmington, MA 01887
USA

AMICON CORP.
21 Hartwell Avenue
Lexington, MA 02173
USA

ASAHI GLASS CORP.
2-1-2, Marumouchi,
Chyoda-ku, Tokyo
JAPAN

BERGHOF Gmbh.
P.O. Box 1523
7400 Tübingen 1
WEST GERMANY

D.D.S., INC.
P.O. Box 149
4900 Nakskov
DENMARK

DESALINATION SYSTEMS, INC.
1107 West Mission Avenue
Escondido, CA 92025
USA

DORR-OLIVER
77 Havemeyer Lane
Stamford, CT 06904
USA

KALLE
Postfach 3540
6200 Wiesbaden 1
WEST GERMANY

MILLIPORE CORP.
Ashby Road
Bedford, MA 01730
USA

NITTO ELECTRIC INDUSTRIAL CO., LTD
1-1-2 Shimohozumi, Ibaraki
Osaka
JAPAN

NUCLEPORE CORP.
7035 Commerce Cir.
Pleasanton, CA 94566
USA

OSMONICS, INC.
5951 Clearwater Drive
Minnetonka, MN 55343
USA

PATERSON CANDY INTERNATIONAL, LTD.
Laverstoke Mill,
Whitchurch, Hampshire RG28 7NR
ENGLAND

RHÔNE-POULENC SPECIALITES CHIMIQUES
18, avenue d'Alsace
92400 Courbevoie
FRANCE

ROMICON, INC.
100 Cummings Park
Woburn, MA 01801
USA

SARTORIUS Gmbh.
Postfach 19
3400 Göttingen
WEST GERMANY

SOCIETE DE FABRICATION D'ELEMENTS CATALYTIQUES (SFEC)
BP 33
84500 Bollène
FRANCE

WAFILIN
P.O. Box 5
7770 AA Hardenberg
THE NETHERLANDS

DAICEL CHEM. IND., LTD
8-1, 3-chome, Kasumigaseki
Chlyoda-ku, Tokyo 100
JAPAN

TEIJIN, LTD
Membrane project
1-1, Uchisaiwai-cho 2-chome,
Chlyoda-ku, Tokyo 100
JAPAN

REVERSE OSMOSIS*

H.K. Lonsdale

Bend Research, Inc.
64550 Research Road
Bend, Oregon 97701

This chapter provides a survey of reverse osmosis science and technology. It begins with a description of the process fundamentals, the various types of membranes used, and the performance of these membranes. Module and system design and typical operation of plants are considered, along with the problems inherent in the operation and the solutions to these problems. The chapter concludes with an account of the principal current reverse osmosis applications.

P. M. Bungay et al. (eds.), Synthetic Membranes: Science, Engineering and Applications, 307–342.

1. A BRIEF HISTORY OF REVERSE OSMOSIS

Detailed histories have been presented previously (1–3). Modern membrane filtration had its beginnings in this century, with Zsigmondy in Germany. Asymmetric or "skinned" membranes were certainly made as early as the 1930's and probably earlier, in Europe and England. This early work led to a small but viable ultrafiltration and microfiltration industry. Ultrafiltration and microfiltration membranes were developed for the laboratory-scale separation, concentration, and purification of biologicals. The first published works on what we now call reverse osmosis date back to about 1930 (4–6). Those early researchers found that when dilute solutions of salt or nonelectrolytes were forced through cellophane or cellulose nitrate membranes, the permeate was less concentrated than was the feed solution. That early work went largely unnoticed, however, and reverse osmosis lay dormant for another generation.

The work that led to current reverse osmosis processing began in the early 1950's when C.E. Reid and his students at the University of Florida (U.S.A.) began to study reverse osmosis. In an unrelated effort at the University of California in Los Angeles (U.S.A) in the late 1950's, Loeb and

Sourirajan made the first high-performance reverse osmosis membrane from cellulose acetate. The U.S. Interior Department's Office of Saline Water (also called the Office of Water Research and Technology) began to support research and development in the reverse osmosis field shortly thereafter. The early work was centered on a search for alternatives to cellulose acetate. The limitations of this material, particularly its hydrolytic instability, were obvious from the beginning. Improved membranes did emerge after several years. Some of these membranes are superior to cellulose acetate and are now commercially available for single-stage seawater desalination.

Much of the more recent research and development effort has been devoted to module development. Two principal types have evolved to capture virtually the entire market: the spiral-wound module and the hollow-fiber module. Spiral-wound modules are somewhat more expensive, whereas hollow-fiber modules foul very easily. Activity continues today in module development to improve these deficiencies.

In recent years, reverse osmosis has become the key method for water desalination, competing successfully with distillation and electrodialysis. Several dozen companies world-wide, both large and small, are now engaged in developing, manufacturing, and distributing reverse osmosis equipment for uses ranging from laboratory and domestic water supplies to large water desalination plants for industrial and municipal applications. Reverse osmosis has also found important applications in food processing and in pollution control.

2. PROCESS FUNDAMENTALS

Reverse osmosis is a process in which pressure is used to reverse the normal osmotic flow of water across a semipermeable membrane. Referring to Figure 1, the normal direction of water flow across a membrane is down its concentration gradient, i.e., from a solution of lower solute concentration to a solution of higher solute concentration. If a pressure (Δp) is applied to the concentrated solution just equal to the osmotic pressure difference between the two solutions ($\Delta\pi$), water flow ceases, and we have osmotic equilibrium. At a higher pressure ($\Delta p > \Delta\pi$), water will flow from the concentrated to the dilute solution. If the membrane is sufficiently semipermeable, this process can be used to desalt the concentrated solution. Most waters treated by reverse osmosis have a substantial osmotic pressure. For seawater, $\pi \simeq 25$ atm, while for most brackish waters $\pi \simeq 1-4$ atm. Even tap water containing 500 ppm total dissolved solids has an osmotic pressure of about 0.4 atm. Furthermore, as water is removed from these solutions, the osmotic pressure increases. A simple relationship between π and salt concentration is the van't Hoff equation:

$$\pi = CRT, \tag{1}$$

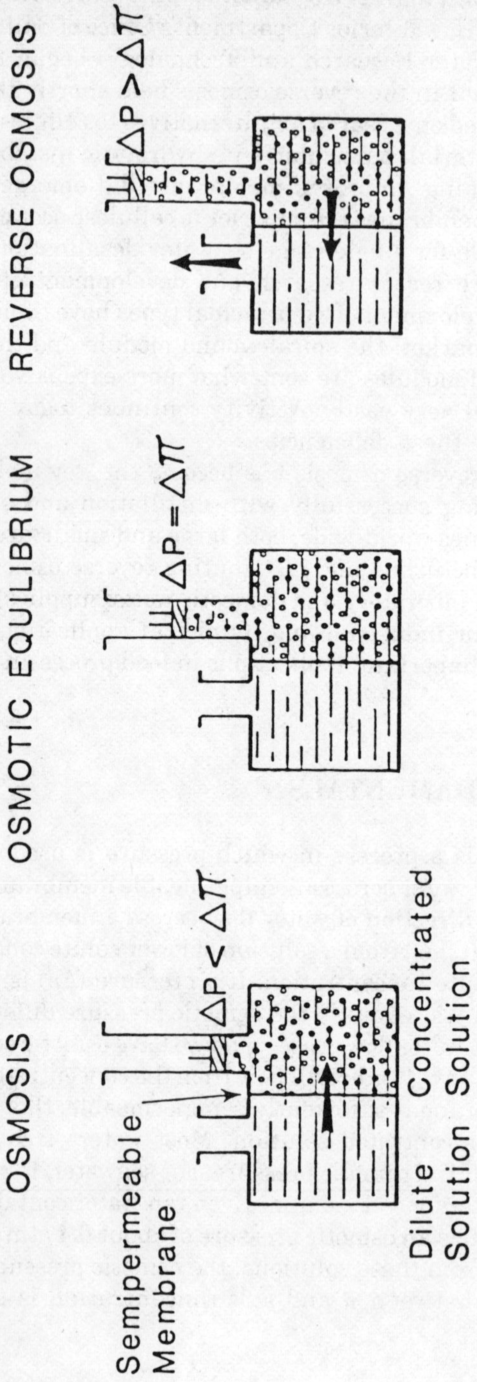

FIGURE 1.—Osmotic processes.

where C is the total number of moles of ions per liter of solution and R and T have their usual meanings. For NaCl solutions,

$$\pi \text{ (atm)} = 8.0 \times \text{wt\% (at 25°C)}. \tag{2}$$

The reversible work for producing pure water from a solution is given by the product of the solution osmotic pressure, π, and the partial molar volume of water, $\tilde{V}_w$. For desalting seawater the reversible work is 2.6 kWh/1000 gal. The actual work of desalting is, of course, considerably higher because the process is carried out irreversibly, at pressures well in excess of the osmotic pressure.

2.1. Water Flux and Solute Retention

The principles of pressure-driven membrane transport have been described in detail in several books (2,3,7), and will not be treated in detail here. The mass flux of water, J_w, through reverse osmosis membranes is given by the expression

$$J_w = \frac{D_w c_w \tilde{V}_w(\Delta p - \Delta\pi)}{RT\Delta x}, \tag{3}$$

where D_w and c_w are the diffusivity and concentration of water dissolved in the membrane, respectively; $\tilde{V}_w$ is the partial molar volume of water in the membrane; Δp and $\Delta\pi$ are the applied pressure difference and the osmotic pressure difference across the membrane, respectively; and Δx is the effective membrane thickness. For a given membrane, many of these quantities are fixed and equation 3 reduces to

$$J_w = A_w(\Delta p - \Delta\pi), \tag{4}$$

where A_w is a constant of the membrane.

Reverse osmosis membranes are also slightly permeable to dissolved salts and other solutes, and the solute molar flux, J_i, is given by:

$$J_i = D_i K_i \frac{c_i' - c_i''}{\Delta x} \tag{5}$$

where D_i is the diffusivity of solute in the membrane; K_i is the distribution coefficient for solute between the membrane and the adjacent solution; and c_i' and c_i'' are the solute concentrations in the feed and permeate solutions, respectively. For a given membrane, equation 5 reduces to

$$J_i = B_i(c_i' - c_i''), \tag{6}$$

where B_i is a solute permeation constant.

Thus, water flux is linear in the "net" pressure, $\Delta p - \Delta\pi$, while solute flux is nearly independent of pressure. The result is that, with increasing pressure, more water passes through the membrane along with a fixed amount of solute; the water is thus purer. The usual measure of solute retention is the "solute rejection," defined as

$$\text{Rejection} = 1 - \frac{c_i'}{c_i''}. \tag{7}$$

Equations 4, 6, and 7 can be combined using $c_i'' = J_i \rho_w / J_w$ to give

$$\text{Rejection} = A_w\,(\Delta p - \Delta\pi)/[A_w(\Delta p - \Delta\pi) + B_i\,\rho_w]. \tag{8}$$

in which ρ_w is the mass density of water. An idealized plot of water flux and solute rejection vs. pressure is presented in Figure 2.

In reverse osmosis the applied pressure is substantial, typically 20–50 atm. This high operating pressure affects the design of both the membrane

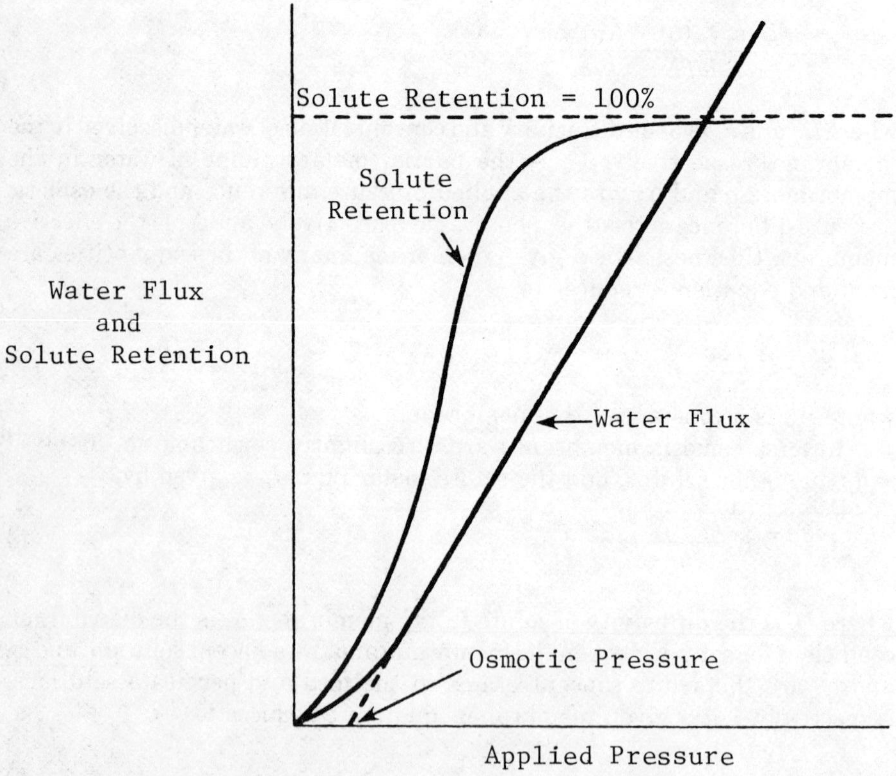

FIGURE 2.—Solute retention and water flux in an idealized reverse osmosis membrane.

module and the remainder of the system. A second conclusion is that the chemical nature of the membrane is important. Reverse osmosis membranes are typically very permeable to water and highly impermeable to most solutes. These permeabilities are determined by the chemistry of the membrane material. The materials from which reverse osmosis membranes are made typically absorb water readily but do not absorb salts or other solutes from aqueous solution. Only a handful of successful reverse osmosis membrane materials are known, although several hundred materials have been evaluated.

It is instructive at this point to compare reverse osmosis with ultrafiltration. In ultrafiltration the solute retention mechanism is different. Ultrafiltration membranes are essentially very fine sieves. Those solutes with a size substantially smaller than the pore diameter of the membrane will pass through readily, whereas those substantially larger will be completely retained. For those materials whose size is comparable to the pore size, the situation is more complex. In general, solute rejection is independent of applied pressure. Because the materials retained by ultrafiltration membranes are of relatively high molecular weights, the osmotic pressure difference across the membrane is generally negligible. Notwithstanding these differences, the water flux realized in reverse osmosis processing is comparable to that in ultrafiltration, i.e.: 10–30 gal/ft^2-day or 0.4–1.2 m^3/m^2-day (1 m^3/m^2-day = 24.6 gal/ft^2-day). A summary of the differences between reverse osmosis and ultrafiltration is presented in Table 1.

TABLE 1. Differences between reverse osmosis and ultrafiltration (

	Reverse Osmosis	*Ultrafiltration*
Size of solute retained	Molecular weights generally less than 300	Molecular weights generally greater than 1000
Osmotic pressure of feed solution	Important, can reach several hundred psi	Generally negligible
Operating pressure	To 1000 psi (68 atm)	10 to 100 psi (0.7–7 atm)
Nature of membrane retention	Solution–diffusion transport	Molecular screening
Chemical nature of membrane	Important in affecting transport properties	Unimportant in affecting transport properties if proper pore size distribution is obtained

2.2. Concentration Polarization

In reverse osmosis, the flow of solvent through the membrane carries solutes with it to the membrane surface, where they are typically rejected. These solutes tend to diffuse back into the feed solution, and at the steady state the amount transported to the membrane is just balanced by the diffusive flux away from the membrane:

$$J_V\, c_i' = D_i' \mathrm{d}c_i'/\mathrm{d}x, \tag{9}$$

where J_V is the transmembrane volumetric flux approximated by J_w/ρ_w, c_i' denotes the solute concentration in the feed solution, D_i' is the diffusion coefficient of the solute in the feed solution, and x denotes the direction normal to the membrane surface. The solution to this equation is

$$c'_{i,\mathrm{int}}/c'_{i,\mathrm{bulk}} = \exp\,(J_V\, \delta/D_i'), \tag{10}$$

where $c'_{i,\mathrm{int}}/c'_{i,\mathrm{bulk}}$ is the ratio of the concentration of the solute at the membrane-solution interface to that in the bulk solution, and δ is the thickness of the "boundary layer" or unstirred layer adjacent to the membrane surface. This boundary layer thickness varies approximately inversely with the mean velocity of the feed solution parallel to the membrane surface. In reverse osmosis with most modular configurations (i.e., in spiral-wound, tubular, and plate-and-frame systems), $c'_{i,\mathrm{int}}/c'_{i,\mathrm{bulk}}$ is typically maintained at values of 1.0–1.5. In hollow-fiber systems, $c'_{i,\mathrm{int}}/c'_{i,\mathrm{bulk}}$ is usually larger because of the system design, but is not directly measurable. Note that the polarization effect is enhanced by high transmembrane flux J_V, by a large boundary layer thickness δ, and by a low solute diffusivity D_i'—i.e., by high-molecular-weight solutes. The polarization effect has several negative consequences: first, the local solute concentration at the membrane surface is increased, thus promoting solute flux through the membrane, or lower solute rejection. Second, the osmotic pressure at the membrane–solution interface is increased, thus reducing the water flux, in accordance with equation 4. Finally, if the local concentration of solute becomes too high, the solute can precipitate or gel, thus fouling the membrane surface. These phenomena are illustrated schematically in Figure 3. This polarization problem is handled in reverse osmosis by maintaining a high fluid flow rate parallel to the membrane surface. This tends to reduce the thickness of the boundary layer.

3. MEMBRANE TYPES AND PREPARATION

Reverse osmosis membranes have a basically porous structure with a thin, dense skin on one surface which forms the solute-rejecting barrier. The several membrane types differ in the manner in which the pores and skin are formed. A detailed accounting of the formation mechanism of most types of so-

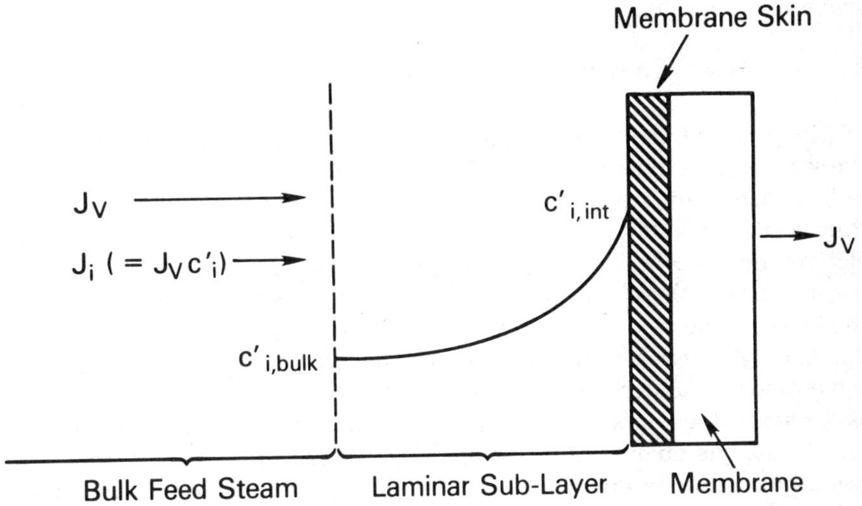

FIGURE 3.—Concentration polarization in reverse osmosis.

called asymmetric membranes has been given by Kesting (9), and we will present here only a short summary.

3.1. Early Microporous Membranes

The microporous membranes made by researchers in the first half of this century were prepared by dissolving a non-water-soluble polymer in an organic solvent—a typical example is cellulose nitrate dissolved in a mixture of ether and alcohol and called “”collodion”—and then casting a film with a “doctor blade.” The film was then either allowed to dry in air or it was immersed in water to precipitate the polymer. These membranes thus underwent a transition from a single phase to a two-phase system in which the polymer forms one continuous phase and the imbibed water or solvent–water mixture forms a second continuous phase. This second phase forms the membrane pores.

When solutions of polymers in volatile solvents are cast into films in air, non-homogeneity of the resulting membrane structure is the rule rather than the exception. If the as-cast membrane is subsequently immersed in water or other nonsolvent for the polymer, gross asymmetry results. Thus, virtually any polymer can be made into an asymmetric microporous membrane. Adjusting the pore size and the membrane microstructure is an operation that remains highly empirical but which is also highly developed by the membrane manufacturers.

3.2. The Loeb–Sourirajan Membrane

In principle, the Loeb–Sourirajan cellulose acetate membrane is made in the same way as those described above (see, for example, References 2, 3, and 9). The membrane itself, however, is distinctly different in that it possesses a thin skin on the air-dried surface. This skin is difficult to detect, even in the electron microscope. Its density can be increased and the solute rejection of the membrane thereby improved by annealing the membrane in water. The thickness of the skin is obtained only inferentially, by comparing the water flux through a Loeb–Sourirajan membrane with that of a dense film of the same material under the same operating conditions. On this basis, typical skin thicknesses appear to be about 0.2 μm, while a typical commercial membrane has an overall thickness of about 100 μm. The Loeb–Sourirajan technique, or variations thereon, has been used to prepare skinned membranes in flat-sheet, tubular, and hollow-fiber forms.

The Loeb–Sourirajan membrane is thus effectively extremely thin, and with this thinness goes very high water flux. At the time of its discovery, this was easily the thinnest free-standing membrane that could be fabricated without imperfections by any known technique. This effective thinness is exceedingly important in reverse osmosis membranes.

3.3. Composite Membranes (10, 11)

These represent a relatively new approach to making effectively very thin membranes. The concept is as follows. A finely microporous substrate is first prepared by the method described in Section 3.1., above. A thin solute-rejecting film, made from a material known to have a high permeability to water and a low permeability to salts and other solutes, is then applied. These thin films are applied either by a film-casting technique such as dip-coating or by interfacial polymerization—as in the case of the FilmTec FT–30 membrane and the Universal Oil Products (UOP) PA–300 membrane. (A listing of the manufacturers of reverse osmosis membranes and modules is presented in the Appendix.) The steps used in preparing one such membrane are shown diagrammatically in Figure 4.

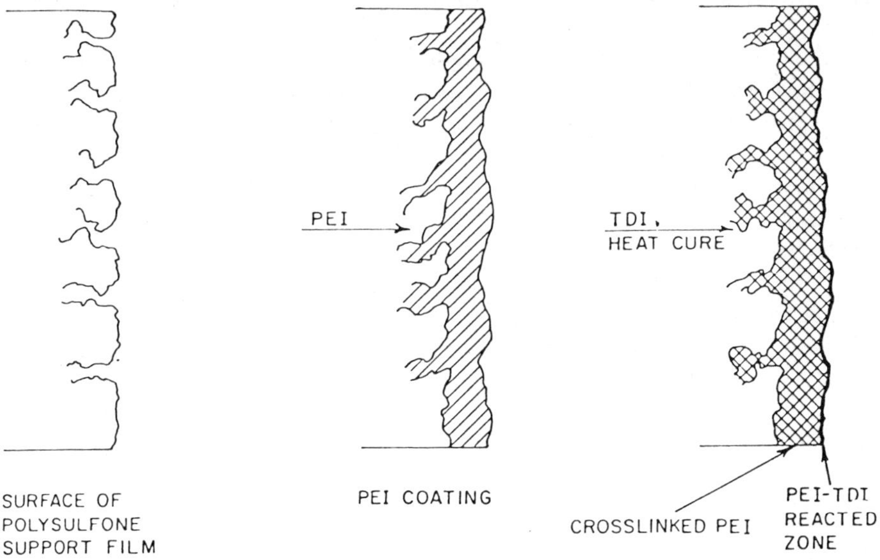

FIGURE 4.—Representation of the formation of the NS–100 membrane (13) (PEI = polyethyleneimine, TDI = toluene diisocyanate).

The advantages of the composite membrane approach are: 1) a material can be used for the thin film that could not be readily prepared into asymmetric membrane form by other techniques such as the Loeb–Sourirajan process; 2) the thickness of the thin film can be reproducibly controlled; and 3) rugged substrate materials can be used, which are less subject to the "compaction" phenomenon usually observed with reverse osmosis membranes. The successful production of these membranes requires a very finely porous substrate to which the thin film adheres well. For adequate film support the pores in the substrate must be smaller in diameter than the thickness of the film to be applied, and the pores must also be rather close together to enable the water permeating the thin film to pass through the substrate unimpeded (12). Typical thin-film thicknesses are 0.05–0.1 μm.

Reverse osmosis membranes of this type have recently been introduced commercially by UOP and by FilmTec. They appear to exhibit superior performance to earlier reverse osmosis membranes on all counts except their resistance to free Cl_2 in solution.

3.4. Dynamically Formed Membranes

The conceptualization and early development of another approach to making high-performance membranes was carried out at Oak Ridge National Laboratory (U.S.A.) (14). In this method, a membrane is formed *in situ* on a finely porous substrate from colloidal material present in a circulating, pressurized feed solution. Typical substrates are finely porous carbon, ceramic, or metallic tubes. Membranes are prepared from hydrous zirconium oxide colloids and polyelectrolytes such as polyacrylic acid. Actually, salt-rejecting membranes have been dynamically formed from a broad spectrum of materials, including the suspended matter already present in sewage effluent, laundry wastes, pulp mill wastes, and so on. Relative to conventional reverse osmosis membranes, these dynamically formed membranes exhibit higher water fluxes (frequently an order of magnitude higher), but lower salt rejections and poor reproducibility.

3.5. Other Membrane Types

Several other approaches have been taken in recent years to making very thin and very permselective membranes. Most notable of these are the "plasma polymerized" membranes (15, 16). Plasma polymerized membranes are prepared by depositing polymers, formed in the vapor state by a glow discharge through a monomer gas, onto a suitable substrate. The deposited film is dense and can be made quite thin. These membranes have exhibited high performance on occasion, but they have not yet been commercialized.

To summarize, all commercial reverse osmosis membranes have an essentially microporous structure with a thin solute-rejecting skin on one surface. Several distinct methods for producing reverse osmosis membranes have been developed. Photomicrographs or sketches of cross sections of the important membrane types are represented in Figures 5 and 6.

3.6. Membrane Configurations

Most of the membranes of interest here can be prepared in either flat-sheet, tubular, or hollow-fiber form. The methods for producing these membranes on a laboratory scale have been described in the literature. Manufacturing methods for flat-sheet membranes have been described in the patent literature (17), while equipment designs for the production of tubular membranes (18), hollow fibers (19), and composite membranes (20) have also been described. The original literature should be consulted for details. Key drawings from these sources are reproduced in Figures 7 to 9. The way in which these membranes are utilized in reverse osmosis modules is described in Section 5, below.

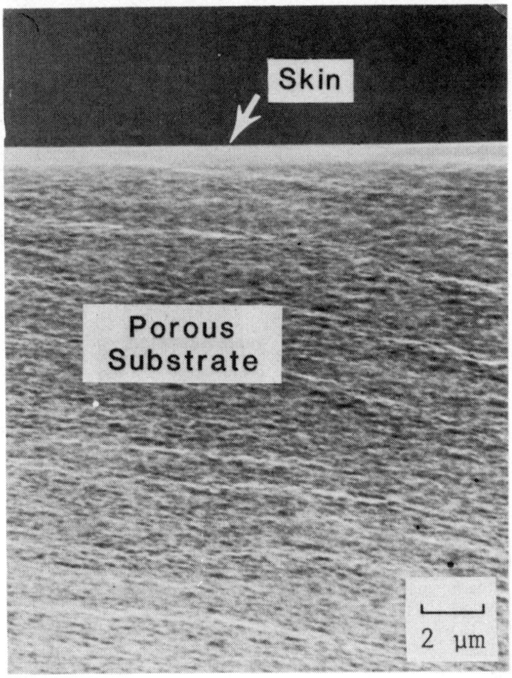

Electron photomicrograph of the cross section of a Leob–Sourirajan cellulose acetate membrane. (Taken from reference 3, p. 137.)

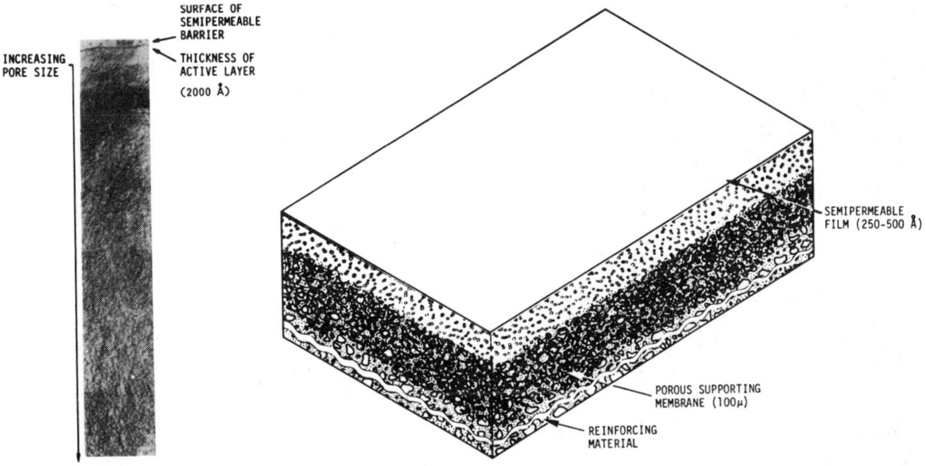

A cross-sectional drawing of a composite membrane. (Taken from reference 20, p. 376.)

FIGURE 5.—Membrane cross sections.

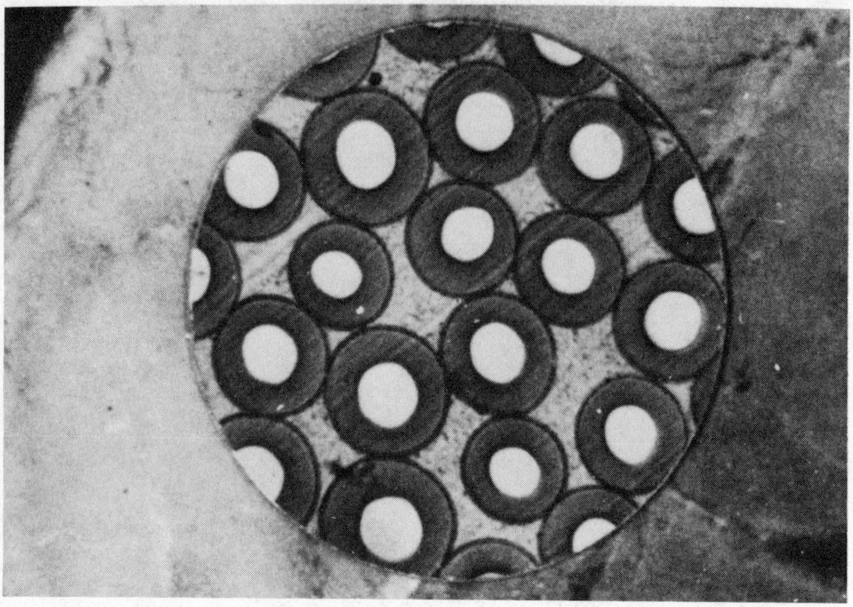

A. Photomicrograph of polyamide hollow fibers (330× magnification).

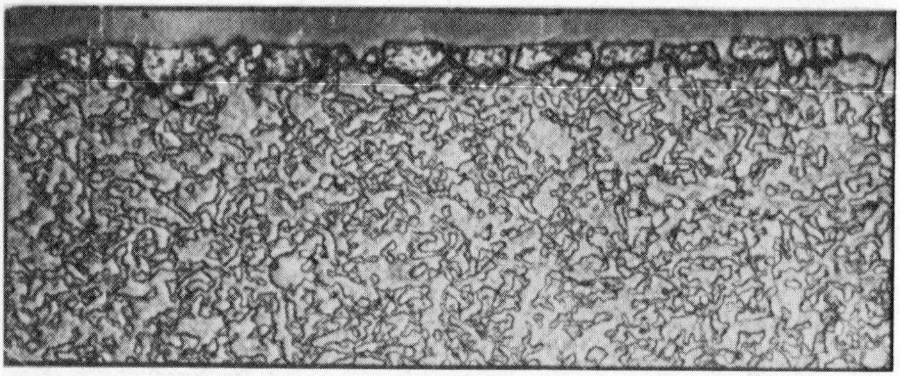

B. Photomicrograph of the cross section of a ceramic tube on which a hydrous Zr (IV) oxide membrane has been dynamically formed. The membrane is shown at the upper surface of the support tube. Photograph courtesy of Dr. J.S. Johnson, Jr., ORNL.

FIGURE 6.—Membrane cross sections.

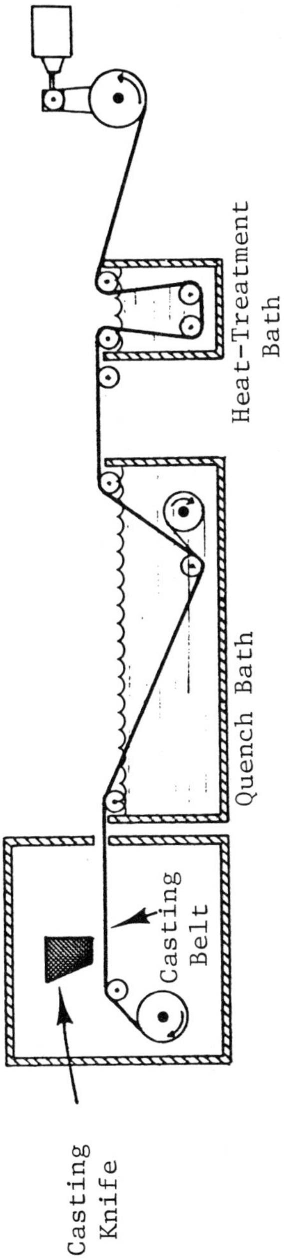

FIGURE 7.—Schematic drawing of a flat-sheet casting machine, from reference 17.

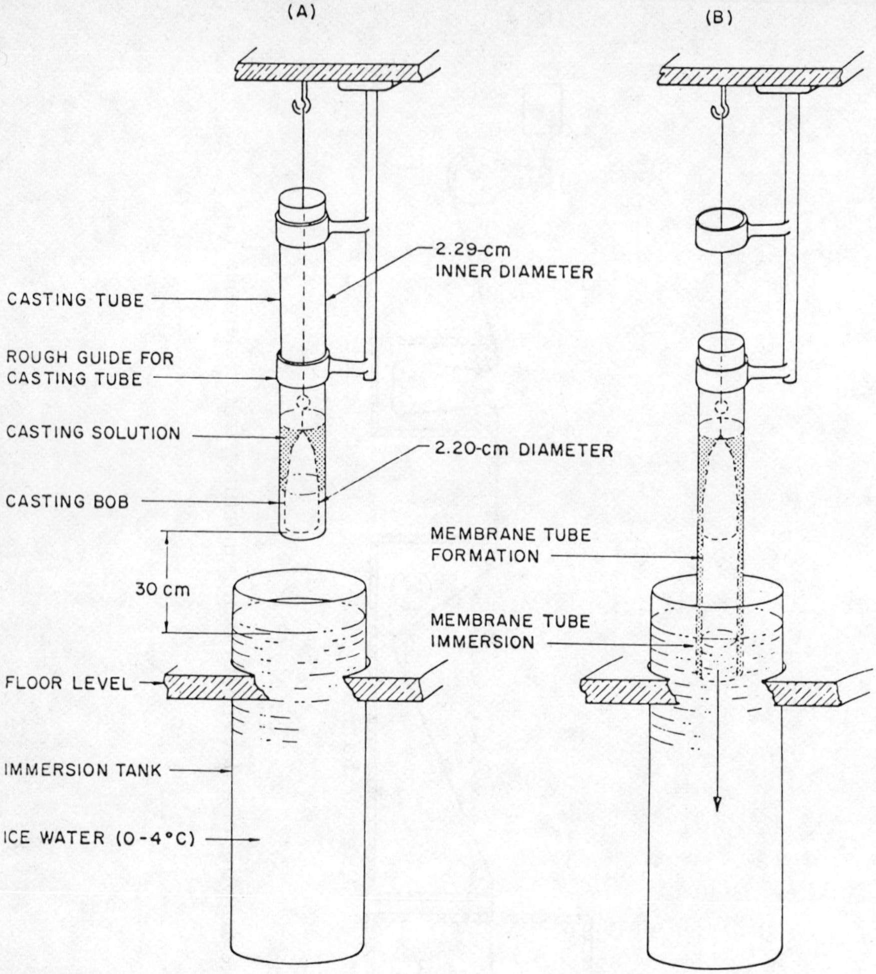

FIGURE 8.—Tubular membrane casting.

4. MEMBRANE AND MODULE PERFORMANCE

The literature on reverse osmosis membrane performance is extensive. We will attempt here only to describe general behavior, and to provide references where more detailed information can be found. There are three commercially important types of reverse osmosis membranes: the Loeb–Sourirajan cellulose acetate membrane, the aromatic polyamide hollow-fiber membrane, and composite membranes.

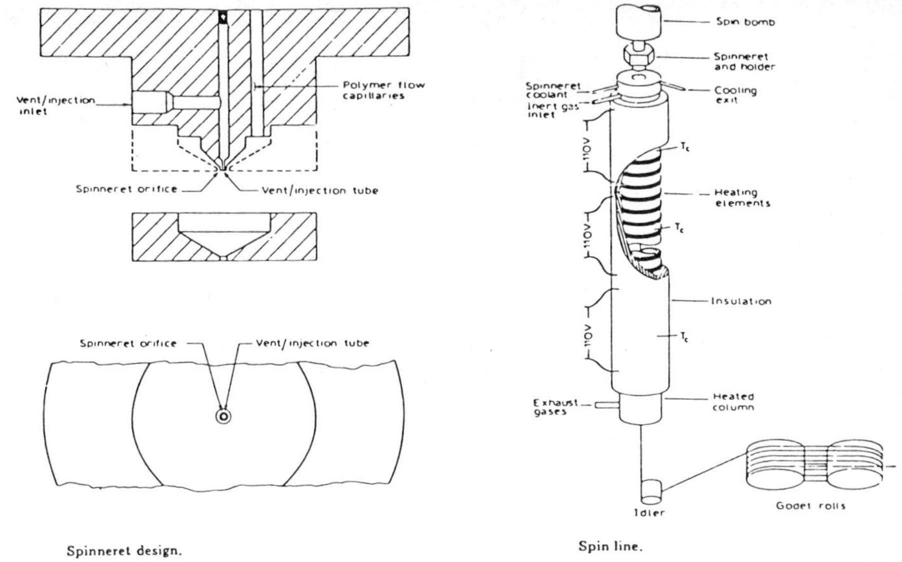

FIGURE 9.—Hollow-fiber spinning, showing a spinneret at the left and a melt spinning line at the right.

4.1. Cellulose Acetate Membranes

These membranes are made in three forms: flat sheets, which are usually assembled into spiral-wound membrane modules; relatively large-diameter tubes (10–25 mm diameter); and hollow fibers. The performance of flat-sheet membranes is comparable to that of tubular membranes, but hollow fibers generally exhibit lower water flux.

The properties of all these membranes depend stongly on the temperature at which the membranes are annealed in water before testing. A plot of water flux and NaCl rejection at 800 psi (54 atm) for flat-sheet or tubular membranes is presented in Figure 10. The data are shown as wide bands because performance varies considerably from supplier to supplier and even from batch to batch. We point out that these are the performance data for the membranes themselves and not for membrane modules. When the membranes are incorporated into modular form the salt rejection usually decreases slightly because of leaks, which commonly occur at the seals.

The characteristic dependence of cellulose acetate membrane performance on applied pressure is presented in Figure 11. A family of curves is obtained for the membrane parameters A_w and B_i at different annealing temperatures. The water flux can be calculated at any condition of applied pressure and osmotic pressure difference using the values of A_w from Figure 11 and equation 4; and the salt rejection can be calculated from B_i and equation 8.

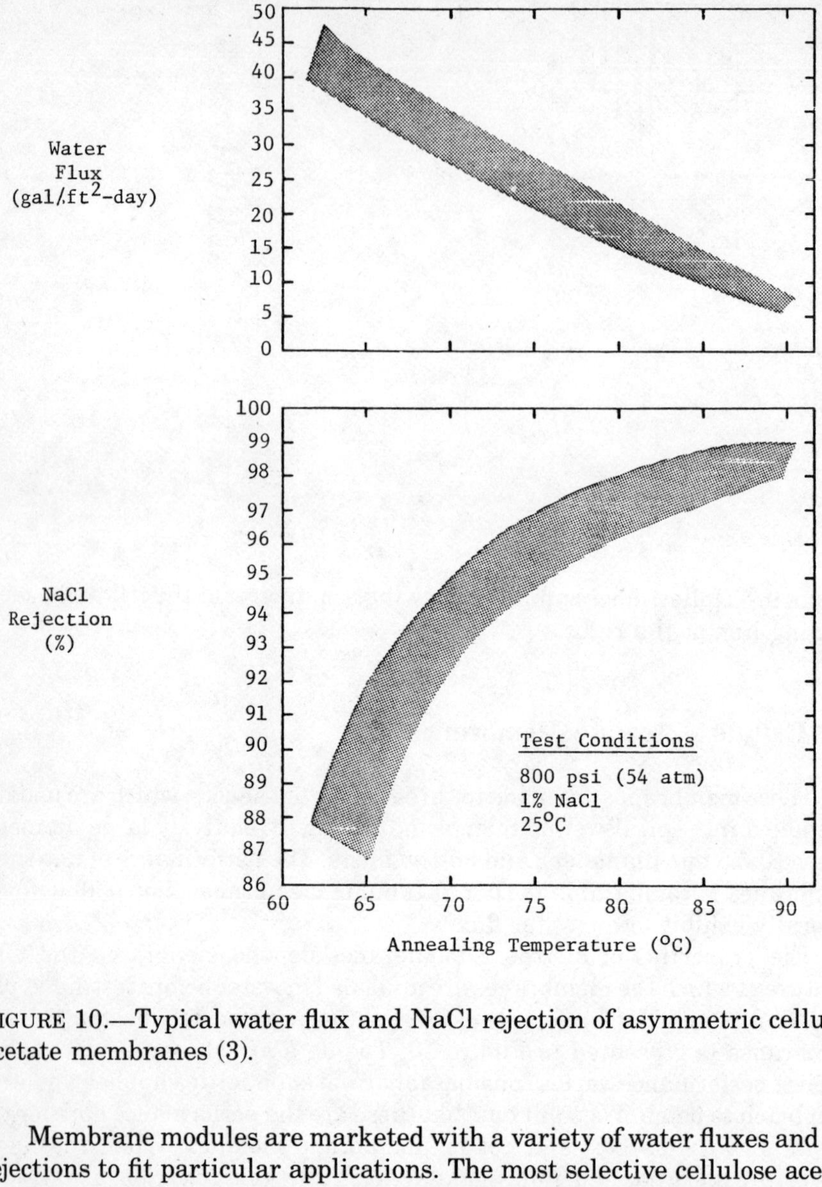

FIGURE 10.—Typical water flux and NaCl rejection of asymmetric cellulose acetate membranes (3).

Membrane modules are marketed with a variety of water fluxes and salt rejections to fit particular applications. The most selective cellulose acetate membrane modules reject an average of 97% NaCl. Note that the observed salt rejection will depend on the fraction of feed water that passes through the membrane, i.e., on the water recovery. As recovery is increased, the feed water salt concentration also increases, and the rejection, based on feed concentration, falls even though the membrane performance *per se* remains unchanged.

Rejections of a variety of salts and non-electrolytes are presented in Table 2. These rejection data refer to spiral-wound modules, and were taken from Osmonics' literature. Still other flux and rejection data for flat-sheet reverse

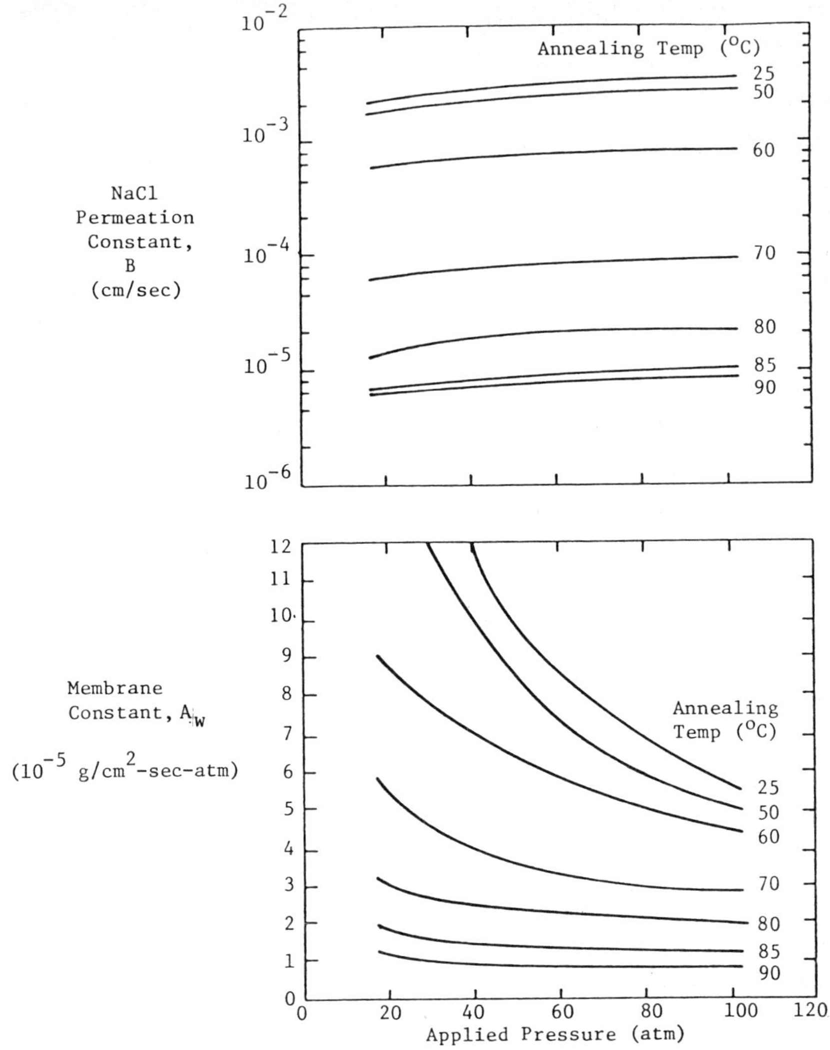

FIGURE 11.—Pressure dependence of salt permeation constant for NaCl and of membrane constant. Loeb–Sourirajan membranes tested at 25°C (3).

osmosis membranes can be found in references 2 and 3, and other references found therein.

Cellulose acetate membranes are known to "compact," i.e., the water flux declines with time, even when the membranes are not fouled. This appears to be an inherent property of all reverse osmosis membranes. It is accentuated by increasing applied pressure. It is customary to depict this phenomenon in terms of a log-log plot of flux against time. A typical plot is presented in

TABLE 2. Typical cellulose acetate membrane rejections at 400 psi (27 atm) and room temperature

SALTS	
CATIONS	Rejection (%)
Sodium	94–96
Calcium	96–98
Magnesium	96–98
Potassium	94–96
Iron	98–99
Manganese	98–99
Aluminum	99+
Ammonium	88–95
Copper	98–99
Nickel	98–99
Zinc	98–99
Strontium	96–99
Cadmium	96–98
Silver	94–96
Mercury	96–98
Hardness	96–98
ANIONS	
Chloride	94–95
Bicarbonate	95–96
Sulfate	99+
Nitrate	93–96
Fluoride	94–96
Silicate	95–97
Phosphate	99+
Bromide	94–96
Borate	35–70*
Chromate	90–98
Cyanide	90–95*
Sulfite	98–99
Thiosulfate	99+
Ferrocyanide	99+

*Dependent on pH

ORGANICS	Molecular Weight	Rejection (%)
Sucrose	342	100
Lactose	342	100
Proteins	≥ 10,000	100
Glucose	180	99.9
Phenol	94	**
Acetic acid	60	**
Formaldehyde	30	**
Dyes	400 to 900	100
Biochemical oxygen demand		90–99
Chemical oxygen demand		80–95
Urea	60	40–60
Bacteria & viruses	50,000–500,000	100
Pyrogens	1000–5000	100

**Permeate is enriched in the solute.

DISSOLVED GASES	
Carbon dioxide (CO_2)	30–50%
Oxygen (O_2)	**
Chlorine (Cl_2)	30–70%

**Permeate is enriched in the solute.

Figure 12. The slope of the lines in such a plot is generally referred to as the "compaction slope." We present below typical values for the slope at three applied pressures, and the effect of compaction slope on the long-term flux through the membrane.

Applied Pressure (psi)	Typical Compaction Slope	$(Flux)_{1\text{-Year}}/(Flux)_{1\text{-Day}}$
500 (34 atm)	−0.03	0.84
1000 (68 atm)	−0.06	0.74
1500 (102 atm)	−0.09	0.59

Another important measure of membrane performance is its long-term stability. Cellulose acetate membranes are subject to hydrolytic attack at extremes of pH, and all manufacturers recommend continual use of their products between the pH limits of 4.0 and 6.0. Brief exposures to extreme pH can be tolerated, and some recommended membrane-cleaning methods require treatment with dilute acids. Cellulose acetate membranes are also somewhat sensitive to free chlorine. It is recommended that ≤ 1.0 ppm free Cl_2 residual be present for long-term use. Short exposure to 10 ppm Cl_2 is also acceptable.

In general, the chemical sensitivity of cellulose acetate membranes is not a serious limitation. pH control is practiced to minimize hydrolytic degrada-

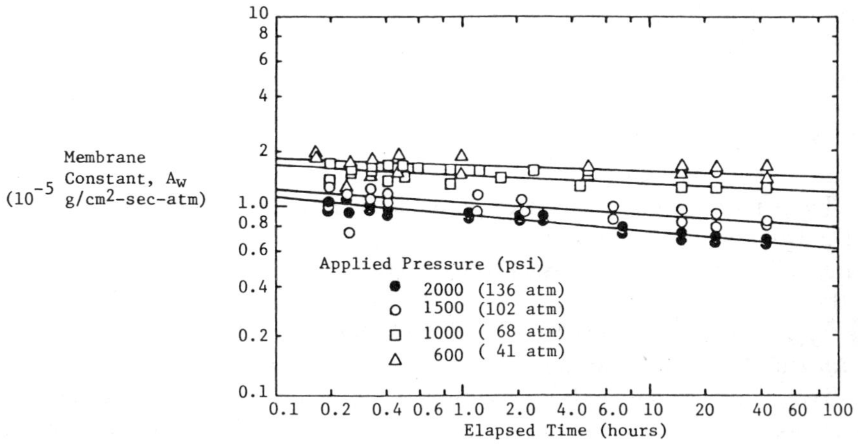

FIGURE 12.—Change in membrane constant, A_w, with time for Loeb–Sourirajan cellulose acetate membranes annealed at 80°C (3).

tion, but pH adjustment is generally recommended in reverse osmosis processing in any case, to minimize membrane fouling caused by precipitation of $CaCO_3$ from solution, for example. Similarly, the Cl_2 sensitivity is rarely a problem. Waters to be treated rarely contain sufficient Cl_2 to be of concern. Sterilization with Cl_2 can also be tolerated, but care must be exercised to minimize exposure. There have been isolated reports of microbial attack on cellulose acetate membranes, but this is generally not a problem. With reasonable care, a membrane lifetime of 2–3 years is usually achieved. The combined effects of hydrolysis and compaction on the long-term performance of cellulose acetate membranes are presented in Table 3.

TABLE 3. Long-term performance* of cellulose acetate reverse osmosis modules (23)

	Time (hr)		
Performance characteristic	1	10,000	25,000
Flux (gal/ft^2-day)	16	12	11
Product quality (ppm)	90	150	250

*At 400 psi (27 atm), 25°C, with 2000 ppm NaCl feed and 50% water recovery

Operating temperature, of course, has an effect on membrane performance. There is about a 3% increase in flux for every 1°C temperature rise near room temperature, and salt rejection declines slightly with increasing temperature.

Cellulose acetate membranes are also produced commercially in hollow-fiber form, both by Dow (21) and Toyobo (22). Little has been reported about inherent transport properties, but salt rejections appear to be comparable to those of flat-sheet membranes, although water fluxes are apparently lower. For further details on membrane performance, the manufacturers' literature should be consulted.

4.2. Aromatic Polyamide Membranes

These membranes are currently sold only in the form of hollow fine fibers by the DuPont Co. through its many distributors. These fibers are typically 80 μm O.D. and 40 μm I.D. (see Fig. 6A). The water flux is not stated, but it is on the order of 1–3 gal/ft^2-day (0.04–0.12 m^3/m^2-day), depending upon the fiber type and operating pressure, and the water flux and salt rejection depend on Δp and $\Delta\pi$ in the manner indicated in equations 4 and 8 and Figure 2.

Two types of hollow-fiber membrane modules are available—the so-called B–9 and B–10 permeators. The latter exhibits higher salt rejection and is sold for applications where high water recovery or high rejection is required, as in single-stage seawater desalination. Typical solute rejection characteristics of a B–9 permeator are presented in Table 4. (The "solute

TABLE 4. Solute passage* for DuPont B–9 permeator

Solute	Passage (%)
Calcium	4
Magnesium	4
Sodium	10
Potassium	10
Phosphate	2
Sulfate	4
Chloride	10
Nitrate	15
Silica	15
Carbon dioxide	100

*At 400 psi (27 atm) with 75% recovery of 1500 ppm feed solution

passage" referred to in the table is 100% − Rejection (%), as previously defined.)

Compaction also occurs with these hollow-fiber membranes, and the flux at one year is about 83% of the initial flux (at 25°C and 400 psi (27 atm) applied pressure, without fouling). This is comparable to the flux decline exhibited by cellulose acetate flat-sheet membranes. The hydrolytic stability of these membranes is superior to that of cellulose acetate, and the recommended pH range is 4–11. However, these membranes are considerably more susceptible to attack by free chlorine. For long-term use, a free chlorine residual below 0.1 ppm is recommended.

Additional information on the DuPont membranes and modules is contained in the "Engineering Design Manual" referred to in the Bibliography.

4.3. Composite Membranes

Two composite membranes are now commercially available. The first of these is the PA–300 membrane from Fluid Systems Div. of UOP (24). It is a poly(ether–amide) membrane which, in spiral-wound module form, exhibits a water flux comparable to that of cellulose acetate membranes, but with 98–98.5% rejection of NaCl at 400 psi (27 atm). One membrane of this type is now being used for single-pass desalination of seawater at 800 psi (54 atm) (25). The membrane is quite sensitive to free chlorine, which must be completely absent from the feed water. An indicator of the superior solute rejection of this membrane can be obtained from Table 5.

More recently, FilmTec Corp. introduced a membrane denoted FT–30, an interfacially polymerized membrane made from m-phenylene diamine and trimesoyl chloride (26). These membranes are somewhat less sensitive to Cl_2 than the PA–300 membrane, and single-pass seawater desalination can be achieved (27).

TABLE 5. Solute rejections of UOP PA–300 thin-film composite membranes (24)

Solute	Concentration (ppm)	pH	Rejection (%)
Sodium nitrate	10,000	6.0	99.0
Ammonium nitrate	9,600	5.7	98.1
Boric acid	280	4.8	65–70
Urea	1,250	4.9	80–85
Phenol	100	4.9	93
Phenol	100	12.0	>99
Ethyl alcohol	700	4.7	90
Glycine	1,400	5.6	99.7
DL-Aspartic acid	1,500	3.2	98.3
Ethyl acetate	366	6.0	95.3
Methyl ethyl ketone	465	5.2	94
Acetic acid	190	3.8	65–70
Acetonitrile	425	6.3	>25
Acetaldehyde	660	5.8	70–75
Dimethyl phthalate	37	6.2	>95
2,4-Dichlorophenoxy acetic acid	130	3.3	>98.5
Citric acid	10,000	2.6	99.9
Alcozyme (soap)	2,000	9.3	99.3
O–Phenyl phenol	110	6.5	>99
Tetrachloroethylene	104	5.9	>93
Sodium silicate	42	8.6	>96
Sodium chromate	1,200	7.8	>99
Chromic acid	870	3.9	90–95
Cupric chloride	1,000	5.0	99.2
Zinc chloride	1,000	5.2	99.3
Trichlorobenzene	100	6.2	>99
Butyl benzoate	220	5.8	99.3

Conditions: 1000 psi (68 atm) applied pressure, 25°C

Other interfacially polymerized composite membranes are under development by U.S. and Japanese firms. One of these, which has only recently become commercially available, is the PEC–1000 membrane developed for single-pass seawater desalination by Toray (28). Salt rejections are reported to be 99.5%, but the water flux is apparently lower than that exhibited by the other composite membranes.

4.4. Other

There are still other reverse osmosis membranes now being marketed. Little has been reported about their chemistry, however. Teijin has recently introduced a polybenzimidazolone (PBIL) membrane, with superior solute-rejection capability, as noted in Table 6. It is said to be resistant to pH extremes of 1–12.

TABLE 6. Solute rejections* of PBIL membrane (29)

Inorganics	Rejection (%)	Organics	Rejection (%)
$NaNO_3$	98.0	Acetic acid	40
Na_2SO_4	99.7	Propionic acid	67
Na_3PO_4	99.8	Sulfosalicylic acid	98
LiCl	97.8	Oxalic acid	90
$CaCl_2$	97	Succinic acid	95
$Ca(NO_3)_2$	96	Tartaric acid	99
$MgCl_2$	96	Citric acid	99
KCN	98.0	Ethanol	61
KF	99.3	Isopropanol	72
CuCN	99.6	Ethylene glycol	80
$CuSO_4$	99.7	Glycerin	97
$Zn(CN)_2$	99.9	Methylethyl ketone	77
$ZnSO_4$	99.9	Acetaldehyde	50
$NiSO_4$	99.6	Triethylamine	87
$NiCl_2$	99.5	Ethylenediamine	83
$SnSO_4$	99.7	Glucose	99
$HgSO_4$	99.7	Lactose	99
$MnSO_4$	99.8	Sucrose	99
$CdSO_4$	99.8	Sorbitol	99
$Al(NO_3)_3$	99.5	Xylose	99
$Al_2(SO_4)_3$	99.8	Glycine	99
CrO_3	98.5	DMF	82
$Na_3[Au(SO_3)_2]$	99.6	DMAc	89
$KAu(CN)_2$	75	NMP	78
$KAg(CN)_2$	90	Urea	65
NH_4Cl	92		
H_2SO_4	99.3		
H_3BO_3	72		

*For 1% aqueous solutions at 50 atm

5. SYSTEM DESIGNS

In reverse osmosis it is vital to control concentration polarization and membrane fouling. It is also necessary to support the membrane against a substantial applied pressure. These requirements affect both the design of modules and the layout of reverse osmosis plants.

5.1. Module Types

There are four designs in use: spiral-wound, hollow-fiber, tubular, and plate-and-frame modules. The first two designs have by far the largest portion of the market. Schematic diagrams of these four configurations are presented in Figure 13. All of these module designs require some type of pressure vessel around the membranes, which is not always shown in the figure.

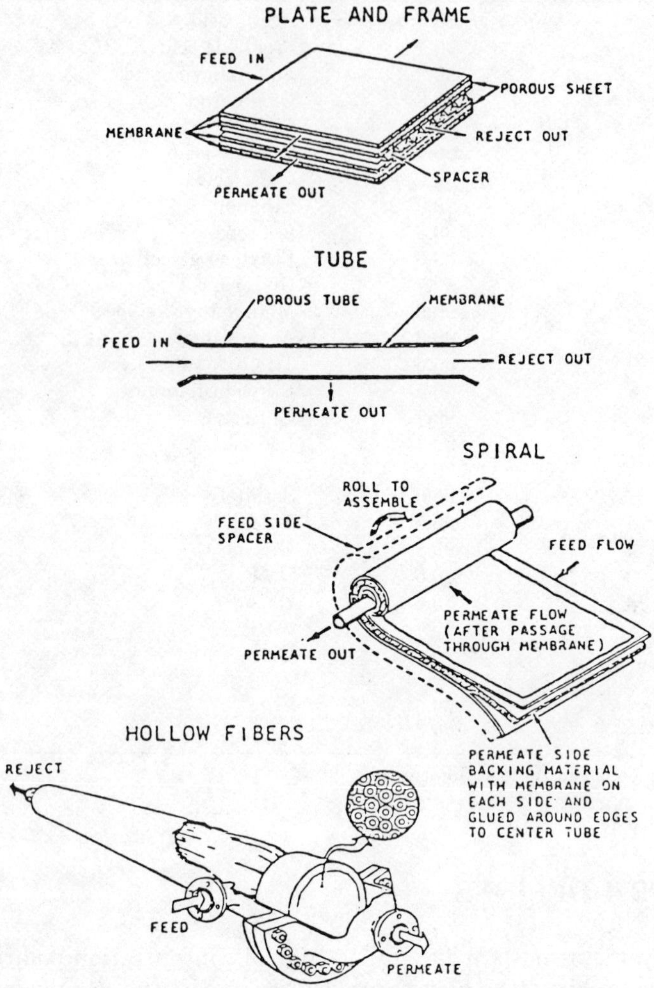

FIGURE 13.—Schematic representations of the four principal membrane module designs.

A good comparison of the relative merits of these module designs is presented in Table 7. As a general rule, hollow-fiber modules are used only on waters that are free of virtually all suspended and colloidal matter, because of the ease with which they foul and the difficulty of cleaning the fibers once they are fouled. Thus, surface waters must be extensively pretreated before being fed to a hollow-fiber unit. Spiral-wound modules have a broader range of

TABLE 7. Reverse osmosis membrane module comparison (30)

Characteristic	Hollow-Fiber	Spiral-Wound	Tubular and Plate-and-Frame
Flux and rejection range	High	High	High
Range of pH tolerance	4–10	4–7	4–7
Production per unit space	Excellent	Good	Fair
Resistance to fouling	Low	Medium	High
Ease of cleaning	Fair	Good	Excellent
Cost per unit capacity	Low	Low	Higher
Operating pressure (psi)	250–800 (17–54 atm)	250–800 (17–54 atm)	400–800 (27–54 atm)
Product pumping required	Rarely	Occasionally	Always

applicability because of the fouling consideration and this design has, in fact, captured about one-half of the world-wide desalting market. Tubular and plate-and-frame systems are competitive only when highly turbid waters are involved, and even then only when pretreatment cannot be readily performed. Plate-and-frame systems are the most expensive of all, their principal virtue being that they are readily cleanable when dissassembled.

In reverse osmosis modules, some provision is made for directing the feed flow over the membrane surface to control concentration polarization. In spiral-wound modules, this is effected by means of a brine-side spacer that becomes an integral part of the module. A porous feed flow distributing tube is used in hollow-fiber modules to direct the flow radially outward across the fibers.

An important consideration in module design is the membrane packing density. When this density is high, the required pressure vessel volume is reduced, tending to lower costs and reduce floor-area requirements. The key parameter, however, is water production rate per unit pressure vessel volume. On this basis, because hollow-fiber membranes exhibit a substantially lower flux than do flat-sheet membranes under the same conditions, hollow fibers offer only a small advantage over spiral-wound modules.

5.2. Plant Layout and Operation

In reverse osmosis processing, pretreatment of the water is invariably used to protect the membranes from fouling and chemical degradation. A precipitation inhibitor such as sodium hexametaphosphate is generally used, and pH control is almost always practiced. Post-treatment of the product water is also sometimes required. A flow diagram of a typical reverse osmosis plant is shown in Figure 14.

The necessity to minimize concentration polarization in reverse osmosis plants results in the tapered plant design shown in Figure 15, in which high

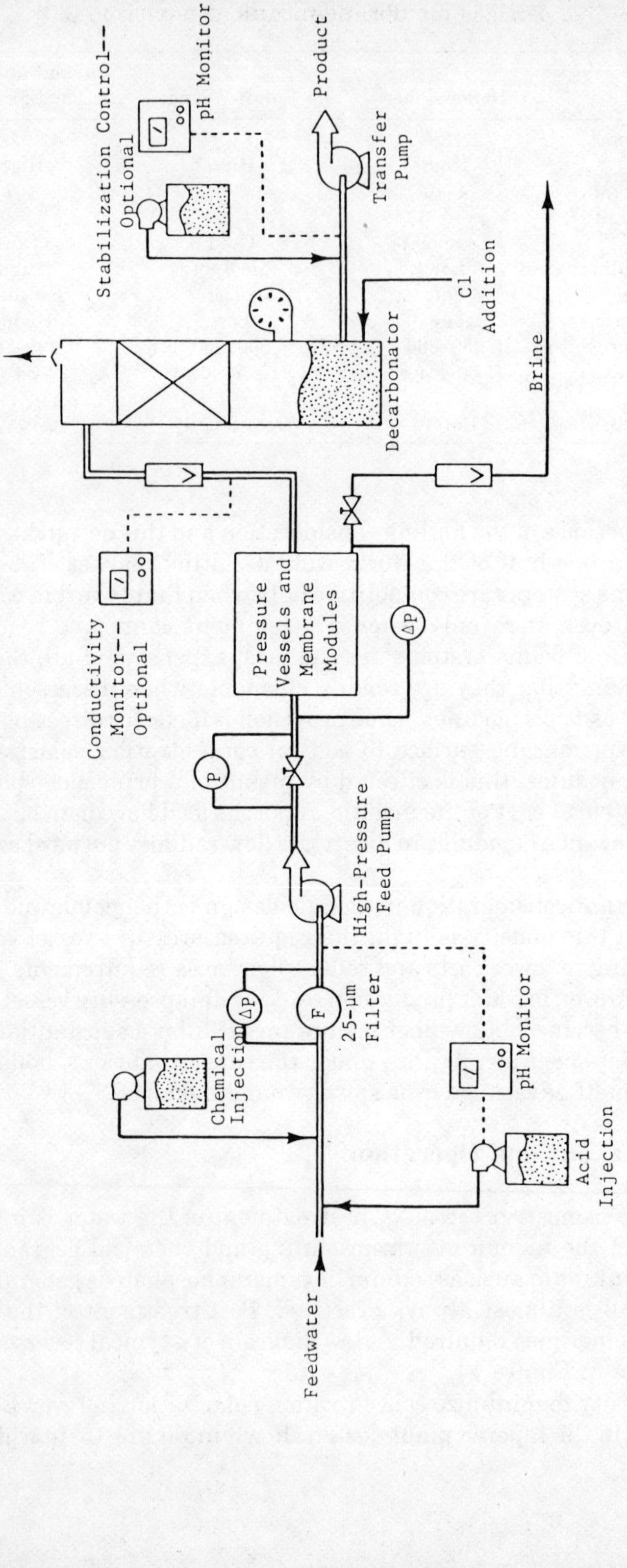

FIGURE 14.—Reverse osmosis flow sheet (31).

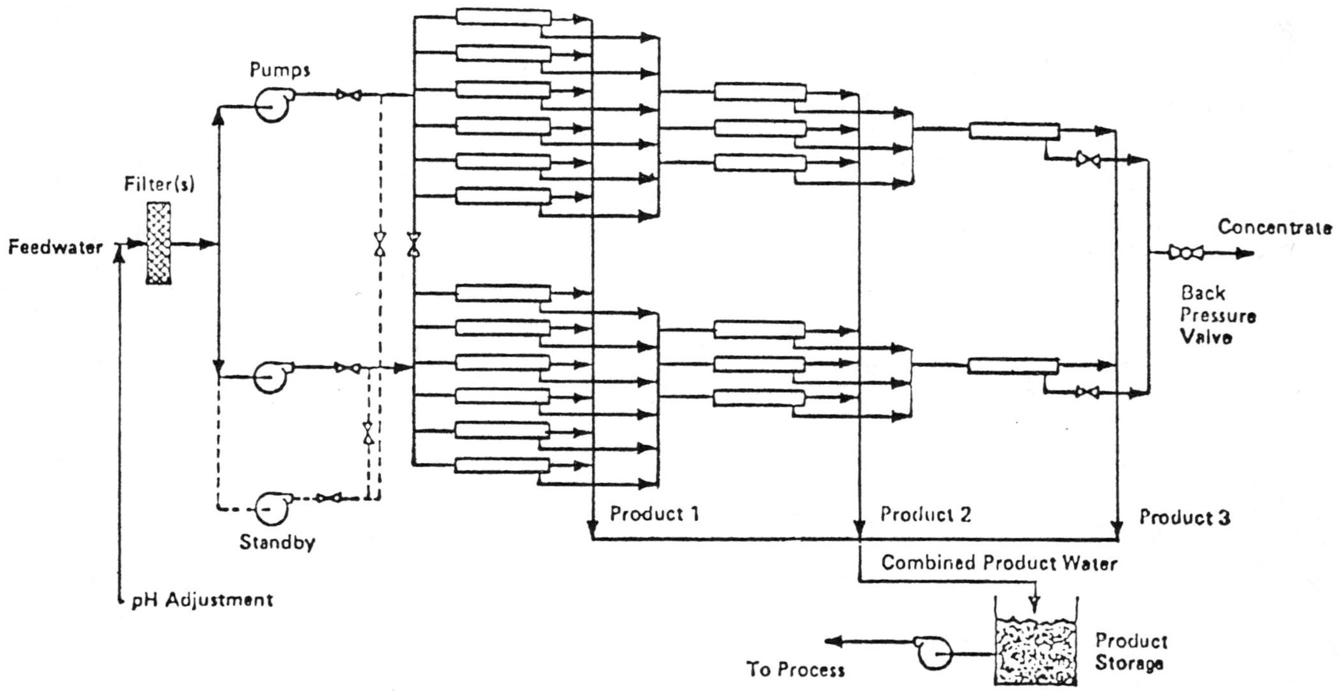

FIGURE 15.—Design of a reverse osmosis module assembly, showing "tapering" of modules to maintain feed velocity (32).

feed velocity and high water recovery can both be achieved. The feed water thus passes through the modules in a series-parallel arrangement.

All pressure-driven membrane processes are subject to membrane fouling. The fouling layer can be of virtually any character: inorganic or organic colloid, precipitates (such as $CaCO_3$ or $CaSO_4$), bacteria, particulates, slimes, etc. A variety of methods have been developed for removing these fouling layers without injuring the membrane. The methods can be classified as either chemical or physical methods. Among the important chemical cleaners are citric acid, EDTA, enzyme detergents, nonionic surfactants, and formaldehyde. The important physical methods include osmotic or hydraulic backflushing and physical scouring.

An important consideration in desalination applications is the energy requirement. Because reverse osmosis is carried out without phase change, it offers a considerable advantage over the various distillation processes. A comparison of these energy requirements is presented in Table 8. In larger reverse osmosis plants, recovery of the energy contained in the pressurized reject brine is economically feasible; as shown in the table, about 30% of the energy can be recovered with a recovery turbine.

TABLE 8. Energy consumption in seawater desalination (33)

	Multi-Flash	Multi-Effect	Vapor Compression	RO without Energy Recovery	RO with Energy Recovery
Maximum temperature	90° C	70° C	100° C	24–32° C	
Energy of vaporization (kWh/m³)	11.5	12.3	–	–	–
Auxiliaries/ compressor (kWh/m³)	4	1.5	11	Incl.	Incl.
Total power consumption (kWh/m³)	15.5	13.8	11	7.9–10	4.7–6.6

6. APPLICATIONS

A broad range of applications have been examined for reverse osmosis, but few have become of commercial significance. There are four principal applications at this time: production of potable water from substandard sources, production of process water for various industries, waste treatment, and food processing. A listing of all the large reverse osmosis plants installed in the world as of June 30, 1980, including their applications, is contained in "Desalting Plants Inventory Report No. 7" (34).

6.1. Potable Water

Many municipalities have been established in areas of substandard (e.g., brackish) water supplies. Reverse osmosis has been used in many of these cases to produce potable water. The source is typically a brackish water of 1500–5000 ppm total dissolved solids, although seawater desalination by reverse osmosis is now coming into general use. When the source is a ground water, membrane fouling is minimal and reverse osmosis has performed very well in this application. The concentrated brine from the reverse osmosis plant is usually rejected to the sea or a river.

6.2. Process Water

Municipal water supplies are frequently of inadequate quality for industrial or commercial processes, and reverse osmosis is now widely used to upgrade these water supplies. The most important single application is the electronics industry, where high-quality rinse water is essential. Reverse osmosis is used there as a pretreatment for mixed-bed deionizers which remove virtually all of the dissolved salts. Reverse osmosis has also found applications in the pharmaceutical, chemical, biological, and food processing industries, as well as in power and heat generation where water hardness can scale pipes and plug boilers.

6.3. Waste Treatment

Reverse osmosis can be used for pollution control and, under favorable circumstances, to recover products from effluents. Some specific applications include treatment of electroplating rinse waters to recover metals, in which the treated water is suitable for reuse; removal of metals from cooling tower blowdown; and purification of pulp and paper industry effluents.

6.4. Food Processing

In recent years, reverse osmosis has been successfully employed in food concentration applications, where the process is being used to partially dewater foods prior to further water removal by distillation or low-temperature evaporation. The advantages of reverse osmosis here are threefold: lower energy requirement; no thermal damage to high-temperature-labile foods, combined, in some cases, with better flavor retention; and lower capital cost, particularly in smaller operations.

There are three key limitations to the use of reverse osmosis in food processing: many foods have very high osmotic pressures (35), requiring very high applied pressures to dewater them (refer, for example, to Figure 16 for the osmotic pressure of orange juice and sucrose solutions); membranes are readily fouled by the constituents of most foods; and the cleaning methods and temperatures commonly used in the food-processing industry are injurious to most classes of membranes.

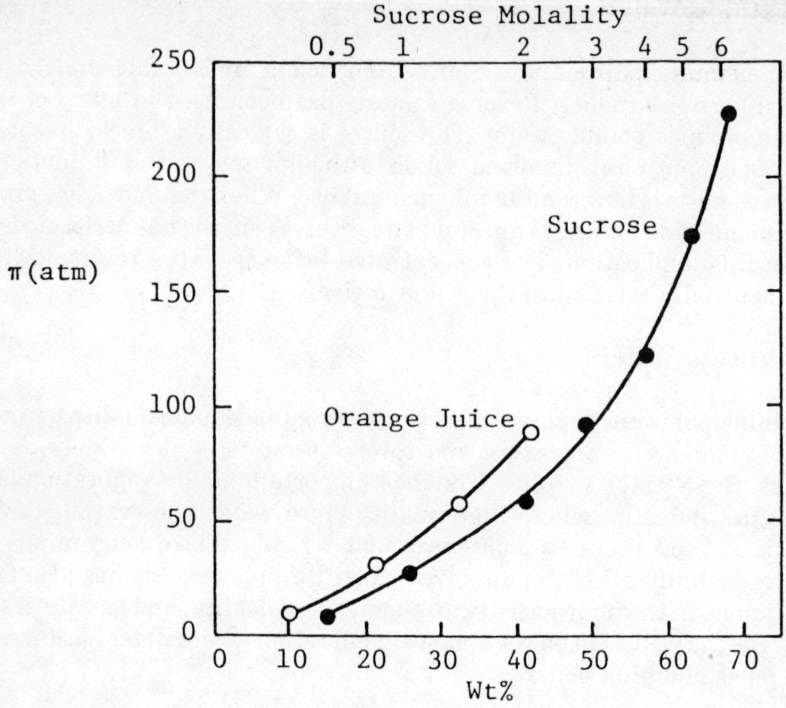

FIGURE 16.—Osmotic pressure of fresh orange juice and aqueous sucrose solutions at 25°C.

In spite of the limitations, several applications of reverse osmosis in the food-processing industry have been found to be economically viable:

—A large tubular system is in use at Foxhol, The Netherlands, to concentrate potato starch effluent from 4% to 8% solids (36). By treating 360,000 pounds of effluent per hour, the reverse osmosis plant reduces the waste volume discharged from the potato processing plant by 500,000 gallons per day. The reverse osmosis plant allows recovery of 50% of the wash water, and also reduces the size and energy consumption required of the evaporator used to concentrate the starch.

—Maple sap is concentrated from 2% to 8% total solids in several small plants utilizing spiral-wound reverse osmosis modules (37). Energy savings are effected in the low-temperature evaporators used in maple syrup production.

—Reverse osmosis is now being used to preconcentrate cheese whey by twofold prior to shipment to a central plant where protein and lactose fractions of the whey are being separated (38). This reduces both the transportation costs and the processing costs in the fractionation plant. Preconcentration of cheese whey by reverse osmosis prior to evaporation has also been shown to be practical in plants producing dried whey for animal feed (39).

—Reverse osmosis is also being effectively utilized to preconcentrate milk prior to the manufacture of cheese, as well as in the production of dried whole milk and skim milk (40).

—Dilute sugar solutions are being concentrated by several candy manufacturers in order to recover sugar for reuse and reduce effluent problems (41,42). Dilute molasses is being similarly concentrated, up to 20–25% solids.

Cleaning and sanitizing the reverse osmosis units is accomplished in these applications by a variety of conventional methods, which have been reviewed (43,44).

REFERENCES

1. Ferry, J.D.: 1936, "Ultrafilter membranes and ultrafiltration." Chem. Rev. 18, p. 373.
2. Merten, U., (Ed.): 1966, "Desalination by Reverse Osmosis," The M.I.T. Press, Cambridge, Mass.
3. Lacey, R.E., and Loeb, S., (Eds.): 1972, "Industrial Processing with Membranes," Wiley-Interscience, New York.
4. Manegold, E.: 1929, Kolloid-Z. 49, p. 372.
5. Michaelis, L.: 1926, Naturwissenschaften 14, p. 33.
6. McBain, J.W., and Kistler, S.S.: 1931, "Ultrafiltration as a test for colloidal constituents in aqueous and nonaqueous systems." J. Phys. Chem. 35, p. 130.
7. Spiegler, K.S., (Ed.): 1966, "Principles of Desalination," Academic Press, New York.
8. Taken from a table in Goldsmith, R.L., R.P. DeFilippi, S. Hossain, and R.S. Timmins: 1971, "Industrial ultrafiltration," in "Membrane Processes in Industry and Biomedicine," M. Bier (ed.), Plenum Press, New York, pp. 267–300.
9. Kesting, R.E.: 1971, "Synthetic Polymeric Membranes," McGraw-Hill, New York.
10. Riley, R.L., Hightower, G., and Lyons, C.R.: 1972, "Preparation, morphology, and transport properties of composite reverse osmosis membranes for seawater desalination," in "Reverse Osmosis Membrane Research," H.K. Lonsdale and H.E. Podall, (eds.), Plenum Press, New York, pp. 437–456.
11. Ward, W.J. III, Browall, W.R., and Salemme, R.M.: 1976, "Ultrathin silicone/polycarbonate membranes for gas separation processes." J. Membrane Sci. 1, pp. 99–108.
12. Lonsdale, H.K., Riley, R.L., Lyons, C.R., and Carosella, D.P., Jr.: 1971, "Transport in composite reverse osmosis membranes," in "Membrane Processes in Industry and Biomedicine," M. Bier (ed.), Plenum Press, New York, pp. 101–122.
13. Rozelle, L.T., Cadotte, J.E., Cobian, K.E., and Kopp, C.V.: 1977, "Nonpolysaccharide membranes for reverse osmosis: NS-100 membranes," in "Reverse Osmosis and Synthetic Membranes," S. Sourirajan (ed.), National Research Council of Canada, Ottawa, Chapter 12.
14. Kraus, K.A., Shor, A.J., and Johnson, J.S. Jr.: 1967, "Hyperfiltration studies. X. Hyperfiltration with dynamically-formed membranes." Desalination 2, pp. 243–266.
15. Yasuda, H., Marsh, H.C., Brandt, E.S., and Reilley, C.N.: 1976, "Preparation of composite reverse osmosis membranes by plasma polymerization of organic compounds. IV. Influence of plasma–polymer (substrate) interaction." J. Appl. Poly. Sci. 20, p. 543.
16. Bell, A.T., Wydeven, T., and Johnson, C.C.: 1975, "A study of the performance and chemical characteristics of composite reverse osmosis membranes prepared by plasma polymerization of allylamine." J. Appl. Poly. Sci. 19, p. 1911.
17. Watson, E.R., Rowley, G.V., and Wunderlich, C.R.: 1969, U.S. Patent 3,432,585.
18. Loeb, S.: "A Composite Tubular Assembly for Reverse Osmosis Desalination." Department of Engineering, UCLA, unpublished.
19. Baum, B., Holley, W. Jr., and White, R.A.: 1976, "Hollow fibers in reverse osmosis, dialysis, and ultrafiltration," in "Membrane Separation Processes," P. Meares, (ed.), Elsevier, Amsterdam, Chapter 5.

20. Riley, R.L., Hightower, G.R., Lyons, C.R., and Tagami, M.: 1974, "Thin-film composite membrane performance in a spiral-wound single-stage reverse osmosis seawater pilot plant," in "Permeability of Plastic Films and Coatings to Gases," H.B. Hopfenberg, (ed.), Plenum Press, New York, pp. 375–388.
21. "Dowex RO-20K Permeator," product brochure, Dow Chemical U.S.A., Midland, Michigan.
22. "Hollowsep," product brochure, Toyobo Co., Ltd., Osaka, Japan.
23. "Reverse Osmosis Principles and Applications," by the Staff, Fluid Systems Division, UOP, San Diego, California, U.S.A.
24. Riley, R.L., Fox, R.L., Lyons, C.R., Milstead, C.E., Seroy, M.W., and Tagami, M.: 1976, "Spiral-wound poly(ether/amide) thin-film composite membrane systems." Desalination 19, pp. 113–126.
25. Al-Gholaikah, A., El-Ramly, N., Jamjoom, I., and Seaton, R.: 1979, "The world's first large sea water reverse osmosis desalination plant at Jeddah, kingdom of Saudi Arabia." The National Water Supply Improvement Assoc. J. 6, pp. 1–11.
26. Cadotte, J.E.: 1981, "Interfacially Synthesized Reverse Osmosis Membrane," U.S. Patent 4,277,344.
27. Larson, R.E., Cadotte, J.E., and Petersen, R.J.: 1981, "The FT-30 seawater reverse osmosis membrane—Element test results." Desalination 38, pp. 473–483.
28. Kurihara, M., Harumiya, N., Kanamaru, N., Tonomura, T., and Nakasatomi, M.: 1981, "Development of the PEC-1000 composite membrane for single-stage seawater desalination and the concentration of dilute aqueous solutions containing valuable materials." Desalination 38, pp. 449–460.
29. Taken from Teijin America literature.
30. Taken from Permutit Co. literature.
31. Taken from UOP, Inc., Fluid Systems Division literature.
32. Buckley, J.: 1975, Consult. Engr. 45:5, pp. 55–61.
33. Sackinger, C.T.: 1980, Middle East Water & Sewage, April/May.
34. El-Ramly, W.A., and Congdon, C.F.: 1981, "Desalting Plants Inventory Report No. 7." The National Water Supply Improvement Association, Ipswich, Massachusetts.
35. Kennedy, T.J., Monge, L.E., McCoy, B.J., and Merson, R.L.: 1973, "Concentrating liquid foods by reverse osmosis: The problems of polarization and high osmotic pressure." AIChE Symp. Ser. 69, pp. 81–86.
36. Pepper, D., and Orchard, A.C.J.: "Starch Effluent Concentration." Paterson Candy International, Ltd., Hampshire, England.
37. Mans, J.: 1981, "Membrane processing—Cost, efficiency spark interest in new technology." Processed Prepared Foods, May, pp. 60–63.
38. "The Vermont Project." Paterson Candy International, Ltd., Hampshire, England.
39. Product bulletin: 1979, "Cheese Whey Concentration by Reverse Osmosis." Paterson Candy International, Ltd., Hampshire, England.
40. Abbot, J., Glover, F.A., Muir, D.D., and Sukudder, P.J.: 1979, "Application of reverse osmosis to the manufacture of dried whole milk and skim milk." J. Dairy Res. 46, pp. 663–672.
41. Spatz, D.D.: 1972, "Reclamation and Reuse of Waste Products from Food Processing by Membrane Processes," paper presented at the AIChE meeting, St. Louis, Missouri, May.
42. Vane, G.W.: 1977, "The applicability of membrane processes in the sugar industry." La Sucrerie Belge 96, pp. 277–282.
43. McDonough, F.E., and Hargrove, R.E.: 1972, "Sanitation of reverse osmosis/ultrafiltration equipment." J. Milk Food Technol. 35, pp. 102–106.
44. Beaton, N.C.: 1979, "Ultrafiltration and reverse osmosis in the dairy industry—An introduction to sanitary considerations." J. Food Protection 42, pp. 584–590.

BIBLIOGRAPHY

Burns and Roe Industrial Services Corp.: 1979, "Reverse Osmosis Technical Manual," a report prepared for the Office of Water Research and Technology, NTIS No. PB 80-186950.

Catalytic, Inc.: 1979, "Desalting Handbook for Planners, Second Edition," a report prepared for the Office of Water Research and Technology, NTIS No. PB 80-202518.

Desalination Journal, Elsevier Scientific Publishing Co., Amsterdam.

"Engineering Design Manual" and "Seawater Reverse Osmosis," Permasep Products, DuPont Co., Wilmington, Delaware.

Fluor Corp.: 1978, "Desalting Plans and Progress—An Evaluation of the State-of-the-Art and Future Research and Development Requirements," Office of Water Research and Technology Contract No. 14-34-0001-7707. Available from NTIS as PB 290 786, Springfield, Virginia.

Journal of Membrane Science, Elsevier Scientific Publishing Co. Amsterdam.

Keller, P.R.: 1976, "Membrane Technology and Industrial Separation Techniques," Noyes Data Corp., Park Ridge, New Jersey.

Lonsdale, H.K.: 1982, "The growth of membrane technology." J. Membrane Sci. 10, pp. 81–181.

Meares, P. (Ed): 1976, "Membrane Separation Processes," Elsevier Scientific Publishing Co., Amsterdam.

Merten, U. (Ed.): 1966, "Desalination by Reverse Osmosis," MIT Press, Cambridge, Massachusetts.

Parrett, T.: 1982, "Membranes: Succeeding by separating." Technology 3, pp. 16–29.

"Reverse Osmosis Principles and Applications" by the Staff, Fluid Systems Division, UOP, Inc., San Diego, California.

Sourirajan, S. (Ed.): 1977, "Reverse Osmosis and Synthetic Membranes," National Research Council of Canada, Ottawa.

Spiegler, K.S., and Laird, A.D.K. (Eds.): 1980, "Principles of Desalination," Academic Press, New York.

Strathmann, H., Kock, K., Amar, P., and Baker, R.W.: 1975, "The formation mechanism of asymmetric membranes." Desalination 16, pp. 179–203.

"Technical Proceedings of the WSIA 10th Annual Conference and Trade Fair," July 25–29, 1982, Honolulu, Hawaii; available through the Water Supply Improvement Association, Ipswich, Massachusetts.

Turbak, A.F. (Ed.): 1981, "Synthetic Membranes: Volume I, Desalination" and "Volume II, Hyper- and Ultrafiltration Uses." ACS Symposium Series 153 and 154, American Chemical Society, Washington, D.C.

APPENDIX

MANUFACTURERS OF REVERSE OSMOSIS MEMBRANES AND MODULES

ABCOR, INC.
850 Main Street
Wilmington, MA 01887

CARRE, INC.
P.O. Box 1555
Seneca, SC 29678

CULLIGAN U.S.A.
1 Culligan Parkway
Northbrook, IL 60062

DAICEL CHEMICAL INDUSTRIES, LTD.
1, Teppo-cho
Sakai, Osaka-pref. 590
Japan

THE DANISH SUGAR CO., INC. (DDS)
c/o Niro Atomizer, Inc.
9164 Rumsey Road
Columbia, MD 21045

DESALINATION SYSTEMS, INC. (DSI)
1107 West Mission Avenue
Escondido, CA 92025

DOW CHEMICAL U.S.A.
Walnut Creek Center
2800 Mitchell Drive
Walnut Creek, CA 94598

DuPONT De NEMOURS & CO.
1007 Market Street
Wilmington, DE 19898

ENVIROGENICS SYSTEMS CO.
9255 Telstar Avenue
El Monte, CA 91731

FILMTEC CORP.
7200 Ohms Lane
Minneapolis, MN 55434

GASTON COUNTY DYEING MACHINE CO.
P.O. Box 308
Stanley, NC 28164

HYDRONAUTICS
P.O. Box 1068
Goleta, CA 93017

KALLE
Niederlassung der Hoechst AG
P.O. Box 1068
6200 Wiesbaden
West Germany

MILLIPORE CORP.
Ashby Road
Bedford, MA 01730

NITTO ELECTRIC INDUSTRIAL CO., LTD.
1-1-2 Shimohozumi, Ibaraki
Osaka, Japan

OSMONICS, INC.
5951 Clearwater Drive
Minnetonka, MN 55343

PATERSON CANDY INTERNATIONAL, LTD.
Laverstoke Mill, Near Whitchurch
Hampshire RG28 7NR, England

SUMITOMO CHEMICAL CO.
7-9, 2-chome, Nihonbashi
Chuo-ku, Tokyo 103, Japan

TEIJIN LTD.
1-1, Uchisaiwai-cho
2-chome, Chiyoda-ku
Tokyo, 100, Japan

TORAY INDUSTRIES, INC.
2-2 Nehonbashimu Romachi
Chuo-ku, Tokyo, Japan

TOYOBO CO., LTD.
Katata Research Center
1300-1, Honkatata
Otsu, Shiga 520-02, Japan

UNIVERSAL OIL PRODUCTS, INC.
Fluid Systems Division
10124 Old Grove Road
San Diego, CA 92131

WAFILIN
P.O. Box 5
Bruchterweg 88, 7770 AA Hardenberg
The Netherlands

SELECTIVITY IN MEMBRANE FILTRATION*

Gunnar Jonsson

Instituttet for Kemiindustri
Technical University of Denmark
DK-2800 Lyngby, Denmark

The paper discusses, from a practical point of view, the main factors that determine membrane selectivity in ultrafiltration (UF)/reverse osmosis (RO) processes. Retention characteristics of UF membranes are delineated through comparisons of the combined viscous flow and frictional model with the Ferry–Faxen equation. The molecular weight cut-off concept is discussed using calculations of how sharp cut-off curves one can expect of UF membranes with uniform pore sizes and how heteroporosity will affect such curves. Retention characteristics of RO membranes are presented for non-electrolytes and electrolytes. Retention variations in mixed-solute systems are also discussed. The influence of the porous sublayer on the integral membrane properties in different membrane processes is further discussed.

*Part of this chapter has been reprinted with permission from Desalination *35* (1980), pp. 21–38.

P. M. Bungay et al. (eds.), Synthetic Membranes: Science, Engineering and Applications, 343–366.

1. INTRODUCTION

When proposing theories to describe membrane transport, one can either use a purely thermodynamic description in which the membrane is treated as a "black box" or introduce a physical model of the membrane. In the first case a quite general description is obtained, which can be used in both reverse osmosis and ultrafiltration. However, as it is model independent it sheds no light on flow- and separation-mechanisms. In the second case more information on the flow- and separation-mechanisms is obtained, but the correctness of the information depends on the chosen model.

Semipermeable membranes are most commonly discussed as materials with either porous or nonporous structures. Real pores as shown in Figure 1a are only found in certain microfiltration membranes, and real nonporous membranes are hardly found in even the tightest reverse osmosis membranes. Most membranes can rather be characterized as having structures of either the nodular or sponge-like types (Figure 1 c–d). Macroscopically, these membranes are homogeneous but on the microscopic level they are two-phase systems where the transport of water and solute takes place in the water-filled interstices between the nodules or the polymer chains.

Most membrane theories have been derived for single-layer membranes. However, commercial membranes are normally asymmetric or composite materials, which have at least two different layers. Figure 2 shows a schematic presentation of an asymmetric cellulose acetate (CA) membrane with the highly selective skin layer, the intermediate layer where the selectivity decreases to zero and the non-selective porous sublayer. Because the porous sublayer is much thicker than the skin layer, one can normally expect an appreciable pressure gradient in this layer. This will influence the total hydraulic permeability as given by the relation:

$$\frac{1}{L_p} = \frac{1}{(L_p)_{\mathrm{sl}}} + \frac{1}{(L_p)_{\mathrm{il}}} + \frac{1}{(L_p)_{\mathrm{pl}}} \tag{1}$$

However, it will only give an indirect effect on the measured retention (1). On the other hand, the influence of a selective intermediate layer depends very much on the thickness of this layer (2).

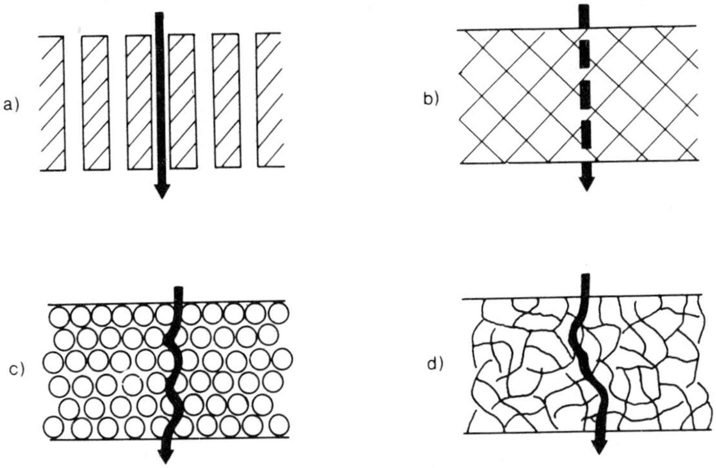

FIGURE 1.—Schematic presentation of different membrane structures: a) porous membrane, b) nonporous membrane, c) nodular structure, consisting of close packed spheres, d) sponge like structures. Used with permission from reference (29).

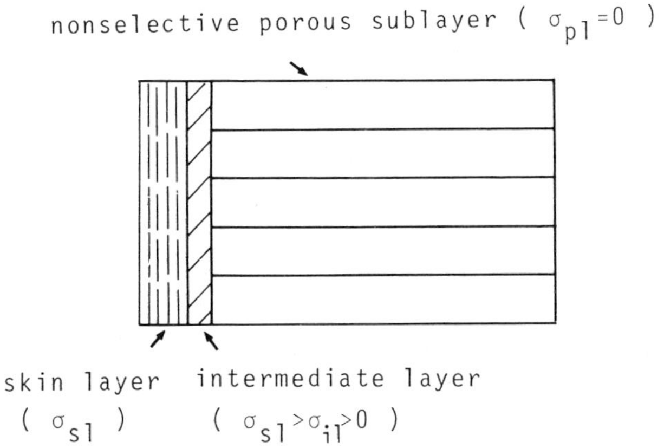

FIGURE 2.—Schematic presentation of an asymmetric CA membrane. Used with permission from reference (29).

2. MEMBRANE TRANSPORT MODEL

In this paper, we will use a combined viscous flow and frictional model, originally proposed by Merten (3) and presented in detail by Jonsson and Boesen (4), because it gives some physically understandable parameters and it can be used quite generally independent of the retention level of the membrane (5,6). The model combines the concept that flow within the "pores" occurs by both viscous flow and diffusion with the concept of frictional interactions within the pores, as proposed by Spiegler (7). The flux of solute per unit of pore area is given by:

$$J_{ip} = -\frac{RT}{f_{iw}+f_{im}}\frac{dc_{ip}}{d\xi} + \frac{c_{ip}\,v}{1+f_{im}/f_{iw}} \tag{2}$$

Using the following relations:

$$K_{ip} = c_{ip}/c_i' \tag{3}$$

$$b = 1 + f_{im}/f_{iw} \tag{4}$$

$$\xi = \theta\, x \tag{5}$$

$$D_{iw} = RT/f_{iw} = D_{i\infty} \tag{6}$$

$$J_V = v\,\epsilon \tag{7}$$

$$J_i = J_{ip}\,\epsilon \tag{8}$$

the flux per unit of membrane area may be written:

$$J_i = -\frac{D_{iw}K_{ip}\epsilon}{b\,\theta}\frac{dc_i'}{dx} + \frac{K_{ip}c_i'}{b}J_V \tag{9}$$

This equation can be integrated with the appropriate boundary conditions to give:

$$\frac{c_i'}{c_i''} = \frac{b}{K_{ip}} + \left(1-\frac{b}{K_{ip}}\right)\exp\left(-\frac{\theta h}{\epsilon}\frac{J_V}{D_{iw}}\right) \tag{10}$$

Here b is a friction factor and K_{ip} is the distribution coefficient of solute between pore fluid and bulk solution. The "effective skin layer thickness," $\theta h/\epsilon$, is a membrane parameter that should be independent of the solute used.

The volume flux per unit membrane area is derived from Poiseuille's equation taking into account that the effective driving force per volume of pore fluid is the pressure gradient minus the frictional force between the solute and pore wall:

$$J_V = L_p \left(1 + \frac{L_p}{\epsilon} \cdot \frac{f_{im}c_i''}{\tilde{M}_i}\right) \frac{\Delta p}{\theta h} \tag{11}$$

where $L_p \equiv \epsilon d_p^2/32\ \eta$ is the hydraulic permeability.

Spiegler and Kedem (8) derived from a friction model the following expression for the reflection coefficient:

$$\sigma = 1 - \frac{K_i}{K_w} \frac{1 + (f_{wm}/f_{iw})(\tilde{V}_i/\tilde{V}_w)}{1 + (f_{im}/f_{iw})} \tag{12}$$

which in our notation can be transformed to:

$$\sigma = 1 - K_{ip} \left(\frac{1}{b} + \frac{D_i\tilde{V}_i}{D_w\tilde{V}_w}\right) \tag{13}$$

Thus two terms influence to what extent a membrane is selective for a given solute: The exclusion term, which is determined by the ratio of solute to water uptake in the membrane and the kinetic term, which is determined by the sum of the drag factor and the pressure diffusion factor.

3. ULTRAFILTRATION

The selectivity of UF membranes is determined primarily by the ratio between the hydrodynamic diameter of the solute and the apparent pore diameter. Thus the retention characteristics of a given membrane are usually presented as the retention versus the molecular weight of different macromolecules. The molecular weight cut-off is said to be that $\tilde{M}$-value which is almost totally rejected. However, for most UF membranes the retention-$\tilde{M}$ curve is not very sharp, indicating a relatively heteroporous membrane structure. Also, factors such as the shape and dissociation of the macromolecules influence the retention.

Most ultrafiltration data are so highly influenced by the concentration polarization that a determination of the membrane transport properties is irrelevant. Very often, one finds that the flux is independent of pressure and that the retention decreases with pressure (9), which is inconsistent with all transport theories that consider only the membrane. However, if the membrane system design is such that a correction for the concentration polarization can be calculated, models for transport within the membrane can be validated.

3.1. Homoporous Model

In one such study, Jonsson and Boesen (4) applied the combined viscous flow and frictional model to data from an unannealed cellulose acetate membrane possessing a $\tilde{M}$ cut-off value of about 6000. They found that equations 10 and 11 gave a good fit to the experimental data. Specifically, the coefficients were rather constant in the measured pressure and concentration region; thus the model gave a better description of the variation in flux and retention as a function of concentration than did other models.

From the model, the "effective skin layer thickness", $\theta h/\epsilon$, and the pore diameter of the skin layer, d_p, were estimated to be 30 μm and 30 Å, respectively. With reasonable assumptions about the water content and the tortuosity factor, this gives a skin layer thickness of about 1 μm.

The pore distribution coefficient, K_{ip}, and the friction parameter, b, were correlated with the Ferry–Faxen equation (10, 11) in the following way:

$$K_{ip} = \frac{(A_i/A_p)\ \text{steric}}{(A_w/A_p)\ \text{steric}} \tag{14}$$

and

$$b = \frac{(A_w/A_p)\ \text{friction}}{(A_i/A_p)\ \text{friction}} \tag{15}$$

where

$$\left(\frac{A_i}{A_p}\right)_{\text{steric}} = 2\left(1 - \frac{d_i}{d_p}\right)^2 - \left(1 - \frac{d_i}{d_p}\right)^4 \tag{16}$$

and

$$\left(\frac{A_i}{A_p}\right)_{\text{friction}} = 1 - 2.104\left(\frac{d_i}{d_p}\right) + 2.09\left(\frac{d_i}{d_p}\right)^3 - 0.95\left(\frac{d_i}{d_p}\right)^5 \tag{17}$$

When a value of 46 Å was used for the pore diameter, the experimentally determined K_{ip} and b values were in reasonable agreement with equations 14–17, as shown in Figure 3. The difference between the pore diameter of 46 Å as correlated with the Ferry–Faxen equation and that of 30 Å as determined from equation 11 has recently been shown to be caused by a combined resistance to the volume transport in both the skin layer and the porous layer (1). From Figure 3 one can further see that the drag factor contributes more to the membrane selectivity than the exclusion term does with this type of membrane.

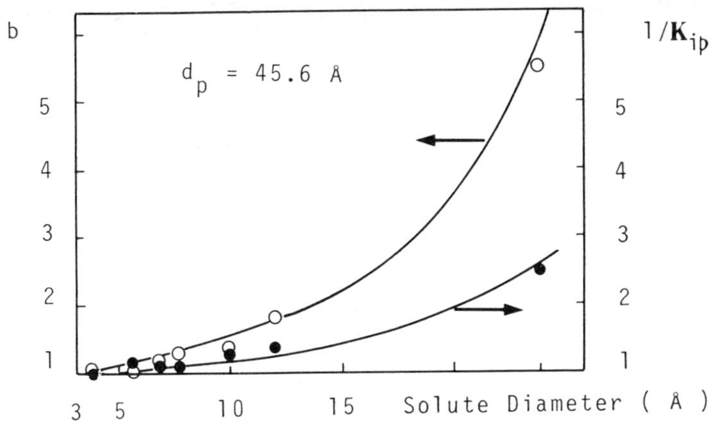

FIGURE 3.—Experimental values of the friction parameter, b (o), and the reciprocal pore distribution coefficient, $1/K_{ip}$ (●), versus the solute diameter. The curves drawn in are calculated from equations 14–17. Used with permission from reference (29).

3.2. The Effect of Heteroporosity on the Retention-Solute Size Curve

From equation 10 the maximum retention corresponding to $J_V \rightarrow \infty$ is given by:

$$\mathcal{R}_{max} = 1 - \frac{K_{ip}}{b} \tag{18}$$

Compared with equation 13, $\mathcal{R}_{max}$ and σ are identical, apart from the pressure diffusion term, which is normally assumed to be negligible in proportion to the drag factor, $1/b$.

Using equations 14–18 it is possible to calculate the maximum retention as a function of solute diameter for different pore diameters. Figure 4 shows the calculated $\mathcal{R}_{max}$ values versus the logarithm of the hydrodynamic solute diameter for three different pore diameters: 50, 100 and 150 Å, respectively. For a heteroporous membrane with a mean pore diameter of 100 Å consisting of two pore sizes—50 and 150 Å—in such a ratio that the permeate flux through the small pores is the same as through the big pores, it can be shown that the mean $\mathcal{R}_{max}$ value, $\overline{\mathcal{R}}_{max}$, is given by:

$$(\overline{\mathcal{R}}_{max})_{100} = \frac{1}{2}\left((\mathcal{R}_{max})_{50} + (\mathcal{R}_{max})_{150}\right) \tag{19}$$

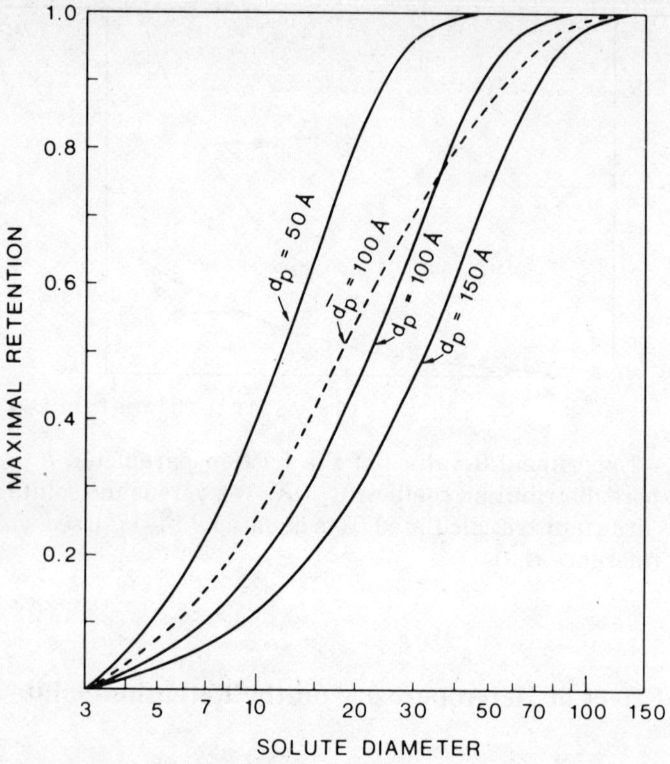

FIGURE 4.—The maximal retention, $\mathcal{R}_{max}$, calculated from equations 14–18 versus the solute diameter for three values of the pore diameter. The dotted curve represents a heteroporous membrane calculated from equation 19.

This is represented by the dotted line shown in Figure 4. Comparing this average curve with the curve for $d_p = 100$ Å in the same figure, we can see that the quite broad selectivity curves for ultrafiltration membranes could be due to the transport mechanism rather than to the variation in pore sizes of the membranes. Thus a very steep selectivity curve cannot be expected even for ultrafiltration membranes with uniform pore sizes.

3.3. The Effect of Pressure on the Retention-Solute Size Curve

Figure 4 only shows the variation with solute size on the intrinsic membrane properties, represented by $\mathcal{R}_{max}$ or σ. In practice, however, the observed retention normally changes with variation in pressure. This is caused by the increase in flux with pressure, thereby increasing the real retention given by equation 10 with $c_i' = c_{int}'$, the concentration at the membrane-solution interface. In addition, the concentration polarization, c'_{int}/c'_{bulk}, which can be calculated from the film model:

$$\frac{c'_{int} - c''_i}{c'_{bulk} - c''_i} = \exp\left(J_V \frac{\delta}{D'_{iw}}\right), \tag{20}$$

also increases with flux. This reduces the observed retention, $\mathcal{R}_{obs}$, which is the parameter of interest for the practical separation process.

Using equations 14–17 together with equations 10 and 20 it is possible to calculate the real membrane retention, $\mathcal{R} = 1 - c''_i/c'_{int}$, and the observed retention, $\mathcal{R}_{obs} = 1 - c''_i/c'_{bulk}$, when D_{iw}, $\theta h/\epsilon$ and δ are known. Here, the diffusion coefficient in the external solution is calculated from the Stokes–Einstein equation:

$$D'_{iw} = \frac{kT}{3\,\pi\,\eta\,d_i} \tag{21}$$

while $\theta h/\epsilon$ and δ are assumed equal to 30 μm and 10 μm, respectively. These are normal values for ultrafiltration membranes (4) and ultrafiltration systems working at high circulation velocities (12).

Figure 5 shows the calculated retentions for three different levels of permeate flux: $2{\cdot}10^{-3}$, $2{\cdot}10^{-4}$ and $2{\cdot}10^{-5}$ cm/s, which corresponds to a high, low and extremely low ultrafiltration flux, respectively. The dotted curves are calculated from equation 10 only, without taking the concentration polarization from equation 20 into account. At high flux ($2{\cdot}10^{-3}$ cm/s), equation 10 gives $\mathcal{R}$-values that are quite close to the maximal retention. But the concentration polarization increases drastically with increasing solute diameter, so that the observed retention is much smaller than $\mathcal{R}_{max}$. With or without concentration polarization the cut-off curve is quite broad, indicating that sharp solute–solute separation is not possible at high flux levels. In addition, gel formation and fouling are often observed even at low bulk concentrations, which further complicate any solute separation. At low flux ($2{\cdot}10^{-4}$ cm/s), equation 10 gives $\mathcal{R}$-values that are much lower than $\mathcal{R}_{max}$, especially for low and medium retaining solutes. As the concentration polarization is quite low for all solute sizes this gives a quite steep cut-off curve. Therefore, solute separation should be much better at low pressures. At extremely low flux ($2{\cdot}10^{-5}$ cm/s), there is no concentration polarization and the cut-off curve is even steeper; however, the flux is so low that it is not economically feasible.

3.4. Comparison of Solute Separation in Ultrafiltration and Dialysis

In dialysis the volume flux through the membrane, J_V, is normally very small. Using equation 9 the solute flux can therefore be written as:

$$J_i = \frac{D_{iw} K_{ip}}{b} \frac{\epsilon}{\theta h} (c'_i - c''_i) \tag{22}$$

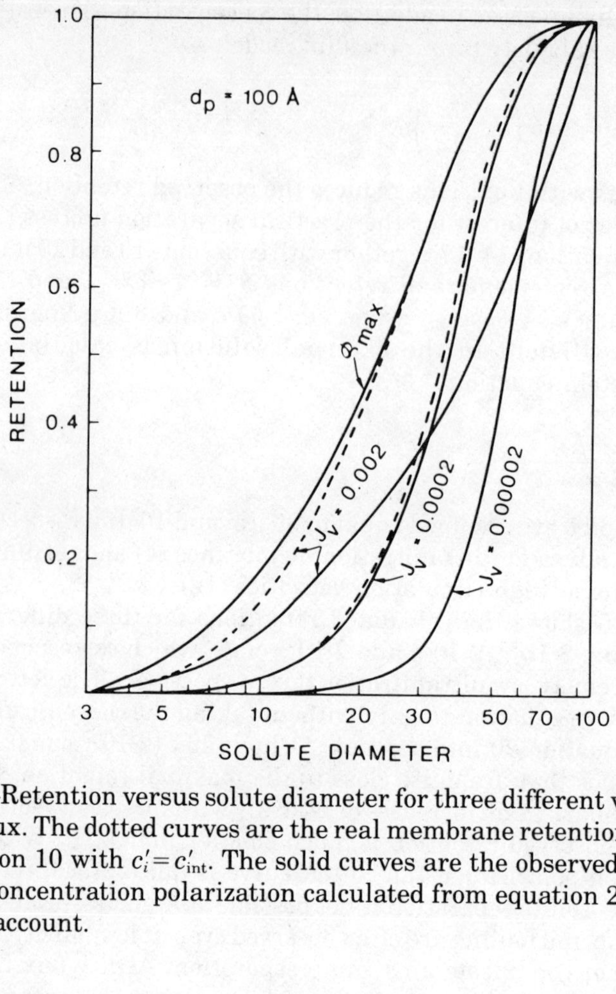

FIGURE 5.—Retention versus solute diameter for three different values of the permeate flux. The dotted curves are the real membrane retention calculated from equation 10 with $c_i' = c_{int}'$. The solid curves are the observed retentions, where the concentration polarization calculated from equation 20 is further taken into account.

Taking a solute with 10 Å in hydrodynamic diameter as reference, the clearance ratio, α, can be expressed as:

$$\alpha = \frac{J_i/c_i' - c_i''}{[J_i/(c_i' - c_i'')]_{10\text{Å}}} = \frac{D_{iw}K_{ip}/b}{(D_{iw}K_{ip}/b)_{10\text{Å}}} \tag{23}$$

On the basis of equations 14–17 and 21, α has been calculated as a function of d_i for a membrane with 100 Å in pore diameter, as shown in Figure 6.

Combining the definition of the observed retention:

$$\mathcal{R}_{obs} = 1 - \frac{c_i''}{c_{bulk}'} \tag{24}$$

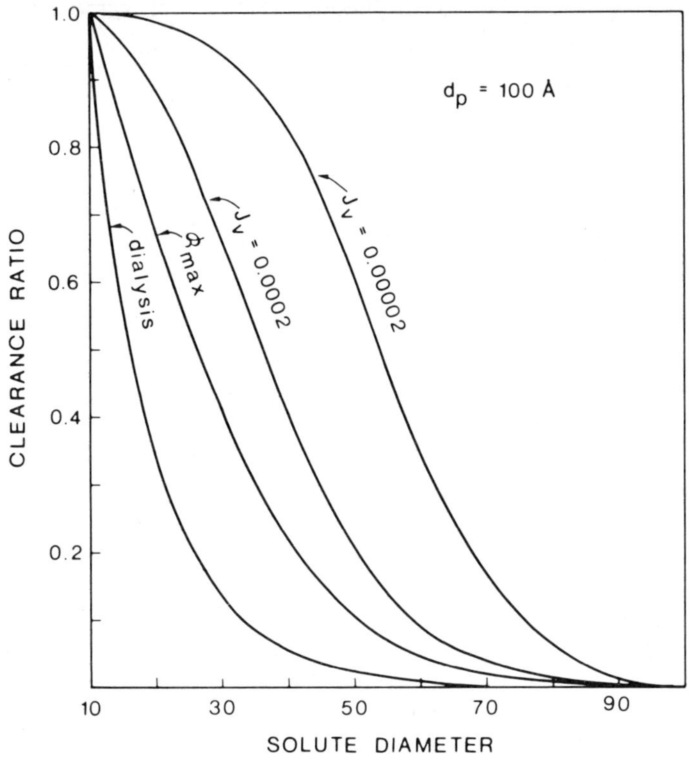

FIGURE 6.—Clearance ratio versus solute diameter. For the dialysis process, α has been calculated from equation 23. For the ultrafiltration process, α has been calculated from equation 26 at different permeate fluxes.

together with the product condition:

$$c_i'' = J_i/J_V \tag{25}$$

the clearance ratio for the ultrafiltration process can be expressed as:

$$\alpha = \frac{J_i/c'_{\text{bulk}}}{(J_i/c'_{\text{bulk}})_{10\text{Å}}} = \frac{1-\mathcal{R}_{\text{obs}}}{(1-\mathcal{R}_{\text{obs}})_{10\text{Å}}} \tag{26}$$

Using the data from Figure 5, α has been recalculated for the three situations shown in Figure 6. This clearly shows that the purely diffusive process, dialysis, is much more dependent on solute size than is the ultrafiltration process, in which convection plays a major part. Separation of solutes close in size should therefore be easier with dialysis than in ultrafiltration. On the other hand, if clearance of solutes up to 5,000 molecular weight at the same rate is necessary in the artificial kidney (13), then a hemofiltration process should be used instead of the normal hemodialysis process.

4. REVERSE OSMOSIS

In contrast to ultrafiltration, the separation in reverse osmosis is not achieved by the size of the solute. The synonym, hyperfiltration, is therefore somewhat misleading, as it is not a filtration process but rather a solution –diffusion process. Thus the retention on a typical CA membrane decreases from 99% to -10% in the following sequence: $CaCl_2 > NaCl \approx$ glycerol $>>$ butanol $>$ hexanol.

The retention of various solutes by asymmetric CA membranes has been examined by a number of groups, both in laboratory and field tests, and a large body of retention data exists. Unlike UF membranes, where the retention is mainly given by the pore size of the membrane, the retention by RO membranes also depends on the membrane material used.

4.1. Retention of Nonelectrolytes

Matsuura and Sourirajan have intensively studied the reverse osmosis separations of several organic solutes and correlated the solute separation with some polar, steric and nonpolar parameters (14). However, as the solute separation is determined by the product of an exclusion term and a kinetic term, the real separation mechanism cannot always be determined by retention data alone.

Table 1 shows some estimated values of the distribution coefficient and the diffusion coefficient of different alcohols determined from desorption

TABLE 1. Estimated values of the distribution coefficient and the diffusion coefficient of different alcohols.

Solute	(1) $\mathcal{R}_{max}$	(2) K_i	(2) $\bar{D}_i \times 10^9$ (cm²/s)	(3) K_i	(3) $\bar{D}_i \times 10^9$ (cm²/s)
Methanol	0.06	0.10	45	0.22	–
Ethanol	0.18	0.21	12		
Propanol	0.22	0.50	7.7		
2-propanol	0.43	0.44	6.0		
Butanol	0.10	1.2	4.3		
2-butanol	0.39	1.0	3.4		
Tert-butanol	0.76	0.8	0.9		
Hexanol	−0.17	14	0.7	0.23	9
Cyclo-hexanol	0.61	5.7	0.14		
Phenol	0.34	30	0.9	0.8	10

(1) Determined from RO experiments on asymmetric CA membranes.
(2) Determined from desorption experiments on homogeneous CA membranes.
(3) As in (2), but with methanol as solvent instead of water.
Used with permission from reference (29).

experiments on homogeneous CA membranes. The distribution coefficient is primarily a function of the aliphatic chain length or the hydrophobic nature of the alcohol molecule. The diffusion coefficient is, on the other hand, essentially a function of the hydrodynamic volume of the molecule, which is both determined by the chain length and the branching of the chain.

Thus, with increasing branching of the butanols the diffusivity decreases much more than the distribution coefficient, which gives rise to the increase in retention by reverse osmosis. For the solutes methanol, butanol, and hexanol, the distribution coefficient increases by a factor of 1 : 12 : 140, while the diffusivity decreases by a factor of 1 : 11 : 64. This explains why methanol and butanol have almost the same positive retention in reverse osmosis, though the mechanism is quite different. It also explains why hexanol has a negative retention: the exclusion term has increased much more than the kinetic term has decreased.

Further, Table 1 also shows a few estimated values of the distribution coefficient and the diffusion coefficient, determined on the same homogeneous CA membranes but with methanol as solvent instead of water. This has a very pronounced effect: For hexanol the distribution coefficient decreases by a factor of 60 to almost the same value as the methanol content in the membrane, whereas the diffusivity increases by a factor of 15. This shows that the main cause of the very strong sorption of higher aliphatic alcohols and similar flavour components on CA membranes is not a solute–membrane interaction, but rather a lack of solute–water interaction.

As discussed by Kozak et al. (15), nonpolar groups have a tendency to adhere to one another in an aqueous environment, known as hydrophobic bonding. This tendency reflects the aversion of these solutes to an aqueous environment. When brought into contact with a membrane, where the polymer phase is hydrophobic and the interstitial water phase is hydrophilic, the solutes will be squeezed into the membrane phase, which acts as a much better solvent medium for the solute than the water phase alone.

4.2. Retention of Electrolytes

The reverse osmosis separation of electrolytes is to a great extent determined by electrostatic forces that cause ions to avoid a region of low dielectric constant. This is easily seen by the fact that the retention of acetic acid is very low whereas the retention of sodium acetate is high. Unlike nonelectrolytes, where the retention-flux curve does not change very much with concentration or the presence of other solutes, variations of this nature are normally found with electrolytes.

Figures 7 and 8 show retention-flux curves for NaCl and $MgSO_4$, respectively, on a rather open cellulose acetate RO membrane, DDS 930. For NaCl the selectivity decreases very much when the concentration increases from 0.0002 to 0.01 eq/L, whereas it is almost constant in the concentration region 0.01–0.2 eq/L. Contrary to this, the selectivity for $MgSO_4$ is almost constant

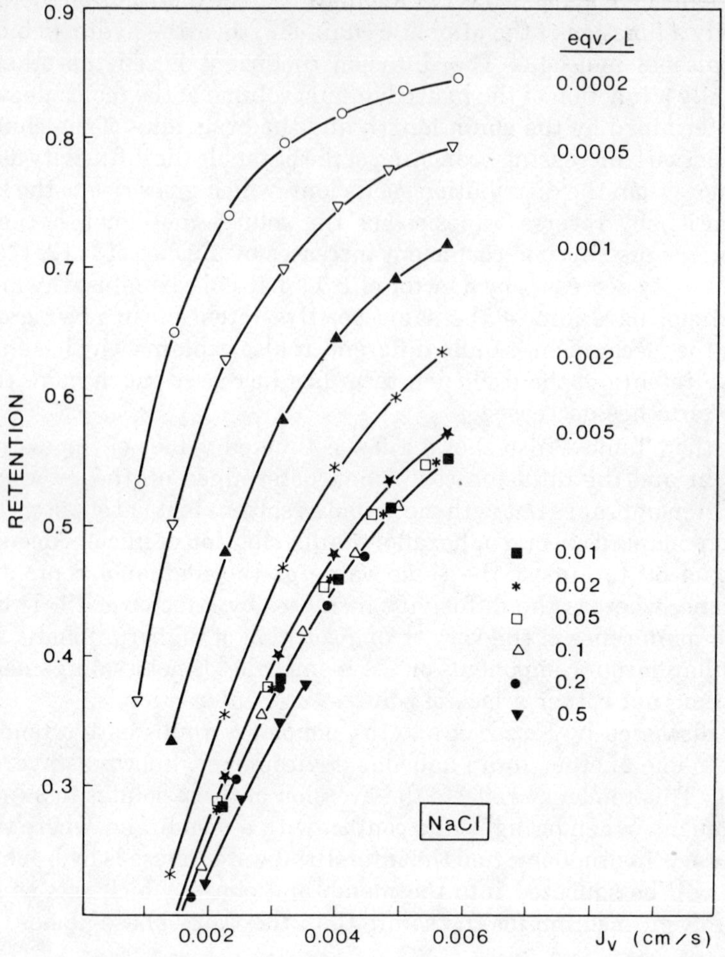

FIGURE 7.—Retention versus permeate flux for different concentrations of NaCl on a DDS 930 membrane. Used with permission from reference (29).

in the concentration region 0.005–0.02 eq/L, and increases very much when the concentration increases from 0.02 to 0.5 eq/L.

An explanation for the decrease in selectivity for NaCl could be a kind of Donnan exclusion at very low concentrations. Normally CA membranes are regarded as neutral membranes; however, it has been shown that they contain a small amount of carboxylic groups, which gives them the character of a weak cation exchange membrane (16). That this is the real mechanism is verified in Figure 9, where the retention-flux curves for a 0.001 eq/L NaCl are shown at different pH-values in the bulk solution. Increasing the pH results

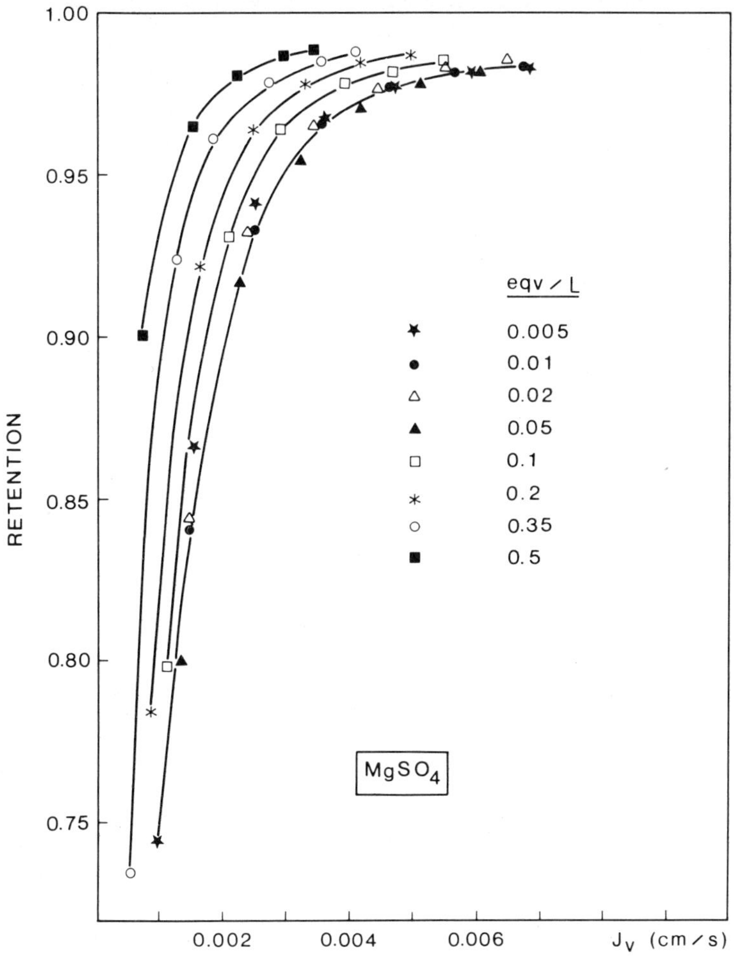

FIGURE 8.—Retention versus permeate flux for different concentrations of $MgSO_4$ on a DDS 930 membrane. Reprinted with permission from reference (29).

in an increasing dissociation of the carboxylic acid groups and therefore an increase in the Donnan exclusion. At low pH-values the carboxylic acid groups are undissociated and the Donnan exclusion disappears.

It is well known that the activity coefficient for 2:2 electrolytes drops drastically with increasing concentration. At the same time the retention is very high, which means that the activity coefficient in the bulk and permeate solutions can be quite different. As the equilibrium at the membrane-solution interface on a RO membrane is a matter of activities rather than concentrations, one could expect that the retention calculated from activity

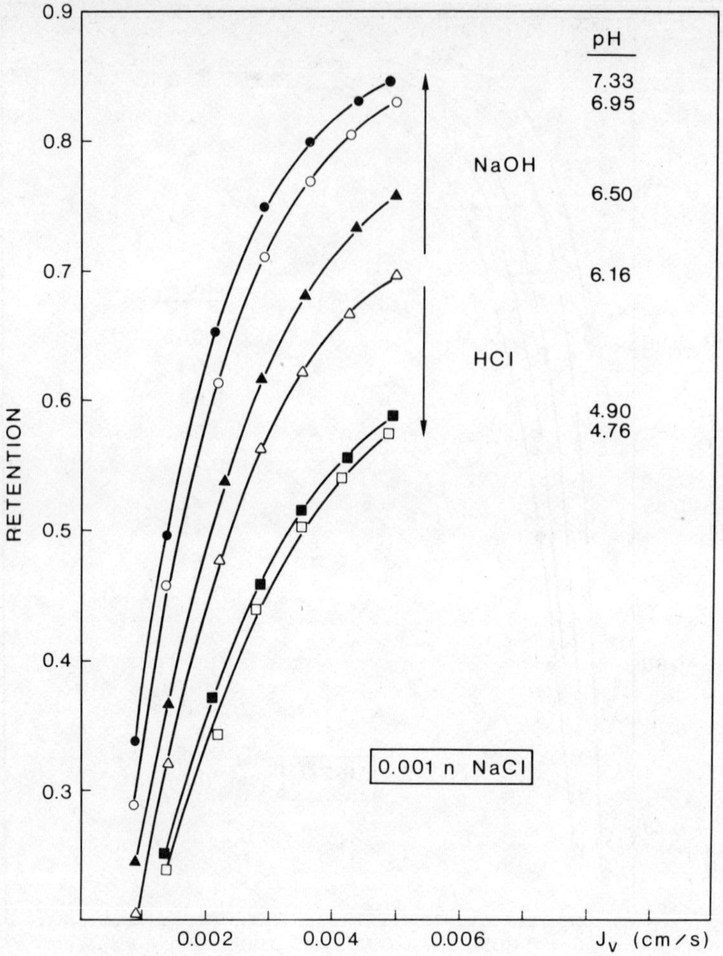

FIGURE 9.—Retention versus permeate flux for 0.001 eq./L NaCl at different pH-values on a DDS 930 membrane. Reprinted with permission from reference (29).

values would be more constant than the retention based on concentrations. In Figure 10 the activity retention, defined as:

$$\mathcal{R}_a = 1 - \frac{a_i''}{a_i'} = 1 - \frac{c_i''}{c_i'} \cdot \frac{\gamma_i''}{\gamma_i'} \qquad (27)$$

is plotted as a function of the permeate flux. Now the selectivity is almost constant in the concentration region 0.05–0.35 eq/L, whereas it increases

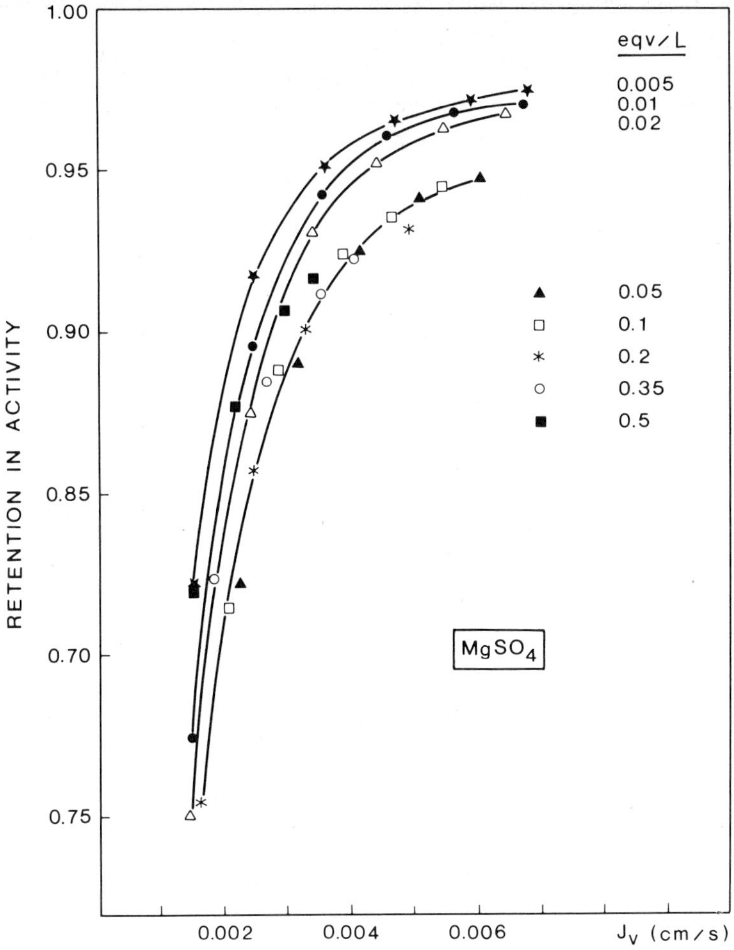

FIGURE 10.—Retention in activity calculated from equation 27 versus permeate flux for different concentrations of $MgSO_4$ on a DDS 930 membrane. Reprinted with permission from reference (29).

when the concentration decreases from 0.05 to 0.005 eq/L. As with NaCl this increase in selectivity at low concentrations is probably caused by a Donnan exclusion. Within the ion exchange literature it is well known that divalent cations such as Ca^{++} and Mg^{++} dissociate the carboxylic acid groups to a much higher extent than Na^+ at neutral pH-values (17). Therefore, it seems likely that the Donnan exclusion is effective at much higher concentrations for $MgSO_4$ than for NaCl.

4.3. Retention Variations in Mixed-Solute Systems

Very few results have been published on mixed-solute systems, from which a possible interaction between nonelectrolytes can be seen. Jonsson and Boesen (4) studied the system glucose–raffinose and found that a "postulated interaction" could be explained from the convection term. This means that the coupling coefficients L_{ij} disappear, and that the change in retention of component i by introducing another solute j is caused by the water–solute coefficients L_{wi} and L_{wj}.

Matsuura et al. (18, 19) studied systems containing mixed alcohols. At low concentrations where the permeate flux was constant, no change in the retention could be seen. At high concentrations a change in retention was observed, which could be calculated from the retentions of the single-solute systems of the same total molality. From the data it is not possible to see if a real coupling exists between the two solutes, or if the convection is responsible for the change in retention.

Boesen and Jonsson (20) recently found a significant change in the retention of *tert*-butanol when either ethanol or phenol was introduced to the feed solution. By measuring the distribution coefficients of *tert*-butanol on homogeneous membranes, they concluded that this change in retention is not caused by changes in the permeability, but rather by a virtual coupling between the flows of the two solutes.

The flux equations for a two-solute system were derived based on non-equilibrium thermodynamics. With some simplifying assumptions the difference between the retention in a single-solute and a two-solute system at the same permeate flux can be expressed by the equation:

$$\Delta\mathcal{R}_i = \overset{o}{\mathcal{R}}_i - \mathcal{R}_i = \frac{P_i^*}{P_j} \cdot \frac{J_V(\sigma_j - \mathcal{R}_j)}{P_i + J_V} \cdot \frac{c_j'}{c_i'} \tag{28}$$

where $\overset{o}{\mathcal{R}}_i$ is the retention of i in the absence of the j and P_i^* is a solute permeability coefficient related to the phenomenological coefficients by:

$$P_i^* = \nu_i RT\bar{c}_i \,(L_{ij} - (L_{wi}L_{wj}/L_{ww}))$$

Here the change in retention at a given permeate flux, $\Delta\mathcal{R}_i$, should increase with increasing c_j'/c_i' ratio, be positive if the interfering solute has negative retention, and negative if the interfering solute has positive retention. The experimental data shown in Figure 11 are qualitatively in agreement with the above equation: with increasing ethanol content the retention of *tert*-butanol decreases roughly proportional to the ethanol concentration at constant J_V. With increasing phenol content the retention of *tert*-butanol increases with the phenol concentration.

Few results have been published on binary salt solutions. The general trend found is that in a mixture of more and less permeable ions, the retention

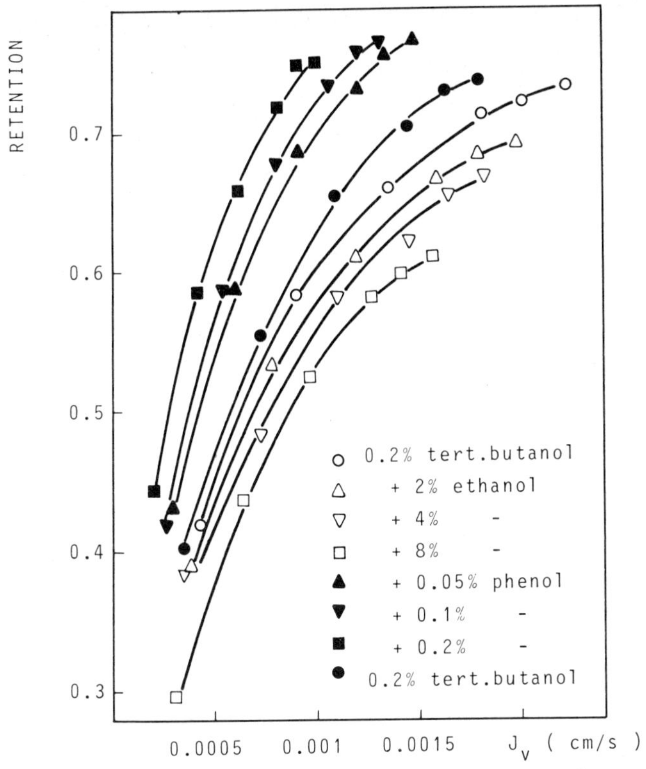

FIGURE 11.—Retention versus permeate flux for 0.2% *tert*-butanol at different concentrations of the interfering solute. Reprinted from reference (20).

of the more permeable ions decreases and that of the less permeable ions increases (21, 22). Heyde and Anderson (23) have studied the influence of added electrolytes upon ion sorption by membranes. They found that membrane sorption of permeable ions is substantially increased by addition of membrane impermeable salts to the bulk solution and vice versa. This change in sorption was explained in terms of constrained phase equilibria, using the ideas developed by Donnan (24). Lonsdale et al. (25) were able to pass from positive to negative retention of Cl^- ions upon addition of membrane impermeable sodium citrate. The experimental results were explained by a solution–diffusion model coupled to a Donnan equilibrium theory.

Recently, Jonsson (26) studied the binary system HCl–$CaCl_2$ and found a very strong coupling between the retentions of HCl and $CaCl_2$ on medium tight CA membranes. The retention curves, shown in Figure 12, have a quite abnormal shape, indicating a rather complex interaction between the fluxes of the individual H^+, Ca^{++} and Cl^- ions. As seen, the retention of the H^+–ion in a 10^{-3} N HCl solution decreases from +65% to −1100% by increasing the

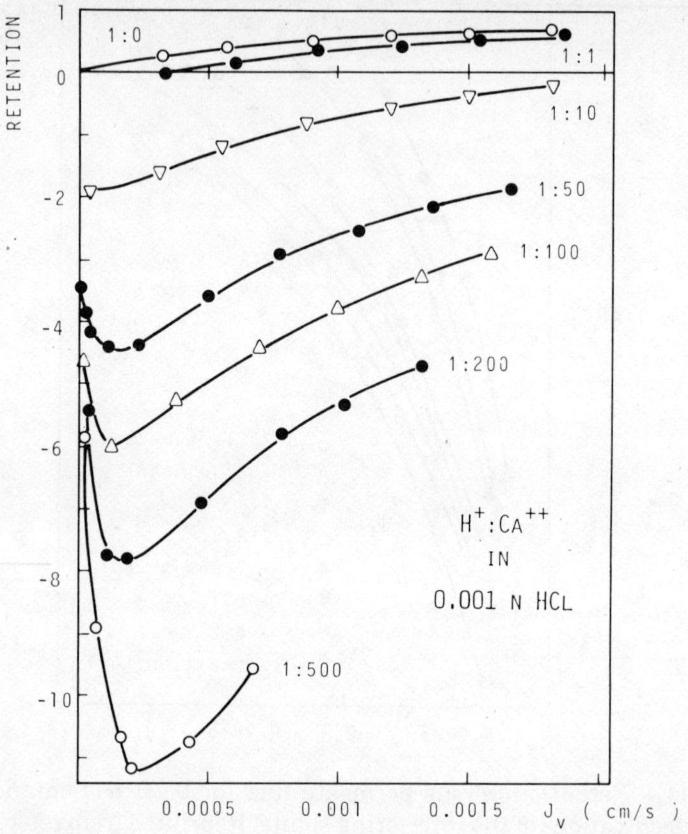

FIGURE 12.—Retention versus permeate flux for the H^+ – ion in a 0.001 N HCl solution at different $CaCl_2$ concentrations. The numbers indicated are the normality ratio H^+: Ca^{++}. Reprinted from reference (26).

concentration of $CaCl_2$ from 0 to 0.5 N. Further the retention curve for the H^+–ion shows a minimum at quite low permeate fluxes. With increasing $CaCl_2$ concentration this minimum goes to higher flux values. The experimental results can be correlated by the combined viscous flow and frictional model with the extension of a flux-dependent electrical potential term together with a boundary potential, causing the distribution coefficients of the individual ions to be concentration dependent.

The reason for the minimum in the retention of the H^+–ion can be explained from the three induced potential terms: The boundary potential on the permeate side has a maximum value corresponding to a pure HCl-solution, which is attained at a quite low permeate flux. At low permeate flux the potential gradient in the membrane is determined mainly by the diffusion potential arising from the diffusion of $CaCl_2$, which gives an increase in the H^+ concentration. With increasing permeate flux the streaming potential arising from the convective motion of HCl partly cancels the diffusion potential, thereby decreasing the H^+ concentration.

From a practical point of view the results show that the retention of the separate ions in multi-salt solutions is strongly dependent on the induced potentials, which are caused by all the ions present in the bulk solution. Therefore, the retention of a single ion can be quite different from the measured retention of the same ion in a single-salt solution. Thus the results explain why pH is normally found to be one pH unit lower in the permeate than in the bulk solution, when using reverse osmosis on natural waters.

5. ASYMMETRY IN MEMBRANES

In reverse osmosis and ultrafiltration the application of asymmetric or composite membranes creates no difficulties if the porous sublayer is mechanically stable and has a high hydraulic permeability, so that the pressure gradient is mainly across the skin layer. The problem arises when these membranes are used in situations in which the concentration at the porous layer side is no longer determined by the ratio of the solute flux to the volume flux, as given by equation 25.

The overall reflection coefficient of an asymmetric membrane is related to the properties of the skin layer (sl) and the porous sublayer (pl) by the relation:

$$\sigma = \sigma_{sl} /(1 + \omega_{sl}/\omega_{pl}) \tag{29}$$

Jonsson (27) has recently shown how the solute permeabilities, ω, of the individual layers can be determined from two sets of reverse osmosis experiments. It was found that the ratio ω_{sl}/ω_{pl} varies greatly with the selectivity of the membrane giving a non-linear relation between σ and σ_{sl}. However, the maximal retention, $\mathcal{R}_{max}$, and σ_{sl} seem to be fairly equally independent of the pressure level, thus confirming the theoretically derived relation $\mathcal{R}_{max} = \sigma_{sl}$. For solutes with low selectivity ($\sigma_{sl} < 0.2$), it was shown that the diffusion resistance in the skin layer was less than 10% of that in the porous sublayer. Therefore the porous layer will often be the diffusion determining layer when asymmetric membranes are used for dialysis purposes.

This phenomenon is one of the reasons why the process called "Pressure Retarded Osmosis" (PRO) often works quite badly. Loeb and Mehta (28) have reported some PRO measurements on two hollow fiber membranes with quite different thicknesses of the porous layer. They found that the volume flux through the membranes could not be expressed by the simple solution–diffusion model but rather by a two-coefficient transport equation such as:

$$-J_V = L_\pi \, \Delta\pi - L_p \, \Delta p \tag{30}$$

where L_π and L_p are the water permeation coefficients from a pure osmotic driving force and a pure hydraulic driving force, respectively. When the pressure exceeds the threshold value, $(L_\pi/L_p)\,\Delta\pi$, the volume flux reverses for the PRO process.

For the B–10 fiber, which has a total wall thickness of 28 μm, L_p/L_π was equal to 1.4. For the FRL fiber, which has a much greater total wall thickness of 83 μm, L_p/L_π was equal to 2.4. Instead of using equation 30 one should use the more general equation:

$$-J_V = L_p\,(\sigma\Delta\pi - \Delta p) \tag{31}$$

where the overall reflection coefficient for the asymmetric fiber is related to the threshold ratio by the relation:

$$\sigma = \frac{L_\pi}{L_p} = \frac{\sigma_{sl}}{1 + \omega_{sl}/\omega_{pl}} \tag{32}$$

Assuming that the reflection coefficient for the skin layer, σ_{sl}, is equal for the two fibers as they have almost equal reverse osmosis properties, the ratio of ω_{sl}/ω_{pl} for the two fibers can be related to the threshold ratios as follows:

$$\frac{(\omega_{sl}/\omega_{pl})_{B-10}}{(\omega_{sl}/\omega_{pl})_{FRL}} = \frac{(L_p/L_\pi - 1)_{B-10}}{(L_p/L_\pi - 1)_{FRL}} = 0.29 \tag{33}$$

This value is quite close to the ratio of the total wall thickness: 28/83 = 0.34. Thus the calculations confirm that the main reason for the poor results with asymmetric membranes in PRO compared with RO is caused by the thickness of the porous sublayer. The only way to overcome this problem is to use a membrane that has a σ_{sl} – value very close to 1, in which case the ratio ω_{sl}/ω_{pl} can become quite low. However, this normally means that the hydraulic permeability will be low, too.

REFERENCES

1. Jonsson, G.: 1978, "The influence of pressure on the compaction of asymmetric cellulose acetate membranes." Proc. 6th Intern. Symp. on Fresh Water from the Sea 3, pp. 203–212.
2. Jonsson, G.: 1983, "Concentration profiles and retention-flux curves for composite membranes in reverse osmosis." J. Membrane Sci. 14, pp. 211–227.
3. Merten, U.: 1966, in Desalination by Reverse Osmosis. Merten, U. (ed.) Cambridge, Mass.: MIT Press, pp. 15–54.
4. Jonsson, G. and Boesen, C.E.: 1975, "Water and solute transport through cellulose acetate reverse osmosis membranes." Desalination 17, pp. 145–165.
5. Jonsson, G.: 1978, "Methods for determining the selectivity of reverse osmosis membranes." Desalination 24, pp. 19–37.
6. Burghoff, H.-G., Lee, K.L., and Pusch, W.: 1980, "Characterization of transport across cellulose acetate membranes in the presence of strong solute–membrane interactions." J. Appl. Polym. Sci. 25, pp. 323–347.
7. Spiegler, K.S.: 1958, "Transport processes in ionic membranes." Trans. Faraday Soc. 54, pp. 1409–1428.
8. Spiegler, K.S., and Kedem, O.: 1966, "Thermodynamics of hyperfiltration (reverse osmosis): Criteria for efficient membranes." Desalination 1, pp. 311–326.
9. Blatt, W.F., Dravid, A., Michaels, A.S., and Nielsen, L.: 1970, in Membrane Science and Technology, Flinn, J.E. (ed.) New York: Plenum Press, pp. 47–97.

10. Ferry, J.D.: 1936, "Statistical evaluation of sieve constants in ultrafiltration." J. Gen. Physiol. 20, p. 95.
11. Faxen, H.: 1922, "Die Bewegung einer starren Kugel längs der Achse eines mit zäner Flüssigkeit gefütten Rohres." Ark. Mat. Astron. Fysik 17, p. 27.
12. Jonsson, G.: 1977, "Concentration polarization in a reverse osmosis test cell." Desalination 21, pp. 1–10.
13. Gotch, F.A.: 1980, "A quantitative evaluation of small and middle molecule toxicity in therapy of uremia." Dialysis and Transplantation 9, pp. 183–194.
14. Sourirajan, S., and Matsuura, T.: 1977, in Reverse Osmosis and Synthetic Membranes. Sourirajan, S. (ed.) Ottawa: National Research Council Canada, pp. 5–43.
15. Kozak, J.J., Knight, W.S., and Kaufmann, W.: 1968, "Solute–solute interactions in aqueous solutions." J. Chem. Phys. 48, pp. 675–690.
16. Demisch, H.-U., and Pusch, W.: 1976, "Ion exchange capacity of cellulose acetate membranes." J. Electrochem. Soc. 123, p. 370.
17. Arden, T.V.: 1968, Water Purification by Ion Exchange. London: Butterworths.
18. Matsuura, T., Bednas, M.E., and Sourirajan, S.: 1974, "Reverse osmosis separation of single and mixed alcohols in aqueous solutions using porous cellulose acetate membranes." J. Appl. Polym. Sci. 18, pp. 567–588.
19. Matsuura, T., and Sourirajan, S.: 1971, "Reverse osmosis separation of some organic solutes in aqueous solutions using porous cellulose acetate membranes." Ind. Eng. Chem. Process Des. Develop. 10, pp. 102–108.
20. Boesen, C.E., and Jonsson, G.: 1978, "Solute–solute interactions by reverse osmosis of three component solutions." Proc. 6th Intern. Symp. on Fresh Water from the Sea 3, pp. 157–164.
21. Erickson, D.L., Glater, J., and McCutchan, J.W.: 1966, "Selective properties of high flux cellulose acetate membranes toward ions found in natural waters." Ind. Eng. Chem. Prod. Res. and Develop. 5, pp. 205–211.
22. Hodgson, T.D.: 1970, "Selective properties of cellulose acetate membranes towards ions in aqueous solutions." Desalination 8, pp. 99–138.
23. Heyde, M.E., and Anderson, J.E.: 1975, "Ion sorption by cellulose acetate membranes from binary salt solutions." J. Phys. Chem. 79, pp. 1659–1664.
24. Donnan, F.G.: 1924, "The theory of membrane equilibria." Chem. Rev. 1, p. 73.
25. Lonsdale, H.K., Pusch, W., and Walch, A.: 1975, "Donnan-membrane effects in hyperfiltration of ternary systems." J. Chem. Soc., Faraday Trans. 71, p. 501.
26. Jonsson, G.: 1980, "Coupling of ion fluxes by boundary-, diffusion- and streaming-potentials under reverse osmosis conditions." Proc. 7th Intern. Symp. on Fresh Water from the Sea 2, pp. 153–163.
27. Jonsson, G.: 1980, "The influence of the porous sublayer on the salt rejection and reflection coefficient of asymmetric CA membranes." Desalination 34, pp. 141–157.
28. Loeb, S., and Mehta, G.D.: 1979, "A two-coefficient water transport equation for pressure-retarded osmosis." J. Membrane Sci. 4, pp. 351–362.
29. Jonsson, G.: 1980, "Overview of theories for water and solute transport in UF/RO membranes." Desalination 35, pp. 21–38.

SYMBOLS

A_i	area available for transport of i
b	friction parameter
K_{ip}	distribution coefficient of solute between pore fluid and bulk solution
α	clearance ratio
ϵ	fractional pore area
ξ	longitudinal coordinate inside pore

Subscripts

il	intermediate layer
m	membrane
p	in the pore
il	intermediate layer
pl	porous layer
sl	skin layer
w	water

CONCENTRATION POLARIZATION IN REVERSE OSMOSIS AND ULTRAFILTRATION

Mark C. Porter

Consultant
3449 Byron Court,
Pleasanton, California 94566

The major limiting factor in pressure-driven membrane processes such as ultrafiltration and reverse osmosis is concentration polarization. This paper discusses its effect on membrane flux and solute retention both theoretically and experimentally. Fluid management techniques effective in reducing concentration polarization are reviewed.

P. M. Bungay et al. (eds.), Synthetic Membranes: Science, Engineering and Applications, 367–388.

1. INTRODUCTION

The development of the anisotropic membrane in the late fifties by Loeb and Sourirajan (1) was the major breakthrough that made reverse osmosis (RO) and ultrafiltration (UF) commercially feasible. The membranes were said to be "skinned," having a very thin retentive film integrally supported by a finely porous substrate (see Figure 1). This gave the membrane sufficient strength while the thin skin reduced the resistance to flow. Further, solutes that penetrate the skin usually pass through the membrane with the filtrate, resulting in virtually no entrapment of solutes within the pore network.

However, the development of anisotropic membranes is only half the battle. The high-flux characteristics of these membranes result in rapid convection of retained solutes to the membrane surface, in turn resulting in the well-known phenomenon of concentration polarization. This accumulation of solute at the membrane interface can severely limit the flux, leading to an apparent fouling of the membrane, which some workers have interpreted as pore blockage. That this is not the case is demonstrated by the easy restoration of the initial flux simply by washing the surface of the membrane.

In RO, where the solutes retained can have a significant osmotic pressure, concentration polarization can result in osmotic pressures considerably higher than those represented by the bulk stream concentration. Higher pressures are required to overcome the osmotic pressure and produce reasonable flux values.

FIGURE 1.—Cross-sectional view of a "skinned" membrane.

For UF, where macromolecular solutes and colloidal species are involved, the concentration at the membrane surface can rise to the point of incipient gel precipitation, forming a dynamic secondary membrane on top of the primary structure. This secondary membrane can offer the major resistance to flow.

Therefore, the second major hurdle to be overcome in the development of a practical industrial unit operation is concentration polarization.

2. REVERSE OSMOSIS

2.1. Theory

Figure 2 is a schematic of the boundary layer formation in RO. The transport of water through the membrane with a volumetric flux J_V carries solute with it to the membrane surface at a transport rate of $J_i = J_V c_i'$ where c_i' is the concentration of solute in feed solution. Because the solute is mostly retained by the membrane, the concentration at the wall ($c'_{i,\mathrm{int}}$) increases until the concentration-dependent back diffusion, $D_i \dfrac{\mathrm{d}c_i'}{\mathrm{d}x}$, plus the solute leakage through the membrane $B_i(c_i' - c_i'')$, is just equal to the transport of solute to the membrane.

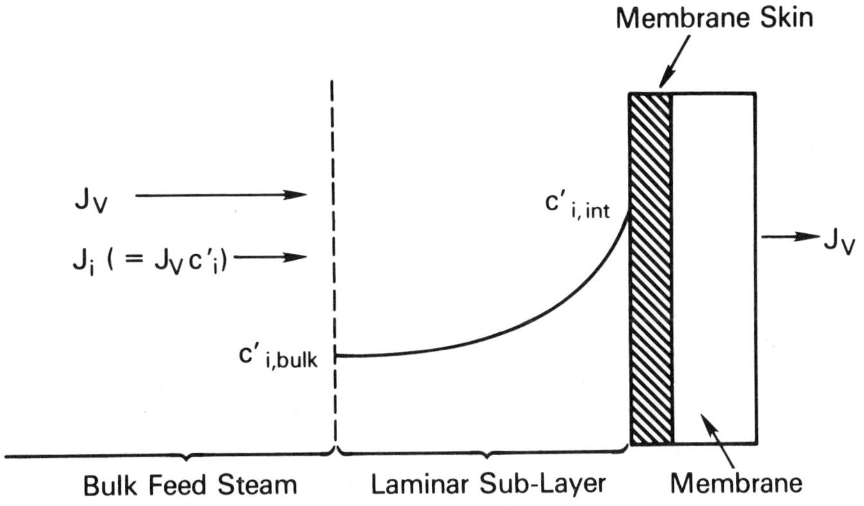

FIGURE 2.—Concentration polarization in reverse osmosis.

In RO, the solvent flux is represented by:

$$J_V = A_V\,(\Delta p - \Delta\pi), \tag{1}$$

where J_V = volumetric flux of solvent
A_V = solvent permeation coefficient
Δp = differential pressure across the membrane
$\Delta\pi$ = differential osmotic pressure across the membrane

On the other hand, the solute flux is represented by:

$$J_i = B_i\,(c_i' - c_i''), \tag{2}$$

where J_i = solute flux,
B_i = permeation membrane constant for a specified solute (takes into account solubility of solute in membrane and diffusibility across membrane).

Therefore, a steady state mass balance for the laminar sub-layer results in the following equations (2):

$$J_V\,c_i' - D_i'\frac{dc_i'}{dx} = B_i\,(c_i' - c_i''), \tag{3}$$

where D_i' = solute diffusion coefficient in the feed solution,
x = distance parameter in solution perpendicular to membrane surface,

and

$$\frac{c'_{i,\,\mathrm{int}}}{c'_{i,\,\mathrm{bulk}}} = \exp\left(\frac{J_V\,\delta}{D_i'}\right), \tag{4}$$

where $c'_{i,\,\mathrm{bulk}}$ = "bulk" solute concentration far from membrane surface,
$c'_{i,\,\mathrm{int}}$ = interfacial or "wall" concentration,
δ = characteristic boundary layer distance in solution.

The ratio, $c'_{i,\,\mathrm{int}}/c'_{i,\,\mathrm{bulk}}$, is often referred to as the "polarization modulus." It is increased by a higher water flux and decreased by a reduction in the boundary layer thickness, δ. Because we want to maximize J_V and we have little control over D_i', fluid management techniques designed to reduce δ and consequently increase the concentration gradient for back diffusion seem to be the most practical approach.

Using the mass transfer–heat transfer analogies, δ may be calculated for turbulent flow in tubular membranes by the following equations:

$$\delta = \frac{2(D_i')^{1/3}\,\nu^{2/3}\,Re^{1/4}}{0.08\,\bar{u}} \qquad (5)$$

$$= \frac{35.3(D_i')^{1/3}\,\nu^{0.42}\,d_t^{1/4}}{\bar{u}^{3/4}}, \qquad (6)$$

where ν = kinematic viscosity,
d_t = tube diameter,
$\bar{u}$ = mean velocity of the feed stream parallel to the membrane,
Re = Reynolds number $= \frac{\bar{u}\,d_t}{\nu}$.

Substituting equation 6 into equation 4 gives

$$\frac{c'_{i,\text{int}}}{c'_{i,\text{bulk}}} = \exp\left[\frac{35.3 J_V\,\nu^{0.42}\,d_t^{1/4}}{(D_i')^{2/3}\,\bar{u}^{3/4}}\right]. \qquad (7)$$

Thus, the polarization modulus may be reduced by decreasing the tube diameter or increasing the cross-flow velocity $\bar{u}$. The latter is more important as it appears to the 0.75 power; there is a weak dependence on tube diameter.

In laminar flow, the solution is more complex because the boundary layer is developing all the way down the tube. The results do not appear in closed form and finite difference methods have been used (3). The essential difference is that tube diameter is much more important in laminar flow. This is demonstrated in Figure 3, which shows the polarization modulus for tubular membranes with a 50% water recovery and a flux of 10 gals/(ft^2 · day). Flow was assumed to be laminar for values of the Reynolds number at the channel inlet less than 2100 and turbulent for values at the channel outlet greater than 2100. Because half the water is removed through the membrane, there is a range representing a factor of 2 in inlet velocity which separates these two regions; in this transition region, a dashed line is drawn to connect the laminar and turbulent flow solutions.

Note that since the water recovery is fixed at 50% in Figure 3, the polarization modulus is independent of the inlet channel velocity $\bar{u}_{\text{In}}$ for laminar flow but decreases with inlet channel velocity in turbulent flow. This curious result is explained by the fact that even though in laminar flow the feed velocity $\bar{u}_{\text{In}}$ affects the polarization at each point along the length of the

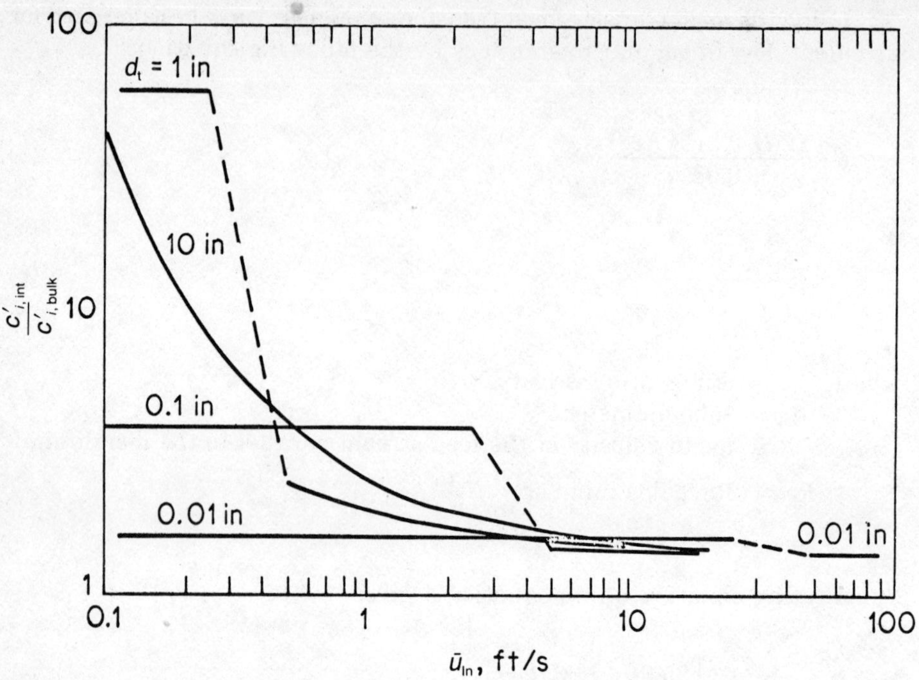

FIGURE 3.—Theoretical average concentration polarization modulus for tubular reverse osmosis membranes with a 50% water recovery and an average flux of 10 gal/(ft^2) (day) as a function of inlet channel velocity.

channel, the length of the channel required for a given recovery increases with increasing velocity, resulting in a constant average polarization modulus.

Also note in Figure 3 that the polarization modulus decreases as the tube diameter decreases in either laminar or turbulent flow, but the effect is generally more pronounced in the laminar-flow regime.

2.2. Effect of Concentration Polarization on Salt Retention

Equation 2 indicates that the solute flux will increase with a higher differential solute concentration across the membrane. This is demonstrated in Figure 4, which shows the effect of brine velocities on Cl^- rejection. Higher velocities increase the rejection by virtue of decreasing the polarization modulus. Higher flux membranes show the most marked improvement.

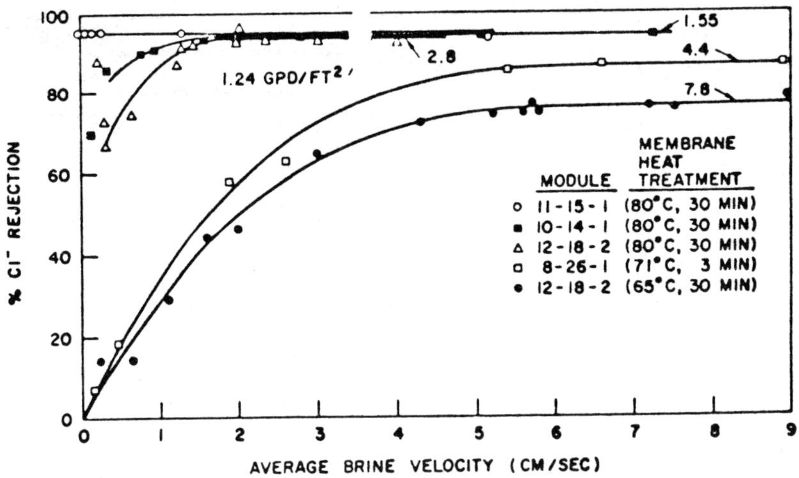

FIGURE 4.—Effect of brine velocities on Cl^- rejection by heat-treated membranes.

2.3. Effect of Concentration Polarization on Membrane Flux

Equation 1 indicates that the solvent flux J_V is decreased by an increased osmotic pressure differential. The osmotic pressure π due to solute species i may be approximated for dilute solutions by the well-known van't Hoff equation

$$\pi = \nu_i c_i RT \tag{8}$$

where c_i = molar concentration of the solute
R = gas constant
T = absolute temperature
ν_i = number of ions formed if the solute dissociates (e.g., for NaCl, $\nu = 2$; for $BaCl_2$, $\nu = 3$)

Inasmuch as the osmotic pressure difference $\Delta\pi$ is associated with the concentration of solutes at the membrane surface, an increased polarization modulus will result in a higher $\Delta\pi$ and a reduced solvent flux J_V. This is demonstrated in Figure 5, which shows the effects of velocity (feed flow rate) on membrane flux. Higher velocities decrease the polarization modulus, decreasing the osmotic pressure difference and increasing the flux.

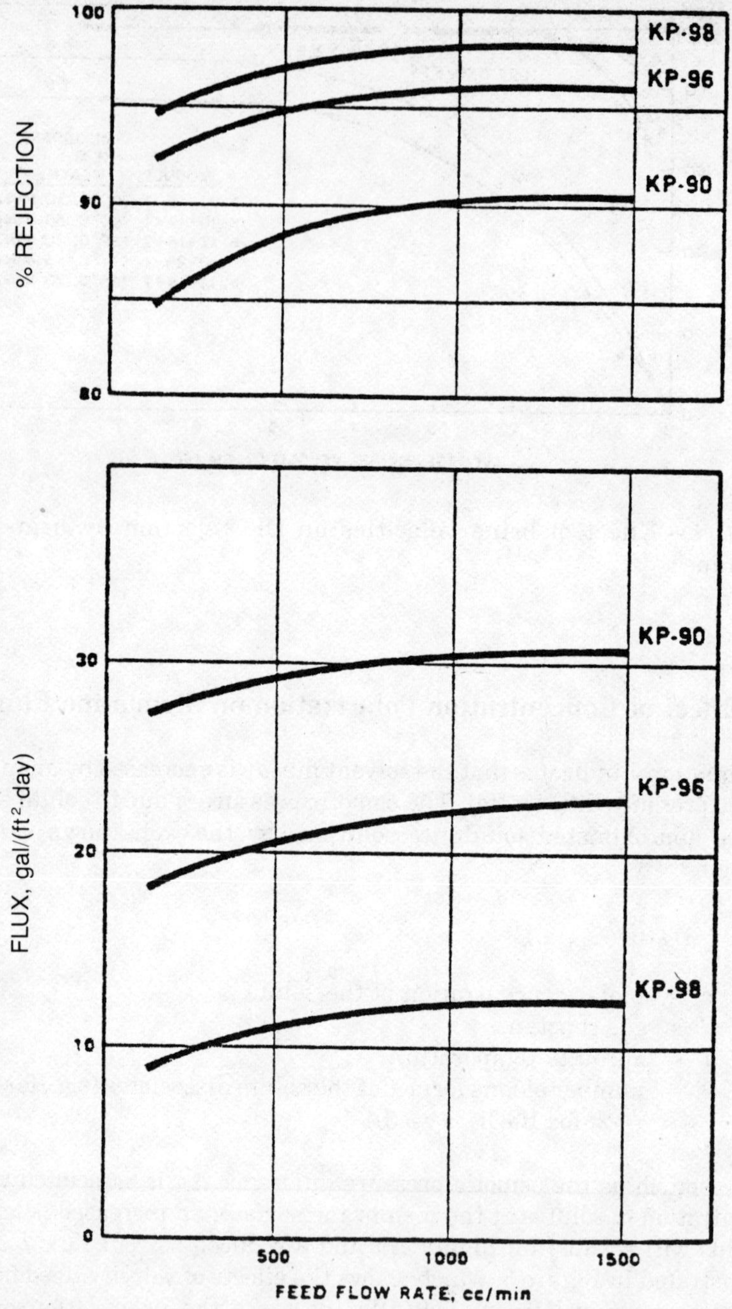

FIGURE 5.—Effect of feed velocity on performance of Eastman Membrane (typical values in Eastman's Test Cell at 600 psi and 78°F.)

2.4. Fluid Management Techniques to Reduce Concentration Polarization

As already mentioned, maintenance of a high axial velocity is important in reducing the polarization modulus. As water is removed from the tube, for a given recovery, the velocity will drop, resulting in increased salt passage and decreased water flux (see Figure 6). To alleviate this effect and maintain a high axial velocity, a "Christmas tree" configuration or "tapered system" is sometimes used (see Figure 15 of chapter on reverse osmosis).

In addition to maintaining high-sweep velocities, turbulence promoters can also be effective in reducing the polarization modulus. Figure 7 shows the effect of a promoter in increasing salt rejection.

Other techniques such as pulsed flow and secondary flow effects have also resulted in improvements.

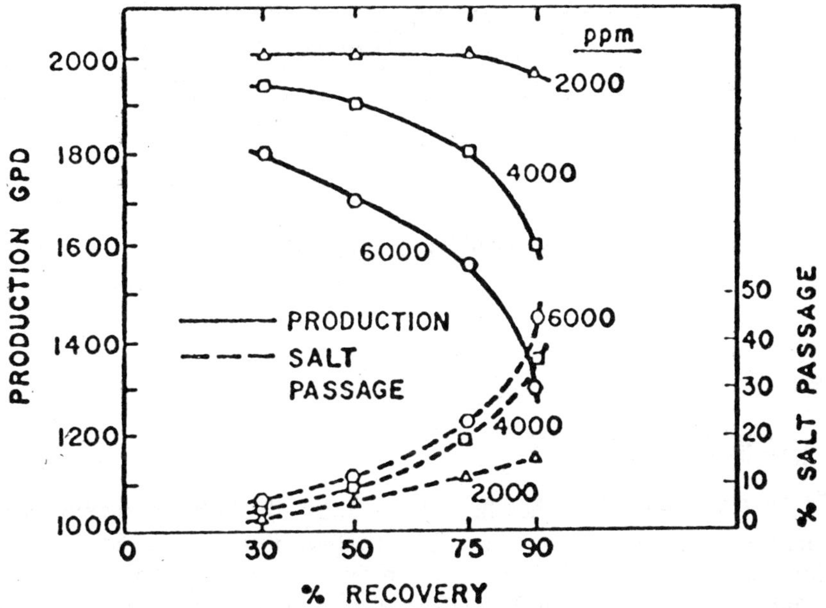

FIGURE 6.—Influence of recovery level on transmembrane flux of water and passage of salt in reverse osmosis.

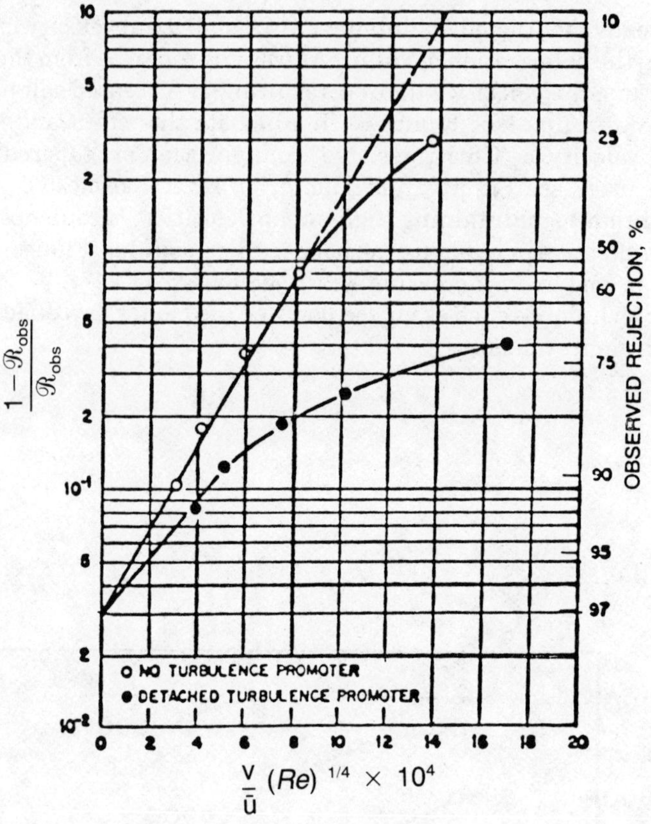

FIGURE 7.—Effect of twisted tape turbulence promoters.

3. ULTRAFILTRATION

For UF, the macromolecular solutes and colloidal species usually have insignificant osmotic pressures. Nevertheless, flux is affected along with retention by concentration polarization and the formation of a gel-layer or secondary membrane. In RO, the precipitation of macromolecules or salts on the membrane surface would be called "fouling." In UF, the higher flux often drives the concentration at the wall above the solubility limit and we are accustomed to operating in the gel-polarized regime (see Figure 8).

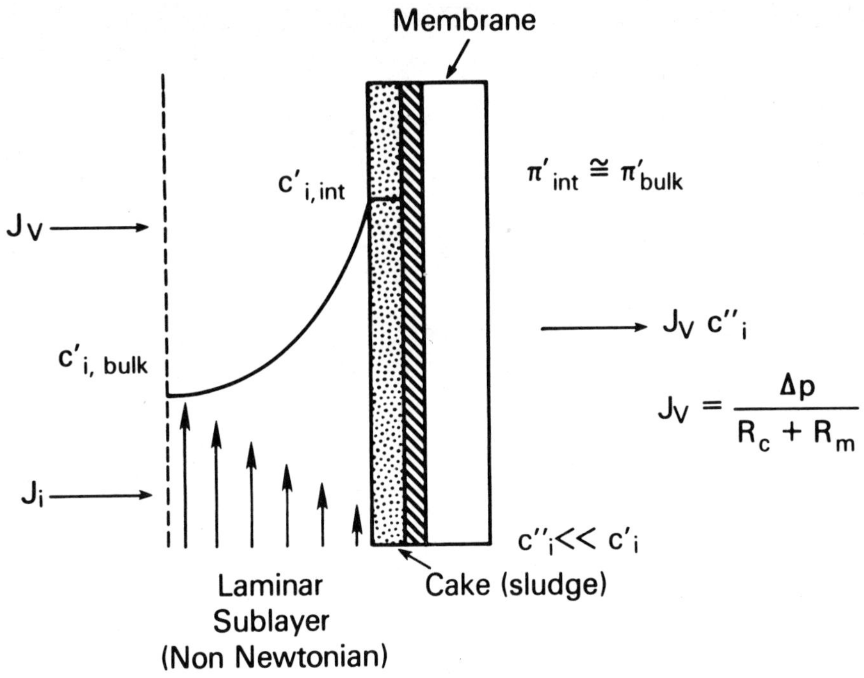

FIGURE 8.—Boundary layer formation in ultrafiltration.

3.1. Effect of Concentration Polarization on Retention

For dilute solutions without gel formation, the rejection of solutes is not unlike the mechanism discussed for RO. As shown in Figure 9, increases in transmembrane pressure increase the flux and the polarization modulus, with a resultant decrease in rejection.

For more concentrated solutions, after incipient gel precipitation, the concentration at the membrane surface is fixed and solute passage occurs at a constant rate irrespective of changes in the feed concentration.

The biggest disappointment is what happens with binary solutes. For example, the pharmaceutical industry is keenly interested in having a UF membrane capable of fractionating albumin/globulin mixtures. In fact, there are membranes available that offer better than 95% retention for gamma globulin and less than 10% retention for albumin. However, as shown in Figure 10, the presence of even small amounts of gamma globulin forms a gel layer on the membrane that is retentive for the smaller albumin molecule. This limits our ability to fractionate various proteins. On a laboratory basis, one can dilute the mixture, perform the fractionation, and then reconcentrate the two fractions. However, this is too cumbersome for a production process.

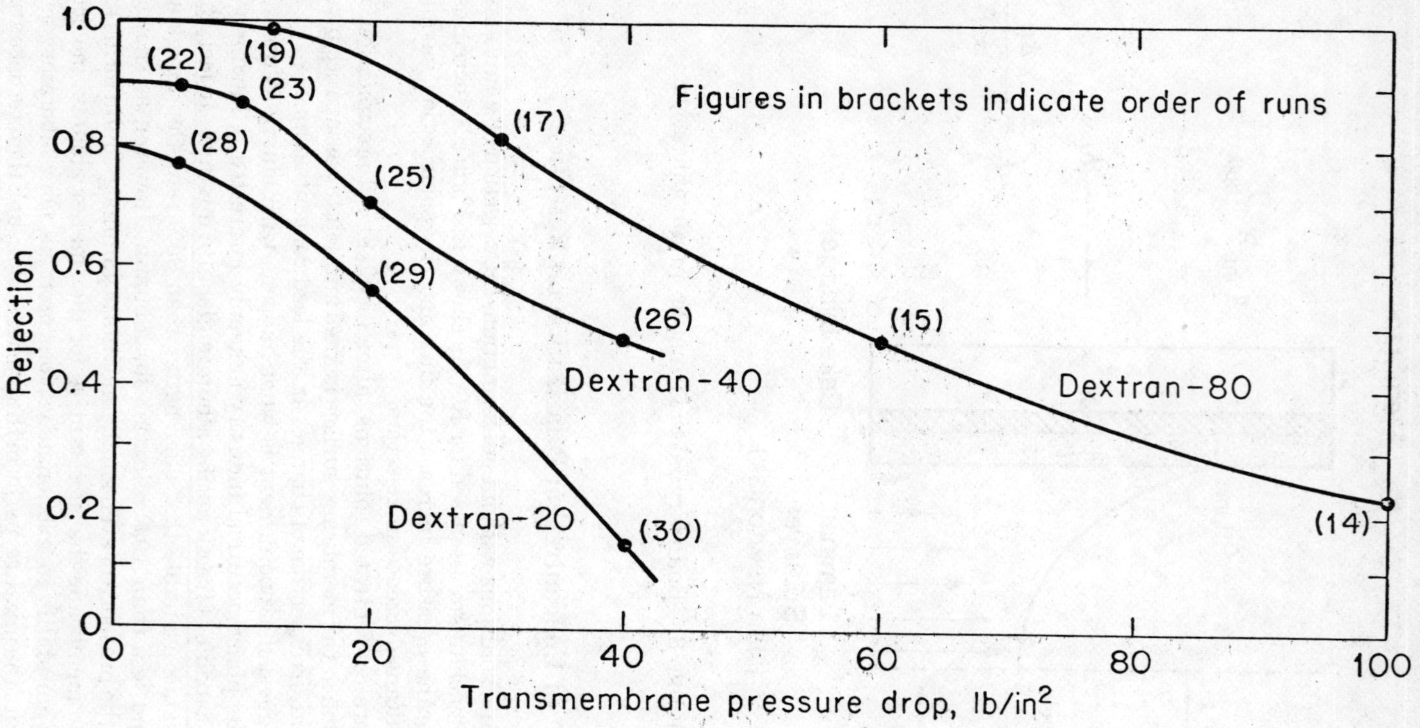

FIGURE 9.—Decrease in dextran rejection with increasing pressure difference across ultrafiltration membrane.

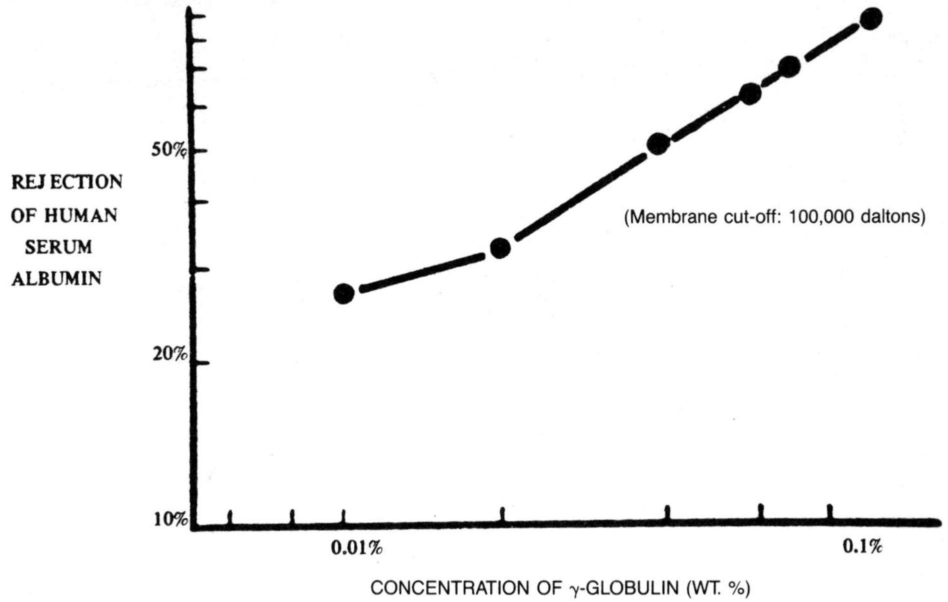

FIGURE 10.—Rejection of albumin by gamma-globulin polarization layer.

3.2. Effect of Concentration Polarization on Membrane Flux

Figure 11 shows flux data as a function of transmembrane pressure drop. The curious thing about these data is that at high enough protein concentrations and pressures the flux is invariant with pressure. Apparently the gel layer resistance R_c increases to offset any increases in Δp

$$J_V = \frac{\Delta p}{R_c + R_m}, \tag{9}$$

where R_m is the membrane resistance.

At steady state and with no leakage of solute, equation 3 applies within the laminar sublayer

$$J_V c_i' - D_i' \frac{d c_i'}{d x} = 0. \tag{10}$$

For any increase in Δp, J_V increases, carrying more solute to the wall. However, the back-diffusive transport is fixed. Therefore, the gel layer increases in thickness or is compacted by the increase in Δp such that the resistance to flow increases, reducing J_V to a level balancing the back-diffusive transport.

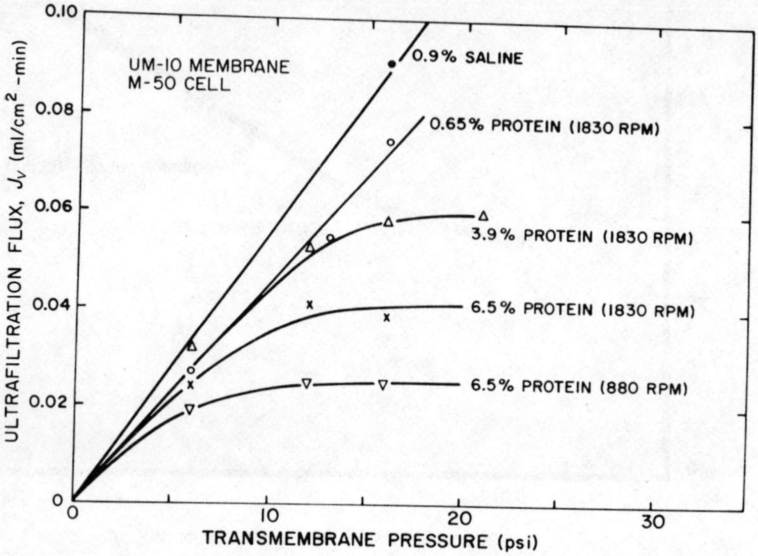

FIGURE 11.—Effect of transmembrane pressure difference on ultrafiltration flux for bovine serum albumin solutions.

Note in Figure 11 that lower protein concentrations have a higher asymptotic flux level and the threshold pressure to achieve pressure independence is higher. This is because the driving force to remove material from the wall is higher. Assuming the gel concentration, $c_g' = c'_{i,\text{gel}}$, is fixed, lower bulk concentrations, $c_b' = c'_{i,\text{bulk}}$, provide a higher concentration gradient $\mathrm{d}c_i'/\mathrm{d}x$.

Likewise, note that higher stirrer speeds also result in higher asymptotic flux levels and threshold pressures. This is because the boundary layer thickness has been decreased, again providing a higher concentration gradient for back-diffusive transport.

Figure 12 demonstrates that the gel layer is indeed the limiting resistance to flow. Three membranes of widely different water permeabilities have essentially the *same* flux with proteins if the transmembrane pressure drop is sufficiently high. The decrease in flux with the XM-100 membrane may be due to membrane compaction.

The gel polarization model enables us to fix the boundary conditions and solve the differential equation 10

$$J_V = k_i \ln (c_g'/c_b'), \tag{11}$$

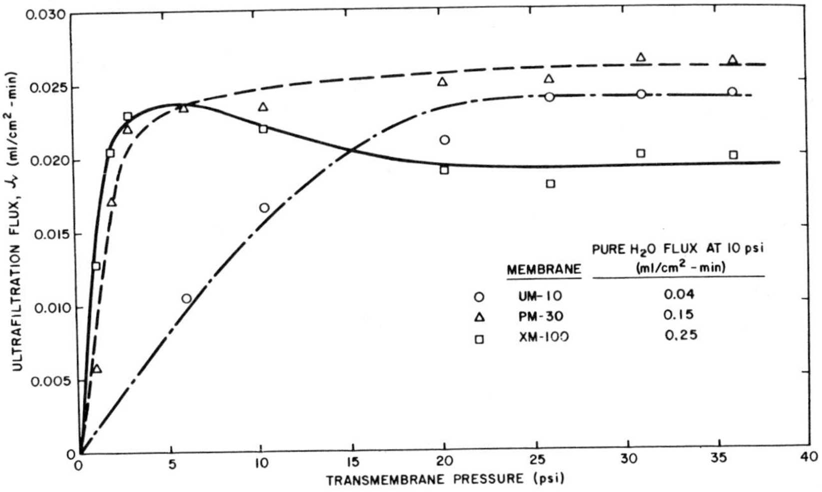

FIGURE 12.—Flux-pressure relationships for human plasma in stirred cell by use of membranes of differing water permeabilities.

where k_i is the solute mass transfer coefficient,

$$k_i = D_i'/\delta. \tag{12}$$

Equation 11 indicates that under conditions where the gel polarization model holds, the membrane flux is invariant with respect to the transmembrane pressure drop or membrane permeability and is dependent only on the solute characteristics (D_i' and c_g') and the boundary layer thickness δ. Thus fluid management techniques must be directed toward decreasing the boundary layer thickness, i.e., toward increasing the mass transfer coefficient k_i.

The validity of equation 11 has been demonstrated for a large number of macromolecular solutes and colloidal species. The data of Figure 13 show the semi-logarithmic variation of flux with concentration for two proteins and two colloidal suspensions. Note that the gel concentration c_g' may be determined from these plots as the intercept with the abscissa (where the flux drops to zero and $c_b' = c_g'$); it is higher for colloidal suspensions (60–70%) than for protein solutions (25–45%). This is not without significance: in the case of colloidal suspensions, the gel layer would be expected to resemble a layer of close-packed spheres having 65–75% solids by volume.

The data of Figure 14 show the increase in the slope of the J_V *vs* ln c_b' plot (i.e., the mass transfer coefficient k_i) for higher recirculation rates (tangential velocities). Higher velocities would be expected to decrease the boundary layer thickness—increasing k_i.

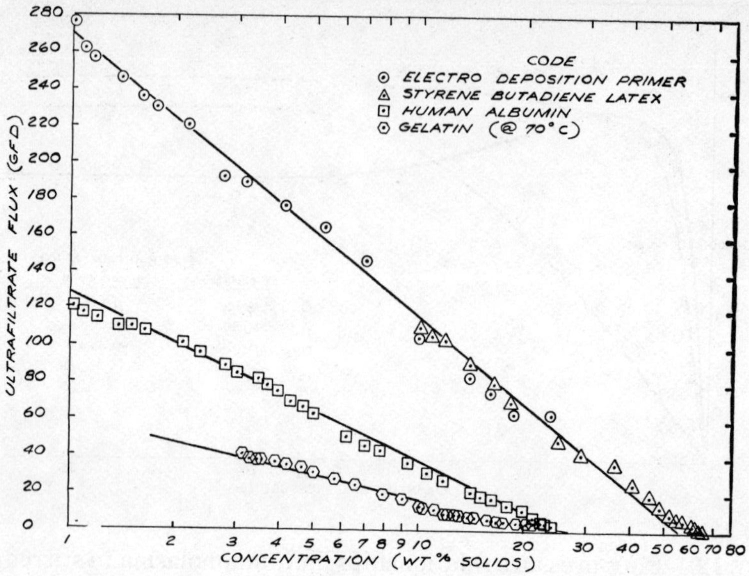

FIGURE 13.—Variation of flux with concentration for several solutions and suspensions by use of thin-channel ultrafiltration.

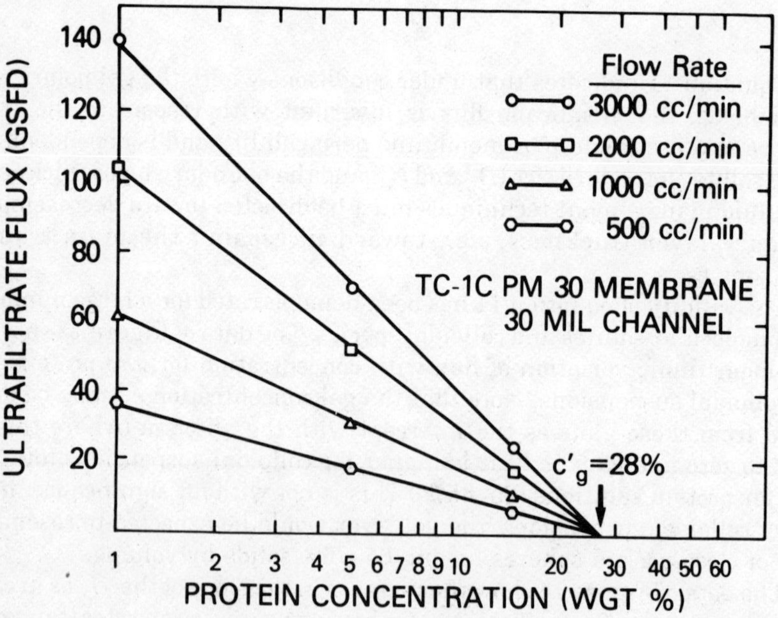

FIGURE 14.—Cross plot of flux vs. recirculation rate data (TC–1C, PM30 membrane, 30-mil channel).

3.3. Evaluation of the Mass-Transfer Coefficient

The mass transfer/heat transfer analogies well-known in the chemical engineering literature make possible an evaluation of the mass-transfer coefficient k_i and provide insight into how membrane geometry and fluid-flow conditions can be specified to optimize flux. A detailed derivation, using the Graetz and Lévêque solutions for convective heat transfer in laminar flow and the Dittus-Boelter correlation for fully developed turbulent flow, is presented in reference (4). The final solutions are given here:

For laminar flow:

$$k_i = 1.62 \left[\frac{\bar{u}\,(D_i')^2}{d_h\,L} \right]^{0.33}, \tag{13}$$

where d_h = equivalent hydraulic diameter, ($d_h = 2\lambda$ for flat rectangular channels of height λ)

L = channel length,

or

$$k_i = 0.816 \left[\frac{6\,Q\,(D_i')^2}{\lambda^2\,wL} \right]^{0.33}, \tag{14}$$

where Q = feed stream volumetric flow rate,

w = channel width.

Figure 15 shows the 0.33 power dependence on the wall shear rate (proportional to $\bar{u}/d_h$) for a variety of channel configurations and dimensions.

For turbulent flow:

$$k_i = 0.023 \frac{\bar{u}^{0.8}\,(D_i')^{0.67}}{d_h^{0.2}\,\nu^{0.47}} \tag{15}$$

or

$$k_i = 0.02 \frac{Q^{0.8}\,(D_i')^{0.67}}{\lambda\,w^{0.8}\,\nu^{0.47}} \tag{16}$$

Equations 13–16 show that for both laminar and turbulent flow, the mass transfer coefficient may be increased by increasing the channel velocity and decreasing the channel height. The flux is dependent on channel length in laminar flow because the boundary layer is still developing even at the end of the channel. Naturally, in turbulent flow there is no dependence on channel length because the boundary layer is developed rapidly at the channel inlet.

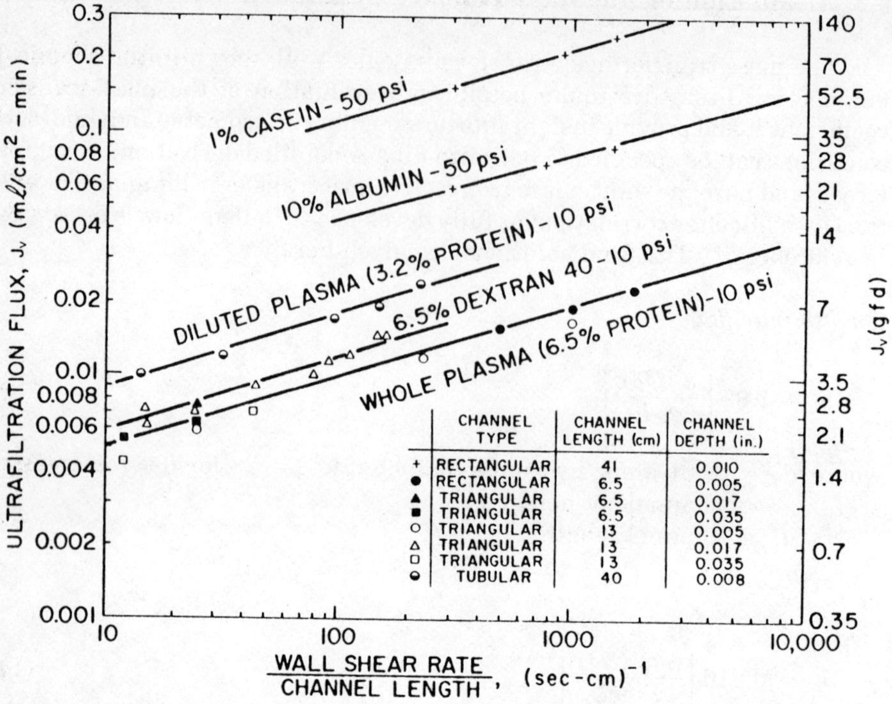

FIGURE 15.—Flux dependence on wall shear rate in laminar flow.

Equations 13–16 can be used satisfactorily to calculate quantitatively the flux for macromolecular solutions, but they are inaccurate for colloidal suspensions. Figure 16 shows excellent agreement between theory and experiment for human albumin solutions, but Figure 17 shows calculated flux values for latex that are more than an order of magnitude too low (4). An evaluation of more than 40 different colloidal suspensions in the author's laboratory have indicated that the diffusion coefficient calculated from the flux is generally from one to three orders of magnitude higher than the theoretical Stokes-Einstein diffusivity for colloidal suspensions.

It was first suggested by the author (4) in 1972 that the augmented colloidal particle mass transfer coefficients could be explained by the so-called "tubular-pinch effect". The lateral movement of suspended particles flowing down a tube toward the center-line of the tube was first observed in 1836, when Poiseuille noted that the region immediately adjacent to the walls of a capillary tended to be free of blood corpuscles. In 1961 and 1962, Segre and Silberberg (5) noted that rigid, spherical, neutrally buoyant particles carried along in Poiseuille flow also migrate radially. The lateral migration of rigid particles arises from fluid inertial effects (6).

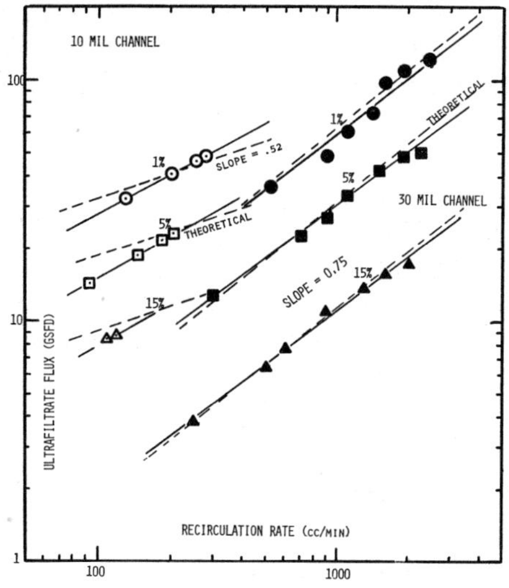

FIGURE 16.—Ultrafiltration of human albumin (TC–1C, PM30 membrane).

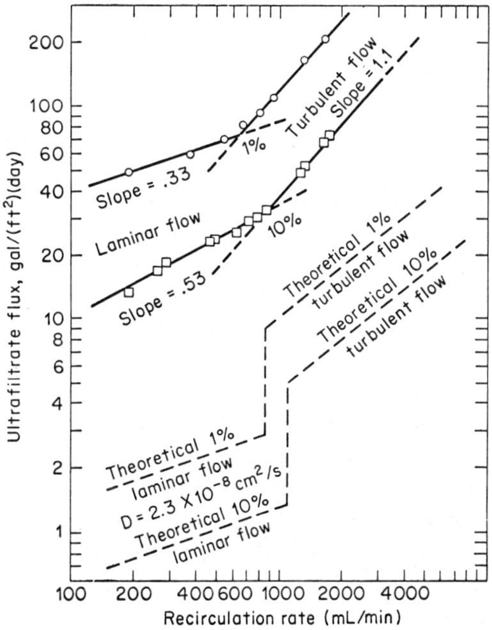

FIGURE 17.—Ultrafiltration of styrene butadiene latex suspension in a module with a 30-mil channel and PM30 membrane at an average transmembrane pressure difference of 40 psi.

The lateral migration velocity, v_L, is given by an expression of the form:

$$v_L = \bar{u}\, Re \left(\frac{b}{r_t}\right)^n F\left[\frac{r}{r_t}\right], \tag{17}$$

where
b = particle radius,
r_t = tube radius,
n = positive integer,
r = radial position of the particle in the tube,
$F[r/r_t]$ = function of the ratio, r/r_t, particle buoyancy and tube orientation.

Thus for colloidal suspensions, equation 10 must be modified to read:

$$J_V\, c_i' - D_i' \frac{d\, c_i'}{d\, x} - v_L\, c_i' = 0, \tag{18}$$

Integrating the above using equations 12 and 17 gives

$$J_V = \frac{F\, \bar{u}^2\, d_t}{\nu}\left(\frac{b}{r_t}\right)^n + k_i \ln\left(\frac{c_g'}{c_b'}\right). \tag{19}$$

Belfort (7) has obtained agreement between theory and experiment using the data of Porter (4) and taking into account the tubular-pinch effect.

These results suggest another technique besides fluid management and membrane geometry to reduce concentration polarization. The introduction of particulates to a solution may tend to transport adsorbed macromolecules away from the membrane surface thereby enhancing the flux. Dorr-Oliver has done this in commercial plants and Bixler obtained a patent (8) for improving the ultrafiltrate flux by introducing solid particulate materials into the process stream (see Figure 18).

A review of equations 13–16 might lead one to surmise that hollow fibers (see Figure 19) represent the optimum membrane geometry. Inside diameters of 0.2 to 2 mm can be spun easily and there is no parasite drag—i.e., there is no hydraulic friction against an inactive surface. Unfortunately, hollow fibers seldom operate in the gel-polarized regime because of the inherent limitation of the burst strength. This limits the inlet pressure to the fiber and thereby limits the maximum attainable velocity. Flux values are often below those obtainable in gel-polarized UF and are pressure dependent. New techniques for supporting and/or making stronger hollow fibers should overcome this problem.

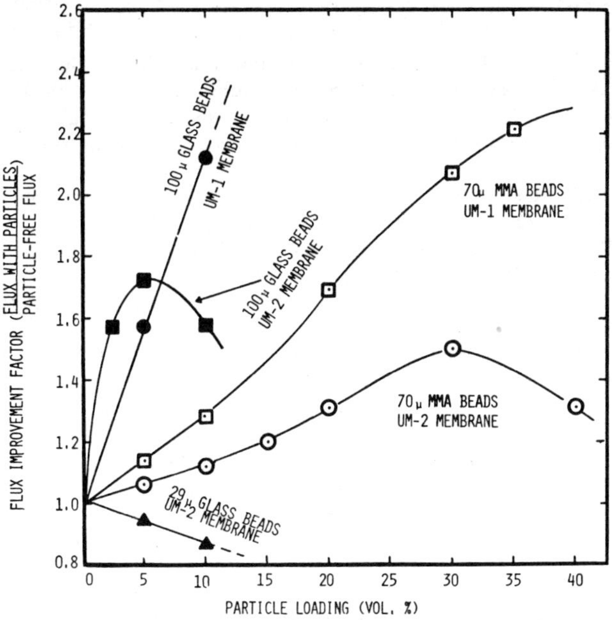

FIGURE 18.—Stirred cell flux improvement factor for various loadings of glass and MMA beads.

Type A hollow fiber cross section

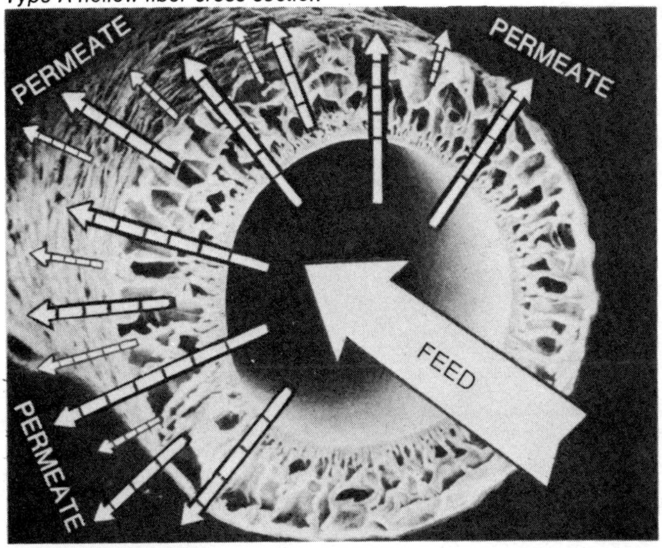

Hollow fiber fluid flow

FIGURE 19.—Hollow-fiber fluid flow. Type A hollow-fiber cross-section.

REFERENCES

1. Loeb, S.: 1981, "The Loeb-Sourirajan membrane: How it came about." Synthetic Membranes: Desalination (A.T. Turbak, ed.) Vol 1, ACS Symposium Series 153, ACS, Washington, D.C.
2. Lonsdale, H.K.: 1972, "Theory and practice of reverse osmosis and ultrafiltration." Industrial Processing with Membranes, R.E. Lacey and S. Loeb (Eds), John Wiley, New York, pp. 123–178.
3. Porter, M.C.: 1979, "Membrane filtration." section 2.1, Handbook of Separation Techniques for Chemical Engineers, (Ed. P.A. Schweitzer), McGraw-Hill, N.Y.
4. Porter, M.C.: 1972, "Concentration polarization with membrane ultrafiltration." I&EC Product Research and Development, vol 11, pp. 234–248.
5. Segre, G., and Silberberg, A.: 1962, "Behavior of macroscopic rigid spheres in Poiseuille flow." J. Fluid Mech. 14, pp. 115–157.
6. Brenner, H.: 1966, "Hydrodynamic resistance of particles at small Reynolds numbers." Advances in Chemical Engineering 6, pp. 287–438.
7. Belfort, G., and Green, G.: 1980, "Fouling of ultrafiltration membranes; Lateral migration and the particle trajectory model." Desalination 35, pp. 129–147.
8. Bixter, H.J., Rappe, G.C.: (Nov. 17, 1970). U.S. Patent 3,541,006.

SYMBOLS

F	particle radial migration function, equation 17
L	channel length
$\bar{u}$	feed stream mean velocity parallel to membrane
v_L	lateral migration velocity

Subscripts

b, bulk	interior of stream outside of boundary layer
g, gel	gel layer

MEMBRANE GAS SEPARATIONS—WHY AND HOW

William J. Ward III

General Electric Company
Schenectady, New York 12301

Significant advances are being made in the areas of membrane fabrication and packaging, and as a result it is reasonable to expect an increasing number of commercial membrane gas separation processes will be available. Therefore process engineers need to have an understanding of this area. To this end, the subjects to be covered in this chapter are:

(1) An analysis showing how the basic process variables of pressure difference across the membrane, pressure ratio across the membrane, and membrane thickness determine system performance.

(2) A discussion of available types of membranes, and membrane packages.

(3) An introduction to carrier transport membranes. This is a new class of membranes, not yet commercially available, which promises much higher performance than conventional polymeric membranes.

1. INTRODUCTION

2. THE PROCESS VARIABLES (p', p'', Δp, h)

3. GAS SEPARATIONS WITH CONVENTIONAL MEMBRANES AND PACKAGES

4. HIGH-PERFORMANCE MEMBRANES AND SYSTEMS

5. FACILITATED TRANSPORT

P. M. Bungay et al. (eds.), Synthetic Membranes: Science, Engineering and Applications, 389–401.

1. INTRODUCTION

A schematic membrane gas separation process is shown in Figure 1. A mixture of gases is fed to one surface of a membrane, assumed to be more permeable to species A than B. A pressure difference is maintained across the membrane by compressing the feed gas and/or evacuating the low pressure side of the membrane. The product streams are the low pressure extract, rich in A, and the high pressure raffinate, rich in B. Such a separation process is inherently attractive:

1. It is passive except for the compressor. No regeneration or replacement of components is required as it would be, for example, in an absorption process.
2. It may be extremely energy efficient.

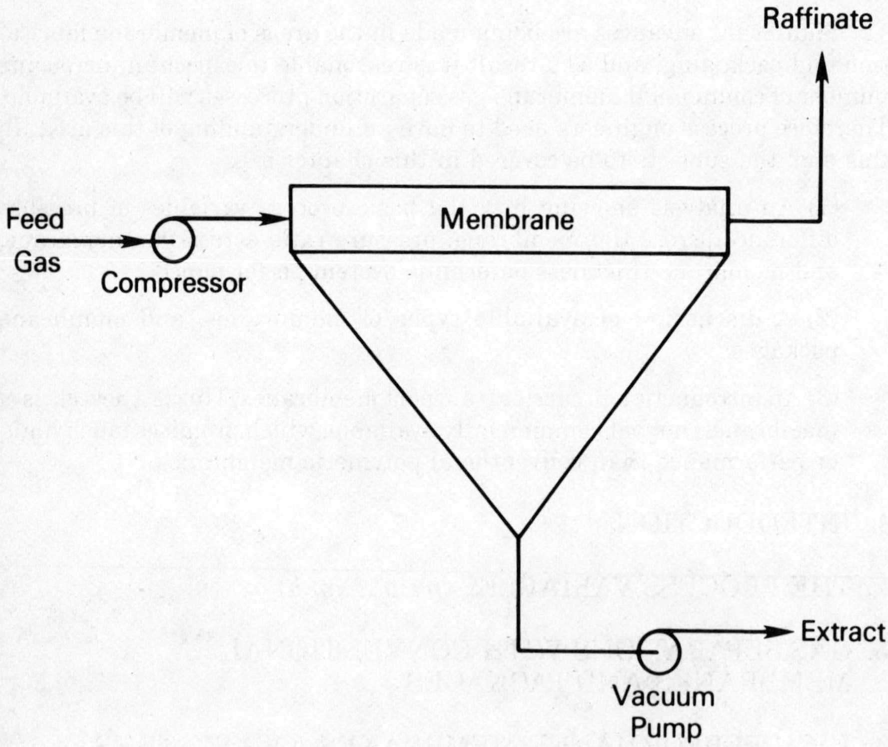

FIGURE 1.—Membrane gas separation process.

3. It may be highly compact, especially if the membrane is provided in the form of a bundle of hollow fibers.

These desirable features have been appreciated for many years and patents on membrane separations date back 100 years. Obviously there have been problems, because practical membrane separations are rare, and just a few years ago there were none. The difficulties have been of two types. First, the separation capability and the production capability of "commercial" polymeric films of "conventional" thickness are completely inadequate for commercially attractive processes. Second, the fragility of thin polymeric membranes has made it difficult to build reliable systems of high surface area that work under realistic operating conditions for a long time.

A great deal of work in many countries is being done to overcome these difficulties, and progress is being made. Ten years ago the first commercial membrane separation system was introduced. The system is a home appliance that produces a continuous flow of 40% oxygen for patients in need of oxygen-enriched air. At the heart of the system, there is approximately 30 ft^2 of 1×10^{-5} cm-thick membrane and a vacuum pump. The second commercial gas separation system was the Prism® separator developed by Monsanto for hydrogen recovery in various industrial processes, including ammonia synthesis. The heart of this system is a bundle of asymmetric hollow fibers.

It seems reasonable to expect that more commercial gas separation processes will soon be available. In a sense, membranes are moving out of the research laboratory and into commercial use. It thus becomes the responsibility of the process engineer to have some understanding of this area.

2. THE PROCESS VARIABLES (p'/p'', Δp, h)

In this discussion of gas permeation through a membrane, it will be assumed that the gas solubility is proportional to the gas partial pressure (Henry's law), and the diffusion coefficient is independent of time, position, and concentration. In this case Fick's first law characterizes transport in the membrane, and as shown in the chapter on Transport Principles,

$$P = Ds. \tag{1}$$

Typical units for permeability are

$$\frac{\text{cm}^3\ (\text{STP}) \cdot \text{cm thickness}}{\text{cm}^2 \cdot \text{sec} \cdot \text{cm Hg}\ \Delta p}.$$

The total volumetric flow rate of gas (j) across a membrane of thickness (h) is

$$Q = \frac{P\,A\,(p_j' - p_j'')}{h}, \tag{2}$$

where p_j' and p_j'' are the partial pressures of gas j on the high (′) and low (″) pressure sides of the membrane, respectively. Equation 2, which is intuitively obvious, simply says that flow rate across the membrane is proportional to area and pressure difference, and inversely proportional to thickness.

Now consider a gas mixture of components (i) and (j), which is allowed to permeate through a membrane across which a pressure difference is maintained. It is assumed that the feed gas flow is sufficiently large that its composition does not change as it passes parallel to the membrane. In other words, the raffinate and feed streams have the same composition. The flow rate of each component through the membrane is written as is done in equation 2. The fraction of component (i) in the product gas is simply

$$x_i'' = \frac{Q_i}{Q_i + Q_j}. \tag{3}$$

Substituting equation 2 for each component into equation 3, and solving gives

$$x_i'' = \frac{1}{2}\,\frac{(\Gamma_{ij}-1)(x_i'+\phi) + 1}{(\Gamma_{ij}-1)\,\phi} - \sqrt{\left[\frac{(\Gamma_{ij}-1)(x_i'+\phi) + 1}{(\Gamma_{ij}-1)\,\phi}\right]^2 - \frac{4\,x_i'\,\Gamma_{ij}}{(\Gamma_{ij}-1)\,\phi}}, \tag{4}$$

where Γ_{ij}, known as the selectivity coefficient, is the ratio of permeabilities of species i and j in the membrane, respectively, and ϕ is the ratio of product to feed gas pressure. This is a rather bulky equation, but an interesting conclusion is evident. The composition of the product gas (x_i'') is a function of the pressure ratio across the membrane, and not the pressure difference across the membrane.

As the value of ϕ approaches zero, equation 4 becomes

$$x_i'' = \frac{\Gamma_{ij}\,x_i'}{1 - x_i'(1-\Gamma_{ij})}. \tag{5}$$

Equation 5 is very useful. It represents, in a simple way, the maximum enrichment that can be obtained in one stage for a given selectivity coefficient (Γ_{ij}).

To summarize, we have seen that the production rate of a membrane separation system is proportional to the pressure difference and surface area, and inversely proportional to thickness. The separation of a gas mixture that can be achieved is a function of the selectivity coefficient of the membrane, and the pressure ratio across it.

3. GAS SEPARATIONS WITH CONVENTIONAL MEMBRANES AND PACKAGES

Now that we know the basic process variables, let us consider what we could have accomplished with membranes up until approximately ten years ago. At that time all membranes for gas separations were polymeric and the thinnest practical membranes were the order of tens of microns. Membrane packages consisted of a stack of flat sheets similar to a plate-and-frame filter press.

Assume that we wish to make oxygen-enriched air, and we consult the literature to see what the optimum membrane material would be. Table 1 is a collection of oxygen permeabilities and oxygen/nitrogen selectivity coefficients for some common polymers. Based on the high permeability of silicone rubber, let us consider this material in more detail. Figure 2 shows, for a system based on silicone rubber, the percent oxygen in the product gas as a function of pressure ratio across the membrane. This is a plot of equation 5 and assumes no depletion of oxygen in the feed gas. If we account for this depletion, and if we assume a product composition of 30% oxygen and a pressure ratio of 0.1 across the membrane, then the power required to make this mixture is five times that required to produce oxygen in a cryogenic air separation plant. If we assume a membrane thickness of 25 μm, 1000 ft^2 of membrane is required to make 1 ton/day of the gas mixture. If the membrane were provided as flat sheet, the investment cost would be many times that for a cryogenic plant. Obviously we could have chosen a more selective but less permeable membrane. In this case the power would be less, but the investment cost would have been even higher because of the lower oxygen permeability.

Many brief analyses of this type have been done, and all come to the same conclusion, i.e., that no practical gas separation can be performed with conventional membranes in a conventional package. It also has been shown that staging the process to achieve higher separation never makes sense. Both the energy and membrane area required increase enormously as stages are added.

TABLE 1. Selected oxygen permeabilities and oxygen/nitrogen selectivity coefficients

Polymer	Oxygen Permeability $\left[\frac{\text{cc (STP), cm}}{\text{sec, cm}^2\text{, cm Hg } \Delta p}\right] \times 10^9$	Oxygen/Nitrogen Selectivity Coefficient
Dimethyl silicone rubber	60	2.2
Natural rubber	2.4	2.7
Ethyl cellulose	2.1	3.1
Polyethylene	0.8	2.8
BPA polycarbonate (Lexan)	0.16	6.7
Butyl rubber	0.14	4.1
Polystyrene	0.12	7.6
Cellulose acetate	0.08	2.5
Methyl cellulose	0.07	2.9
Polyvinyl chloride	0.014	3.0
Nylon 6	0.004	3.8
Mylar	0.0019	6.0
Kel F	0.001	4.3
Polyvinylidene chloride (Saran)	0.0005	5.0

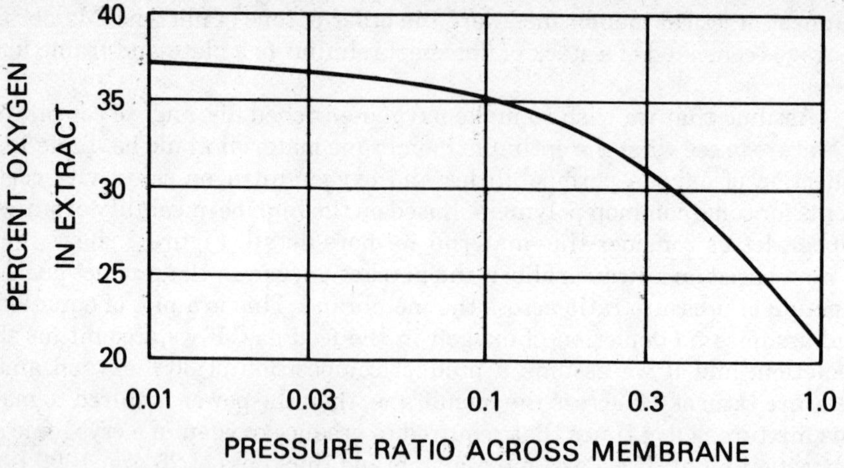

FIGURE 2.—Percent oxygen in extract vs. pressure ratio across silicone rubber membrane.

4. HIGH PERFORMANCE MEMBRANES AND SYSTEMS

For membranes to be of interest for gas separations, new approaches are clearly needed both for membrane preparation and packaging, and it is in these areas that dramatic progress is being made. An example of progress in producing ultrathin membranes is illustrated schematically in Figure 3 (1). Teflon-coated rods, shown in cross section, rest on Teflon walls of a tray filled with water almost to the point of overflowing. The "well" next to the Teflon rod is filled with a solution of polymer in a water-immiscible solvent. The rod is moved down the length of the tray creating new surface upon which the casting solution spreads. When the solvent evaporates, a membrane, which can be as thin as 5 nm, is left on the water surface. The membrane is transferred to a porous support and is ready for packaging. It is this technology that is used to manufacture the home oxygen enrichment appliance mentioned earlier. Because the membrane is so thin, sufficient flow rate can be achieved simply by evacuating the extract side of the membrane. Inasmuch as a compressor on the feed side is not required, the feed air is virtually free and can be provided at a flow rate such that it is not significantly depleted in oxygen. In this way, the enrichment of oxygen is maximized and the power requirement is minimized.

The Prism separator from Monsanto (2) is also based on an ultrathin membrane, but it is packaged in a different and very clever way. The separating layer is a thin skin on an asymmetric walled hollow fiber. Industrial-scale separations will undoubtedly be done with hollow fiber systems rather than

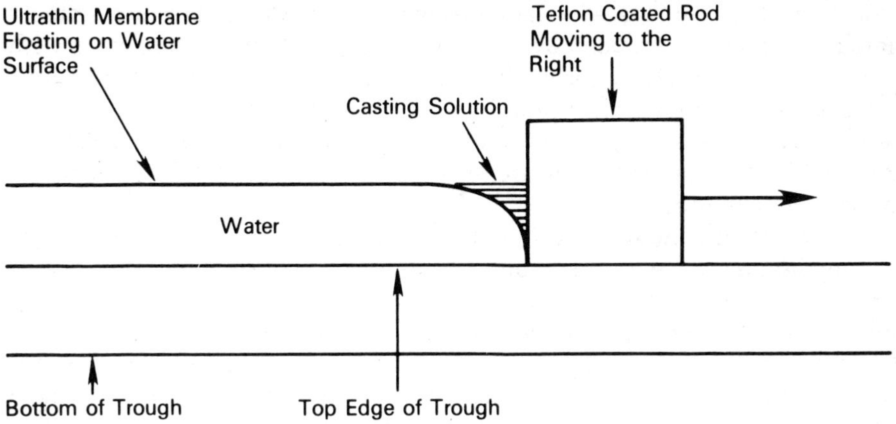

FIGURE 3.—Apparatus for producing ultrathin polymeric membranes.

flat sheets. Hollow fibers are self supporting, even with very large pressure differences between the feed stream on the exterior and the extract collected from the interior of the fibers. In addition, they have a very high membrane surface to system volume ratio ($\sim$5000 ft^2/ft^3), and they are relatively inexpensive per unit surface area. They do suffer the disadvantage that vacuum operation on the extract is probably not practical because of excessive pressure drop along the fiber interior.

These two approaches to improved membrane systems are based on maximizing what can be done with conventional polymeric membrane materials. Another approach is to improve the membrane itself, and in particular, its separation factor. Before considering how to do this, let us see how advantageous it could be. Assume again that we wish to extract oxygen from air. The most selective polymeric membrane for this application has an oxygen/nitrogen selectivity coefficient of approximately 6. Suppose instead it were 60! In this case with a pressure ratio of 0.1 and a large excess of feed air, the product gas would be 90% O_2. Based on use of commercially available vacuum pumps on the extract side, and depending on the scale of the process, the power required to produce this gas would be 100 to 200 kilowatt-hours/ton of mixture. This compares with 250 kilowatt-hours/ton of mixture to produce 90% oxygen with a cryogenic plant. Producing oxygen with such efficiency would be a genuine breakthrough and would profoundly affect many industrial processes. Obviously no such "super" oxygen membrane exists, although serious efforts are underway to develop this and other very high performance membranes. An approach to high performance is "facilitated transport."

5. FACILITATED TRANSPORT

Because of the inherent limitations of polymers as membrane materials, Robb (3) in 1967 considered a new class of materials for membranes, namely liquids. In general, liquids have large diffusion coefficients compared with polymers, and for certain gases they can have enormous solubilities. Robb's concept was to immobilize liquid films by filling the pores of a solid microporous film, or by gelling the liquid to give it solid-like properties. Thus he obtained the mechanical properties of solid membranes, and the permeation properties of liquids.

In 1967 Ward and Robb (4) found that perhaps the most important aspect of using liquids as membrane materials is that across liquid films one can do facilitated transport, which can increase permeability and selectivity by orders of magnitude. To date there are no commercial-scale applications of facilitated transport, but at least several industrial and government groups are known to be pursuing this approach. Here, the basic ideas of facilitated transport will be presented and simplified mathematical treatments will be covered, which should help provide an intuitive feeling for the process.

The mathematical treatment that follows is based on a paper by Ward (5). The system is shown schematically in Figure 4. Gas (A) is present at known pressures on both sides of a thin liquid film. The liquid film has dissolved in it nonvolatile material (B), and (A) and (B) react reversibly to form (AB). Schematic concentration profiles are shown for (A), (B), and (AB) for two cases, (1) fast reaction, solid lines, and (2) slow reaction, dotted lines. The major features of the concentration profiles can be deduced as follows. It is assumed that (A) in the gas phase is in equilibrium with dissolved (A) at the boundaries, and the equilibrium is unaffected by the rate of the reaction. Therefore, because we know the pressure of (A) on either side of the membrane, we know the boundary concentrations, and they are the same for fast and slow reactions. We specified that (B) and (AB) are nonvolatile and do not leave the membrane. Thus the concentration gradients of these species at the boundaries must be zero, as shown. If the rate of reaction is fast compared with diffusion, then clearly a concentration difference in (A) will cause concentration differences in (B) and (AB) across the film. Near the high pressure side of the membrane the reaction will be driven to the right and the concentration of (AB) will be high and (B) will be low. The same reasoning indicates that (AB) will be low and (B) high at the low pressure side. Knowing the magnitude (qualitatively) of the concentrations and the slopes for both (AB) and (B) at the boundaries, we can now sketch in the concentration profiles for fast reaction. As the reaction slows down, in the limit to an insignificant rate compared with diffusion, there would be no concentration differences in (B) and (AB) across the film, so the slow rate profiles are drawn with smaller concentration differences across the membrane, and still with zero gradients at the boundaries. At very slow reaction rate, there would be a linear concentration profile for (A).

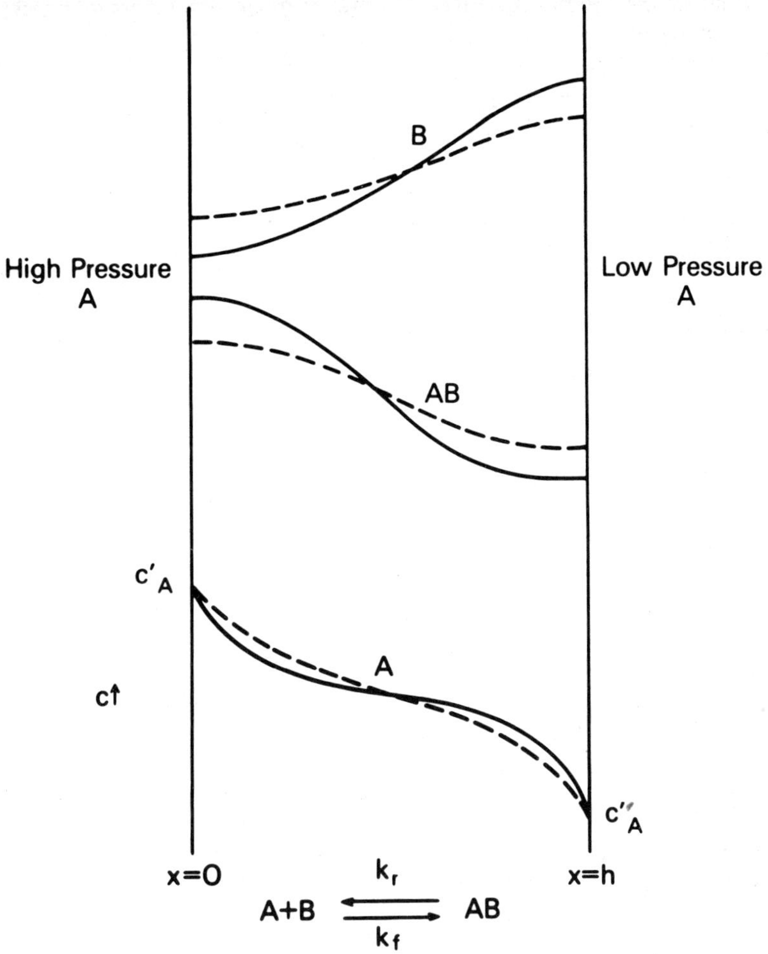

FIGURE 4.—Facilitated transport: concentration profiles for fast (__) and slow (---) reaction.

Qualitatively it should be clear that the transport of (A) across the membrane is increased due to the presence of (B). This chemical augmentation is referred to as facilitated transport, and was known to biologists studying cell membrane transport long before it was applied to gas separations.

In applying facilitated transport to gas separations, one needs to be able to calculate the flux of (A) across the membrane. The purpose of the following analysis is to do just this. As was done above for the case of gas diffusion in rubbery polymers, we begin with differential mass balances on (A), (B), and (AB) at any point within the film:

$$D_A \frac{d^2 c_A}{dx^2} = k_f c_A c_B - k_r c_{AB}, \tag{6}$$

$$D_B \frac{d^2 c_B}{dx^2} = k_f c_A c_B - k_r c_{AB}, \tag{7}$$

$$D_{AB} \frac{d^2 c_{AB}}{dx^2} = - k_f c_A c_B + k_r c_{AB}. \tag{8}$$

The boundary conditions that apply are

$$c_A(x=0) = c_A',$$

$$c_A(x=h) = c_A'',$$

$$\int_0^h (c_B + c_{AB})\, dx = n_B^T \tag{9}$$

$$\left.\frac{dc_{AB}}{dx}\right|_{x=0} = \left.\frac{dc_{AB}}{dx}\right|_{x=h} = 0.$$

There is no general closed solution to equations 6–8 from which the total flux of (A) could be calculated. However, two sets of simplifying solutions that do apply in certain real systems enable a straightforward calculation of the total flux of (A).

The first assumption is that the chemical reaction is very fast and that all species are present in equilibrium concentration. In this case, add equations 6 and 8:

$$D_A \frac{d^2 c_A}{dx^2} + D_{AB} \frac{d^2 c_{AB}}{dx^2} = 0. \tag{10}$$

The solution of equation 10 is

$$D_A c_A + D_{AB} c_{AB} = a_1 x + a_2, \tag{11}$$

where a_1 and a_2 are constants. The total flux of (A) is

$$J_A^T = - D_A \frac{dc_A}{dx} - D_{AB} \frac{dc_{AB}}{dx}. \tag{12}$$

Equation 12 is valid at all points in the film, and it is seen from equation 11 that

$$J_A^T = -a_1. \tag{13}$$

If D_B and D_{AB} are equal, then

$$c_{AB} + c_B = c_T, \tag{14}$$

where c_T is a constant. Equation 14 plus the assumption of chemical equilibrium among all reacting species leads to the expression for the total flux of (A):

$$J_A^T = \frac{D_A}{h}(c_A' - c_A'') + \frac{D_{AB} K_{eq} c_T}{h(1 + K_{eq} c_A')(1 + K_{eq} c_A'')}(c_A' - c_A''). \tag{15}$$

The first term is the flux for no facilitation; the second term is the augmented flux due to reaction. This term can be very large compared with the first, but clearly there is an optimum set of values for c_A', c_A'', and K_{eq} which will maximize this term. This is an important feature of facilitated transport systems, regardless of reaction rate, of which the engineer must be cognizant. It is quite possible to have a system involving a fast reversible reaction, but to select c_A' and c_A'' such that virtually no facilitated transport will occur. If c_A'' is too high, the carrier (B) will be saturated throughout the membrane, and if c_A' is too low, the carrier will remain unreacted throughout. No facilitated transport will occur in either case.

The other assumption that results in an expression for the total flux of (A) is that the reaction is sufficiently slow that the concentrations of (B) and (AB) are essentially constant throughout the film. In this case a closed solution for the total flux of (A) can be derived (5):

$$J_A^T = -B_1 L_1, \tag{16}$$

where

$$L_1 = \sqrt{D_A/K_1}$$

$$K_1 = k_f \bar{c}_B$$

$$B_1 = \frac{K_1 c_A'' - K_2 - B_2 \cosh(h/L_1)}{\sinh(h/L_1)} \tag{17}$$

$$K_2 = k_r \bar{c}_{AB}$$

$$B_2 = K_1 c_A' - K_2$$

In the above expressions the overbar indicates quantities averaged across the membrane.

Although less useful than equation 15 in trying to obtain a qualitative understanding of facilitated transport, equation 16 does describe the transport of carbon dioxide across a thin film of concentrated $HCO_3^-/CO_3^=$ solution (14).

It remains to be seen whether or not facilitated transport will find practical commercial applications. A general problem with immobilized liquid membranes is their fragility, particularly where large pressure differences across the membrane must be sustained. Volatility is another problem with most liquid membrane systems. In an attempt to deal with both problems, LeBlanc et al (6) used an ion exchange membrane as the base membrane, and used counterions to facilitate transport. For example, silver ions impregnated in a cation exchange membrane were used to facilitate ethylene transport, which was made possible by the reversible reaction of Ag^+ and ethylene. The silver ions and silver ethylene complex ion were highly mobile within the membrane, and yet were confined to it by electrostatic charge. A ratio of ethylene to ethane permeability of 290 was demonstrated, which is the order of 100 times higher than could be achieved by any polymeric membrane.

REFERENCES

1. Ward W.J., Browall W.R., and Salemme R.M.: 1976, "Ultrathin silicone/polycarbonate membranes for gas separation processes." J. Membrane Science 1, p. 99.
2. Bollinger W.H., MacLean D.L., and Narayan R.S.: 1982, "Separation systems for oil refining and production." Chem. Eng. Progress, pp. 27–32.
3. Robb W.L., Reinhard D.L.: 1967, "Gas separation by differential permeation." U.S. Patent 3, 335, 545.
4. Ward W.J., Robb W.L.: 1967, "Carbon dioxide–oxygen separation: Facilitated transport of carbon dioxide across a liquid film." Science 156, p. 1481.
5. Ward W.J.: 1970, "Analytical and experimental studies of facilitated transport." AIChE Journal 16, p. 405.
6. LeBlanc O.H., Ward W.J., Matson S.L., and Kimura S.G.: 1980, "Facilitated transport in ion-exchange membranes." J. Membrane Science 6, pp. 339–343.

SYMBOLS

a_1	constant, equation 11
a_2	constant, equation 11
B_1	constant, equation 17 (mol cm^{-3} s^{-1})
B_2	constant, equation 17 (mol cm^{-3} s^{-1})
$\bar{c}$	concentration averaged across the membrane (mol cm^{-3})
J_A^T	total flux of (A) (mol cm^{-2} s^{-1})
k_f	forward reaction rate constant (cm^3 mol^{-1} s^{-1})
k_r	reverse reaction rate constant (s^{-1})
K_{eq}	equilibrium constant (cm^3 mol^{-1})
K_1	constant, equation 17 (s^{-1})

K_2	constant, equation 17 (mol cm^{-3} s^{-1})
L_1	constant, equation 17 (cm)
$n_B{}^T$	Total moles of (B), equation 9 (mol cm^{-2})
P	permeability $\left(\frac{cm^3\ (STP) \cdot cm\ thickness}{cm^2 \cdot s \cdot cm\ Hg\,\Delta p}\right)$
p	pressure (cm Hg)
Q	transmembrane volumetric flowrate (cm^3/s)
s	Henry's Law solubility $\left(\frac{cm^3\ (STP)}{cm^3 \cdot cm\ Hg}\right)$
Λ_{ij}	selectivity coefficient
ϕ	ratio of product to feed gas pressure

Superscripts

$'$	high pressure (feed) side of membrane ($x=0$)
$''$	low pressure (extract) side of membrane ($x=h$)

Subscripts

i	component i
j	component j
A	component A
B	component B
AB	component AB

PERVAPORATION

Philippe Aptel

Laboratoire de Génie Chimique (CNRS L.A. 192)
Université Paul Sabatier
118, route de Narbonne
31062 Toulouse CEDEX, France

Jean Néel

Laboratoire de Chimie-Physique Macromoléculaire (CNRS E.R.A. 23)
Ecole Nationale Supérieure des Industries Chimiques
1, rue Grandville
54042 Nancy CEDEX, France

The basic principles of the pervaporation process are presented in an overall view intended as an introduction to this unconventional technique. The mass transfer within the membrane for pure liquid and for liquid mixtures is discussed. Limitations due to concentration polarization and temperature drop at the membrane interface are considered. Basic calculations are given for cross-flow continuous pervaporation and some potential applications are explained. Possible improvements of both membrane and process design are reviewed.

P. M. Bungay et al. (eds.), Synthetic Membranes: Science, Engineering and Applications, 403–436.

1. INTRODUCTION

Pervaporation is a separation process in which a liquid mixture is in direct contact with one side of a membrane and in which the permeated product, the "pervaporate", is removed in the vapour state from the other side. The mass flux is brought about by maintaining the downstream partial pressure lower than the saturation pressure. This is usually achieved by creating a vacuum or by employing a carrier gas (Figure 1).

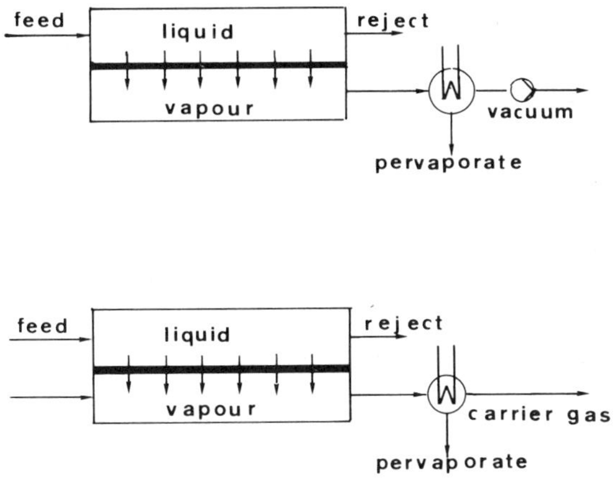

FIGURE 1.—Pervaporation and carrier-gas pervaporation.

In this process, the membrane acts as a thin solvent layer and the pervaporate composition is mainly governed by the preferential solvation of the polymeric barrier material. As a result, the liquid–vapour equilibrium is greatly perturbed as shown in Figure 2.

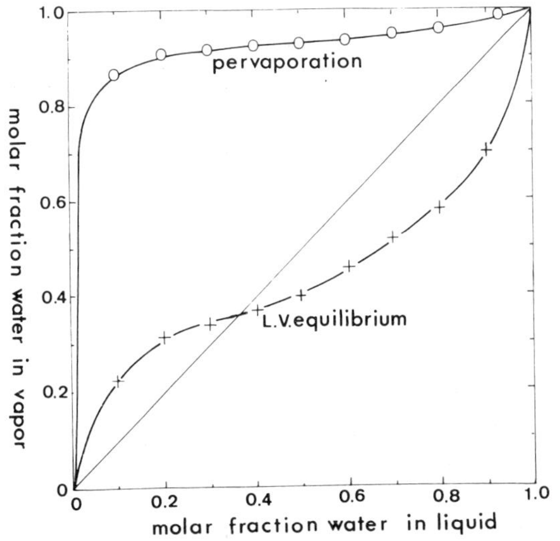

FIGURE 2.—Comparative separation of water-dioxane mixtures by distillation and by pervaporation. Reprinted by courtesy of Elsevier Science Publishers. (25)

1.1. Brief History

The evaporation of water through the walls of a collodion membrane was called "pervaporation" by Kober in a short note published in 1917 (1). However, the first separation of a liquid mixture, reported as early as 1906 (2), was performed by Kahlenberg using a rubber membrane. Half a century later (3) Schwob published a thesis on the separation of water–alcohol mixtures through a cellophane membrane. Later, Heisler et al. (4) studied the effect of the presence of a solute on the pervaporation of a water–ethanol mixture through a similar membrane.

The extensive work of Binning et al. (5,6) between 1958 and 1962 provided the starting point for several investigations conducted by different groups of workers (7–76).

The potential of this technique has been recognized by numerous workers in several countries. Pervaporation is one of the topics of the ten-year research and development program on membranes, launched by the Japanese in 1981. And a 1982 workshop on pervaporation attracted eight groups of investigators, from Western Europe alone, who are working in this field.

1.2. Potential Applications

Pervaporation is potentially applicable to mixtures that are difficult to separate by more conventional techniques, such as azeotropic mixtures, or mixtures of close-boiling components. Other applications could include the separation of heat-sensitive products, the concentration of fruit juices, the elimination of traces of impurities, the enrichment of organic pollutants for quantitative detection, etc.

In spite of the fact that the first industrial trials date from the late forties (3), pervaporation is not yet employed on an industrial scale. The main reason is that, unlike other membrane processes, this separation technique requires the vaporization of a part of the liquid charge. The input of the necessary enthalpy of vaporization poses two serious problems: the additional energy cost and the difficulty of designing compact permeator modules having sufficient membrane area for industrial use. Moreover, transport phenomena in pervaporation are considerably less well understood than in gas permeation or other membrane processes. This means that at present there are no established criteria for tailoring the best membrane and for designing the best separator unit.

2. MASS TRANSPORT WITHIN THE MEMBRANE AND MEMBRANE SELECTION

The main criteria for assessing the usefulness of a membrane process are the selectivity and the permeation flux. In pervaporation, the flux is usually quoted as a mass flux J ($kg \cdot m^{-2} \cdot s^{-1}$ or $kg \cdot m^{-2} \cdot hr^{-1}$). In practice, one can usually expect to obtain a higher value for J in the pervaporation mode than in the reverse osmosis mode for the same membrane (72, 73).

One measure of selectivity is the separation factor defined as:

$$\alpha_{AB} = \frac{w''_A / w''_B}{w'_A / w'_B} \tag{1}$$

where w'_A, w'_B, w''_A, and w''_B denote the weight fraction of component A and B in the feed solution and in the pervaporate, respectively. A is the species that is preferentially pervaporated. A more convenient expression for the selectivity is given by the enrichment factor β:

$$\beta = \frac{w''_A}{w'_A} \tag{2}$$

The flux (J) and the selectivity (α or β) are governed by the mass transport through the membrane. It is commonly accepted that pervaporation through a nonporous polymer membrane takes place according to the following sequence of three steps:

— solution of permeating molecules at the liquid side of the membrane;
— diffusion of these molecules through the membrane;
— evaporation from the vapour side of the membrane.

It is important to note that the vapour side of the membrane is kept in a "dry" state by vacuum pumping or by a carrier gas, whereas the sorption on the liquid side produces a substantial swelling of the membrane matrix. This anisotropic swelling is a characteristic feature of pervaporation and the membrane can be viewed as a stack of thin films in which the interactions between polymer and penetrants lead to nonlinear expressions for solubility and diffusion coefficients.

2.1. Pervaporation of Pure Liquids

Ideal case The first problem we will consider is the pervaporation of a pure liquid through a non-porous homogeneous polymer membrane. If we assume that transport occurs by a solution–diffusion–desorption mechanism, the steady state flux is described by Fick's first law:

$$J_V = -D \frac{d\phi}{dx} \tag{3}$$

where J_V is the volume flow per unit area time ($m.s^{-1}$), ϕ the fractional volume of penetrant within the membrane, D the mutual diffusion coefficient ($m^2.s^{-1}$) and distance x is measured from the incoming side of the membrane.

If the membrane with a thickness h is uniform and D is independent of ϕ (ideal case), the flux per unit area of the membrane is given by:

$$J_V = D \frac{\phi' - \phi''}{h} \tag{4}$$

The superscript prime relates to the upstream (liquid) and the superscript double prime to the downstream (vapour) boundary of the membrane.

The gradient in the concentration of the liquid within the membrane is induced by the chemical potential difference, which is maintained across the system:

$$\mu' - \mu'' = RT \ln \frac{p'}{p''} \tag{5}$$

p' and p'' are the equilibrium vapour pressure of the liquid and the partial pressure of the vapour, respectively.

The dependence of the concentration of the liquid in the membrane on the vapour pressure is required to express J as a function of the pressures p' and p''. In the case of an ideal liquid membrane pair, the simple relationship is given by Henry's law:

$$\phi = sp \tag{6}$$

where s is a constant solubility coefficient that depends only on the temperature and the nature of the system.

If we assume there is always equilibrium at the interfaces, equations 4 and 6 give:

$$J_V = D\, s \frac{p' - p''}{h} \tag{7a}$$

or $$J_V = P \frac{p' - p''}{h} \tag{7b}$$

where $P = D\, s$ is the permeability coefficient.

Non-ideal cases In general, D will depend significantly on ϕ and Henry's law is not obeyed. The integration of equation 3 will depend on the functional forms of $D(\phi)$ and $s(p)$.

1) $s\ (p)$ = constant

Empirical expressions have been proposed to express the increase of D with the concentration (19, 24, 37, 48, 49, 77):

$$D(\phi) = D(0) \exp g\, \phi \tag{8a}$$

$$D(\phi) = D(0)\,(1 + g\, \phi^n) \tag{8b}$$

Assuming a steady state transport, equation 8a yields:

$$J_V = \frac{D(0)}{gh}\left[\exp g\, \phi' - \exp g\, \phi''\right] \tag{9}$$

or
$$J_V = \frac{D(0)}{gh}\left[\exp g\, sp' - \exp g\, sp''\right] \tag{10}$$

In these expressions $D(0)$ is the diffusion coefficient in the limit $\phi \rightarrow 0$, and g is an empirical constant that depends on the nature of the penetrant–polymer system and the temperature. The term g reflects the plasticization factor, i.e., the effect of the presence of the penetrant on polymer segmental motion.

2) $D(\phi)$ and $s(p)$ variable

This general case has been successfully treated by Greenlaw et al. (48), using the data of Rogers et al. (79) on the sorption and diffusion of hexane vapour through polyethylene. However, the need to determine six experimental constants (three in the expression for $D(\phi)$ and three in the expression for s (p)) limits the usefulness of the model.

We should emphasize here the need for theoretical studies on the mechanism of pervaporation : the above binary system (hexane–polyethylene) intuitively seems one of the simplest systems to analyse because of the chemical similarity in the repeating unit -CH_2- in the two components. In spite of this fact, six empirical parameters are necessary.

Recent progress in the characterization of vapour transport in and through rubbery and glassy polymers has been analysed by Stern (79). Suitable molecular models are now available for gas transport in rubbery polymers and we can expect the development of similar models for vapor transport in glassy polymers. Before such a complete elucidation, models based on the free-volume theory for diffusion will be useful for practical purposes (14, 81).

Experimental validity of the solution–diffusion mechanism Our aim here is to show that experimental studies are in good agreement with the above phenomenological equations 3 to 10.

Effect of the thickness of the membrane on flux.—The validity of the thickness dependence of flux in the form of equation 7 or 10 was confirmed experimentally (6, 23).

Effect of the upstream and downstream pressure on pervaporation flux.—Because the solubility of liquid in a polymer is not very sensitive to hydrostatic pressure, the upstream pressure has no significant influence on the flux as has been shown by several investigations (5, 23, 48, 56, 66). Note that this independence of permeation rate on upstream pressure is expected only for an upstream pressure greater than the saturation pressure and insofar as it does not exceed a limit beyond which the swollen membrane is physically modified.

On the other hand, the downstream pressure is a determining parameter since it governs the activity of the pervaporant at the membrane–vapour interface, and according to equations 9–10, pervaporation flux is progressively reduced when the permeate pressure is increased (Figure 3; 19, 48, 53, 56, 69).

Concentration profile through the membrane.—Integration of Fick's first and second laws, using the functional form of $D(\phi)$ given by equation 8a, leads to the value of the concentration at any distance x in the membrane:

$$\phi(x) = \frac{1}{g} \ln \left[\exp g\,\phi' - \frac{x}{h}\left(\exp g\,\phi' - \exp g\,\phi''\right)\right] \quad (11)$$

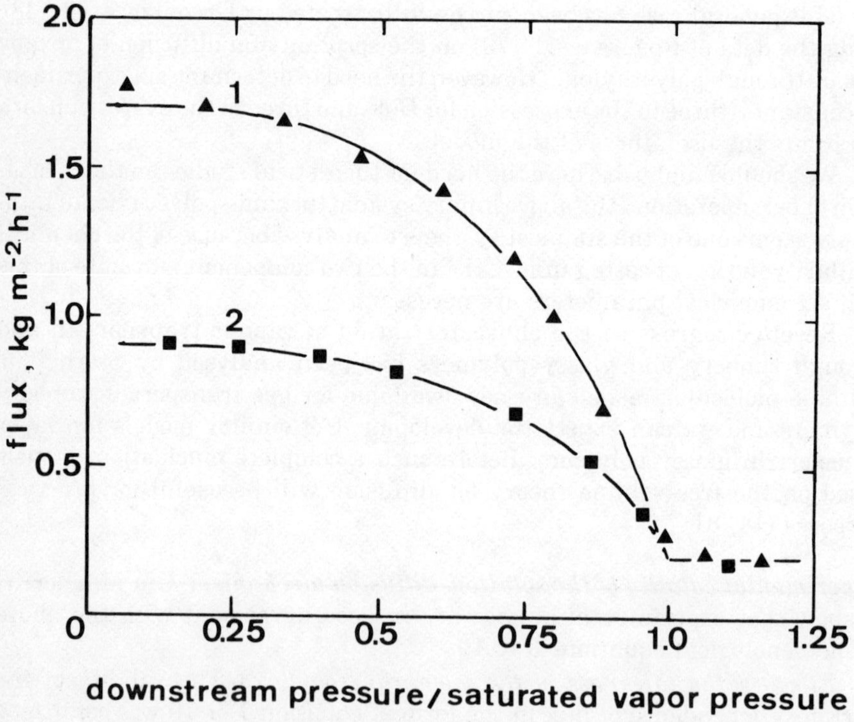

FIGURE 3.—Variation of pervaporation flux with downstream pressure. 1: Benzene. 2: Cyclohexane. Polyethylene membrane. (53)

Kim et al. (19) and Aptel et al. (24) measured the actual concentration profile in a stacked film and were able to fit the measured profiles successfully to equation 11, as shown in Figure 4.

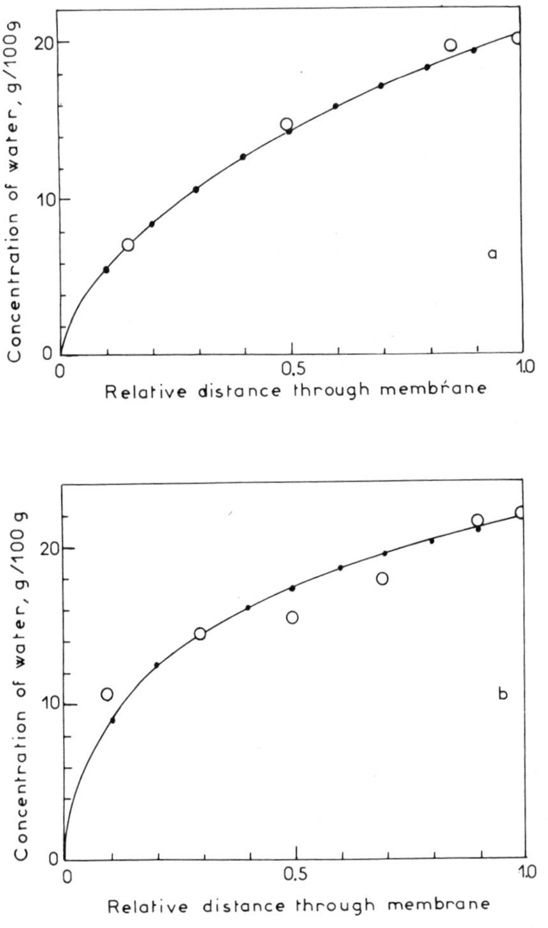

FIGURE 4.—Concentration gradient of water in PTFE-P4VP membranes. a) Stack of three membranes (PTFE thickness 17 μm; 150% grafting); b) Stack of five membranes (PTFE thickness 50 μm; 160% grafting); (○) experimental average values; (●) calculated values from equation 11: a, g = 10; b, g = 16. Reprinted by courtesy of John Wiley & Sons. (24)

2.2. Pervaporation of Liquid Mixtures

While in gas permeation, the selectivity of the separation of a binary mixture can be estimated from the ratio of the experimental permeabilities of the individual components, it is well established that in pervaporation, it is not valid to assume that the dissolved penetrants do not interact.

The absence of suitable models for the pervaporation of pure liquids is the main reason why so few attempts have been made to develop models for binary mixtures (14, 49, 50, 53, 56). Though these models could describe the performance of a given membrane toward a given binary mixture under different operating conditions, they contain numerous empirical parameters and so could not be used to predict the selectivity of a specific membrane toward a given binary mixture.

From a practical point of view, however, it is necessary to have at least some basis (even a qualitative one) for making this sort of prediction. Hence, the second part of this chapter includes a summary of various data that could be used as a guide in developing efficient pervaporation membranes.

Effect of pressure, temperature and concentration on pervaporation flux and selectivity It has been reported that an increase in downstream pressure decreases the selectivity (23, 45, 60), but this is not necessarily true in all cases and the reverse situation has been observed (49, 50).

As in the case of pure liquid, the downstream pressure is a determining parameter in fixing the flux, whereas the upstream pressure has little effect.

When the temperature is raised, the pervaporation flux increases according to an Arrhenius-like relationship. The activation energy can be as high as 50 kJ/mole (23, 34, 35), which corresponds to an increase of flux of 20 per cent per K. This effect is of great practical importance because the selectivity is often only slightly reduced by an increase in temperature (23, 33, 39, 66); in fact, a higher selectivity has been reported (14, 57, 66).

Because the solubility and the diffusivity of species A and B are functions of the concentrations of both components of the binary system, a complex "coupled" transport occurs; flux and selectivity are strongly dependent on the composition of the feed. In general, the flux decreases when the mixture becomes poorer in A—the species A refers to the more rapidly permeating species—and loses its swelling properties; the selectivity simultaneously increases (Figure 5). However, a decrease in selectivity has also been observed (70, 75).

Criteria for developing efficient pervaporation membranes

Analogies with liquid–liquid extraction.—One way to increase the fractionation efficiency consists of inserting an agent used in extractive-distillation or solvent-extraction processes into a relatively inert polymer film. For example, a vinylidene fluoride film plasticized with 3-methylsulfolene was found to be an effective permselective membrane for the extraction of benzene

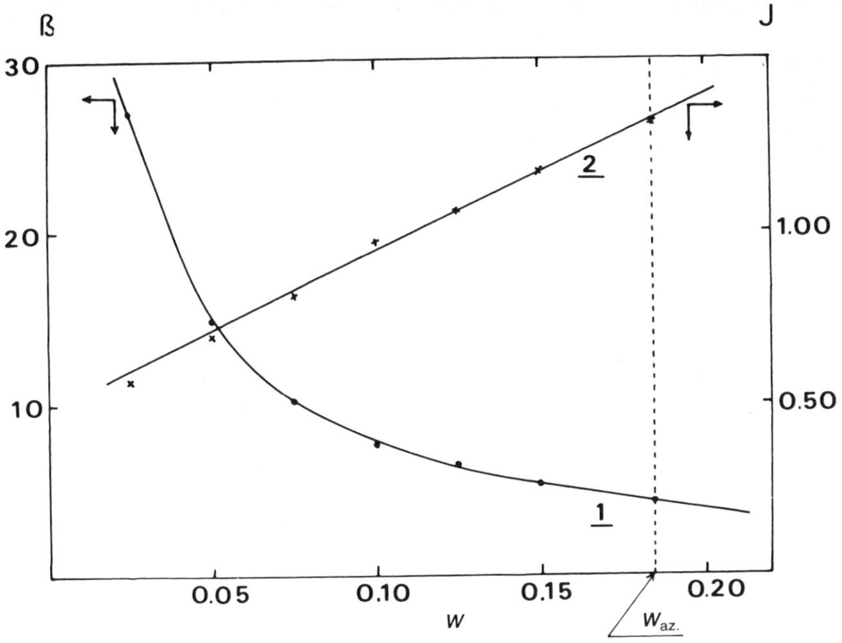

FIGURE 5.—Variation of pervaporation flux (J) and enrichment factor (β) with concentration (w) of water in water-dioxane mixtures. Concentration is expressed in weight fraction and flux in kg $m^{-2}hr^{-1}$. PTFE-PVP membrane. Reprinted by courtesy of Association Entropie. (31)

from cyclohexane. The addition of 17% 3-methylsulfolene resulted in an approximately 15-fold increase in the flux, while the enrichment factor β decreased only slightly from 1.9 to 1.8 (60).

Other examples have been studied by Larchet and Brun who inserted a nitrile functional group into polybutadiene by co-polymerisation with acrylonitrile. Another promising modification of membranes is to include Werner complexes so as to mimic clathration processes (62).

We should note that the membranes prepared from cellulose acetate–polyphosphonate blends and the membranes obtained by grafting polyvinylpyrrolidone onto polytetrafluoroethylene films could also be considered as films containing substituents chemically similar to additives commonly used in extractive processes.

Hildebrand–Hansen solubility parameters.—Cabasso (40, 41, 43) and Mulder (74) used Hansen's three-dimensional solubility parameter to predict the separation capabilities of polymeric membranes toward liquid mixtures. Hansen (81) suggested representing every organic compound by three dimensional coordinates: δ_d, δ_p and δ_h, which reflect London forces, polar forces and hydrogen bonds, respectively. Similarly, every polymer can be represented by a volume bounded by the coordinates of all those compounds that act as solvents for the polymer. Thus, organic compounds whose coordinates are not within the boundaries of a polymer's solubility volume are not solvents for the polymer. In practice, it should be sufficient to consider the two coordinates δ_p and δ_h and draw only a solubility area for each polymer (Figure 6).

For the separation of a given mixture A-B, Cabasso suggested selecting a polymer soluble in A and insoluble in B. The selectivity is expected to be highest when A is closest to the center of the solubility area and B is located far from the solubility boundary. To prevent the dissolution of such a membrane, the "active" polymer can be alloyed with another polymer, or grafted onto such a polymer. Figure 6 shows the solubility areas for an alloy mem-

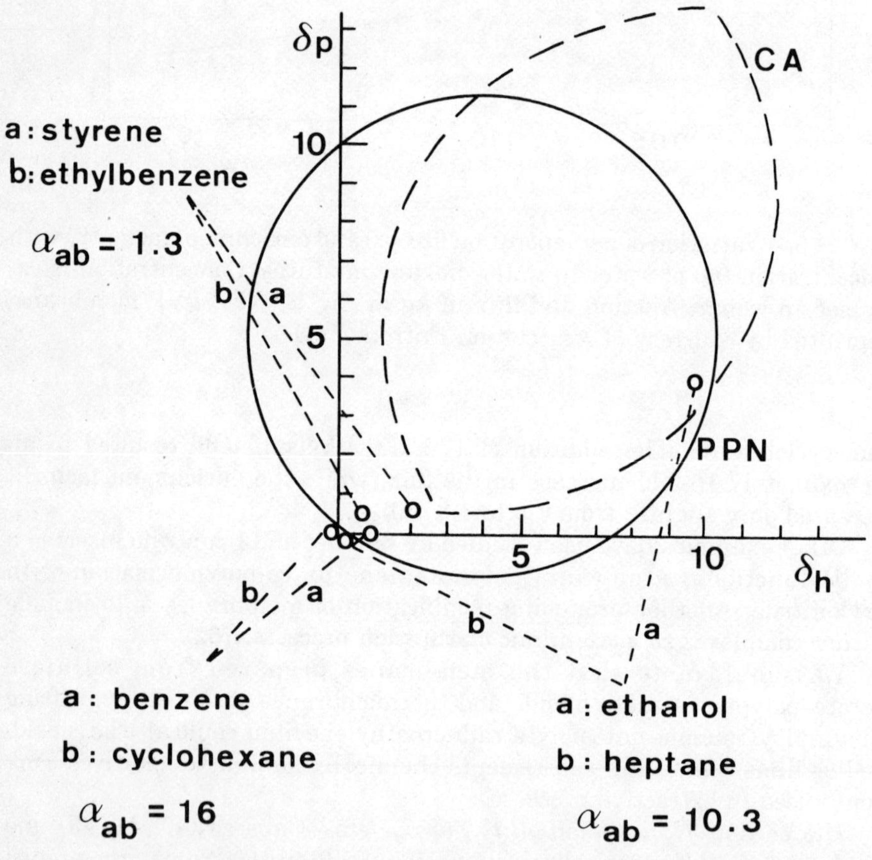

FIGURE 6.—Solubility parameter diagram of PPN/CA alloy membrane. α_{ab} is the separation factor for the azeotropic composition. CA: Cellulose acetate; PPN: Polyphosphonate. Reprinted by courtesy of I. Cabasso (41).

brane composed of cellulose acetate and polyphosphonate. This membrane is very efficient in separating an azeotropic mixture such as benzene–cyclohexane ($\alpha = 16$) or ethanol–heptane ($\alpha = 10.3$). The separation factor is significantly smaller for the mixture styrene–ethylbenzene ($\alpha = 1.3$), because the two components are both within the solubility area of the polyphosphonate and because of the small distance between their locations in this area.

The same qualitative trends can be detected in an examination of the selectivities obtained with a PTFE–PVP membrane, which is prepared by grafting N-vinylpyrrolidone onto a polytetrafluoroethylene film. However, while some binary mixtures composed of a solvent and a non-solvent of PVP are easily separated (i.e. alcohol + saturated hydrocarbons), others (alcohol + unsaturated hydrocarbons) show a separation factor close to 1 (25, 26).

A more quantitative approach has been proposed by Mulder (74), who correlated experimental data with the distance between the points representing polymer and solvents in δ_p, δ_d, δ_h space. Though swelling properties are in fact well predicted, it was concluded that selectivity cannot be predicted by this δ-parameter concept.

Preferential solvation coefficient.—The above discussion shows that a simple analysis that takes into account only the interactions existing in the two binary systems, component A/membrane and component B/membrane, can only be successfully applied to the separation of "ideal" binary liquids. Here, the term "ideal" means that the permeability of one species is not influenced by the presence of the other: i.e., there are no coupling or plasticization effects. To describe the actual situation, it is necessary to look for an approach taking into account all the interactions existing in the ternary system A/B/membrane. Aptel et al. (26) have shown that the preferential solvation coefficient displayed by a polymer toward various mixed solvents provides a powerful tool for predicting selectivity. For example, Figure 7 shows that a very close correlation exists between preferential solvation of polyvinylpyrrolidone (PVP) in solution and the selectivity of a PVP-grafted membrane. Note that where an inversion occurs in the preferential solvation at a given concentration, an inversion of selectivity is also observed in pervaporation at about the same feed composition. In the same way, the high selectivity of the above membrane toward a water/dioxane mixture is also in good agreement with the observation of a large excess of water molecules in the vicinity of PVP chains in solution (Figure 8).

It is likely that further progress in the understanding of preferential solvation phenomena will furnish tabulated data that will be of direct interest in the development of efficient pervaporation membranes.

Positive and negative azeotropic mixtures.—Evidence for the importance of the penetrant–penetrant interactions in pervaporation is also provided by data concerning the separation of typical azeotropic mixtures (Tables 1 and 2).

All of the mixtures in Table 1 share the feature of belonging to the class of positive azeotropic mixtures (characterized by a minimum boiling temperature). Positive azeotropism indicates a molecular segregation in the A-B

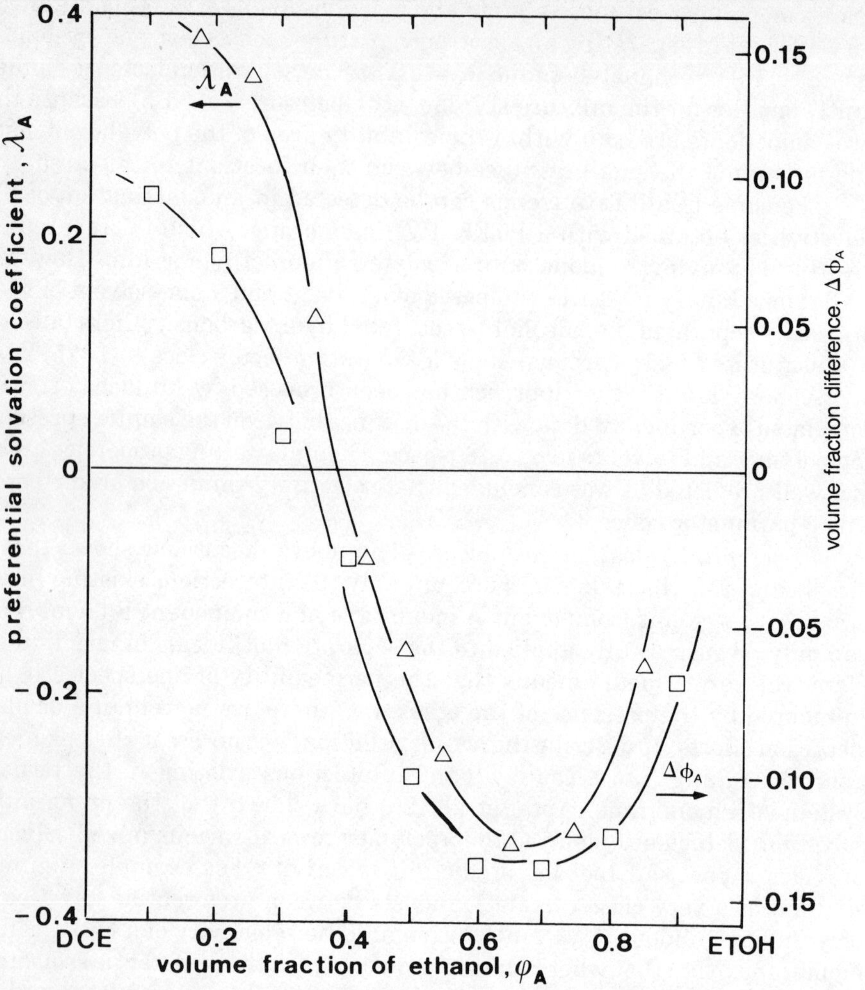

FIGURE 7.—Variation of the preferential solvation coefficient λ and the volume fraction difference, $\Delta\phi_A$, with the composition of the liquid mixture ethanol (A)/1,2 − dichloroethane (B). λ_A was measured by gel permeation chromatography and represents the volume of A in excess in the vicinity of the polyvinylpyrrolidone (PVP) chain per gram of polymer:

$$\lambda_A = \frac{V}{M}(U_A \phi_A)$$

V and *M* are the volume and the mass of the monomer unit, respectively. *U* and ϕ_A are the volume fraction of A in the vicinity of the polymer chain and in the bulk solution, respectively. The ordinate on the right is the difference between the volume fraction of A in the liquid (ϕ_A') and in the vapor (ϕ_A'') as measured in pervaporation through a PTFE-PVP membrane.

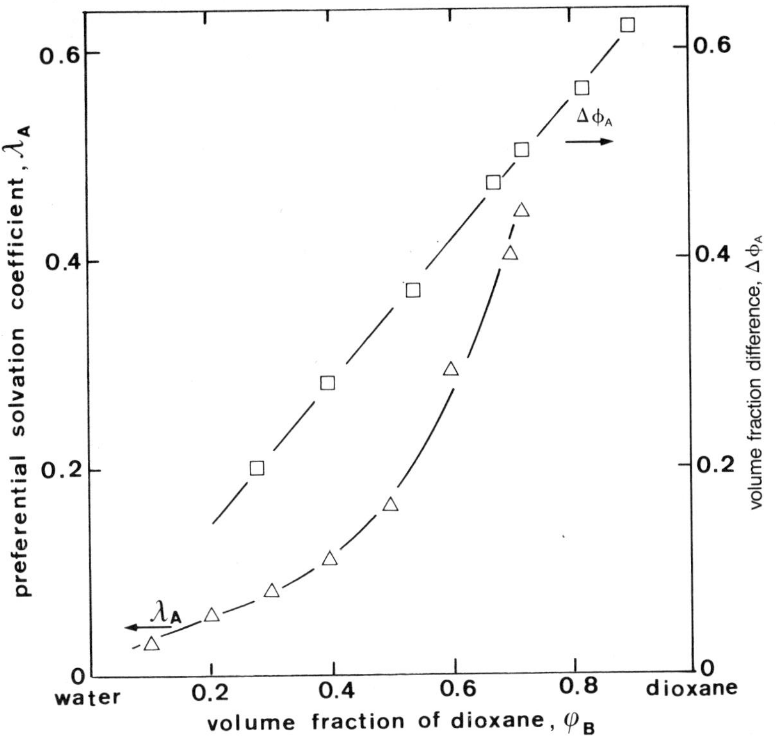

FIGURE 8.—Variation of the preferential solvation coefficient, λ_A, and the volume fraction difference, $\Delta\phi_A$, with the composition of the liquid mixture water(A) / dioxane(B). λ_A and $\Delta\phi_A$ are defined in Figure 7.

TABLE 1. Pervaporation of positive binary azeotropic mixtures across a PVP-PTFE membrane (T = 25°C): A = Faster permeant.

Components		T_b (°C)	Azeotrope T_b (°C)	Azeotrope w (%w-w)	β	J (Kg.m.$^{-2}$h.$^{-1}$)	$J(\beta-1)$
A	Water	100	63.8	5.7	9.24	0.94	7.75
B	THF	65.4		94.3			
A	Water	100	87.8	18.4	4.36	1.33	4.47
B	Dioxane	101.3		81.6			
A	Water	100	78.2	4.4	2.68	2.20	3.70
B	Ethanol	78.5		95.6			
A	Ethanol	78.5	58.7	21.0	3.24	1.10	2.46
B	Hexane	69.0		79.0			

TABLE 2. Pervaporation of negative binary azeotropic mixtures across a PVP-PTFE membrane (T = 25°C): A = Faster permeant.

Components	T_b (°C)	Azeotrope T_b (°C)	Azeotrope w (%w/w)	β	J (Kg.m.$^{-2}$h.$^{-1}$)	J(β − 1)
A Butanol B Pyridine	117.7 115.3	118.7	71.0 29.0	1.09	1.25	0.112
A Chloroform B Acetone	61.2 56.2	64.7	80.0 20.0	1.10	0.85	0.085
A Formic acid B Pyridine	100.7 115.3	150.0	63.5 36.5	1.31	0.22	0.068
A Water B Formic acid	100 100.7	107.1	22.5 77.5	1.00	2.74	0

liquid phase, in which A-A and B-B interactions prevail over A-B heteromolecular forces. On the other hand, the mixtures in Table 2 are negative azeotropic mixtures. Whereas efficient separations are observed for the former class, pervaporation is practically incapable of fractionating negative azeotropic mixtures because, in that case, the mutual solvation of both components favours their simultaneous and unselective transport.

Effect of the network structure.—The above discussions are concerned with the qualitative criteria to be used as a guide in selecting the chemical nature of the polymer. There is considerable evidence to show that the morphology of the polymer matrix also plays a part in the transport process; this is quite consistent with the solution–diffusion mechanism.

It was first established that for a homologous series of compounds, the permeability decreases as the size is increased; it is thus easier to separate water from isopropanol or *tert*-butanol than from methanol or ethanol.

It has also long been recognized that in a semi-crystalline polymer, transport occurs essentially within the intercrystalline amorphous region. The polymer can be considered as a molecular "sieve," wherein the amorphous regions constitute the "holes" and the interconnected crystalline elements, the "mesh" (7). Clearly, this two-phase picture is only an approximation to reality and the boundaries between amorphous and crystalline regions are not really distinct. As such a structure is dependent on past history (method of preparing the membrane, heat annealing, solvent treatment, etc.), a number of investigators have studied the influence of membrane conditioning on the transport properties (7, 33–35, 51, 52).

Another promising way to obtain a two-phase membrane is to graft an amorphous "active" polymer onto a highly crystalline "inert" film (27,70). Because grafting proceeds in the amorphous region surrounding the crystalline elements, the distribution of the diffusion paths and their density in "active" sites will depend essentially on the initial structure of the crystalline polymer. It appears that the best results are obtained with membranes prepared by grafting those films that originally displayed the highest crystallinity (Figure 9).

From this result, it could also be expected that by blending two polymers together, satisfactory membranes would be obtained. Moreover, the performances would be improved if some phase separation occurred during the membrane formation!

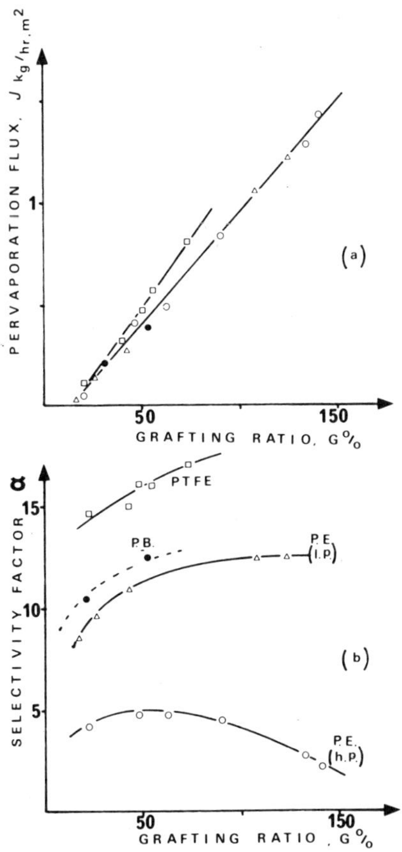

FIGURE 9.—Effect of grafting ratio on flux(a) and separation factor(b) for membranes prepared by grafting PVP into films of different crystallinity. Pervaporation of water (A) / dioxane(B) azeotropic mixture. (○) 15μm thick high-pressure polyethylene (50% crystalline phase); (△) 17μm thick low-pressure polyethylene (85% crystalline phase); (●) 20μm thick polybutadiene (90% crystalline phase); (□) 18 – μm thick PTFE (95% crystalline phase). Reprinted by courtesy of John Wiley & Sons. (27)

3. DESIGN OF A PERVAPORATION UNIT

To improve the separation efficiency of a membrane process, two factors should be examined thoroughly: 1) the transport mechanism through the membrane, and 2) the effect of boundary-layer phenomena on the mass transport. In pervaporation, a third factor has to be considered: the temperature drop due to the phase change from liquid to vapour.

In section 2 we have shown that even in the absence of a suitable model for the transport mechanism through the membrane, we have some useful criteria for designing a satisfactory membrane, and we have a number of semi-empirical equations that describe the effect of temperature and pressure on process performance.

We will now examine the effect of boundary-layer phenomena on mass transfer and evaluate the importance of the temperature drop due to the phase change from liquid to vapour. With this in mind, we will present the operating equations for a cross-flow continuous pervaporator. Possible improvements of both the membrane and the apparatus design will then be discussed.

3.1. Concentration Polarization

Pervaporation fluxes are generally less than 2 kg m^{-2} hr^{-1}, which corresponds to a transmembrane velocity of 6.10^{-5} $cm.s^{-1}$, i.e., one order magnitude lower than in reverse osmosis. Hence it can be expected that for feed flow conditions similar to those used in reverse osmosis, the boundary-layer effect at the liquid–membrane interface should be negligible in pervaporation.

However, for laminar flow in the liquid phase (at low Reynolds number), the situation is different. For example, assuming that the Sherwood number is given by:

$$Sh = 1.62 \left[Re\ Sc \frac{2r_t}{L} \right]^{1/3}, \tag{12}$$

LeBlanc (33) calculated that the concentration at the membrane interface could differ by a factor of 5 from that existing in the bulk solution (tubular pervaporator $r_t = 0.5$ cm and $L = 1$ m; feed mean velocity of 1 $cm.s^{-1}$). This result shows that for a continuous laminar flow mode of operation it is necessary to use a thin-channel module ($r_t = 0.2 - 0.5$ mm), so as to make sure that concentration polarization at the liquid–membrane interface does not affect the overall transfer phenomena.

On the vapour side, the pervaporate stream is drawn into a high vacuum; consequently, concentration polarization is practically absent.

3.2. Temperature Drop at the Membrane Interface

Due to the phase change from liquid to vapour of a part of the feed, an unavoidable local temperature drop at the membrane interface occurs, which in turn, affects the permeation flux. Rautenbach et al. (53) have shown that this drop could reach 12 K at low Reynolds number during pervaporation of water through a cellulose acetate membrane.

Néel et al. (32) discussed the case of a laminar-flow operation under adiabatic conditions and deduced a simple expression for the temperature drop in the feed mixture between the inlet and the outlet of the pervaporator:

$$\Delta T = \theta \frac{\Delta \hat{H}_{V,B} + (\Delta \hat{H}_{VA} - \Delta \hat{H}_{V,B}) \langle w''_A \rangle}{C_B + (C_A - C_B)\, w'_{A,0}} \qquad (13)$$

In this expression, $\langle w''_A \rangle$ and $w'_{A,0}$ are the average mass fraction of A in the pervaporate and the initial mass fraction of A in the feed, respectively. In most practical cases, the cut ratio θ lies between 0.1 and 0.3. The latent heat of vaporisation, $\Delta \hat{H}_v$, of most liquids (expressed in $kJ.kg^{-1}$) is generally 500 times as large as their heat capacity (C in $kJ.kg^{-1}.K^{-1}$). So, to a first approximation:

$$\Delta T \sim \theta \frac{\Delta \hat{H}_{v,B}}{C_B} \sim 50 - 150 \text{ K} \qquad (13a)$$

This large decrease in temperature can be avoided by a correct design of the process. The simplest means appears to be to operate with a sequence of short modules with intermediate heat exchangers (32, 65); alternative suggestions consist of furnishing the enthalpy of vaporisation on the vapour side of the membrane: heated inert carrier-gas (71) or carrier-liquid (39) and inert condensable carrier-vapour (8).

3.3. Cross-Flow Continuous Pervaporation

There is a general agreement that for obvious economic reasons, the possibility of utilizing pervaporation will be essentially limited to the extraction of the minor component (A) contained in an azeotropic mixture.

In order to reduce the original impurity content ($w'_{A,0}$) to a definite level $w'_{A,r}$, an attractive method would be to operate continuously in a single-staged membrane module (32, 65).

The low-pressure (pervaporate) stream is assumed to be drawn into a vacuum such that the flow is mainly in the direction normal to the membrane surface-cross-flow mode of operation.

Moreover, in the following, the liquid stream is presumed to be in plug flow (Figure 10).

Simple analysis First, consider the simple case where the enrichment factor β and the transmembrane flux J are both constant within the range $w'_{A,0} - w'_{A,r}$. A set of operating equations can easily be deduced from a mass-balance analysis. These equations involve the purification ratio:

$$\kappa = \frac{w'_{A,0}}{w'_{A,r}} \qquad (14)$$

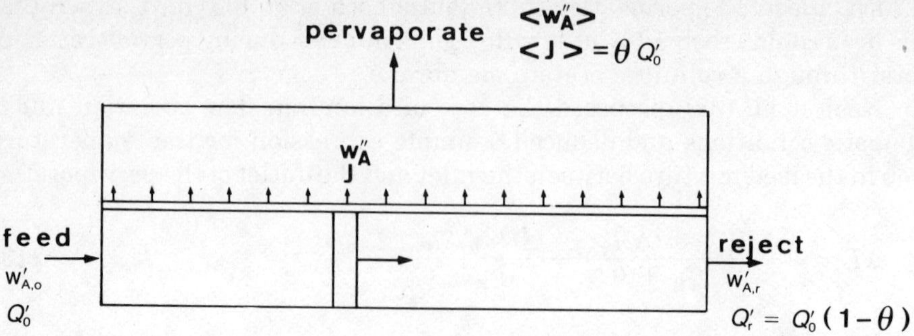

FIGURE 10.—Cross-flow continuous pervaporation.

and another dimensionless parameter X:

$$X = \kappa^{(\beta-1)^{-1}} \tag{15–1}$$

X is a function of both the selectivity of the membrane, β, and the required purification ratio, κ. The knowledge of X and J are then sufficient to describe the operation.

The production capacity of a pervaporation unit could be expressed as the flow rate of reject per unit area of membrane Q'_r (kg.m^{-2}.hr^{-1}):

$$Q'_r = \frac{J}{X - 1} \tag{16–1}$$

or as the flow rate per unit membrane area Q'_0 (kg.m^{-2}.hr^{-1}) of the feed to be purified:

$$Q'_0 = \frac{XJ}{X - 1} \tag{17–1}$$

The recovery yield of B in the purified reject will be:

$$Ry = \frac{Q'_r\,(1 - w'_{A,r})}{Q'_0\,(1 - w'_{A,0})} = \frac{1}{X}\left(\frac{1 - w'_{A,r}}{1 - w'_{A,0}}\right) \tag{18–1}$$

For highly selective membranes (β >> 1):

$$X \cong 1 + \frac{1}{\beta - 1}\ln\kappa \tag{19}$$

and $$Q'_r \cong \frac{(\beta - 1)\,J}{\ln\kappa} \tag{20}$$

Note that in equation 20, the production capacity Q'_r appears to be proportional to the term $(\beta - 1)\,J$. Equation 20 is thus in good agreement with the idea that the magnitude of the term $(\beta - 1)\,J$ can be considered as a suitable criterion for classifying different membranes.

An alternative formulation of operating equations The overall mass-balance equation is the following:

$$J\,(1 - \langle w''_A \rangle) = Q'_0\,(1 - w'_{A,0}) - Q'_r\,(1 - w'_{A,r}) \tag{21}$$

where $\langle w''_A \rangle$ is the mean mass fraction of A in the pervaporate.

Substitution for Q'_0 and Q'_r, using expressions 16–1 and 17–1, gives X as a function of $w'_{A,0}$, $w'_{A,r}$ and $\langle w''_A \rangle$:

$$X = \frac{\langle w''_A \rangle - w'_{A,r}}{\langle w''_A \rangle - w'_{A,0}} \tag{15–2}$$

An alternative set of operating equations for a pervaporation unit is thereby obtained.

Production capacity:

$$Q'_r = J\,\frac{\langle w''_A \rangle - w'_{A,0}}{w'_{A,0} - w'_{A,r}} \tag{16–2}$$

Treatment capacity:

$$Q'_0 = J\,\frac{\langle w''_A \rangle - w'_{A,r}}{w'_{A,0} - w'_{A,r}} \tag{17–2}$$

Recovery yield of B in the reject:

$$Ry = \frac{\langle w''_A \rangle - w'_{A,0}}{\langle w''_A \rangle - w'_{A,r}} \cdot \frac{1 - w'_{A,r}}{1 - w'_{A,0}} \tag{18–2}$$

General analysis A subsequent more exact analysis will take into account the variation of both β and J when the composition of the liquid mixture progressively decreases from $w'_{A,0}$ to $w'_{A,r}$. For a sequence of elementary cells of the same length in which the values of β and J are assumed constant, average values, $\langle J \rangle$ and $\langle w''_A \rangle$, can be calculated numerically from experimental measurements of J and β as functions of w'_A. (see Fig. 5) It follows that we obtain a set of equations similar to equations 16–2 to 18–2.

Production capacity:

$$Q_r' = \langle J\rangle \frac{\langle w_A''\rangle - w_{A,0}'}{w_{A,0}' - w_{A,r}'} \tag{16–3}$$

Treatment capacity:

$$Q_0' = \langle J\rangle \frac{\langle w_A''\rangle - w_{A,r}'}{w_{A,0}' - w_{A,r}'} \tag{17–3}$$

Recovery yield of B:

$$Ry = \frac{\langle w_A''\rangle - w_{A,0}'}{\langle w_A''\rangle - w_{A,r}'} \cdot \frac{1 - w_{A,r}'}{1 - w_{A,0}'} \tag{18–3}$$

Energy cost of pervaporation The major part of the energy consumed in a pervaporator corresponds to the heat required to vaporize the permeate. The production of a unit weight of purified liquid reject therefore requires the energy input:

$$E = \frac{\langle J\rangle \langle \Delta\hat{H}_{v,p}\rangle}{Q_r'} \tag{22}$$

where $\langle \Delta\hat{H}_{v,p}\rangle$ is the mean latent heat of vaporization of the permeate:

$$\langle \Delta\hat{H}_{v,p}\rangle = \Delta\hat{H}_{v,B} + (\Delta\hat{H}_{v,A} - \Delta\hat{H}_{v,B})\,\langle w_A''\rangle \tag{23}$$

$\Delta\hat{H}_{v,A}$ and $\Delta\hat{H}_{v,B}$ are the latent heats of vaporization of A and B, respectively, Referring to equation 16-3, we obtain:

$$E = \frac{w'_{A,0} - w'_{A,r}}{\langle w''_A\rangle - w_{A',0}} \quad [\Delta\hat{H}_{v,B} - (\Delta\hat{H}_{v,A} - \Delta\hat{H}_{v,B})\langle w''A\rangle] \tag{24}$$

Numerical applications

Dehydration of the water–dioxane azeotrope by pervaporation through a PVP–PTFE membrane.—As an example, we will now examine the dehydration of the water–dioxane azeotrope, which contains a water mass fraction $w_{A,0} = 0.184$. The variation of J and β as functions of $w_{A'}$ is shown in Figure 5 when a PVP–PTFE membrane is used. To specify the value of the purification ratio κ, we will assume that the residual water-content in the reject is required to be less than 1 percent ($w_{A,r}' = 0.01$). The corresponding κ ratio then equals 18.4.

Average values $\langle J \rangle$ and $\langle w''_A \rangle$ are evaluated from experimental data shown in Figure 5:

$$\langle J \rangle = 0.76 \text{ kg.m}^{-2} \text{ hr}^{-1} \tag{25}$$

$$\langle w''_A \rangle = 0.72 \tag{26}$$

Referring to equations 16-3 – 18-3, it is then possible to estimate Q'_r, Q'_0 and Ry:

$$Q'_r = 2.3 \text{ kg.m}^{-2} \text{ hr}^{-1} \tag{27}$$

$$Q'_0 = 3.1 \text{ kg.m}^{-2} \text{ hr}^{-1} \tag{28}$$

As long as the 8.6% loss ($Ry = 91.4\%$) in dioxane is acceptable, no recycling is needed and the energy cost of the process is reduced approximately to the heat of vaporization of the permeate (equation 24). From the thermodynamic characteristics of water and dioxane ($\Delta \hat{H}_{v,W} = 2456$ kJ.kg^{-1} and $\Delta \hat{H}_{v,D} = 431.3$ kJ.kg^{-1}) we deduce:

$$E = 615 \text{ kJ.kg}^{-1} \text{ (dioxane 99\% purity)} \tag{29}$$

Equation 13 makes it possible to estimate the temperature reduction that the liquid would undergo if the apparatus were used under adiabatic conditions.

Substituting for C_A and C_B their tabulated values:

$$C_A = C_W = 4.2 \text{ kJ.kg}^{-1}.\text{K}^{-1} \tag{30}$$

$$C_B = C_D = 1.736 \text{ kJ.kg}^{-1}.\text{K}^{-1} \tag{31}$$

we obtain:

$$\Delta T = 212 \text{ K} \tag{32}$$

It becomes apparent that the pervaporator could not work without a compensating heat supply. In order to keep the liquid at its original temperature, it will be necessary to provide approximately 615 kJ for each kg of purified dioxane retained. Since a unit pervaporator produces 12.3 kg of refined solvent per hour, it must be equipped with a 400-watt heater. In the case of a metal-plate module, the heat could be readily transferred to the liquid. On the other hand, technical problems would arise if a hollow-fiber or a spiral-wound device were used. To overcome that difficulty, it would be possible to split the system into several sub-units, each one achieving a partial purification of the liquid. Within each step, the temperature drop could then be reduced to an acceptable value and any cooling could be compensated by passing the liquid through intermediate heat-exchangers inserted between the successive sub-units.

Production of pure ethanol by pervaporation through a polyvinylalcohol membrane.—Preliminary results from a pilot plant for the production of power alcohol from biomass have been reported recently (65); in this process, pervaporation is coupled with distillation. The ethanol content in the fermentor is 10% w/w. In a first distillation step the alcohol concentration is brought up to 80%; pervaporation is then used to produce an alcohol containing less than 0.4% of water.

A composite membrane is used to eliminate water selectively; this membrane is prepared by depositing a layer of cross-linked polyvinylalcohol (~ 10 μm thick) onto the surface of a porous polyacrylonitrile support.

The enrichment factor β, which is around 5 at the initial feed concentration ($w'_{A,0} = 0.2$), reaches a value of above 15 at the end of the purification ($w'_{A,r} = 0.04$). The average pervaporate flux is of the order of 2 $kg.m^{-2}.hr^{-1}$.

Due to the temperature drop, four spiral-wound modules, each with 3 m^2 of active surface, are installed in series. The feed enters each module at 80°C and leaves at a temperature of 35°C. Heat exchangers are inserted between the condenser of the distillation tower and each module.

3.4. Improvement of the Membrane: Asymmetric and Composite Membranes

The above discussion of the design of a pervaporator shows the importance of the flux in determining the production capacity (equation 16). Hence, besides the chemical nature and the morphology of the polymer matrix which have been discussed in section 2, an intriguing question is "Could pervaporation be improved by employing asymmetric membranes in the same way as in reverse osmosis or in gas permeation?"

Rautenbach and Albrecht (54) recently studied the pervaporation of a water–isopropanol mixture through symmetric and asymmetric cellulose acetate membranes. If the flux of asymmetric membranes is found to be higher than that of symmetric membranes of identical total thickness, the improvement is far less pronounced than might be anticipated from equation 4 or 7. Moreover, the increase in flux is accompanied by a decrease in selectivity.

It is interesting to note that large differences were observed when the orientation of the membrane was reversed; flux was higher and selectivity lower when the porous sublayer faced the feed (Table 3). The efficiency coefficient for the separation $J(\beta - 1)$ is higher, then, when the porous sublayer of the membrane faces the feed ("skin down" position). In terms of reverse osmosis, the membrane is installed upside down!

TABLE 3. Effect of the structure of the membrane and of the direction of pervaporate flux on performances of cellulose acetate membranes. J is expressed in kg/m^{2}hr.

Membrane		water/isopropanol(54) $w_{water} = 0.4$			ethanol/cyclohexane $w_{EtOH} = 0.3$		
		J	β	$J(\beta-1)$	J	β	$J(\beta-1)$
Asymmetric	"Skin up"	1.5	1.9	1.3	2.6	2.8	4.7
	"Skin down"	3.5	1.7	2.5	5.2	2.6	8.3
Homogeneous		0.6	2.5	0.9	—	—	—

These intriguing results merit discussion; they could be explained qualitatively as a result of mass and heat transfer resistances at the interfaces:

1) "Skin down" The compounds are transported to the dense skin as a liquid by slow viscous flow through the porous substructure; because of the selectivity of the transport across the skin, a concentration gradient will be built up. Moreover, the temperature will decrease as a result of the phase change. Because these two events occur inside the pores of the membrane, an increase in feed flow rate is therefore only of a limited efficacy in lowering the resistance of the boundary layers.

2) "Skin up" When the dense skin is facing the feed, the boundary layer resistance at the upstream interface is similar to the value obtained with a symmetric membrane. In flowing through the porous support layer, the vapours can undergo a noticeable pressure loss, which in turn increases the activity of the pervaporate at the downstream side of the skin. The local pressure may even be larger than the capillary pressure and the membrane will operate in liquid–liquid permeation.

The value of the boundary layer resistance will depend strongly on the structure of the porous part of the film. Therefore, a quantitative analysis (54) requires a precise knowledge of this structure. For this purpose, composite membranes appear to be of greater interest than asymmetric membranes.

When considering asymmetric or composite membranes it should also be emphasized that depending on the concentration (or swelling) profile in the dense layer, a large difference in swelling between the two sides of the skin could create cracks if the skin is too thin.

3.5. Process Design Improvements

Because the downstream side of the membrane must be maintained at very low vapour pressure, most investigations have used a vacuum to remove the pervaporate and a low-temperature trap to condense the vapours in order to avoid any loss of product.

However, other modes of operation have received some attention: Yuan and Schwartzberg (71) used an air sweep stream, Cabasso et al. (39) an inert liquid carrier, Aptel et al. (25) a temperature gradient. More recently, Hoover and Hwang (17) have extended the "continuous membrane column" concept to the pervaporation process and Henis has discussed the use of vapor permeation (82).

Air-heated pervaporation (71) Heated air can be used to carry away the pervaporate while at the same time providing the latent heat of vaporization. The system is particularly simple when the pervaporate can be discarded, because eliminating the need for a vacuum line and condenser lowers the capital cost. However, the added mass transfer resistance due to the air may somewhat counterbalance these advantages.

Osmotic distillation (39) The air or gas carrier can be replaced by an inert liquid, as shown in Figure 11. The process was called "osmotic distillation" because of the difference in the osmotic pressure between the two compartments. This difference is revealed by the build-up of a hydrostatic pressure in the downstream compartment. If the liquid does not swell the membrane, it serves only to maintain an essentially zero activity of the penetrants at the downstream interface. In this way, a selectivity as high as pervaporation can be achieved. Moreover, the absence of vaporization eliminates the temperature drop and higher fluxes are observed.

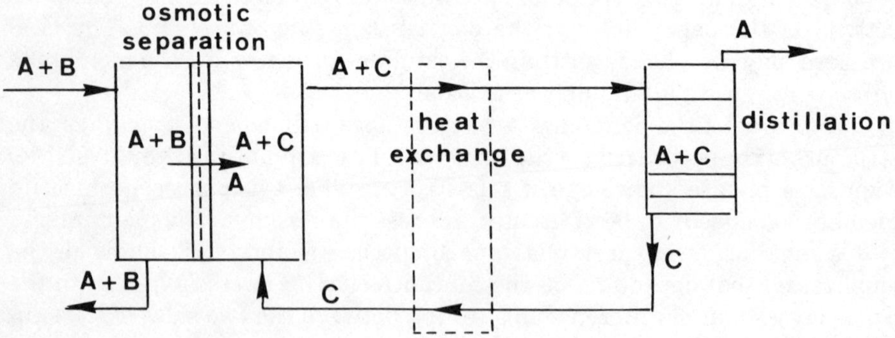

FIGURE 11.—Schematic representation of an "osmotic distillation" setup. (39)

Thermopervaporation (25) Conventional pervaporation, like the two previous modes of operation, needs an external means (vacuum, gas or liquid flow) for removing the pervaporated species from the membrane. As a result, we get a very diluted product and we must concentrate it in a second step outside the module: condensation (in conventional and air-heated pervaporation) or distillation (in osmotic distillation).

By slightly modifying the conventional apparatus, it is possible to condense the pervaporate on a cold wall inside the module. The temperature of the cold wall can be fixed lower or higher than the ambient temperature when the downstream compartment is insulated from the vacuum line (Figure 12).

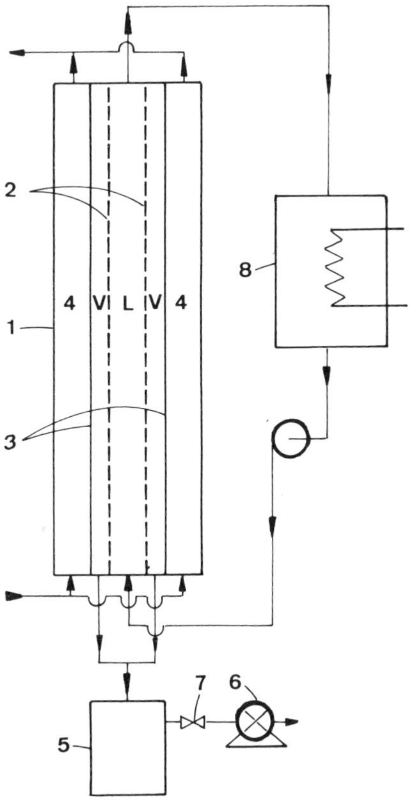

FIGURE 12.—Schematic representation of a "thermopervaporation" apparatus. 1 = thermopervaporation cell; 2 = membranes; 3 = cold wall; 4 = cooling chamber; 5 = pervaporate container; 6 = vacuum pump; 7 = stopcock; 8 = hot feed container; L = feed compartment; V = pervaporate compartment. The vacuum pump is only used to evacuate the air from compartment (V) before starting the process. Reprinted by courtesy of Elsevier Science Publishers. (35)

This mode of operation was named "thermopervaporation" because the driving force for the transport is directly related to the temperature difference between the feed and the cold wall facing the downstream side of the membrane. When a reasonable temperature difference is maintained (i.e., 40°C for a water–dioxane mixture), selectivities and fluxes are of the same order of magnitude as in conventional pervaporation.

"Thermopervaporation" presents several advantages over conventional pervaporation: the pervaporate is condensed at ambient temperature (which implies a low energy consumption) and a liquid effluent is directly obtained and might possibly serve as the feed for a second stage of pervaporation.

Proper design of such a process results in recovery of the condensation enthalpy to pre-heat the feed. The major drawback is the loss of heat by conduction, in a stack of successively hot and cold chambers.

Continuous membrane column (17) A recent breakthrough in the area of gas separation is Hwang's "continuous membrane column," which has been recently extended to the pervaporation process. A sketch of the column is presented in Figure 13. By circulating the liquid feed mixture through the

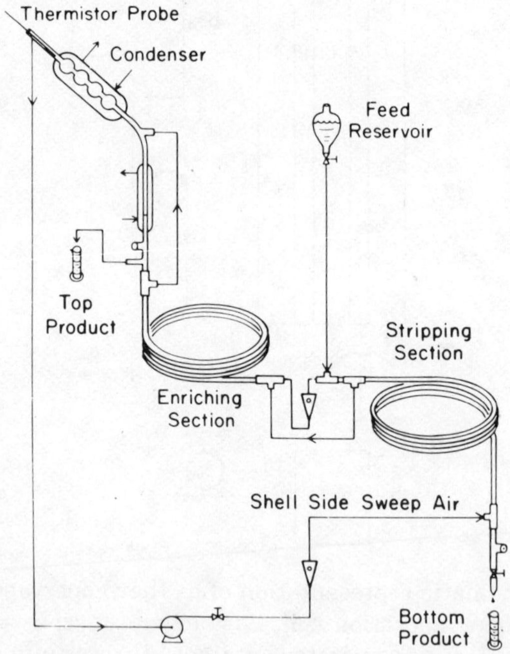

FIGURE 13.—Continuous membrane column for pervaporation. Reprinted by courtesy of Elsevier Science Publishers. (17)

column continuously and bleeding off only a small fraction of the most permeable (A) and least permeable (B) liquids at the top and bottom of the column, respectively, a clear separation can theoretically be achieved with membranes that do not have inherently high selectivities. In the apparatus shown in Figure 13, a closed circuit sweep air stream was used to remove the pervaporate from the shell side of the tubular membrane and transfer it to the condenser.

Despite the difficulties encountered in the first experiments, this novel mode of operation could be more attractive than conventional cascade enrichments for the complete separation of azeotropic mixtures. However, because of the need to recirculate an inert gas containing a low concentration of vapour, the energy consumption could be higher than in cross-flow pervaporation when extracting the minor component from an azeotropic mixture.

Vapor permeation (82) To avoid any temperature drop due to vaporization, the membrane system can be combined with a distillation tower and the vapor stream heated directly. Under these conditions, for the azeotropic ethanol–water mixture, the energy savings associated with the use of a membrane is of the order of 80%, compared with conventional azeotropic distillation.

4. CONCLUSION

Pervaporation is on the verge of becoming an industrial process and the first field of application will be the extraction of the minor component contained in an azeotropic mixture.

Up to now the main efforts have been devoted to studying dense membranes and some progress has been made in understanding the relationship between chemical composition and selectivity. The next step in this development may be theoretical and experimental research into the effect of using asymmetric and composite membrane structures.

The aim of developing a membrane with optimum chemical properties and physical structure for a particular application must not be considered independently from the optimization of the module and the overall system. The design of membranes, modules and plants must be performed as a single effort in order to solve the two main problems: keeping the pressure losses low on the pervaporate side and supplying the minimum energy for evaporation.

Moreover, pervaporation will generally constitute only one concentration step in a separation process. Finding the best combination with other conventional unit operations in an overall flow-sheet is another essential condition for industrial success.

REFERENCES

1. Kober P.A.: 1917, "Pervaporation, perstillation and percrystallization." J. Am. Chem. Soc. 39, p. 944.
2. Kahlenberg L.: 1906, "On the nature of the process of osmosis and osmotic pressure with observations concerning dialysis." J. Phys. Chem. 10, p. 141.
3. Schwob Y.: 1963, "Sur l'hémiperméabilité à l'eau des membranes de cellulose régénérée" (Thesis, Toulouse (France), May 23 1949). C.R. Séances Acad. Sci., 256, p. 3459.
4. Heissler E.G., Hunter A.S., Siciliano J. and Treadway R.M.: 1956, "Solute and temperature effects in the pervaporation of aqueous alcoholic solutions. Science 124, p. 77.
5. Binning R.C. and James F.E.: 1958, "New separate by membrane separation." Pet. Ref. 37, p. 214.
6. Binning R.C., Lee R.J., Jennings J.F., Martin E.C.: 1961, "Separation of liquid mixtures by permeation." Ind. Eng. Chem. 53, p. 45.
7. Michaels A.S., Baddour R.F., Bixler H.J., and Choo C.Y.: 1962, "Conditioned polyethylene as a permselective membrane." Ind. Eng. Proc. Des. Develop. 1, p. 14.
8. Michaels A.S. and Bixler H.J.: 1968, "Membrane permeation: theory and practice," in E.S. Perry (Ed.) "Progress in separation and purification". Wiley, New York.
9. Huang R.Y.M. and Lin V.J.C.: 1968, "Separation of liquid mixtures using polymer membranes. I—Permeation of binary organic liquid mixtures through polyethylene." J. Appl. Polym. Sci. 12, p. 2615.
10. Huang R.Y.M. and Fels M.: 1968, "Separation of organic liquid mixtures by the permeation process using modified membranes." Chem. Eng. Prog. Symp. Ser. 65, p. 52.
11. Huang R.Y.M. and Jarvis N.R.: 1970, "Separation of liquid mixtures by using polymer membranes. II—Permeation of aqueous alcohol solutions through cellophane and poly(vinyl-alcohol)." J. Appl. Polym. Sci. 14, p. 2739.
12. Fels M. and Huang R.Y.M.: 1970, "Diffusion coefficient of liquids in polymer membranes by a desorption method." J. Appl. Polym. Sci. 14, p. 523.
13. Fels M. and Huang R.Y.M.: 1970, "The effect of styrene grafting on the diffusion and solubility of organic liquids in polyethylene." J. Appl. Polym. Sci. 14, p. 537.
14. Fels M. and Huang R.Y.M.: 1971, "Theoretical interpretation of the effect of mixture composition on separation of liquids in polymers." J. Macromol. Sci. Phys. B5 (1) p. 89.
15. Fels M.: 1972, "Permeation and separation behavior of binary organic mixtures in polyethylene." A.I.Ch.E. Sym. Ser. 68, p.49.
16. Hwang S.T. and Kammermeyer K.: 1975, "Pervaporation" in "Membranes in Separations." Techniques of Chemistry, Vol. VII, A. Weissberger, Editor, J. Wiley, New York.
17. Hwang S.T. and Hoover K.C.: 1982, "Pervaporation by a continuous membrane column." J. Memb. Sci. 10, p. 253.
18. Kammermeyer K. and Hagerbaumer D.H.: 1965, "Membrane separations in the liquid phase." A.I.Ch.E. J. 1, p. 215.
19. Kim S.N. and Kammermeyer K.: 1970, "Actual concentration profiles in membrane permeation." Sep. Sci 5, p. 679.
20. Aptel P., Fries R. and Néel J.: 1969, "Transferts sélectifs à travers des membranes obtenues à partir de polybases de Lewis. Membranes à perméabilité sélective." Ed. du CNRS, pp. 119-129.
21. Aptel P., Cuny J., Jozefonvicz J., Morel G., and Néel J.: 1972, "Liquid transport through membranes prepared by grafting of polar monomers onto polytetrafluoroethylene films. Part 1—Some fractionations of liquid mixture by pervaporation." J. Appl. Polym. Sci. 16, p. 1061.
22. Aptel P., Cuny J., Jozefonvicz J., Morel G., and Néel J.: 1973, "Pervaporation à travers des films de polytétrafluoréthylène modifiés par greffage radiochimique de N-vinylpyrrolidone." Europ. Polym. J. 9, p. 877.
23. Aptel P., Cuny J. Jozefonvicz J., Morel G., and Néel J.: 1974, "Liquid transport through membranes prepared by grafting of polar monomers onto polytetrafluoroethylene films. Part II—Some factors determining pervaporation rate and selectivity." J. Appl., Polym. Sci. 18, p. 357.

24. Aptel P., Cuny J., Jozefonvicz J., Morel G., and Néel J.: 1974, "Liquid transport through membranes prepared by grafting of polar monomers onto polytetrafluoroethylene films. Part III—Steady state distribution in membrane during pervaporation." J. Appl. Polym. Sci. 18, p. 365.
25. Aptel P., Challard N., Cuny J. and Néel J.: 1976, "Application of the pervaporation process to separate azeotropic mixtures." J. Memb. Sci. 1 (3), p. 271.
26. Aptel P., Cuny J., Jozefonvicz J., Morel G., and Chaufer B.: 1978, "Perméabilité sélective et solvatation préférentielle." Europ. Polym. J. 14, p. 595.
27. Morel G., Jozefonvicz J., and Aptel P.: 1979, "Pervaporation membranes prepared by radiochemical grafting of N-vinylpyrrolidone onto different base-films. Influence of the grafting conditions and nature of the base-film." J. Appl. Polym. Sci. 23, p. 2397.
28. Morel G. and Jozefonvicz J.: 1979, "Structure and morphology of poly(tetrafluoroethylene)-poly(N-vinylpyrrolidone) copolymer membranes." J. Appl. Polym. Sci. 24, p. 771.
29. Morel G.: 1979, "Synthèse et propriétés de membranes greffées permsélectives." Thesis, Paris XIII.
30. Neel J. and Fries R.: 1965, "Transfert sélectif à travers des membranes actives." J. Chim. Phys. 6, p. 494.
31. Neel J. and Aptel P.: 1982, "La pervaporation. I—Principe de la technique." Entropie 104, p. 15.
32. Néel J., Nguyen Q.T., LeBlanc L. and Clement R.: 1982, "La pervaporation. II—Fractionnement en continu d'un mélange liquide binaire." Entropie 104, p. 27.
33. LeBlanc L.: 1982, "Evaluation de la pervaporation en tant que procédé continu pour fractionner les mélanges liquides." Thesis, Nancy, France (June 22, 1982).
34. Li N.N.: 1969, "Plasticizing effect of permeates on the selectivity of polymeric membranes." Ind. Eng. Chem. Prod. Res. Devel. 8, p. 282.
35. Li N.N. and Long R.B.: 1969, "Permeation through plastic films." Am. Inst. Chem. Eng. J. 15, p. 73.
36. Li N.N. and Long R.B.: 1972, "Separation by permeation through polymeric membranes." Recent Develop. Separ. Sci. 2, p. 107.
37. Long R.B.: 1965, "Liquid permeation through plastic films." Ind. Eng. Chem. Fund. 4, p. 445.
38. Cabasso I., Jaguar-Grodzinski J. and Vofsi D.: 1974, "Polymeric alloys of polyphosphonates and acetyl cellulose. I—Sorption and diffusion of benzene and cyclohexane." J. Appl. Polym. Sci. 18, p. 2117.
39. Cabasso I., Jaguar-Grodzinski J. and Vofsi D.: 1974, "A study of permeation of organic solvents through polymeric membranes based on polymeric alloys of polyphosphonates and cellulose acetate. II—Separation of benzene, cyclohexane and cyclohexene." J. Appl. Polym. Sci. 18, p. 2137.
40. Cabasso I., Jaguar-Grodzinski J. and Vofsi D.: 1974, "Organic separation with polymeric alloy membranes." Am. Chem. Soc. 168th meeting, Atlantic City, N.J., (Sept. 1974).
41. Cabasso I., and Leen H.: 1975, "Liquid mixture separation by flat-sheet and hollow-fiber membranes." AIChE 80th National Meeting, Boston, Massachusetts (Sept. 1975).
42. Cabasso I., Eyer C. and Smith J.K.: 1977, "Water diffusion membrane. Pervaporation and heat rejection through composite membranes." Final report. NASA 27650151967 (March 1977).
43. Cabasso I.: 1983, "Organic liquid mixtures separation by permselective polymer membranes. 1—Selection and characteristics of dense isotropic membranes employed in the pervaporation process." Ind. Eng. Chem. Prod. Res. Dev. 22, p. 313.
44. Kubica J., Kucharski M. and Stelmaszek J.: 1968, "Separation of liquid mixtures by the permeation method. III—Separation of aqueous solution." Int. Chem. Eng. 8, p. 81.
45. Kucharski M. and Stelmaszek J.: 1967, "Separation of liquid mixtures by permeation." Int. Chem. Eng. 7, p. 618.
46. Stelmaszek J.: 1977, "Dehydration of isopropyl alcohol by means of a permeation method. Comparison with other ones." Proc. 3rd Conf. Appl. Chem. Unit Oper. Process, Veszprem, Hungary, pp. 481-486.
47. Stelmaszek J., Borlai O., Nagy E.: 1981, "Application of permeation process for separating alcohol-water mixtures." Chisa 81, Praha, Czechoslovakia.

48. Greenlaw F.W., Prince W.D., Shelden R.A. and Thomson E.V.: 1977, "Dependence of diffusive permeation rates on upstream and downstream pressures. I—Single component permeant." J. Memb. Sci. 2, p. 349.
49. Greenlaw F.W., Shelden R.A. and Thompson E.V.: 1977, "Dependence of diffusive permeation rates on upstream and downstream pressures. II—Two component permeant." J. Memb. Sci. 2, p. 334.
50. Shelden R.A. and Thompson E.V.: 1978, "Dependence of diffusive permeation rates on upstream and downstream pressures. III—Membrane selectivity and implications for separation processes." J. Memb. Sci. 4, p. 115.
51. Borlai O., Nagy E. and Lijhiday A.: 1977, "Dehydration of alcohol-water azeotropic mixtures by permeation," in Proceedings of the 3rd Conference on Applied Chemistry Unit Operations and Processes. Veszprem, Hungary, pp. 347-352.
52. Nagy E., Borlai O. and Ujhioy A.: 1980, "Membrane permeation of water-alcohol binary mixtures." J. Memb. Sci. 7, p. 109.
53. Rautenbach R. and Albrecht R.: 1980, "Separation of organic binary mixtures by pervaporation," J. Memb. Sci. 7, p. 203.
54. Rautenbach R. and Albrecht R.: 1982, "On the behaviour of asymmetric membranes in pervaporation." European Workshop on Pervaporation, Nancy, France (Sept. 21-22).
55. Brun J.P., Bulvestre G., Kergreis A. and Guillou M.: 1974, "Hydrocarbons separation with polymer membranes. 1-3-butadiene-isobutene separation with nitrile rubber membranes." J. Appl. Polym. Sci. 18, p. 1663.
56. Brun J.P.: 1981, "Etude thermodynamique du transfert sélectif par pervaporation à travers des membranes élastomères d'espèces organiques dissoutes en milieu aqueux." Thesis, Paris XII (June 26, 1981).
57. Larchet C., Brun J.P., Guillou M.: 1978, "Etude de la pervaporation de mélanges d'hydrocarbures aromatiques et aliphatiques à travers différentes membranes élastomères." C.R. Acad. Sci. Paris 287, Ser C., p. 31.
58. Vasse F.: 1974, "Contribution à l'étude de l'extraction du 1,3-butadiene des coupes pétrolières par pervaporation à travers des membranes permsélectives." Thesis, Paris VI. (Dec. 16, 1974).
59. Vasse F., Brun J.P. et Guillou M.: 1975, "Etude de la pervaporation de mélanges binaires 1,3-butadiène-isobutène à travers des copolymères 1,3-butadiène-acrylonitrile: influence des conditions opératoires." C.R. Acad. Sci., Paris, 281, Ser. C, p. 1073.
60. McCandless F.P.: 1973, "Separation of aromatics and naphthenes by permeation through modified vinylidene fluoride films." Ind. Eng. Chem. Proc. Des. Develop. 12, p. 354. (1973)
61. McCandless F.P., Alzheimer D.P., Hartman R.B.: 1974, "Solvent membrane separation of benzene and cyclohexane." Ind. Eng. Chem. Proc. Des. Develop. 13, p. 310.
62. Sikonia J.G. and McCandless F.P.: 1978, "Separation of isomeric xylenes by permeation through modified plastic films." J. Memb. Sci. 4, p. 229.
63. Zavaleta R. and McCandless F.P.: 1976, "Selective permeation through modified polyvinylidene fluoride membranes." J. Memb. Sci. 1, p. 333.
64. Blackadder D.A. and Keniry J.S.: 1972, "The measurement of the permeability of polymer membranes to solvating molecules." J. Appl. Polym. Sci. 16, p. 2141.
65. Bruschke H.E.A., Schneider W.H. and Tusel G.F.: 1982, "Pervaporation membrane for the separation of water and oxygen-containing simple organic solvents." European Workshop on Pervaporation, Nancy, France (Sept. 21-22, 1982).
66. Carter J.W. and Jagannadhaswamy J.: 1964, "Liquid separations using polymer films." Proceedings of the Symposium on the Less Common Means of Separations, Inst. Chem. Engrs. 35.
67. Eustache H. and Histi G.: 1981, "Separation of aqueous organic mixtures by pervaporation and analysis by mass-spectrometry or a coupled gas chromatograph mass spectrometer." J. Membr. Sci. 8, p. 105.
68. Kimura S. and Nomura T.: 1982, "Pervaporation of alcohol-water mixtures with silicone rubber membrane." Membrane 7, p. 353.

69. Peeters H., Vanderstraeten P. and Verhoeye L.: 1979, "The permeation of hydrocarbons through a polyethylene membrane." J. Chem. Techn. Biotechnol. 29, p. 581.
70. Tealdo G.C., Canepa P. and Munari S.: 1981, "Water-ethanol permeation through radiation grafted PTFE membranes." J. Memb. Sci. 9, p. 191.
71. Yuan S. and Schwartzberg H.G.: 1972, "Mass transfer resistance in cross membrane evaporation into air." A.I.Ch.E. Symp. Ser. 68, p. 41.
72. Paul D.R. and Ebra-Lima O.M.: 1971, "Mechanism of liquid transport through swollen polymer membranes." J. Appl. Polym. Sci. 15, p. 2199.
73. Paul D.R. and Paciotti J.D.: 1975, "Driving force for hydraulic and pervaporative transport in homogeneous membranes." J. Polym. Sci. Polym. Phys. Ed. 13, p. 1201.
74. Mulder M.H.V., Kruitz F. and Smolders C.A.: 1982, "Separation of isomeric xylenes by pervaporation through cellulose ester membranes." J. Memb. Sci. 11, p. 349.
75. Takizawa A., Kinoshita T., Sasaki M. and Tsujita Y.: 1980, "Solubility and diffusion of binary water-methylalcohol vapor mixtures in cellulose acetate membrane." J. Memb. Sci. 6, p, 265.
76. Stannet V. and Yasuda H.: 1963, "Liquid versus vapor permeation through polymer films." J. Polym. Sci., Part B, 1, p. 289.
77. Stern S.A. and Saxena V.: 1980, "Concentration-dependent transport of gases and vapors in glassy polymers." J. Memb. Sci. 7, p. 47.
78. Stern S.A. and Frisch H.L.: 1984, "Selective permeation of gases through polymers." Annual Review of Materials Science. In press.
79. Rogers C.E., Stannett V. and Szwarc M.: 1960, J. Polym. Sci. 45, p. 61.
80. Fujita H.: 1968, in "Diffusion in polymers" ed. J. Crank and G.S. Park, Acad. Press, London.
81. Hansen C. and Beerbower A.: 1971, Kirk-Othmer, Ency. of Chem. Tech., Supplement Vol. p. 889.
82. Henis J.M.S.: 1983, "Aspects of organic vapor and gas stream dehydration using membranes." 4th Symposium on Synthetic Membranes in Science and Industry, Tübingen, Germany (September 6-9, 1983).

SYMBOLS

Values are given in SI units unless otherwise indicated: length (m), mass (kg), time (s), temperature (K), energy (J), pressure (Pa).

E	energy consumption ($kJ.kg^{-1}$)
g	empirical constant, equation 8
G	grafting ratio
n	empirical exponent, equation 8
P	permeability coefficient ($m^2.s^{-1}.Pa^{-1}$)
Q'_0, Q'_r	inlet and outlet flow rates per square meter of membrane ($kg.m^{-2}.s^{-1}$ or $kg.m^{-2}.hr^{-1}$)
Ry	recovery yield, equation 18
X	dimensionless parameter, equation 15-1
δ	Hildebrand-Hansen solubility parameters (δ_d, δ_h, δ_p)
θ	cut ratio ($\langle J \rangle / Q'_0$)
κ	purification ratio, equation (14)
λ	preferential solvation coefficient
ϕ	fractional volume of penetrant within the membrane

Superscripts

$'$	upstream (liquid)
$''$	downstream (vapour)

Subscripts

A	more rapidly permeating component
B	less rapidly permeating component
0	feed
r	reject

Other

$\langle\,\rangle$	average quantity

MEMBRANES IN ENERGY CONSERVATION PROCESSES

Richard W. Baker

Membrane Technology and Research, Inc.
Menlo Park, California 94025

The increase in the price of energy following the oil crisis of 1973 has encouraged the use of membranes in a number of energy conversion, recovery, and conservation processes. Membranes are often considered for use in these processes because of their low energy consumption compared with conventional techniques.

Coal gasification and liquifaction and energy production from biomass sources represent significant potential applications of membranes in energy production processes. Energy production from salinity gradients is another application, but the economics of this process do not appear favorable with existing membranes.

Important applications of membranes in energy recovery and conservation processes are bleed stream gas recovery systems, pervaporation, and hydrometallurgical separations.

P. M. Bungay et al. (eds.), Synthetic Membranes; Science, Engineering and Applications, 437–455.

1. INTRODUCTION

Every thermodynamics textbook includes an example of an ideal semi-permeable membrane. In principle, this membrane is able to separate mixtures with the thermodynamic minimum expenditure of energy. Despite this inherent advantage of membrane processes, few membrane separation plants were in operation until the early 1970's. Membranes suffered from three problems that prohibited their widespread use. They were too slow, too expensive, and too unselective. Partial solutions to these problems have now been developed, and as a result there is a surge of interest in membranes as a separation technique. The increase in the price of oil from \$1–2 per barrel in 1970 to \$25–30 per barrel in 1985 has further contributed to this interest, particularly in energy conversion and recovery processes.

In this paper, a number of these membrane applications are surveyed. The applications illustrate the opportunities for membrane processes in the coming decade. Based on this survey, the developments required to make membranes a widely used separation process can be determined.

2. RECENT DEVELOPMENTS IN MEMBRANE TECHNOLOGY

The problem of slow permeation rates through membranes was largely overcome by the development of imperfection-free asymmetric membranes in the late 1960's and early 1970's. These membranes consist of a permselective surface film supported by a much thicker microporous substrate. Because the permselective surface film is very thin, asymmetric membranes exhibit high fluxes. The microporous substrate provides mechanical strength. A scanning electron micrograph of an asymmetric membrane is shown in Figure 1. In practice, the permselective layer is typically less than 0.2 μm in thickness. In this example, the permselective film has deliberately been made thicker than normal to better show the dense surface structure.

The Loeb–Sourirajan method was the first method developed for the production of effectively very thin membranes (1). It was first applied to

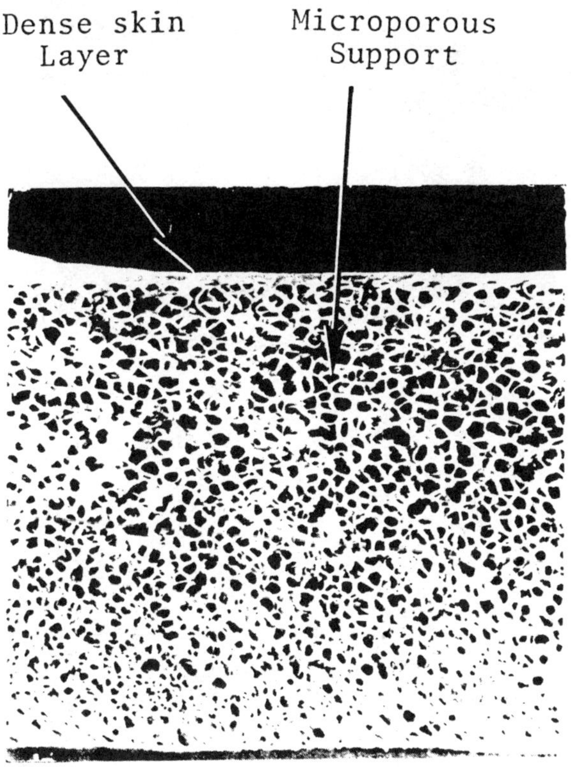

FIGURE 1.—Scanning electron micrograph of an asymmetric membrane.

cellulose acetate membranes, and the development of these high-flux reverse osmosis membranes spawned the reverse osmosis industry in the 1960's. In recent years, new approaches to making asymmetric membranes appear to produce membranes with even thinner permselective layers than those produced by the Loeb–Sourirajan method. Barrier layers only a few hundred Angstrom units in thickness have been claimed for these so-called thin film composite membranes, and it appears that these barrier layers are effectively free of imperfections (2,3).

The problem of producing low-cost membrane modules was also solved during the development of modules for reverse osmosis membranes. There are now two principal module designs in use: the spiral wound module (4) and the hollow fiber module (5). Diagrams of both designs are presented in Figure 2. Both configurations are considerably less expensive than the older plate and frame and tubular systems.

Progess has also been made in making membranes more selective. One approach has been to carefully tailor the chemistry of the polymer films to improve the intrinsic selectivity of the membrane. In this way, for example,

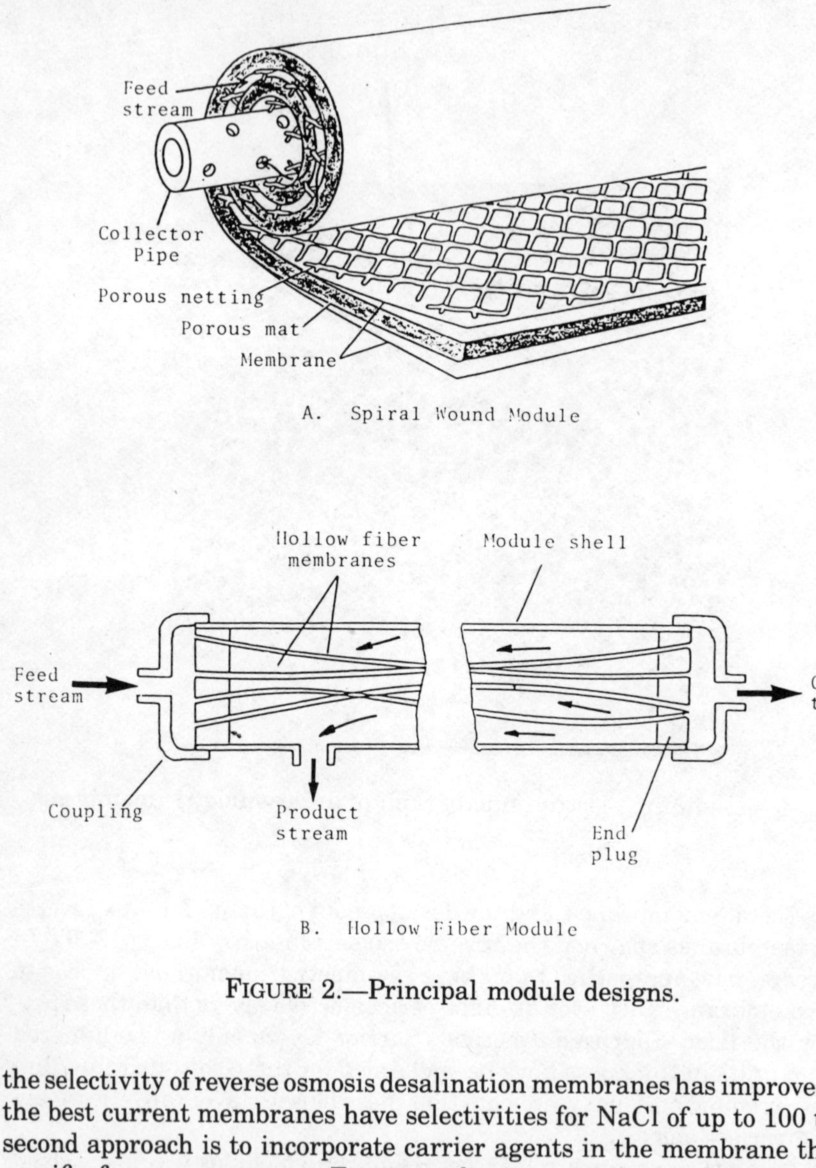

FIGURE 2.—Principal module designs.

the selectivity of reverse osmosis desalination membranes has improved until the best current membranes have selectivities for NaCl of up to 100 to 1. A second approach is to incorporate carrier agents in the membrane that are specific for one permeant. For example, in coupled transport, liquid ion exchange reagents have been used that selectively transport Cu^{++} across the membrane, but not Fe^{++} or other cations (6, 7). Similarly, in the separation of oxygen from nitrogen by facilitated transport, oxygen-specific carriers are used that can lead to membrane selectivities of oxygen over nitrogen of 50 to 1 (8, 9). A final approach to improving the selectivity of membrane systems is the use of multi-stage membrane systems (10) or innovative recycling modules (11). These processes are now practical because of the relatively low cost of modern membrane modules.

3. MEMBRANES IN ENERGY PRODUCTION PROCESSES

Membranes are used in a number of conventional ways in energy production processes, for example, the production of boiler feed water or treatment of evaporator blow-down water by reverse osmosis. Rather than discuss these conventional processes, however, we will concentrate on new applications of membranes, in particular coal gasification and liquification processes and biotechnological processes. These applications could become major applications of membranes in the next decade.

3.1. Coal Gasification and Liquifaction

South Africa is currently the only country to use coal resources on a large scale for hydrocarbon synthesis (12), but several authors believe that coal-derived synthesis gas, a mixture of carbon monoxide and hydrogen, will be widely used to produce methane, liquid fuels, and oxochemicals in the future.

A simplified production scheme for synthesis gas is shown in Figure 3. In coal gasification plants of this type, between 30 and 40% of the capital costs and operating expenses are related to gas separation problems (13, 14). Coal gasification usually requires the production of oxygen from air, and the oxygen plant is approximately 20% of the total cost. Removal of the acid gases CO_2 and H_2S represents another 15 to 20%. Finally, the composition of the mixture of CO and H_2 must be adjusted depending on the final use of the gas. This adjustment is now performed by a catalytic shift converter, which represents another 10% of the plant cost. In principle, however, this adjustment could also be made by a membrane separation unit. We will briefly discuss each of these separations.

O_2/N_2 Separations: The production of O_2-enriched air using simple ultrathin silicone-based membranes is being studied by a number of groups (9, 15). Silicone polymers are used because of their intrinsically high oxygen permeability, but the selectivity of the membrane for oxygen over nitrogen is only 2. Thus, the concentration of oxygen in the product gas after a single pass through the membrane is approximately 30–35%. More selective polymeric membranes are available, for example, cellulose triacetate, with a selectivity of 6 (16), and an oxygen content of almost 60% can be achieved with these membranes in a single pass. However, the permeability of cellulose triacetate membranes is very low compared with silicone polymer membranes, and the economics of the process do not seem viable. Thus, although oxygen-enriched air from a single silicone membrane separation unit may be used in medical applications or in processes such as combustion, it is unlikely to be used in a coal gasification plant.

An alternative method capable of producing high purity oxygen is to use an oxygen carrier complex in a facilitated transport process. The concept is illustrated in Figure 4. This concept was first demonstrated by Scholander (17, 18) using hemoglobin, and later by Basset and Schultz (8) using cobaltodihistidine. These early carriers both suffered from relatively rapid

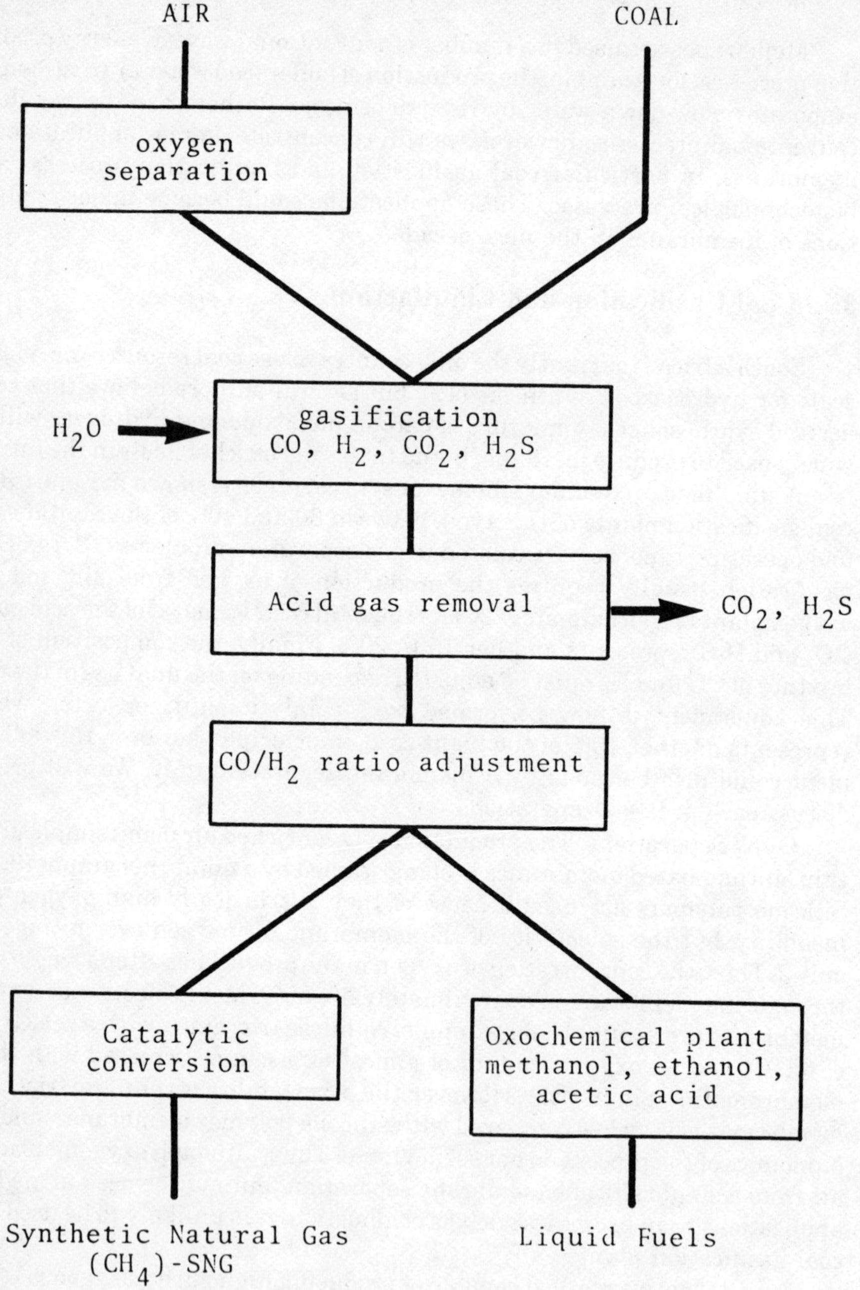

FIGURE 3.—A simplified flow scheme for one proposed synthesis route to SNG or liquid fuels.

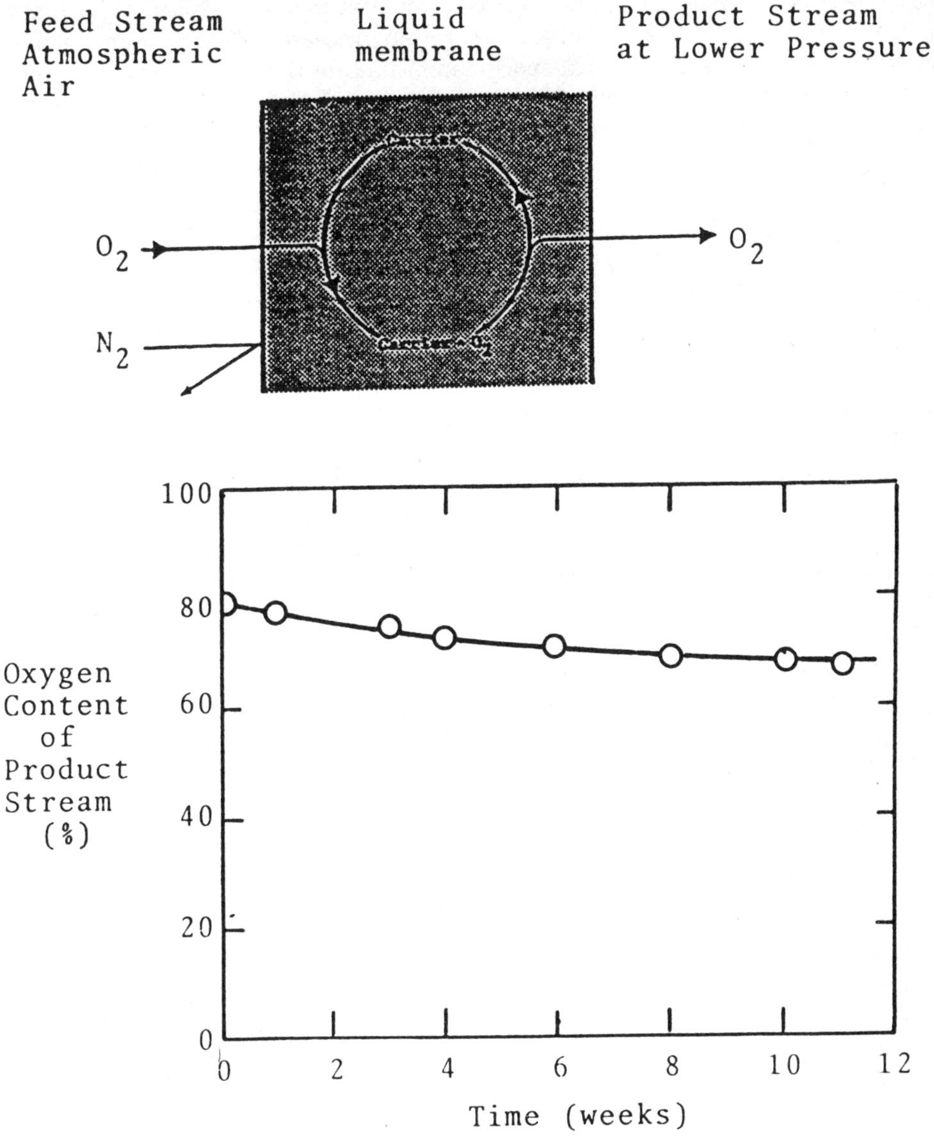

FIGURE 4.—Production of oxygen-enriched air via carrier-facilitated transport.

degradation of the carrier and rather slow carrier reaction kinetics. However, a large number of oxygen carriers are now known (19, 20). Carriers have been reported with stabilities in excess of a month, and their rapid oxygen kinetics result in oxygen to nitrogen selectivities of nearly 30 (9). With these membranes, 88% product oxygen can be produced, although, because the membranes used are relatively thick, oxygen fluxes are low.

Acid/Gas Separations: Facilitated transport membranes are also an attractive system for use in acid gas separation processes. Ward produced some of the very first facilitated transport membranes for the separation of H_2S and CO_2 with carriers such as $NaHCO_3$ (21, 22). Separation of CO_2/H_2S from CO/H_2 mixtures with conventional polymeric membranes is more difficult, principally because of the high permeability of H_2, which tends to permeate the membrane together with the acid gases.

CO/H_2 Separations: Adjustment of the CO/H_2 ratio in synthesis gas is required for production of oxochemicals. Because several membranes are substantially more permeable to H_2 than CO, this separation is relatively easily achieved. The Prism (Monsanto Corp., St. Louis, MO) hollow fiber gas separation system has been suggested for this application (23).

3.2. Energy From Biomass Sources

Biomass feed stocks are being studied for use as a source of liquid and gaseous fuels and industrial chemicals. Membranes could find a number of applications in these processes. An interesting comparison between a conventional fermentation process and an idealized membrane-moderated plant is shown in Figure 5 (24).

As Figure 5 shows, the recovery of a product from a fermentation reaction represents a significant portion of the cost of the process. Membranes are currently used in a number of these processes to recover the product. Because the fermentation broths are relatively dilute aqueous solutions (usually 1 to 5%), ultrafiltration (UF), reverse osmosis (RO), and electrodialysis (ED) can all be easily applied. In some cases, the products of the fermentation are gases, for example, CO_2 and CH_4 produced by aerobic digestion of biomass sources. Typically these gases contain 40 to 60% CH_4, which it is desirable to recover. Although conventional absorption techniques could be used, the economics are not attractive. Both conventional polymer membranes and facilitated transport membranes could be used in these membrane processes. The group at General Electric (10, 22) has worked with both systems, and several small pilot plants based on ultrathin silicone carbonate membranes have been installed. It appears that this type of membrane process will be used on a fairly large scale in a few years.

3.3. Salinity Gradient Energy Conversion

Production of energy from saline sources is a novel application of membranes that experienced a brief vogue from 1977 to 1981. A vast amount of energy is irreversibly dissipated when fresh river waters mix with sea waters. In principle, this energy could be recovered by using a suitable osmotic membrane system to convert the osmotic pressure of sea water into hydrostatic pressure, which could in turn be used to drive a turbine and generate electricity. Such a scheme was proposed by Normann (25), and independently

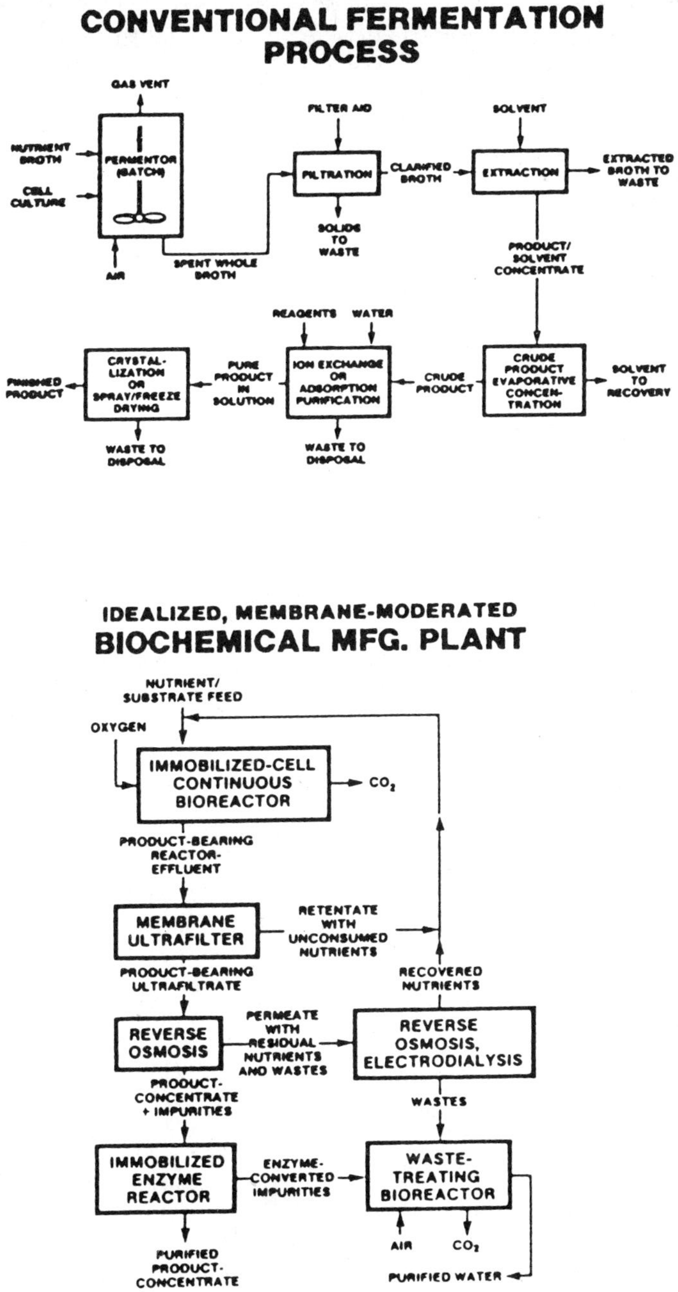

FIGURE 5.—Comparison between a conventional fermentation process and an idealized membrane-moderated biochemical manufacturing plant.

by Jellinek (26) and Loeb (27). The process is illustrated in Figure 6. Loeb has coined the term "pressure-retarded osmosis" (PRO) for this process. A similar process, called reverse electrodialysis (RED), is shown in Figure 7 (28). In

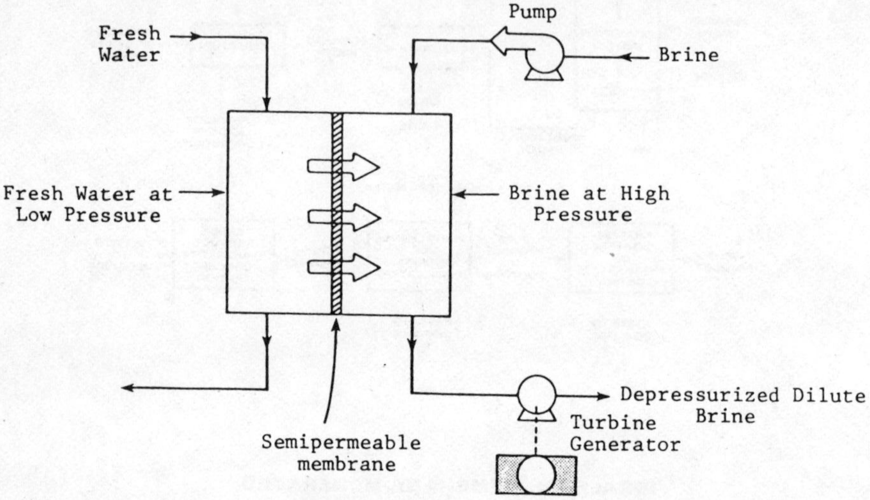

FIGURE 6.—Conceptual representation of an energy production scheme based on pressure-retarded osmosis (PRO).

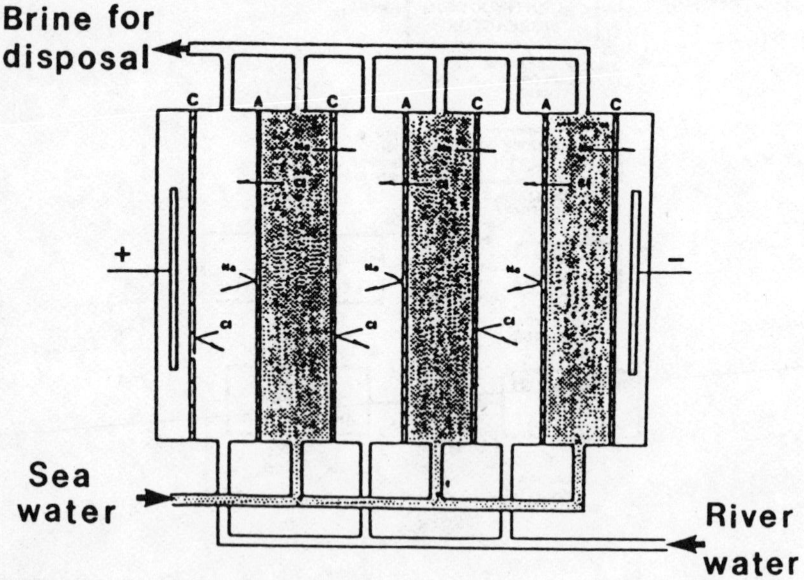

FIGURE 7.—Reverse-electrodialysis stack. Only a few cells are shown here. (A and C refer to anionic and cationic membranes.)

RED, the diffusion of ions from sea water to fresh water across the membranes in a normal electrodialysis stack is used to develop an electrical potential from which energy can be recovered.

Both PRO and, to a lesser extent, RED have been evaluated in the laboratory (29, 30). Both processes appear to require order-of-magnitude improvements in membrane system costs and performance before the processes will be economically viable methods of producing energy.

Concentration polarization is the major problem of PRO. The phenomenon is illustrated in Figure 8. External concentration polarization in the liquid boundary layers on either side of the membrane could in principle be controlled in properly designed modules. However, internal concentration polarization occurs in the porous substrate underneath the permselective skin of the asymmetric membrane. Internal polarization cannot be controlled by circulation of the fluids on either side of the membrane. As a result, internal concentration polarization sharply reduces the fluxes obtained under PRO conditions, compared with the fluxes predicted from normal RO experiments.

Based on a simple economic analysis (29), it appears that when sea water is used as the salt solution and river water as the fresh water source, a water

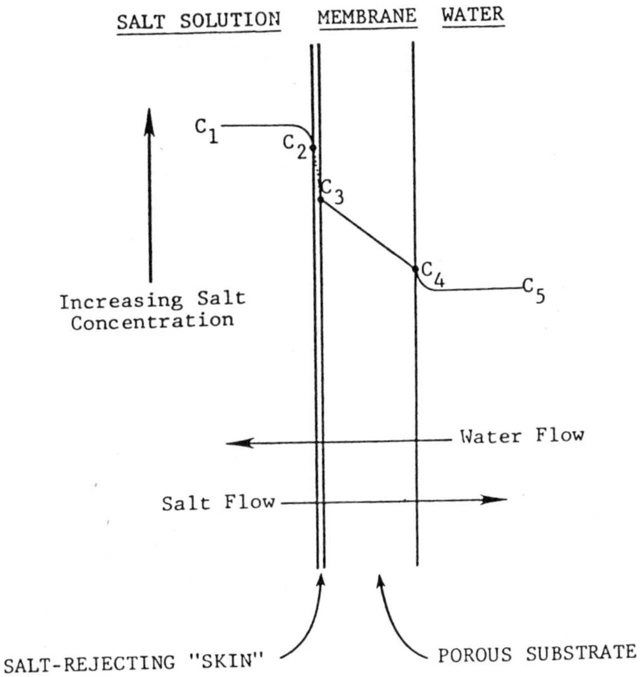

FIGURE 8.—A schematic representation of concentration polarization across a skinned membrane in PRO.

flux of 200 gal/ft^2·day is required to make the process economically viable. PRO fluxes under these conditions are on the order of 1–2 gal/ft^2·day, and thus a substantial improvement in membrane performance is required. Economics of PRO systems using brines and fresh water are more favorable. However, brines normally exist in deserts where there is limited fresh water. Brines from salt domes have been suggested as a resource, but pose difficult effluent disposal problems (31).

4. MEMBRANES IN ENERGY RECOVERY AND CONSERVATION PROCESSES

As well as finding an application in new energy production processes, membranes are being employed in energy recovery and conservation applications. An example of an energy recovery application is separation of H_2, CH_4, and CO from industrial bleed streams. By recycling these recovered gases to the process, the overall energy consumption and the cost of the processes can be reduced. An example of an energy conservation process is the use of membranes in reverse osmosis or pervaporation as a low-energy alternative to simple distillation.

4.1. Bleed Stream Gas Recovery Systems

The bleed gas stream from a typical industrial process contains the unreacted feed gas together with impurities and by-products that would otherwise accumulate in the process reactor. In the past, when energy and feed stock costs were low, bleed streams were usually flared. Higher energy costs now make it economically attractive to use a membrane separation process, illustrated in Figure 9, to recover the feed stock, which can be recycled to the reactor.

Hydrogen-containing bleed streams are a common example of this type of process. These gas streams are a particularly attractive candidate for recovery because of the high permeability of most membranes to hydrogen relative to other gases. Typical hydrogen-containing streams are H_2/CH_4 (32) and $H_2/CH_4/N_2$ streams produced by refining hydrodesulfurizers and hydrocrackers, and H_2/Ar, N_2 (33) bleed streams produced in Harber process ammonium synthesis plants. The economics of H_2 recovery from all of these streams is favorable, particularly because the streams are usually produced at very high pressures which largely eliminate the cost of compressing the gas.

Other similar applications include CO recovery from reducing furnace waste gases. Methane from CO_2-contaminated natural gas wells is another valuable product that can be recovered from a low grade resource. Similarly, more than 200 million cubic feet per day of CH_4 are lost in exhausts from bituminous coal miners (34,35). Almost all of this gas could be collected for use instead of being emitted into the mine atmosphere.

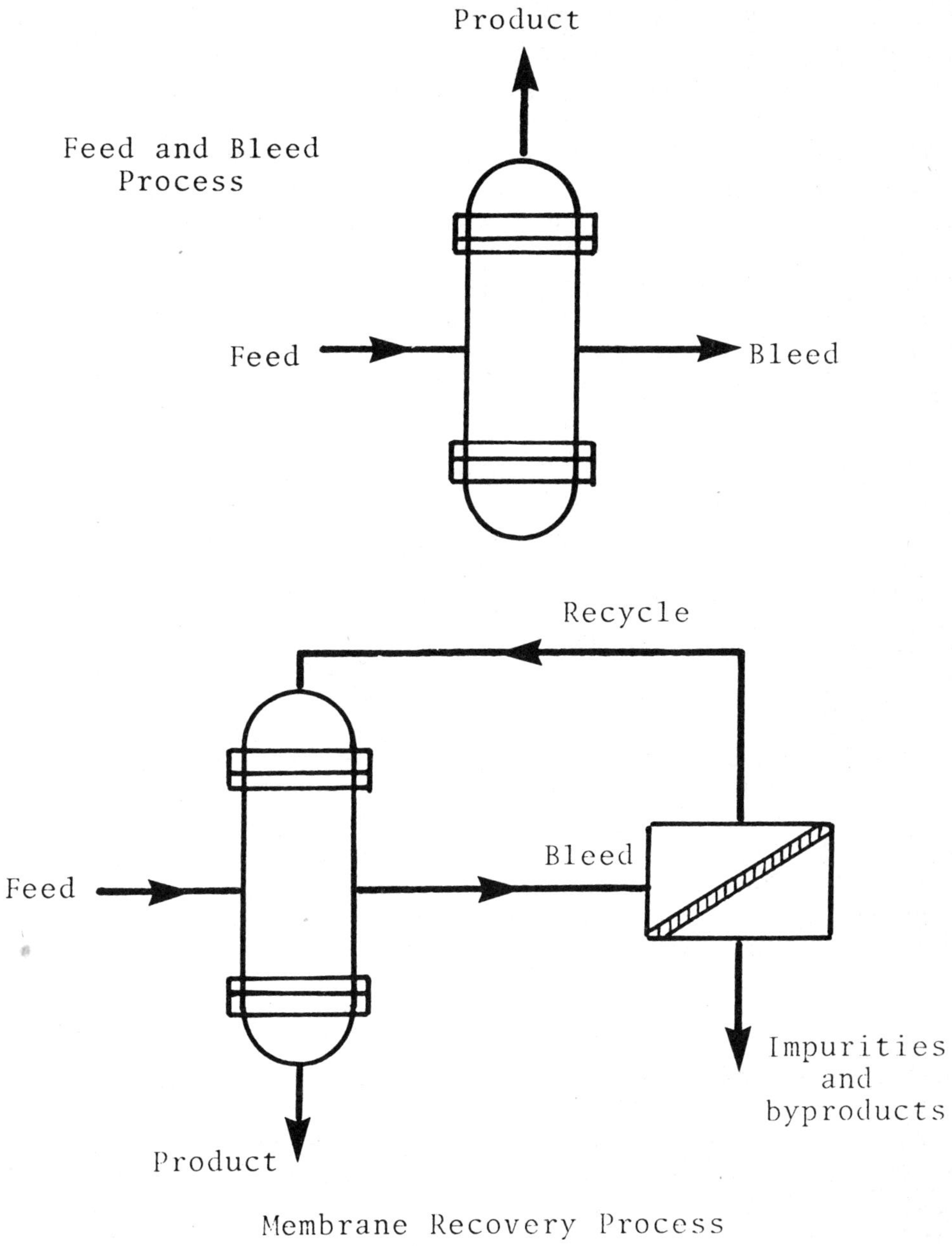

FIGURE 9.—Schematic of feed and bleed recovery process.

4.2. Alternatives to Distillation

Reverse osmosis Between 1 and 2% of the total energy consumed in the United States is used in various distillation processes (36). Because of their isothermal nature, membrane processes such as reverse osmosis (RO) or electrodialysis (ED) are often touted as low energy alternatives to distillation, particularly in the production of potable water. These claims are valid for small water desalination plants, and RO and ED are now being widely used in plants with capacities up to 5 million gallons/day. However, distillation by multi-stage flash (MSF) remains the preferred process for large plants. Many of these plants are dual-purpose systems designed to produce electricity and water. In these plants, use of MSF avoids the thermodynamic inefficiencies of converting heat to electricity, and some of the low grade waste heat produced is utilized by the power plant. As a result, dual-purpose MSF plants use almost 50% less energy than membrane plants (37).

An example of an application in which RO is more competitive with distillation is the concentration of fruit juices and sugar solutions. The nature of these solutions requires that energy-inefficient distillation systems be used. Degradation of fruit juice flavor elements is also a problem. In the past, RO could not be applied to these solutions because the available cellulose acetate membranes were insufficiently selective and allowed alcohol aldehyde and ester flavor elements to permeate the membrane. Also, some of these solutions are viscous, and must be processed at temperatures of 80 to 100°C. This was not possible with cellulose acetate membranes. However, since the development of high-flux, high-rejection composite membranes, a number of small plants are being installed (38).

High osmotic pressure is a problem with these solutions. Because RO systems are limited to operating pressures of 1000 to 1500 psi, concentrations achievable with a single pass membrane system are limited. A number of alternative concentration modes to RO could be employed, as shown in Figure 10. For example, a multi-pass RO system employing partial recycling of the product is possible. This system would have a relatively high membrane cost, but might be used for products with a high value. The use of an osmotic sink to dewater the solution by pressure-assisted dialysis is also possible (39). However, this system would have the same problems of concentration polarization found in PRO. The diluted osmotic sink solution would have to be evaporated for recycling, and thus the energy savings would be negligible. Despite these drawbacks, the process might find an application with solutions such as some fruit juices that lose flavor elements during normal evaporative concentration.

Pervaporation Pervaporation is another membrane process that appears to have promise as an alternative to distillation. The process is shown in Figure 11. As in simple evaporation, material moves from a liquid to a vapor phase.

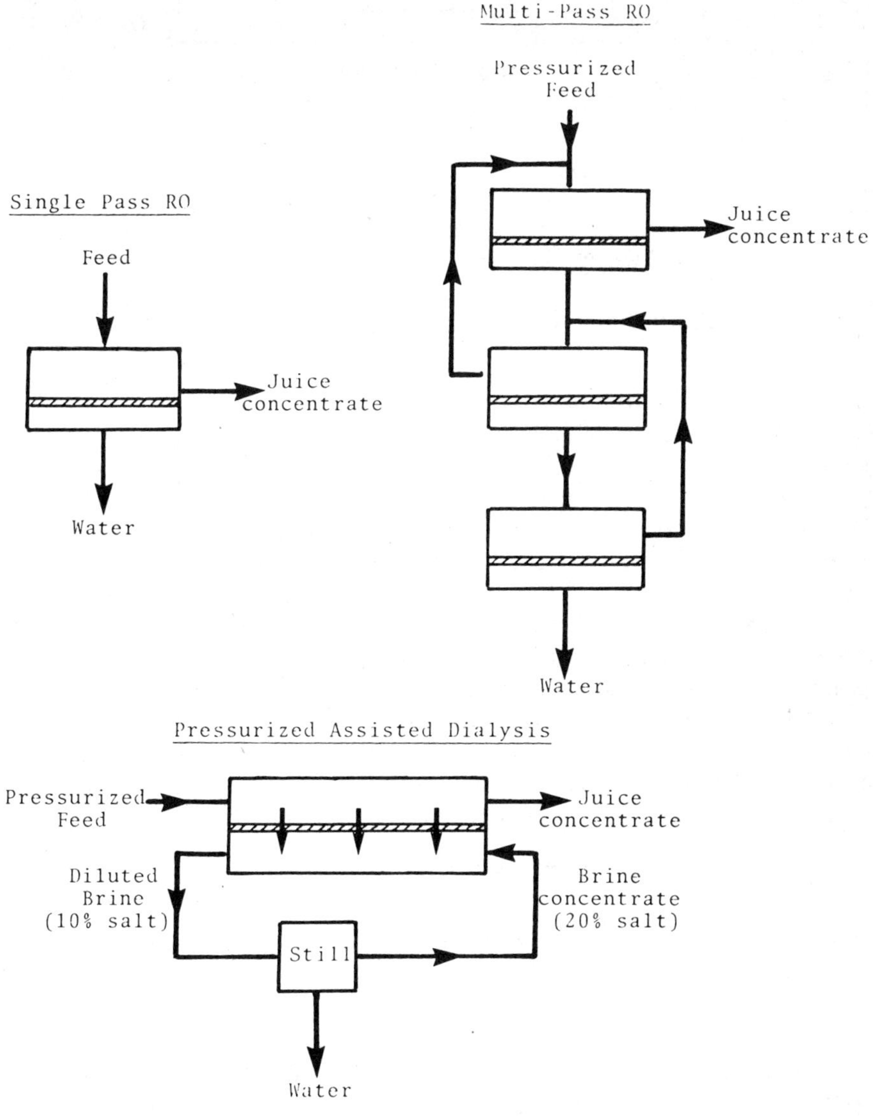

FIGURE 10.—Schematics of alternative concentration processes.

In contrast to evaporation, where the concentration of the components in the vapor phase is determined by the concentration and volatility of the components in the liquid feed, in pervaporation a membrane is placed between the two phases. The selectivity of the membrane, therefore, largely determines the vapor phase composition.

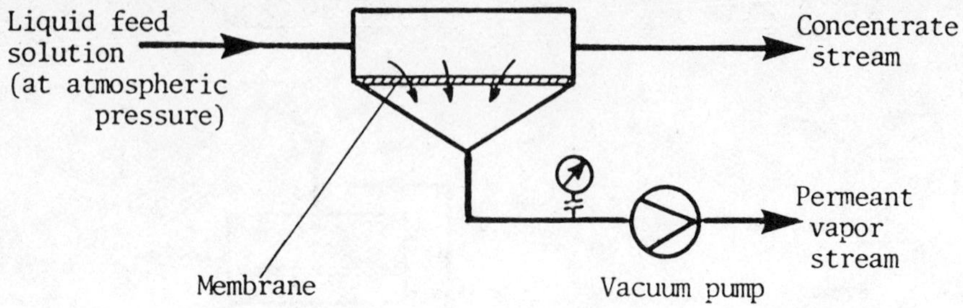

FIGURE 11.—Schematic of pervaporation process.

The low selectivity of pervaporation membranes appears to be the principal problem to be solved before pervaporation can be economically used to separate organic mixtures. As a rule of thumb, a separation factor of at least 20 is required to develop a commercial process. A separation factor of 20 is acceptable even though a pure product is required because of the high value of organic liquids. This allows the liquids to be economically treated in a two-stage process.

The first report of a pervaporation process appears to have been a paper and a series of patents by Binnings and Lee, 1959–1962 (40). However, until recently, interest in the process was low. Important early research groups were Aptel and Néel in France (41, 42) and Cabasso (43) in the United States. A promising application of pervaporation is the breaking of azeotropes such as alcohol–water. The first pilot plant, designed to remove water from an 80% alcohol/20% water solution, is now being installed in Brazil.

5. FUTURE DEVELOPMENTS

It is clear from the summary of applications described in the preceding sections that membrane processes are beginning to establish themselves in a number of energy conversion and recovery processes. Gas separation with membranes is a particularly attractive new area of application and RO and ED are likely to gradually expand their market during the coming decade. Commercial applications of coupled transport, facilitated transport, pervaporation, and liquid bubble membranes are still speculative.

It appears that despite the advances made since Loeb's and Sourirajan's discovery that began the modern membrane era, new breakthroughs in membrane systems are still required before these processes will be economically justifiable.

5.1. New Module Designs

Better module designs are required. In many cases the costs of currently available membrane modules are too high to allow membranes to compete with other separation techniques. Currently, spiral wound modules can be produced for \$2–3/ft^2 of membrane. Significant reductions in the cost of either spiral wound or hollow fiber module designs are unlikely. Although hollow fiber membranes are significantly less expensive than spiral wound modules, this cost advantage is offset by the generally lower fluxes of hollow fiber membranes and their susceptibility to fouling, which requires expensive feed pretreatment systems. In addition, the most widely used module systems were developed for reverse osmosis and ultrafiltration, which only require control of concentration polarization on the feed side of the membrane. In some of the newer membrane processes, such as gas separation, coupled transport, and PRO, control of concentration polarization on both sides of the membrane is required. Current modules can be modified to allow concentration polarization control on both sides of the membrane, but these are only marginally acceptable.

5.2. Better Membranes

Improvements in the techniques used to prepare ultrathin membranes are still possible. Membranes have been made with claimed thicknesses as low as 500 Å, but commercial membranes generally have thicknesses on the order of 2000 to 5000 Å. Development of new composite membrane techniques would appear to be a profitable research area. Also, the supported liquid membranes used in facilitated and coupled transport have very high permeabilities, but they are very thick, 50 to 100 μm. Techniques for making thinner and more stable liquid membranes are required.

5.3. Driving Forces

Pressure and concentration gradients are the most commonly used driving forces in commercial membrane systems. Processes driven by electrical potential driving forces appear to be a particularly attractive area of research. Electrodialysis is a very old technique, but compared with other membrane processes, it has been largely ignored in recent years. Improvements in electrodialysis membranes and membrane modules are possible.

REFERENCES

1. Loeb, S. and Sourirajan, S.: 1963, "Sea water demineralization by means of an osmotic membrane," in Saline Water Conversion II, Advances in Chemistry Series, No. 38, American Chemical Society, Washington, D.C., pp. 117–132.
2. Kremen, S.: 1977, "Technology and engineering of ROGA spiral-wound reverse osmosis membrane modules," in Reverse Osmosis and Synthetic Membranes, Sourirajan, S. (Ed.), National Research Council Canada, Ottawa, pp. 371–385.
3. Riley, R.L., Lonsdale, H.K., Lyons, D.R., and Merten, U.R.: 1967, "Preparation of ultrathin reverse osmosis membranes and the attainment of theoretical salt rejection." J. Appl. Poly. Sci. 11, p. 2143.

4. Rozelle, L.T., Cadotte, J.E., Cobian, K.E., and Copp, Jr., C.V.: 1977, "Nonpolysaccharide membranes for reverse osmosis: NS-100 membranes," in Reverse Osmosis and Synthetic Membranes, Sourirajan, S. (Ed.), National Research Council Canada, Ottawa, pp. 249–261.
5. Baum, B., Holley, Jr., W., and White, R.A.: 1976, "Hollow fibres in reverse osmosis, dialysis, and ultrafiltration," in Membrane Separation Processes, Meares, P. (Ed.), Elsevier Scientific Pub. Co., Amsterdam, pp. 187–228.
6. Baker, R.W., Tuttle, M.E., Kelly, D.J., and Lonsdale, H.K.: 1977, "Coupled transport membranes I. Copper separations." J. Membrane Sci. 2, p. 213..
7. Largman, T. and Sifniades, S.: 1978, "Recovery of copper (II) from aqueous solutions by means of supported liquid membranes." Hydrometallurgy 3, p. 153.
8. Bassett, R.J. and Schultz, J.S.: 1970, "Nonequilibrium facilitated diffusion through membranes of aqueous cobaltodihistidine." Biochem. Biophys. Acta. 211, p. 194.
9. Baker, R.W., Roman, I.C., Smith, K.L., and Lonsdale, H.K.: 1982, "Liquid membranes for the production of oxygen-enriched air," in Proceedings of the Ninth Energy Technology Conference, Hill, R.F. (Ed.), Government Institutes, Inc. (Reprinted in Industrial Heating, pp. 16–18, July 1982.)
10. Kimura, S.G. and Walmet, G.E.: 1980, "Fuel gas purification with permselective membranes." Sep. Sci. & Tech. 15, p. 1115.
11. Hwang, S.T. and Yuen, K.H.: 1980, "Gas separation by a continuous membrane column." Sep. Sci. & Tech. 15, p. 1069.
12. Shires, M.J. and Gellender, M.: 1980, "Gasified coal—Starting point for chemical manufacture." Chem. Int'l., No. 6, p. 5.
13. Fossil Energy Program Summary Document. 1980, U.S. Department of Energy, DOE/FE-0006, May.
14. National Academy of Engineering–National Research Council–Division of Engineering. Committee on Air Quality Management–Committee on Pollution Abatement and Control. Ad Hoc Panel on Evaluation of Coal–Gasification Technology: 1973, Evaluation of Coal–Gasification Technology: Part II—Low- and Intermediate-BTU Fuel Gases. R & D Report No. 74, Interim Report No. 2, U.S. Government Printing Office, Washington, D.C.
15. Ward III, W.J., Browall, W.R., and Salemme, R.M.: 1976, "Ultrathin silicone/polycarbonate membranes for gas separation processes." J. Membrane Sci. 1, p. 99.
16. Yasuda, H. and Stannett, V.: 1975, "Permeability coefficients," in Polymer Handbook, Brandup, J., Immergut, E.H., and McDowell, W. (Eds.), John Wiley & Sons, New York, p. 249.
17. Scholander, P.F.: 1960, "Oxygen transport through hemoglobin solutions." Science 131, p. 585.
18. Kreuzer, F.: 1970, "Facilitated diffusion of oxygen and its possible significance; a review." Respir. Physiol. 9, p. 1.
19. Basolo, F., Hoffman, B.M., and Ibers, J.A.: 1975, "Synthetic oxygen carriers of biological interest." Accounts of Chemical Research 8(11), p. 384.
20. Martell, A.E. and Calvin, M.: 1952, "Catalytic effects of chelate compounds," in Chemistry of the Metal Chelate Compounds, Martell, A.E. and Calvin, M. (Eds.), Prentice-Hall, Englewood Cliffs, N.J., p. 336.
21. Ward III, W.J. and Robb, W.L.: 1967, "Carbon dioxide–oxygen separation: Facilitated transport of carbon dioxide across a liquid film." Science 156, p. 1481.
22. Matson, S.L., Herrick, C.S., and Ward III, W.J.: 1977, "Progress on the selective removal of H_2S from gasified coal using an immobilized liquid membrane." Ind. Eng. Chem., Process Des. Dev. 16(3), p. 370.
23. Henis, J.M.S. and Tripodi, M.K.: 1983, "The developing techniques of gas separating membranes." Science 220, p. 4592.
24. Michaels, A.S.: 1980, "Membrane technology and biotechnology." Desalination 35, pp. 329–351.
25. Norman, R.S.: 1974, "Water salination: A source of energy." Science 186, p. 350.
26. Jellinek, H.H.: 1975, "Osmotic work I. Energy production from osmosis on fresh water systems." Kagaku Kojo 19, p. 87.

27. Loeb, S.: 1976, "Production of energy from concentrated brines by pressure-retarded osmosis I. Preliminary technical and economic correlations." J. Membrane Sci. 1, p. 49.
28. Weinstein, J.N. and Leitz, F.B.: 1976, "Electric power from differences in salinity: The dialytic battery." Science 191, p. 557.
29. Lee, K.L., Baker, R.W., and Lonsdale, H.K.: 1981, "Membranes for power generation by pressure-retarded osmosis." J. Membrane Sci. 8, pp. 141–171.
30. Mehta, G.D. and Loeb, S.: 1978, "Internal polarization in the porous substructure of a semipermeable membrane under pressure-retarded osmosis." J. Membrane Sci. 4, p. 261.
31. Wick, G.L. and Issacs, J.D.: 1978, "Salt domes: Is there more energy available from their salt than their oil?" Science 199, p. 7436.
32. Maciula, E.A.: 1980, "High H_2 purity is key in new refining era." Oil & Gas Journal, May 26, pp. 63–68.
33. Gardner, R.J., Crane, R.A., and Hannan, J.F.: 1977, "Hollow fiber permeator for separating gases." Chem. Eng. Progr. 73, pp. 76–78.
34. Duel, M. and Kim, A.: 1978, "Methane drainage—An update." Mining Congress Journal, July, pp. 38–42.
35. Duel, M. and Skow, W.: 1975, "Speeding coal mining operations by recovering and utilizing methane from coal beds." Coal Age, July, p. 104.
36. Rush, Jr., F.E.: 1980, "Energy saving alternatives to distillation." Chem. Eng. Progress, July.
37. Cook, B.: 1980, "Trends in desalting." Newsletter International Desalination and Environmental Association, p. 10.
38. Department of Energy, Office of Industrial Programs: 1981, Agriculture and Food Processes Branch Program Summary Document, June.
39. Loeb, S. and Bloch, M.R.: 1973, "Countercurrent flow osmotic processes for the production of solutions having a high osmotic pressure." Desalination 13, p. 207.
40. Binning, R.C., Lee, R.J., Jennings, J.F., and Martin, E.C.: 1961, "Separation of liquid mixtures by permeation." Ind. & Eng. Chem. 53, p. 45.
41. Aptel, P., Cuny, J., Jozefowiczz, J., Morel, G., and Neel, J.: 1972, "Liquid transport membranes prepared by grafting polar monomers onto poly(tetrafluorethylene) films: I. Some fractions of liquid mixtures by pervaporation." J. Appl. Poly. Sci. 16, p. 1061.
42. Aptel, P., Challard, N., Cuny, J., and Neel, J.: 1976, "Application of the pervaporation processes to separate azeotropic mixtures." J. Membrane Sci. 1, p. 271.
43. Cabasso, I.: 1983, "Organic liquid mixtures separation by permselective polymer membranes. 1. Selection and characteristics of dense isotropic membranes employed in the pervaporation process." Ind. & Eng. Chem., Prod. Res. & Dev. 22(2), p. 313.

PROCESS DESIGN AND OPTIMIZATION

Robert Rautenbach

Institut für Verfahrenstechnik
Rheinisch-Westfälische Technische Hochschule Aachen
D-5100 Aachen, West Germany

The potential of membrane processes will be by far overestimated if only the transport properties of the membrane are taken into consideration. Concentration polarization and pressure losses in the module, i.e., the membrane arrangement, will result in sometimes much lower figures for capacity and product quality. Therefore, the first part of this chapter will discuss the major aspects which have to be considered in module design—based on the equations governing the local mass transport at and in the membrane. The next step will be a discussion of how modules have to be arranged in a process in order to achieve the desired result. The last part of this chapter will deal with the application of membrane processes, the economics of membrane processes compared with conventional separation techniques and with hybrid processes, i.e., the combination of a membrane process and a conventional separation process such as distillation.

P. M. Bungay et al. (eds.), Synthetic Membranes: Science, Engineering and Applications, 457–521.

1. INTRODUCTION: MASS TRANSPORT AT THE MEMBRANE SURFACE

1.1. Local Mass Transport

Consideration of only the transport mechanisms in membranes will, in general, lead to an overestimation of the specific permeation rates in membrane processes. Formation of a concentration boundary layer in front of the membrane surface or within the porous support structure reduces the permeation rate and, in most cases, the product quality as well. For reverse osmosis, Figure 1 shows how a concentration boundary layer forms as a result of membrane selectivity. At steady state conditions, the retained components must be transported back into the bulk of the liquid. As laminar flow is present near the membrane surface, this backflow is diffusive in nature, i.e., is based on a concentration gradient. At steady state conditions, the concentration profile is calculated from a mass balance as

$$\frac{c_2 - c_3}{c_1 - c_3} = e^{v_m/k}, \tag{1}$$

where k is the mass transfer coefficient, which can be assumed, at least with a good approximation, to be independent of the permeate flux v_m. For this reason, the analogy between heat and mass transfer is valid and k can be calculated using the well-known heat transfer equations. This has been tested experimentally for a number of cases (1).

In Tables 1 and 2, mass transfer equations for important cases of forced and free convection are listed.

In the case of reverse osmosis, for example, the concentration boundary layer (concentration polarization) reduces the product quality on two accounts: 1. the driving force for the solvent ($\Delta p - \Delta\pi$) is reduced because of increased osmotic pressure difference $\Delta\pi$; 2. the driving force for the dis-

TABLE 1. Forced convection: mass transfer equations

	Forced convection Mass transfer equations	$\mathrm{Re} = \frac{d_h \bar{v}_x}{\nu}$ $\mathrm{Sc} = \frac{\nu}{D}$
Tube diameter d $d = d_h$	$\overline{\mathrm{Sh}} = \frac{\bar{k}\, d}{D} = 1.62 \left(\mathrm{Re} \cdot \mathrm{Sh} \frac{d}{L}\right)^{1/3}$	Laminar flow
	$\mathrm{Sh} = \frac{k\, d}{D} = 0.0023\, \mathrm{Re}^{7/8}\, \mathrm{Sc}^{1/4}$ or $\mathrm{Sh} = 0.04\, \mathrm{Re}^{3/4}\, \mathrm{Sc}^{1/3}$	Turbulent flow
Channel height 2b $d_h = 4b$	$\overline{\mathrm{Sh}} = 1.85 \left(\mathrm{Re}\, \mathrm{Sc} \frac{d_h}{L}\right)^{1/3}$	Laminar flow $\left(\mathrm{Re}\, \mathrm{Sc} \frac{d_h}{L}\right) > 10^2$
	$\mathrm{Sh} = \mathrm{Sh}_{tube}$	Turbulent flow

TABLE 2. Free convection: mass transfer equations

	Free convection Mass transfer equations	$\mathrm{Ra} = \mathrm{Gr} \cdot \mathrm{Sc}$ $\mathrm{Gr} = \frac{\rho_2 - \rho_1}{\rho_1\, \nu^2}\, g L^3$ $\mathrm{Sc} = \nu / D$ $\mathrm{Sh} = \frac{k \cdot L}{D}$
Vertical wall height L	$\overline{\mathrm{Sh}} = C(\mathrm{Sc}) \cdot \mathrm{Ra}^{1/4}$ $C(\mathrm{Sc} = 1000) = 0.663$	Range of laminar flow $\mathrm{Ra} < 10^9$
	$\mathrm{Sh} = 1.08 + 0.41\, \mathrm{Ra}^{1/4} + 0.04\, \mathrm{Ra}^{1/3}$	Transition stage $10^2 \leq \mathrm{Ra} \leq 10^{12}$
	$\mathrm{Sh} = 0.10 \cdot \mathrm{Ra}^{1/3}$ or $\mathrm{Sh} = 0.025 \cdot \mathrm{Gr}^{2/5} \cdot \mathrm{Sc}^{7/15} (1 + 0.5\, \mathrm{Sc}^{2/3})^{2/5}$	Range of turbulent flow $\mathrm{Ra} > 2 \cdot 10^9$
Horizontal tube d	The equations of the vertical wall remain valid with $L = 2.76 \cdot d$	

solved component Δc increases. Both effects increase the concentration of the undesirable component in the product:

$$c_3 = \frac{\dot{m}_i\, \rho_3}{\dot{m}_i + \dot{m}_w} \approx \rho_w \frac{\dot{m}_i}{\dot{m}_w} \approx \frac{J_i}{J_V} \tag{2}$$

Figures 2 and 3 indicate the order of magnitude of concentration polarization for laminar and turbulent flows through tubular membranes. The diagrams illustrate the dependence of the concentration boundary layer on flow conditions along the membrane (*Re*) and on the permeation flux (Pe_m). According to Figures 2 and 3, turbulent flow is more advantageous.

Furthermore, these diagrams explain why, in the case of gas permeation, in contrast to separation processes in the liquid state, the resistance caused by a boundary layer can be neglected: the diffusion coefficients of gases, for example O_2/N_2, are 10^4 times larger than those of dissolved components in liquids such as $NaCl/H_2O$ which, in combination with the permeation rates reported in the literature, leads to *Pe* numbers of less than unity.

As demonstrated, the concentration profile depends not only on hydrodynamics and material properties but on flux itself. For this reason the (local) flux of a membrane unit can be calculated in general only by iteration, i.e., numerically, even in simple cases without additional mass transfer resistances like a gel-layer or, eventually, within the support layer of asymmetric membranes.

In the case of reverse osmosis, for example, the following set of equations has to be solved with respect to permeate flux v_m and product quality c_3:

$$\frac{c_2 - c_3}{c_1 - c_3} = e^{v_m/k} \tag{3}$$

$$v_m = A_V [\Delta p - b_\pi (c_2 - c_3)] \tag{4}$$

$$c_3 = \frac{B (c_2 - c_3)}{v_m} . \tag{5}$$

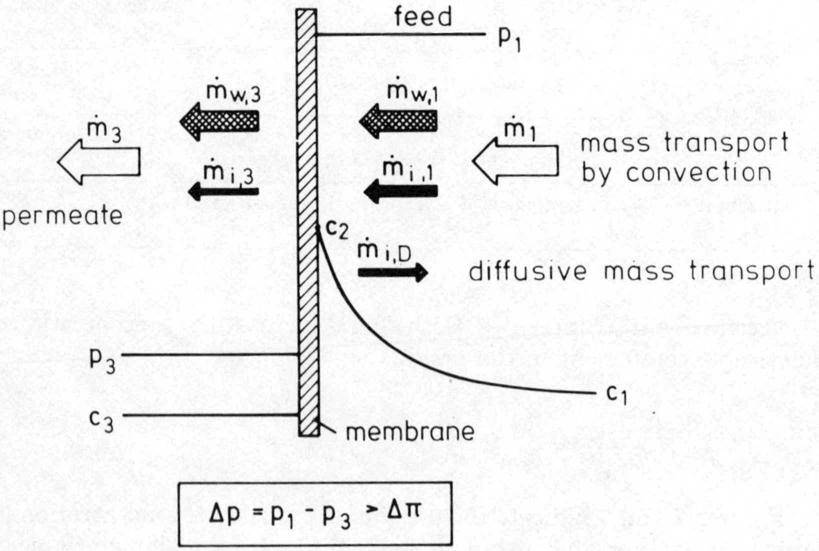

FIGURE 1.—Concentration polarization.

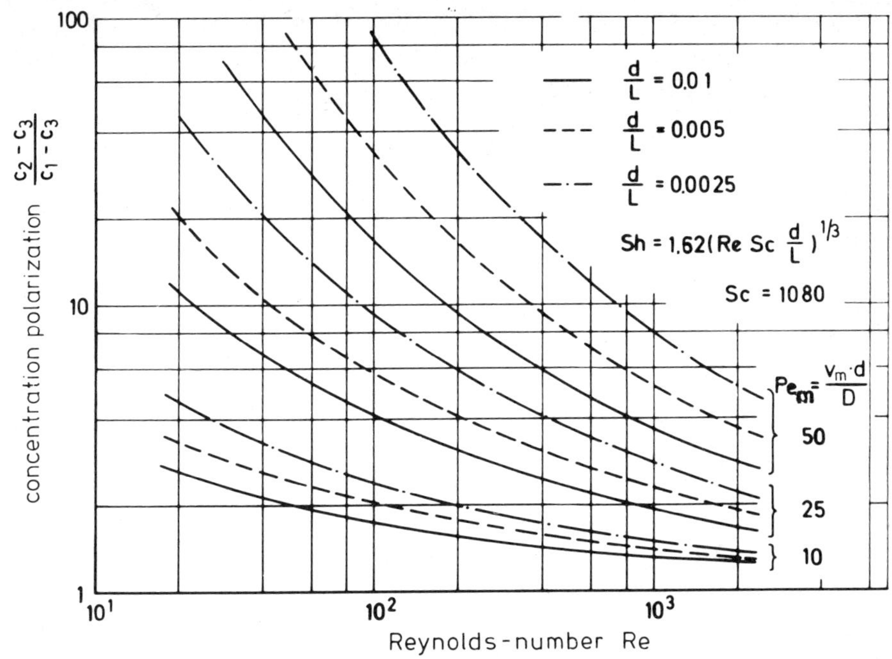

FIGURE 2.—Concentration polarization in laminar flow.

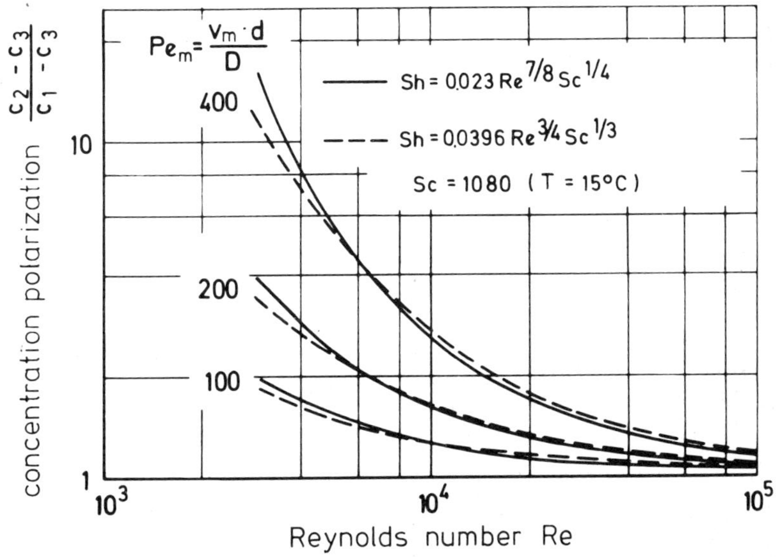

FIGURE 3.—Concentration polarization in turbulent flow.

Using equation 3 to eliminate c_2 from equations 4 and 5 yields

$$v_m = A_V \left[\Delta p - b_\pi \left(c_1 - \frac{B\, c_1 \cdot e^{v_m/k}}{v_m + B \cdot e^{v_m/k}} \right) \cdot e^{v_m/k} \right] \tag{6}$$

$$c_3 = \frac{B\,(c_1 - c_3)\, e^{v_m/k}}{v_m} \tag{7}$$

For some special cases an analytical solution can be found:

1. a) No concentration polarization ($e^{v_m/k} = 1$). In this case, equation 6 is reduced to

$$v_m^2 - v_m \left[A_V (\Delta p - b_\pi c_1) - B\right] = B \cdot A_V \cdot \Delta p \tag{8}$$

and finally

$$v_m = \frac{A_V (\Delta p - b_\pi c_1) - B}{2} + \sqrt{\left(\frac{A_V (\Delta p - b_\pi c_1) - B}{2}\right)^2 + B \cdot A_V \cdot \Delta p} \tag{9}$$

1. b) If, additionally, the membrane rejection is extremely good ($B \to 0$, $c_3 \to 0$), equation 9 is reduced further to

$$v_m = A_V (\Delta p - b_\pi c_1) \tag{10}$$

2. Low concentration polarization ($e^{v_m/k} = 1 + \frac{v_m}{k} + \cdots$) and, at the same time, a high membrane rejection ($B \to 0$, $c_3 \to 0$).

In this case, equation 6 is reduced to

$$v_m = A_V \left[\Delta p - b_\pi c_1 \left(1 + \frac{v_m}{k}\right) \right] \tag{11}$$

and finally

$$v_m = \frac{A_V (\Delta p - b_\pi c_1)}{1 + \frac{A_V b_\pi c_1}{k}} \tag{12}$$

In general, however, equation 6 must be solved graphically or numerically.

1.2. Influence of the Asymmetric Structure of Membranes

Although it is possible to enhance mass transfer in the boundary layer by improving the flow conditions, this is impossible if a concentration profile exists within the porous support structure.

For reverse osmosis, Figure 4 shows the predicted concentration profile in the porous structure of an asymmetric membrane. According to Figure 4, in liquid systems, the concentration c_3, which affects permeate flux and product quality, cannot be influenced significantly by concentration of the liquid on the product side.

Conditions are nearly always favourable for the permeate to flow unhindered. Even in the extreme case of local concentration on the permeate side being equal to that of the liquid on the high pressure side of the membrane ($c_1 = c_5$), the error made in calculating the solvent flux v_m and the salt flux J_i is only 2% or 5%, respectively, if the effect of c_5 is neglected.

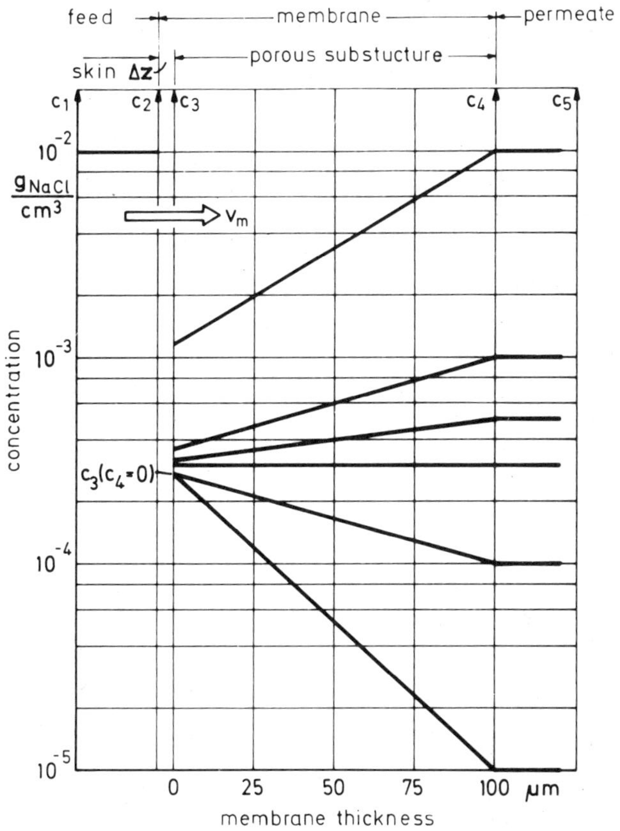

FIGURE 4.—Concentration profile in porous substructure of asymmetric membranes.

The situation is different in the case of gas permeation. Here, the local concentration affects the (local) permeate fluxes of the mixture components and for this reason the type of flow for gas permeation on the permeate side is much more important than for liquid separation processes such as reverse osmosis. Figure 5 shows theoretical results of Blaisdell and Kammermeyer (2). Mole fraction of oxygen is plotted against the yield during fractionation of air for different flow configurations and cases of mixing. The diagram demonstrates that 1) countercurrent flow provides the best results, and 2) marked differences exist between perfect mixing and plug flow.

1.3. Change of Conditions *Along* the Membrane

As a consequence of the permeate flux, the pressure gradient as well as the mean velocity and concentration will vary along the membrane surface—in principle at the feed side as well as at the product side. Depending on module design, this has to be taken into account for one or both sides.

In any case it should be kept in mind that all the equations derived so far are only valid for a membrane element. If, for example, the membrane is of

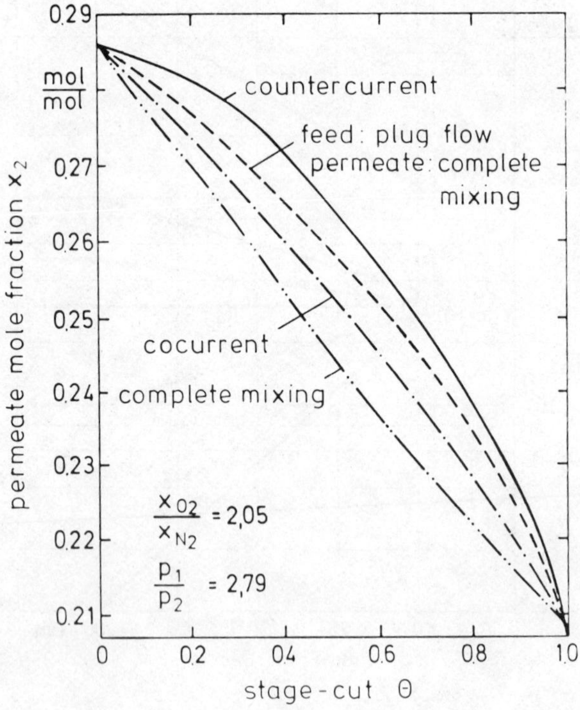

FIGURE 5.—Effect of flow configuration in gas permeation.

tubular design with feed-flow inside, the following should be expected along the tube-axis:

1) Decrease of the transmembrane pressure-difference as a consequence of the drop in pressure caused by friction (→ decrease of permeate flux).
2) Increase of the mean concentration of the feed solution (→ increase of osmotic pressure → decrease of permeate flux and increase of salt flux).
3) Decrease of the mean velocity (→ increase of concentration polarization → decrease of permeate flux and increase of salt flux).

For these reasons, the overall permeation rate and the product quality of a membrane channel have to be calculated by integrating over the channel length, employing, in addition to the equations discussed above,

1. the mass balance
2. the material balance
3. the energy balance.

The equations necessary for the calculation of tubular RO-membrane channels (flow of permeate unhindered, no influence of the porous support-layer) are:

mass transfer at the membrane surface

concentration polarisation $\frac{c_2 - c_3}{c_1 - c_3} = e^{v_m/k}$

equation for mass transfer $Sh = \frac{k\,d}{D} = Sh(Re, Sc)$

mass transfer in the membrane

permeate flux $v_m = A \cdot [\Delta p - b(c_2 - c_3)]$; $\Delta p = p_n - p_{3,n}$

salt flux $J_i = B \cdot (c_2 - c_3)$

salt concentration $c_3 = \frac{J_i}{v_m}$

balance for a length element

mass balance $\bar{v}_{x,n+1} = \bar{v}_{x,n} - \frac{4\Delta x}{d} v_{m,n}$

material balance (salt) $\bar{v}_{x,n+1} \cdot c_{1,n+1} = \bar{v}_{x,n} \cdot c_{1,n} - \frac{4\Delta x}{d} \cdot v_{m,n} \cdot c_{3,n}$

energy balance (ρ = const) $p_{n+1} = \frac{\bar{v}_{x,n}}{\bar{v}_{x,n+1}} [p_n + \frac{1}{2}\rho \bar{v}_{x,n}^2 (1 - \xi \frac{\Delta x}{d})] - \frac{4\Delta x \cdot v_m}{d\,\bar{v}_{x,n+1}} \cdot p_n$

$$- \frac{1}{2}\rho \bar{v}_{x,n+1}^2 - \Delta p_v$$

in which,

$$\xi = \frac{2\,\Delta p \cdot d}{L \cdot \rho\,\bar{v}^2} = 4f\,(f = \text{Fanning friction factor}).$$

This set of equations has to be solved numerically. Figure 6 portrays the results of such a numerical calculation of a long tubular RO-membrane channel for seawater desalination. It clearly demonstrates the decrease of permeate (mass) flux J_w and, at the same time, the increase of local and mean product concentration c_3, $\bar{c}_3$, respectively.

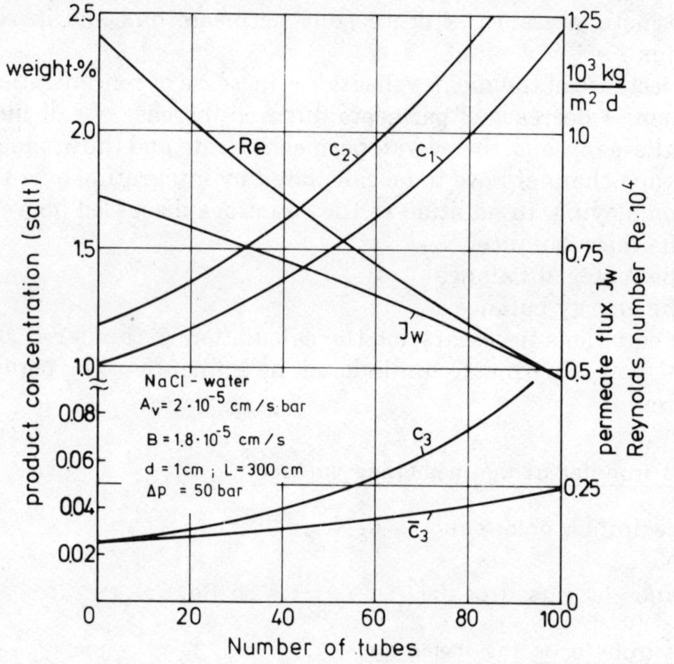

FIGURE 6.—Calculation of a tubular module.

2. MODULE CONCEPTS AND DESIGN

The module, i.e., the industrial configuration of membranes, has to meet a variety of requirements that are sometimes even contradictory. Points of major importance are: 1) flow conditions along the membranes, 2) ratio of membrane area to pressurized vessel volume, 3) price of module and 4) at least for a number of applications, the possibility of cleaning the membranes.

Depending on the process application, one or the other of these requirements is of primary importance and, for this reason, a number of different modules have been designed. Figures 7–11 show some of the more important designs, i.e., tubular, plate, spiral and hollow fiber modules. Optimization procedure and some of its results will be discussed for only one module configuration, namely the hollow fiber module. This module is employed for reverse osmosis and gas permeation.

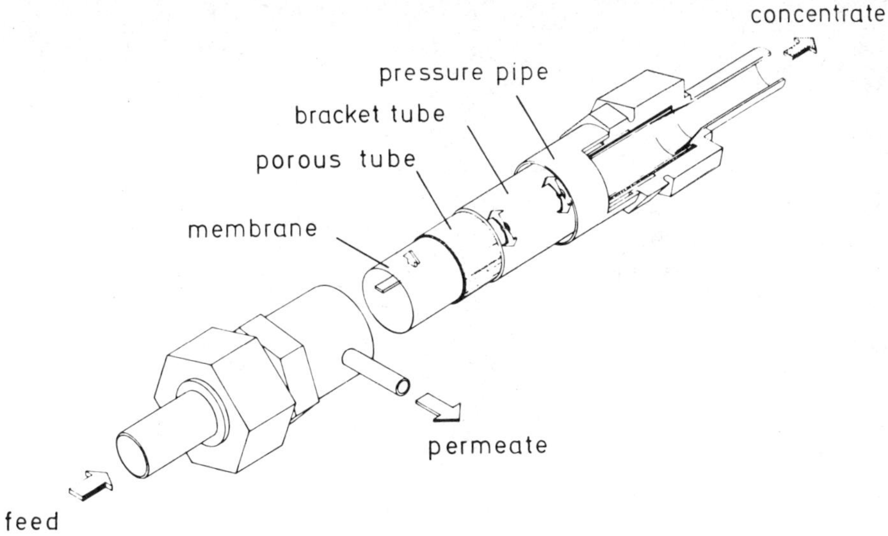

FIGURE 7.—Tubular membrane module.

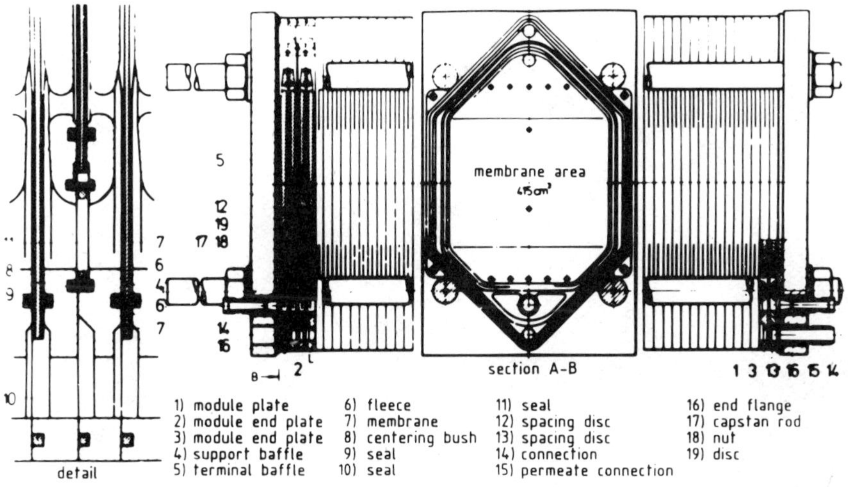

FIGURE 8.—Plate module (Courtesy of Steinmüller).

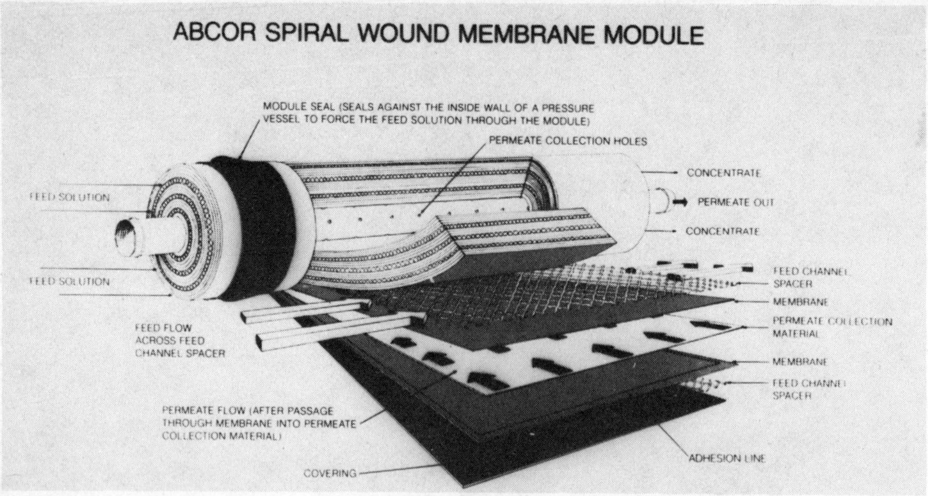

FIGURE 9.—Spiral module (Courtesy of Abcor).

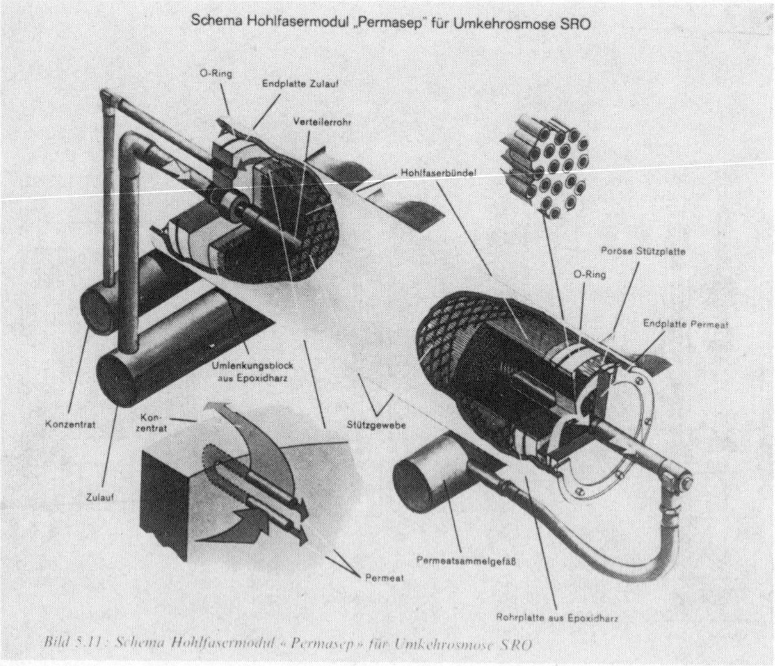

FIGURE 10.—Hollow-fiber module (Courtesy of Hager & Elsässer).

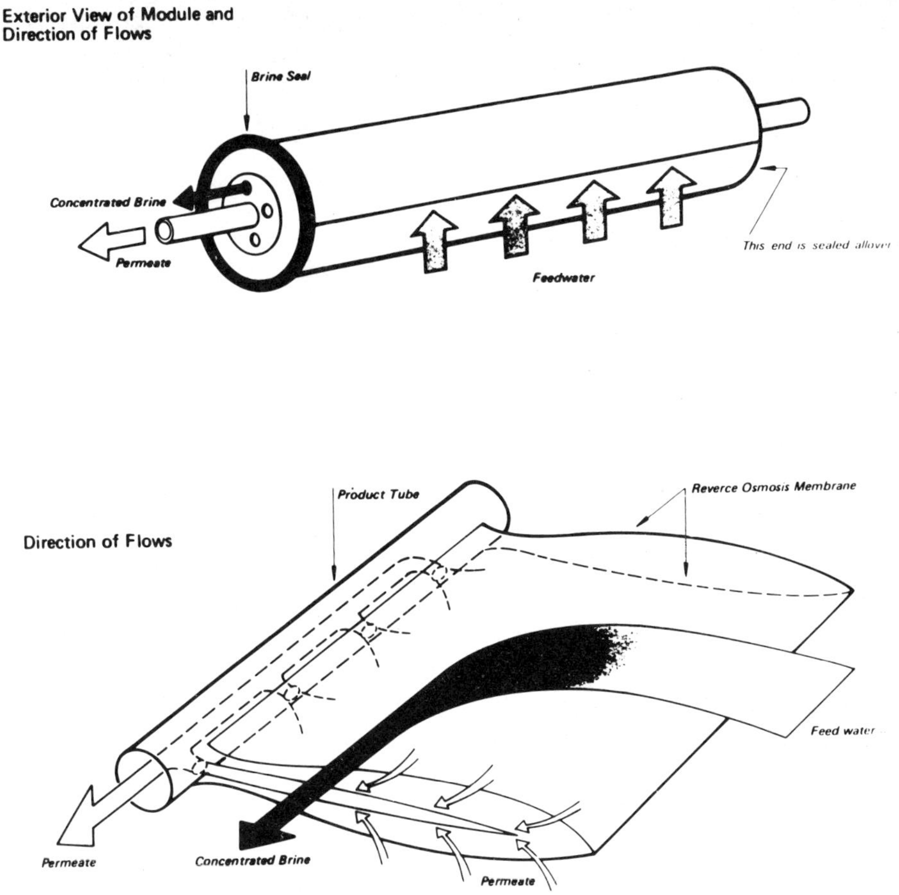

FIGURE 11.—Spiral module (Courtesy of Toray).

In sea-water desalination for example, spiral-wound and hollow fiber modules are generally employed. However, a plate-and-frame module is claimed to be competitive, despite its higher specific investment costs, for two reasons, both leading to a very high possible water recovery (WCF):

1) plate-and-frame systems show excellent mass transfer coefficients as a consequence of their excellent hydrodynamics;
2) plate-and-frame systems can be operated at higher transmembrane pressure differences (up to 100 bar).

At present, hollow fiber modules for sea-water desalination contain symmetric cellulose acetate fibers or asymmetric polyamide fibers (Table 3).

TABLE 3. Characteristics of several commercial hollow fiber modules (1)

Manufacturer	Du Pont		Dow Chemical	Toyobo	
Name of Module	Permasep		Dowex	Hollosep	
Type	B–9	B–10	4K, 20K	Low/Middle Press. Type	High Press. Type
Material of membrane	PA	PA	CTA	CA	CTA
Diameter (μ)	85	90	250	225	165
Filling density (%)	60	60	70	50	55
Spacing (μm)	12	13	15	57	32
Arrangement	Parallel	Parallel	Parallel	Cross-lapped	Cross-lapped
Supporting cloth	required	required	required	unrequired	unrequired
Main usage	Brackish water Pure water	Sea water Single pass (Partial dual pass)	Brackish water Pure water	Pure water Medical water Brackish water	Sea water (one pass)

2.1. The Hollow-Fiber Module

As shown in Figure 10 and Table 3, hollow fiber modules contain very fine fibers forming asymmetric or symmetric membranes and capable of withstanding pressure differences (high-pressure on the outside) up to 70 bar. The fiber bundle is U-shaped and its open ends are sealed in an epoxy resin plate. The active layer of the asymmetric membranes is on the outside of the fiber; the feed solution occupies the shell side.

In principle, feed flow can be parallel to the fibers or in the radial direction. For parallel flow, feed and permeate may flow co- or countercurrently. Gill and Bansal (2) demonstrated that countercurrent flow is (slightly) superior to cocurrent flow. Radial flow, however, is even better, especially for long fiber-bundles. In this case, the feed is distributed by a porous tube in the center of the bundle (Figure 12).

In this paper, only the hollow-fiber arrangement with radial flow will be analyzed. Assumptions:

1) Concentration polarization can be neglected because of the low water flux ($c_2 = c_1$). This assumption is certainly valid for the high pressure hollow fibers employed in sea-water desalination.
2) Pressure, concentration and mean velocity of the feed are only a function of the radial position:

$$p_1 = p_1(r),\ c_1 = c_1(r),\ \bar{v} = \bar{v}(r).$$

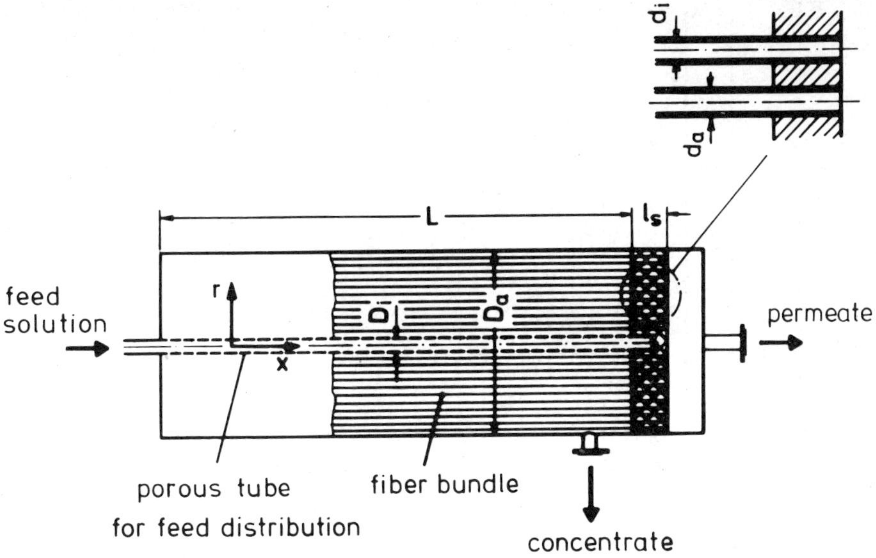

FIGURE 12.—Schematic representation of a hollow-fiber module.

3) Mass transfer in the membranes can be characterized by the sorption-diffusion model, i.e., by the solvent permeation coefficient A_V and the solute permeation coefficient B. In part of the calculations, however, a constant rejection coefficient $\mathcal{R} = 1 - \frac{c_3}{c_1}$ will be employed instead of the coefficient B of the sorption-diffusion model. In this case, analytical solutions of the design equations can be derived.

With respect to design, the hollow fiber module is characterized by

1) the dimensions of the fiber bundle (length of fibers L, bundle diameter D_a, inner diameter D_i; Figure 12)
2) seal length l_s
3) inner and outer diameters of the fibers d_i and d_a
4) a measure of the packing density of the bundle:
   a) membrane area / bundle volume, A/V
   b) number of fibers / cross-sectional area, n_f
   c) porosity ϵ

These three measures are correlated according to:

$$1 - \epsilon = n_f \frac{\pi}{4} d_a^2 = \frac{A}{V} \cdot \frac{d_a}{4} \tag{13}$$

The capacity of a hollow fiber module is based on the capacity of the single fibers. In every fiber, the permeate-*flow* is increasing along the axis as a consequence of the permeate *flux* across the membrane (Figure 13). As a

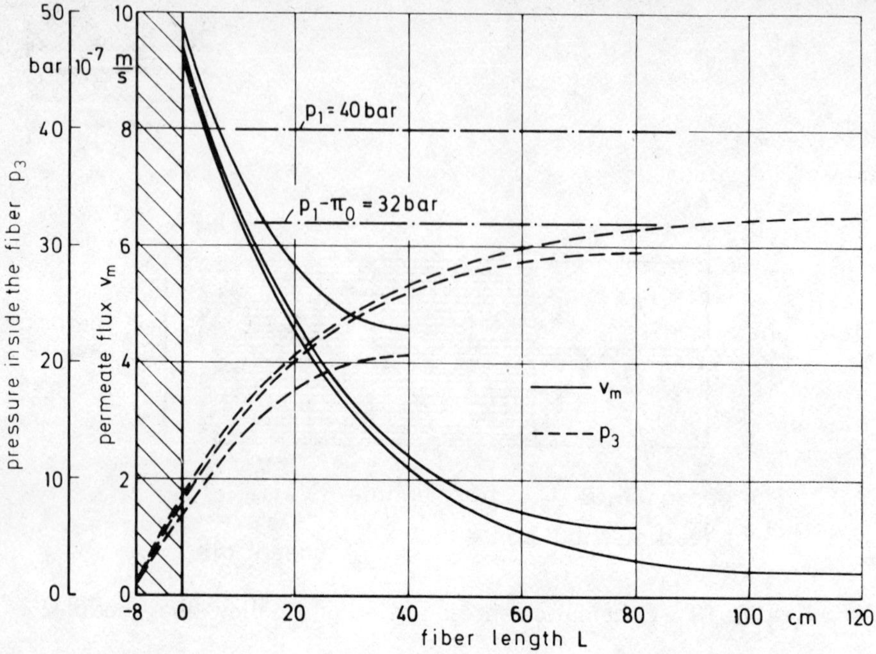

FIGURE 13.—Permeation behavior of hollow fibers (see Figure 14 for module parameter values.

consequence, pressure gradient as well as the transmembrane pressure difference will vary along the fiber-axis. Assuming that the Hagen-Poiseuille-equation is valid for every cross-section of the fiber (meaning that acceleration effects can be neglected), the following set of equations has to be solved:

pressure gradient: $$\frac{\mathrm{d}p_3}{\mathrm{d}x} = -\frac{64}{Re}\cdot\frac{\rho}{d_i\cdot 2}\cdot\bar{v}_{x3}^2 = -\frac{32\eta}{d_i^2}\,\bar{v}_{x3}(x) \tag{14}$$

mass balance: $$\frac{\mathrm{d}\bar{v}_{x3}}{\mathrm{d}x} = 4\,\frac{d_a}{d_i^2}\,v_m \tag{15}$$

sorption-diffusion model: $$v_m = A_V\,[p_1(r)-p_3(x)-b_\pi(c_1(r)-c_3(x))] \tag{16}$$

boundary conditions: $$x = 0 : \frac{\mathrm{d}p_3}{\mathrm{d}x} = 0$$

$$x = L : p_3 = p_{3L}$$

pressure loss in the seal: $$\Delta p_s = p_{3L}-p_o = \frac{32\eta}{d_i^2}\,l_s\,\bar{v}_{x3o} \tag{17}$$

"yield" of a single fiber: $$Q_f = \int_0^L v_m(x)\pi\, d_a\, dx = \frac{\pi}{4} d_i^2\, \bar{v}_{x3o} \tag{18}$$

The set of equations can be solved analytically if the influence of the permeate flux on the (local) concentration c_3 is neglected, or if A_V and $\mathcal{R}$ are employed instead of the parameters A_V and B of the sorption-diffusion model.

With

$$c_3 = (1-\mathcal{R})\, c_1$$

instead of

$$c_3 = \frac{B\, c_1}{v_m + B}$$

the integration of equations 14–18 leads to

$$Q_f = \frac{\pi d_a}{H} \cdot A_V \cdot (p_1 - p_o - \mathcal{R} b_\pi c_1) \frac{\tanh\,(HL)}{1 + H \cdot L_s \cdot \tanh\,(HL)} \tag{19}$$

where H is defined as

$$H = \sqrt{\frac{128\, \eta\, d_a\, A_V}{d_i^4}}$$

In some cases, such as the production of ultrapure water from fresh water, equation 19 will be sufficient. For the design of modules for the production of fresh water from sea- or brackish water, however, the sorption–diffusion model must be employed and in this case, the set of equations can only be solved numerically.

Figures 13 and 14 are the result of such a numerical calculation. They illustrate the negative effect of excessive fiber length. With reverse osmosis, only the front sections of long fibers are effective, at best. For the rear section, the transmembrane pressure difference and consequently, the flux, are very low, on account of pressure drop within the fiber. The product quality becomes even worse because the flux of the dissolved (and undesirable) components depends on the concentration difference and is not affected by the pressure difference across the fiber wall. Therefore, a high permeability would only be feasible if the inner diameter of the fiber could be increased, without losing its pressure stability, or if very short modules could be economically assembled.

In theory, larger fiber diameters should increase the flux, i.e., the performance of a single fiber. However, at the same time, the membrane area per unit volume decreases. Use of some simplifying assumptions facilitates the discussion of how fiber diameter and fiber length have to be chosen in order to

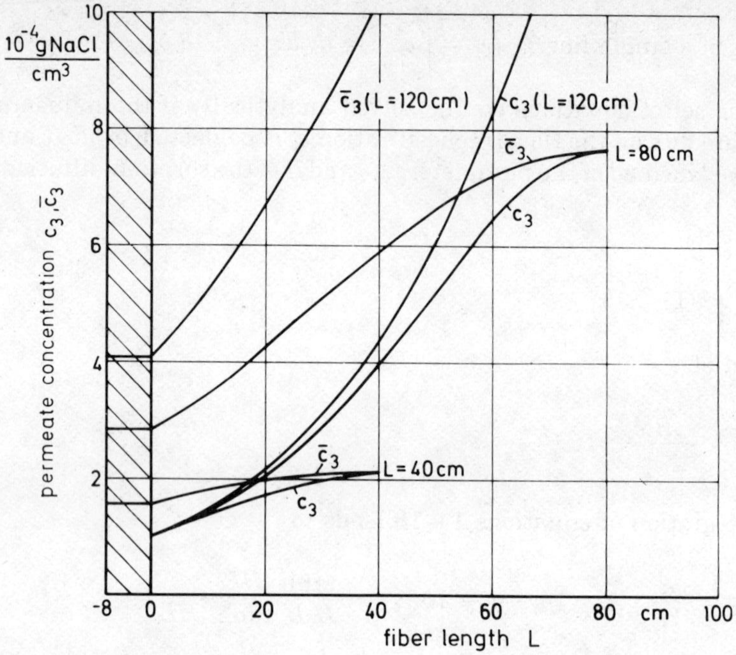

Module characteristics:

Fiber geometry	Material properties (NaCl–H_2O)	Membrane
$l_s = 8$ cm	$c_1 = 0.01 \dfrac{\text{gNaCl}}{\text{cm}^3}$	$A_V = 4 \cdot 10^{-6} \dfrac{\text{cm}}{\text{s} \cdot \text{bar}}$
$d_a = 4 \cdot 10^{-3}$ cm	$\eta = 0.01$ g/cm sec	
$c_i = 2 \cdot 10^{-3}$ cm	$b_\pi = 800$ bar · cm³/g	$B = 2 \cdot 10^{-6} \dfrac{\text{cm}}{\text{s} \cdot \text{bar}}$

Operating data

$p_1 = 40$ bar $\quad p_o = 1$ bar

FIGURE 14.—Concentration profiles in hollow fibers

maximize the "yield" per unit volume of the bundle. Starting with equation 19 and neglecting

1) the radial pressure losses in the feed, and
2) the influence of osmotic pressure on flux,

the specific yield of the fiber bundle is given by:

$$\frac{Q}{V} = \frac{Q_f}{A_f} \cdot \frac{A}{V} = \frac{Q_f}{\pi d_a L} \cdot \frac{4(1-\epsilon)}{d_a}$$

$$\frac{Q}{V} = \frac{4(1-\epsilon)}{d_a} \cdot \frac{A_V \cdot \Delta p}{H \cdot L} \cdot \frac{\tanh(HL)}{1 + H \cdot l_s \cdot \tanh(HL)}. \tag{20}$$

For a fixed ratio of $\frac{d_a}{d_i} = 2$, equation 20 is plotted in Figure 15.

According to Figure 15, an optimal fiber length exists for a given fiber diameter. Figure 13 clearly indicates that hollow fiber modules for reverse osmosis should be made as short as possible. Note that a simple correlation between yield of the single fiber and yield of the bundle can only be derived if the concentration and the pressure in the shell (outside of the bundle) are considered to be constant! In principle, the concentration of the feed increases between entrance and outlet while the pressure of the feed decreases. A numerical solution taking this into account can be achieved if the bundle is considered as a continuum containing sinks (Figure 16).

In this case, the following equations have to be solved:
mass balance (assuming constant liquid density):

$$\frac{\mathrm{d}}{\mathrm{d}r}\left(\frac{Q}{2\pi}\right) = \frac{\mathrm{d}}{\mathrm{d}r}(r\ \bar{v}_1(r)) = \frac{r\ n_f Q_{fR}}{L} \tag{21}$$

material balance:

$$\frac{\mathrm{d}}{\mathrm{d}r}\left(\frac{Qc_1}{2\pi}\right) = \frac{\mathrm{d}}{\mathrm{d}r}(r\ \bar{v}_1(r)c_1(r)) = \frac{n_f Q_{fR} c_{30} r}{L} \tag{22}$$

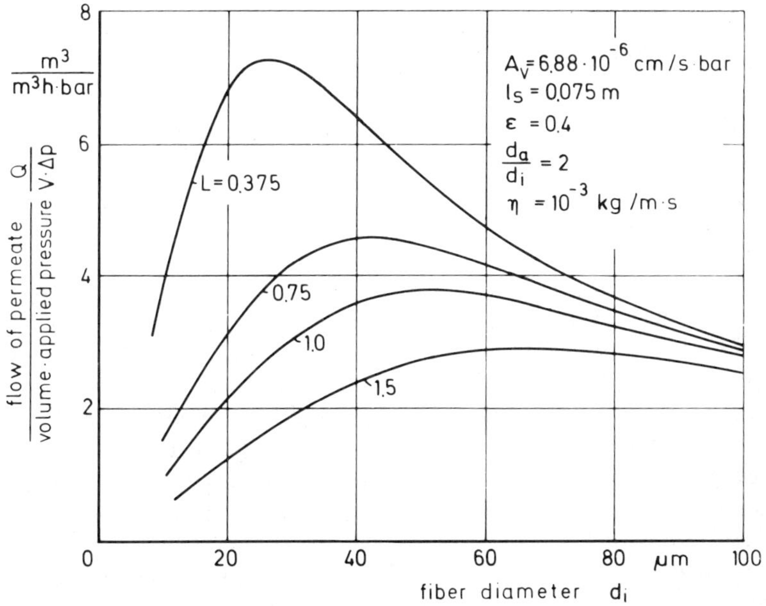

FIGURE 15.—Optimum design of hollow fibers.

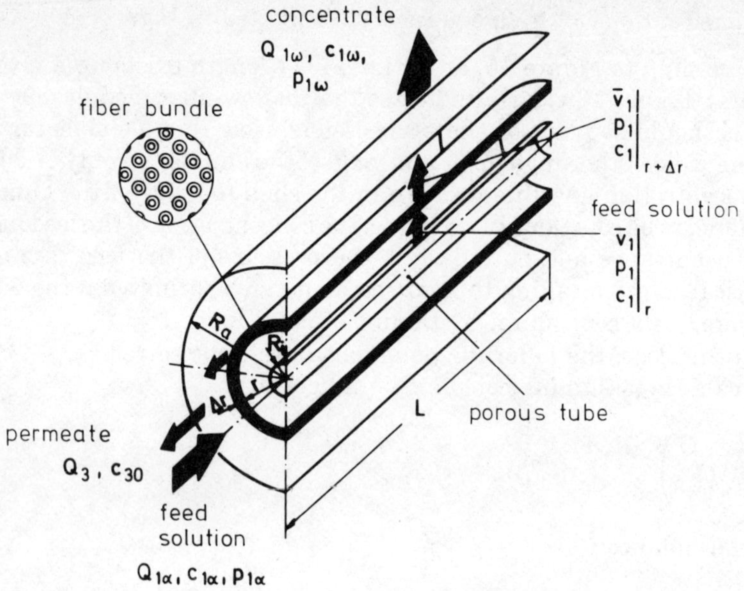

FIGURE 16.—Fiber bundle control volume for conservation equation.

energy balance:

$$\frac{dp_1(r)}{dr} = -K_o\,\eta\,(A/V)^2\,\frac{(1-\epsilon)^2}{\epsilon^3}\,\bar{v}_1(r) \tag{23}$$

in which K_o is a proportionality coefficient that must be varied according to the fiber arrangement. $K_o = 30$ is valid for fibers in parallel.

In the case of reverse osmosis, the relative flow configuration does not affect the performance to any large extent. As already indicated, the situation is quite different for gas permeation. While countercurrent flow only slightly improves the separation efficiency of hollow fiber modules in reverse osmosis, as far as gas permeation is concerned, countercurrent flow produces the best results. A consequent application of the countercurrent principle led to the concept of Hwang (3), i.e., the membrane column (Figure 17). This column consists of a rectification and a stripping section and permits a binary mixture to be fractionated into its components. For a given geometry of the membrane fractionating column (length of each section, membrane area per unit length), the product quality is determined by the flow rate of the gas being recirculated.

The modules employed in a membrane fractionating column must be different from those used in cascades; in hollow fiber modules, for example, the fibers cannot be u-shaped and must be open at both ends (4).

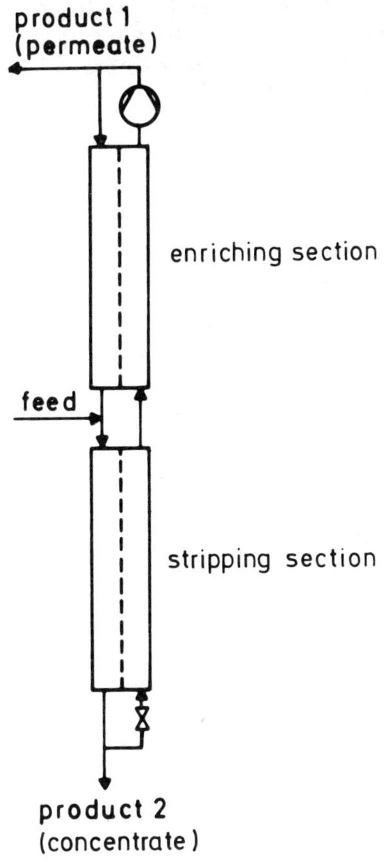

FIGURE 17.—Membrane column for gas separation.

3. CASCADES

There are cases where it is impossible to fractionate a given mixture in one stage to the desired degree of purity—the separation process will necessarily consist of several stages. These stages can be realized in several ways, for example, by employing the countercurrent flow principle as much as possible within one unit (e.g., in packed towers for absorption or rectification), or by combining definite separation units in the form of a cascade. The best known examples for this case are the fractionating towers employing trays. In fractionating towers, the trays are still incorporated in one vessel, i.e., part of one unit. In other cases, e.g., cascades of hydrocyclones, gas centrifuges and—last but not least—membrane modules, the unit itself will be the stage of the cascade.

In membrane technology, processes consisting of more than one stage are known in gas permeation and for the production of boiler feed water from seawater. Furthermore, cascades will be necessary if organic mixtures are to be separated by membrane processes only. The best known example of a membrane cascade consisting of a high number of stages is the Oak Ridge plant for enriching U^{235} in the gaseous phase (employing porous membranes).

Cascades—whether the stages are sedimentation ponds, membrane modules or gas centrifuges—do have one degree of freedom more than the "standard" tray column employed for distillation: while the (molar) reflux is constant from tray to tray in a column (for equal molar evaporation enthalpies of the components and provided that no sidestreams are extracted from the column), in general the reflux can be adjusted individually for every stage.

3.1. Definitions

In the literature, terms such as "stage", "yield," etc., sometimes have a different meaning. Therefore, the terms used here for the description and design of cascades will be defined as follows:

Separation unit. The elements of a separation process. Separation units of a separation process could be, for example,

- a gas centrifuge
- a membrane module
- a tray of a distillation column
- the evaporator of a multiple effect plant.

Remember, however, that in plate-and-frame modules as well as in modules of the spiral-wound type, usually every block or pressure vessel contains more than one separation unit!

Stage. The element of a cascade will be named the "stage," which might contain several separation units connected in series and/or in parallel. Separation units are considered part of a stage as long as they are connected only at the feed- and/or retentate-side.

The elements of a stage can be arranged in a "tapered" fashion (Figure 18) or in a "squared-off" fashion (Figure 19). A tapered arrangement is advantageous with respect to mass transfer across the boundary layer because the volume flow on the feed side decreases as the feed progresses along the "axis" of a stage.

Cascade. A combination of stages where the permeate of a stage will be the feed of the next stage.

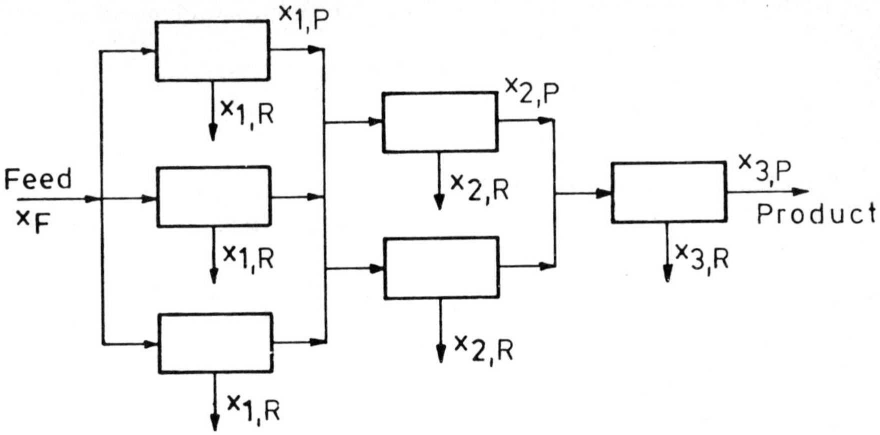

FIGURE 18.—"Tapered" fashion.

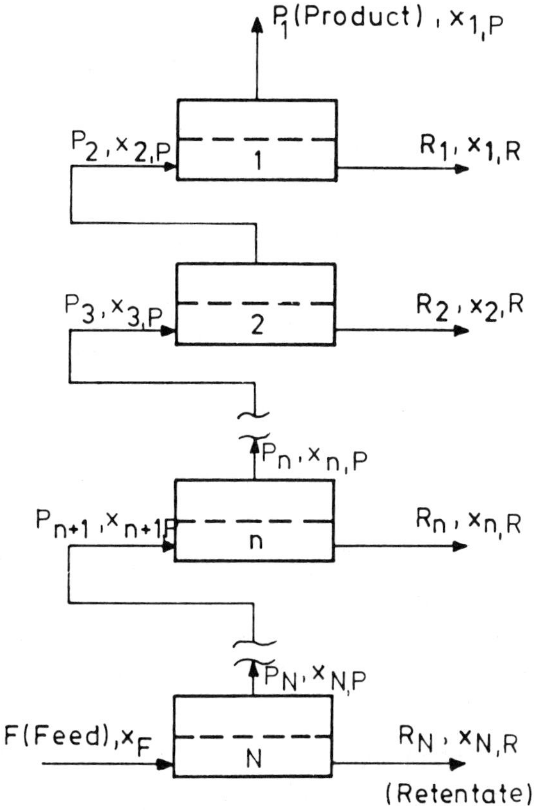

FIGURE 19.—Cascade without reflux.

Yield. Yield is defined as the ratio of a certain (key) component A in the product and in the feed:

$$Y = \frac{\dot{m}_{AP}}{\dot{m}_{AF}} = \frac{P_A}{F_A} . \tag{24}$$

Selectivity. A convenient measure of selectivity is the separation factor defined in terms of partial pressures p_i or mole fractions x_i as

$$\alpha_{molar} \equiv \frac{(p_A/p_B)_{Prod}}{(p_A/p_B)_{Feed}} = \frac{(x_A/x_B)_{Prod}}{(x_A/x_B)_{Feed}} \tag{25a}$$

or in terms of mass fractions w_i as

$$\alpha_{mass} = \frac{(w_A/w_B)_{Prod}}{(w_A/w_B)_{Feed}} \tag{25b}$$

For the permeation of ideal gases, where

$$\dot{n}_i = \mathcal{P}_i (p_1 x_{i1} - p_2 x_{i2})$$

describes the transport mechanism (in sorption-diffusion) membranes, equation 25a is reduced to

$$\alpha_{molar} = \frac{\mathcal{P}_A}{\mathcal{P}_B} \tag{25c}$$

(in cases, where $p_2 x_2 << p_1 x_1$ and consequently

$$\dot{n}_i \sim \mathcal{P}_i \, p_1 x_{i1}$$

is valid). In the above, subscript "1" refers to the feed stream and subscript "2" to the product stream.

Sometimes, selectivity is related to both products, for example,

$$\lambda \equiv \frac{(x_A/x_B)_{Permeate}}{(x_A/x_B)_{Retentate}} \tag{25d}$$

For a binary mixture and for low concentrations of the key component in both streams,

$$\lambda \sim \frac{(x_A)_{Per}}{(x_A)_{Ret}}$$

follows from equation 25d because of

$$x_B = 1 - x_A \sim 1$$

"Cut-rate" (split-factor). The cut-rate, sometimes called "split-factor" and "stage-cut," is defined as the ratio of permeate flow and feed flow. However, there will be a difference between the cut-rate defined for molar flows and for mass flow.

$$\theta_{\text{mass}} = \frac{\dot{m}_P}{\dot{m}_F} \tag{26a}$$

or

$$\theta_{\text{molar}} = \frac{P}{F} \tag{26b}$$

Operating line and "equilibrium" curve. Both terms are of importance for the graphical solution of a separation problem, i.e., for the graphical determination of the number of stages of a cascade. This method, which has been developed for the design of distillation columns by McCabe and Thiele, should be well known. For all cases, the *operating line* represents the *mass and material balances*. In distillation, the equilibrium curve represents the thermodynamic vapor/liquid equilibrium. For an ideal binary system, the equilibrium curve can be calculated from Raoult's law and the saturation-pressure curves of the pure components of the mixture.

In all other cases, however, including all membrane processes, the equilibrium curve does not represent a thermodynamic equilibrium at all but will represent the separation characteristics of the stage.

One-dimensional cascades. Only one-dimensional cascades will be discussed, i.e., cascades for the separation of a binary mixture. In practice, of course, these cascades will be employed for the separation of a multi-component mixture as well. In the latter case the multi-component mixture has to be regarded as a binary mixture and the fractionation will be calculated for two key components.

3.2. Cascades Without Reflux

Such cascades serve a purpose only in cases where the retentate is practically worthless. Norsk-Hydro, for example, is operating an electrolysis heavy water production plant where the deuterium is separated in a cascade without reflux (at least in the lower part of the cascade). Figure 19 illustrates the principle of a cascade without reflux.

Molar balance

$$P_{n+1} - P_n - R_n = 0 \tag{27a}$$

and material balance

$$P_{n+1}\, x_{Pn+1} - P_n\, x_{Pn} - R_n\, x_{Rn} = 0 \tag{27b}$$

in connection with the molar cut-rate

$$\theta_n = \frac{P_n}{P_{n+1}},\ 0 < \theta < 1 \tag{27c}$$

and the selectivity

$$\lambda_n = \frac{x_{Pn}}{x_{Rn}} \tag{27d}$$

result in the equation of the operating line

$$x_{Pn+1} = [1 + \theta_n(\lambda_n - 1)]\, x_{Rn} \tag{27e}$$

Feed flow rate and feed concentration as well as the desired product quality are known entities. Accordingly, the number of stages, product flow and the membrane area of every stage can be calculated:

$$P_n = F \cdot \prod_{k=n}^{N} \theta_k \tag{27f}$$

$$x_{Pn} = x_F \frac{\prod_{n}^{N} \lambda_k}{\prod_{n}^{N} [1 + \theta_k\,(\lambda_k - 1)]} \tag{27g}$$

Equation 27f demonstrates that permeate flow is rapidly decreasing from stage to stage.

For a constant cut-rate and a constant selectivity, the equations 27e–27g are reduced to

$$x_{Pn+1} = [1 + \theta(\lambda - 1)]\, x_{Rn} \tag{27h}$$

$$P_n = F\theta^{(N+1-n)} \tag{27i}$$

$$x_{Pn} = x_F \left[\frac{\lambda}{1 + \theta(\lambda - 1)}\right]^{N+1-n} \tag{27j}$$

The condition $\lambda \equiv \dfrac{x_{Pn}}{x_{Rn}}$ = constant remains to be discussed. In principle, this condition cannot be valid between $x_P = 0$ and $x_P = 1$ because of its linear character. Of practical importance are the following two cases:

1) $\lambda \rightarrow 1$ (very low selectivity and, accordingly, a high number of stages if larger differences in product and feed quality have to be achieved; and
2) low concentrations in the whole range of operating conditions.

In the first case the operating line will be the diagonal and the "equilibrium curve" will be a nearly straight line very close to it (Figure 20). In the second case the "equilibrium curve" can be replaced by a straight line because the whole operating range is small (Figure 20).

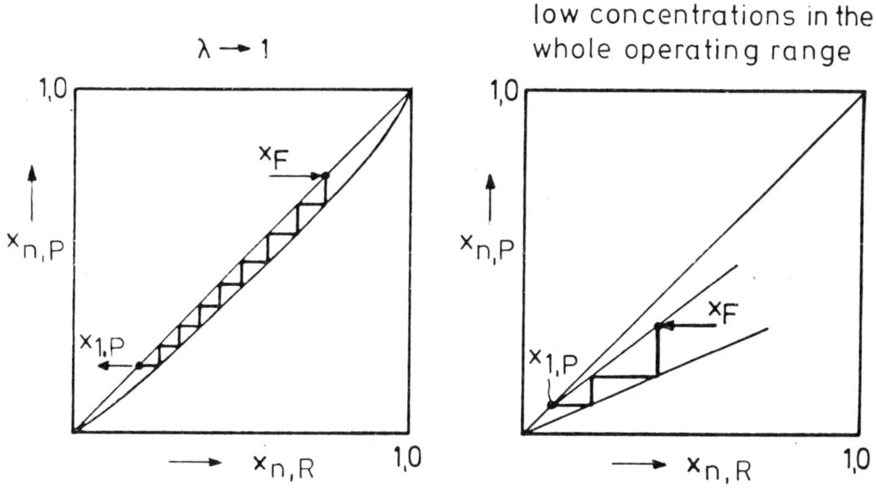

FIGURE 20.—McCabe-Thiele diagram.

3.3. Reflux Cascades

The inevitable losses of product in the retentate of cascades without reflux can be avoided by recycling the retentate, according to Figure 21. This advantage, however, must be weighed against the higher capital costs of reflux cascades. Only a detailed calculation of the specific separating costs will lead to a sound decision whether a separation process should operate with or without reflux.

In principle, there are four possible modes of operation:

1) constant reflux $\frac{R}{P_1}$
2) constant cut rate and variable reflux
3) variable cut rate and variable reflux
4) "ideal" cascade operation ($x_{n-1,R} = x_{n+1,P}$).

Only the first two cases will be discussed here, because a graphical solution is of value. The last two cases can be solved using a computer.

Constant reflux In general, the operating line of a cascade follows from a mass/molar balance and a component material balance between the top and a certain stage "n" of the cascade.

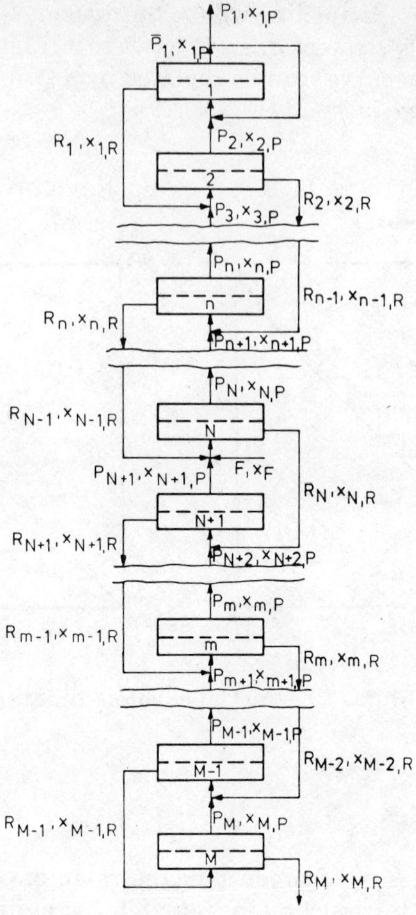

FIGURE 21.—Cascade with reflux.

For constant reflux: R = constant.

$$\frac{R}{P_1} = v = \text{constant},$$

the molar balance $$P_{n+1} - R_n - P_1 = 0 \qquad (28a)$$

and the material balance $$P_{n+1}\, x_{Pn+1} - R_n x_{Rn} - P_1\, x_{P1} = 0 \qquad (28b)$$

result in

$$x_{pn+1} = \frac{v}{v+1}\, x_{Rn} + \frac{1}{v+1}\, x_{P1} \qquad (28c)$$

The operating line is a straight line with the slope $\frac{v}{v+1}$.

If the "equilibrium curve" is known, the number of stages necessary to achieve a certain product quality can be easily found, according to Figure 22, by a step-by-step graphical procedure.

Constant cut rate, variable reflux When the molar balance

$$P_{n+1} - R_n - P_1 = 0 \tag{29a}$$

is developed for several stages, a pattern can be noticed if the (constant) cut rate is introduced:

$$P_2 = P_1 + R_1 = P_1\,(1+\gamma) \tag{29b}$$

Here $\gamma = \frac{R}{P}$ is used instead of $\theta = \frac{P}{F}$.

θ and γ are related by the molar balance of a stage: $\gamma = \frac{1-\theta}{\theta}$.

$$P_3 = P_1 + R_2 = P_1\,(1+\gamma\,(1+\gamma))$$

$$P_n = P_1(1+\gamma+\gamma^2 + \ldots \gamma^{n-1}) \tag{29c}$$

$$P_n = P_1 \frac{1-\gamma^n}{1-\gamma}$$

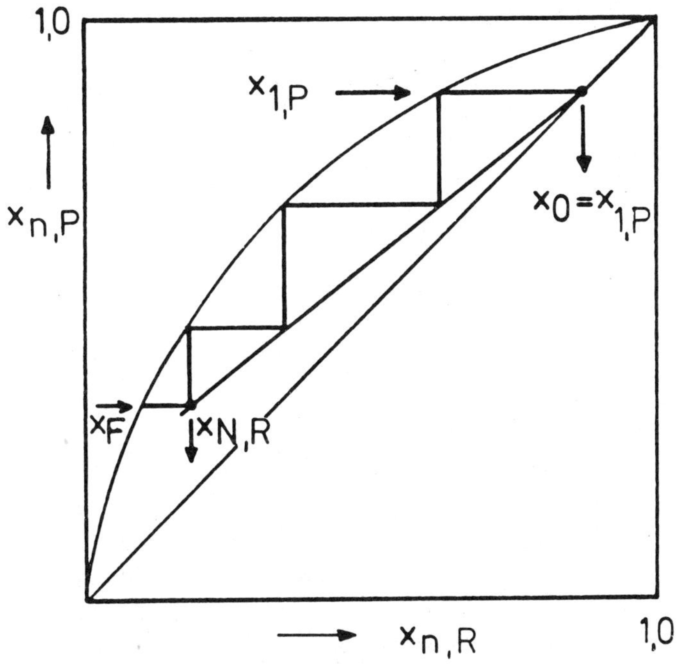

FIGURE 22.—Constant reflux.

With the material balance

$$x_{Pn+1} = \frac{R_n}{P_{n+1}} x_{Rn} + \frac{P_1 x_{P1}}{P_{n+1}} \tag{29d}$$

the operating line follows:

$$x_{Pn+1} = \frac{\gamma - \gamma^{n+1}}{1-\gamma^{n+1}} x_{Rn} + \frac{1-\gamma}{1-\gamma^{n+1}} x_{P1} \tag{29e}$$

Quite obviously, the slope of the operating line differs from stage to stage.

The number of stages follows according to Figure 23. Note that all operating lines will intersect at $x_{P1} = x_{R1}$.

Once the necessary number of stages is known, the product rate P_1/F can be calculated according to

$$\frac{P_1}{F} \equiv \frac{P_1}{P_{N+1}} = \frac{1-\gamma}{1-\gamma^{N+1}} \tag{29f}$$

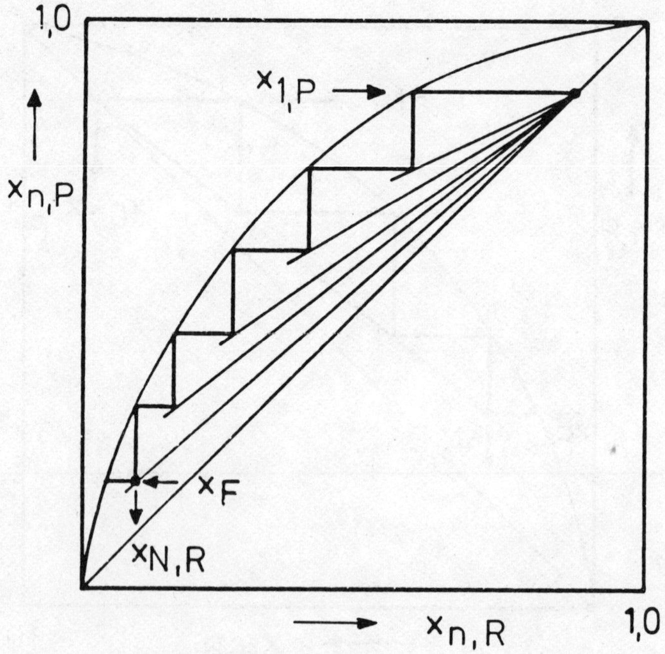

FIGURE 23.—Constant cut rate, variable reflux.

3.4. The "Equilibrium Curve"

As mentioned above, the "equilibrium curve" in the case of membrane cascades is not at all a thermodynamic equilibrium but will represent the separation characteristics of the stage. This separation characteristic is in general a function of:

1) the selectivity of the membrane
2) the fluid dynamics in the module
3) the driving force
4) the concentration level (because the membrane selectivity is, or might be, concentration dependent)
5) the flow pattern in the module, for example, co- or counter-current flow.

With respect to flow pattern, five different patterns are possible:

1) complete mixing at both sides of the membrane [Weller Steiner Case 1 (5)]
2) Cross-plug flow, i.e., plug flow at the feed side and permeate flow orthogonal to the membrane without mixing [Weller Steiner Case 2 (5), Naylor–Backer (6)]
3) Parallel plug flow, cocurrent flow [Blaisdell–Case (7)]
4) Parallel plug flow, countercurrent flow
5) Partial (incomplete) mixing of both feed and permeate [Breuer Case (8)].

The first two cases will be discussed here. A detailed discussion of the other cases can be found elsewhere (9, 10).

Complete mixing within feed and permeate streams In this case (Figure 24), the material balances for a binary mixture are:

$$P_n\, x_{Pn} = \mathcal{P}_A(p_1\, x_{Pn} - p_2\, x_{Pn}) \tag{30a}$$

$$P_n\,(1 - x_{Pn}) = \mathcal{P}_B\,[p_1(1 - x_{Rn}) - p_2(1 - x_{Pn})] \tag{30b}$$

Combining both equations and introducing $r = p_1/p_2$ and $\alpha = \dfrac{\mathcal{P}_A}{\mathcal{P}_B}$

leads to:

$$\frac{x_{Pn}}{1 - x_{Pn}} = \alpha\,\frac{r\,x_{Rn} - x_{Pn}}{r(1 - x_{Rn}) - (1 - x_{Pn})} \tag{30c}$$

As equation 30c and Figure 25 indicate, the equilibrium curve in this case is influenced by the separation factor α and the pressure ratio r.

For large pressure ratios, the equilibrium curve simplifies to

$$x_{Pn} = \frac{\alpha\,x_{Rn}}{1 + (\alpha - 1)x_{Rn}}. \tag{30d}$$

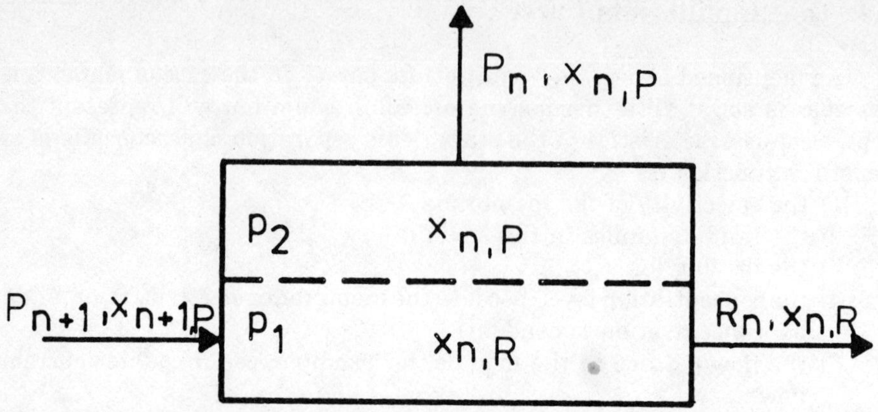

FIGURE 24.—Complete mixing.

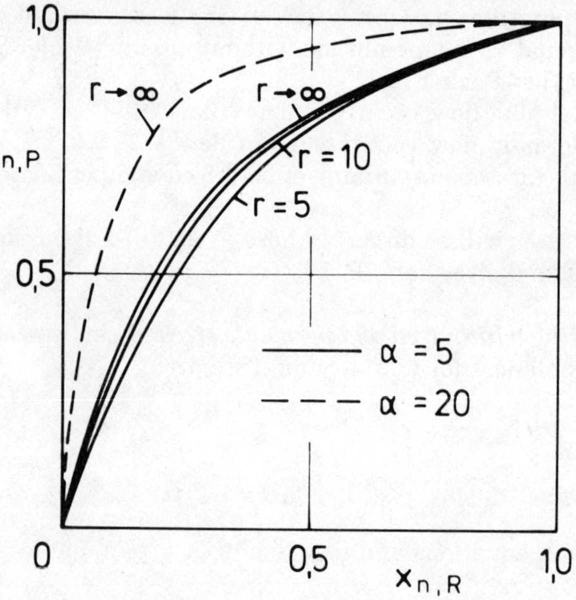

FIGURE 25.—Equilibrium curves (complete mixing).

Cross-plug flow In this case, the molar and material balances have to be formulated for a differential element of the membrane channel (Figure 26). Because of the assumption of plug flow in the membrane channel, the module in this case is looked upon as sum of an infinite number of infinitesimally small modules of the complete mixing case.

The molar balance

$$P - (P + dP) - dP_n = 0 \tag{31a}$$

and the material balance

$$(dP_n)\, x'_P = -\, d(P \cdot x'_R) = -\, x'_R\, dP - P\, dx'_R \tag{31b}$$

have to be solved in combination with equation 30c, which is valid for local values.

The set of equations can be integrated easily in case of large pressure ratios and for constant local separation factors α':

$$\ln \frac{R_n}{P_{n+1}} = \int_{x_{Pn+1}}^{x_{Rn}} \left(\frac{1}{x'} + \frac{\alpha'}{1 - x'} \right) \frac{dx'}{\alpha' - 1} \tag{31c}$$

The material and molar balances around the stage relate the feed concentration x_{Pn+1} to the permeate and retentate concentrations of the stage (module):

$$x_{Pn+1} = \theta\, x_{Pn} + (1-\theta) x_{Rn} \tag{31d}$$

Integration of equation 31c gives the "equilibrium curve" in an implicit form:

$$\frac{x_{Rn}}{\theta x_{Pn} + (1 - \theta)\, x_{Rn}} = \left[\frac{1 - x_{Rn}}{1 - \theta x_{Pn} + (1 - \theta)\, x_{Rn}} \right]^{\alpha'} (1 - \theta)^{\alpha - 1} \tag{31e}$$

It is of interest, whether the overall separation factor α of the stage is higher or lower than the local separation factor α'. This problem can be

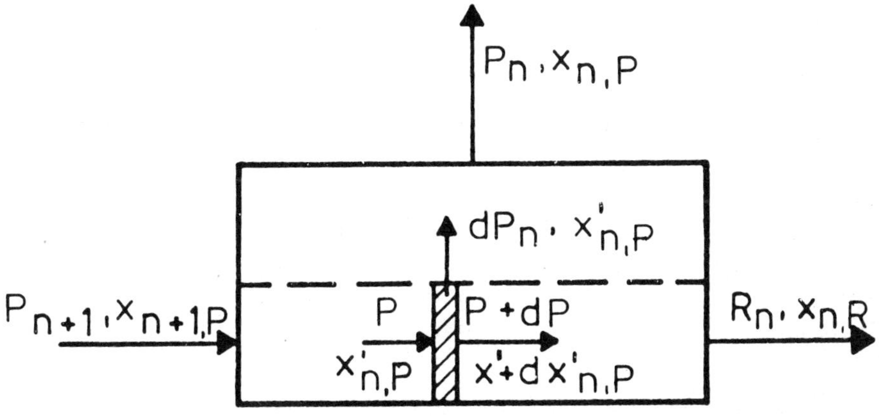

FIGURE 26.—Cross-plug flow.

discussed easily for small local separation factors ($\alpha' - 1 << 1$), which is the case for the above-mentioned separation of U^{235}/U^{238}. In this case, the (overall) separation factor of the stage is given by

$$\frac{\alpha - 1}{\alpha' - 1} = -\frac{\ln(1-\theta)}{\theta} \tag{31f}$$

According to equation 31f, the overall separation factor is larger than the local separation factor. Although equation 31f is valid only for small local separation factors, the general statement $\alpha > \alpha'$ is valid for larger separation factors as well.

Note that in the discussion of these cases with respect to the equilibrium curve, concentration polarization has not been taken into account. As pointed out, this assumption is in general valid for gas permeation, but not for the separation of liquid systems.

3.5. Membrane Column

The membrane column can be an alternative to membrane cascades (3, 11, 12). The membrane column consequently utilizes the countercurrent-flow principle (Figure 27), and is thus restricted to cases where concentration polarization is negligible. (It has been suggested that the membrane-column concept might be applied to the separation of liquid mixtures. However, this concept fails here, at least for asymmetric membranes, because of the concentration polarization within the support layer.) Membrane columns utilize maximal driving forces because of the countercurrent flow at both sides of the membrane.

The permeating component of higher separation factor is depleted on the high pressure side in the direction of the flow. This component can be withdrawn at the top of the column, and the slower permeating component of the mixture at the bottom.

At present, all modules for membrane columns described in the literature employ hollow fiber membranes.

The design equations for a membrane column are:

1) the equation for species transport in the membrane,
2) mass, material and energy balances (pressure losses), formulated for the differential element of the membrane channel.

Transport of species i across the membrane:

$$\dot{n}_i = \mathcal{P}_i(x_i\, p_1 - y_i\, p_2) \tag{32a}$$

Mass (molar) balance (M = constant, combined with equation 32a):

$$\frac{d\dot{N}}{dz} = 2\pi r\left[\,\mathcal{P}_i\,(x_i\, p_1 - y_i\, p_2) + \mathcal{P}_j\,[(1-x_i)\,p_1 - (1-y_i)\,p_2]\right] \tag{32b}$$

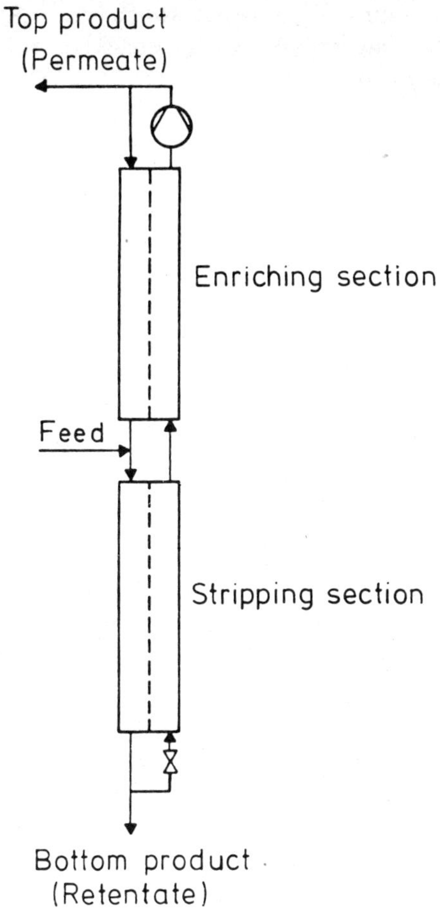

FIGURE 27.—Membrane column.

Material balance:

$$\frac{d^2x_i}{dz^2} + \frac{\dot{N}\,RT}{p_1 D_{ij}\pi r^2}\,\frac{dx_i}{dz} = \frac{2\pi r\,RT\,\mathcal{P}_i}{D_{ij}\,\pi r^2}\left[x_i - \frac{p_2}{p_1}\,y_i - x_i\left(x_i - y_i\frac{p_2}{p_1}\right)\left(1 - \frac{\mathcal{P}_j}{\mathcal{P}_i}\right) - x_i\frac{\mathcal{P}_j}{\mathcal{P}_i}\left(1 - \frac{p_2}{p_1}\right)\right] \qquad (32c)$$

For a membrane column it is essential that axial backmixing by diffusion (dispersion) be negligible compared with convective transport. In this case, equation 32c is reduced to

$$\frac{d}{dz}(x_i \dot{N}) = 2\,\pi\, r\, \mathcal{P}_i\,(x_i p_1 - y_i\, p_2) \tag{32d}$$

Combining equation 32d with equation 32a (transport in the membrane) leads to

$$\frac{dx_i}{dz} = \frac{2\pi r \mathcal{P}_i p_1}{\dot{N}}\left[\, x_i - \frac{p_2}{p_1} y_i - x_i \left(x_i - y_i \frac{p_2}{p_1}\right)\left(1 - \frac{\mathcal{P}_j}{\mathcal{P}_i}\right) - x_i \frac{\mathcal{P}_j}{\mathcal{P}_i}\left(1 - \frac{p_2}{p_1}\right)\right] \tag{32e}$$

Energy balance:

For the calculation of pressure losses, one must take into account not only the compressibility of gases, but also the fact that hollow fibers are elastically deformed (widened) by the higher pressure inside the fiber. According to Hwang and Thorman (3), in the case of a deforming fiber the pressure losses can be calculated by

$$\frac{dp}{dz} = \frac{K_1\, \eta \dot{N}\, RT}{\pi\, r^4\, p_1}\left(\frac{1}{K_2} - \frac{4\, Re_m z}{Re_z\, r}\right) - \frac{4}{r}\,\frac{1}{\dfrac{Re_z}{3p_1} - \dfrac{\pi r^3\, p_1}{2\eta\, \dot{N}RT}} - \frac{8\eta\, \dot{N}RT}{\pi\, r^4 p_1} \tag{33}$$

with

$$Re_m = \frac{\rho v_m r_z}{\eta}; \quad Re_z = \frac{\rho v_z r_z}{\eta}$$

$$K_1 = 8\,(1 + 0.75\, Re_m - 0.0407\, Re_m^2 + 0.0125\, Re_m^3)$$

$$K_2 = 1 + 0.056\, Re_m - 0.0153\, Re_m^2$$

The set of equations 33 has to be solved numerically. For the special case of total reflux, however, an analytical solution for the concentration profile along the membrane column can be derived if the pressure losses are neglected.

In the case of total reflux, for every section of the column (Figure 28)

$$x_i = y_i \tag{34}$$

is valid.

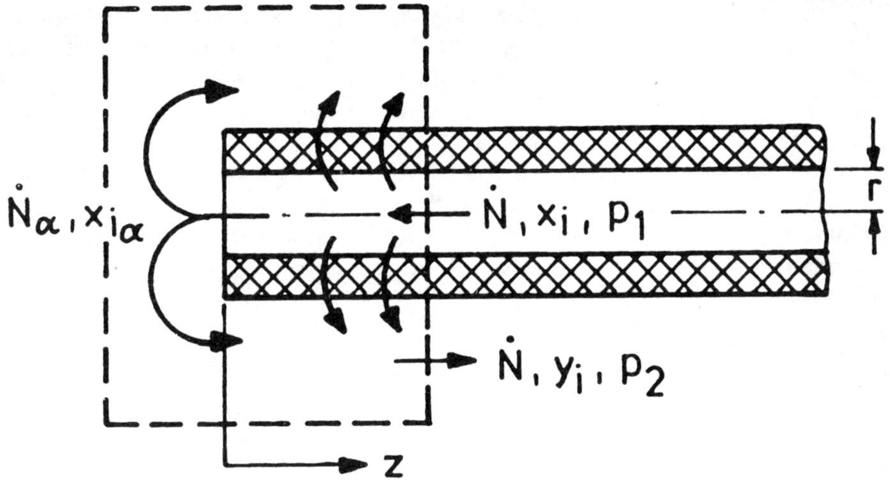

FIGURE 28.—Membrane column for the case of total reflux showing control volume for conservation equations.

The boundary conditions at $z=0$ are

$$x_i = x_{i\alpha}, \text{ and } \dot{N} = \dot{N}_\alpha, \tag{34a}$$

in which the subscript α indicates evaluation at the entrance to the column. With the above, equation 32d can be integrated:

$$z = \frac{\dot{N}_\alpha \cdot x_{i\alpha}\left(\dfrac{1 - x_{i\alpha}}{x_{i\alpha}}\right)^{\frac{1 + \mathcal{P}_j/\mathcal{P}_i}{1 - \mathcal{P}_j/\mathcal{P}_i}}}{2\pi\, r\mathcal{P}_j p_1\left(1 - \dfrac{p_2}{p_1}\right)} \left[\left\{\left(\frac{x_i}{x_{i\alpha}}\frac{1 - x_{i\alpha}}{1 - x_i}\right)^{\frac{\mathcal{P}_j/\mathcal{P}_i}{1 - \mathcal{P}_j/\mathcal{P}_i}} - 1\right\}\right.$$

$$\left. + \frac{x_{i\alpha}}{1 - x_{i\alpha}}\left\{\left(\frac{x_i}{x_{i\alpha}}\cdot\frac{1 - x_{i\alpha}}{1 - x_i}\right)^{\frac{1}{1 - \mathcal{P}_j/\mathcal{P}_i}} - 1\right\}\right] \tag{34b}$$

$$\dot{N} = \dot{N}_\alpha\left(\frac{x_i}{x_{i\alpha}}\right)^{\frac{\mathcal{P}_j/\mathcal{P}_i}{1 - \mathcal{P}_j/\mathcal{P}_i}}\left(\frac{1 - x_{i\alpha}}{1 - x_i}\right)^{\frac{1}{1 - \mathcal{P}_j/\mathcal{P}_i}} \tag{34c}$$

The separation process is determined (among other parameters) by the reflux ratio. Similar to fractionating by distillation, the calculation with total reflux leads to the minimal necessary column length. For a given column length, on the other hand, equations 34b and 34c indicate the maximum possible purities of top and bottom product. For these reasons, a discussion of the simple equations 34b and 34c is useful for a first estimate of the column length and the operating conditions.

Column length depends on the recycled molar flux $\dot{N}_\alpha$. Figure 29 indicates that a specific molar flux (molar flow per fiber) as low as possible should be chosen. The lower limit is set by the condition that diffusive axial backmixing must be negligible with respect to the convective mass transport. On the other hand, the volume-specific yield (yield per unit-volume of fiber bundle) should be high.

In general, the optimization of a membrane column must consider molar flux $\dot{N}_\alpha$, reflux ratio and the pressure losses. The mutual dependencies, which have been only briefly discussed here, explain why hollow fiber modules for the membrane column are relatively long ($L/r \sim 20000$), as compared with hollow fiber modules for the separation of liquid mixtures. Figure 30 shows the concentration profile of a membrane column as calculated by equations 34b, 34c (total reflux).

The concentration profile is very similar to the profile calculated by a numerical solution of the set of equations 33 for the separation of CO_2/N_2 (Figure 31), taking into account the pressure losses and a finite reflux ratio.

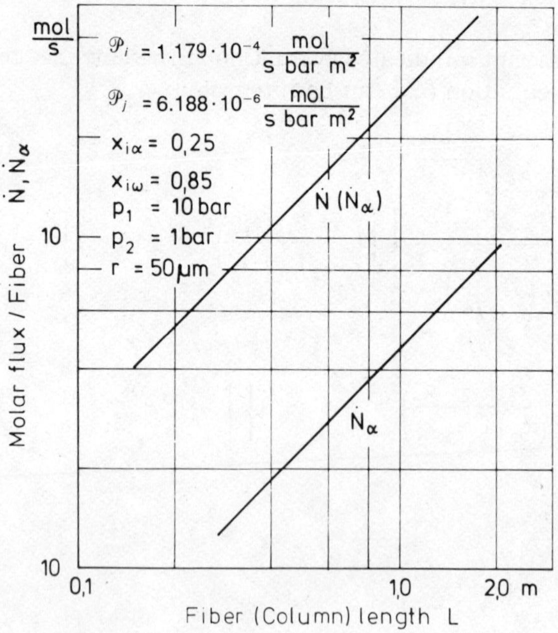

FIGURE 29.—Influence of the recycle flux on the length of the membrane column.

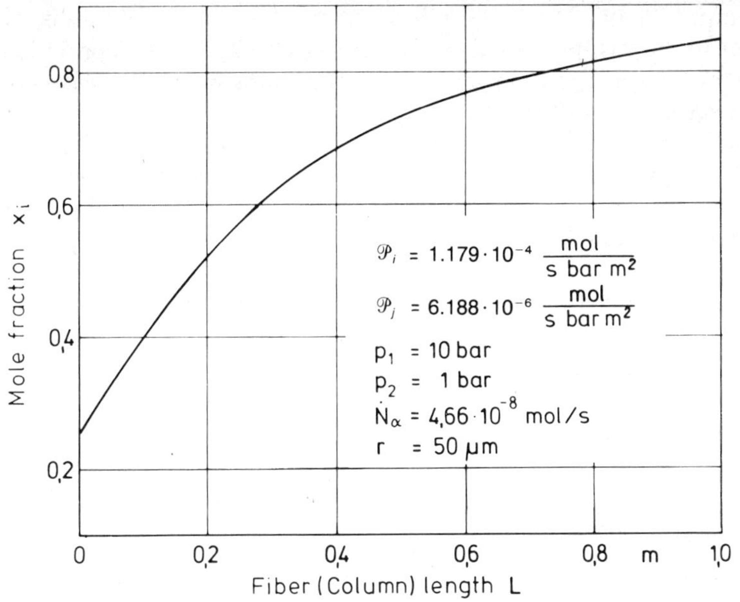

FIGURE 30.—Concentration profile of the membrane column for total reflux.

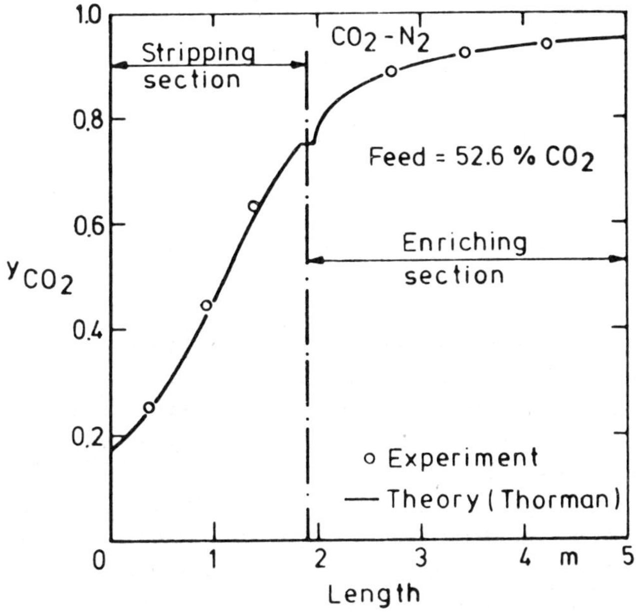

FIGURE 31.—Separation of CO_2/N_2 by membrane column.

Figures 30 and 31 clearly indicate that a membrane column is especially effective in the stripping section. For the enriching section, especially for high concentrations of the preferably permeating component, the necessary membrane area and the reflux ratio (pressure losses) are increasing rapidly. In this range of concentrations, a cascade will be superior. The advantages of a combination of membrane column and membrane cascade, as compared with a cascade or a membrane column has been discussed by Werner and coworkers (13) (Table 4).

TABLE 4. Comparison of the module arrangements used for membrane rectification

Module arrangement	Number of modules	Number of compressors	Total amount of compressed gases [mol/h]	Total membrane area [m^2]
Cascade	11	10	87.2	7.6
Single staged membrane column enriching section	1	1	346	6.0
Membrane column enriching and stripping section	4	2	80	2.3
Membrane column (stripping section) +2 staged Cascade	4	2	48.2	1.9

4. HYBRID PROCESSES

4.1. Prime Energy Demand of Separation Processes—Evaporation versus Reverse Osmosis

It seems to be the general impression that membrane processes require smaller amounts of energy than do alternate processes. Although in the majority of cases this is true, it is worth analyzing the above statement more substantially. The discussion will be confined to only one illustration, the desalination of sea- or brackish water.

In principle, there are three methods by which one can achieve the separation of fresh water from solutions: vapor phase (i.e., evaporation), solid phase (i.e., crystallization of pure ice) and membranes.

In our opinion, which is based on experiments and a thorough process evaluation (14), crystallization processes do not merit consideration for desalination of seawater.

Evaporation processes with a low specific energy consumption are multiple-effect evaporation (ME), multi-stage flash evaporation (MSF) and vapor compression (VC). Membrane separation processes, especially reverse osmosis (RO), are *the* alternative to evaporation processes. The quality of energy needed by those various processes is totally different. Although MSF and ME utilize heat of a relatively low temperature level, mechanical vapor compression as well as reverse osmosis use mechanical work (power). The principle of the ME process, as illustrated in Figure 32, is that the vapor generated in the first effect is condensed in the second, thus producing more vapor, which is condensed in the third and so on. The principle of VC is illustrated in Figure 33. By increasing the pressure of the vapor, the condensation temperature is increased accordingly. As a consequence, the vapor can be condensed utilizing its latent heat for the evaporation of the brine.

Figure 34 is a flow-sheet of a RO plant, including water pretreatment.

Theoretical development In comparing such different processes, one must use a measure common to all of them: thermal availability. This measure weighs both the quantity as well as the quality of any energy consumed.

Instead of using availability, which is a somewhat abstract measure, we will employ the term "primary energy demand." The primary energy demand

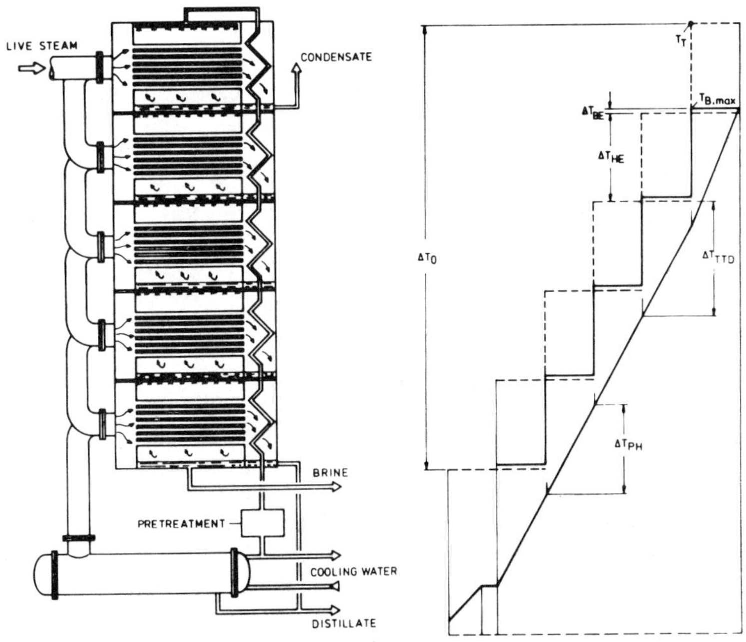

FIGURE 32.—Flow scheme of a horizontal tube multiple effect (HTME) desalination plant.

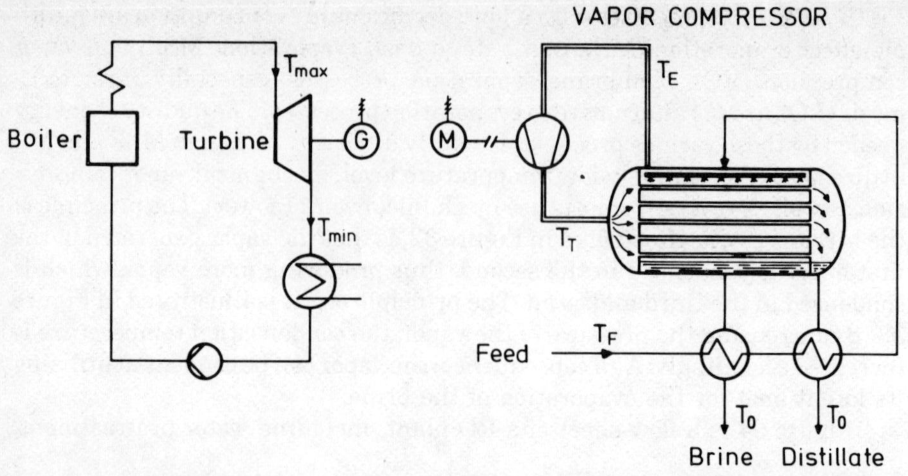

FIGURE 33.—Vapor compression and its primary energy demand.

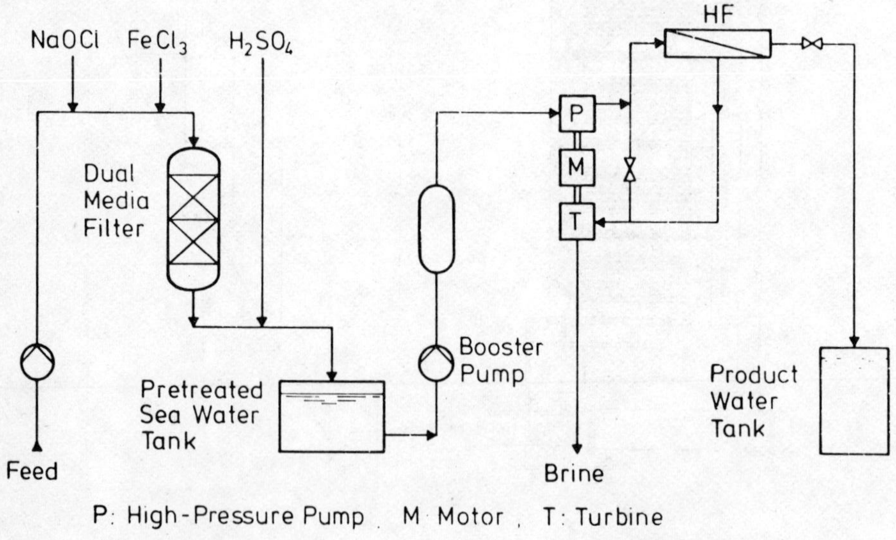

FIGURE 34.—Flow scheme for a reverse osmosis desalination plant.

is much easier to visualize and, at least with respect to desalination processes, just as sound. Employing primary energy demand instead of availability is possible because the heat content of our fossil fuels is equivalent to their availability content within an accuracy of ± 3% (15,16). What part of this can be utilized depends in principle on the conversion process.

For combustion as a conversion process (via boiler or via internal combustion engine), the extent to which the availability of a fossil fuel is utilized is very much determined by the maximum and minimum temperature of the working fluid employed. The maximum temperature should be as high as possible and in general, current construction materials limit these temperatures to about 600°C. It is completely impossible to operate evaporation processes in desalination at such temperatures. Because of the fast-rising vapor pressure of water and because of the serious scaling and corrosion problems at temperatures above 100°C, the maximum operating temperature of seawater evaporators is about 120°C.

Therefore, a combination of a power plant and an evaporation process is the best solution—utilizing the availability of the fuel in the entire (technically possible) temperature range. Of course, this has been well known since the days of S. Carnot—all major desalination plants are designed as dual purpose plants (Figure 35).

When comparing vapor compression and dual purpose plants, it has to be taken into account that the power needed for compression is generated in a

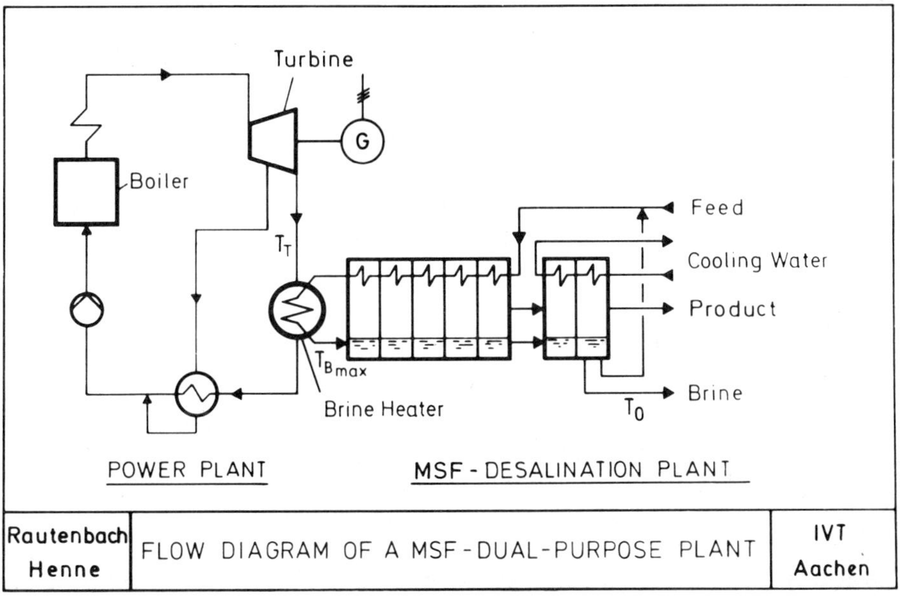

FIGURE 35.—Dual purpose plant for improved utilization of primary energy.

power plant, preferably a condensing type plant with a highly sophisticated flow scheme (high overall efficiency) (Figure 33), or by diesels with high overall efficiencies. The same holds true for RO.

Vapor compression Essentially, the vapor compressor must compensate the boiling point elevation and provide a certain thermal driving force. The power consumption of a vapor compressor is therefore a function of the temperature difference

$$\Delta T_{VC} = T_T - T_E \tag{35}$$

between the evaporation and the condensing side of the evaporator. Assuming polytropic compression, the specific power consumption is given by

$$\frac{N}{\dot{m}_D} = R_D\, T_E \frac{n}{n-1}\left[\left(\frac{p_T}{p_E}\right)^{\frac{n-1}{n}} - 1\right]\frac{1}{\eta_{VC}} \tag{36}$$

The pressures p_T and p_E are related by the vapor pressure curve of water

$$\frac{p_T}{p_E} = \exp\left[-\frac{\Delta\hat{H}_v}{R_D}\left(\frac{1}{T_T} - \frac{1}{T_E}\right)\right] = \exp\left[\frac{\Delta\hat{H}_v}{R_D}\,\frac{\Delta T_{VC}}{T_T(T_T - \Delta T_{VC})}\right] \tag{37}$$

where the temperature difference ΔT_{VC} is defined as

$$\Delta T_{VC} = \Delta T_{BE} + \Delta T_{HE} \tag{38}$$

This specific power consumption is related to the specific primary energy demand by the overall efficiency η_{pp} of a condensing power plant or the overall efficiency η_{Dies} of a diesel engine:

$$N = \eta_{pp}\, \dot{E}_F = \dot{E}_F \frac{T_{max} - T_{min}}{T_{max}}\, \eta_B\, \eta_T \tag{39}$$

or

$$N = \eta_{Dies}\, \dot{E}_F \tag{39a}$$

Combining equations 36, 37, and 39 leads to the specific primary energy demand of vapor compression, which is plotted in Figure 36 as a function of the top temperature T_T. Included in Figure 36 is the average primary energy demand of the auxiliary pumps (17). It should be noted that expanding

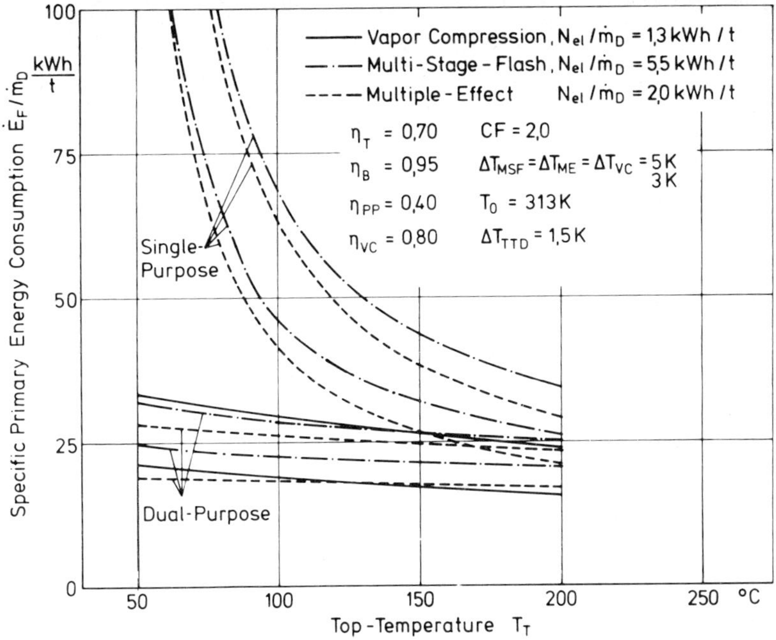

FIGURE 36.—Primary energy consumption of evaporation processes.

equation 37 into a series and employing only the first term in the series will be sufficient in most cases. This leads to the very simple equations

$$\frac{\dot{E}_F}{\dot{m}_D}\left(=\frac{E_F}{m_D}\right)=\frac{\Delta\hat{H}_v}{T_T}\,\Delta T_{VC}\,\frac{1}{\eta_{pp}\,\eta_{VC}} \tag{40}$$

or

$$\frac{\dot{E}_F}{\dot{m}_D}\left(=\frac{E_F}{m_D}\right)=\frac{\Delta\hat{H}_v}{T_T}\,\Delta T_{VC}\,\frac{1}{\eta_{Dies}\,\eta_{VC}}. \tag{40a}$$

Multiple-effect evaporation—single purpose Multiple-effect as well as multi-stage flash processes, operating between a top temperature T_T and a discharge temperature T_o, have to be assessed in a different way.

The specific heat consumption of a (once-through) ME-process (Figure 32) is approximated by

$$\frac{qA}{\dot{m}_D}=\frac{\Delta\hat{H}_v}{T_T-T_o}\,\Delta T_{ME}+\frac{\dot{m}_F}{\dot{m}_D}\,C_{PF}\,\Delta T_{PH} \tag{41}$$

with

$$\Delta T_{ME} = \Delta T_{BE} + \Delta T_{HE} \tag{42}$$

and

$$\Delta T_{PH} = \Delta T_{BE} + \Delta T_{TTD} \tag{43}$$

If the overall temperature differences of the last preheater ΔT_{PH} and of the effects of ΔT_{ME} are equal, which is at least a possible mode of operation, equation 41 becomes

$$\frac{qA}{\dot{m}_D} = \Delta T_{ME}\left(\frac{\Delta\hat{H}_v}{T_T - T_o} + \frac{C_F}{C_F - 1}\, C_{PF}\right) \tag{44}$$

with the concentration factor C_F

$$C_F = \frac{\dot{m}_F}{\dot{m}_F - \dot{m}_D} \tag{45}$$

For a single-purpose plant, the heat consumption qA is related to the consumption of primary energy $\dot{E}_F$ simply by the efficiency of the boiler η_B. In addition to this, the shaft power needed for the pumps has to be considered, which amounts to about

$$\frac{N_{el,ME}}{\dot{m}_D} = 2.0\ \frac{\text{KWh}}{\text{t}}$$

for once-through ME-plants (17).

Taking this into account, the specific primary energy demand of a single purpose ME-plant is given by

$$\frac{\dot{E}_F}{\dot{m}_D}\left(=\frac{E_F}{m_D}\right) = \left(\frac{\Delta\hat{H}_v}{T_T - T_o} + \frac{C_F}{C_F - 1}\, C_{PF}\right)\Delta T_{ME}\,\frac{1}{\eta_B} + \frac{N_{el,ME}}{\dot{m}_D \eta_{pp}} \tag{46}$$

Multiple-effect evaporation—dual purpose The consumption of primary energy of the same evaporation process incorporated in a dual-purpose plant can be calculated because essentially this primary energy consumption is related to the difference in power generation between a condensing turbine, operating between T_{max} and T_{min}, and a back pressure turbine, operating between T_{max} and T_T.

The above-mentioned difference in power generation is calculated by

$$N = qA\,\frac{T_T - T_{min}}{T_T}\,\eta_T . \tag{47}$$

Because of

$$N = \dot{E}_F \eta_{pp} = \dot{E}_F \frac{T_{max} - T_{min}}{T_{max}} \eta_B \eta_T \tag{48}$$

the specific primary energy consumption of a dual-purpose plant ME (once-through) process can be calculated according to equations 10, 13, & 14

$$\frac{\dot{E}_F}{\dot{m}_D} \left(= \frac{E_F}{\dot{m}_D} \right) = \frac{1}{\eta_{pp}} \Delta T_{ME} \left(\frac{\Delta \hat{H}_v}{T_T - T_o} + \frac{C_F}{C_F - 1} C_{PF} \right) \frac{T_T - T_{min}}{T_T} \eta_T + \frac{N_{el,ME}}{\dot{m}_D} \tag{49}$$

including the term $\frac{N_{el,ME}}{\dot{m}_D} = 2 \frac{\text{KWh}}{\text{t}}$ for the specific power consumption of the pumps (4).

Primary energy consumption of evaporation processes In Figure 36 (18) the specific primary energy demand of the various processes is plotted against the maximum process temperature T_T with ΔT as a parameter (VC: equation 40; single purpose ME: equation 46; dual purpose ME: equation 49).

The most important conclusions to be drawn from Figure 36 are:

1. VC and dual-purpose plant processes, especially ME, are comparable thermodynamically.
2. Any increase in maximum process temperature is much less important for VC and for distillation processes connected to a back pressure turbine than for single-purpose plants, producing heating steam in a fossil-fueled boiler.

Evaporation vs. membrane separation In reverse osmosis the transmembrane pressure difference, Δp, must exceed the difference of the osmotic pressures $\Delta \Pi$. Additionally, the pressure losses by friction within the (membrane) modules must be compensated:

$$\Delta p = \Delta p_{RO} + \Delta \Pi + \Delta p_{LOSS} \tag{50}$$

The osmotic pressures are proportional to the corresponding concentrations of solubles in the water (van't Hoff). Disregarding concentration polarization, the equation

$$\Delta \Pi = b_\pi (w_B - w_P) \sim b_\pi w_B \tag{51}$$

is valid, where w_B is the mass fraction of solubles in the discharged brine and w_P the mass fraction of solubles in the product.

If a turbine for energy recovery is included in the system, the power consumption of reverse osmosis will be

$$N = \frac{\dot{m}_F \, \Delta p}{\eta_P \, \rho_L} - \dot{m}_B \, \eta_T \, \frac{\Delta p - \Delta p_{LOSS}}{\rho_L}. \tag{52}$$

Considering the balance for the solubles

$$\dot{m}_F \left(1 - \frac{w_F}{w_B}\right) - \dot{m}_D = 0, \tag{53}$$

the mass balance

$$\dot{m}_F - \dot{m}_B - \dot{m}_D = 0 \tag{54}$$

and the concentration factor

$$C_F = \frac{w_B}{w_F}$$

results in the specific power consumption of reverse osmosis:

$$\frac{N}{\dot{m}_D} = \frac{1}{\rho} \left[\frac{\Delta p_{RO} + \Delta p_{LOSS} + b_\pi \, w_F \, C_F}{1 - \dfrac{1}{C_F}} \frac{1}{\eta_p} - \frac{\Delta p_{RO} + b_\pi \, w_F \, C_F}{C_F - 1} \eta_T \right] + \frac{N_{el,RO}}{\dot{m}_D} \tag{55}$$

By introducing the overall efficiency of a condensation power plant η_{pp}, the specific primary energy demand of reverse osmosis is given by

$$\frac{\dot{E}_F}{\dot{m}_D} \left(= \frac{E_F}{m_D} \right) = \frac{1}{\eta_{pp}} \cdot \left\{ \frac{1}{\rho} \left[\frac{\Delta p_{RO} + \Delta p_{LOSS} + b_\pi \, w_F \, C_F}{1 - \dfrac{1}{C_F}} \frac{1}{\eta_p} - \frac{\Delta p_{RO} + b_\pi \, w_F \, C_F}{C_F - 1} \cdot \eta_T \right] + \frac{N_{el,RO}}{\dot{m}_D} \right\} \tag{56}$$

Figure 37 displays the specific primary energy consumption of reverse osmosis for two different driving forces Δp_{RO}, assuming 65% and 85% for pump and turbine efficiency, respectively, and including an average figure of

$$\frac{N_{el,\,RO}}{\dot{m}_D} = 0.5 \text{ KWh/t}$$

shaft power for the pretreatment (17). We chose Δp_{RO} as a variable because constant Δp_{RO} plants of the same quality, i.e., the same specific membrane area, will be compared.

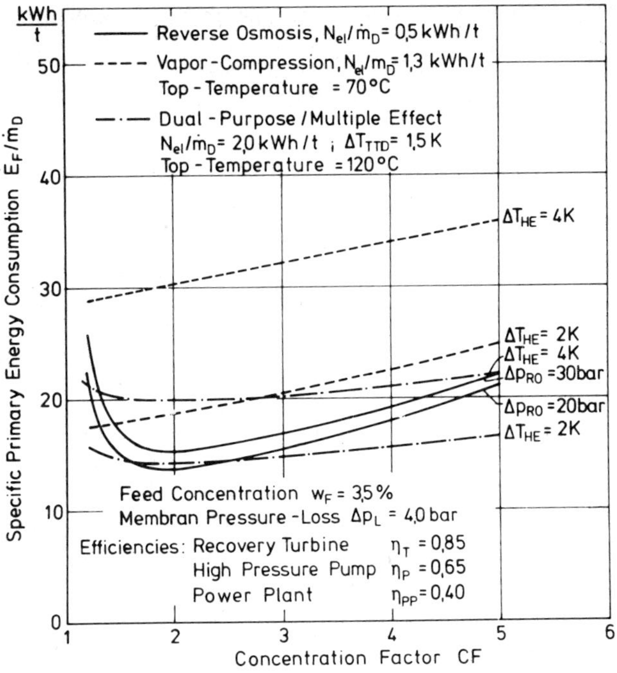

FIGURE 37.—Primary energy consumption of reverse osmosis and evaporation processes.

One might wonder why RO plants are displaying an optimum of primary energy consumption with regard to the concentration factor. This optimum is a direct consequence of the pump and turbine efficiencies being always smaller than unity. At low concentration factors, especially, large amounts of brine have to be pumped for a given fresh water production—leading to high specific friction losses.

For comparison, vapor compression as well as a dual-purpose ME process are included in Figure 37. Plotting the specific primary energy demand against the concentration factor clearly indicates the increase of primary energy demand of all evaporation processes resulting from the increase in boiling point elevation. Furthermore, it illustrates the fact that the increase in boiling point elevation in evaporation corresponds to the increase in osmotic pressure in RO.

Figure 37 indicates that RO is very economical with respect to its primary energy demand, but at the same time, evaporation processes in combination with a power plant are also competitive.

It is well understood that the detailed economic analysis of any process must consider the investment cost as well. Nonetheless, as energy consumption tends to increase in importance because of the rising cost of primary energy, process analysis with respect to energy consumption is becoming a prime managerial tool.

4.2. Example Hybrid Processes

The decision to utilize membrane processes commercially depends ultimately on the economics compared with proven separation processes such as distillation. Strictly speaking, membrane processes will not replace conventional processes; instead, a combination of membrane processes and conventional processes is likely to be the optimal solution, as illustrated by the following examples.

Removal of nitrate from well water As a consequence of intensive fertilizing, the nitrate concentration of certain well waters is well above the standards permitted by law. At places where blending with other water sources is impossible, this nitrate concentration has to be reduced. RO is a relatively simple and dependable process for the removal of nitrate. However, in cases where it is impossible to discharge the concentrate of the RO process into a river or canals, the economic treatment of this concentrate will be a major problem. Even for yields of about 90%, which can be achieved by RO in this case, the costs of evaporation of the concentrate are prohibitively high. Storing and recycling of the concentrate as a liquid fertilizer component proves to be no alternative (19).

A hybrid process combining essentially RO, electrodialysis (ED) and evaporation (Figure 38) seems to be the optimal solution with respect to total cost (20). Because of the very high specific cost of evaporation, the yield of the membrane process should be as high as possible. It is impossible, however, for a number of reasons (e.g., increase in osmotic pressure), to increase the yield of a RO stage much above 90%. Concentrations of about 15% total dissolved solids (TDS), corresponding to a total yield of the membrane processes of about 99.5%, can be achieved by electrodialysis.

Thus, by combining RO and ED, the amount of brine fed to the evaporator is reduced to only 0.5% of the RO feed (compared with 10% in the case of a simple combination RO–evaporation). For the concentration of brines up to 20% before evaporation, ED is already employed in Japan in the production of salt from seawater with processing costs considerably below the comparative costs of evaporation (21).

At the above-mentioned concentrations, the limit of solubility would be exceeded for the hardening components ($CaCO_3$, $Mg(OH)_2$, $CaSO_4$), causing membrane scaling. Therefore, these components have to be removed partially by selective cation-exchange. Regeneration of the ion exchange resin can be achieved by the softened brine, when a certain amount of NaCl is added to meet the stoichiometric balance. The added salt does not affect the energy consumption of the process, as the quantity of water to be evaporated remains the same (the prime energy consumption will be increased slightly as the boiling point elevation is increased—heating steam of higher temperature is

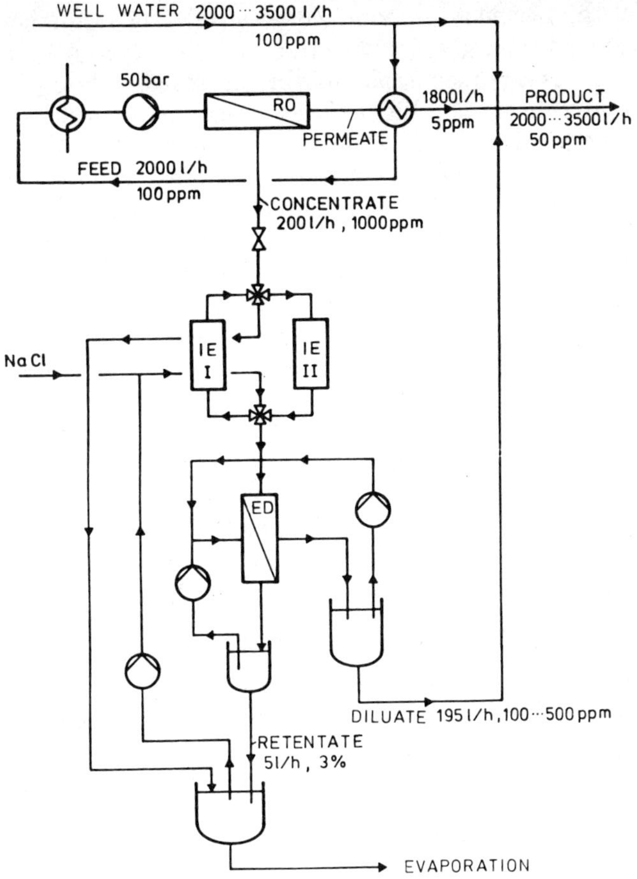

FIGURE 38.—Simplified flow diagram of the pilot plant with flow rates and respective nitrate concentrations.

necessary). The diluate of the ED is blended with the RO-permeate while the concentrate is fed to an agitated thin film evaporator where the crystallizing of the salts occurs. A crystal slurry will be the only waste, which can be easily disposed of.

Combination of ultrafiltration, cross-flow microfiltration and evaporation for recycling of detergents and process water in the automotive industry The final step in the production of automotive parts like pistons, axles, etc., is usually a rinse. Besides detergents, the spent wash water of such a process contains cutting oil and suspended matter and it is impossible to discharge it without treatment. Use of ultrafiltration (UF) allows one to fractionate the waste water into a concentrate containing the oil and the suspended particles and into a permeate containing practically no oil. The detergent concentration of the permeate is high, however, leading to further treatment costs if it is to be discharged. On the other hand, a recycling of the permeate for reuse in the rinse step seems possible but in this case, the oil concentration of the permeate can be even higher than in case of a discharge. Therefore, cross-flow microfiltration (MF) should be considered because of its higher average flux and—even more important—its higher yield of detergents.

With UF and MF, there is always the chance of formation of a gel layer. In many cases the formation of a gel layer is reversible; flux is reduced, but the gel layer can be removed by cleaning. The thickness of the gel layer can be controlled by hydrodynamics. However, for reasonable flow conditions the permeate flux is almost totally controlled by the gel layer—not by the membrane itself. Figure 39 clearly indicates that the permeate flux for steady-state conditions is nearly the same for UF membranes (A,B) and an MF membrane.

In general, periodic cleaning and a higher average flux have to be weighed against the time intervals necessary for cleaning. With respect to down-time for cleaning, MF is advantageous, as compared with UF. MF membranes, which are available in the form of tubes withstanding outer or inner pressure of several bar, can be easily cleaned by reversing the flux for a short time.

For this reason, the average flux of MF can be kept at a high level. Of course the oil rejection of UF is better than that of MF. According to pilot-plant studies (22), the optimal process is a combination of cross-flow MF and

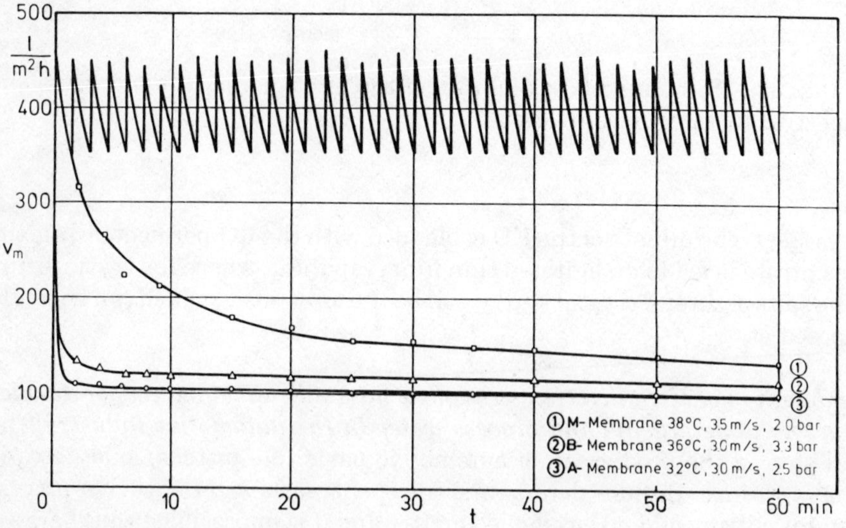

FIGURE 39.—Permeate flux of ultrafiltration membranes (A,B) and microfiltration membranes (M).

UF connected in series, with a (central) evaporation of the concentrate (Figure 40). The recovery of detergents is, on the average, 85%. It should be noted that the permeabilities of the individual components of a detergent differ—the upgrading of the recycled permeate must be based on the selectivity of the process.

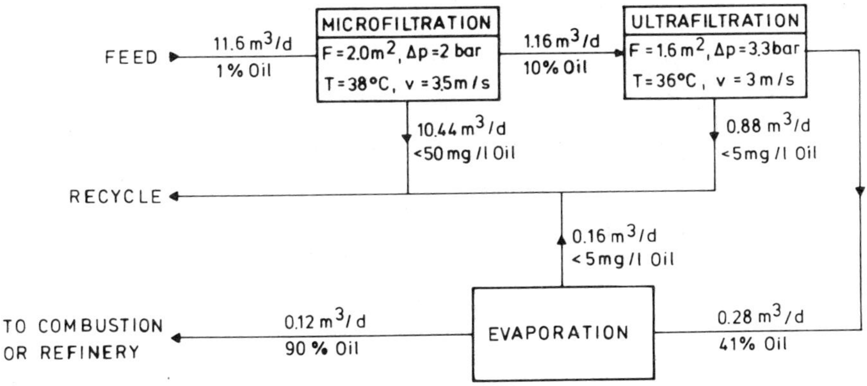

FIGURE 40.—Flow scheme—MF/UF with evaporation.

Gas permeation With a few exceptions, gas permeation on a technical scale employs membranes of the sorption–diffusion type. In this case, the flux of a permeating component is proportional to the difference of the partial pressures at both sides of the membrane

$$J_i = \mathcal{P}_i (p_1 x_{i1} - p_2 x_{i2}), \tag{57}$$

at least as long as the conditions of the gases are well above the critical point. On a commercial basis, gas separation started with the recovery of H_2 from the bleed of (high pressure) synthesis loops, employing in most cases a composite membrane "silicone/polysulfone" in the form of hollow fibers.

Asymmetric phase inversion membranes such as those employed in RO are difficult to prepare, as gas permeation is much more sensitive to micropores than RO, given the much higher diffusion coefficients of gases. For the same reason, the composite membrane (Figure 41) differs from RO composite membranes: in gas permeation, the top layer of the asymmetric support structure is responsible for the separation, whereas it is the sole duty of the coating to plug the micropores. Consequently, the material chosen for the coating (silicone rubber) has a high permeability but a low selectivity, and the membrane material (polysulfone) has a high selectivity (and a much lower) permeability (Figures 42, 43).

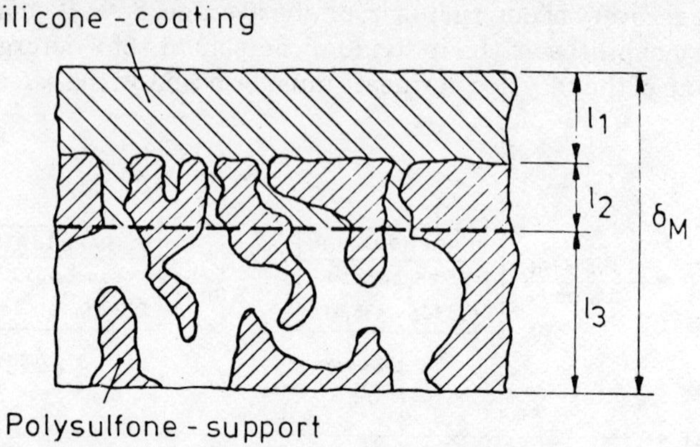

FIGURE 41.—Schematic structure of polysulfone/silicone composite membranes used for gas separation.

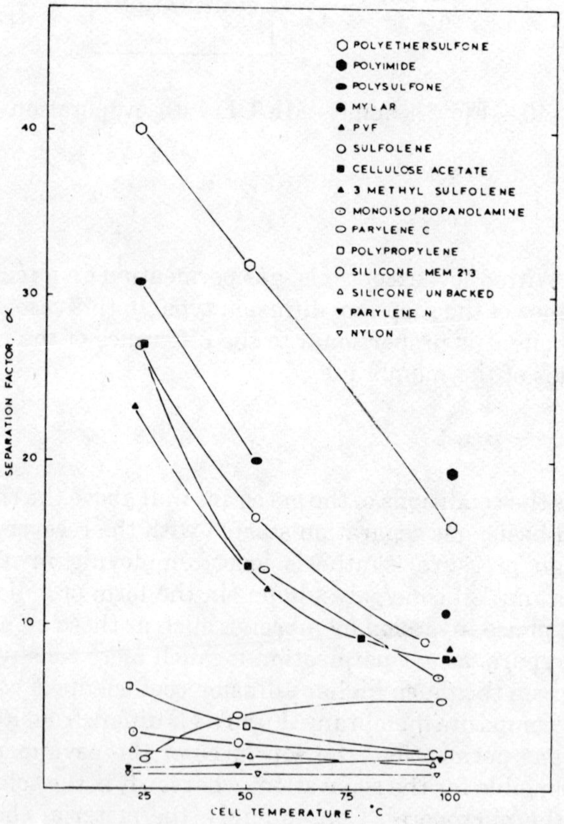

FIGURE 42.—Separation factor as a function of cell temperature. $\alpha = x_2 \cdot (1-x_1)/x_1(1-x_2)$. Reprinted from Ellig, et al.: 1980, J. Membrane Sci. 6, pp. 259–263.

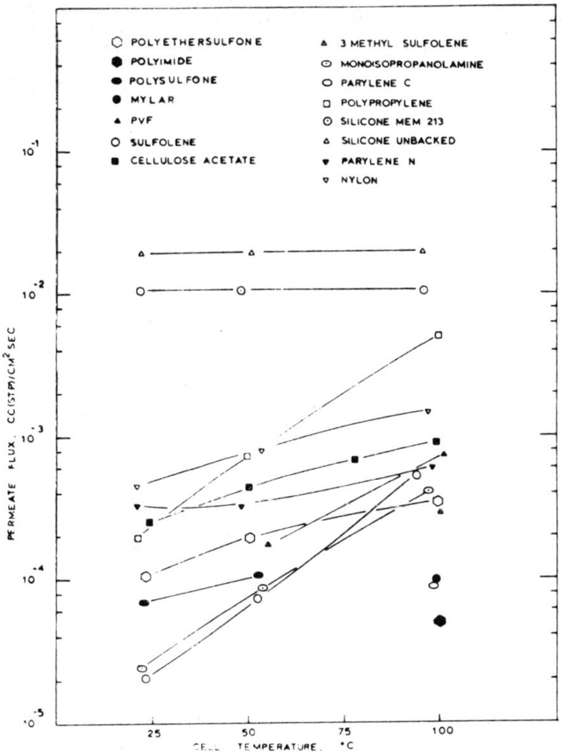

FIGURE 43.—Permeate flux as a function of cell temperature. Reprinted from Ellig, et al.: 1980, J. Membrane Sci. 6, pp. 259–263.

Figure 44 shows the flow sheet for the recovery of H_2 from the bleed of the synthesis loop of ammonia plants. In ammonia synthesis, a recycling of the unreacted components is mandatory because of its low equilibrium conversion rate. In all technical processes, however, a bleed is necessary: otherwise, the concentration of impurities would increase to an intolerable level. Normally, the bleed is utilized for heating purposes in the reforming stage.

Here, the bleed is fed to a conventional separation unit beginning with a scrubber for recovering ammonia. Behind the scrubber, the modules for the recovery of H_2 are arranged in a "one stage–two unit" form. The first unit, consisting of eight hollow fiber modules (total feed capacity, 3800 Nm^3/h), is operated with a transmembrane pressure difference of 60 bar, the permeate leaving at a pressure of about 70 bar. At this pressure, the permeate can be fed to the second stage of the synthesis feed-compressor.

The retentate of the first unit is fed to the second unit (and for this reason it cannot be considered a two-stage cascade), the permeate leaving at 25 bar as an additional feed to the first stage of the compressor. The retentate is utilized for heating purposes (23).

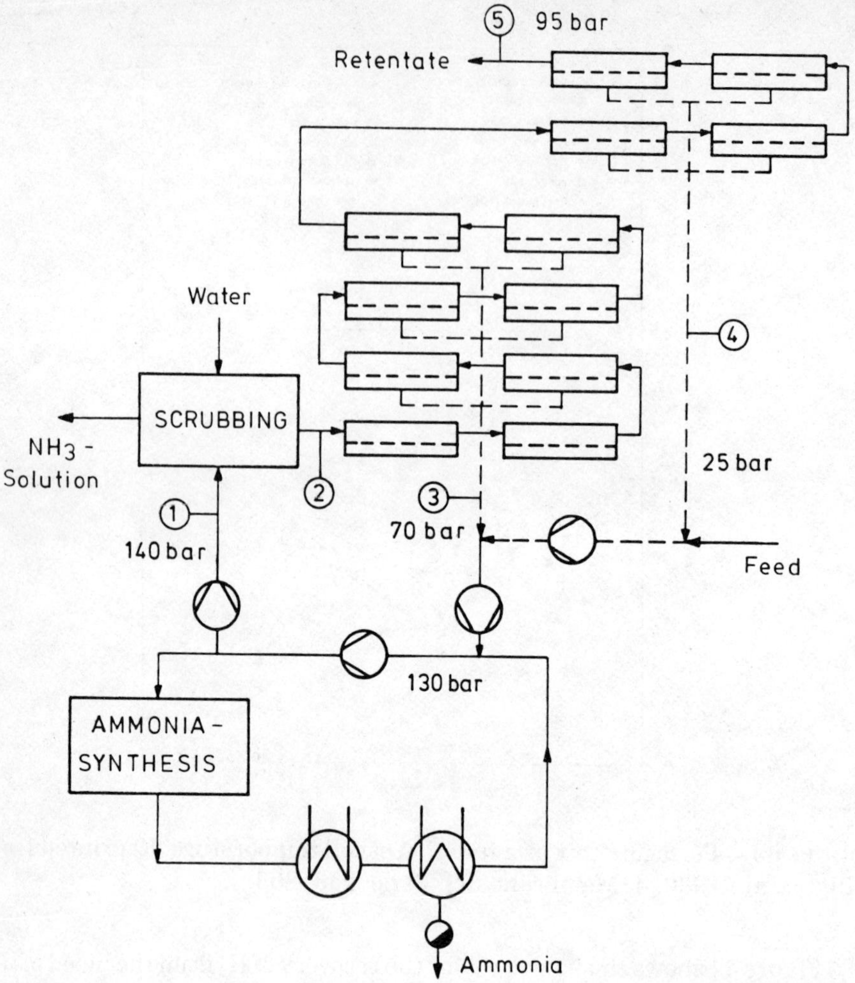

	1	2	3	4	5
H_2 [mol-%]	61,0	62,2	87,3	84,8	20,8
N_2 [mol-%]	20,5	20,9	7,1	8,4	43,8
CH_4 [mol-%]	10,5	10,7	3,6	4,3	22,5
Ar [mol-%]	6,0	6,1	2,0	2,5	12,9
NH_3 [mol-%]	2,0	< 0,02	< 0,02	< 0,02	< 0,02
Flow [m_N^3/h]	3868	3821	1591	820	1410

FIGURE 44.—Flow diagram for Monsanto-Louisiana demonstration plant.

Figure 45 shows another example of H_2 recovery, here from the tail gas of a high-pressure synthesis (UOP–Butamer process). In this case, the tail gas leaving the conventional fractionating system contains about 70% H_2. The membrane unit placed behind the fractionating system recovers about 90%, with a purity of >96% (25). This process seems to be interesting for two reasons:

1) spiral-wound modules are employed with dry asymmetric phase inversion (cellulose-acetate) membranes; and
2) the feed contains small amounts of HCl.

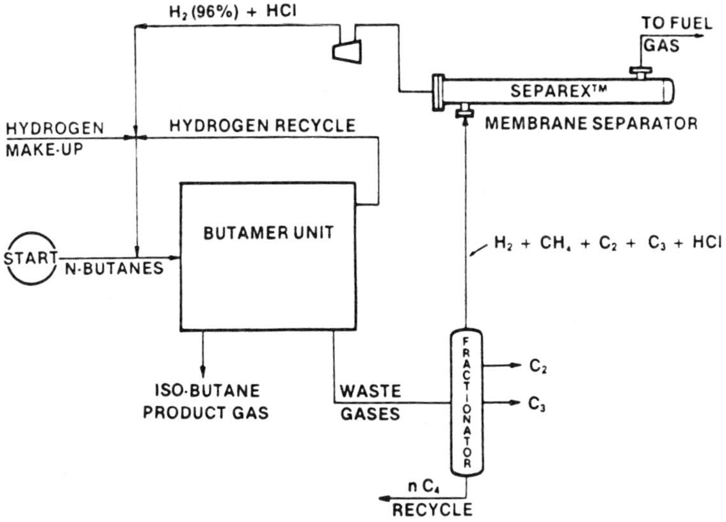

FIGURE 45.—Hydrogen recovery schematic.

Figure 46 shows a two-stage cascade for the purification and dehydration of sour gases, mainly separating CO_2 and CH_4 (24). Again, spiral-wound modules with asymmetric cellulose acetate membranes are employed. Note that in this case, as in all other cases discussed here, no compressors had to be installed. This is the main reason why these applications show excellent payback times!

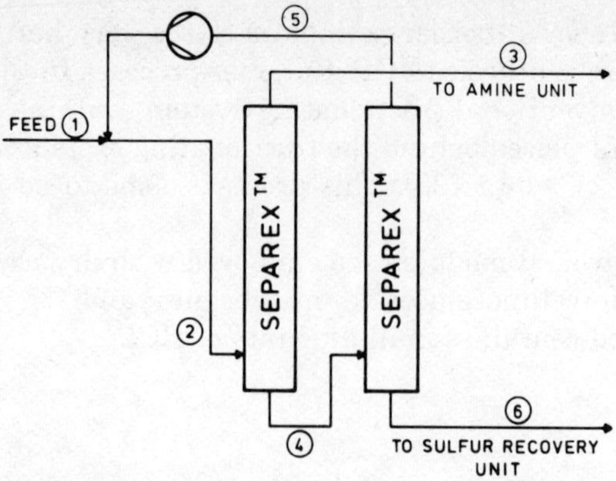

STREAM / COMPONENT	1	2	3	4	5	6	
CO_2	10	11,5	2	16	13,7	23	
CH_4	64,9	64,2	86,49	48,8	63,19	4,8	
C_2	5	4,2	8,6	2,4	3,1	0,1	
H_2S	20	20	3	32,7	20	71,6	ξ_{CH_4}
H_2O	0,1	0,1	0,01	0,1	0,01	0,5	98%
VOLUME FLUX [Nm^3/h]	17110	29830	12625	16872	12742	4130	
MOLAR FLUX [kmol/h]	763	1331	563	753	568	184	
PRESSURE [bar]	66,5	66,5	65,5	18,2	17,2	1,35	

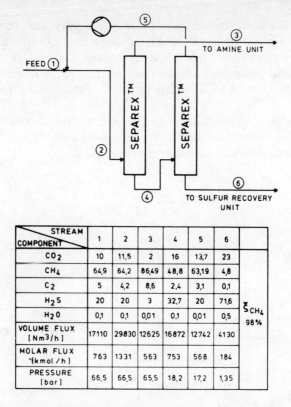

FIGURE 46.—Offshore natural gas dehydration.

Pervaporation Pervaporation differs from all other membrane processes because of the phase change of the permeate. Mass transport across the membrane is not achieved by raising the pressure at the feed side, as in RO and gas permeation, but by lowering the activity of the permeating component at the permeate side. Usually this is achieved by applying a vacuum of about 100 mbar at the permeate side. Phase change occurs because the partial pressure of the permeating components is lower than the saturation pressure.

In general, the selectivity of pervaporation is high (25), as demonstrated in Figure 47 for the system "benzene–cyclohexane, polyethylene membrane." The separation characteristics are shifted to much more favourable figures, as compared with the thermodynamic equilibrium curve; the azeotropic point is suppressed.

Pervaporation will always be a relatively expensive process, as compared with other membrane processes, for two reasons:

1) The process requires heat transfer surfaces because the heat of evaporation necessary for the phase change of the permeate has to be supplied by a stepwise heating of the liquid feed (and has to be rejected by condensation of the permeate).

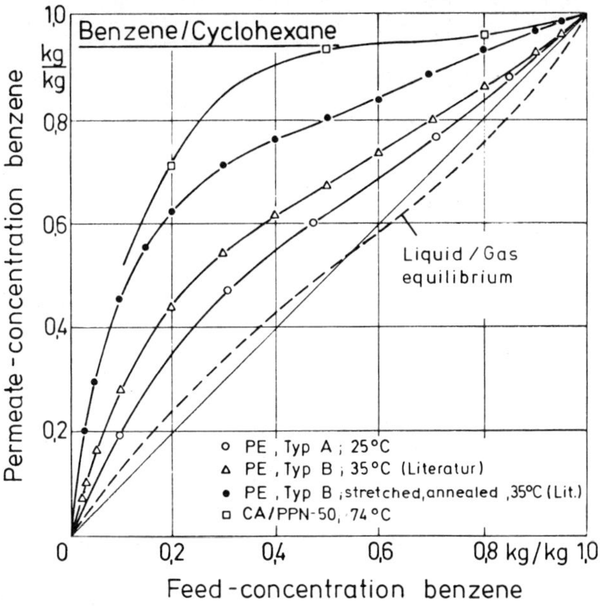

FIGURE 47.—Thermodynamic equilibrium vs. equilibrium curves for pervaporation, using various membrane materials.

2) The modules must be designed for a low pressure drop at the permeate side despite the increasing volume of the permeate caused by the phase change, as the principle of pervaporation is very sensitive to such pressure losses.

There are already applications of pervaporation on a technical scale. A hybrid process (Figure 48) for the production of pure alcohol, combining distillation and pervaporation, proved superior to the conventional approach (extractive distillation) with respect to specific energy consumption (1 kg/kg low pressure steam against 3 kg/kg). This is another example of the above-mentioned combination of membrane processes and conventional processes. According to Figure 48, the alcohol concentration is increased from about 80% to 99.5% by only one pervaporation stage (26). Note that for this range of concentrations, membranes have to be employed where water is the preferable permeating component.

Although the high separation potential of permeation has been demonstrated for a number of systems in laboratory experiments (27), pervaporation seems to be economical only in cases where high product purities are requested—and in combination with a conventional separation process. This statement is based at present only on a detailed analysis of the separation of benzene–cyclohexane and corroborated by the above mentioned production of

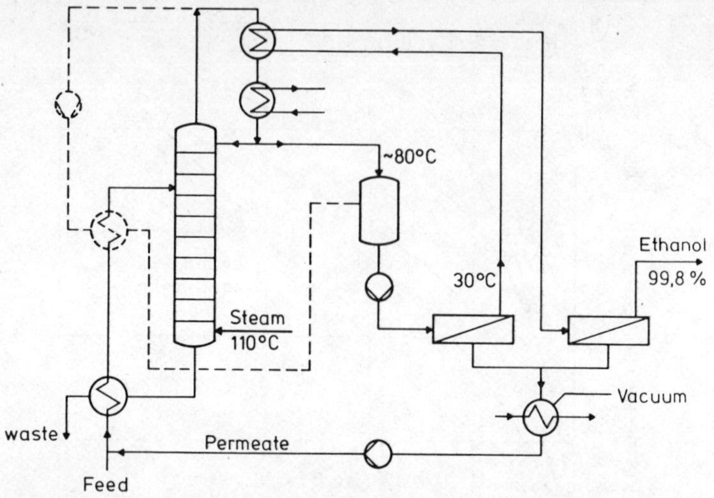

FIGURE 48.—Hybrid plant for production of pure alcohol.

	1 Stage	2. Stage	3. Stage
$\bar{J}$ [kg/m²h]	2,83	1,43	0,16
$\bar{S}_{ij}$ —	7,48	8,94	7,92
ϑ [°C]	70,9	70,5	70,8
F_M [m²]	621	4940	6998
n_{WAT} —	6	26	10
F_{WAT} [m²]	9,5	41	6,2

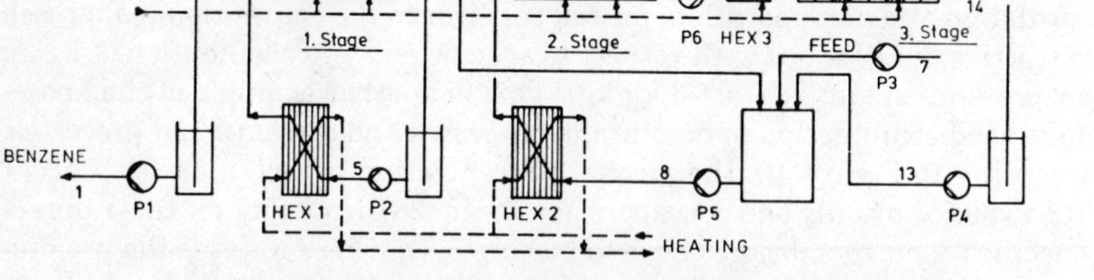

	1	2	3	4	5	6	7	8	9	10	11	12	13	14
$\dot{m}$ [kg/h]	1760	1760	7040	5280	7040	7040	3520	9905	9905	2865	2865	1105	1105	1760
w_i [Gew.-%]	98	98	88,2	85	88,2	88,2	50	67,5	67,5	16,4	16,4	39,3	39,3	2
ϑ [°C]	29,1	70,9	75	68,1	27,5	70,5	75	66,9	75	67,5	75	70,8	25,9	67,7

FIGURE 49.—Flow sheet of an optimized cascade for the separation of a 50 wt % benzene-cyclohexane mixture, product quality 98 wt % benzene, cyclohexane, respectively.

pure alcohol; but there can be no doubt that the results are valid for many other systems as well.

The fractionating of a 50% benzene–cyclohexane mixture into products of 98% purity is normally achieved by extractive distillation with furfural as a carrier. If pervaporation is to be employed instead, at best a three-stage cascade with the refluxes indicated in Figure 49 is needed (28). A comparison of total cost, based on a capacity of 3.5 t/h feed, reveals that pervaporation will not be competitive at all (Figure 50). Even if the costs of the installed membranes were zero, the conventional extractive distillation would be superior!

The situation is different, however, if product purities of about 99.5% are required. In this case, the energy consumption of the extractive distillation is very high because of the high reflux ratio (in the extractive distillation tower, the effectiveness of the carrier is rapidly decreasing in the section above the feed tray for the carrier and as a consequence, benzene will "enrich" in this section). As shown in Figure 51, a hybrid process, with a one-stage pervapora-

		spec. costs $\frac{DM}{t\,product}$	
		extractive distillation	pervaporation
Investment costs	sieve-tray column	6,33	-
	module	-	10,50
	evaporator/heat exchanger	4,38	1,02
	condensor	1,40	6,19
	pumps	0,46	0,74
	piping	1,89	2,76
Operating costs	membranes (DM 100/m²)	-	85,93
	steam	18,48	11,65
	cooling water	1,34	14,43
	power	0,10	0,04
	total costs	34,38	133,76

FIGURE 50.—Extractive distillation vs. pervaporation; cost analysis.

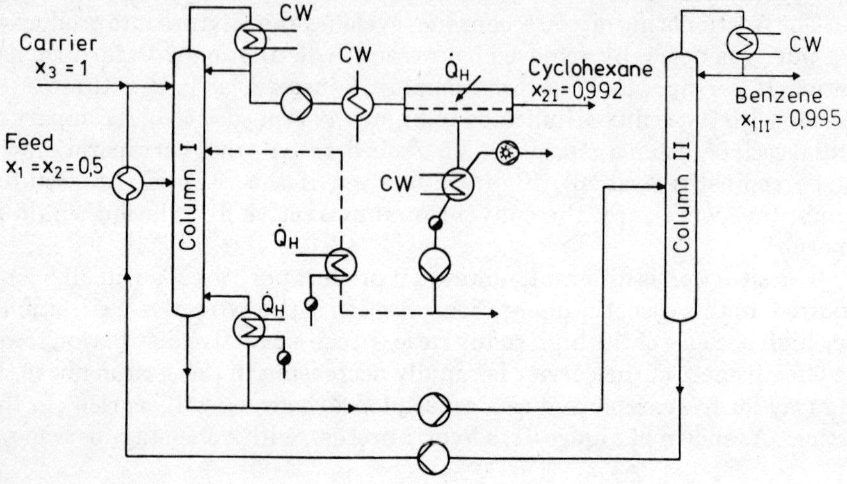

FIGURE 51.—Hybrid process consisting of extractive distillation and pervaporation.

tion for the separation of these low amounts of benzene, has a cost advantage of 20%, even for very conservative assumptions.

REFERENCES

1. Mitsubishi International GmbH. Hollosep for Seawater Desalination.
2. Gill, W.N., Bansal, B.: 1973, "Hollow fiber reverse osmosis systems analysis and design." AIChE J. 19, pp. 823–831.
3. Hwang, S.T., Thorman, J.M.: 1980, "The continuous membrane column." AIChE J. 26, pp. 558–566.
4. Kammermeyer, K.: 1976, "Technical gas permeation processes." Chemie Ing. Technik 48, pp. 672–675.
5. Weller, S., Steiner, W.A.: 1950, J. Appl. Phys. 21, p. 279.
6. Naylor, R.W., Backer, P.O.: 1955, AIChE J. 1, p. 95.
7. Blaisdell, C.T., Kammermeyer, K.: 1973, Chem. Eng. Sci. 28, p. 1249.
8. Breuer, M.E., Kammermeyer, K.: 1967, Separation Sci. 2, p. 319.
9. Hwang, S.T., Kammermeyer, K.: 1965, "Membranes in separations." In Techniques of Chemistry, vol. 7, New York, John Wiley & Sons.
10. Walawender, W.P., Stern, S.A.: 1972, "Analysis of membrane separation parameters. II. Countercurrent and co-current flow in a single permeation stage." Separation Sci. 7, pp. 553–584.
11. Thorman, J.M.: 1979, "Engineering aspects of capillary gas permeators, and the continuous membrane column." Dissertation, University of Iowa.
12. Hwang, S.T., Ghalchi, S.: 1982, "Methane separation by a continuous membrane column." J. Membrane Sci. 11, pp. 187–198.

13. Michels, H., Schulz, G., Werner, U.: 1981, "Schaltungen zur Gasrektifikation mit Membranen." Chem.-Ing.-Tech. 53, pp. 206–207 (Synopse 887).
14. Rautenbach, R., Seide, A.: 1978, "Technical problems and economics of hydrate processes." Proceedings of the 6th Symp. on Fresh Water from the Sea 4, pp. 43–51.
15. Grassman, P.: 1970, Grundlagen der Verfahrenstechnik. Sauerländer, Aarau und Frankfurt.
16. Riekert, L.: 1980, Ber. Bunsengesellschaft Phys. Chemie 84, pp. 964–973.
17. Finnegan, D.R., Wagner, W.M.: 1981, "Process selection guide to seawater desalination." Proceedings of the International Congress on Desalination and Water Re-Use (Bahrain), vol. 1. pp. 29–48.
18. Rautenbach, R., Arzt, B.: 1983, "Assessment of present and future desalination processes with respect to their primary energy demand." Arabian J. Sci. and Engin.
19. Rautenbach, R., Henne, K.H.: 1980, "Denitration von Grundwasser mittels Umkehrosmose." BMFT Forschungsbericht T, pp. 90–155.
20. Kawahara, T., Suzuki, K.: 1981, "Utilization of the waste concentrated seawater in the desalination plants." Water, the Essence of Life, (Bahrain) vol. 1. pp. 499–507.
21. Peters, T., van Opbergen, G., et al.: 1983, "Reduction of nitrate concentration in drinking water by a hybrid process with zero discharge based on reverse osmosis." First World Congress on Desalination and Water Re-Use (Florence, Italy; May 23–27).
22. Paul, H., Schock, G., Rautenbach, R.: 1983, "Ölhaltige Industrieabwässer mit Mikrofiltration oder Ultrafiltration aubereiten." Maschinenmarkt 89, pp. 1785–1788.
23. Monsanto Communications. 1982. Symposium in Scheveningen, The Netherlands, March 24, 1982.
24. Schell, W.J., Houston, C.D.: 1982, "Use of spiral-wound gas permeators for purification and recovery."
25. R. Rautenbach, R. Albrecht: European Workshop on Pervaporation, Nancy (F) 21.–22.9.1982, Journal of Membrane Science (in print)
26. H. Brüschke, W. Schneider, G.F. Tusel: European Workshop on Pervaporation, Nancy (F) 21.–22.9.1982
27. P. Aptel: European Workshop on Pervaporation, Nancy (F) 21.–22.9.1982
28. R. Albrecht: Ph.D.–Thesis, RWTH Aachen, W.-Germany, 1983

SYMBOLS

A	membrane, area in bundle
A_f	membrane area per fiber
b	channel half width
b_π	osmotic (van't Hoff) coefficient (~RT)
C_F	concentration factor
C_P	specific heat capacity
d_i (d_a)	fiber inner (outer) diameter
D_a	bundle diameter
D_i	diameter of feed distribution tube
$\dot{E}_F$	primary energy consumption
F	molar flow rate of feed
Gr	Grashof number
H	coefficient defined in equation 19
k	mass transfer coefficient
K_o	Proportionality coefficient in equation 23
l_s	seal length
L	length of sheet, tubular or hollow fiber membrane
$\dot{m}$	mass transfer rate

n	polytropic exponent
$\dot{n}_i$	transmembrane molar flow rate of species i
n_f	number of fibers per cross-sectional area of bundle
N	number of stages
N	power consumption; shaft power
$\dot{N}$	total molar flow rate per fiber
Δp_{LOSS}	pressure losses due to friction
Pe_m	Peclet number based on permeation through membrane, $v_m\ d/D$
$\mathcal{P}_i$	permeability coefficient
P_n	molar flow rate of product from the nth stage
r	ratio of retentate to permeate total pressures
Ra	Rayleigh number
R_D	gas constant
R_n	molar flow rate of retentate from the nth stage
Sh	Sherwood number, $k\ L/D$, dimensionless
ΔT_{VC}	temperature difference of vapor compression
ΔT_{BE}	boiling point elevation
T_E	evaporation temperature
T_{max}	top temperature of the Rankine cycle, T_{max} = 798 K
T_{min}	minimal temperature of the Rankine cycle, T_{min} = 313 K
T_o	brine discharge temperature
ΔT_o	gross temperature difference
ΔT_{TTD}	terminal temperature difference
x_i	mole fraction of species i (in feed and retentate)
y_i	mole fraction of species i (in permeate)
γ	ratio of retentate to product molar flow rates, R/P
η_B	efficiency of boiler
η_{Dies}	efficiency of diesel engine
η_{pp}	overall efficiency of a condensation power plant
η_T	efficiency of turbine
η_{VC}	efficiency of vapor compressor
θ	cut-rate, split-factor, or stage-cut
λ	permeate to retentate separation factor, equation 25d

Superscript

local quantity

Subscripts

A	key component
B	brine, secondary component
D	distillate
E	evaporation side
F	feed
L	value at x = L

o	discharge
P	product or permeate
R	retentate
T	top
BH	brine heater
el	electrical
HE	heat exchanger
MSF	multi-stage flash
ME	multiple-effect
PH	preheater
RO	reverse osmosis
VC	vapor compression
α	inlet
ω	outlet
0	value at $x = 0$
1	bulk value in feed or retentate solution
2	membrane–solution interface
3	value in permeate solution or downstream side of dense skin

CARRIER-MEDIATED TRANSPORT

Jerome S. Schultz

Chairman
Department of Chemical Engineering
University of Michigan
Ann Arbor, Michigan 48109

This chapter describes the concept and potential of carrier-mediated transport, in which a specific carrier species is added to the membrane phase to selectively bind specific solutes present in the bathing solutions. The carrier provides an additional mechanism of transport through the membrane and, in so doing, generally accelerates the transport of specific permeants. Several modes are discussed—accelerated diffusion, coupled transport, and exchange transport—to illustrate how carrier-mediated transport membranes can be operated to achieve very high separations and purifications. Important factors to be considered in membrane design, such as reaction kinetics and diffusion rates, are reviewed. The creation of asymmetric transport properties by means of imposed spatial variations in membrane structure is demonstrated. The chapter concludes with an overview of a number of approaches that couple external energy sources (electrical and light) to these membranes and thereby allow modulation of transport rates.

P. M. Bungay et al. (eds.), Synthetic Membranes: Science, Engineering and Applications, 523–566.

1. INTRODUCTION

By carrier-mediated membrane transport we mean that a carrier is confined to the membrane phase and interacts with solutes on both sides of the membrane phase. The carrier provides an additional mechanism of transport through the membrane and in general, accelerates the transport of specific solutes across the membrane phase.

The first system described that could be easily studied experimentally was that of Scholander (1), who showed that oxygen diffusion through a filter paper membrane containing a hemoglobin solution was greatly accelerated. Oxygen diffusion through a membrane that had been soaked in methemoglobin, which has no oxygen-carrying capacity, showed the same lower response as would be expected for simple diffusion of oxygen through water. However, in the presence of hemoglobin an additional amount of oxygen is carried through the membrane by means of the diffusion of a hemoglobin–oxygen complex within the membrane phase (2). This increase in oxygen transport due to the action of hemoglobin as a carrier is termed the "facilitation effect." In Figure 1, it can be seen that the facilitation effect reaches a constant value

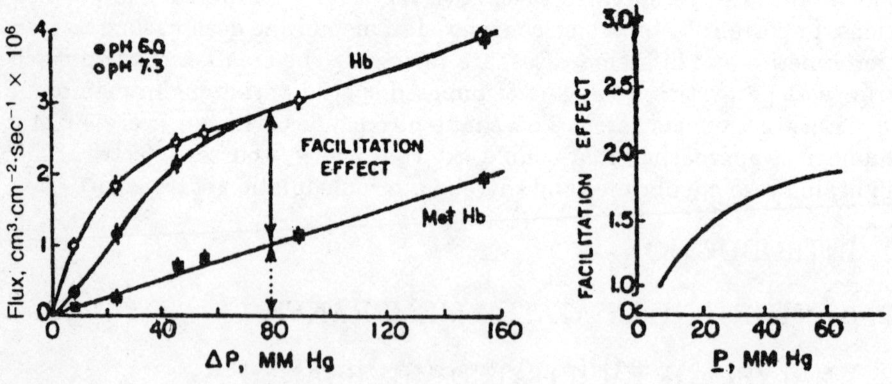

FIGURE 1.—Typical nonlinear flux patterns for carrier-mediated facilitated diffusion: a. Oxygen diffusion through hemoglobin solution, measured by O_2^{18} (2); b. Effect of hemoglobin saturation on facilitated oxygen flux.

at higher partial pressures of oxygen on the upstream side. This is due to the saturation of the carrier hemoglobin with oxygen on the downstream side (3). If one plots the incremental increase in oxygen diffusion through the membrane, that is, the difference between the upper and lower flux line in Figure 1A, shown in Figure 1B, it can be seen that the facilitated oxygen flux saturates at approximately the pressure at which hemoglobin is saturated with oxygen.

The basic mechanism for this effect is the reversible reaction of a permeable solute, which can enter the membrane phase from either side of the membrane, with a carrier, which is confined to the membrane phase. This is shown schematically in Figure 2, where the symbol B represents the "carrier molecule," A is the "permeant," and AB is the "complex of carrier and permeant." An important characteristic of carrier-mediated diffusion through membranes is that the reaction must be reversible. Otherwise, either the carrier or the permeant will be consumed in the reaction, which would limit the operational life of the membrane in this mode of operation.

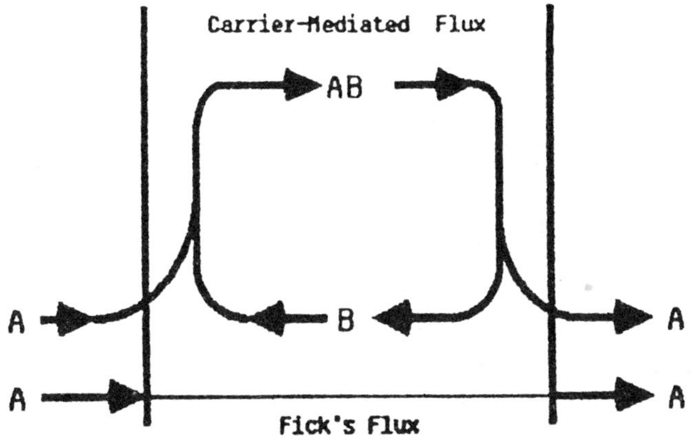

FIGURE 2.—Schematic representation of mechanism of carrier-mediated transport.

The selectivity of carrier-mediated diffusion membranes is based on the fact that the permeant, in reacting with the carrier, has a higher flux across the membrane than other substances, which cannot react with the carrier. For example, in Figure 3, we show some data obtained on the diffusion of oxygen through a solution of cobaltohistidine, a chelate that reacts reversibly with oxygen to form a complex (4). Therefore, in the presence of cobaltohistidine, oxygen diffuses through the membrane in two forms, as molecular oxygen and as the complex with cobaltohistidine.

The lowest line in Figure 3 shows the Fickean flux of nitrogen through a solution of cobaltohistidine. The middle line shows the Fickean flux of oxygen through water, i.e., a water-soaked membrane. The ratio of these two (the

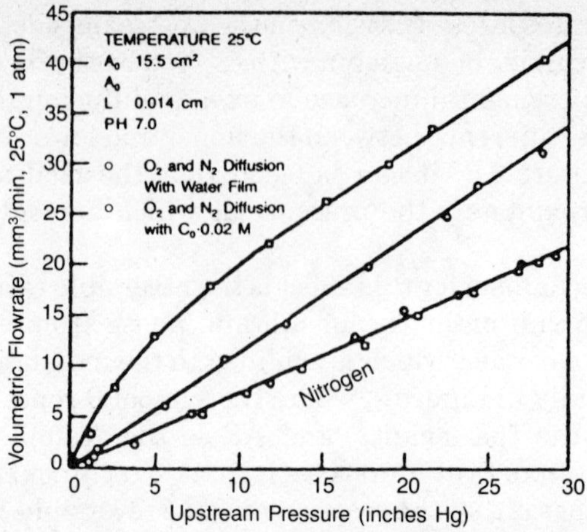

FIGURE 3.—O_2 and N_2 fluxes in cobaltohistidine solutions at 25°C (4).

slopes of these two lines) gives the separation factor of oxygen to nitrogen in the absence of a carrier. For the same partial pressure driving forces, the flux of oxygen is roughly 50% more than nitrogen through a water film. This is due to differences in the solubilities and diffusivities of oxygen and nitrogen in water.

The upper curve shows the flux of oxygen in the presence of the cobaltohistidine carrier. Because of the nonlinear characteristics of this flux, due to the presence of the carrier, the ratio of oxygen flux to nitrogen flux is approximately 9 to 1 at the lower oxygen partial pressures. Of course, at higher oxygen partial pressures the ratio of oxygen flux to nitrogen flux decreases because the carrier becomes saturated. Consequently, a higher degree of purification or separation of oxygen from nitrogen can be obtained at lower partial pressures. Thus, one can see that the mechanism for separating different solutes by a carrier-mediated transport membrane is based on their relative affinities for the carrier. In the particular case of oxygen and nitrogen in the cobaltohistidine system, nitrogen has no affinity for the carrier and passes through the membrane simply by diffusion as molecular nitrogen.

Some appreciation for the potential in selectivity of carrier-mediated membranes that can be obtained for particular solutes is exemplified by the behavior of some biological systems. Table 1 shows binding constants for a variety of sugars with the carrier in red blood cell membranes that is responsible for the transport of sugars. Binding constants of the carrier are very specific for certain structural characteristics and vary over a range of roughly 1000. Thus, the protein carrier that resides in the membrane of red blood cells can distinguish between the two isomeric forms of glucose, D-glucose and L-glucose. At low concentrations the D-glucose will have a much higher affinity

TABLE 1. Carrier–complex dissociation constants for sugar transport into red blood cells

Sugar	Affinity Constant (mol/L)
L-Glucose	>3
L-Galactose	>3
L-Xylose	>3
L-Fucose	2.5
D-Fucose	0.25
D-Xylose	0.06
D-Galactose	0.04
D-Glucose	0.007

(From LeFevre (25)).

to the carrier than L-glucose. Under the circumstances of very low concentrations, D-glucose will be transported roughly 1000 times faster than L-glucose. Thus, from a mixture of D- and L-glucose the cell will accept almost pure D-glucose by this mechanism.

2. SYNTHETIC FACILITATED MEMBRANES

Given that the basic mechanism for the effectiveness of carrier-mediated diffusion is the binding of permeants to carriers that are contained within the membrane phase, a number of different examples of carrier-mediated diffusion can easily be ascertained. Some of these mechanisms are shown in Figure 4. The first example is that mentioned earlier of a simple "carrier-mediated" diffusion where one solute interacts with a single carrier and the carrier complex provides an additional path for the transport of the permeant. In this example, oxygen is being transported by hemoglobin.

Another category is a "coupled mediated" transport. In this case, the permeant, an ionic species, cannot enter the membrane because the membrane consists of a hydrophobic organic solvent, which rejects ions because they have very low solubility. But, if the membrane contains a carrier that can bind to both the positive ion and negative counterion by the formation of an ion pair complex, then an additional flux of the ions through the membrane is obtained by means of this neutral complex. In the case shown, the carrier has a selective affinity for the cation but also can accommodate the counterion, the anion, to provide a neutral complex. This complex can diffuse freely within the organic phase of the membrane, providing an additional path for the transport of the metallic ion through the membrane. Usually, there will be virtually no transmembrane flux of the ions in the absence of the carrier because of the limited solubility of naked ion pairs.

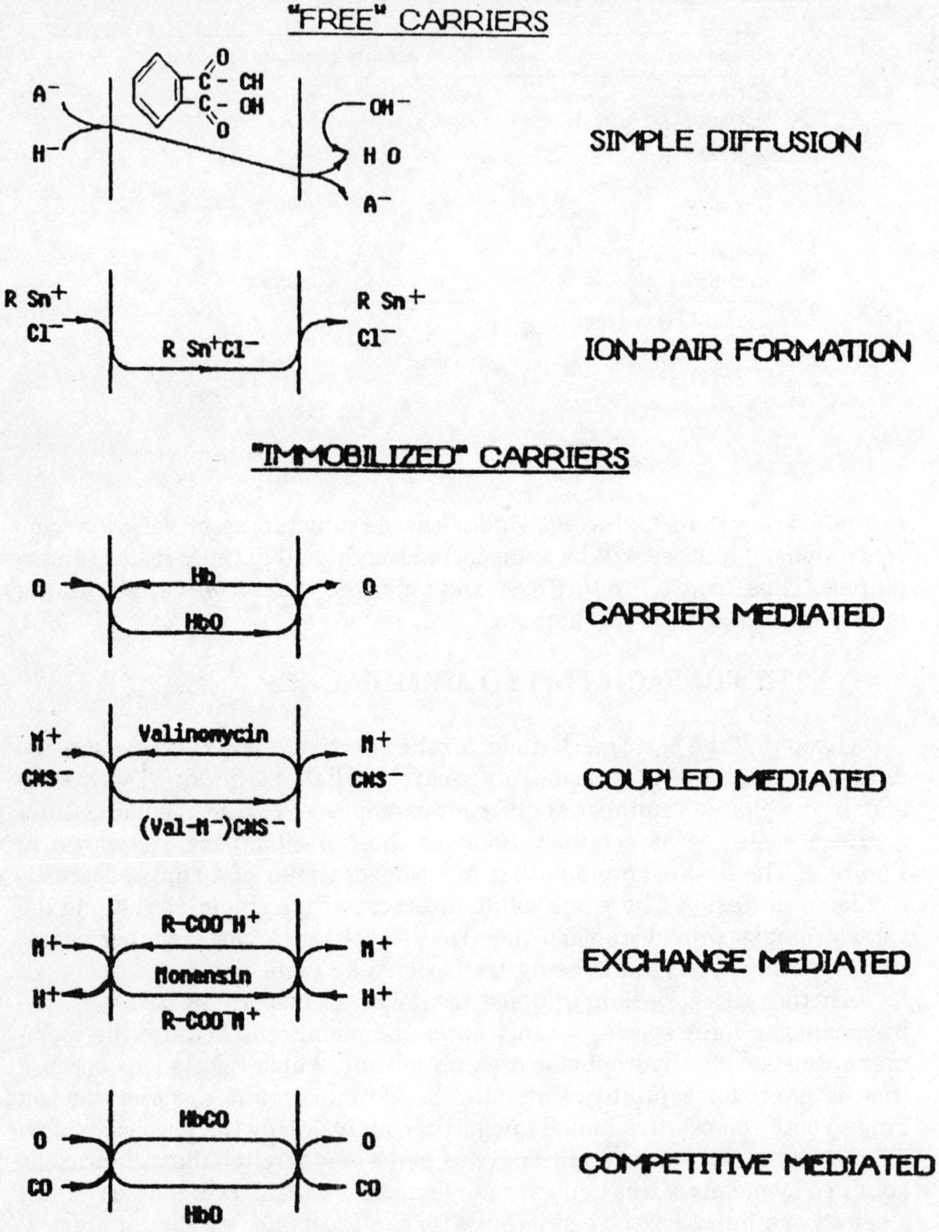

FIGURE 4.—Types of carrier-mediated diffusion.

Another type of transport can be termed "exchange-mediated diffusion." In this case the carrier in the organic membrane phase is a dissociable ion pair as is found in ion exchange resins. But, in the organic phase, the carrier cannot have a free charge; it can only exchange one charged species for another of similar charge at the membrane solution interface. Here the carrier is mobile within the phase, and thus, based on the relative amounts of the two competing cations for the carrier in the membrane, one can achieve the transport of a metallic ion in one direction while another ion, for example, the less valuable hydrogen ion, diffuses in the other direction.

Another type of mechanism that can be used is called "competitive mediated diffusion." Here, two permeants can react independently and compete for the carrier molecule. In the example shown, both oxygen and carbon monoxide can react with hemoglobin. In this case, one can use one gas to drive the other gas across the membrane. If there were a low partial pressure of oxygen in the gas sweeping past the left side of the membrane and a high partial pressure of carbon monoxide on the right side of the membrane, the hemoglobin on the left side of the membrane would pick up some of the oxygen and diffuse across the membrane. On the right side of the membrane the higher concentration of carbon monoxide would effectively displace the oxygen from the hemoglobin and thereby release the oxygen into the right hand chamber. Not only can this type of membrane cause the removal of oxygen from the left side and transport it to the right side, but under the proper circumstances, the concentration of oxygen on the right side of the membrane can be forced to be much higher than the concentration on the left side of the membrane.

Also shown in Figure 4 are some examples of what might be termed "free carrier membranes." In these cases, there are no specific chemicals or carriers that are confined to the membrane phase alone. Actually, reactants, usually ionic in the external phases, can form complexes that have a high solubility in the membrane phase, and thus, the reactants or permeants can transport through the organic membrane phase in the form of a complex. The magnitude of the facilitated diffusion of these permeants through the membrane would depend upon the solubility of the ion pair complexes. However, in the sense of this discussion these membranes would not be considered as carrier-mediated membranes.

The distinction between carrier-mediated diffusion, free diffusion and reactive membranes is shown more clearly in Figure 5. Some simple rules have been developed which can be used to determine the classification of a carrier-mediated transport membrane. Basically, if the invariant "I" for a membrane transport process is greater than zero, then the membrane would be considered a carrier-mediated transport membrane. If the invariant for the membrane is equal to zero then it is a free carrier membrane, and if the invariant is less than zero, then the membrane is really a reactor and not a carrier transport membrane. These different cases are shown in diagrammatic form in Figure 5.

	F	R	I
↑A + B ⇌ C	2	1	1
↑A + ↑B ⇌ C	1	1	0
↑A + ↑B ⇌ ↑C	0	1	-1

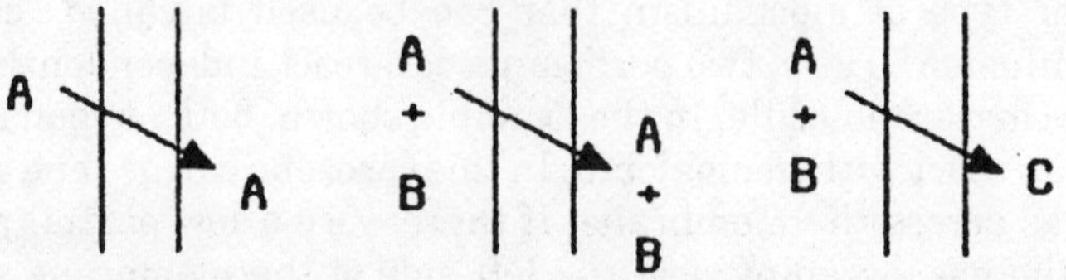

F = Number of non-permeant species

R = Number of reactions

I = Invariant = F − R

FIGURE 5.—Invariant and membrane properties.

3. ANALYSIS OF FACILITATED MEMBRANES

3.1. Mathematical Model

In order to describe the flux of a permeant through a carrier-mediated membrane quantitatively, one has to account for the coupled reaction-diffusion processes that are occurring within the membrane. These processes are shown schematically in Figure 6. First, there is a possibility of an interfacial reaction of the permeant with the carrier at the interface between the membrane and the solution phase. Interfacial reactions are particularly important if the permeant is not soluble in the membrane phase. Secondly, if the permeant can dissolve in the membrane phase it will react with the carrier to form a complex. The complex moves freely throughout the membrane phase as do the free carrier and the free permeant. For most of the discussion in this paper, we will assume that the permeant is soluble in the membrane so that the process can be modelled as a homogeneous reaction–diffusion problem. Also indicated in Figure 6 is the possibility that there may be some heterogeneity in the membrane. The heterogeneity may occur because of a number

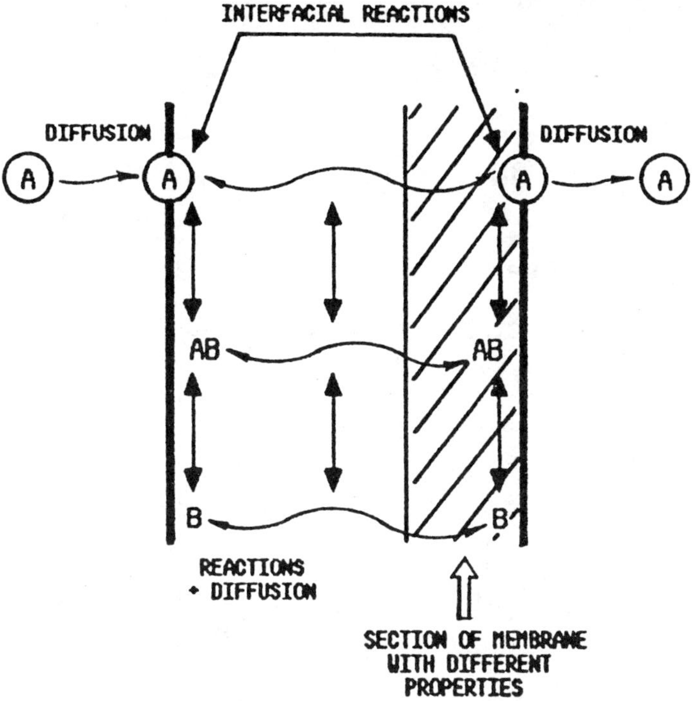

FIGURE 6.—Schematic for coupled reaction–diffusion processes occurring in a carrier-mediated membrane.

of factors such as variation in the membrane structure across the membrane, or various differences in chemical reaction rates in regions of the membrane. We will discuss these possibilities below.

The equations for the simplest model of a permeant A reacting with a carrier B within the membrane phase to form a complex AB are shown in Figure 7. Only neutral reactants are considered to obviate consideration of electrical effects, and further, we will only treat the steady state situation (in some situations this assumption may be inappropriate because of thick membranes or slow reaction rates). With these restrictions the mathematical problem is to solve a coupled set of second order differential equations, given two point boundary conditions on permeant concentrations at the two interfaces and also the requirement that the carrier in either the free or complex form cannot leave the membrane phase, i.e., the total amount of carrier in the system is conserved.

DIFFUSION-REACTION EQUATIONS IN MEMBRANES

$$A + B \rightleftharpoons AB$$

$$D_A \frac{d^2 c_A}{dx^2} = r_A = k_f c_A c_B - k_r c_{AB}$$

$$\text{B.C.} \quad C_{A0}, \; C_{AL}, \; \left.\frac{dC_B}{dx}\right|_{0,L} = 0, \; \left.\frac{dC_{AB}}{dx}\right|_{0,L} = 0$$

$$r_A = r_B = -r_{AB}, \quad C_B + C_{AB} = C_T$$

$$J_A = \frac{D_A(C_{A0} - C_{AL})}{L} + \frac{D_{AB}(C_{AB0} - C_{ABL})}{L}$$

$$\text{Equilibrium}: \quad C_{AB} = C_T\left[\frac{C_A}{K + C_A}\right]$$

$$\text{Non-equilibrium}: \quad \frac{d^2\hat{C}_A}{dy^2} = \frac{C_T k_f L^2}{D_A}\left[\hat{C}_A\hat{C}_B - \frac{\hat{C}_{AB}}{K C_T}\right]$$

FIGURE 7.—Diffusion–reaction equations in membranes.

For this simple case, the flux of the permeant A through the membrane is shown to be given by the sum of two terms. The first term represents the Fick's Law contribution for free permeant diffusion through the membrane (D_A is the diffusion coefficient and L is the thickness of the membrane). The second term represents the facilitated portion of permeant transport. Here the transport rate is proportional to the difference in complex concentration between the two surfaces of the membrane (an assumption is made that the diffusivity of the carrier is the same whether the permeant is complexed or not). Thus, to predict the transport rate of the permeant through the membrane, one needs to know the concentration of the permeant–carrier complex on each side of the membrane. Under some special conditions, which we will call here "equilibrium conditions," these concentrations can be estimated by assuming that the carrier is in equilibrium with the surface concentrations of the permeant. Under these circumstances the equilibrium concentration of the carrier–complex is given by the simple algebraic relation as shown.

More generally, this equilibrium relation between the carrier and permeant will not be valid and then the complete set of simultaneous differential equations must be solved. Because of the non-linear nature of these equations, no general algebraic or closed-form solution exists. One then must resort to either approximate analytical solutions of the differential equations or exact numerical solutions.

If the differential equation is put into a dimensionless form, as shown at the bottom of Figure 7, one can identify a dimensionless parameter (the Damköhler number), which is a key to the behavior of the membrane. This parameter, which in this simple case is defined by the forward reaction rate constant, $k_f\,C_T$, multiplied by the square of the membrane thickness, L, divided by the diffusion coefficient D_A, is a dimensionless characterization of the relative degree of diffusion or reaction control of the transport process.

The normalized performance of the membrane as a function of the Damköhler number is shown in Figure 8. Here the facilitated flux divided by the Fick's flux is plotted against the Damköhler number. At high values of the Damköhler number the membrane can be regarded to be in the equilibrium regime. In this regime the amount of carrier complexed with permeant at each side of the membrane can be estimated by simple equilibrium considerations as mentioned above. In the other extreme, when the Damköhler number

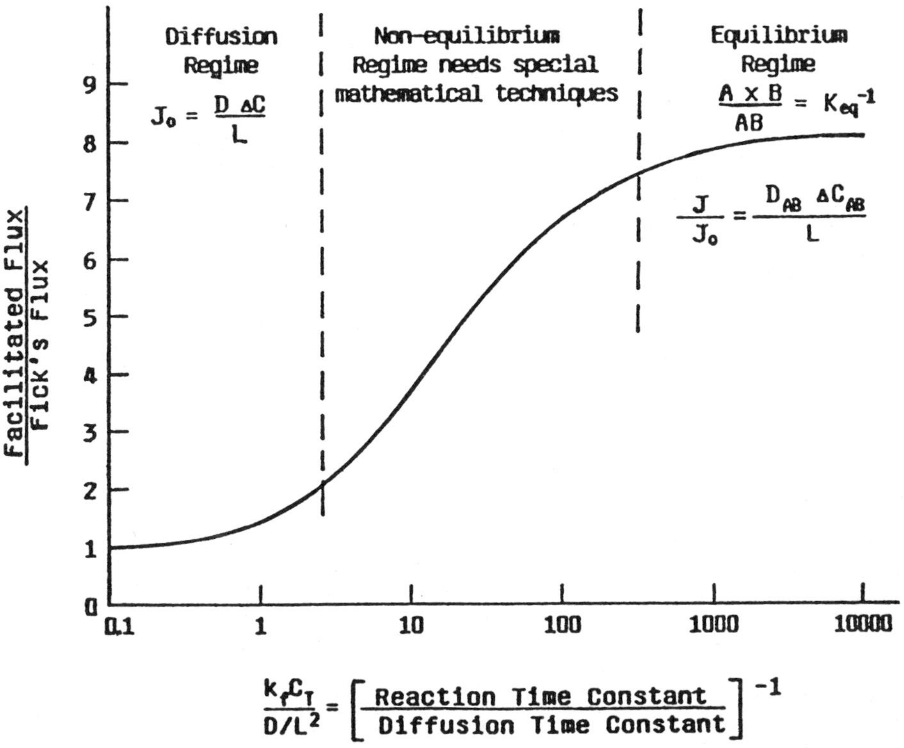

FIGURE 8.—Effect of Damköhler number on membrane performance.

is very small, the carrier does not have sufficient opportunity or time to react with the permeant and therefore, the flux through the system is dominated by simple Fick's Law diffusion with very little extra facilitation taking place.

There is a large reaction–diffusion transition zone between the so-called diffusion regime and the equilibrium regime. In this transition zone one must solve the differential equations completely by numerical methods in order to determine the flux through the membrane.

3.2. Verification of the Model

An experimental verification that the Damköhler number is the important characteristic parameter in these problems is demonstrated in Figure 9 (5). The data represent the flux of carbon dioxide through a membrane containing sodium hydroxide. This system falls into the category of carrier-mediated diffusion because the carbon dioxide can react with hydroxide to form carbonate and bicarbonate. The carbonate and bicarbonate are effectively the carrier–complex species.

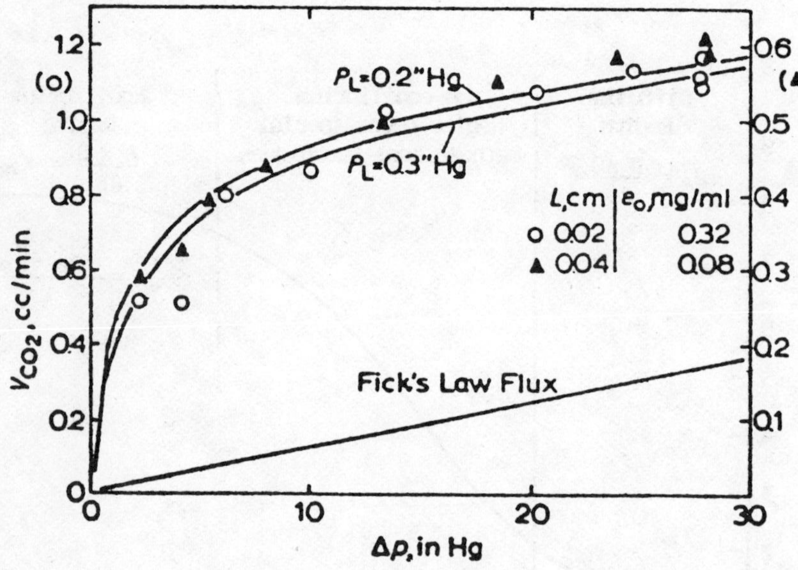

FIGURE 9.—Flux of CO_2 at a fixed Damköhler number (constant $e_o L^2$) illustrating that the reaction rate increases with enzyme concentration (5).

In this study we controlled the rate at which carbon dioxide reacted with hydroxide by the amount of catalyst (carbonic anhydrase) added to the system. We also measured the flux through membranes of different thicknesses. In this case the forward rate constant is proportional to enzyme concentration; thus, the Damköhler number is proportional to enzyme content multiplied by the thickness of the membrane squared. This product was experimentally adjusted to be numerically equal for the two curves shown in this graph. As expected, the net normalized flux across this particular membrane system is the same for two different combinations of length and enzyme concentration that were used.

Note that in this system the diffusion of carbon dioxide through the membrane-containing base was about 8 times greater than if no base were in the membrane. This facilitation of CO_2 transport is one mechanism by which CO_2 removal from gas streams can be achieved.

An appreciation of the complexity of these carrier-mediated transport systems can be obtained by examining the calculated concentration profiles of reactants, as shown in Figure 10. In the lower panel, the calculated concentration profile of CO_2 through the membrane is presented. The CO_2 profile through the membrane is obviously non-linear and non-symmetrical. The net rate of reaction is shown in the upper level. Here it is seen that net reaction occurs primarily at the interface surfaces on both sides of the membrane and that in the bulk of the membrane (in the center) very little reaction

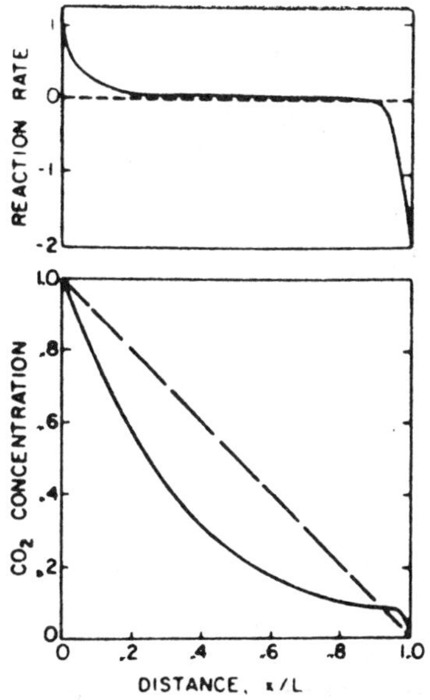

FIGURE 10.—Nonsymmetrical profiles for CO_2 diffusion through bicarbonate solutions: a. Reaction rate more intense at downstream boundary; b. Concentration of CO_2 changes rapidly at membrane surfaces. Conditions: CO_2 partial pressure, upstream = 1 atm; downstream = 0.01 atm, 1N $NaHCO_3$ in membrane; membrane thickness = 0.5 cm (3).

occurs. This pattern is characteristic in carrier-mediated transport membranes and has led to a number of simplified mathematical treatments of membrane behavior, based on a conceptual division of the membrane into three regions: two boundary layer phases where reaction is occurring, and then the central portion or core where the reaction is presumed to be approximately at equilibrium (6, 7).

There are a few approximate analytical expressions for estimating the behavior of carrier-mediated transport membranes of type

$$A + B \underset{k_r}{\overset{k_f}{\rightleftharpoons}} AB \tag{1}$$

A) Boundary-layer analysis—Intermediate to thick films (3)

$$\frac{1 + \Phi}{1 + \Phi_{eq}} = \left[1 + Z_L^2\, \delta(Z_L) + Z_0^2\, \delta(Z_0)\right]^{-1} \tag{2}$$

where

$$Z = \frac{\sigma}{1 + KC_A} \qquad \delta(Z) = \left[\frac{\sigma}{\zeta}\frac{Z}{1 + Z^2}\right]^{1/2} \qquad \Phi = \frac{J_A - J_{A_0}}{J_{A_0}} \tag{3–5}$$

$$\sigma^2 = KC_T\frac{D_{AB}}{D_A} \qquad \zeta = \frac{k_f C_T L^2}{D_A} \qquad K = \frac{k_f}{k_r} \tag{6–8}$$

B) Large excess of carrier "B" (8)

$$1 + \Phi = \frac{1 + \sigma^2}{1 + [\sigma^2/\phi]\tanh\phi} \tag{9}$$

where

$$\phi = \frac{1}{2}\left[\zeta\left(\frac{1 + \sigma^2}{\sigma^2}\right)\right]^{1/2} \tag{10}$$

More complete and exact predictions of the behavior of specific cases are obtainable by numerical methods.

If operating conditions are chosen that place the membrane Damköhler number in the transition range, then the flux tends to be more sensitive to the downstream conditions than the upstream conditions. This behavior is shown in Figure 11, where the CO_2 flux is plotted against the downstream concentration of CO_2, p_L. There is quite a significant reduction in flux, with modest changes in the downstream concentration of permeant.

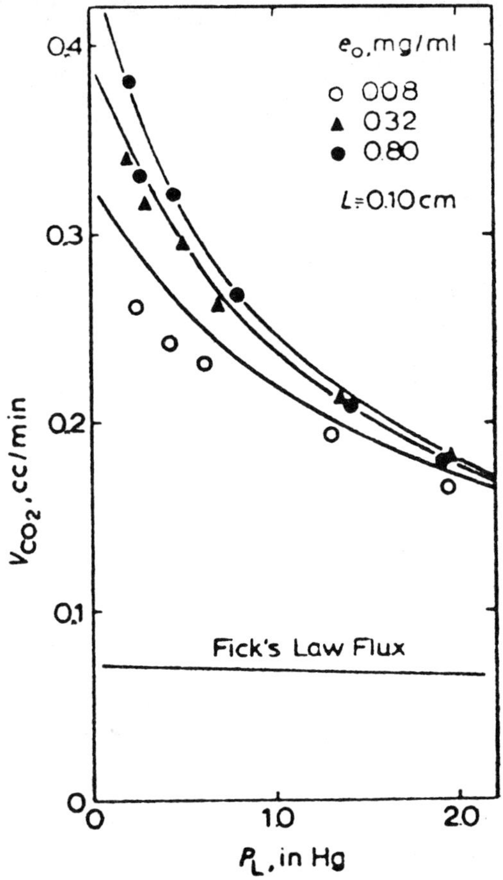

FIGURE 11.—Flux of CO_2 across a 0.10-cm membrane containing 1 M $NaHCO_3$ and carbonic anhydrase as a function of downstream partial pressure of CO_2 (5).

Another manifestation of this effect is shown in Figure 12. Here, the calculated facilitated flux of O_2 across a membrane containing myoglobin as a carrier decreases with very high upstream concentrations of O_2. This unexpected "inhibition" of the flux arises from an increase in the downstream concentration of O_2 from within the membrane. In Figure 10, it is seen that there is a sharp change in the permeant concentration at the downstream boundary. As the upstream concentration is increased further, the downstream boundary layer becomes thinner, the permeant concentration near the boundary increases and the carrier is less able to "unload" the permeant.

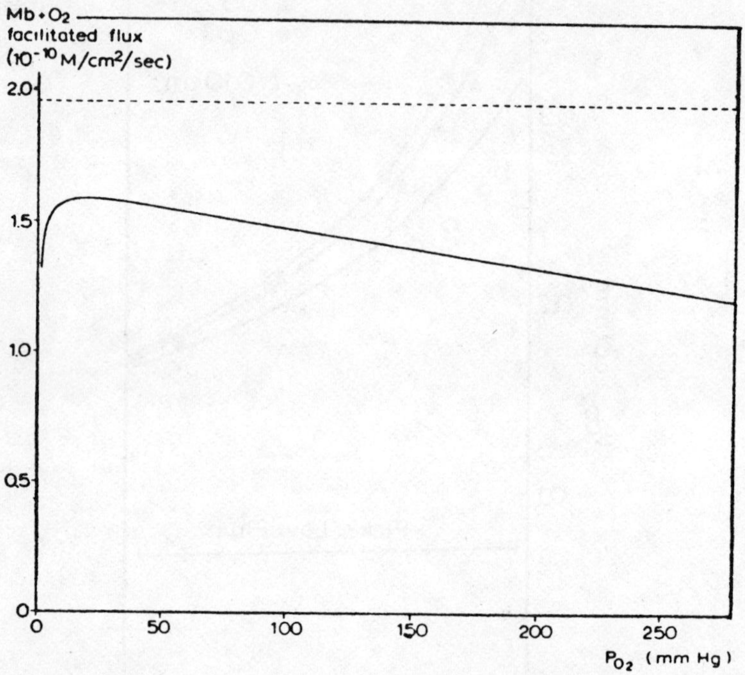

FIGURE 12.—Effect of upstream concentration on facilitated flux of oxygen (6).

For the most part, the performance of carrier-mediated transport membranes is a complex function of kinetic and diffusional parameters and the flux is not controlled by any particular step. However, under specially arranged conditions, the permeant flux is dominated by the reaction rate, and under these conditions, the kinetics of the reaction of permeant with the carrier can be directly inferred from the flux data. Donaldson and Quinn (7) showed that when the carrier is in excess, the equations given above can be recast in a form to linearize the flux data so that kinetic constants can be extracted. Figure 13 shows the facilitation of CO_2 across solutions containing a catalyst, carbonic anhydrase. The data can be plotted as reciprocals, as shown in Figure 14, to provide linearization. Then from the slope and intercept of the line, the kinetic constants can be determined.

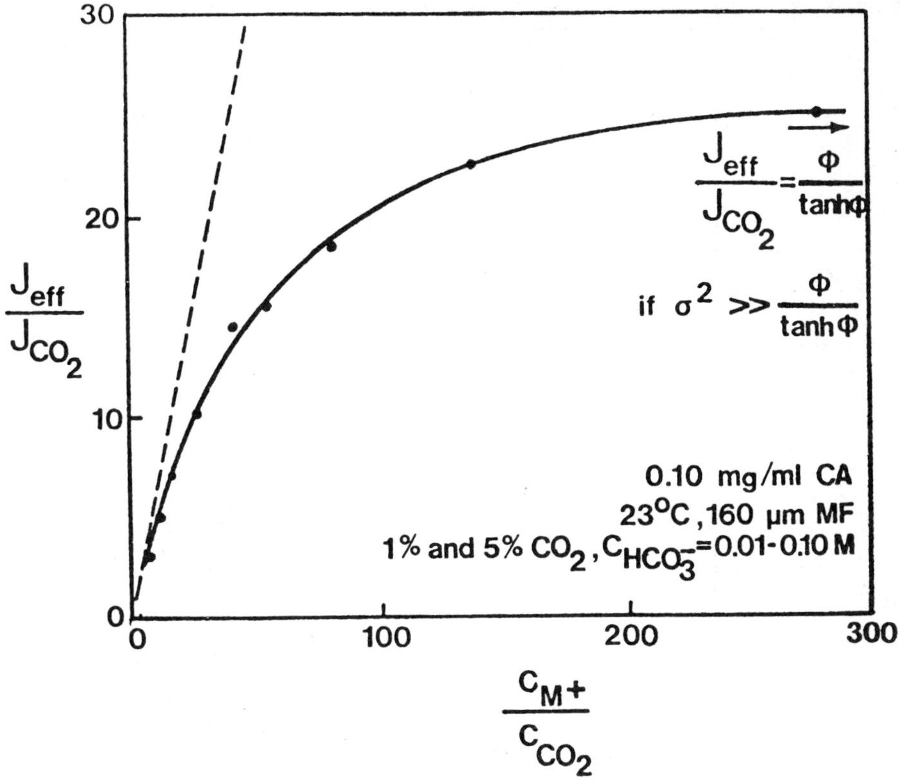

$$\frac{J_{eff}}{J_{CO_2}} = \frac{1 + \sigma^2}{1 + [\sigma^2/\Phi] \tanh \Phi}$$

$$\sigma^2 = \frac{D_{HCO_3^-}}{D_{CO_2}} KC_T = \frac{D_{HCO_3^-}}{D_{CO_2}} \frac{C_{M^+}}{C_{CO_2}} \quad \text{if carrier is in large excess}$$

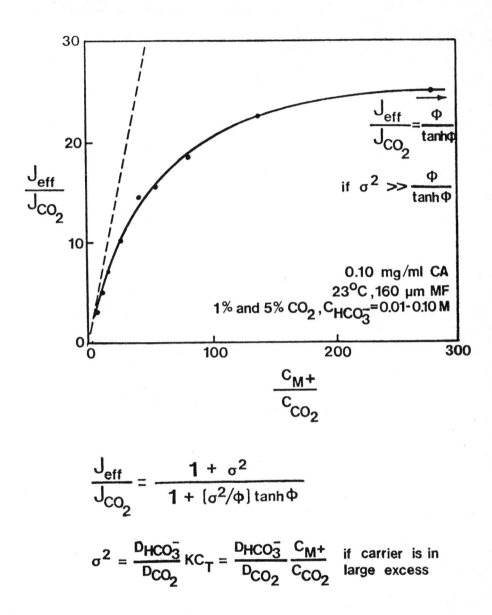

FIGURE 13.—Effective diffusivity of CO_2 across a membrane containing carbonic anhydrase (CA) (8).

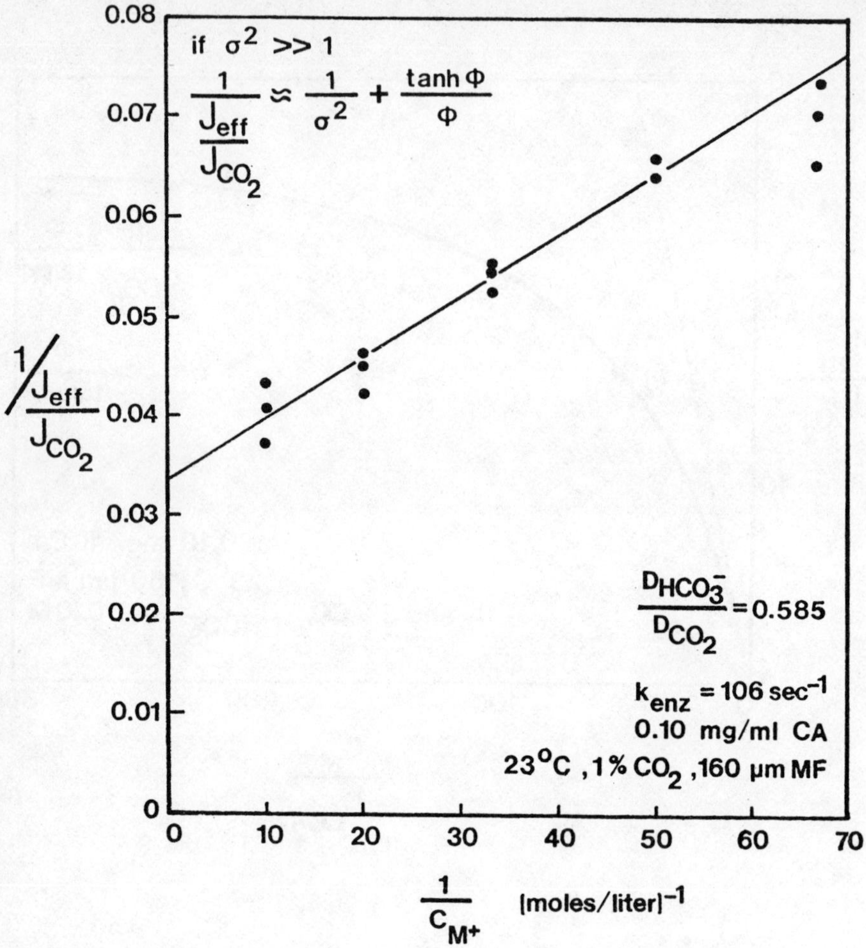

FIGURE 14.—Double reciprocal plot of data from Figure 13 (8).

3.3. The Equilibrium Regime

Many of the characteristics of carrier-mediated transport membranes can be inferred rather simply by consideration of membrane behavior in the equilibrium regime.

Of particular importance in the design of a membrane system for the separation of permeants is the choice of the carrier species. Here the relative affinity of carrier for different permeants will determine the selectivity characteristics of the membrane. On the other hand, if the binding constant between permeant and carrier is too high, then the carrier may have poor "unloading" characteristics on the downstream side of the membrane, and net permeant flux across the membrane will be reduced. This behavior is shown in Figure 15, where the facilitation factor is plotted against the equilibrium binding constant between permeant and carrier.

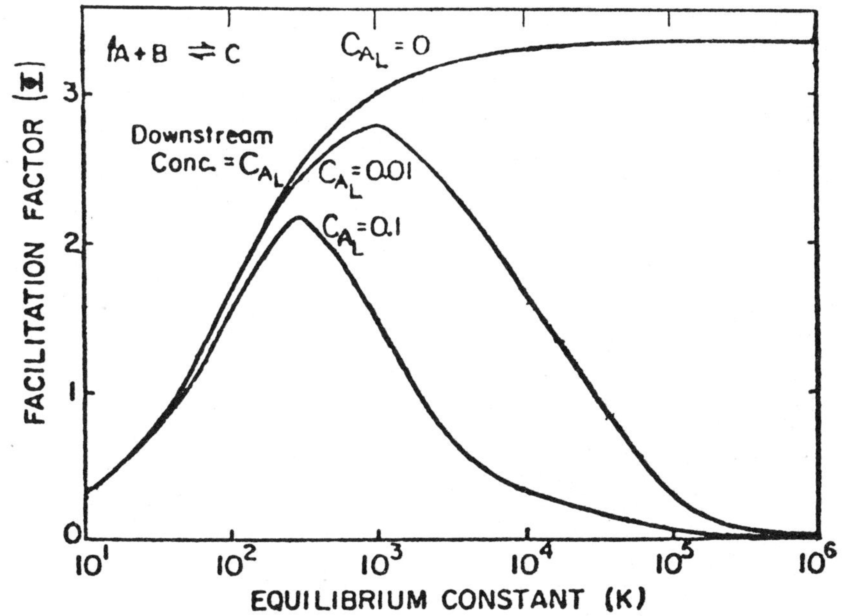

FIGURE 15.—Effect of equilibrium binding and downstream pressure concentration on carrier-mediated facilitation in equilibrium regime (3).

Except for the unusual case where the downstream permeant concentration is maintained at absolute zero, there is an optimum binding constant, which is a function of permeant concentration on both sides of the membrane

$$K_{opt} = (C_{A_0} C_{A_L})^{1/2} \quad (11)$$

Experimental verification for this effect was given by Lamb et al. (9), who measured the binding constant of many crown ether carriers with cations. Figure 16 shows some typical results (10). The selectivity of 18-crown-6 is particularly high for cations which fit into the donut-shaped cavity of the cyclic ether. However, when a comparison of the flux of cations across a series of membranes with different crown ethers was made, an optimum binding constant was found in each case (Figure 17).

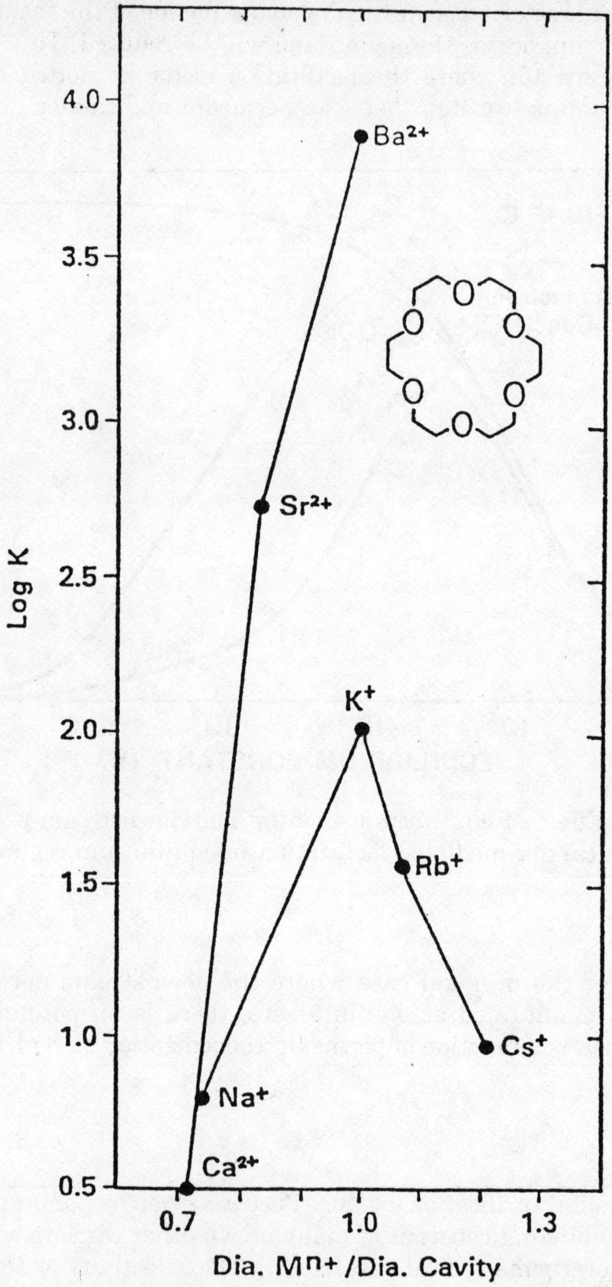

FIGURE 16.—Variation of complex stability constant in water for the reaction of 18-crown-6 with alkali and alkaline earth metal ions (plotted according to ratio of metal ion size to crown cavity size) (10).

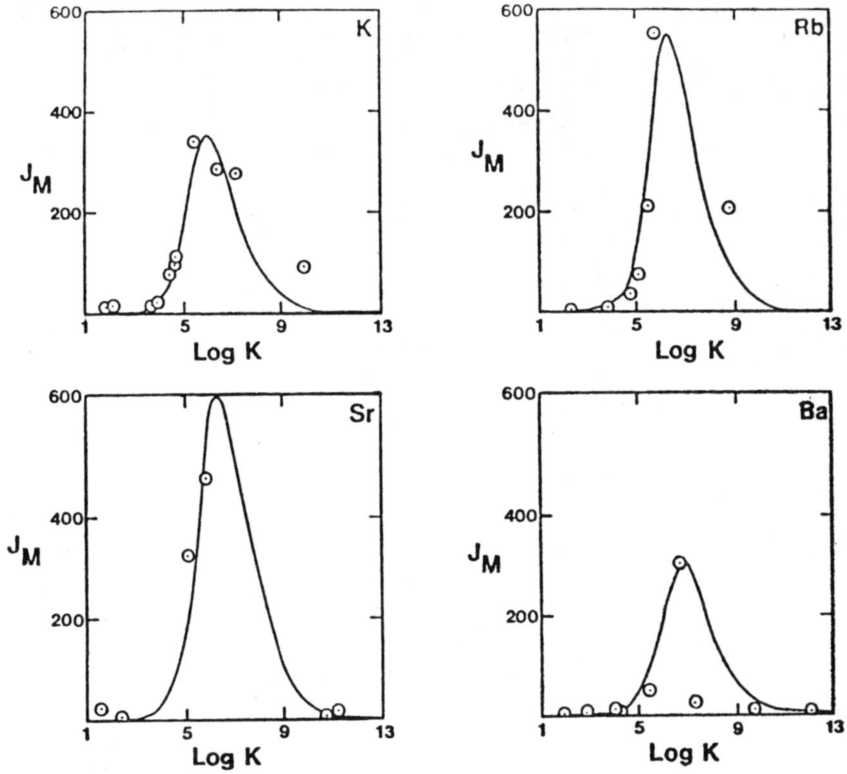

FIGURE 17.—Moles of metal ion × 10^7 transported vs. log K in methanol (9).

In optimizing a carrier-mediated membrane for separations, a compromise may have to be made in the selection of carrier between net flux of the desired permeant and selectivity between the solutes in the feed mixture.

3.4. Effect of Temperature

Temperature affects a variety of properties which will influence the performance of carrier-mediated membranes. Usually the equilibrium binding constant decreases with increasing temperature. This factor alone might be expected to reduce the facilitated flux with increasing temperature. However, at the same time the kinetic rates in the reaction will increase, so that if the membrane is being operated in the reaction-limited regime an increase in facilitation may be the net result (Figure 18).

Other effects of temperature that should be considered are the changes in the stability of carrier, the solubility of carrier, the diffusion coefficients, and the viscosity of the membrane solvent.

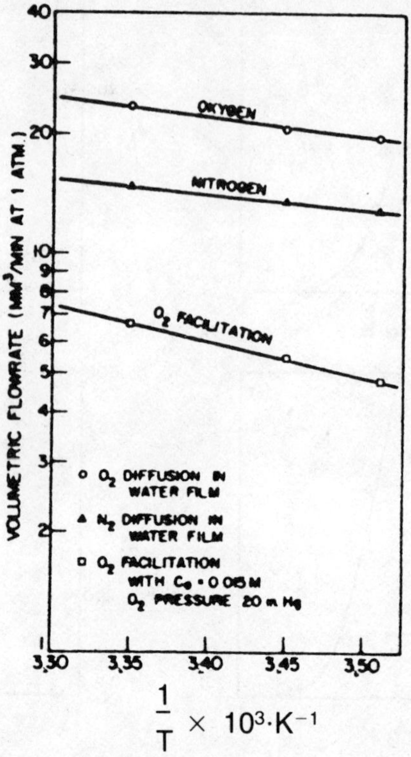

FIGURE 18.—Effect of temperature on facilitation (4).

3.5. Effect of Membrane Thickness

Increasing the thickness of the membrane will generally decrease the net flux of the various permeants through the membrane, as shown in the equations in Figure 7, where membrane thickness is in the denominator in both terms. On the other hand, an increase in membrane thickness increases the Damköhler number and tends to put the membrane in the equilibrium regime, increasing the facilitation effect (Figure 8). Even so, the net effect of increasing thickness is to reduce the total flux (Figure 19).

In some cases, the reaction between permeant and carrier is controlled by the interfacial reaction at the membrane surface, and under these circumstances there may be a regime where the flux is independent of membrane thickness (11). This type of behavior is shown in Figure 20 (curve marked pH 3.0).

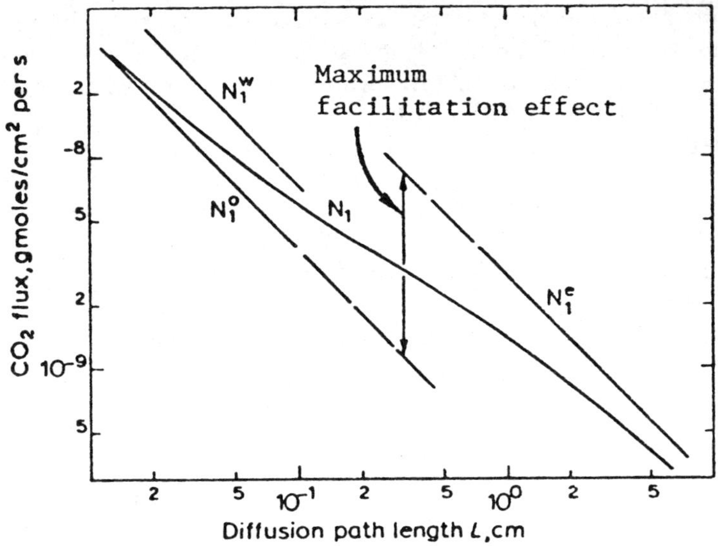

FIGURE 19.—Expected total CO_2 flux across a membrane (of thickness L) containing 1 M $NaHCO_3$. Calculations are for CO_2 partial pressures at the two boundaries of 1.0 and 0.01 atm. (5).

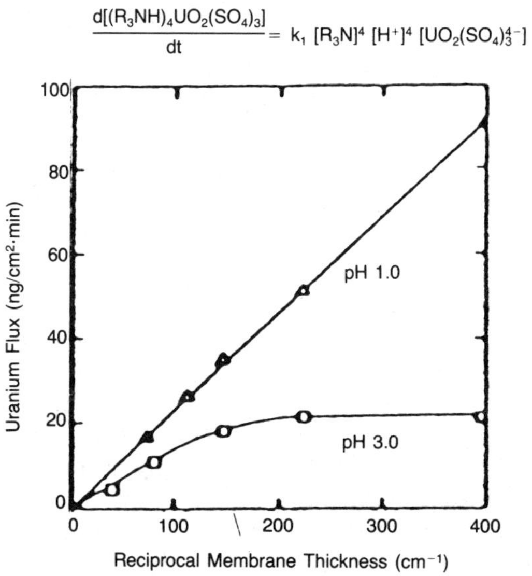

FIGURE 20.—Effect of membrane thickness on uranium flux for two experiments with different feed solution pHs (11).

Often it is assumed that if the flux is inversely proportional to membrane thickness, then the membrane is operating in the equilibrium regime. The curve marked pH 1.0 shows that the uranium flux is inversely proportional to membrane thickness, and the authors interpret this behavior to imply that the reactions are at equilibrium within the membrane. However, we have shown that one can observe nearly inverse behavior between membrane thickness and facilitated flux and still be in the non-equilibrium regime (Figure 21). Here the facilitated O_2 flux through a cobaltodihistidine-loaded membrane was inversely proportional to the .92 power of membrane thickness, yet other data on the binding equilibrium constants established that the membrane was operating in the non-equilibrium regime.

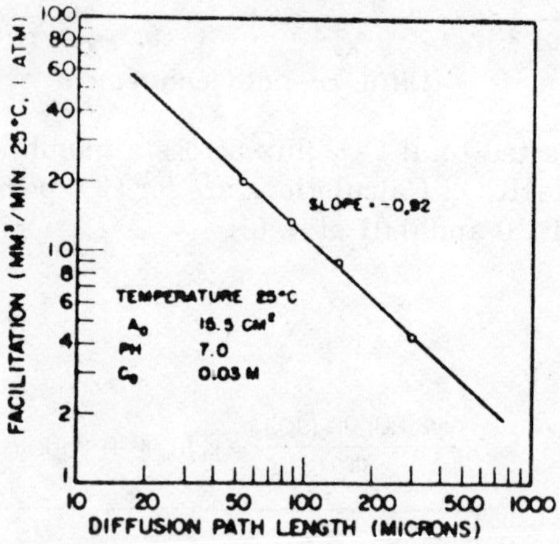

FIGURE 21.—Effect of path length on facilitation (4).

3.6. Carrier Solubility Considerations

In the equilibrium regime, the facilitated flux will be proportional to carrier concentration in the membrane (Figure 7). Thus, increasing carrier concentration will increase the separation of two permeants if one of the permeants does not bind to the carrier. However, there may be a limitation to the extent to which this selectivity can be accomplished, given the limits of carrier solubility in the membrane solvent. This is shown for the O_2–cobaltohistidine system in Figure 22. We found that the net flux of O_2 across the membrane did not continue to increase with increasing carrier loading. The solubility limitation is on the oxygen-complexed carrier, as shown in Figure 23. Thus, although more concentrated cobalt histidine solutions could be prepared, at high concentrations the oxygenated complex precipitated out and did not contribute to an increase in flux.

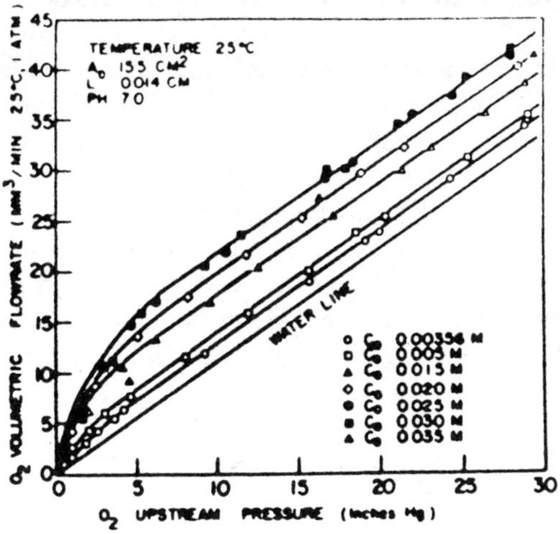

FIGURE 22.—Effect of carrier concentration on O_2 diffusion at 25°C (4).

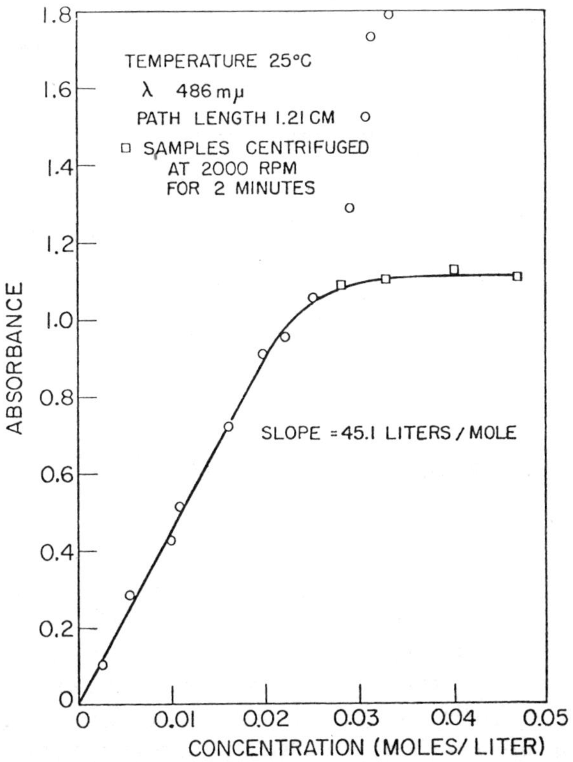

FIGURE 23.—Solubility study of cobaltohistidine (4).

3.7. Screening of Candidate Carrier Compounds

It is relatively easy to test candidate carrier materials for carrier-mediated membranes to be used in separating selective species from aqueous solutions. The behavior of a carrier-mediated membrane is very similar to an extraction stage on the upstream side and a stripping stage on the downstream side. Thus, in a preliminary way, candidate carrier compounds can be screened with the apparatus shown in Figure 24. Here the transport of Rb^+ from one aqueous phase to the other was initiated when nigiricin was added to the bottom solvent phase. Stirring of the solutions insures that diffusion through the rather thick "liquid membrane" is not a limited factor.

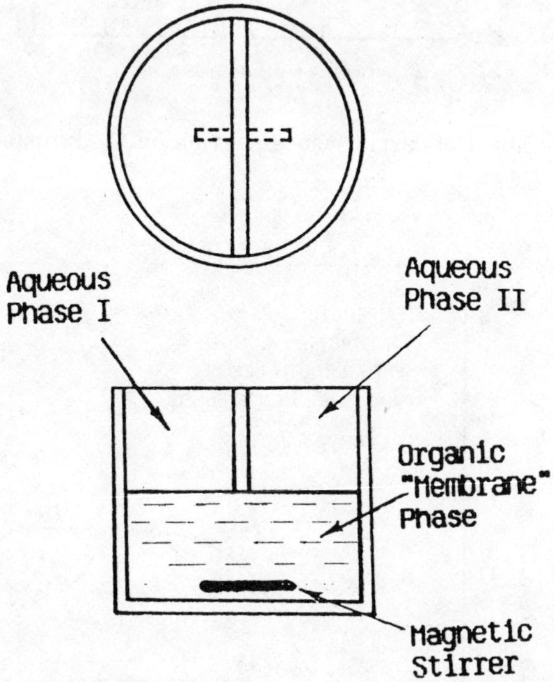

FIGURE 24.—Schematic of apparatus for screening candidate carrier compounds.

Another configuration for using "liquid membranes" (12) is by the dispersion of an aqueous phase within an organic phase (Figure 25). This approach is very similar to traditional extraction practice because the internal stripping solution is not readily accessible and the emulsion has to be repeatedly broken and reestablished for a continued process. A large amount of surface area can be attained without the need for supported membrane structures, thus providing some commercial interest in this approach. An example of the performance of this type of liquid membrane is shown in Figure 26 (13).

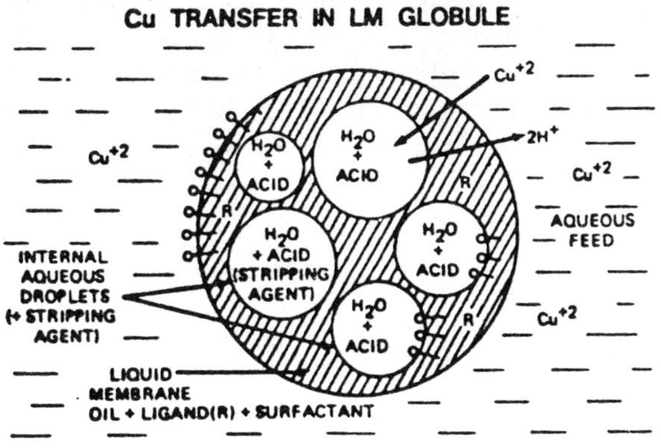

FIGURE 25.—Schematic diagram of liquid membrane capsule for Cu^{+2} extraction (12).

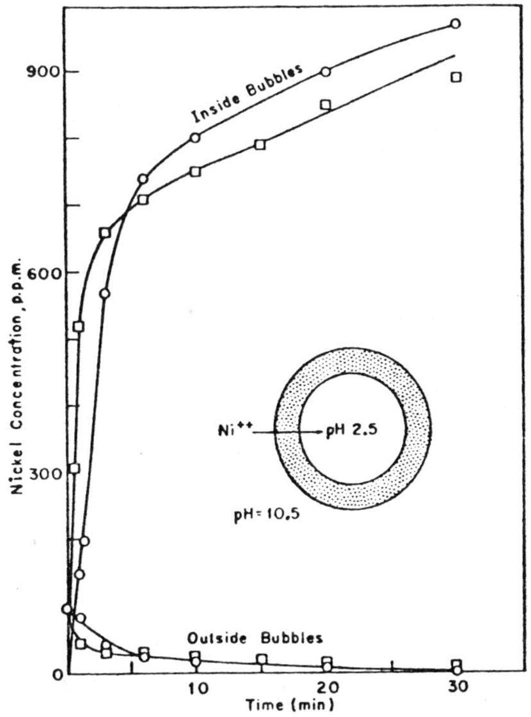

FIGURE 26.—Nickel separations using liquid surfactant membranes. The membranes in this case are mixtures of mineral oil and hexachloro-1, 3-butadiene stabilized by sorbitan mono-oleate. When the membranes contain a high concentration of the carrier LIX-64N, the initial separation is faster (□). When the carrier concentration is lower, the final separation is more complete (○) (13).

The use of counter-transporting approaches has been very fruitful in the separation of ions. The example, shown in Figure 27a (14), is for the concentration of uranium by a pH gradient. The basic mechanism is shown in Figure 27b (11), where a gradient in pH can be used to concentrate uranium 2000-fold. But note that the initial pH ratio across the membrane for this device is on the order of 10^{13}.

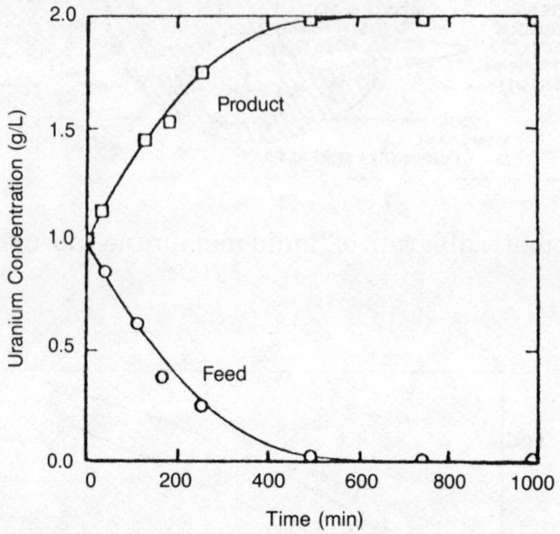

FIGURE 27a.—Uranium concentration in the feed and product solutions vs. time, using a sodium carbonate-buffered product solution. Membrane: 30 vol % tertiary amine. Feed solution: 1.0 g/l uranium, pH 1.0. Production solution: 1.0 g/l uranium, 200 g/l sodium carbonate (14).

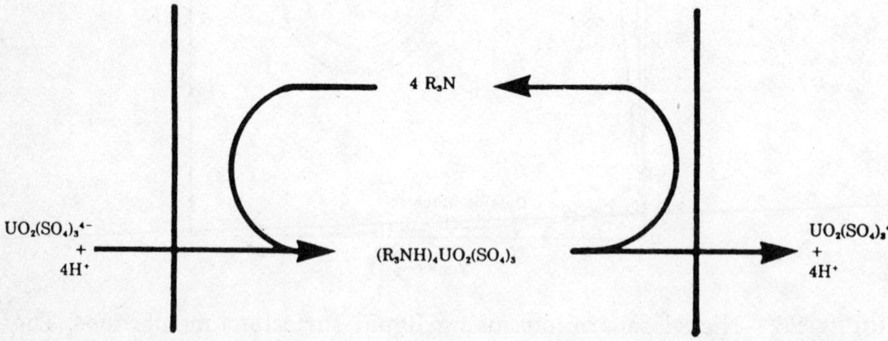

FIGURE 27b.—Coupled mediated transport of uranium across a membrane containing a tertiary amine carrier (11).

3.8. Effects of Asymmetry in Membrane Performance

As indicated in Figure 6, regions of the membrane may vary with respect to carrier concentration, kinetic properties, or other factors. Such variations in properties can cause a directionality to membrane performance.

In Figure 28, a non-carrier containing transport resistance is shown in series with the carrier-mediated membrane. This resistance may be present for a number of reasons. In liquid systems there will be diffusional boundary layer resistances on both sides of the membrane at the interface because of laminar flow conditions near solid surfaces. These resistances may be reduced by higher flow rates or turbulence, but never eliminated. Other situations

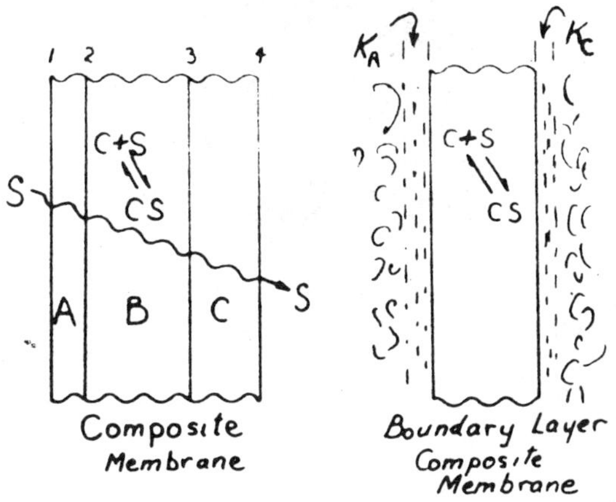

FIGURE 28.—Types of asymmetric membranes.

that may produce composite membrane behavior are cases in which the membrane is coated with a sealing polymer to retain the carrier solution within the membrane, or those that are caused by the asymmetric structure of the membrane material itself, as shown in Figure 29. Here there is a variation in the pore structure of the membrane. Depending on how much carrier solution is placed in the membrane and the properties of the solvent, only partial regions of the membrane will contain the carrier.

These variations in membrane properties can be modeled as membranes in series. Some calculated results are shown in Figure 30 (15) for a hypothetical membrane where the carrier binding characteristics are similar to hemoglobin (Figure 1). The asymmetry ratio is defined as the ratio of the permeant fluxes in each direction at the same concentration driving force. These calculations show that very large asymmetries may be developed when the permeability of the non-carrier film is small (β small) as compared with the mean permeability of the carrier layer.

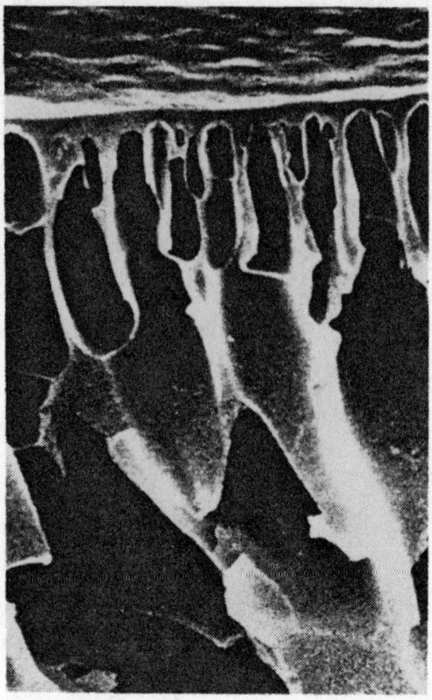

FIGURE 29.—Cross-section of a Nuclepore™ asymmetric membrane.

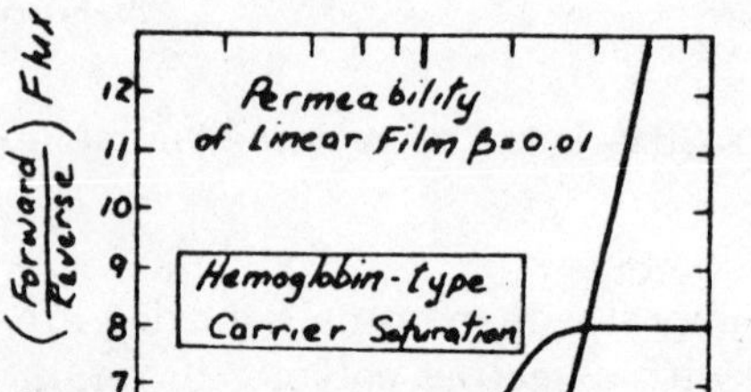

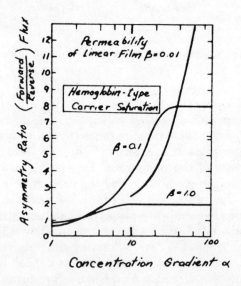

FIGURE 30.—Asymmetric flux calculated for a two-membrane system, one membrane with hemoglobin-type transport characteristics, and the other obeying linear transport (15).

Another factor that can produce asymmetry in the performance of carrier-mediated membranes is a spatial variation in the kinetics of the reaction between carrier and permeant within the membrane. Such variations may result in parts of the membrane having different local Damköhler numbers and effectively different facilitation factors (Figure 8). One mechanism to control the local kinetics within a membrane is by varying the concentration of an immobilized catalyst. Some illustrative calculations were performed based on CO_2 transport through a membrane containing the catalyst carbonic anhydrase.

In Figure 31, the flux of CO_2 is shown as a function of enzyme concentration (i.e., Damköhler number) and fraction of the membrane occupied by the catalyst (16). The conditions were chosen so that the membrane was operating in the non-equilibrium regime. At any given enzyme concentration, the flux drops significantly whenever the catalyst is not uniformly distributed in the membrane.

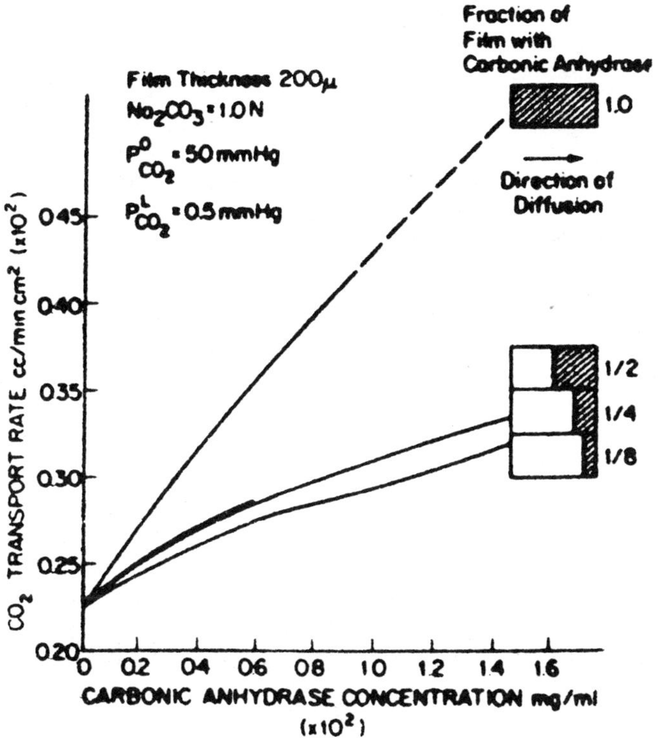

FIGURE 31.—Effect of carbonic anhydrase concentration within the membrane on facilitated transport of CO_2 (16).

The effect of direction of diffusion with respect to enzyme location is shown in Figure 32. The maximum asymmetry effect appears to occur when about one-half of the membrane contains the catalyst.

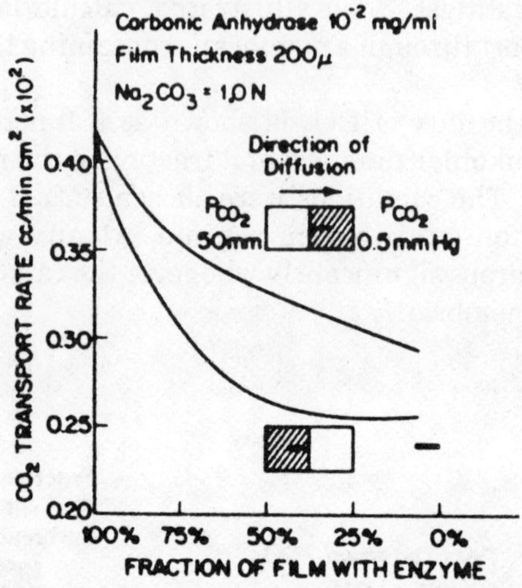

FIGURE 32.—Effect of relative spatial distribution of carbonic anhydrase in a membrane on facilitated transport of carbon dioxide, at a given concentration of carbonic anhydrase (16).

4. ENERGY-COUPLED CARRIER-MEDIATED MEMBRANES

A number of mechanisms have been suggested for coupling external energy sources to carrier-mediated membranes as a means of controlling and/or accelerating the flux of permeants.

Ward (17) has shown that if the carrier can be made as part of an oxidation–reduction couple, then the flux of permeant can be influenced by producing a gradient in the oxidation—reduction state of the carrier across the membrane (Figure 33). The converse has also been demonstrated, that is, a concentration difference in permeant across such a membrane can cause the production of a voltage gradient across the membrane.

Light energy can be coupled to a carrier-mediated membrane if the reaction of the carrier with permeant is light sensitive. The binding of carbon monoxide is an example of such a system as shown in Figure 34. Carbon monoxide is induced to dissociate from myoglobin by visible light, and the effect is proportional to light intensity over a wide range in intensities (18).

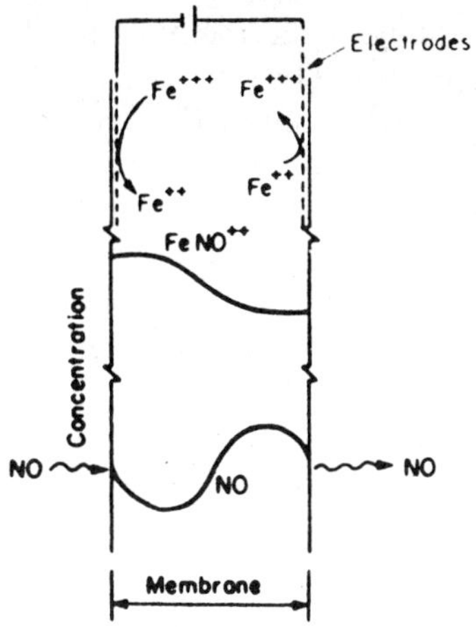

FIGURE 33.—Anomalous concentration profiles (NO) with equal concentrations at the boundaries for an energy-coupled system of carrier-mediated membrane (17).

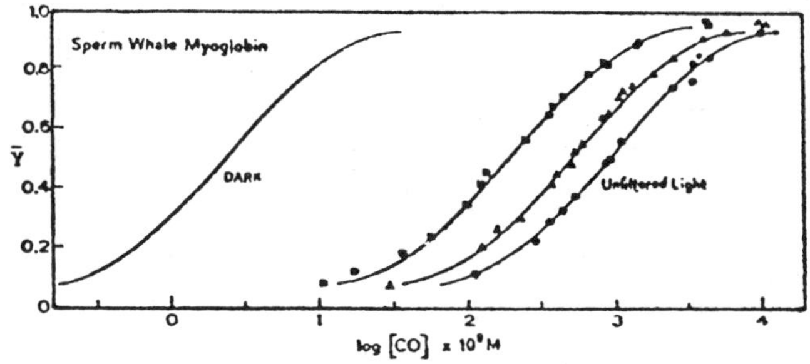

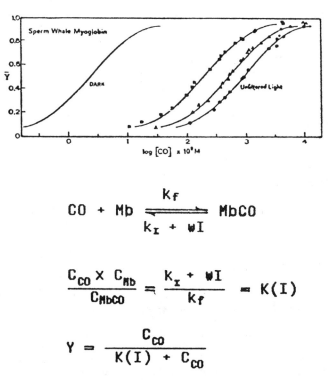

$$CO + Mb \underset{k_I + \psi I}{\overset{k_f}{\rightleftharpoons}} MbCO$$

$$\frac{C_{CO} \times C_{Mb}}{C_{MbCO}} = \frac{k_I + \psi I}{k_f} = K(I)$$

$$Y = \frac{C_{CO}}{K(I) + C_{CO}}$$

FIGURE 34.—Carbon monoxide binding curves for sperm-whale myoglobin for relative light intensities of 1.0 (○), 0.5 (Δ) and 0.21 (□) (18).

One method for coupling light energy to the carrier-mediated membrane is to illuminate only one side of the membrane while the other side of the membrane remains in the dark. The effect then is to shift the equilibrium of the reaction in the illuminated region so that the affinity of the carrier for the permeant is decreased.

The equilibrium transport equations for this situation are shown in Figure 35, and an experimental system for verifying the light coupling effect is shown in Figure 36. In the actual experiments, hemoglobin was used as the carrier instead of myoglobin because it is more readily available and more stable.

Dark Region : $\frac{C_{AB}}{C_A C_B} = \frac{k_f}{k_r} = K$; Light Region : $\frac{\underline{C_{AB}}}{\underline{C_A}\,\underline{C_B}} = \frac{\underline{k_f}}{\underline{k_r}} = \underline{K}$

$$J_A = \frac{D_A}{(d + \underline{d})}(C_{Ad} - \underline{C_{Ad}}) + \frac{D_{AB}\, C_{BT}}{(d + \underline{d})}\left[\frac{C_{Ad}}{K_d^{-1} + C_{Ad}} - \frac{\underline{C_{Ad}}}{\underline{K_d}^{-1} + \underline{C_{Ad}}}\right]$$

Maximum Pumping : $C_{Ad} = \underline{C_{Ad}} = C_A$

$$J_A = \frac{D_{AB}}{(d + \underline{d})}\, C_{BT}\, C_A\, \frac{(\underline{K_d}^{-1} - K_d^{-1})}{(K_d^{-1} + C_A)(\underline{K_d}^{-1} + C_A)}$$

Maximum Concentration Difference : $J_A = 0$

$$\frac{\underline{C_{Ad}}}{C_{Ad}} = \frac{D_A + D_B\, C_{BT}\, K_d}{D_A + D_B\, C_{BT}\, \underline{K_d}}$$

FIGURE 35.—Equations for photodissociation membranes.

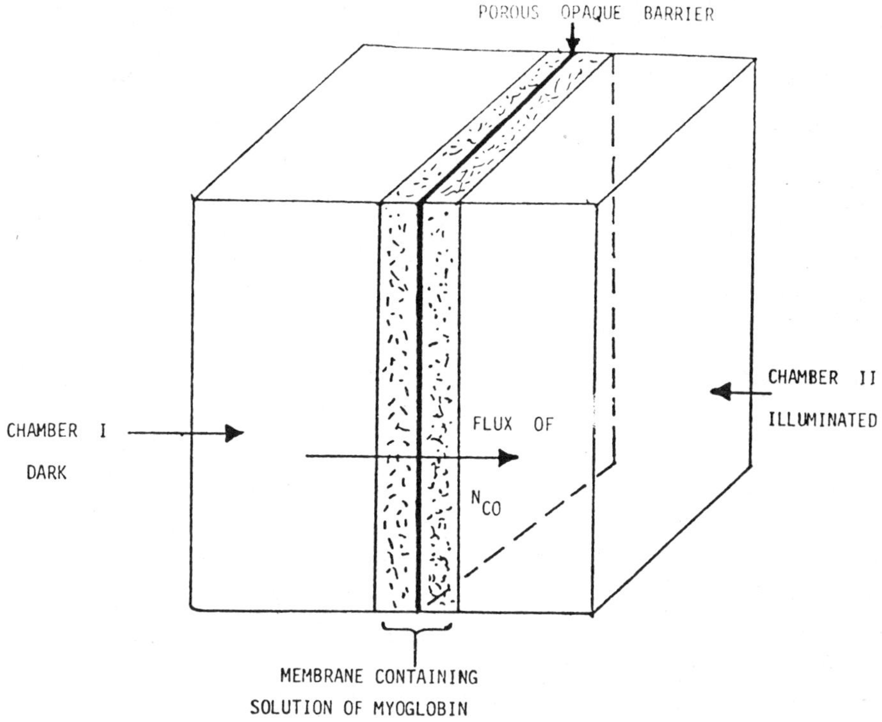

FIGURE 36.—Schematic representation of an experimental system for studying light-coupled transport across a membrane.

The mathematical analysis shows that not only can an increase in permeant flux be achieved by light modulation of the reaction kinetics, but also an enhancement of permeant concentration can be achieved in the illuminated chamber. This effect is demonstrated in Figure 37, where a light-side/dark-side concentration ratio of three-fold was achieved, which reverted toward unity when the light was removed (19).

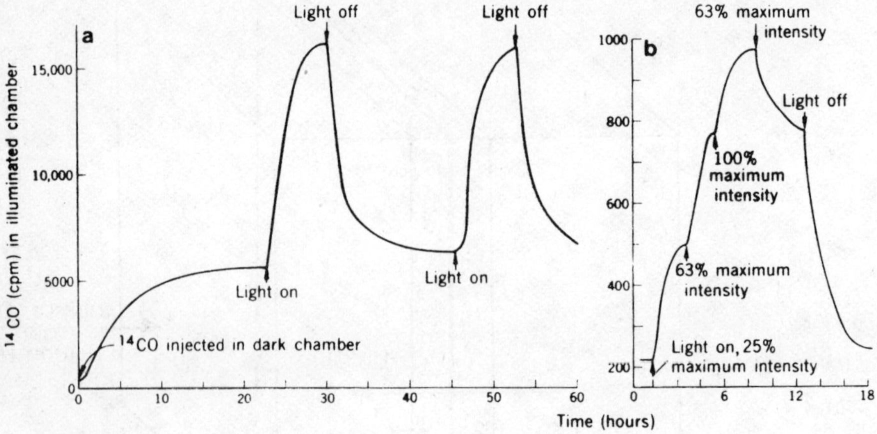

FIGURE 37.—Photodiffusion of CO across a hemoglobin-solution membrane. (a) [^{14}C] Carbon monoxide accumulation of chamber I first reaches equilibrium with the ^{14}CO content of chamber II. When chamber I is illuminated ^{14}CO accumulates above the initial equilibrium level, and the process is reversible as shown by subsequent cycles. (b) The concentration of ^{14}CO in chamber I is directly related to the light intensity on this side of the membrane. The light intensity was changed by interposing different neutral density filters in the illuminating beam. Experimental conditions: membrane sandwich consisting of two microporous filters (Gelman) separated by a porous silver filter (Selas); diameter, 3 cm; thickness, 0.03 cm; solution, 15 percent human hemoglobin; CO partial pressure in nitrogen, 0.125 mm-Hg; chamber I, 4 ml; chamber II, 20,000 ml. In these experiments the Geiger counter for detection of ^{14}C was placed in a bypass loop of chamber I. The circulation of gas through the bypass was very slow and limited the response of the radiation-detecting system, which accounts for the prolonged periods required to reach steady state. Subsequent experiments with improved circulation showed that steady state is reached in less than 30 minutes. The light source was a 100-watt xenon halide lamp, and the light was passed through a $CuSO_4$ solution to remove heat-producing infrared radiation before it was focused onto the membrane surface. The temperature rise in the membrane due to illumination was less than 1°C, as measured by a thermocouple placed within the membrane sandwich (19).

A complete mathematical analysis of this system including the non-equilibrium regime shows that the efficacy of light modulation and coupling is diminished if the membrane is operated in the non-equilibrium regime (Figure 38). Also, in the equilibrium regime the degree of permeant concentration that can be achieved is closely related to the ratio of the apparent affinity constant in the light to the true affinity constant in the dark, $\underline{K}/K$.

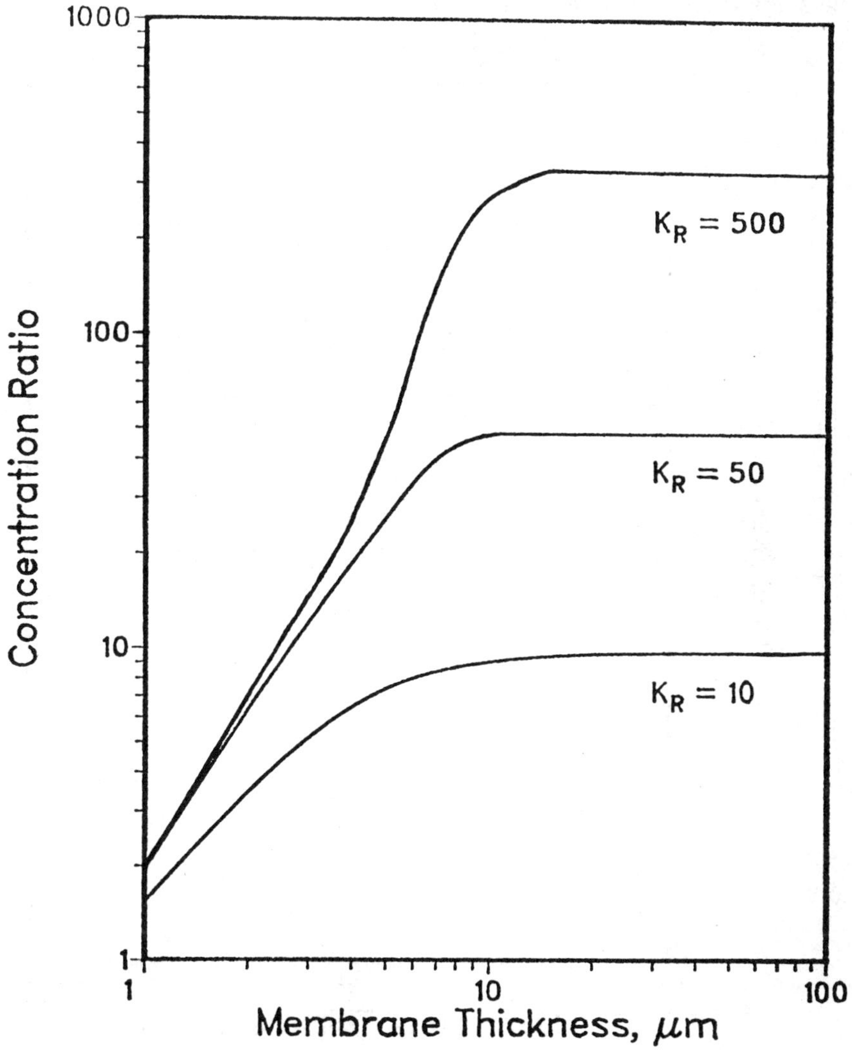

FIGURE 38.—Effect of light intensity of CO enrichment, with hemoglobin as the carrier, as a function of membrane thickness (upstream concentration = 10^{-11} mol/ml; percent illumination = 50) (20, 21).

Since it has been shown that the saturation curve for CO-myoglobin can be shifted about 500-fold at high light intensities, approximately a 500-fold concentration of CO by the coupling of light energy to the membrane is not an unreasonable possibility.

We examined alternatives for the most effective use of light for concentrating the permeant. One question concerns the optimal light intensity profile across the film. Calculations for a step pattern in light profile and a linear pattern are shown in Figure 39. Clearly the step is more effective for the same total light energy absorbed.

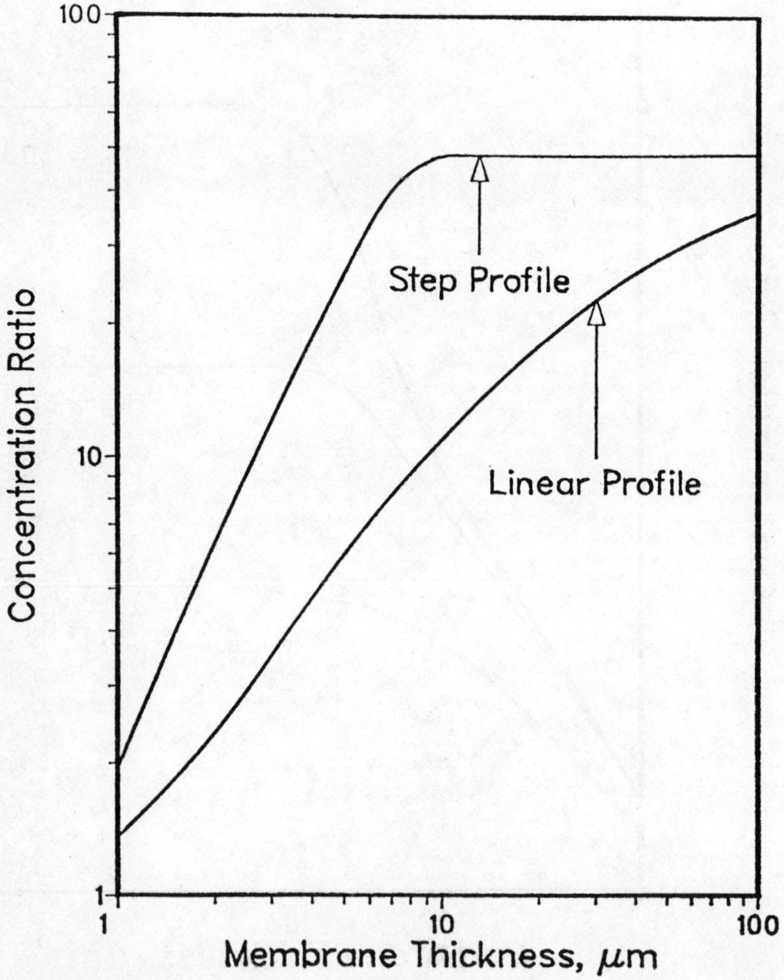

FIGURE 39.—Effect of light intensity profile on CO enrichment as a function of membrane thickness (upstream concentration = 10^{-11} mol/ml; percent illumination = 50) (20, 21).

Another interesting characteristic of these photodiffusion membrane systems is that their effectiveness as concentrating devices is best when the upstream concentration is low. This is demonstrated in Figure 40, where the concentration ratio is shown as a function of dimensionless upstream concentration C_A^0.

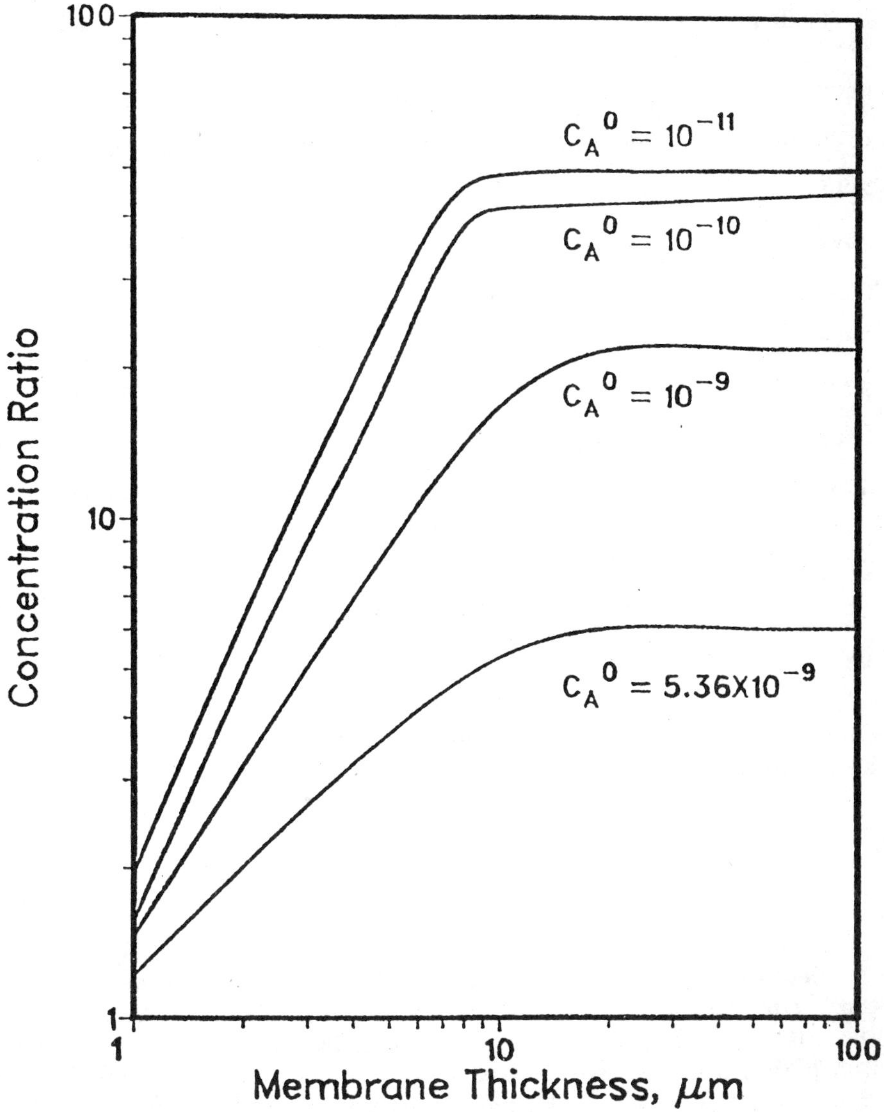

FIGURE 40.—Effect of upstream concentration on CO enrichment as a function of membrane thickness (K_R = 50, percent illumination = 50, C_A^0 in mols/ml) (20,21).

There are many photodissociation reactions that could be implemented as photodiffusion membranes; the main requirement is that the binding reaction must be reversible. Figure 41 shows some of the reaction systems reported in the literature which appear to meet the prerequisites.

OVERALL REACTION			EFFECTIVE WAVELENGTH (nm)
HbCO	$\rightleftharpoons$	Hb + CO (water)	400
$Cr(CO)_6$	$\rightleftharpoons$	$Cr(CO)_5$ + CO (in pyridine)	UV
2HBr	$\rightleftharpoons$	$Br_2 + H_2$	254
$2SO_3$	$\rightleftharpoons$	$O_2 + 2SO_2$	276
$COCl_2$	$\rightleftharpoons$	$CO + Cl_2$	275
$2NO_2$	$\rightleftharpoons$	$2NO + O_2$ (in CCl_4)	435
2NOCl	$\rightleftharpoons$	$2NO + Cl_2$ (in CCl_4)	637
2AgBr(s)	$\rightleftharpoons$	$2Ag(s) + Br_2$	460
SO_2Cl_2	$\rightleftharpoons$	$SO_2 + Cl_2$	300
$2Fe^{+++} + I_3^-$	$\rightleftharpoons$	$2Fe^{+++} + 3I^-$ (in H_2O)	546
$I_2 + NO_2^- + H_2O$	$\rightleftharpoons$	$NO_3^- + 2HI$ (in H_2O)	576
$Ce^{+++} + H^+$	$\rightleftharpoons$	$Ce^{++++} + 1/2\ H_2$	254

FIGURE 41.—Photochromic reactions adaptable to photodiffusion membranes.

Recently, investigators have demonstrated other photochromatic effects that can be utilized to control carrier-mediated transport processes. Figure 42 shows that the configuration of an azo crown ether can be altered by illumination. In this case the crown ether binds to K^+ only in the *cis* form. Shinkai and co-workers (22) showed that a membrane system can be constructed where the transport of K^+ across the membrane can be controlled by illumination. Other examples of photoregulated ion binding (Figure 43) imply that this approach may offer interesting opportunities for new membrane separation techniques (23).

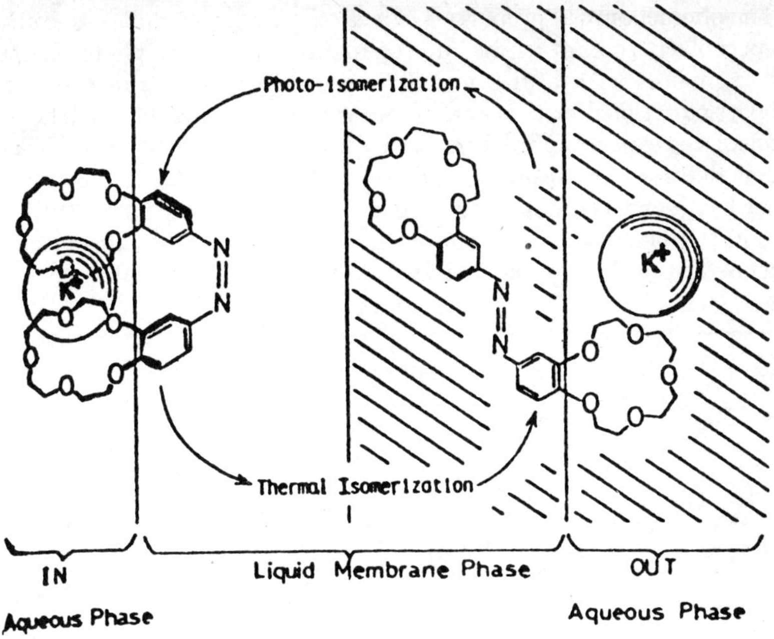

FIGURE 42.—Schematic representation of light-driven ion transport (22).

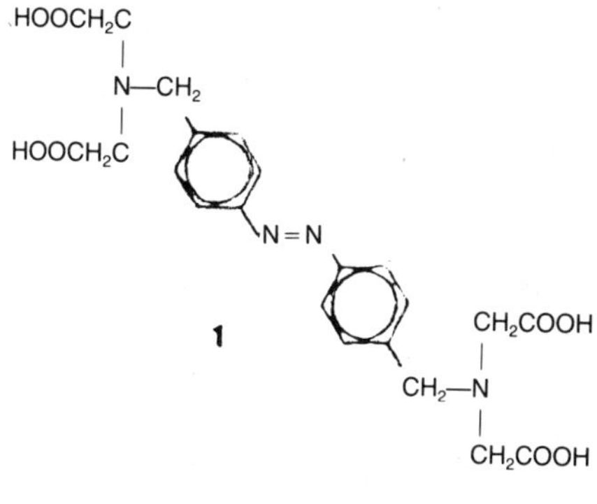

$$\text{Trans } 1 \rightleftharpoons \text{Cis } 1$$

$$\text{Cis } 1 + Zn^{++} \rightleftharpoons \text{Cis } 1 . Zn^{++}$$

$$K \sim 10^{-5} M$$

FIGURE 43.—Structure of 4,4′-bis(-iminodiacetic acid) azotoluene (1), and equations for photoregulated ion binding (23).

The photochemical properties of a carrier can be exploited to control the transport of a permeant whose binding to the carrier is not photosensitive but which competes with a photosensitive permeant. The equilibrium regime transport equations for this system are given in Figure 44, and the effect is demonstrated experimentally in Figure 45 (24). In the latter case, CO binding to hemoglobin is photosensitive, whereas O_2 is not. Because of the higher affinity between CO and hemoglobin, no carrier sites are normally available for O_2 in the presence of CO. However, when the system is illuminated, CO dissociates from the carrier and O_2 facilitation can then take place.

$$Hb + O_2 \rightleftharpoons HbO_2 \qquad \text{NOT LIGHT SENSITIVE}$$

$$Hb + CO \rightleftharpoons HbCO \qquad \text{LIGHT SENSITIVE}$$

$$J_{O_2} = \frac{D_{O_2}}{L}(\underline{C}_{O_2} - C_{O_2}) + \frac{D_{Hb}\, C_{HbT}}{L\, K_{O_2}} \left[\frac{\underline{C}_{O_2}}{1 + \frac{\underline{C}_{O_2}}{K_{O_2}} + \frac{\underline{C}_{CO}}{\underline{K}_{CO}}} - \frac{C_{O_2}}{1 + \frac{C_{O_2}}{K_{O_2}} + \frac{C_{CO}}{K_{CO}}} \right]$$

FIGURE 44.—Equations for indirect photo-diffusion membrane system with hemoglobin as the carrier compound.

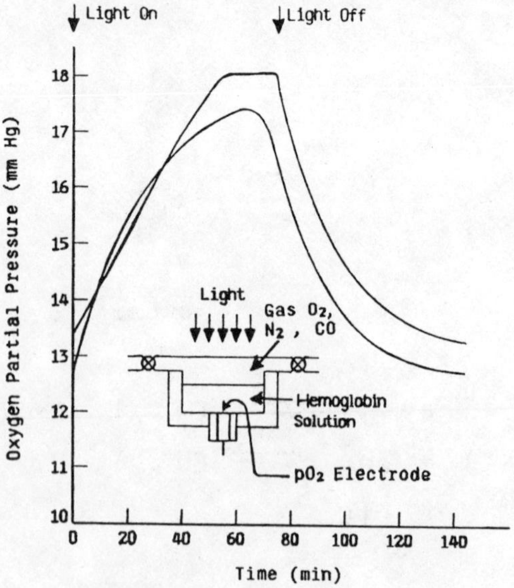

FIGURE 45.—Typical concentration profiles for oxygen transport by indirect photodiffusion across a membrane of hemoglobin solution (24).

REFERENCES

1. Scholander, P. F.: 1960, "Oxygen transport through hemoglobin solutions." Science 131, p. 585.
2. Hemmingsen, E. A.: 1962, "Accelerated exchange of oxygen-18 through a membrane containing oxygen-saturated hemoglobin." Science 135, p. 733.
3. Schultz, J. S., Goddard, J. D., and Suchdeo, S.R.: 1974, "Facilitated transport via carrier-mediated diffusion in membranes." AIChE J. 20, p. 417.
4. Bassett, R. J.: 1969, "Fractionation of oxygen–nitrogen gas mixtures by non-equilibrium facilitated diffusion in liquid membranes." Ph.D. Thesis, University of Michigan.
5. Suchdeo, S. R., and Schultz, J. S.: 1974, "Mass transfer of CO_2 across membranes: Facilitation in the presence of bicarbonate ion and the enzyme carbonic anhydrase." Biochim. Biophys. Acta. 352, p. 412.
6. Kreuzer, F., and Hoofd, L. J. C.: 1972, "Factors influencing facilitated diffusion of oxygen in the presence of hemoglobin and myoglobin." Resp. Physiol. 15, p. 104.
7. Smith, K. A., Meldon, J. K., and Colton, C. K.: 1973, "An analysis of carrier-facilitated transport." AIChE J. 19, p. 102.
8. Donaldson, T. L., and Quinn, J. A.: 1975, "Carbon dioxide transport through enzymatically active synthetic membranes." Chem. Eng. Sci. 30, p. 103.
9. Lamb, J. D., Christensen, J. J., Oscarson, J. L., Nielsen, B. L., Asay, B. W., and Izatt, R. M.: 1980, "The relationship between complex stability constants and rates of cation transport through liquid membranes by macrocyclic carriers." J. Am. Chem. Soc. 102, p. 6820.
10. Christensen, J. J., Izatt, R. M., and Lamb, J. D.: 1980, "Selective transport of metal ions through liquid membranes containing macrocyclic carriers." Presented at the 73rd Annual AIChE Meeting, Chicago, Illinois.
11. Babcock, W. C., Baker, R. W., Lachapelle, E. D., and Smith, K. L.: 1980, "Coupled transport membranes. III. The rate-limiting step in uranium transport with a tertiary amine." J. Membrane Sci. 7, p. 89.
12. Frankenfeld, J. W., and Li, N. N.: 1977, "Waste water treatment by liquid ion exchange in liquid membrane systems," in Li, N.N. (Ed.): "Recent Developments in Separation Science." Vol. III B. CRC Press, Cleveland, Ohio.
13. Cussler, E. L., and Evans, D. F.: 1980, "Liquid membranes for separation and reactions." J. Membrane Sci. 6, p. 113.
14. Babcock, W. C., Baker, R. W., Lachapelle, E. D., and Smith, K. L.: 1980, "Coupled transport membranes. II. The mechanism of uranium transport with a tertiary amine." J. Membrane Sci. 7, p. 71.
15. Schultz, J. S.: 1971, "Passive asymmetric transport through biological membranes." Biophys. J. 11, p. 924.
16. Schultz, J. S.: 1980, "Facilitation of CO_2 through layers with a spatial distribution of carbonic anhydrase," in Bauer, C., Gros, G., and Bartels, H. (Eds.): "Biophysics and Physiology of Carbon Dioxide." Springer-Verlag, Berlin.
17. Ward, W. J. III: 1970, "Electrically induced carrier transport." Nature 227, p. 162.
18. Bonaventura, C., Bonaventura, J., Antonini, E., Brunori, M., and Wyman, J.: 1973, "Carbon monoxide binding by simple heme proteins under photodissociating conditions." Biochem. 12, p. 3424.
19. Schultz, J. S.: 1977, "Carrier-mediated photodiffusion membranes." Science 197, p. 1177.
20. Jain, R.: 1984, "Light-assisted transport of gases across supported liquid membranes." Ph.D. Thesis, University of Michigan.
21. Jain, R., and Schultz, J. S.: 1983, "An analysis of carrier-mediated photodiffusion membranes." J. Membrane Sci. 15, p. 63.
22. Shinkai, S., Nakaji, T., Ogawa, T., Shigematsu, K., and Manabe, O.: 1981, "Photoresponsive crown ethers. 2. Photocontrol of ion extraction and ion transport by a bis(crown ether) with a butterfly-like motion." J. Am. Chem. Soc. 103, p. 111.
23. Blank, M., Soo, L. M., Wasserman, N. H., and Erlanger, B. F.: 1981, "Photoregulated ion binding." Science 214, p. 70.

24. Flessner, M.: Master's Project, University of Michigan.
25. Lefevre, P. G.: 1962, "Rate and affinity in human red blood cell sugar transport." Am. J. Physiol. 203, p. 286.

SYMBOLS

C	molar concentration, mols/cm^3
$\hat{c}$	dimensionless concentration, C/C_T
e_o	enzyme concentration, mg/ml
J	flux of permeating species, mols/cm^2 sec
J_0	flux in absence of facilitation, mols/cm^2 sec
K_R	ratio of equilibrium constant in light to that in dark, $\underline{K}/K$
L	membrane thickness, cm
r	reaction rate, mols/cm^3 sec
$\dot{V}$	volumetric flow rate, cm^3/sec
x	flux coordinate
y	dimensionless coordinate, x/L
Z	dimensionless group, equation 3
α	dimensionless concentration gradient
β	permeability of linear film
δ	reaction layer dimension, equation 4
ζ	Damköhler number, equation 7
σ	parameter defined by equation 6
ϕ	dimensionless group, equation 10
Φ	dimensionless group, equation 5

Subscripts

A	transported species
B	carrier species
AB	complexed carrier species
eff	effective value
enz	enzyme
L	conditions at x = L
0	conditions at x = 0
opt	optimum value
T	total

Diacritical marks

—	Underline distinguishes values for conditions of illumination by light from conditions in dark.

LIQUID MEMBRANES

D. Bargeman and C.A. Smolders

Department of Chemical Technology
Twente University of Technology
P.O. Box 217, 7500 AE Enschede
The Netherlands

A new method for developing high-performance membranes is to use liquids as membrane materials. The liquid membrane generally separates two miscible liquids or gases and controls the mass transfer between the two phases.

Two types of liquid membranes are discussed, firstly the emulsion or liquid surfactant type membrane and secondly, the immobilized or supported liquid membrane where the separating liquid is present in the pores of a microporous membrane.

Recent literature, especially work reported since 1979, is reviewed and the possible application of the technique for the separation of organic compounds, waste-water treatment, the hydrometallurgical recovery of metal ions and the separation of gases is discussed. In particular the use of macrocyclic polyethers as carriers is described and a short survey of literature data of the mathematical modeling of liquid membrane transport is given.

P. M. Bungay et al. (eds.), Synthetic Membranes: Science, Engineering and Applications, 567–579.

1. INTRODUCTION

A new approach in developing high-performance membranes is to use a liquid as membrane material. Such a liquid membrane consists of a thin liquid film which generally separates two miscible liquids or gases and which controls the mass transfer between these two phases.

Two types of liquid membranes can be distinguished: the emulsion type, first described by Li in 1968 (1), and the immobilized liquid membrane, where the liquid is immobilized by impregnation in a porous substrate. The emulsion type membranes are normally formed by first making an emulsion of two immiscible phases followed by dispersing the emulsion in a third phase. The liquid membrane phase refers to the continuous phase of the emulsion prepared in the first step. A schematic representation of the two forms of liquid membranes is given in Figure 1.

The mass transport selectivity and transfer rate of liquid membranes are primarily limited by the solubilities and diffusion coefficients of the permeants in the membrane phase. The driving force for the transport process is the gradient of the chemical potential, which in most cases is nearly proportional to the concentration gradient.

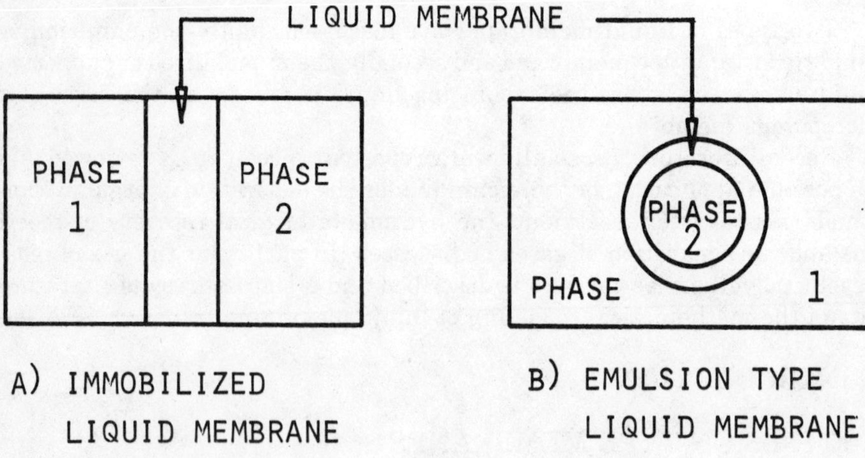

FIGURE 1A—Immobilized liquid membrane. FIGURE 1B—Emulsion type liquid membrane.

Several review papers on liquid membranes have appeared in the literature during the last few years. The theoretical progress and various numerical methods for solving the transport models were reviewed by Schultz *et al.* (2,3) and are further discussed by Schultz in Chapter 00. A summary of experimental and theoretical work since 1974 is presented by Smith *et al.* (4), a general review of liquid membranes is given by Halwachs and Schügerl (5); Marr and Kopp (6) and Way *et al.* (7). Kimura *et al.* (8) discuss potential industrial applications.

2. EMULSION TYPE LIQUID MEMBRANES

The emulsion type liquid membrane separation technique is a highly selective method of separating organics, inorganics and metal ions. Contrary to conventional solvent extraction, liquid membrane permeation is a simultaneous extraction-stripping process. The method was discovered by Li (1) in 1968. The first work dealt with the separation of a binary mixture of aromatic and paraffinic hydrocarbons (9).

Liquid membranes are formed by making an emulsion of two immiscible phases. The continuous phase of the emulsion forms the liquid membrane and the disperse droplets in the emulsion form the so-called "inner phase". The emulsion can be either oil-in-water or water-in-oil. The liquid membrane phase normally contains surfactants and additives, which are used to control the stability, the permeability and selectivity of the membrane. The emulsion then is dispersed in a third phase (the continuous outer phase), which is normally miscible with the inner phase. Figure 2 gives an example of a hydrophilic liquid membrane (10). The inner phase consists of a 50/50 toluene/heptane mixture which has to be separated. The membrane is composed of a saponin, glycerol and water mixture. Large droplets of the emulsion are dispersed in a kerosene solvent, which forms the continuous outer phase.

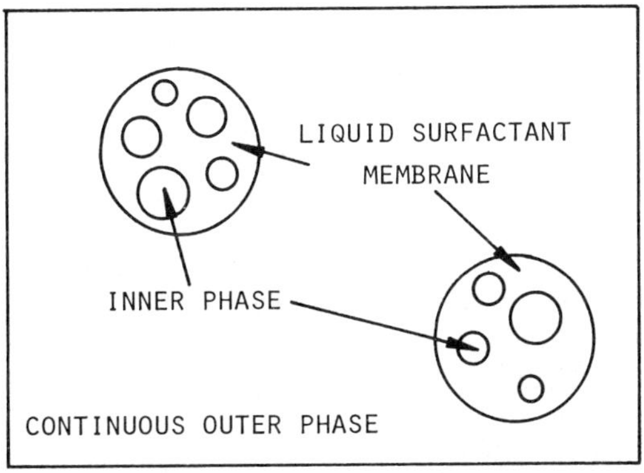

FIGURE 2—Schematic representation of an emulsion liquid membrane.

Figure 3 is a flowsheet for a hypothetical industrial-scale separation process (5). The heart of the plant is the mixer, which is fed with waste water. For the removal of weak acids (e.g., phenol, hydrogen-sulfide, acetic acid, mercaptans, etc.), caustic soda solution should be introduced as the inner phase. For the removal of weak bases (e.g., ammonia, amines, etc.), dilute sulfuric acid can be chosen for encapsulation. After the actual separation of the materials, the multi-emulsion is passed to a settler unit in which the liquid membrane phase and the continuous aqueous phase are separated because of the density difference. The recovered emulsion phase must be broken in the coalescing unit and the dispersed inner phase recuperated or discarded. The membrane phase, being the most valuable component, is recycled to the emulsifier, where fresh inner phase is introduced.

Since the introduction of the emulsion liquid membrane process by Li (1), there have been many reports of various liquid membrane formulations for separations of organics, inorganics and metal ions. Mass transfer studies in these systems have been concerned mostly with the influence of process parameters such as surfactant concentration, complexing agent concentration, feed stream composition and agitation rate.

2.1. Separation of Organic Compounds

A large number of hydrocarbon mixtures have been investigated by Li (9), e.g., n-hexanebenzene, n-heptanetoluene, n-octane-trimethylpentane, etc. Separation factors can be large; for instance, a separation factor $\alpha = 106$ was found for the benzene/n-hexane mixture. The separation factor is given by $\alpha = (c_1/c_2)_d/(c_1/c_2)_r$, in which d indicates solvent phase and r receiving phase; 1 and 2 are the components to be separated. It was shown that the

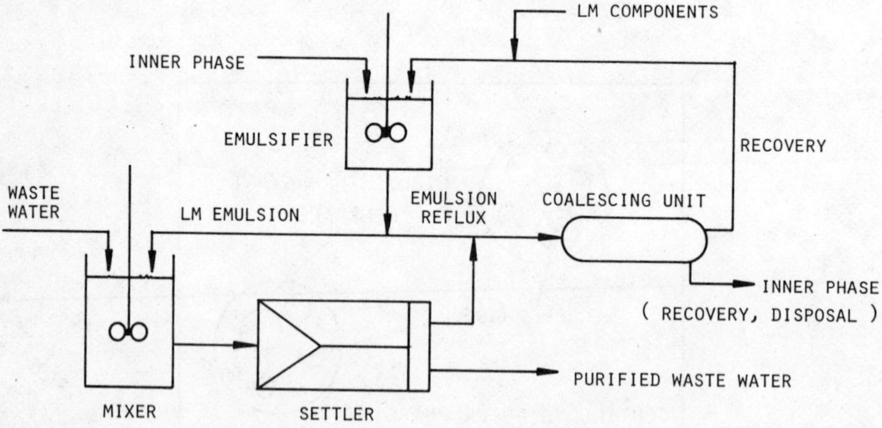

FIGURE 3—Flowsheet for wastewater purification with emulsion liquid membrane system.

selectivity depends on the solubility of the permeant in the water phase and on the diffusivity of the permeant through the liquid membrane.

The separation of mixtures of xylene isomers with liquid membranes was investigated by Kikic *et al.* (11). As a solvent in the outer phase they used n–pentane, N_2 or N_2 saturated with n–pentane. Gas stripping permeation seems to be better than permeation with a liquid solvent and competitive with other techniques under some conditions. Halwachs *et al.* (12) studied the permeation of n–hexane/benzene mixtures in a single-drop extraction column and in a multi-emulsion. Membrane viscosity, type and concentration of surfactant and ionic strength in the membrane were the most important parameters governing the selective permeation of benzene and n–hexane through liquid aqueous glycerol/water membranes. Benzene transport selectivities of up to 7.5 were observed at high surfactant concentrations (0.2 M) and 20 vol % glycerol.

Emulsion liquid membranes are also used in waste-water treatment. The removal of phenol from waste water has been investigated by a number of authors, including Li and Shrier (13) and Cahn and Li (14). Here the waste water feed constitutes the continuous outer phase. These investigators reported results of batch and continuous experiments in which factors affecting the removal of phenol from dilute solutions were examined. These factors were the degree of mixing of emulsion and feed, surfactant concentration, membrane to internal phase ratio and reagent concentration (caustic) in the internal phase. Halwachs *et al.* (12) removed phenol from an aqueous solution (pH = 7) by means of a silicone oil membrane in a multi-emulsion. Up to 95% removal of phenol from dilute phenol solutions (<1mM) could be achieved in some 20 minutes at a waste water/emulsion ratio of 12:1. More efficient phenol removal in a shorter mixing time was obtained by Terry, Li and Ho (15). More than 99% of phenol can be extracted in less than 1 minute when a membrane phase of a 4 wt.% polyamine-derivative surfactant (16) in a paraffin solvent is used. The feed was 5000 ppm phenol by weight and the inner phase was a 10% by weight NaOH solution. The membrane was mixed with internal phase on a 1:1 weight ratio and the feed/emulsion ratio was 5:1. Experimental data on the extraction of phenol from waste water were used to check model descriptions of mass transport through liquid membranes (17,18).

2.2. Metal Ion Separations

The use of emulsion liquid membranes can be of considerable interest for the hydrometallurgical recovery of metal ions. In particular, liquid membrane formulations for Cu^{2+} recovery are described in the literature. In all cases water-in-oil emulsions are dispersed in the feed solution and carrier molecules are added to the membrane phase (oil). The transport of the copper ions is coupled with a counter transport of protons. Hence a pH gradient across the membrane can cause copper ion permeation into the inner phase even against the copper concentration gradient. The transport is given schematically by

$$Cu^{2+} + 2RH \rightleftarrows CuR_2 + 2H^+ \tag{1}$$

Table 1 gives a survey of the research in the field of metal ion separation with the emulsion type liquid membranes. Factors influencing mass transfer such as membrane composition, pH, acid content of the feed and internal phase, agitation rate, bulk viscosity, carrier concentration, feed-emulsion ratio, complex-formation rate, etc., were investigated. From the results it follows that the application of the emulsion liquid membrane process (as compared with other processes such as liquid–liquid extraction) is especially promising for waste water treatment, the recovery of substances from dilute solutions and in hydrometallurgy. An important advantage of the method is that the carrier concentration can be significantly lower than that in liquid–liquid extraction and therefore a smaller loss of carrier agent can be expected. A comparison of pertraction through liquid membranes and double liquid–liquid extraction is given by Schlosser and Kossaczky (30).

TABLE 1.

Metal ion	Counter ion or co-ion	Carrier	Membrane phase	Reference
Zn^{2+}, Cu^{2+}, Cd^{2+}, Pb^{2+}	H^+	fatty acid ($C_{14}-C_{20}$)	kerosene	19
Cu^{2+}, Ni^{2+}	H^+	LIX 64 N^0	mineral oil	20
NH_4^+, Cu^{2+}, Cd^{2+}, Hg^{2+}	H^+	LIX 64 N	oil	21
Zn^{2+}	H^+	Di-2-ethylhexyl-phosphoric acid	paraffin oil	22
Tl^+, Pb^{2+}		crown ether	toluene	23
Pb^{2+}		crown ether	oil	24
Hg^{2+}	H^+	dibutylbenzoyl-thiourea	hexane decane	25
Hg^{2+}	OH^-	trioctylamine	xylene	26
Cu^{2+}	H^+	Shell SME 529	?	27
Cu^{2+}	H^+	LIX 64 N	kerosene	28
Cu^{2+}	H^+	benzoylacetone	toluene	29

3. SUPPORTED LIQUID MEMBRANES

These liquid membranes are obtained by impregnating a microporous polymer support with a liquid containing a carrier molecule. The published reports on immobilized or supported liquid membranes have dealt with the transport and separation of olefins, metal ions and gases. The majority of the investigations demonstrated facilitated transport in membranes with a carrier system under controlled laboratory conditions.

3.1. Separation and Transport of Ions

Many authors have recently reported several examples of coupled transport separations. With the proper selection of carrier agent and working conditions, very selective separations of metal ions can be made and the ion of interest can be concentrated against its concentration gradient. Baker *et al.* (31) discussed copper transport and separation from a mixture of copper and iron ions. Celgard 2400, a microporous polypropylene film with a porosity of about 35% and a pore size of 0.1 μm or less and filled with LIX 64 N was used as the liquid membrane. Copper was concentrated against a 4000-fold concentration difference and separated from iron with a separation factor greater than 1000. A counter transport of protons, due to a difference of two pH-units, acts as the driving force. From a short economic assessment based on the flux they conclude that the process should be competitive with the best current technology. The mechanism of copper transport was investigated by Kataoka *et al.* (32), who used a Teflon millipore filter with a thickness of 125 μm, a porosity of 68% and a pore size of 10 μm. They found that the permeation rate of copper ion depends on the diffusion process in the membrane for a chelating complex between copper and carrier and by the succeeding reaction at the interface between membrane and stripping phase.

Babcock *et al.* (33, 34) studied the transport of uranyl ions complexed with sulfate. Porous polypropylene, polyethylene and teflon support with pore diameters ranging from 0.02–5.0 μm were impregnated with a hydrocarbon solvent containing a tertiary amine as carrier molecule. Both co-transport of protons and counter-transport of bisulfate ions were used. Based on the results obtained, co-transport is the preferred mechanism for concentrating uranium. The uranium flux is affected principally by the concentrations of uranyl and coupled ion in the aqueous solutions, but the base strength of the tertiary amine is also an important parameter. The size of the pores in the support also affects the flux; interaction with the pore walls in membranes with small pores hinders diffusion. The transport and separation of acids were investigated by Imato *et al.* (35). High molecular weight amines were used as carriers in the liquid membrane. The permselectivity to the anions was correlated with the diffusivity of ion pairs in the membrane and the relative ion exchange constant at the membrane–solution interface. The concentration profile of ion pairs under steady state conditions in a stack of liquid membranes was linear with the distance.

Macrocyclic polyethers have been extensively studied for their ability to bind and transfer ions across liquid membranes. The majority of the work was done in modified U-tubes where the feed and product phase were separated by a bulk oil layer acting as a liquid membrane. The transport of ammonium and alkylammonium cations by macrocyclic carriers was investigated by Izatt *et al.* (36). A large number of authors have investigated transport rates of single monovalent and divalent cations and of binary cation mixtures using macrocyclic crown ether ligands (37–45).

Christensen *et al.* (37) correlated the alkali cation transport rates with equilibrium constant values for cation-macrocyclic interaction in methanol. Pannell *et al.* (38) reported on the variation in transport rates of Na^+ and K^+ across a bulk chloroform membrane as a function of substituted dibenzo-18-crown-6 ionophores and temperature. Metal complexation by the ionophore proved to be the controlling feature for this range of ionophores. Yoshida and Hayano (39) showed that the fluxes of cations through the membranes and cation distribution ratios between aqueous solution and membrane (dibenzo-18-crown-6 in chloroform) were strongly dependent on the anions present. Hiratani *et al.* (40), who used noncyclic polyether ionophores, found that the amounts of alkali ions transported and the transport rates depend upon the kind of anion species in the receiving phase. The amount of cation transported increased when an acid could not be counter-transported in the membrane system. Strzelbicki and Batsch (41) investigated competitive alkali metal transport from an alkaline source to an acidic receiving phase through a chloroform membrane. Nine different crown ethers with pendant carboxylic groups were used. Increase of the lipophilicity of the carrier enhanced the transport rate but did not affect the selectivity when the polyether cavity size remained constant. Lamb *et al.* (42) selected carriers for selective transport of various cations (Na^+ or Cs^+ or Sr^{2+}). Relative fluxes from binary cation mixtures were rationalized in terms of macrocycle cavity size, donor atom type and ring substituents. These authors (43) also reported on the use of 18-crown-6 to transport europium after first reducing it to the bivalent oxidation state. The best carrier for ion transport is a ligand that gives a moderately stable rather than a very stable complex (44), because a very stable complex that rapidly extracts ions into the membrane phase does not release the ion efficiently from the complex to the receiving phase. Therefore, Shinkai *et al.* (45) synthesized a bis (crown ether) with an azo linkage so that the stability of the carrier complex could be influenced by light. Shimidzu and Yoshikawa (46) also used light as a driving force for transport of ions across an octanol liquid membrane with spiropyron compounds as transporting carriers.

3.2. Separation and Transport of Gases

The work on facilitated transport started by considering biologically important systems, e.g., the transport of O_2 through thin films. At an early stage Scholander (47) measured the steady-state flux of O_2 through microporous cellulose acetate membranes impregnated with an aqueous hemoglobin solution. Most of the work for the development of industrial applications deals with CO_2, H_2S, NO_x and the separation of unsaturated hydrocarbons. Review papers are given by Kimura, Matson and Ward (8) and by Meldon, Stroeve and Gregoire (48).

One of the most difficult problems in attaining a practical membrane system is the effective immobilization of the liquid membrane. Hydrophobic membranes with high gas permeabilities (silicone rubber membranes) or non-wetting porous polymer membranes (microporous fluorocarbon) are used to sandwich the liquid membrane, which consists of a porous membrane impregnated with aqueous solution.

CO_2 transport in HCO_3^-/CO_3^{2-} solutions depends on a number of reactions. In the pH range of industrial significance at pH 9 to 10 the rate controlling reaction for CO_2 transport is

$$CO_2 + OH^- \rightleftarrows HCO_3^- \tag{2}$$

The total transport rate could be increased markedly by using effective catalysts for the reaction:

$$CO_2 + H_2O \rightleftarrows H_2CO_3 \rightleftarrows HCO_3^- + H^+ \tag{3}$$

Enns (49) demonstrated experimentally that the CO_2 flux increased by catalyzing this reaction with carbonic anhydrase. Ward and Robb (50) used sodium arsenite. Because equilibrium for reaction is far to the right, the contributions of undissociated carbonic acid to the total CO_2 content in solution and the transport of the gas are negligible. Ward and Robb (50) achieved a separation factor of 4100 for the removal of CO_2 from a CO_2/O_2 mixture. The separation of CO_2 from H_2S-free natural gas was studied by Kimura *et al.* (8); they found most effective catalysts to be TeO_3^{2-} and AsO_2^-, with estimated savings of 30–50%, compared with gas scrubbing with hot potassium carbonate or cold methanol systems. Particularly high CO_2 selectivities were obtained by LeBlanc *et al.* (51) with cation exchange membranes containing ethylene-diamine (EDA). A strong acid cation exchange membrane was soaked in aqueous EDA solution and adjusted to pH 11 with HCl in order to introduce EDA monopositive cation as the counter ion. The unprotonated amine group reacts reversibly with CO_2 to form a carbamate. A CO_2/N_2 permeability ratio of 600 was obtained.

The removal of H_2S from gasified coal was investigated by Matson *et al.* (52). The liquid membrane consisted of a carbonate solution, which has a high H_2S permeability and greater selectivity than conventional hot carbonate scrubbers. They also studied membrane life time. Within 10^3 hours no appreciable decrease in membrane permeability was observed. Kimura *et al.* (8), who measured H_2S permeabilities in CO_2/H_2S mixtures, noted that H_2S selectivity over CO_2 could be improved by introducing gas gaps in multilayer membranes. The flux of H_2S was little affected by the gas barrier, whereas forcing the CO_2 reactions to occur an additional number of times decreased the CO_2 flux.

The transport of NO through a liquid layer immobilized between two thin polymeric films was investigated by Ward (53, 54). Fe^{2+} ions were used as carriers. The numerical solution given for the coupled diffusion-reaction equation did fit the experimental data. Bdzil *et al.* (55) presented a mathematical analysis of electrically induced carrier transport in the NO/Fe^{2+} system.

The potential application of liquid membranes to the separation of olefinic hydrocarbons from gas mixtures is described in a patent by Steigelman and Hughes (56). High solubilities of unsaturated hydrocarbons can be achieved in silver solutions.

Facilitated transport of ethylene was investigated by LeBlanc *et al.* (51). A cation-exchange membrane of sulfonated polyphenylene oxide was converted to the Ag^+ form by soaking in a silver nitrate solution. The immobilized silver ions were used as carriers for the transport of ethylene. The ethylene permeability at 25°C was about 300 times the ethane permeability under corresponding conditions.

4. MATHEMATICAL MODELING OF LIQUID MEMBRANE TRANSPORT

A number of analyses based on modeling theory exist for liquid membrane permeation. In general, the treatment of liquid membrane transport based on Fick's law for each permeant and the corresponding permeant complexes requires a numerical treatment of the coupled partial differential equations and is hence only applicable under certain conditions. Schultz *et al.* (2) reviewed approximate solutions for the requisite differential equations and gave criteria for evaluating whether a membrane is in the diffusion or equilibrium regime. They also reviewed the mathematical analysis of facilitated transport (3). Goddard (57) gives an extensive review of facilitated transport theory with applications in the region near reaction equilibrium. Kremesec (58) reviewed two approaches for modeling of the dispersed-emulsion separation system. Both methods yield expressions that must be curve fitted to the data to obtain the effective mass transfer rate parameters. Ho *et al.* (18) formulated a model for the calculation of solute extraction rates in the presence of a diffusion-controlled reaction by uniform emulsion droplets dispersed in a well-mixed fluid. A regular perturbation solution to the model equations was developed and experimentally tested using the batch extraction of phenol from water by aqueous caustic solutions. Donaldson and Lapinas (59) recently analyzed mass transfer enhancement by carrier transport for two independent carrier species in parallel. Additional secondary enhancement or depression was shown to occur. Folkner and Noble (60) presented a graphical solution for the transient flow of a permeate under facilitated transport for various geometries. This permitted determination of the time to reach steady-state and the steady-state flux of the solute.

REFERENCES

1. Li, N.N.: 1968, US Pat. 3 410 794.
2. Schultz, J.S., Goddard, J.D., and Suchdeo, S.R.: 1974, "Facilitated transport via carrier-mediated diffusion in membranes. Part 1. Mechanistic aspects, experimental systems and characteristic regimes." AIChE J. 20, pp. 417–444.
3. Schultz, J.S., Goddard, J.D., and Suchdeo, S.R.: 1974, "Facilitated transport via carrier-mediated diffusion in membranes. Part 2. Mathematical aspects and analysis." AIChE J. 20, pp. 625–639.
4. Smith, D.R., Lander, R.J., and Quinn, J.A.: 1977, in "Recent developments in separation science." Vol. 3, Li, N.N. (Ed.), CRC Press, Cleveland, Ohio.
5. Halwachs, W., and Schügerl, K.: 1980, "The liquid membrane technique—a promising extraction process." Int. Chem. Eng. 20, pp. 519–528.
6. Marr, R., and Kopp, A.: 1980, "Flüssigmembran-Technik: Übersicht über Phänomene, Transportmechanismen und Modellbildungen." Chem. Eng. Tech. 52, pp. 399–410.
7. Way, J.D., Noble, R.D., Flynn, T.M., and Sloan, E.D.: 1982, "Liquid membrane transport: A survey." J. Membrane Sci. 12, pp. 239–259.
8. Kimura, S.G., Matson, S.L., and W.J. Ward III: 1979, in "Recent developments in separation science." Vol. 5, Li, N.N. (Ed.), CRC Press, Cleveland, Ohio, pp. 11–25.
9. Li, N.N.: 1981, "Permeation through liquid surfactant membranes." AIChE J. 17, pp. 459–463.
10. Cahn, R.P., and Li, N.N.: 1976, "Separations of organic compounds by liquid membrane processes." J. Membrane Sci. 1, pp. 129–146.
11. Kikic, J., Alessi, P., and Orlandi Visalberghi, M.: 1978, "Liquid membrane permeation for the separation of xylenes." Inst. Chem. Eng. Symp. Ser. 54, pp. 153–164.
12. Halwachs, W., Flaschel, E., and Schügerl, K.: 1980, "Liquid membrane transport—a highly selective separation process for organic solutes." J. Membrane Sci. 6, pp. 33–44.
13. Li, N.N., and Shrier, A.L.: 1972, "Liquid membrane water treating." Recent Devel. Sep. Sci. 1, pp. 163–174.
14. Cahn, R.P., and Li, N.N.: 1974, "Separation of phenol from waste water by the liquid membrane technique." Sep. Sci. 9, pp. 505–519.
15. Terry, R.E., Li, N.N., and Ho, W.S.: 1982, "Extraction of phenolic compounds and organic acids by liquid membranes." J. Membrane Sci. 10, pp. 305–323.
16. Li, N.N.: 1981, US Patent 4 259 183, March 31.
17. Gladek, L., Stelmaszek, J., and Szust, J.: 1982, "Modeling of mass transport with a very fast reaction through liquid membranes." J. Membrane Sci. 12, pp. 153–167.
18. Ho, W.S., Hatton, T.A., Lightfoot, E.N., and Li, N.N.: 1982, "Batch extraction with liquid surfactant membranes: A diffusion-controlled model." AIChE J. 28, pp. 662–670.
19. Boyadzhiev, L., Kyuchoukov, G.: 1980, "Further development of carrier-mediated extraction." J. Membrane Sci. 6, pp. 107–112.
20. Evans, D.F., Duffey, M.E., Lee, K.H., and Cussler, E.L.: 1976, in "Colloid and Interface Science," Vol. V, Kerker, M. (Ed.), Acad. Press, New York, pp. 119–132.
21. Kitagawa, T., Nishikawa, Y., Frankenfeld, J.W., and Li, N.N.: 1977, "Wastewater treatment by liquid membrane process." Environ. Sci. Technol. 11, pp. 602–605.
22. Boyadzhiev, L., and Bensenshek, E.: 1982, "Zink-Rückgewinnung aus wässrigen Lösungen mittels Extraktion mit einer Carrier-Phase." Chem. Ing. Tech. 54, pp. 506–508.
23. Izatt, R.M., Biehl, M.P., Lamb, J.D., and Christensen, J.J.: 1982, "Rapid separation of thallium (+) and lead (2+) from various binary cation mixtures using dicyclohexano-16-crown-6 incorporated into emulsion membranes." Sep. Sci. Technol. 17, pp. 1351–1360.

24. Biehl, M.P., Izatt, R.M., Lamb, J.D., and Christensen, J.J.: 1982, "Use of a macrocyclic crown ether in an emulsion (liquid surfactant) membrane to effect rapid separation of Pb^{2+} from cation mixtures." Sep. Sci. Technol. 17, pp. 289–294.
25. Weiss, S., Grigoriev, V., and Mühl, P.: 1982, "The liquid membrane process for the separation of mercury from waste water." J. Membrane Sci. 12, pp. 119–129.
26. Schiffer, D.K., Choy, M., Evans, D.F., and Cussler, E.L.: 1974, "More membrane pumps." AIChE J. Symp. Ser. 70, pp. 150–156.
27. Martin, T.P., and Davies, G.A.: 1977, "The extraction of copper from dilute aqueous solutions using a liquid membrane process." Hydrometallurgy, pp. 315–334.
28. Völkel, W., Halwachs, W., and Schügerl, K.: 1980, "Copper extraction by means of a liquid surfactant membrane process." J. Membrane Sci. 6, pp. 19–31.
29. Kondo, K., Kita, K., Korda, J., and Irie, J.: 1979, "Extraction of copper with liquid surfactant membranes containing benzoylacetone." J. Chem. Eng. Japan 12, pp. 203–209.
30. Schlosser, S., Kassaczky, E.: 1980, "Comparison of liquid pertraction through liquid membranes and double liquid–liquid extraction." J. Membrane Sci. 6, pp. 83–105.
31. Baker, R.W., Tuttle, M.E., Kelly, D.J., and Lonsdale, H.K.: 1977, "Coupled transport membranes. I. Copper separations." J. Membrane Sci. 2, pp. 213–233.
32. Kataoka, T., Nishika, T., and Ueyama, K.: 1982, "Mechanism of copper transport through a diaphragm-type liquid membrane." Bull. Chem. Soc. Japan 55, pp. 1306–1309.
33. Babcock, W.C., Baker, R.W., LaChapelle, E.D., and Smith, K.L.: 1980, "Coupled transport membranes. II. The mechanism of uranium transport with a tertiary amine." J. Membrane Sci. 7, pp. 71–87.
34. Babcock, W.C., Baker, R.W., LaChapelle, E.D., and Smith, K.L.: 1980, "Coupled transport membranes. III. The rate-limiting step in uranium transport with a tertiary amine." J. Membrane Sci. 7, pp. 89–100.
35. Imato, T., Yabu, H., Morooka, S., and Kato, Y.: 1982, "Evaluation of permselectivity of a liquid anion-exchange membrane by diffusion dialysis." J. Membrane Sci. 10, pp. 21–33.
36. Izatt, R.M., Nielsen, B.L., Christensen, J.J., and Lamb, J.D.: 1981, "Membrane transport of ammonium and alkylammonium cations using macrocyclic carriers." J. Membrane Sci. 9, pp. 263–271.
37. Christensen, J.J., Lamb, J.D., Brown, P.R., Oscarson, J.L., Izatt, R.M.: 1981, "Liquid membrane separations of metal cations using macrocyclic carriers." Sep. Sci. Technol. 16, pp. 1193–1215.
38. Pannell, K.H., Rodrigues, B.J., Chiocca, S., Jones, L.P., Molinar, J.: 1982, "Dibenzo-crown facilitated transport across a $CHCl_3$ liquid membrane." J. Membrane Sci. 11, pp. 169–175.
39. Yoshida, S., and Hayano, S.: 1982, "Kinetics of partition between aqueous solutions of salts and bulk liquid membranes containing neutral carriers." J. Membrane Sci. 11, pp. 157–168.
40. Hiratani, K., Nozawa, I., Nakagawa, T., and Yamada, S.: 1982, "Synthesis and properties of noncyclic polyether compounds. Part V. The influence of acid and concentration of cations on the uphill transport rates of cations and selectivities through liquid membranes by synthetic, noncyclic polyether ionophores." J. Membrane Sci. 12, pp. 207–215.
41. Strzelbicki, J., and Bartsch, R.A.: 1982, "Transport of alkali metal cations across liquid membranes by crown ether carboxylic acids." J. Membrane Sci. 10, pp. 35–47.
42. Lamb, J.D., Brown, P.R., Christensen, J.J., Bradshaw, J.S., Garrick, D.G., and Izatt, R.M.: 1983, "Cation transport at 25°C from binary Na^+–M^{n+}, Cs^+–M^{n+} and Sr^{2+}–M^{n+} nitrate mixtures in a H_2O–$CHCl_3$–H_2O liquid membrane system containing a series of macrocyclic carriers." J. Membrane Sci. 13, pp. 89–100.
43. Brown, P.R., Izatt, R.M., Christensen, J.J., and Lamb, J.D.: 1983, "Transport of Eu^{2+} in a H_2O–$CHCl_3$–H_2O liquid membrane system containing the macrocyclic polyether 18-crown-6." J. Membrane Sci. 13, pp. 85–88.
44. Kobuke, Y., Hanji, K., Horiguchi, K., Asada, M., Nakayama, Y., and Furukawa, J.: 1976, "Macrocyclic ligands composed of tetrahydrofuran for selective transport of monovalent cations through liquid membrane." J. Am. Chem. Soc. 98, pp. 7414–7419.

45. Shinkai, S., Nakaji, T., Ogawa, T., Shigematsu, K., and Manabe, O.: 1981, "Photoresponse crown ethers. 2. Photocontrol of ion extraction and ion transport by a bis(crown ether) with a butterfly-like motion." J. Am. Chem. Soc. 103, pp. 111–115.
46. Shimidzu, T., and Yoshikawa, M.: 1983, "Photo-induced carrier-mediated transport of alkali metal salts." J. Membrane Sci. 13, pp. 1–13.
47. Scholander, P.F.: 1960, "Oxygen transport through hemoglobin solutions." Science 131, pp. 585–590.
48. Meldon, J.H., Stroeve, P., and Gregoire, C.E.: 1976, "Facilitated transport of carbon dioxide: A review." Chem. Eng. Commun. 16, pp. 263–300.
49. Enns, T.: 1976, "Facilitation by carbonic anhydrase of carbon dioxide transport." Science 155, pp. 44–47.
50. Ward, W.J., and Robb, W.L.: 1967, "Carbon dioxide–oxygen separation: Facilitated transport of carbon dioxide across a liquid film." Science 156, pp. 1481–1484.
51. Leblanc, O.H., Ward, W.J., Matson, S.L., and Kimura, S.: 1980, "Facilitated transport in ion-exchange membranes." J. Membrane Sci. 6, pp. 339–343.
52. Matson, S.L., Herrick, C.S., and Ward III, W.J.: 1977, "Progress on the selective removal of H_2S from gasified coal using an immobilized liquid membrane." Ind. Eng. Chem. Proc. Des. Develop. 16, pp. 370–374.
53. Ward, W.J.: 1970, "Analytical and experimental studies of facilitated transport." AIChE J. 16, pp. 405–410.
54. Ward, W.J.: 1970, "Electrically induced carrier transport." Nature 227, pp. 162–163.
55. Bdzil, J., Carlier, C.C., Frisch, H.L., Ward III, W.J., and Breiter, M.W.: 1973, "Analysis of potential difference in electrically induced carrier transport systems." J. Phys. Chem. 77, pp. 846–850.
56. Steigelman, E.F., and Hughes, R.D.: 1973, US Patent 3758603, September 11.
57. Goddard, J.D.: 1977, "Further applications of carrier-mediated transport theory—A survey." Chem. Eng. Sci. 32, pp. 795–809.
58. Kremesec, V.J.: 1981, "Modeling of dispersed-emulsion separation systems." Sep. and Purif. Methods 10, pp. 117–157.
59. Donaldson, T.L., and Lapinas, A.T.: 1982, "Secondary flux enhancement in two-carrier facilitated transport." Chem. Eng. Sci. 37, pp. 715–718.
60. Folkner, C.A., and Noble, R.D.: 1983, "Transient response of facilitated transport membranes." J. Membrane Sci. 12, pp. 289-301.

CONTROLLED RELEASE DELIVERY SYSTEMS

Richard W. Baker

Membrane Technology Research Inc.
Menlo Park, California 94025

and

Lynda M. Sanders

Syntex Research Corporation
Palo Alto, California 94304

Controlled release systems offer significant advantages over conventional systems for the delivery of biologically active compounds in both the pharmaceutical and agrichemical fields. The advantages and disadvantages of various types of system design are discussed from the perspective of each of these major areas of application. Both erodible and non-erodible systems, providing release of agents by various mechanisms with characteristic rate profiles, are covered. The choice of type of system will depend heavily on its intended use and the rate (and uniformity of rate) and duration of release of compound required. The properties of release profiles provided by various designs of system are qualitatively and quantitatively described, with respect to areas of current and potential application.

P. M. Bungay et al. (eds.), Synthetic Membranes: Science, Engineering and Applications, 581–624.

1. INTRODUCTION

In its broadest sense, the concept of sustained or prolonged release of biologically active agents has existed for decades. In the pharmaceutical field, sustained release has been widely used in oral medication since the early 1950's. Enteric coating of such dosage forms as tablets with pH-sensitive materials has been and is very common. Similarly, encapsulated pellets or beads have been used, as have sparingly soluble salts, complexed systems, and porous insoluble tablets containing dispersed drug.

In the field of agriculture, slow release fertilizers have been available for a similar length of time, with various methods being used to achieve the desired release profile. They include the use of materials that are slowly activated by microbial attack, complexation of the active agent with an ion-exchange resin, and the use of compounds that are very slowly water-soluble.

Membrane-regulated formulations, a more sophisticated approach in which the active agent is released by diffusion through a surrounding membrane, have also been used in this field.

The majority of these early products can be called "sustained release" systems, in that the release of the active agent, although slower than that of conventional formulation, is still substantially affected by the external environment into which it is released. For example, sustained release fertilizers, which rely on the slow dissolution of the active component, are subject to considerable variation in delivery rate due to variation in environmental factors such as rainfall and atmospheric humidity. Controlled release systems, with which this paper is concerned, provide predictable release kinetics largely independent of such external factors, being controlled by the design of the system itself.

1.1. Delivery Rates

The pattern of release achieved by a controlled release system can vary over a wide range, but most release profiles can be categorized into three types. In the simplest of these, the release rate remains constant until the device is essentially exhausted of active agent.

Mathematically, the release rate $\frac{dM_t}{dt}$, from this device is given as

$$\frac{dM_t}{dt} = k_0 \tag{1}$$

where k_0 is constant. This type of release pattern is commonly called zero order release, a term suggested by chemical kinetics.

The second common type of release kinetics is first order release. The release rate in this case will be proportional to the mass of active agent, M_t, released from the device up to time t. The release rate is then given as

$$\frac{dM_t}{dt} = k_1(M_0 - M_t) \tag{2}$$

which on integration and rearrangement gives

$$\frac{dM_t}{dt} = k_1 M_0 \exp(-k_1 t) \tag{3}$$

where M_0 is the mass of agent in the device at $t = 0$. In first order release, therefore, the rate declines exponentially with time, approaching a release rate of zero as the device approaches exhaustion.

The third common release pattern is release proportional to the square root of time. The rate in this case again falls with time, proportional to $1/\sqrt{t}$.

The release rate is given as

$$\frac{dM_t}{dt} = \frac{k_d}{\sqrt{t}} \tag{4}$$

In contrast to first order release, the release rate here will remain finite as the device approaches exhaustion. The release pattern for each of these classes of device is illustrated in Figure 1.

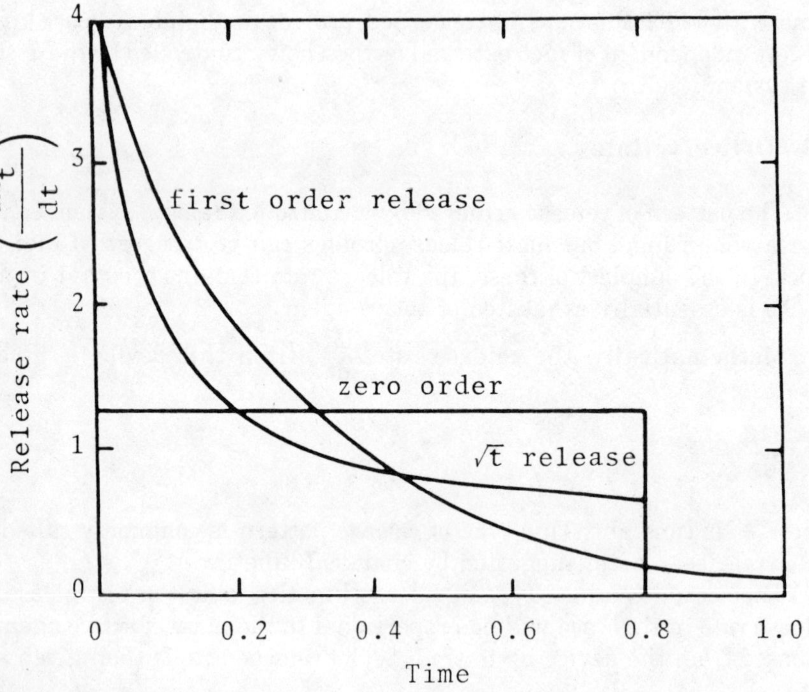

FIGURE 1.—Zero order, first order, and $\sqrt{t}$ release patterns from delivery containing the same initial amount of active agent.

1.2. The Advantages of Controlled Release

Controlled release delivery systems offer five principal advantages over conventional formulations which deliver the active agent in a bolus.

More constant levels of agent One advantage of controlled delivery can be appreciated by examining Figure 2. It illustrates the concentration of material in the local environment as a function of time after delivery in a conventional manner.

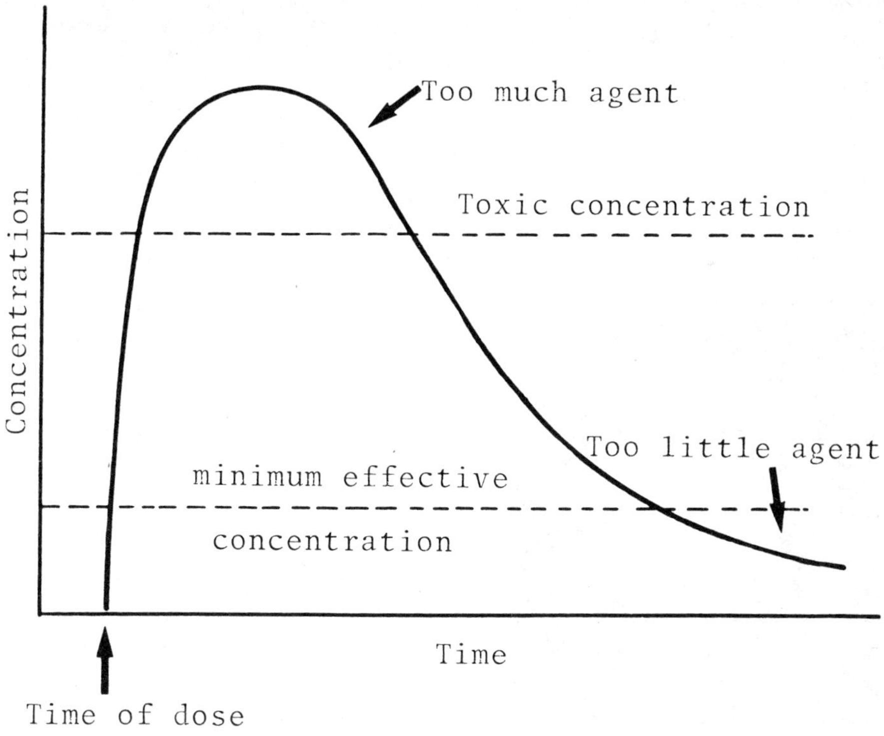

FIGURE 2.—Typical time course of concentration of an active agent in the environment from a conventional delivery formulation showing the potential for over- and under-dosing.

Typically, the concentration rises rapidly to a maximum after the agent is applied. The level then slowly falls as the material is metabolized, excreted, or degraded. Figure 2 illustrates two important concentration levels: the minimum effective concentration, and the toxic concentration above which undesirable "side effects" occur. For example, too little of an insecticide will not kill the target species while too much can kill both pests and plants and endanger humans. It is important, therefore, to maintain the concentration of the active agent between the minimum effective and toxic levels. This is inherently difficult with conventional formulations that tend first to overdose and then underdose the local environment, but is more readily attained with a controlled release device. The Ocusert®, (Alza Corporation, Palo Alto, CA) a controlled reservoir device to deliver pilocarpine to the eye, illustrates this capability (1–3). The device is shown in Figure 3.

More efficient utilization of the agent As a result of the approximately constant levels of the active agent achieved with a controlled release device, considerably less material is usually required to produce a given duration of

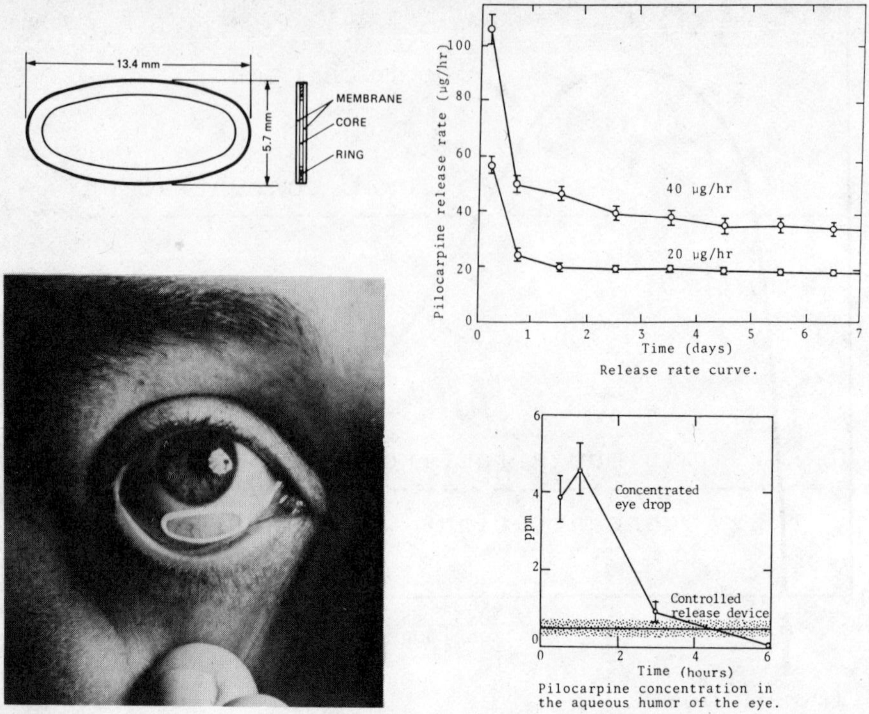

FIGURE 3.—The Ocusert® pilocarpine system is a thin multi-layer membrane device. The central sandwich consists of a core containing the drug pilocarpine. The device is placed in the eye, where it releases the drug at a continuous rate for seven days. Devices with release rates of 20 or 40 μg/hr are used. Controlled release of the drug eliminates over- and under-dosing observed with conventional eyedrop formulations.

effect than with a conventional system. It can be shown (4) that to produce a linear increase in duration of the effective levels of a compound delivered in a conventional form, exponentially greater quantities of material need to be applied. However, when the agent is applied in a controlled release form that releases the compound at a rate just sufficient to maintain the agent at the effective level M_e it can be shown that the required release rate is:

$$\frac{dM_t}{dt} = kM_e = \frac{M_e \ln 2}{t_{1/2}} \tag{5}$$

where $t_{1/2}$ is the half-life of the agent. Thus a dose M_0 applied in an ideal controlled release form will last a period t_e:

$$t_e = \frac{t_{1/2}\,(M_0 - M_e)}{M_e \ln 2} \tag{6}$$

Plots of the effective durations of controlled release and conventional formulations are shown in Figure 4. In this figure the time axis has been normalized for agents with different decay constants by plotting time in terms of half-lives of degradation. This plot shows that controlled delivery is always more efficient than a single dose, particularly if the half-life of the agent is short compared with the desired treatment period. The potential result is improved economics of use and decreased hazard to the user.

Site of action delivery A third important benefit of controlled release is that it may be possible to deliver the agent locally and to contain it at the site of action. In medicine, this reduces the dosage of drug required and consequently the possibility of side effects. In the agricultural area, this allows the localization of insecticides and herbicides so that spreading by wind or

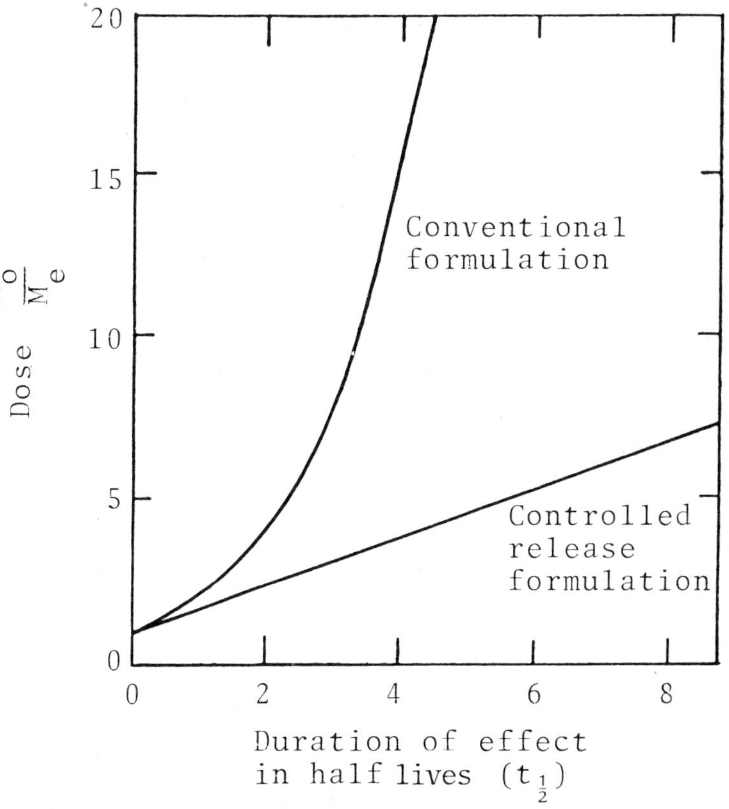

FIGURE 4.—Dosages required to maintain effective environmental concentration with conventional formulation and controlled release formulation vs. half-life. Because the controlled release formulation meters out replacement active agent at the rate it is consumed or degraded, much smaller amounts of active agent are required to produce a long-lasting effect (4).

runoff is avoided thereby reducing the exposure to non-target species. This property of controlled release systems is utilized by drug-releasing intrauterine devices such as the Progestasert® (Alza Corporation, Palo Alto, CA) shown in Figure 5 (5,6).

Less frequent administration Another advantage of controlled delivery is that fewer applications of the material will be required. In medicine, this will

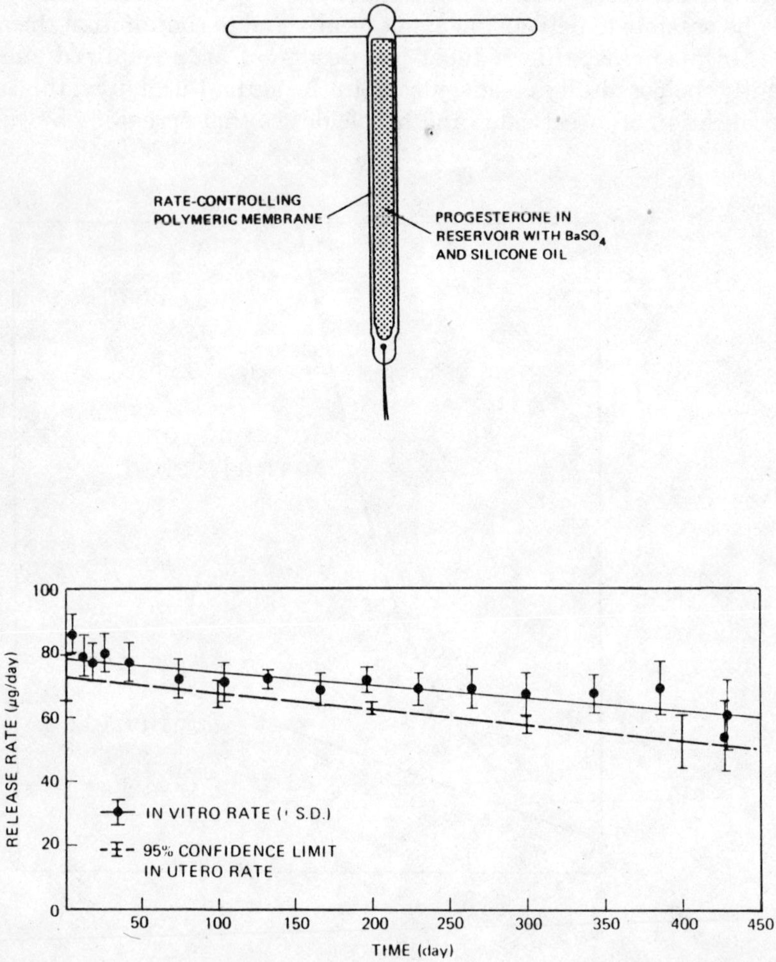

FIGURE 5.—The Alza Progestasert® system is a polyethylene vinyl acetate T-shaped IUD. When placed in the uterus of a woman, the device releases the steroid, progesterone, at an average rate of 65 μg/day for more than one year. This release rate is much lower than the normal progesterone secretions of the adrenal and pituitary glands. However, the site of action delivery of the IUD produces a contraceptive effect.

result in greater convenience and therefore compliance by a patient for the dosing regimen, and consequently a higher response rate to the medication. In agriculture it is economically highly advantageous to minimize the numbers of necessary applications of a pesticide or fertilizer, and this is most effectively realized by use of controlled release preparations.

Qualitatively different response Until recently, controlled release preparations have been seen as useful in adjusting the quantitative aspects of administration of medicinal or agricultural materials. However, it may also be possible to bring about a qualitatively different response by this technology. For example, a luteinising hormone-releasing hormone (LHRH) analog, when given intermittently, is effective in delaying the progesterone surge in the female rhesus monkey, and thereby suppressing ovulation, but the luteinising hormone (LH) surge is not depressed. When the LHRH analog is administered continuously, in a biodegradable system (7, 8), the levels of both progesterone and LH remain depressed, even on bolus injection of a dose of an agonistic LHRH analog as challenge. Similarly, intermittent administration of the analog to men reduces but does not completely inhibit spermatogenesis; it is possible that this will only be achievable with a continuous administration of the compound with a controlled release system.

A number of factors must be weighed against these substantial benefits provided by controlled release when selecting the type of formulation for a particular purpose. For example, development of a controlled release device to the manufacturing stage can be lengthy and complex and result in an expensive product. For some materials that are inexpensive, have a high therapeutic ratio and a long inherent half-life, it is unnecessary and inappropriate. A device designed to function over a relatively long period of time necessarily contains a large quantity of active material, and the potential for device failure resulting in "dumping" must be addressed. Finally, in order to be commercially viable, the consumer market must be educated in the features and advantages of controlled release.

1.3. Methods of Achieving Controlled Release

There are five principal types of controlled release systems.

1. Membrane controlled reservoir systems.
2. Diffusion controlled monolithic systems.
3. Biodegradable systems.
4. Osmotic systems.
5. Mechanically driven (pumping) systems.

The choice of system for any particular application depends upon a number of factors including cost, the nature of the active agent, the loading level required of the device, the environment of use, and the requirement for biodegradability. Perhaps the most critical factor is the release rate required. Diffusion and biodegradation controlled release processes are relatively slow and are generally useful only when release rates of less than 1 to 2 mg/day cm^2 of device are required. Osmotic devices, on the other hand, have much higher release rates and are useful in the region of 2–10 mg/day cm^2. Each of these categories of device is briefly described below.

2. MEMBRANE CONTROLLED RESERVOIR SYSTEMS

The simplest diffusion controlled system is the reservoir device, in which the agent to be released is enclosed within an inert membrane, through which it diffuses at a finite controllable rate. A reservoir device has the advantage of inherent zero order release capability over a substantial portion of its lifetime. The loading level of active material may also be higher than most other systems, giving economy of materials and minimizing costs. However, the methods of encapsulating materials in reservoir systems are usually more expensive than preparing alternative types of systems. Also, the release rate is critically dependent on thickness, area and permeability of the membrane. These parameters, while giving broad flexibility to the design of a reservoir system, must be carefully controlled. Finally, the device is subject to catastrophic failure should the surrounding membrane fail.

There are two principal types of reservoir device. If the reservoir contains excess solid material, the internal concentration of the solute within the reservoir is constant. The concentration gradient across the surrounding membrane and therefore the release rate, is then constant. In the absence of excess solid, the concentration gradient across the membrane drops exponentially with time providing an approximate first order rate of release. The constant release system has obvious advantages and is the form usually preferred.

2.1. Constant Activity Source Devices

The use of excess solid to maintain a saturated solution and constant concentration gradient dates to Fick's original paper but the deliberate application of this property to controlled release devices seems to date to Polin (9) and Lehmann (10). The Ocusert (Figure 3) and the Progestasert (Figure 5) are typical examples of such reservoir systems.

The rate of release of an active material from a reservoir device will be controlled by the permeability of the membrane and by the configuration of the device. For a device containing drug at unit activity the appropriate form of Fick's law for the slab or sandwich geometry, is

$$\frac{\mathrm{d}M_t}{\mathrm{d}t} = \frac{ADKc_s}{h} \tag{7}$$

where M_t is the mass of drug released at time t and hence $\mathrm{d}M_t/\mathrm{d}t$ is the steady state release rate at time t. A is the total surface area of the device (edge effects being ignored), D is the diffusivity of the drug in the membrane, K is the equilibrium coefficient for distribution of the drug between the membrane and the drug solution, c_s is the saturated solution concentration, and h is the membrane thickness. For the cylinder, the steady state release rate (ignoring end effects) is given by (11,12):

$$\frac{dM_t}{dt} = \frac{2\pi lDKc_s}{\ln (r_o/r_i)} \tag{8}$$

where r_o and r_i are the outside and inside radius of the cylinder, respectively, and l is the length of the cylinder. The analogous equation for the sphere is

$$\frac{dM_t}{dt} = \frac{4\pi DKc_s\, r_o\, r_i}{r_o - r_i} \tag{9}$$

Figure 6 illustrates the slab, cylinder and sphere geometries of these devices.

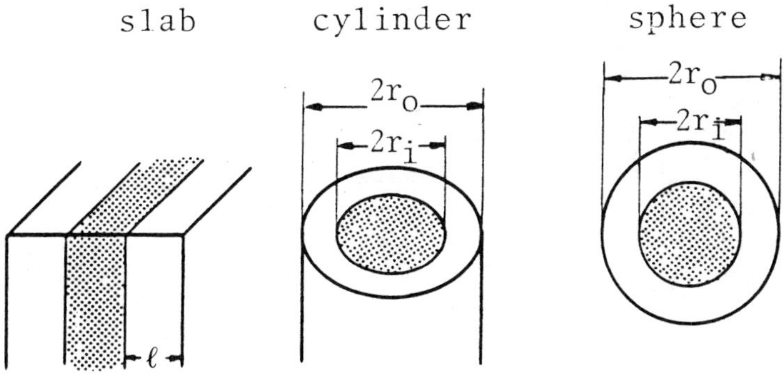

FIGURE 6.—Common membrane reservoir controlled release system geometries.

2.2. Non-Constant Activity Source Devices

It may be impractical to maintain a saturated solution within the reservoir, to provide unit thermodynamic activity of the drug. A liquid agent, for example, may become diluted with imbibed water, and a previously saturated solution that has become exhausted of excess solid will also, of course, provide a non-constant activity source. It can then be shown that the release rate from this type of device is given by the first order expression (11,12).

$$\frac{dM_t}{dt} = \frac{M_0ADK}{Vh} \exp\left(- \frac{ADKt}{Vh}\right) \tag{10}$$

2.3. Time Lag and Burst Effects

The initial release rate of a reservoir device after exposure to the target environment will depend upon its prior history. For example, a freshly prepared reservoir device will require time to establish an equilibrium concentration gradient across the external membrane and will provide a time lag. Similarly, a device aged by storage will have a membrane saturated with the solute, which will release at an initially high rate when placed in a desorbing environment, giving a burst effect. The mathematical analysis of lag times has been reviewed elsewhere (11,12).

More complex transient effects are also possible, for example, if the drug reservoir is only partially equilibrated with the membrane during storage, or if the device is stored at one temperature and then used at another. An example of a release curve characteristic of this effect is shown in Figure 7. Other examples of the same effect are known (13,14).

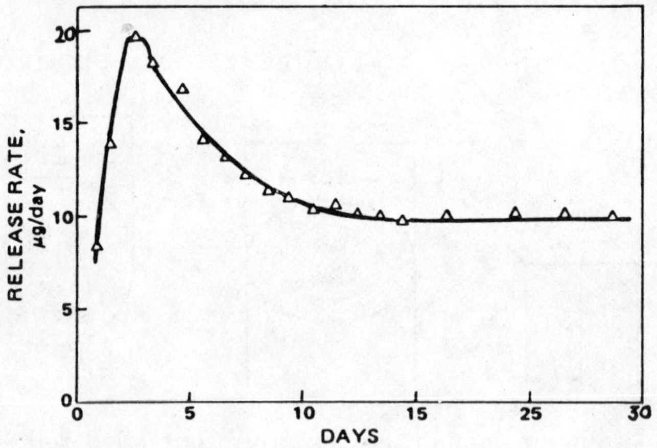

FIGURE 7.—Initial estriol release rate curve from an IUD device stored at room temperature but used at 37° C. The complex pattern of release is a mixture of a time lag and burst effect (13).

3. MONOLITHIC DEVICES

A monolithic device is one in which the material to be released is uniformly dispersed throughout the rate controlling medium. The release profile will be determined by the loading level of the material, the nature of the components of the device, and its geometry. Defects such as thin spots, pin holes and the like, which can be serious problems with reservoir systems, do not substantially affect the performance of monolithic devices. This, together with their relative ease of fabrication (by milling and extruding, for example), make for lower costs. These advantages will often outweigh the less desirable inherent feature of declining release rate.

There are two principal types of monolithic devices: monolithic solutions, in which the agent is present as a solid solution in the polymeric carrier medium, and monolithic dispersions, where the agent has a limited solubility and is partially in solution and partially in suspension in the carrier. The 'No Pest' strip and most domestic animal flea collars are examples of monolithic solutions, which are often formed when the active agent is a liquid and the polymeric carrier, for example, poly(vinyl chloride), can readily dissolve up to 20% of the agent. Release kinetics from monolithic solutions are the simpler to describe mathematically, and will be described first.

3.1. Monolithic Solution Systems

A monolithic device containing dissolved active material may frequently be prepared by equilibrating it with the material, for example, by soaking it in a neat liquid or a concentrated solution. Desorption of such material from a slab may be expressed by either of the two series, derived in the text of Crank (11) and given here for completeness.

$$\frac{M_t}{M_0} = 4\left(\frac{Dt}{h^2}\right)^{1/2}\left(\pi^{-1/2} + 2\sum_{n=1}^{\infty}(-1)^n \operatorname{ierfc}\frac{nh}{2\sqrt{Dt}}\right) \tag{11}$$

and

$$\frac{M_t}{M_0} = 1 - \sum_{n=0}^{\infty}\frac{8\exp(-D[2n+1]^2\pi^2 t/h^2)}{(2n+1)^2\pi^2} \tag{12}$$

These expressions fortunately reduce to two close approximations, reliable to better than 1%, valid for different parts of the desorption curve. The early time approximation, which holds over the initial portion of the curve, is derived from equation 11:

$$\frac{M_t}{M_0} = 4\left(\frac{Dt}{\pi h^2}\right)^{1/2}, \text{ for } 0 \leq \frac{M_t}{M_0} \leq 0.6, \tag{13}$$

and the late time approximation, which holds over the final portion of the desorption curve, is derived from equation 12:

$$\frac{M_t}{M_0} = 1 - \frac{8}{\pi^2}\exp\left(\frac{-\pi^2 Dt}{h^2}\right), \text{ for } 0.4 \geq \frac{M_t}{M_0} < 1.0 \tag{14}$$

These are plotted in Figure 8, which illustrates their different regions of validity. In general, the drug release rate at any time is of more interest than

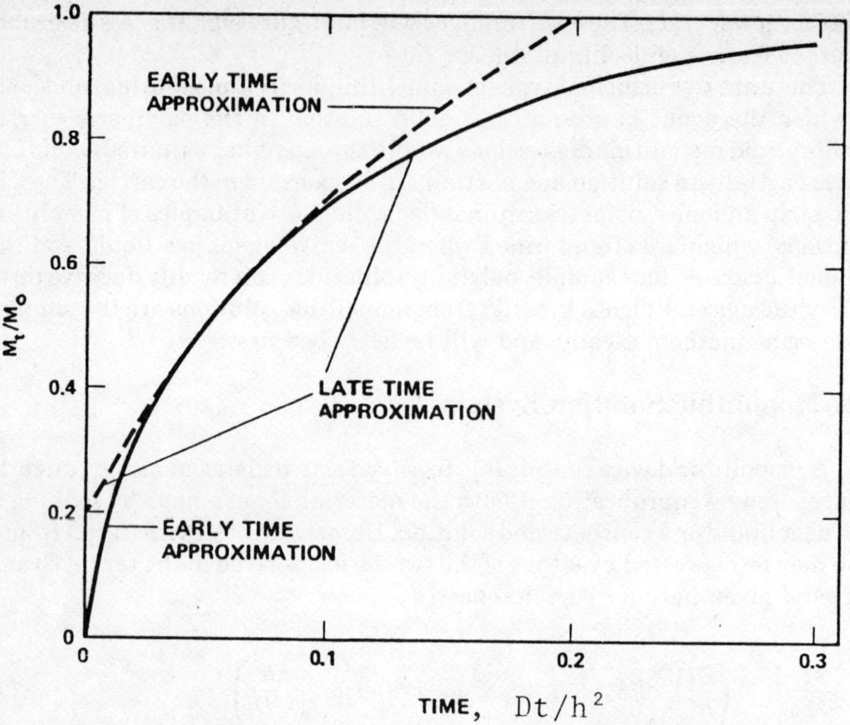

FIGURE 8.—Plots of the fraction of agent desorbed from a slab as a function of time using the early time and late time approximations. The full line shows the portion of the curve over which the approximations are valid.

the accumulated total drug release. This is easily obtained by differentiating equations 13 and 14, which give:

$$\frac{dM_t}{dt} = 2M_0 \left(\frac{D}{\pi h^2 t}\right)^{1/2} \tag{15}$$

and

$$\frac{dM_t}{dt} = \frac{8DM_0}{h^2} \exp\left(-\frac{\pi^2 Dt}{h^2}\right) \tag{16}$$

for the early and late time approximations, respectively. Figure 9 shows a plot of these two approximations against time. The release rate decreases as $t^{-1/2}$ over the first 60% of the release history, after which it decays in an exponential manner following equation 16.

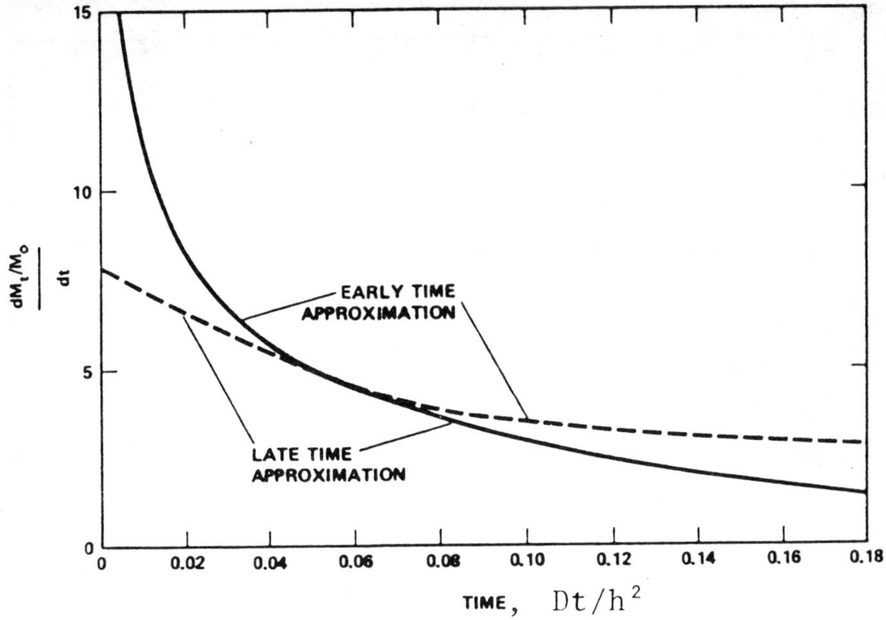

FIGURE 9.—Plots of the release rate of agent initially dissolved in a slab as a function of time, using the early time and late time approximations. The full line shows the portion of the curve over which the approximations are valid.

The principal difficulty with systems containing dissolved material, apart from the relatively nonconstant release rates unacceptable in some applications, is the low loading of agent achievable because of the limited solubility of most agents in polymers. However, some systems exist, especially with non-polar organic liquids such as pesticides, where relatively high loadings can be achieved. An example of the performance of such a device, the Shell No-Pest® strip, is shown in Figure 10 (15). Mathematical solutions for desorption of solutes from devices of other geometries can be found elsewhere (11,12).

3.2. Monolithic Dispersions

A more widely used type of monolithic system comprises a dispersion of solid through the rate-controlling medium. The mechanism of release from these monolithic dispersion devices depends upon the volume fraction of dispersed solids. For low loading levels of agent the release of the compound involves dissolution of the agent in the polymer medium followed by diffusion to the surface of the device. We will call these devices simple monolithic dispersions.

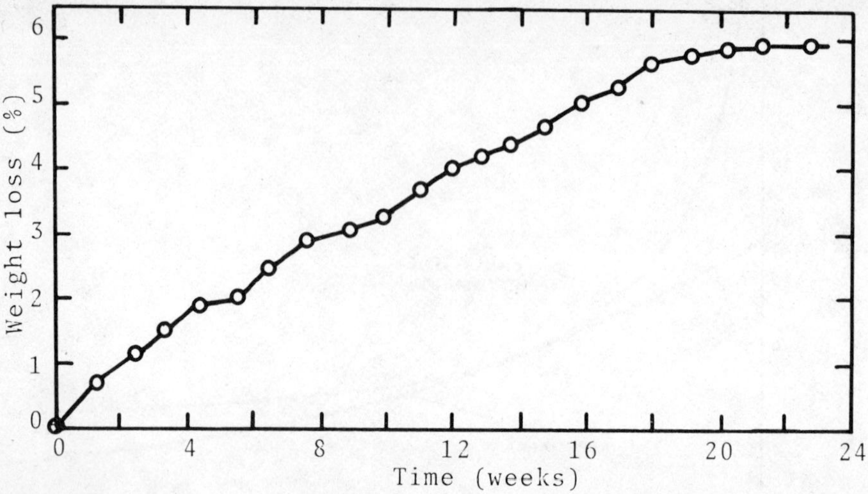

FIGURE 10.—The release of dichlorvos from a Shell No-Pest Strip® (15). This controlled-release device was the first commercially important controlled-release product. The device consists of 10–20% of the pesticide dissolved in a dioctylphthalate/polyvinyl chloride matrix. Release rates are varied by changing the pesticide and dioctylphthalate concentrations. The cat and dog flea collars and various insect pheromone-releasing systems use the same technology.

At slightly higher loading levels the release mechanism is more complex, because the cavities remaining from the loss of material near the surface are filled with fluid imbibed from the external environment. The cavities then provide preferred pathways for the escape of material remaining within the device. At these intermediate loadings the cavities are not connected to form pathways to the surface, but may increase the apparent permeability of the agent in the device. These devices are known as complex monolithic dispersions.

Finally, when the loading of dispersed agent is increased further the cavities left by loss of material are sufficiently numerous to form a continuous channel to the surface of the matrix. In this case the majority of all of the active agent is released by diffusion through these channels. These devices are known as monolithic matrix systems.

Simple monolithic systems At low loading levels the release rate from these systems can be described by a simple model first proposed by T. Higuchi (16). The model is shown schematically in Figure 11. In this model it is assumed that the solid dispersed material dissolves in the polymeric carrier and diffuses from the surface. Exhaustion of the excess solid occurs as the moving

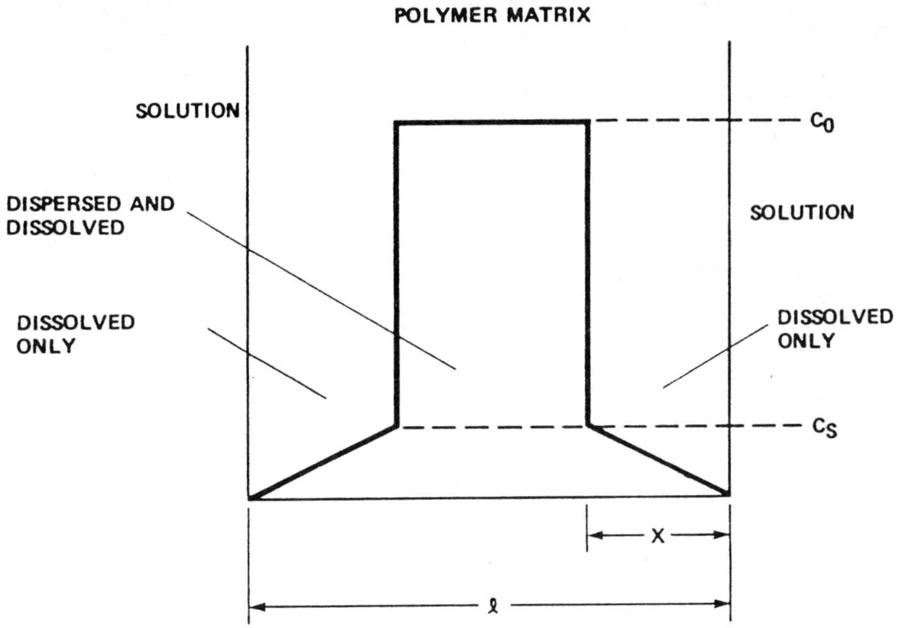

FIGURE 11.—Schematic representation of a cross-section through a polymer matrix initially containing dispersed solid agent. The interface between the region containing dispersed agent and the region containing only dissolved agent has moved a distance x from the surface (16).

front between regions containing only dissolved, and dissolved and dispersed material recedes to the interior of the device. The validity of the Higuchi model has been experimentally demonstrated numerous times. In addition, the movement of a dissolving front was monitored under the microscope by Roseman and W. Higuchi (17). The proof of T. Higuchi is straightforward and leads to the equation

$$M_t = A[DKtc_s(2c_0 - c_s)]^{1/2} \tag{17}$$

$$\simeq A(2DKtc_sc_0)^{1/2}, \text{ for } c_0 >> c_s. \tag{18}$$

The release rate at any time is then given by:

$$\frac{dM_t}{dt} = \frac{A}{2}\left[\frac{DKc_2}{t}(2c_0 - c_s)\right]^{1/2} \tag{19}$$

$$\simeq \frac{A}{2}\left(\frac{2DKc_sc_0}{t}\right)^{1/2}, \text{ for } c_0 >> c_s. \tag{20}$$

The Higuchi model is an approximate solution in that it assumes a "pseudo-steady state" in which the concentration profile from the dispersed drug front to the outer surface is assumed linear. Paul and McSpadden (18) have shown that the correct expression can be written as

$$M_t = A\,[2DKc_s\,(c_0 - Kc_s)]^{1/2} \tag{21}$$

which is almost identical to equation 18 and reduces to it, when $c_0 >> c_s$. It is clear that the release rate is proportional to the square root of the loading and can thus be easily varied by incorporating more or less agent. Furthermore, although the release rate is by no means constant, it varies over a narrower range than would be the case if the agent were merely dissolved, rather than dispersed, in the matrix. An example of the release rate of the drug from an ethylene vinyl acetate slab containing dispersed antibiotic chloramphenicol is presented in Figure 12. The amount of drug released is seen to increase

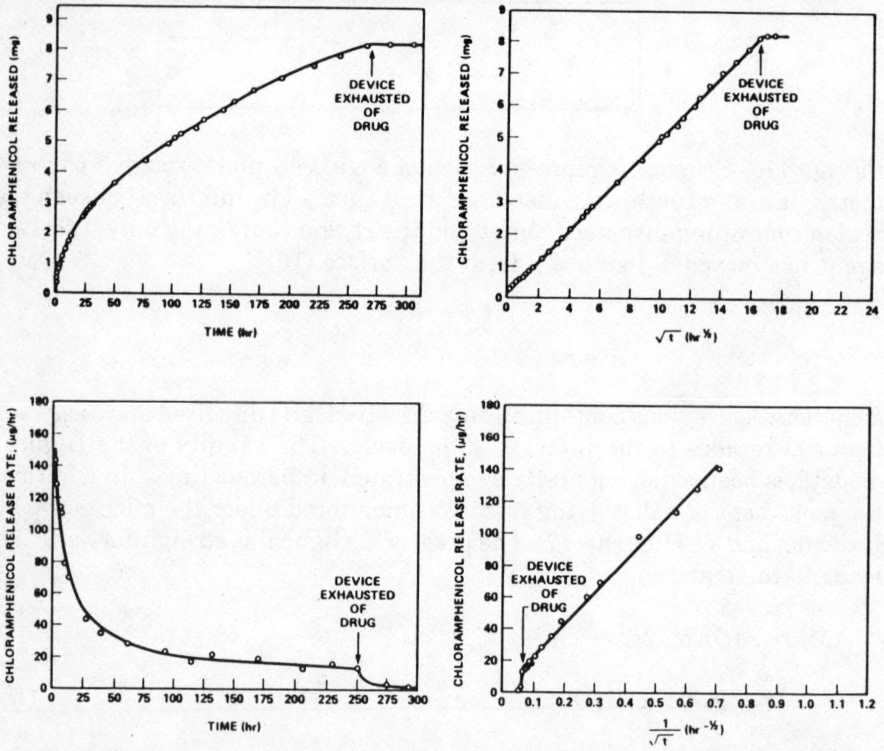

FIGURE 12.—Complete release curves from a simple monolithic dispersion device. The data are plotted as release rate vs. time and $1/\sqrt{t}$, and as total release vs. time and $\sqrt{t}$. From the slope of the total release vs. $\sqrt{t}$ plot or the release rate vs. $1/\sqrt{t}$ plot, the apparent permeability of the agent through the polymer matrix can be calculated using the Higuchi model (19).

with the square root of time in accordance with equation 21. Numerous other examples of this type of behavior can be found.

The equivalent solution for the release kinetics for cylindrical geometry when $c >> c_s$ has been given by Roseman and W. Higuchi (17). In this case, the expressions do not reduce to a simple form as in the case of the slab. Thus, the release rate at time t is:

$$\frac{\mathrm{d}M_t/M_0}{\mathrm{d}t} = \frac{4DKc_s}{r_o^2 c_0 \ln(1 - M_t/M_0)}, \quad (22)$$

where r_o is the radius of the cylinder.

A comparison of release kinetics for the three important geometries is presented in Figure 13. The curves have been normalized so that all the agent is released at $t_\infty = 1$, i.e., the area under all three curves is identical. It was assumed in all cases that $c_0 >> Kc_s$. The three curves are initially similar, the most important differences developing near to the point of device exhaustion.

Complex monolithic dispersions The Higuchi model is generally a good predictor for monolithic polymer dispersions containing low levels (<5%) of active material. However, at higher loadings, deviations from the expected release profile are seen. The rate of release is still proportional to $\sqrt{t}$, but has a higher value than predicted by the model. This, as mentioned earlier, is due to the presence of fluid-filled cavities created by dissolution and diffusion of particles from near the surface, which increases the system's permeability to some substances.

Some experimental results illustrating this phenomenon are shown in Figure 14. In these experiments the release of the drug chloramphenicol from a series of ethylene vinyl acetate monolithic dispersions with various dry loadings was measured. Equation 20 indicates that the release from similar devices with the same area but different loadings can be normalized into a straight line by a plot of $M_t/(c_0)^{1/2}$ vs. $\sqrt{t}$ with a slope of $A\,(2DKc_s)^{1/2}$. As Figure 15 shows, a plot of $M_t/(c_0)^{1/2}$ vs. $\sqrt{t}$ gives a family of straight lines with slopes steadily increasing with loading. Therefore, the membrane permeability apparently increases as the loading increases, an effect not predicted by the simple Higuchi model.

Permeabilities (DK) calculated from plots of the type shown in Figure 15 are shown versus loading in Figure 16. At low loadings, the permeability extrapolates to the value obtained by independent membrane permeability experiments. However, at high loadings of material the apparent permeability is increased because of the high permeability of the cavities left behind by the released active agent. Barrer (19) has reviewed the theory of diffusion through such complex systems; i.e., those consisting of cavities having a permeability $P_{(a)}$ within a continuous medium of permeability $P_{(o)}$. Different equations have been derived for dispersions of different shaped

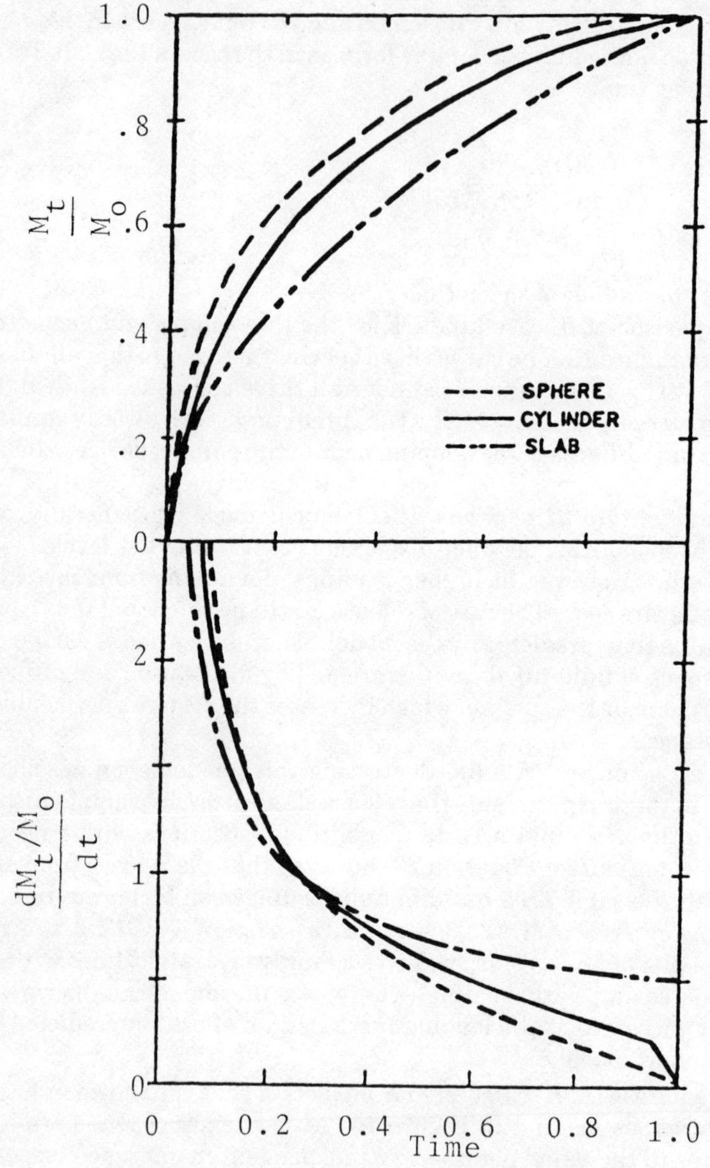

FIGURE 13.—Theoretical fractional release and release rate vs. time for a dispersed agent in a slab, cylinder, and sphere.

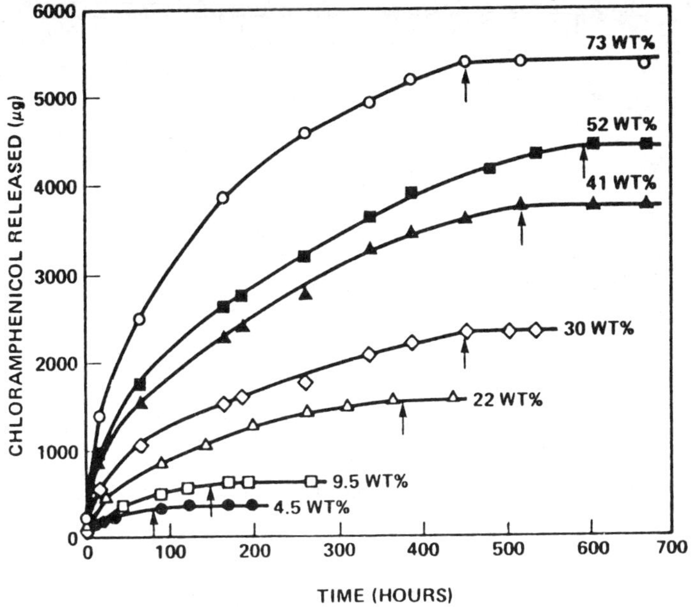

FIGURE 14.—Plots of release vs. time for the same-sized monolithic dispersed-agent devices containing the loadings of agent shown on each plot. The arrows indicate the time of exhaustion of the device (19).

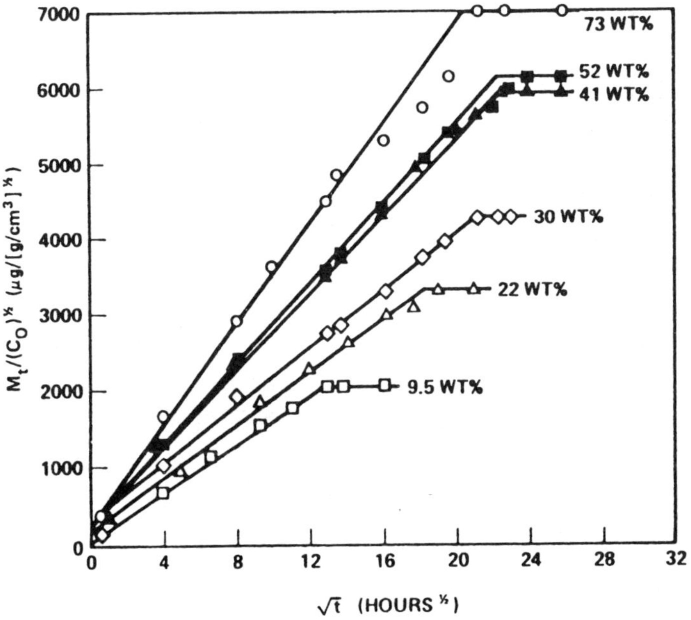

FIGURE 15.—Plots of agent released normalized for drug loading vs. $\sqrt{t}$. The break in the curve is the time of exhaustion of the device. The increase in slope of the plots with increased loading shows an increase in the permeability of the polymer matrix. (19)

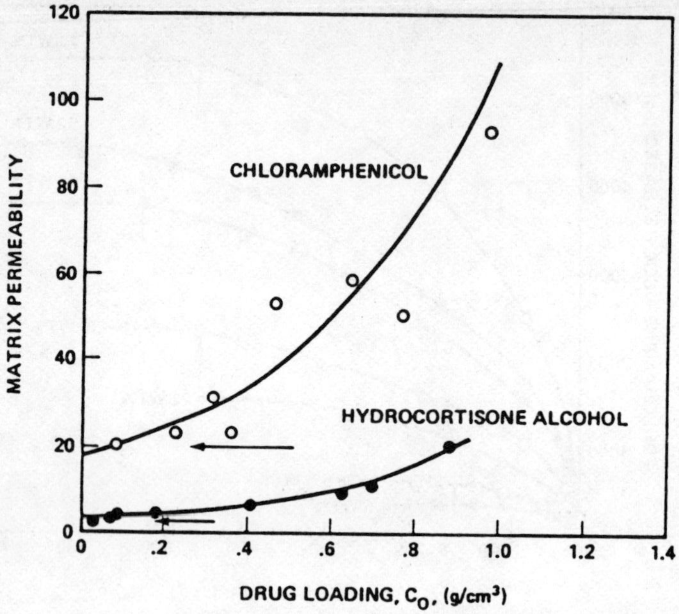

FIGURE 16.—Matrix permeability vs. agent loading. The arrows indicate the matrix polymer permeability (19).

particles and spatial distributions. However, a simple, convenient equation for a dilute random dispersion of spherical particles is that of Maxwell (20).

$$\frac{P}{P_{(o)}} = 3\phi_0 \left(\frac{(P_{(a)}/P_{(o)}) + 2 - \phi_0^{-1}}{P_{(a)}/P_{(o)} - 1} \right) + 1 \tag{23}$$

where P is the limiting flux of the composite system and ϕ_o is the original volume fraction of the dispersed permeant, i.e., $\phi_0 = c_0/\rho_p$ where ρ_p is the density of the permeant. Making the reasonable assumption that $P_{(a)} >> P_{(o)}$ (for the example of the chloramphenicol shown in Figure 15, the aqueous permeability is much greater than in the polymer matrix) reduces equation 23 to

$$P = P_{(o)} \frac{1 + 2\phi_o}{1 - \phi_0} \tag{24}$$

or

$$P = P_{(o)} \frac{1 + 2c_0/\rho_p}{1 - c_0/\rho_p} \tag{25}$$

Thus equations 18 and 20 may be modified for the complex monolithic dispersion to:

$$M_t = A\left[2DKc_s c_0 t\left(\frac{1 + 2c_0/\rho_p}{1 - c_0/\rho_p}\right)\right]^{1/2} \tag{26}$$

$$\frac{dM_t}{dt} = \frac{A}{2}\left[\frac{2DKc_s c_0\,(1 + 2c_0/\rho_p)}{t\,(1 - c_0/\rho_p)}\right]^{1/2} \tag{27}$$

The validity of these two equations is verified by calculating the curve of effective permeability (DKc_s) obtained from the slopes of the lines in Figure 14 versus loading using the independently measured permeability of the bulk polymer (21).

Monolithic matrix systems The release mechanism for a monolithic matrix system is illustrated in Figure 17. As this figure shows, the active agent is released by diffusion through the water-filled pores that form as water is imbibed from the surface of its device to replace the material that leaches out.

The mathematical description of release from this type of system exactly matches equations 20 and 21 previously derived for the simple monolithic-dispersion, the only difference being substitution of the appropriate expression for the permeability for DK in these equations. In the matrix leaching case, release is through the pores of a microporous membrane and thus the appropriate substitution for the partition coefficient is

$$K = \epsilon \tag{28}$$

This reflects the fact that although the fluid inside the membrane pores is the same as the surrounding solution, only a volume fraction ϵ of the membrane is filled with the fluid. The appropriate substitution for the diffusion coefficient is

$$D = \frac{D_w}{\Theta} \tag{29}$$

Where D_w is the diffusion coefficient of the agent in the fluid (water) filling the matrix pores and Θ is a term reflecting the extra distance the agent must on average diffuse to escape from the device. Thus the Higuchi model expression for release from a monolithic slab is

$$\frac{dM_t}{dt} = \frac{A}{2}\left(\frac{2\,D_w\epsilon\,c_s c_0}{\Theta\,t}\right)^{1/2} \tag{30}$$

Boundary layer effects It has been assumed to this point that the agent's release is solely determined by its rate of diffusion to the surface of the device,

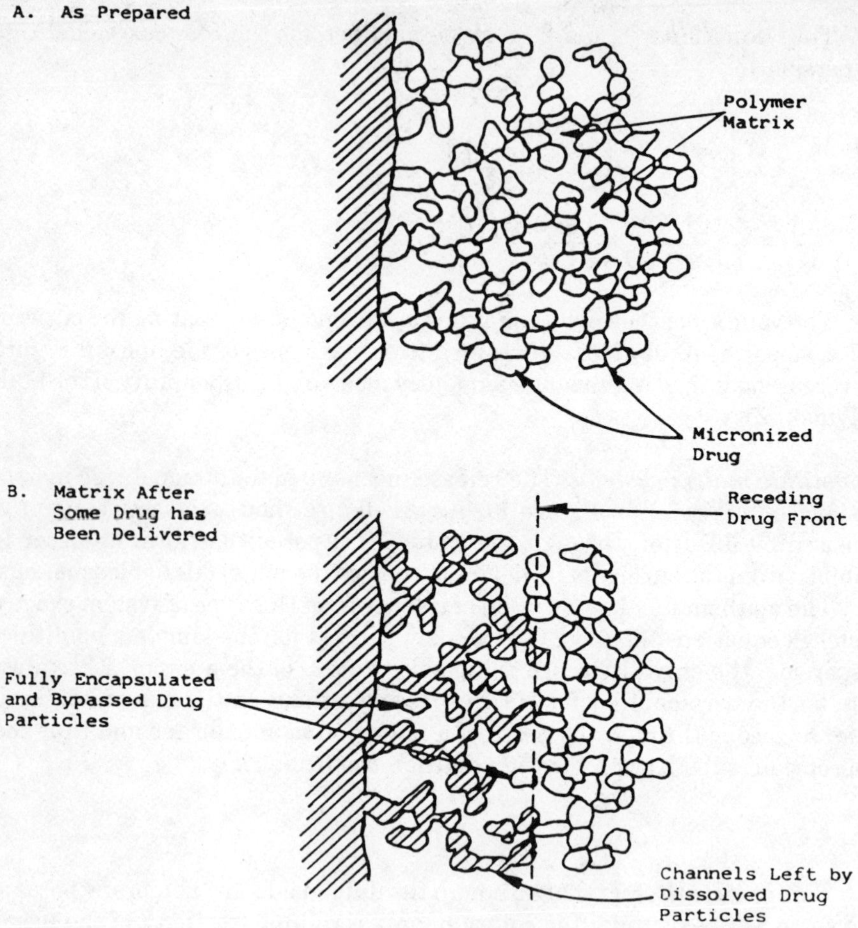

FIGURE 17.—Schematic of the monolithic matrix system.

this surface being maintained at zero concentration essentially for the lifetime of the device. This situation is sometimes not attained in practice because of slow transport of the material from the surface of the device into the surrounding environment. There will then be a finite concentration at the device surface, i.e., non-sink conditions, and in an extreme case the concentration in the fluid immediately surrounding the device, the so-called boundary layer or unstirred layer, can reach the agent's solubility, c_s, at which point the device would cease functioning. Boundary layer effects are most serious with relatively water-insoluble materials, where the concentration gradient, which is the driving force for diffusion across the boundary layer, is low, and saturation will rapidly be approached at the surface of the device. These problems are discussed in more detail elsewhere (12).

4. BIODEGRADABLE SYSTEMS

Biodegradable controlled release devices have found their major application in the pharmaceutical field, particularly in the preparation of controlled release parenteral products. It is anticipated that such injectable or implantable preparations will be widely used in developing areas of the world where patient compliance with conventional dosage forms is a limiting factor in effective treatment, and where medical attention is available infrequently. Non-erodible systems require removal on their exhaustion, and biodegradable systems therefore have a clear advantage in this type of application.

Some early attempts at design of degradable controlled release systems focused on the preparation of monolithic devices containing dispersed materials where the carrier would slowly dissolve at the surface, thereby releasing the material. Such a concept is illustrated in Figure 18. This ideal system would be predictably dependent upon the geometry of the device.

It is now known that this mechanism of release is quite rare, since degradation usually cannot be confined to a narrow boundary at the surface of the polymer, and also the release of material by diffusion prior to degradation is frequently quite significant. These confounding factors proved to make the development of effective biodegradable delivery systems much more complex than originally anticipated.

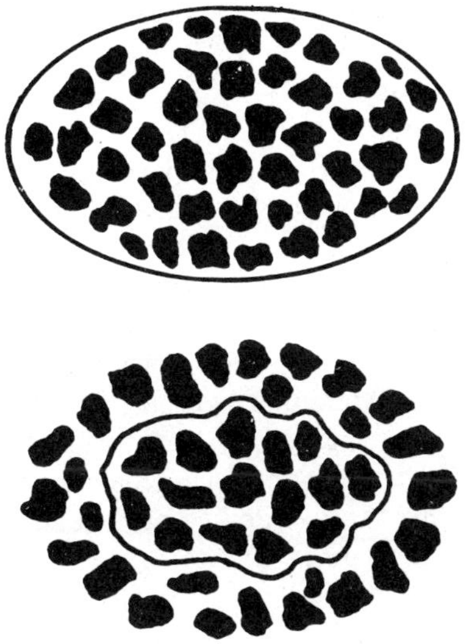

FIGURE 18.—An early concept of drug delivery by surface degradation of the polymer matrix, later proved to be rare.

The mechanism of degradation and material release from degradable controlled release systems can be described in terms of three parameters:

(1) The chemistry of the bioerodible polymer;
(2) The reaction rate kinetics;
(3) The manner in which the active agent is incorporated into the device.

The type of hydrolytically unstable linkage used in the bioerodible polymer and the position of the labile group in the polymer both affect the design of the system. For example, the polymer matrix might contain pendant labile groups which when cleaved would lead to solubilization of the polymer. Another alternative would consist of a hydrogel containing unstable bonds in the polymer backbone. Sufficient cleavage of these bonds would produce a soluble oligomer or monomer. Finally, the active agent could be covalently bonded by labile bonds into the main chain of the polymer. The mechanism of active agent release is then cleavage of the polymer chain. All of these systems have been studied (22,23).

The way in which the biodegradable polymer degrades can substantially affect the performance of the final device. Polymer degradation can occur preferentially at the surface of the device, i.e., heterogeneously, or it can occur uniformly throughout the polymer matrix, i.e., homogeneously. Combinations of these two extremes can also occur. In addition, polymer degradation can proceed at a constant rate or can be autocatalytic, accelerating with time. The interplay between surface and bulk degradation and other kinetic factors has a profound effect on the release kinetics of the device and therefore has a direct bearing on its design and performance.

There are four approaches to the incorporation of an active material within a biodegradable device.

(1) Monolithic systems in which the active agent is essentially immobile until released by degradation of the surrounding polymer matrix (erosion contolled);
(2) Monolithic systems in which the active agent is released by diffusion prior to or concurrent with the degradation of the matrix (diffusion controlled);
(3) Reservoir systems in which the active agent is encapsulated by a rate-controlling membrane through which the agent escapes by diffusion (and the capsule degrades after its delivery role is complete);
(4) Erodible polyagent systems in which the active agent is chemically bonded to the biodegradable polymer by labile covalent bonds and is not released until these bonds are cleaved.

Variations of these four types of device have been constructed and extensively described in the literature.

4.1. Biodegradable Polymers

A number of labile bonds can be used to form biodegradable polymers. Based on the known hydrolysis rates of low-molecular-weight analogs, the relative rates of hydrolysis of these bonds under neutral conditions is

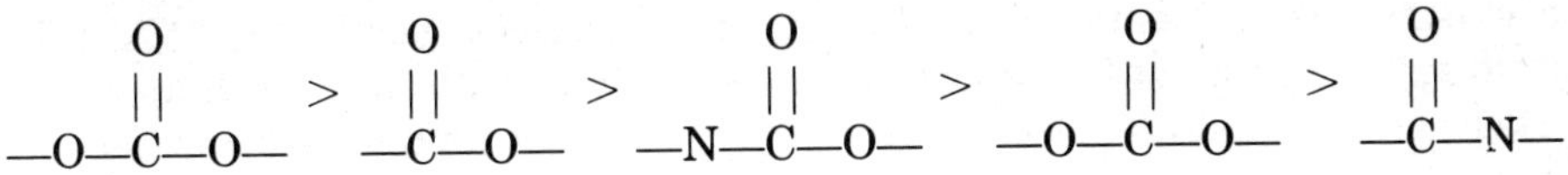

polycarbonate polyester polyurethane polyorthoester polyamide

However, this is very much affected by the morphology of the polymer and the presence of substituent groups. The comparison, therefore, is only an approximate guide.

There are a number of possible conformations of the labile bonds in a biodegradable polymer, as illustrated in Figure 19. In this figure, x denotes an

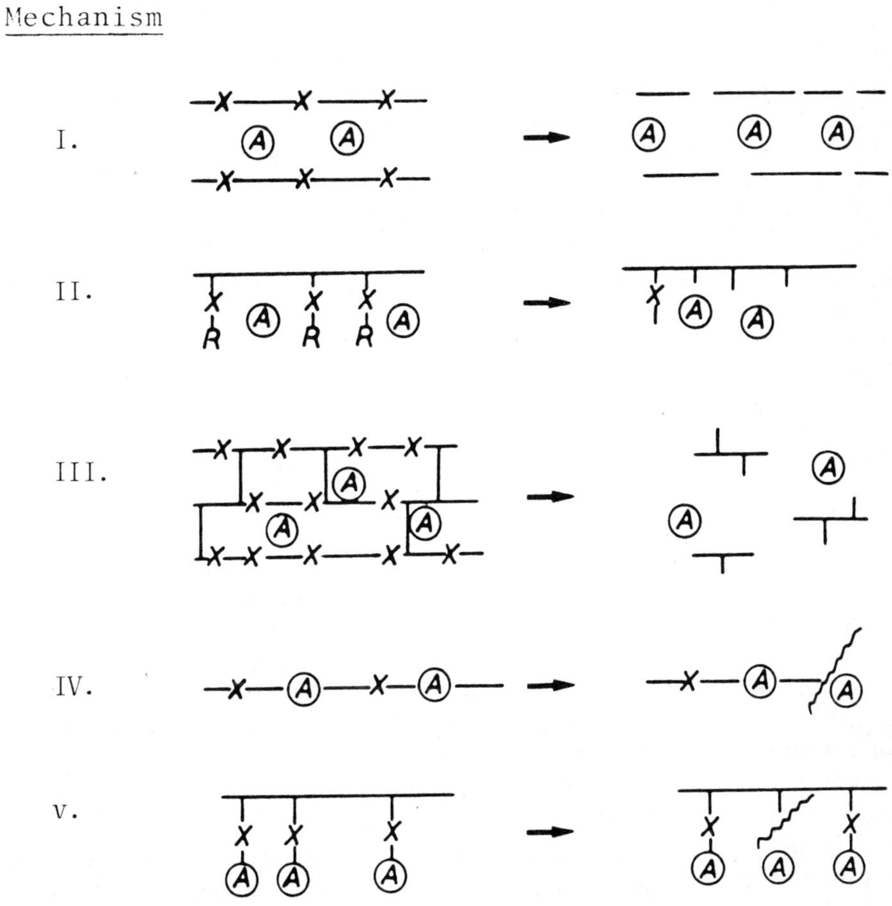

FIGURE 19.—Different approaches to drug delivery systems based on biodegradable polymers.

unstable bond, such as an ester that is subject to hydrolysis. The polymer may become solubilized by a simple decrease in molecular weight of the polymer backbone (mechanism I) with concomitant release of the active agent A, or by cleavage of bonds with a group (R) conferring insolubility (mechanism II). Cleavage of crosslinks or of the backbone of a crosslinked polymer (mechanism III) will liberate polymer fragments where size, and therefore, solubility will depend upon the original crosslink density, while mechanisms IV and V illustrate two means of release of A from direct incorporation within the polymer. These latter two matrix designs are examples of polyagents.

The linear polyesters (22, 24) are by far the most widely studied class of biodegradable polymers. Other polymers have been developed, for example, polyorthoesters (25, 26), polyacetals (26, 27), polyacids and anhydrides (28), and polyamino acids (29). The development and study of these materials is an active research area. However, the bulk of the *in vivo* and reliable *in vitro* data reported in the literature have involved polyesters, and only these polymers will be discussed here.

Polyester degradation rates are influenced by several factors. The morphology of the polymer is perhaps of primary importance. Hard, crystalline polyesters are most resistant to degradation, followed by amorphous glassy polyesters, with amorphous rubbery polyesters being the most easily degraded. The chemistry of the groups adjacent to the polyester bond is also a key factor. Adjacent electron withdrawing groups, for example, destabilize the ester bond, while electron donating groups lead to increased stability (22). Finally, because hydrolysis is pH-sensitive, the molecular weight of the polymer is important, and high molecular weight polymers degrade more slowly than their low molecular weight homologs.

Unfortunately, it is often difficult to change one of these parameters without significantly affecting the other two. For example, a change in the chemistry of the polyester monomer will simultaneously affect both the morphology and molecular weight of the polyester. For this reason, it is quite difficult to predict the degradation rates of different polyesters.

Whatever the chemistry and structure, however, the degradation mechanism for all of the polyesters appears to be similar. The work of Pitt et al. (22, 24) indicates that the reaction is autocatalytic and proportional to the free carboxylic acid concentration in the polymer. Thus the intrinsic viscosity of the polymer decreases exponentially when plotted against time. However, it is not until a critical molecular weight is reached (between 2000 to 5000 MW) that polymer weight loss occurs. These results are consistent with a homogeneous erosion mechanism, and this conclusion is supported by the data of several workers who have shown that samples of the same polymer as film or rod form erode at essentially the same rate, even though this can represent a 10-fold difference in surface-to-volume ratio.

Table 1 illustrates the effect of variations in ester group chemistry on the hydrolysis rate constant for a series of methyl ester analogs (30). It can be seen that the presence of electron withdrawing groups adjacent to the ester

TABLE 1. Hydrolysis rate constants of several esters relative to the rate constant of methyl acetate (30)

Relative Hydrolysis Rate Constant (k/k_{MA})	Compound
0.6	$CH_3—C(=O)—OCH_2CH_3$
1	$CH_3—C(=O)—OCH_3$
761	$Cl—CH_2—C(=O)—OCH_3$
10,000	$CH_3—C(=O)—C(=O)—OCH_2CH_3$
16,000	$Cl—CH(Cl)—C(=O)—OCH_3$
170,000	$CH_3—O—C(=O)—C(=O)—OCH_3$

bond enhances the hydrolysis rate substantially. A carboxyl group, for example, increases the rate of hydrolysis by 3 to 5 orders of magnitude. On the other hand, electron donating groups such as alkyl groups have the opposite effect. Although this effect has not been well documented with polyesters, one would expect that poly(glycolic acid) and lactic acid will degrade more rapidly than polycaprolactone, for example. The relative rates of biodegradation of different aliphatic polyesters are shown in Figure 20.

In order to increase the degradation rates of crystalline homopolymers such as poly(lactic acid), more amorphous copolymers are frequently used. Thus, poly(lactic acid) is often modified by copolymerization with a small percentage of poly(glycolic acid) (15). Similarly, Schindler et al. (24) have studied a series of poly(D,L-lactide-ϵ-caprolactone) polymers. The glassy D,L-lactide and crystalline ϵ-caprolactone homopolymers both degrade slowly, but the random copolymer is more amorphous and rubbery and allows more rapid attack. The copolymer containing 80 mol% lactic acid is the most rapidly degraded, as shown in Figure 21.

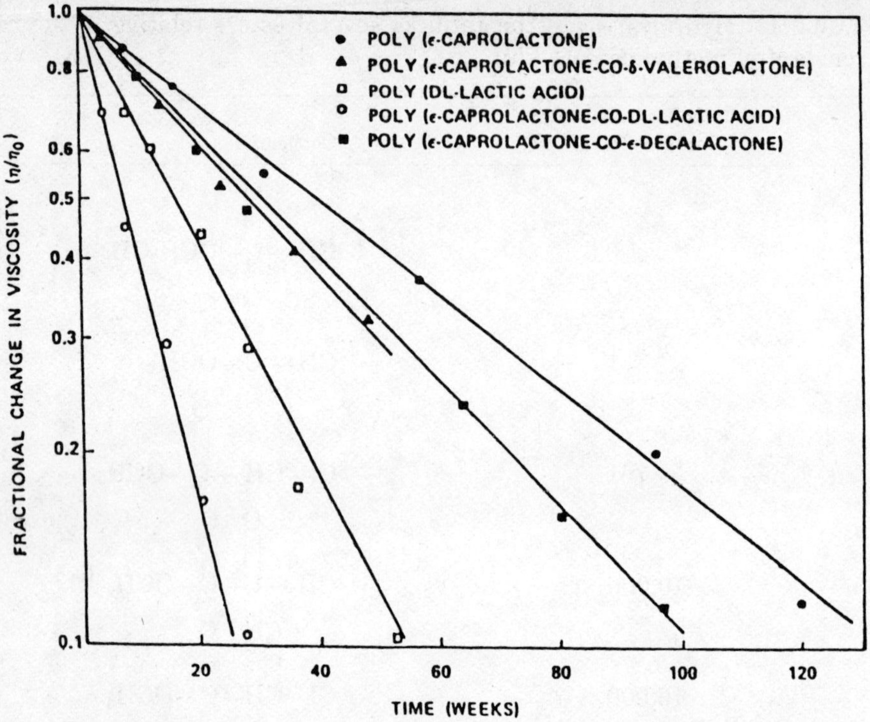

FIGURE 20.—Semi-logarithmic plots of fractional changes in the intrinsic viscosity of different aliphatic polyesters as a function of time *in vivo* (22).

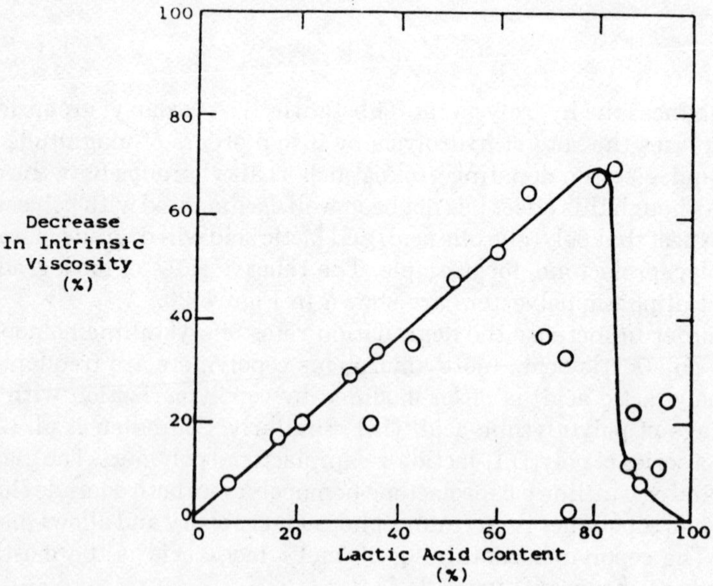

FIGURE 21.—Degradation of poly(lactic acid-caprolactone) copolymers of varying composition after 250 days *in vivo* (22).

4.2. Degradable Polyagents

The final significant type of degradable controlled release system is the poly-agent system in which the active agent is chemically bound to the degradable matrix material. The release of material from this type of system is complex, but the rate-limiting step is usually the rate of hydrolysis of the labile bond attaching the agent to the polymer. The rate of diffusion of the agent from the polymer matrix is usually rapid by comparison. The chemistry of each poly-agent system must be tailor made for each agent. An advantage of degradable poly-agent systems is the relatively high loading of active agent that can be achieved.

A general classification scheme illustrating the approaches available is shown in Figure 22. Perhaps the simplest of these is scheme I, when the active agent, A, contains a readily reactive group, such as -COOH, -OH, or $-NH_2$, which can be chemically linked to a suitable backbone polymer. A good example of this type of agent would be the herbicide 2,4-dichlorophenoxy-acetic acid (2,4-D), which Allan et al. (31) and Harris (32) have studied extensively.

If the active agent does not contain a suitable reactive group, scheme II (Figure 22) can be employed. The active material is first derivativized and then reacted with the polymeric carrier. Petersen et al. (33) have prepared poly-agents of the steroids by this carrier method; for example, the chloroformate derivative of norethindrone is prepared and then reacted with poly-N-(3-hydroxypropyl)-L- glutamine to form the poly-agent. The *in vivo* norethindrone release from this poly-agent is shown in Figure 23.

Alternatively, the active agent can be converted to a polymerizable

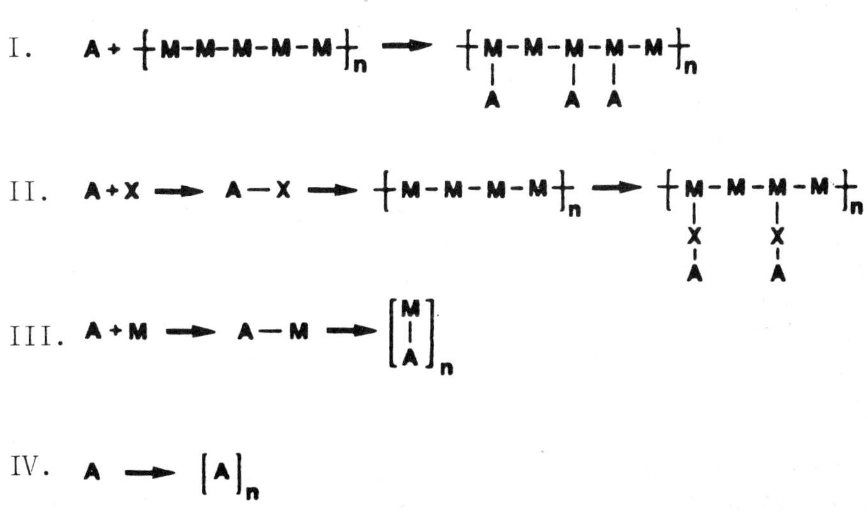

FIGURE 22.—Four different approaches for synthesizing controlled-release poly-agents.

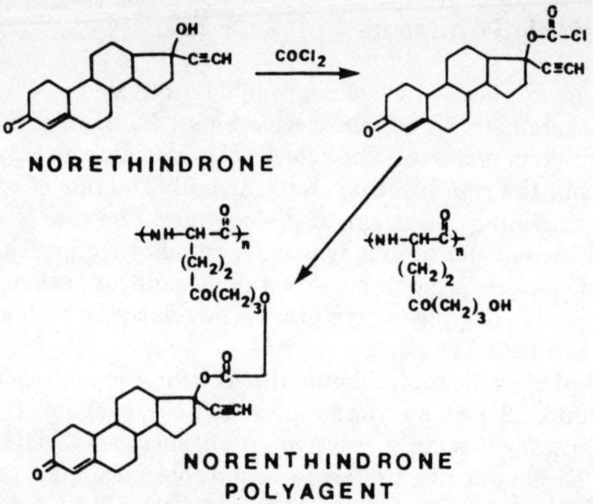

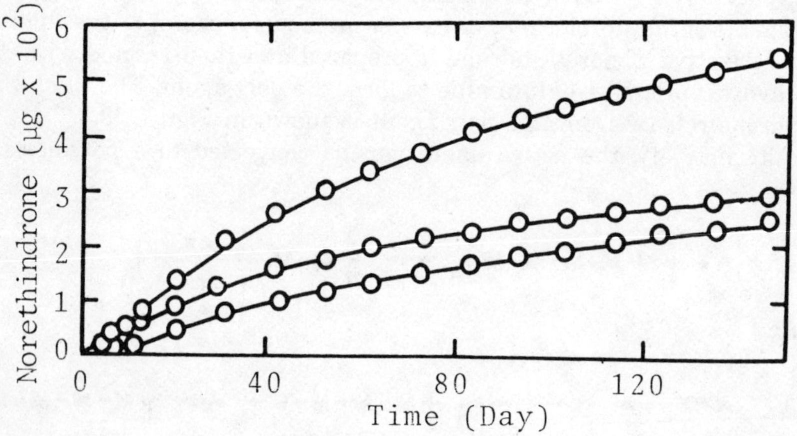

FIGURE 23.—*In vivo* release of a pendant norethindrone polypeptide system. Norethindrone release is changed by varying the degree of substitution of the polymer (33).

derivative, typically by the addition of a vinyl-containing group, which is then homo- or copolymerized to give a wholly synthetic active agent/polymer combination. This approach is shown as scheme III in Figure 22, and has been very widely used both in medicine and with pesticides and herbicides. The principal problem is that the rate of degradation of the poly-agent is often very slow. For example, Harris found that poly (2,4-D) derivatives had only released a few percent of their total 2,4-D content after several months, and this rate of release was only marginally increased when derivatives with rather long pendant agent groups were used. However, by proper choice of the copolymer and the polymerizable monomer, useful release rates can be obtained.

Finally, in a few cases when the active agent contains a difunctional grouping it can be incorporated directly into a suitable homo- or copolymer backbone, as in scheme IV. For example, the herbicide picloram (31) can be homopolymerized thus:

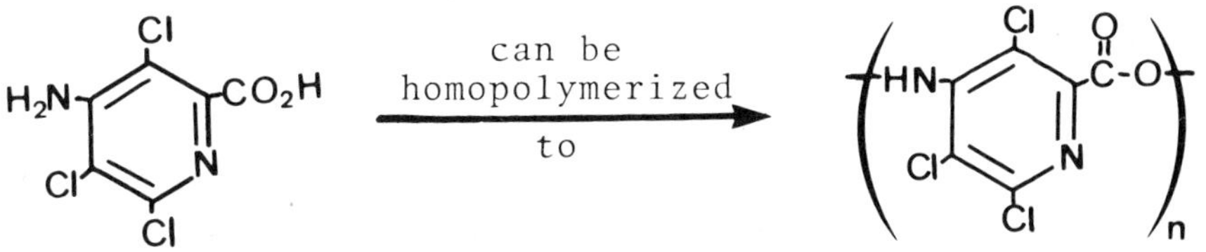

4-amino-3,5,6,-trichloro-picolinic acid (picloram)

The characteristic feature of polymers that degrade by a purely homogeneous (bulk) degradation mechanism is that the molecular weight of the sample exhibits a steady decrease as the degradation proceeds. However, degradation need not be accompanied by any appreciable weight loss until a critical molecular weight or degree of reaction is reached, at which point the polymer will dissolve quite rapidly. The degradation of polymer samples by a homogeneous mechanism will be essentially independent of the surface-to-volume ratio of the samples, since all of the polymer matrix is degrading simultaneously, irrespective of its distance from the surface. Device geometry is therefore a non-critical factor in the design of these systems. The pattern of degradation for a linear polyester containing labile groups along its backbone is shown in Figure 24.

It can be seen that the induction time prior to onset of erosion shows a strong dependence on the initial molecular weight of the polymer. This is something of a paradox, because a typical polymer of this type may have several hundred labile bonds throughout a molecule, and a very small percent degradation will bring about a large drop in molecular weight. This ratio of molecular weight change:% bonds hydrolyzed will increase logarithmically with increase in polymer molecular weight, so that extremely large changes in molecular weight are necessary to affect degradation times. It appears, however, that the observed greater than anticipated dependence of induction time on polymer molecular weight arises due to the autocatalytic nature of the reaction. This is a common feature of this type of homogeneous degradation. It has been determined that the hydrolysis rate increases exponentially with time, and the rate of cleavage of the first few bonds is thus very slow compared with the rate of cleavage of the final bonds. This point is illustrated by the data in Figure 25 from Pitt et al. (22). This plot shows a decrease in the

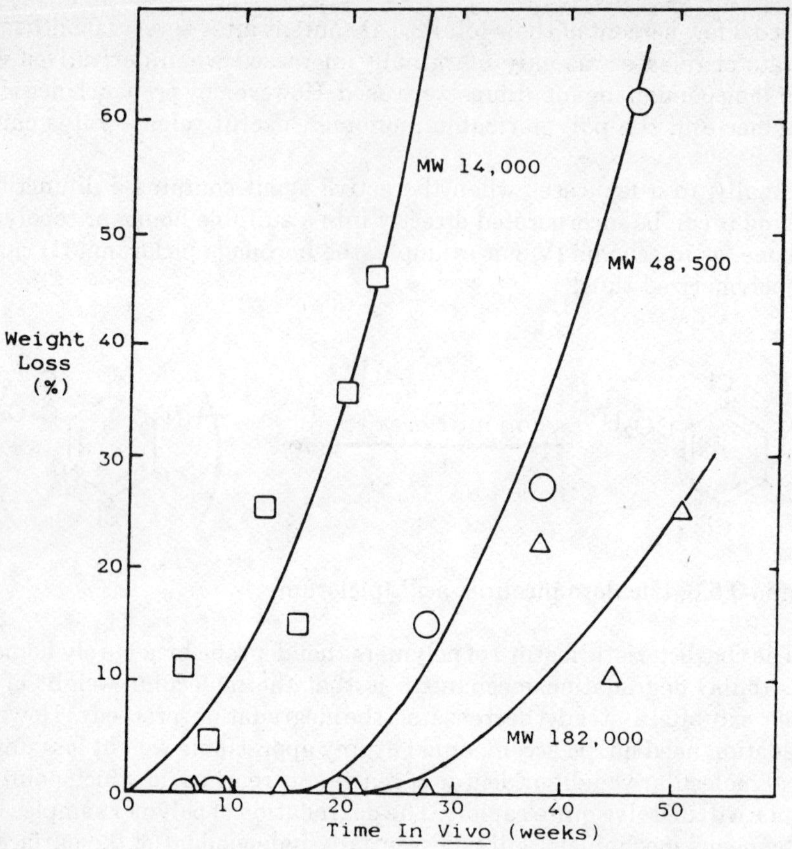

FIGURE 24.—*In vivo* weight loss of poly (D,L-lactic acid) films for samples of different initial molecular weights (22).

average molecular weight of a polycaprolactone *in vivo*. The molecular weight of the polymer is initially 50,000, corresponding to an average of 400 labile ester bonds in each polymer chain. After approximately 100 days *in vivo* the molecular weight had been reduced to 5000, resulting from hydrolysis of 9 of the original 400 labile bonds. At this low molecular weight, however, the sample began to lose many of its polymeric properties, and material was lost from the sample by fragmentation and diffusion. Figure 25 demonstrates two key points. First, the degradation is autocatalytic. Therefore, it takes almost 4 weeks to break the first ester bond in the chain, 2 weeks to break the second, and less than one week to hydrolyze the ninth bond. This autocatalytic behavior is probably due to the hydrophilic nature of the chain ends. The second noteworthy point that is apparent in the figure is that only a small portion of the bonds in the sample degrade prior to disintegration and dissolution of the sample. These factors cause the observed relationship between erosion time and polymer molecular weight.

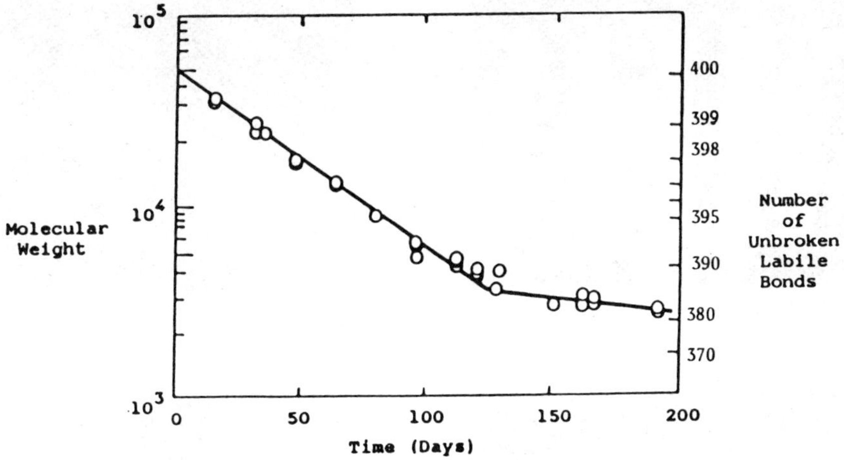

FIGURE 25.—*In vivo* degradation of poly (ε-caprolactone) showing molecular weight and unbroken labile bonds vs. time (22).

Diffusion-controlled reservoir systems constructed from ultimately erodible materials have emerged as one of the more successful drug implant systems. A key advantage of these systems is that drug release kinetics are largely dissociated from the erosion kinetics of the polymer. These systems exhibit generally good drug release profiles, provided the membrane does not deteriorate during the lifetime of the device. Polymers that degrade by a homogeneous mechanism are preferred for this application. Typical release curves for a capsule implant system prepared by Pitt et al. (24), are shown in Figure 26. Photographs showing the course of degradation for one of these devices are shown in Figure 27.

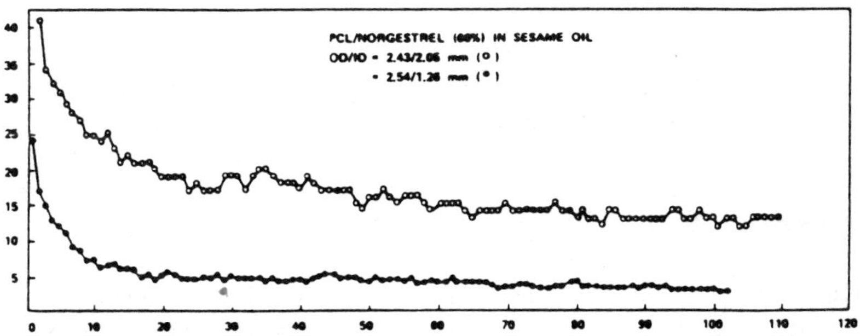

FIGURE 26.—Daily rate of release of Norgestrel from poly (ε-caprolactone) capsules, *in vitro,* per centimeter length of capsule, using two different capsule wall thicknesses (24).

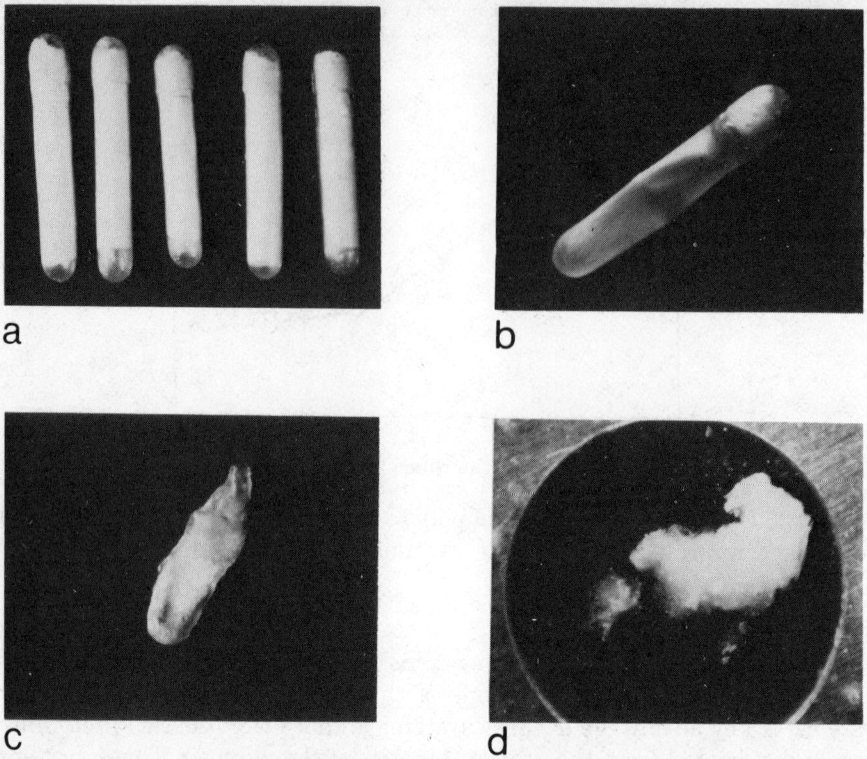

FIGURE 27.—*In vivo* degradation sequence of glutamic acid–ethyl glutamate capsules; (a) capsules prior to implantation, (b), (c), (d) after 30, 60, and 90 days, respectively.

5. OSMOTIC DEVICES

Osmotic effects are often a problem in device design because the imbibition of water causes the device to swell, or dilutes the active material. However, in recent years a series of devices have been designed in which osmotic effects have been used as a driving force for delivery of the active agent.

5.1. The Rose–Nelson Pump

The forerunner of modern osmotic devices is the Rose–Nelson Pump (34), illustrated in Figure 28. The pump consists of essentially three chambers: a drug chamber, a salt chamber containing excess solid salt, and a water chamber. The salt and water chambers are separated by a semi-permeable membrane. Because of the difference in osmotic pressure across the mem-

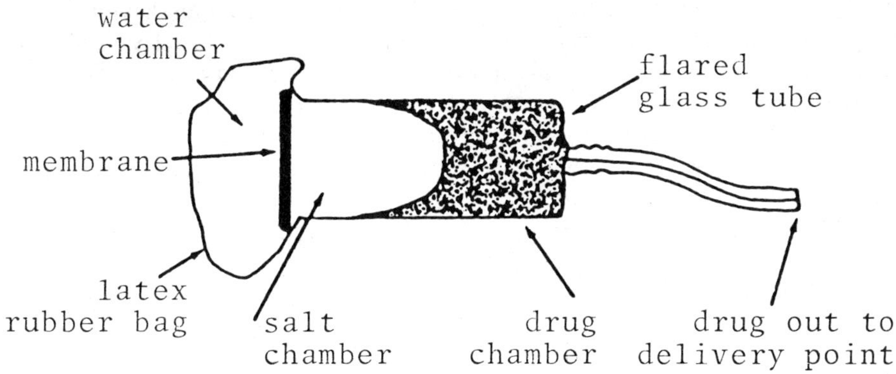

FIGURE 28.—The first osmotic pump device was produced by Rose and Nelson in 1955. In use, water from the latex rubber chamber is osmotically drawn into the salt chamber, which then expands into the drug chamber, pumping the drug from the device (34).

brane, water moves from the water chamber into the salt chamber. In use, the volume of the salt chamber thus increases distending the latex diaphram, thereby pumping drug out of the device. The osmotic pressure of the saturated salt solution is high, on the order of several hundred psi and the small pressure required to pump the suspension of active agent is insignificant in comparison. The permeation rate of water across the semi-permeable membrane and the pumping rate remain constant so long as solid excess salt is present in the salt chamber to maintain saturation.

The osmotic delivery rate of the Rose–Nelson pump is given by the equation:

$$\frac{\mathrm{d}M_t}{\mathrm{d}t} = \frac{Ak(\Delta\pi)c}{h} \tag{31}$$

where k is the osmotic permeability of the membrane, $\Delta\pi$ is the osmotic pressure difference between the salt chamber and the external solution and c is the concentration of active agent in the chamber. The delivery rate is inversely proportional to membrane thickness, h.

Delivery rates from these devices can be relatively high and this is one of their principal advantages.

5.2. The Higuchi and Theeuwes Pumps

Higuchi has proposed a series of variations of the Rose–Nelson pump (35). In this device the water chamber has been removed and the device is designed to be activated by water imbibed from the surrounding media environment, for example, the rumen of cattle in a veterinary product.

Another variant of the Rose–Nelson pump has been developed by Theeuwes (36, 37). As with the Higuchi pump the device is activated by the imbibition of water from the surrounding environment. In this device, however, the membrane forms the outer rigid casing. The device is loaded with any desired drug immediately prior to use. When placed in an aqueous environment this drug is then delivered following a time course determined by the salt used in the salt chamber and the permeability of the outer membrane casing. The device is illustrated in Figure 29.

Small osmotic pumps of this form are sold under the trade name Alzet. They are frequently used as implantable controlled release delivery systems in experimental animal studies of the effects of continuous administration of various drugs.

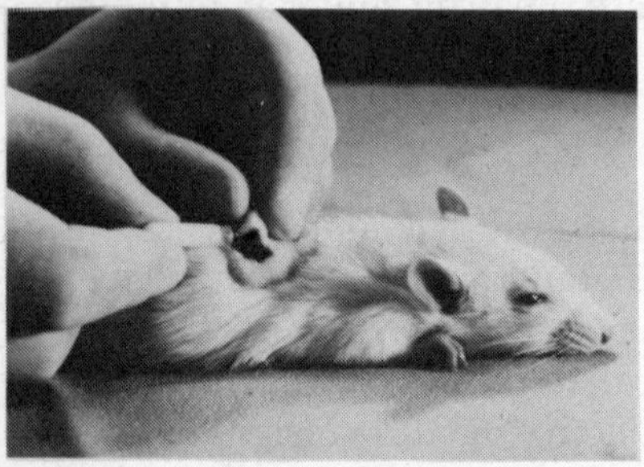

Pump in use.

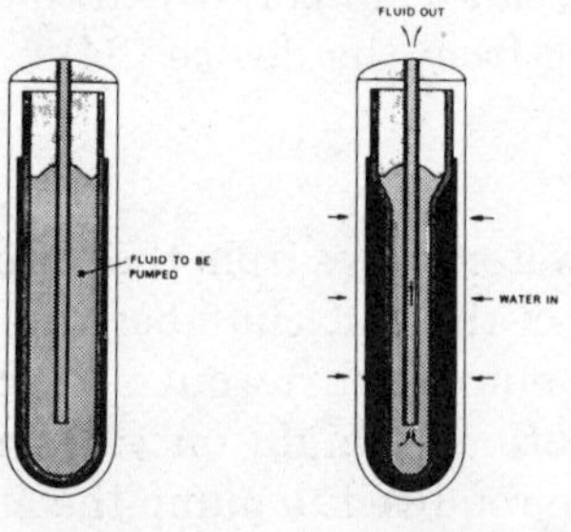

Pump filled and assembled.

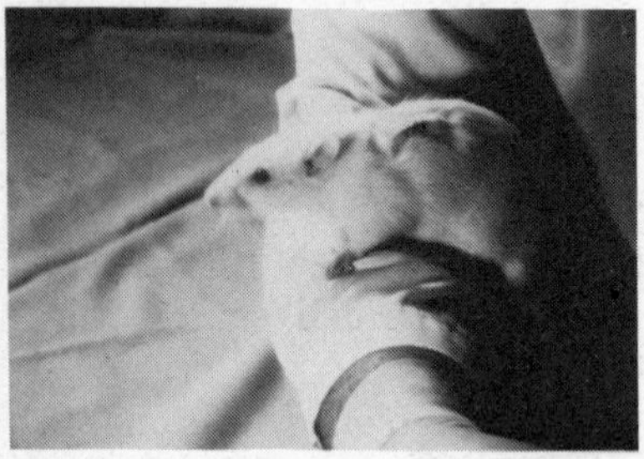

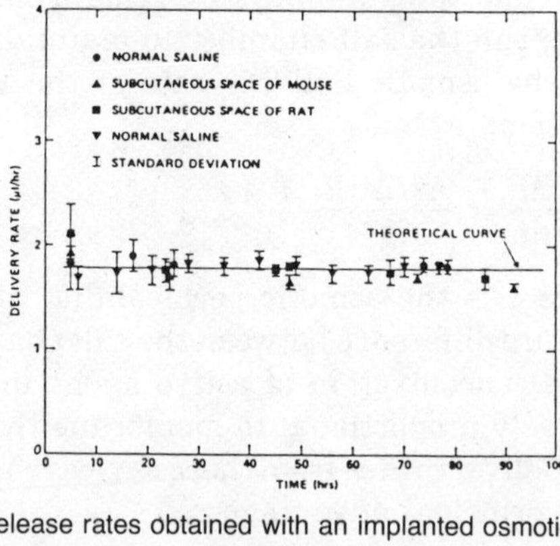

Drug release rates obtained with an implanted osmotic pump.

FIGURE 29.—The Theeuwes osmotic pump. This device has found a considerable use in *in vivo* pharmacokinetic studies. The drug to be studied is loaded into the device which is then implanted in the test animal. The release rate is controlled by the osmotic membrane and the osmotic agent used to power the device.

Figure 29 shows one of the devices implanted in a laboratory rat. Because the delivery rate is fairly low, a rather long delivery orifice is used to minimize diffusional loss of drug from the device.

Yet a further simplification of the original Rose–Nelson concept is the osmotic tablet also developed by Theeuwes (38) and shown in Figure 30. In this device the separate salt chamber is eliminated. The agent, alone or with a

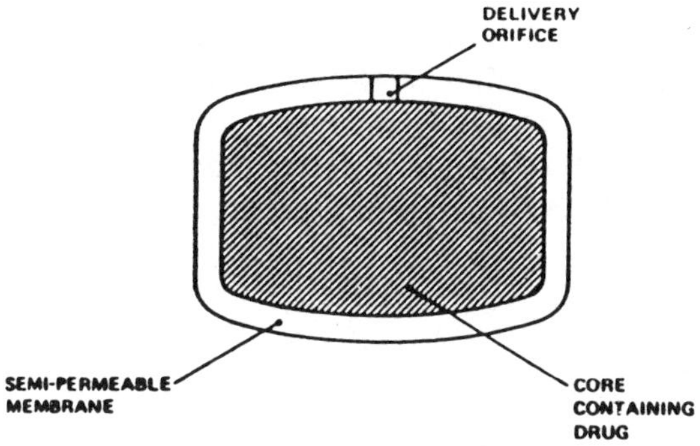

Elementary osmotic pump cross section.

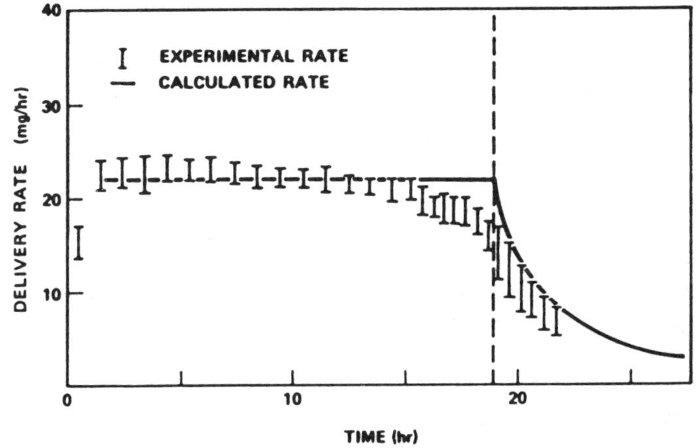

FIGURE 30.—The Theeuwes elementary osmotic pump (38).

diluent of suitable osmotic pressure, is first formed into a tablet. The tablet is then coated with a semi-permeable membrane such as cellulose acetate. A small orifice is drilled through the membrane. When placed in an aqueous environment the osmotic pressure of the agent inside the tablet draws water through the semi-permeable coating forming a saturated aqueous solution inside the device. Because the membrane is non-extensible the saturated solution inside the device is pumped out at a constant rate as water is osmotically imbibed. This process continues until all of the solid drug inside the tablet has been expelled and only a solution-filled shell remains. One constraint in the design of this system is the solubility of the active material compared with the dose rate required. Excipients of suitable water solubility may be included in the tablet core to effect the required osmotic pressure and therefore release rate. In this case, the effects of preferential depletion of the excipient or the active material must also be considered. For constant release until device exhaustion, excess of both materials must be present over the entire lifetime of the device. This may be achieved by incorporating any excipients at a ratio to the active agent equal to the ratio of their water solubility.

5.3. Osmotic Monolithic Devices

In addition to the osmotic pumping devices described above osmosis can also be used as the releasing mechanism for monolithic systems. The principle is illustrated in Figure 31, which shows a simple dispersion of a soluble agent in a polymer matrix (31). Some of the agent can permeate the polymer continuum and escape by simple diffusion as described in section 3. However, if the material has sufficient solubility, and consequently osmotic pressure, water imbibition can also be significant. Water uptake may generate internal pressure sufficient to rupture the walls of vesicles containing a solution of the agent together with excess solid, thereby liberating the agent to the external environment. Initially this process will occur at the outside layer of the polymer matrix and then in serial fractions towards the interior of the matrix. The principle of controlling release in this fashion was first proposed by Marson to explain the release of copper salts from anti-fouling paints (39). The approach has been used by a number of workers since and in particular by Gale et al. (49).

Figure 32 shows the pattern of release that can be obtained by this means from films impregnated with soluble salts. There is an initial high release of the salt from exposed salt particles at the surface of the film. This burst is then followed by a prolonged period of relatively constant release after which the release decreases as the device becomes exhausted. The release rate controlling step appears to be the rate of osmotic imbibition of water into the top layer of salt particles. Thus the release rate depends on the difference in osmotic

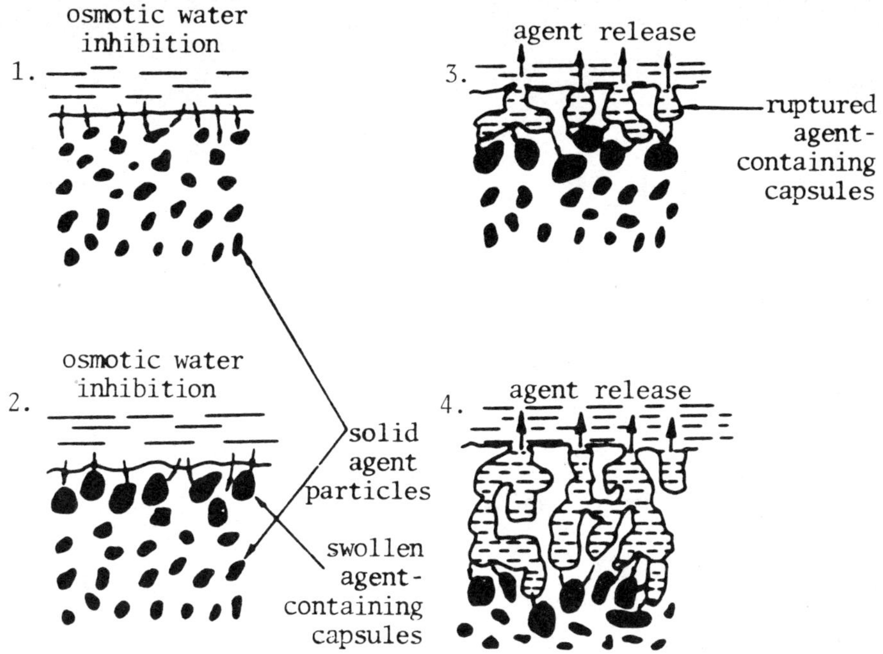

FIGURE 31.—The mechanism of active agent release by osmotic rupture. Water is drawn into the polymer matrix by the water-soluble active agent. This water swells the agent-containing capsules, eventually leading to their rupture and escape of the active agent to the outside solution. The next layer of agent particles is then exposed to the aqueous surrounding media and begins to imbibe water and the process is repeated. This mechanism of serial rupture can be used to produce prolonged release of water-soluble active agents.

pressure ($\Delta\pi$) between the external medium (π_{ext}) and a saturated solution of the active agent (π_s):

$$\Delta\pi = \pi_s - \pi_{ext}. \tag{32}$$

This expression indicates that when the osmotic pressure of the external medium is increased by addition of dissolved salts the release rate will fall, and in the limit when the external osmotic pressure equals the osmotic pressure within the device the release rate should approach zero. This is indeed the case. The data are a convincing demonstration of the validity of the osmotic model.

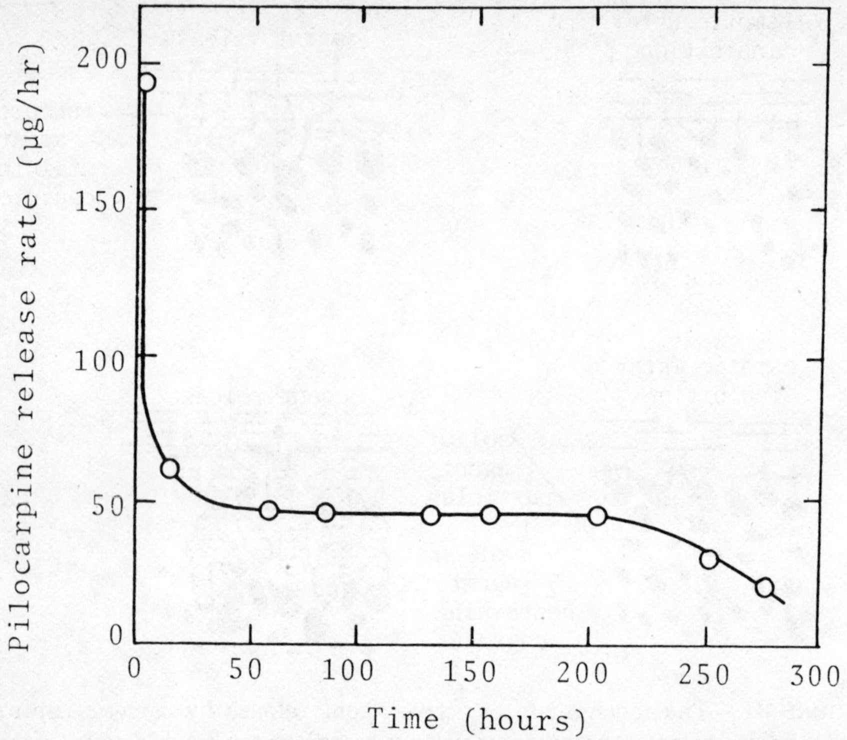

FIGURE 32.—Release of pilocarpine nitrate from a matrix containing 27% by volume of the osmotic agent dispersed in polyethylene vinylacetate (40).

The osmotic release mechanism begins to fail when more than 20% to 30% of active agent is incorporated into the device, at which point there is a significant contribution of simple leaching of agent from the device to the release mechanism.

6. CONCLUSIONS

Although the use of sustained release systems can be traced back to the early fifties, the use of truly controlled release technology has emerged into prominence only in the last decade. During this period the basic principles governing the design and performance of these systems have been elucidated, and have been reviewed in this paper. Although further advances will no doubt be made, particularly in the field of biodegradable systems, the thrust of future progress will center around the application of these principles to commercial products, both pharmaceutical and agrichemical. The first generation products have already emerged (Progestasert, Transderm, Osmosin,

Penncap M, No Pest Strip, etc). The field of application for this sophisticated technology is rapidly expanding.

REFERENCES

1. Sendelbeck, L, Moore, D. and Urquhart, J.: 1975, "Comparative distribution of pilocarpine in ocular tissues of the rabbit during administration by eyedrop or by membrane-controlled delivery systems." Am. J. Ophthalm. 80 (2), pp. 274–283.
2. Alza Corporation: 1974, "Ocusert® (Pilocarpine) Ocular Therapeutic System. A Monograph." Palo Alto, CA.
3. Shell, J.W., and Baker R., W.:, 1974, "Diffusional systems for controlled release of drugs to the eye." Ann. Ophthalm. *6*(10), pp. 1037–1043.
4. Allan, G.G., Chopra, C.S. Friedhoff, J.F., Gara, R.I., Maggi, M.W., Neogi, A.N., Robers, S.C. and Wilkins, R.M.: 1973, "Pesticides, pollution and polymers." Chemtech 4, p. 171.
5. Tilson, S.A., Marion, M. Hudson, R., Wong, P., Pharriss, B.B., Aznar, R. and Martinez-Manautou, J. : 1975, "The effect of intrauterine progesterone on the hypothalamic-hypophyseal-ovarian axis in humans," Contraception 11, p. 179.
6. Pharris, B.B., Erickson, R., Bashaw, J., Hoff, S., Place, V.A., and Zaffaroni, A., : 1974, "Progestasert: A uterine therapeutic system for long-term contraception: I. Philosophy and clinical efficacy." Fertil. Steril. 25, p. 915.
7. Sanders, L.M., Kent, J.S., McRae, G.I., Vickery, B.H., Tice, T.R., and Lewis, D.H. : 1984, "Controlled release of an LHRH analogue from poly (D, L lactide-co-glycolide) microspheres." J. Pharm. Sci. 73, p. 1294.
8. Sanders, L.M., McRae, G.I., Kent, J.S., and Vickery, B.H.: 1983, "Design and characterization of a biodegradable controlled release system of an LHRH analogue." Proc. 10th Int. Symp. on Controlled Release of Bioactive Materials.
9. Polin, H.S. : 1958, "Devices for dosage control." U.S. patent 2,846,057.
10. Lehmann, K. : 1967, "Acrylic resin coatings for the manufacture of depot preparations of drugs." Drugs Made in Germany 10, p. 115.
11. Crank, J.: 1956, The Mathematics of Diffusion, Oxford University Press, London.
12. Baker, R.W. and Lonsdale, H.K. : 1974, "Controlled release: Mechanisms and rates," in Controlled Release of Biologically Active Agents. Tanquary, A.C. and Lacey, R.E. (eds.) Plenum Press, New York, pp. 15–71.
13. Baker, R.W., Tuttle, M.E., Lonsdale, H.K. and Ayres, J.W.: 1979, "Development of an estriol-releasing intrauterine device." J. Pharm. Sci. 68(1), p. 20.
14. Shippy, R.L., Hwang, S.T. and Bunge, R.G. : 1973, "Controlled release of testosterone using silicone rubber." Biomed. Mater. Res. 7, p. 95.
15. Roseman, T.J. and Cardarelli, N.F.: 1980, "Monolithic polymer devices," in Controlled Release Technologies: Methods, Theory, and Applications, Vol. I. A. G. Kydonieus (Ed.), CRC Press, Boca Raton, FL, p. 21.
16. Higuchi, T. : 1961, "Rate of release of medicaments from ointment bases containing drugs in suspension." J. Pharm. Sci. 50, p. 874.
17. T. J. Roseman and Higuchi, W.I.: 1970, "Release of medroxyprogesterone acetate from a silicone polymer." J. Pharm. Sci. 59, pp. 353–356.
18. Paul, D.R. and McSpaden, S.K.: 1976, "Diffusional release of a solute from a polymer matrix." J. Membrane Sci. *1*, 33.
19. Barrer, R.M.: 1968, "Diffusion and permeation in heterogeneous media," in Diffusion in Polymers. J. Crank and G.S. Pard (Eds.), Academic Press, London, p. 165.
20. Maxwell, C.: 1873, Treatise on Electricity and Magnetism. Vol. I, Oxford University Press, London, p. 365.
21. Baker, R.W., Lonsdale, H.K. and Gale, R.M.: 1974, "Membrane controlled delivery systems." Proceedings of the Controlled Release Symposium, Akron, Ohio (Chairman, Nate Cardarelli).

22. Pitt, C.G. and Schindler, A.: 1980, in Biodegradables and Delivery Systems for Contraception, Hafez, E.S.E. and Van Os, W.A.A. (Eds.), MTP Press, Lancaster, England.
23. Heller, J. and Baker, R.W.: 1980, "Theory and practice of controlled drug delivery from erodible polymers," in Controlled Release of Bioactive Materials. Baker, R.W. (Ed.), Academic Press, New York.
24. Schindler, A., Jeffcoat, R., Kimmel, G.L., Pitt, C.G., and Zweidinger, R. : 1977, in Contemporary Topics in Polymer Sciences. Pierce, E.M. and Schaefgen, J.R. (Eds.), Plenum Press, New York.
25. Heller, J., Penhale, D.W.H., Helwing, R.F.: 1980, "Preparation of poly (ortho esters) by the reaction of diketene acetals and polyols." J. Polymer Science: Polymer Letters Edition, 18, p. 619.
26. Heller, J., Penhale, D.H., Helwing, R.F., Fritzinger, B.K., Baker, R.W.: 1981, "Release of norethindrone from polyacetals and poly (ortho esters)." AIChE Symposium Series No. 206, Chandrasekaren, S.K. (Ed.). Vol. 77, p. 28.
27. Heller, J., Penhale, D.W.H., and Helwing, R.F.: 1980, "Preparation of polyacetals by the reaction of divinyl ethers and polyols." J. Polymer Science: Polymer Letters Edition 18, p. 293.
28. Heller, J., Baker, R.W., Gale, R.M., and Rodin, J.O.: 1978, "Controlled drug release by polymer dissolution I. Partial esters of maleic anhydride copolymers - properties and theories." Journal of Applied Polymer Science 22, p. 1991.
29. Sidman, K.R., Schwope, A.D., Steber, W.D., Rudolph, S.E., and Poulin, S.B.: 1980, "Biodegradable, implantable sustained release systems based on glutamic acid copolymers." J. Membrane Science 7, p. 277.
30. Heller, J.: SRI International, Palo Alto, California. Private communication.
31. Allan, G.G., Beer, J.W., Cousin, M.J., and Mikels, R.A.: 1980, "The biodegradative controlled release of pesticides from polymeric substrates" in Controlled Release Technologies—Methods, Theory, and Applications, Vol. II, Kydonieus p. 7. A.F. (Ed.). CRC Press, Boca Raton, FL.
32. Harris, F.W.: 1980, "Polymers containing pendent pesticide substituents" in Controlled Release Technologies—Methods, Theory, and Applications. Vol II, Kydonieus, A.F. (Ed.), CRC Press, Boca Raton, FL, p.63.
33. Petersen, R.V., Anderson C.G., Fang, S.M., Gregonis, D.E., Kim, S.W., Feijen, J., Anderson, J.M., and Mitra, S.: 1980, "Controlled release of progestins from poly (α amino acid) carriers" in Controlled Release of Bioactive Materials. Baker R. (Ed.), Academic Press, NY, p. 45.
34. Rose, S. and Nelson, J.F.: 1955, "A continuous long-term injector." Austral. J. Exp. Biol. 33, p. 415.
35. Higuchi, T. U.S. Patent 3,760,805 Sept. 1973.
36. Theeuwes, F., and Yum, S.J.: 1976, "Principles of the design and operation of generic osmotic pumps for the delivery of semisolid or liquid drug formulations." Ann. Biomedical Engineering 4, p. 343.
37. Theeuwes, F., and Eckenhoff, B.: 1980, "Applications of osmotic drug delivery" in Controlled Release of Bioactive Materials. Baker, R. (Ed), Academic Press, NY, p. 61.
38. Theeuwes, F.: 1975, "Elementary osmotic pump," J. Pharm. Sci. 64, p. 1987.
39. Marson, F.: 1969, "Anti-fouling paints. I. Theoretical approach to leaching of soluble pigments from insoluble paint vehicles." J. Appl. Chem. 19, p. 93.
40. Gale, R., Chendrasekaran, S.K., Swanson, D. and Wright, J.: 1980, "Use of osmotically active therapeutic agents in monolithic systems." J. Membrane Sci. 7, p. 319.

DIALYSIS

Gunnar Jonsson

Instituttet for Kemiindustri
Technical University of Denmark
DK 2800 Lyngby, Denmark

The basic principles of the dialysis process are presented. The relationship between dialysance and mass transfer parameters for different flow geometries is derived. A brief survey of dialyzer design is given. A summary of permeability properties for different commercial membranes is given and the relative mass transfer resistance in the three phases—membrane, feed solution and dialysate as a function of molecular weight—is discussed.

Commercial and potential applications of the dialysis process are presented. This includes hemodialysis, Donnan dialysis, ion-exchange dialysis and alcohol reduction in beer. For the last application a comparison with a similar diafiltration process is given, based on the necessary membrane area and dialysate volume.

P. M. Bungay et al. (eds.), Synthetic Membranes: Science, Engineering and Applications, 625–645.

1. INTRODUCTION

Dialysis is a membrane process by which various solutes having widely different molecular weights may be separated by diffusion through semipermeable membranes. Dialysis requires that the membranes separating the two liquids permit diffusional exchange between at least some of the solutes while effectively preventing any convective mixing between the concentrated and dilute solutions. Because the driving force in dialysis is a concentration gradient, membranes should be thin and the transmembrane concentration difference should be large, if high fluxes are to be achieved. Further, the material from which the membranes are made must be stable enough when in its operating environment to permit suitably long service life times.

Originally, dialysis was used to separate solutes with large differences in size. Low molecular weight substances were separated from colloids, e.g., in 1861 by Graham, who is generally conceded to be the discoverer of dialysis (1). In recent years smaller differences in solute size have been found adequate to achieve significant separations and in some cases, size was deemed an inadequate parameter by which to judge the potential of dialysis for effecting the separations.

Even though dialysis was the first membrane separation process to be discovered and studied, it has never really become industrially important. With the development of new membranes in the 1960's other membrane processes, such as ultrafiltration, reverse osmosis and electrodialysis, have been more thoroughly exploited. These processes are normally operated at higher fluxes than dialysis, because they utilize external driving forces to establish the flux.

Because capital costs increase nearly linearly with the membrane area and the amortized costs of the membrane itself are relatively high, overall processing costs have often been found to be lower with high flux processes despite their higher energy costs per unit of product. Dialysis has persisted as a competitive process where the intrinsic driving force is large or where external driving forces are not effective. The latter may occur because of concentration polarization phenomena or because the external forces are damaging to the fluid being processed.

2. MASS TRANSFER IN DIALYZERS

Figure 1 shows schematically the inlet and outlet flow rates and concentrations in a dialyzer. The solution to be depleted of solute is called the feed

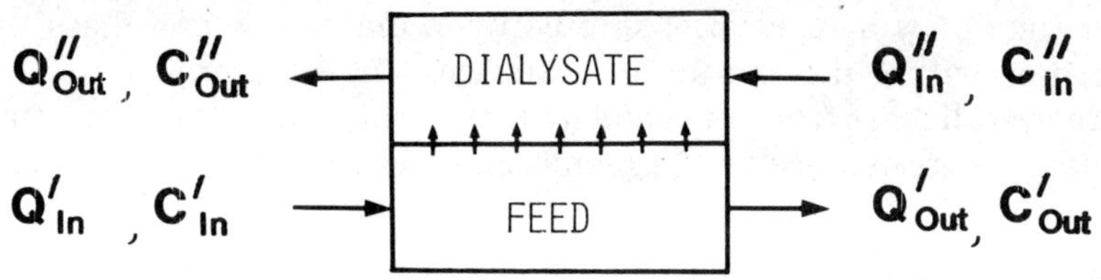

FIGURE 1.—Schematic presentation of the inlet and outlet flow rates and concentrations in a dialyzer operating in counter-current flow.

and the solution receiving the solute is termed the dialysate. The overall efficiency of a dialyzer is determined by two independent factors: the ratio of the flow rates of the two solutions, and the rate constant for solute transport between the solutions, which is determined by the properties of the membrane, the fluid channel geometry, and the local fluid velocities.

In the diagram, Q represents the volumetric flow rates and c the solute concentrations. Subscripts "In" and "Out" refer to inlet and outlet conditions, respectively. Since mass is neither created nor destroyed in the device, the rate of mass transfer, $\dot{M}$, in moles per unit time, is obtained from a mass balance:

$$\dot{M} = Q'_{\mathrm{In}}\, c'_{\mathrm{In}} - Q'_{\mathrm{Out}}\, c'_{\mathrm{Out}} = Q''_{\mathrm{Out}}\, c''_{\mathrm{Out}} - Q''_{\mathrm{In}}\, c''_{\mathrm{In}} \tag{1}$$

The efficiency of a dialyzer is commonly expressed in terms of its dialysance, D, defined as:

$$D = \frac{\dot{M}}{c'_{\mathrm{In}} - c''_{\mathrm{In}}} \tag{2}$$

The extraction ratio, E, defined as:

$$E = D/Q' \tag{3}$$

is a more meaningful quantity than dialysance alone, inasmuch as it represents the fraction of maximum attainable solute depletion in the feed that can actually be achieved in the dialyzer.

Another common term is the clearance, Cl, defined as:

$$Cl = \frac{\dot{M}}{c'_{\mathrm{In}}} \tag{4}$$

It may be viewed as the volumetric rate at which the feed is completely cleared of solute. When $c''_{\mathrm{In}} = 0$, which is the case for most single-pass dialysate systems, clearance and dialysance are equal.

In the following a relationship between dialysance and mass transfer parameters will be derived for a counter-current dialyzer.

An overall mass transfer coefficient, k_o, is calculated from the sum of the mass transfer resistance in the membrane and the two boundary layers:

$$\frac{1}{k_o} = \frac{1}{P} + \frac{1}{k'} + \frac{1}{k''} \tag{5}$$

It is assumed that the local overall mass transfer coefficient, k_o, is constant along the dialyzer.

Because k' and k'' will change with position under laminar flow conditions, a length-averaged overall mass transfer coefficient, $\bar{k}_o$, will be used.

It is further assumed that the volumetric flow rate across the dialysis membrane is negligible, compared with the flow rate of the feed and dialysate, so that the Q are constant in these two phases.

The rate of mass transfer of a given solute across a differential element of area dA is given by:

$$\mathrm{d}\dot{M} = \bar{k}_o\,(c' - c'')\,\mathrm{d}A \tag{6}$$

Differential mass balances on the feed and dialysate streams yield:

$$\mathrm{d}\dot{M} = -Q'\,\mathrm{d}c' \tag{7}$$

$$\mathrm{d}\dot{M} = -Q''\,\mathrm{d}c'' \tag{8}$$

The concentration difference between the two liquids is:

$$\Delta c = c' - c'' \tag{9}$$

By differentiating equation 9, one obtains:

$$\mathrm{d}(\Delta c) = \mathrm{d}c' - \mathrm{d}c'' \tag{10}$$

Substituting the concentration differentials from equations 7 and 8 herein gives:

$$\mathrm{d}(\Delta c) = -\left(\frac{1}{Q'} - \frac{1}{Q''}\right)\mathrm{d}\dot{M} \tag{11}$$

This equation can immediately be integrated to give:

$$\Delta c_{\mathrm{In}} - \Delta c_{\mathrm{Out}} = \left(\frac{1}{Q'} - \frac{1}{Q''}\right)\dot{M} \tag{12}$$

Inserting the mass flow from equation 6 into equation 11 gives the relation:

$$\frac{d(\Delta c)}{\Delta c} = -\left(\frac{1}{Q'} - \frac{1}{Q''}\right)\bar{k}_o \, dA \tag{13}$$

This is integrated across the entire dialyzer area to give:

$$\ln\left(\frac{\Delta c_{Out}}{\Delta c_{In}}\right) = -\bar{k}_o\left(\frac{1}{Q'} - \frac{1}{Q''}\right)A \tag{14}$$

Combining equations 12 and 14 one obtains for the mass flow through the entire membrane area the relation:

$$\dot{M} = \bar{k}_o \, A \, (\Delta c)_{lm} \tag{15}$$

where the log–mean concentration difference is defined by:

$$(\Delta c)_{lm} = \frac{\Delta c_{In} - \Delta c_{Out}}{\ln\left(\frac{\Delta c_{In}}{\Delta c_{Out}}\right)} \tag{16}$$

Δc_{In} and Δc_{Out} are the concentration differences between feed and dialysate at the inlet and outlet ends of the dialyzer, respectively. Equation 15 holds also for other flow geometries using the correct $(\Delta c)_{lm}$ value. For example,

$$\Delta c_{In} = c'_{In} - c''_{In} \tag{17}$$

for co-current and cross-flow geometries, and

$$\Delta c_{In} = c'_{In} - c''_{Out} \tag{18}$$

for counter-current and well-mixed dialysate flows. Similarly,

$$\Delta c_{Out} = c'_{Out} - c''_{Out} \tag{19}$$

for co-current, cross-flow and well-mixed dialysate, and

$$\Delta c_{Out} = c'_{Out} - c''_{In} \tag{20}$$

for counter-current flow.

Equations 1, 2, 15 and 16 may be combined to give the overall performance equations that are summarized in Table 1 for the four flow geometries

TABLE 1. Equations relating extraction ratio to mass transfer parameters

Flow geometry	Extraction ratio
Co-current	$E = \dfrac{1 - \exp[-N_T(1 + Z)]}{1 + Z}$
Counter-current	$E = \dfrac{1 - \exp[-N_T(1 - Z)]}{1 - Z\exp[-N_T(1 - Z)]}$
	If $Z = 1$: $E = \dfrac{N_T}{N_T + 1}$
Well-mixed dialysate	$E = \dfrac{1 - \exp[-N_T]}{1 + Z(1 - \exp[-N_T])}$
Cross-flow	$E = \dfrac{1}{N_T Z} \sum_{n=0}^{\infty} (S_n[N_T]\, S_n[N_T Z])$
	where $S_n[X] = 1 - \exp[-X] \sum_{m=0}^{n} \left(\dfrac{X^m}{m!}\right)$

that are relevant for dialyzers. From the equations the extraction ratio can be calculated from two dimensionless parameters defined by:

$$N_T = \frac{\bar{k}_o A}{Q'} \tag{21}$$

$$Z = \frac{Q'}{Q''} \tag{22}$$

The number of transfer units, N_T, is a measure of the mass transfer size of the dialyzer. The equations shown in Table 1 derive originally from the heat transfer literature (2) and they have been variously introduced into the dialysis literature (3, 4).

Figures 2a–2d are plots of the extraction ratio versus the number of transfer units for each contacting geometry. The Z-value is varied from 0 to 2 and shown as a discrete parameter in the figures. From these figures an estimate of the expected dialyzer performance can be deduced. From the curves it can be seen that with all other parameters fixed the extraction ratio decreases in the order counter-current flow > cross flow > co-current flow > well-mixed dialysate. It is further seen that for a low number of transfer units ($N_T < 0.5$) or low values of the feed-to-dialysate flow rate ratio ($Z \approx 0$), the extraction ratio is independent of the dialyzer geometry.

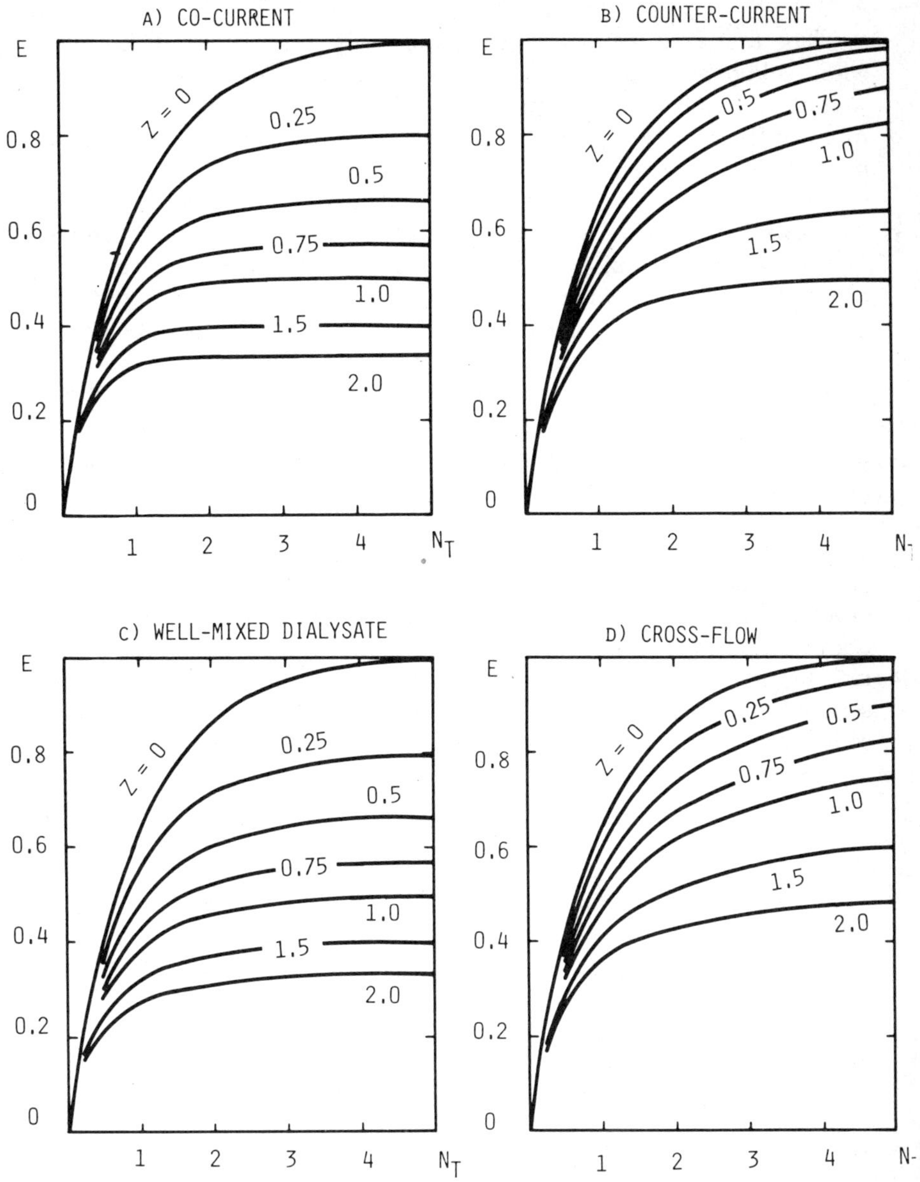

FIGURE 2.—Extraction ratio versus number of transfer units with the Z-value as discrete parameter. Calculated from the equations for the four flow geometries given in Table 1: a) co-current flow, b) counter-current flow, c) well-mixed dialysate, d) cross-flow.

3. DIALYSIS SYSTEMS

The first significant application of dialysis occurred in the mid 1930's when cellophane was first used to remove salts and other low molecular weight solutes from serum proteins and vaccines. At that time it also came into use for the recovery of caustic soda from hemicellulose solutions in the manufacture of rayon.

For these industrial dialyzers a simple plate-and-frame device was used. As shown in Figure 3, it consists of an assembly of frames, with a membrane between each adjoining pair, held together by a clamping device. The goal of any dialysis system is to install a maximum of membrane area in a minimum volume and to provide optimal flow control in the cells, which normally contain plastic screens as spacers and turbulence promoters. A dialyzer normally contains a few hundred single compartments filled in alternating sequence with the feed solution and the dialysate. The two solutions are mostly operated in counter-current flow with the liquid of higher density flowing upwards and that of lower density flowing downwards.

The dominant use of dialysis today is in the treatment of patients with kidney failure to remove urea, uric acid, creatinine and other products of protein metabolism from their blood. The first operating artificial kidney, developed in 1944 by Kolff and Berk (5), consisted of a cellophane sausage casing wrapped around a drum. In this "rotating drum dialyzer," the dialysis membrane was rotated through a bath of dialyzing fluid. The blood was screwed along the tubing by gravity forming a thin film on the wall of the cellophane. Kolff and coworkers also introduced the Twin Coil dialyzer in

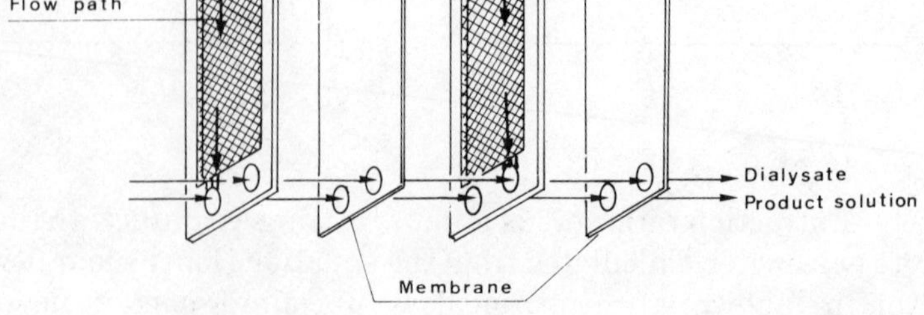

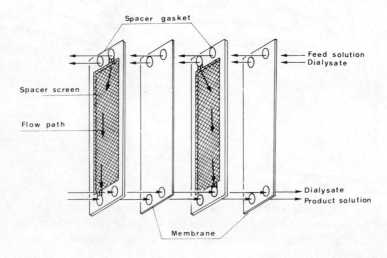

FIGURE 3.—Schematic diagram of a plate-and-frame type dialyzer. Taken with permission from H. Strathmann, J. Membrane Sci., *9* (1981), pp. 121–189.

1956 (6). The coil design, shown in Figure 4, consists of a continuously wrapped roll of flattened cellophane tubing through which blood flows. Supporting layers of plastic screening are spaced to allow dialysate flow crosswise to the coil and at right angles to the flow of blood. The entire coil is submerged in a large bath, and dialysate is recirculated through the coil at a high flow rate. In 1960 Kiil introduced the first plate-and-frame hemodialyzer (7) using the same flow principles as the early industrial dialyzers shown in Figure 3.

A breakthrough in artificial kidney design occurred in 1967 with the report of the hollow fiber artificial kidney (8) originally developed for water desalting by reverse osmosis. The hollow fiber dialyzer, shown in Figure 5, consists of a bundle of many capillary membranes through which blood flows. The fibers are potted into tube sheets at each end of the device, thereby forming headers for the inlet and outlet blood flow paths. Dialysate flows on the outside of the fibers, and the overall design is similar to a shell-and-tube heat exchanger.

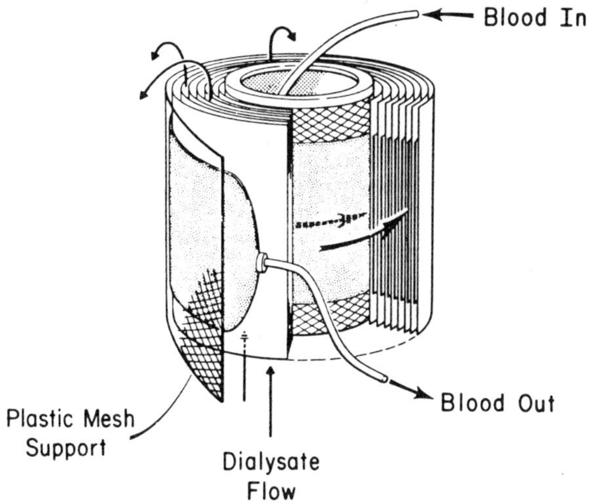

FIGURE 4.—Schematic diagram of a coil type dialyzer. Taken with permission from reference (4).

4. HEMODIALYSIS

The main components of an artificial kidney system (Figure 6) include the hemodialyzer and a dialysate delivery system. The delivery system functions to mix a solution of electrolytes, the composition of which is designed to resemble plasma water. A concentrated solution is mixed with an appropriate quantity of filtered and purified tap water to yield the desired concentrations. Most modern artificial kidney machines incorporate a proportioning pump for continuous manufacture of dialysate at a rate of 200–500 ml/min for

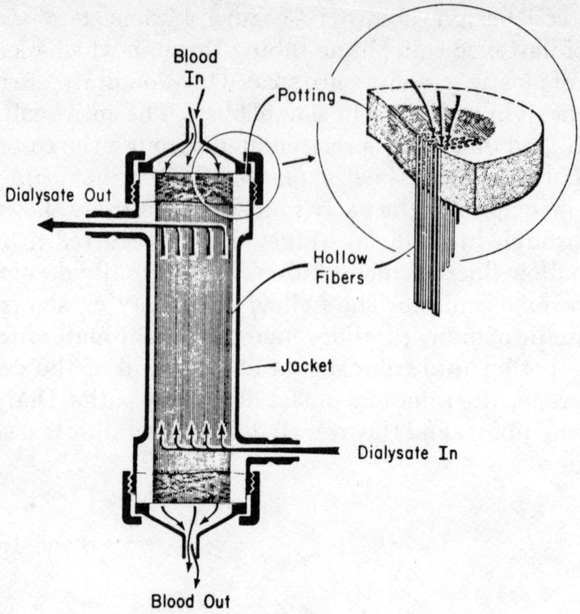

FIGURE 5.—Schematic diagram of a hollow-fiber type dialyzer. Taken with permission from reference (4).

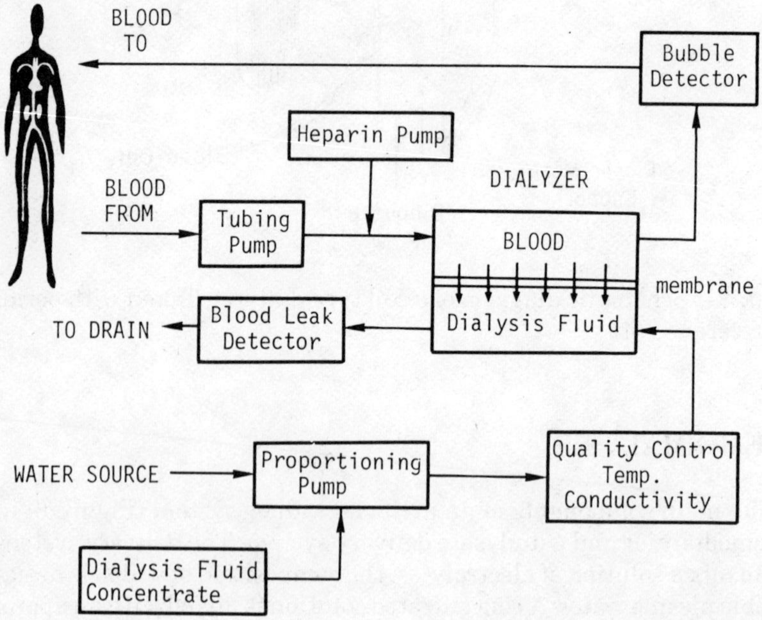

FIGURE 6.—Schematic presentation of the main components of a typical hemodialysis system.

delivery to the dialyzer. Before reaching the dialyzer, the dialysate is heated to 38°C, freed of dissolved gases, and checked both for composition and temperature. The dialysate outlet stream passes through a blood leak detector to protect the patient against rupture or leaks of the dialysis membrane. Blood is usually pumped through the dialyzer at a flow rate up to 300 ml/min. Heparin is added to the blood inlet stream to prevent clotting, and the blood outlet is equipped with a bubble detector. Most conventional hemodialyzers are disposable, require a low blood-priming volume and have about 1 m^2 of membrane area. Although a number of noncellulosic membrane materials have been studied, the vast majority of membranes in actual use are cellulosic, either the more common regenerated cellulose or, in the case of certain hollow fibers, saponified cellulose acetate.

The first long-term treatment of renal patients on the artificial kidney was reported in 1960 (9) and since that time a large number of people—130,000 currently worldwide—have been maintained on the artificial kidney (10). A typical hemodialysis procedure lasts from 4 to 6 hours, during which time the patient is usually confined to bed, and chronic uremia sufferers typically undergo this procedure twice each week.

Some of the molecular species cleared by the human kidney are known to be of a higher molecular weight than those removed in artificial kidneys. It has been speculated that these "middle molecules" are in some way responsible for the lack of complete well-being suffered by artificial kidney users (11). This suspicion has opened up the way for a new hemodialysis membrane, polycarbonate, which has a somewhat higher clearance rate to solutes in the 1,000–2,000 mol wt range, than do the various grades of regenerated cellulose (12). The middle molecule hypothesis has also triggered a major study of a different approach to the removal of toxins from blood: hemodiafiltration (13). Here, the blood is ultrafiltered rather than dialyzed, and small- as well as medium-size molecules are removed with the ultrafiltrate. To maintain appropriate blood volume, the blood is diluted, either before or after the ultrafiltration step. While hemodiafiltration is not yet in significant clinical use, favorable results have been reported with both polyacrylonitrile and polysulfone membranes.

4.1. Permeability Properties of Hemodialysis Membranes

The transport properties of the principal hemodialysis membranes have now been rather well explored (14,15). In the work of Colton et al. (14) the diffusive permeabilities of commercial, modified commercial and laboratory cast regenerated cellulose membranes were measured using 15 solutes ranging in molecular weight from 58 to 68,000. All commercial cellophanes were similar in sodium chloride permeability on a unit thickness basis and were significantly less permeable than hydrophilic wet gel membranes. This difference was attributed to the irreversible collapse of membrane structure upon drying. As the solute characteristic size increased, the effective permeability decreased more sharply with commercial cellophane than with the wet gel cellulose, the ratio between the two becoming an order of magnitude for large solutes. Taking the measured partition coefficients into account, Colton et al. calculated the true membrane diffusion coefficient. Plotting the ratio of the diffusion coefficient in the membrane and in free solution, versus the critical solute radius (Figure 7), they found reasonable agreement with a straight line on semilogarithmic coordinates. When extrapolated to zero radius, the lines yield intercepts of 0.36 and 0.50 for cuprophane and wet gel, respectively. These quantities represent reciprocal squared tortuosities of the membranes that take into account that the path length for a permeant molecule through the hydrated polymer matrix is greater than the membrane thickness.

In the work of Klein et al. (15), the convective permeability of water and diffusive permeability of 13 solutes ranging in molecular weight from 20 to 5200 were measured on four hemodialysis membranes. The membranes tested represented three commercially available materials: Cuprophan 150 PM, Rhône Poulenc RP AN-69 (a copolymer of acrylonitrile and methallylsulfonate) and Bard PCM polycarbonate; and a developmental cellulose acetate (CA-2) made by Celanese. Using a simplified pore model coupled with two measurements, tritiated water permeability and hydraulic permeability, they were able to estimate the solute permeability within a factor of 2 of the true value. Further, they found the theoretical permeability values to be a

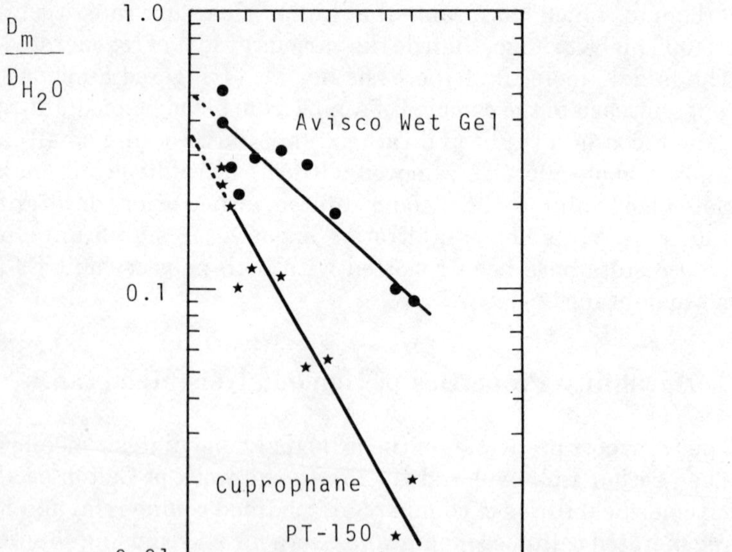

FIGURE 7.—Diffusion coefficient reduction as a function of characteristic molecular radius. Taken with permission from reference (14).

semilog function of the solute radius, in accordance with the experimental findings already shown in Figure 7.

4.2. Simulation of Artificial Kidney Performance

In the work of Colton and Lowrie (4) the mass transfer performance of a hollow fiber dialyzer, the Cordis Dow Model 3, was simulated by estimating the component mass transfer resistances and the whole-blood dialysance under actual clinical conditions. First, they estimated the individual mass transfer resistances in the membrane and the dialysate and blood compartments for urea under in vitro test conditions using data obtained from the literature. Next, they estimated the individual resistances for a solute of arbitrary molecular weight using an observed relation for the membrane resistance versus the molecular weight (16) and general equations relating mass transfer in tubes to the diffusion coefficient in aqueous solution. The results of their calculations are shown in Figure 8, which is a plot of the relative contribution of each phase to the total mass transfer resistance versus the molecular weight. For urea, the membrane accounts for only 45 per cent of the total resistance. With increasing molecular weight the membrane takes on an increasing fraction of the total resistance, because diffusivity in the membrane is a stronger function of solute size than is diffusivity in liquids. Figure 9 is a similar plot of the relative mass transfer resistance in the Kolff Twin Coil dialyzer. Comparison of the two figures shows the very substantial difference in the contribution of the blood mass transfer resistance. The decrease results from the much smaller blood channel thickness

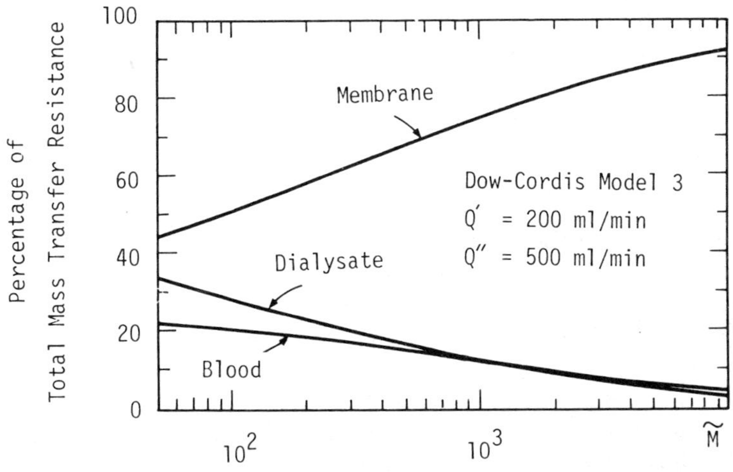

FIGURE 8.—Fraction of total mass transfer resistance in a hollow-fiber dialyzer accounted for by the individual phases. Taken with permission from reference (4).

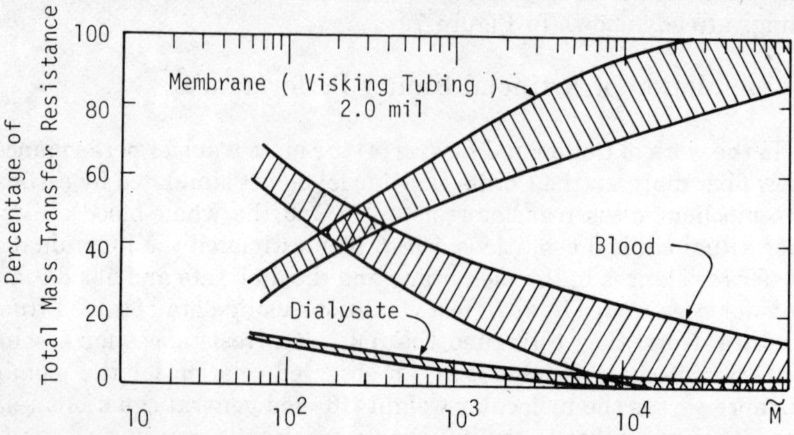

FIGURE 9.—Fraction of total mass transfer resistance in a Kolff Twin Coil hemodialyzer accounted for by the individual phases. Taken with permission from reference (4).

(225 μm versus about 1,000 μm) employed in the hollow fiber device. This improvement, which has also been duplicated in coil and plate-and-frame devices, came about largely following recognition that the large blood channel thickness in early artificial kidneys reduced the mass transfer rates of small solutes compared with the intrinsic capabilities of the membrane.

Whole-blood dialysance was further calculated using the total mass transfer resistance and the equation for counter-current flow in Table 1. The model prediction was compared with clinical dialysance data as shown in Figure 10 and the agreement is quite satisfactory. Both the data and the theoretical prediction show the rather dramatic decrease in dialysance with increasing solute size. Because the membrane constitutes the largest mass transfer resistance, one obvious approach to increasing dialysance is to increase the membrane permeability. The upper curve in Figure 10 was calculated in the same way but with the membrane replaced by a film of water of equivalent thickness. This curve, therefore, provides an upper boundary for the dialysance attainable with hydrophilic gel membranes in a dialyzer of about 1 m^2 area. Although there is a substantial increase of dialysance for solutes with molecular weights above several hundred, the decrease in dialysance with increasing molecular weight remains, because it is inherent in diffusive processes such as dialysis. If dialysance of solutes up to about 5,000 molecular weight should be at nearly the same rate (as is the case with the natural kidney), then other mass transfer processes will have to be used, such as those that rely on convective transport, which more closely emulate the natural kidney.

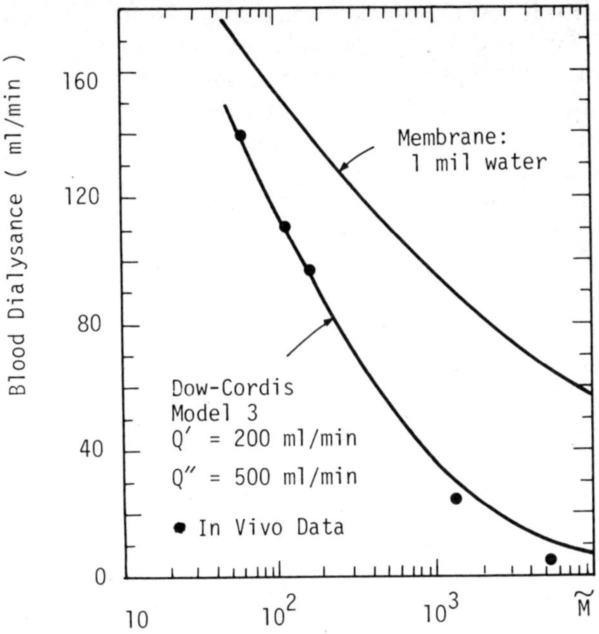

FIGURE 10.—Whole-blood dialysance as a function of solute molecular weight for a hollow-fiber dialyzer. Taken with permission from reference (4).

5. DONNAN DIALYSIS

This process is a special variant of dialysis which combines the Donnan exclusion attainable with ion-exchange membranes with a transmembrane concentration gradient driving force as used in conventional dialysis. It is essentially a continuous ion-exchange process. Its primary use is for recovering valuable cations or anions from dilute solutions by exchanging ions of the same sign from a relatively concentrated solution that is cheaper than the material to be recovered.

Elements of Donnan dialysis of copper ions are illustrated in Figure 11. The cation-selective membrane excludes the anions so that the anion concentration of the two solutions remains constant. The cations, however, will redistribute by diffusion through the membrane until the system approaches the equilibrium dictated by the Donnan relation, which for the actual case is:

$$\frac{a'_{Cu^{++}}}{a''_{Cu^{++}}} = \left(\frac{a'_{H^+}}{a''_{H^+}}\right)^2 \tag{23}$$

Thus, all cations tend to concentrate in the solution containing the higher anion concentration; however, cations of higher charge tend to concentrate preferentially over those of lower charge. For the case shown in Figure 11, Lacey (17) reports that it is possible to concentrate a stream containing 30 ppm Cu^{++} up to 3,000 ppm in the H_2SO_4 solution.

By adding to the stripping solution a complexing agent that can form a complex with a specific cation in the feed stream, it is further possible to separate the ions in a mixed solution, given that equation 23 refers to free ions in solution. Thus the equilibrium of the ions between the two sides of the membrane can be varied by adjusting the composition of the two solutions. A wide variety of concentration and separation processes is therefore possible for two counter-current solutions flowing over opposite surfaces of a perm-selective membrane. Several potential applications of Donnan dialysis have been cited (18):

- removal of ^{137}Cs and mercury from a dilute waste
- removal of ^{90}Sr from a dilute waste with a complexing agent
- separation of silver and copper with a complexing agent
- concentration of uranium from dilute processing solutions
- water softening: removal of calcium and magnesium
- pH adjustment
- removal of zinc from textile waste
- recovery of copper from industrial wastes and low-grade ore
- recovery of nickel, cadmium, and chromium from electroplating wastes
- stripping common pollutants such as mercury, cadmium, copper, and lead from industrial and mining waste.

In a recent paper Ng and Snyder (19) studied the mass transport correlations in a shell-and-tube dialyzer used to extract nickel from dilute $NiSO_4$ solutions with H_2SO_4 as the stripping agent. They found that within the 0–0.0017 molar range of Ni^{++} concentration studied, the boundary layer

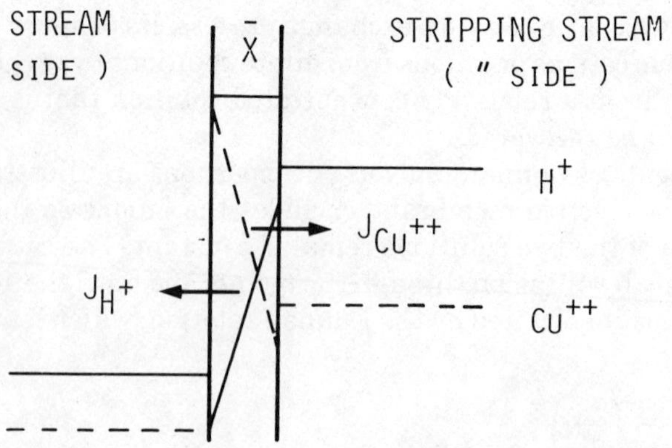

FIGURE 11.—Schematic presentation of the concentration profiles and flux directions in Donnan dialysis.

resistance in the feed stream is the major resistance to the Ni^{++} transport. Only at high flow rates and high concentrations will the membrane resistance be important.

6. ION-EXCHANGE DIALYSIS

This process takes advantage of a phenomenon often considered to be a disadvantage of ion-exchange membranes used for electrodialysis. Anion-exchange membranes, especially the weak-base type, are not very effective in blocking the transport of H^+ ions, although they block other cations effectively. Thus the current efficiency is often found to be as low as 20% for electrodialysis involving highly acid media.

Ion-exchange dialysis is most often used to separate acids from metal salts. Elements of ion-exchange dialysis of a mixed solution of sulphuric acid and nickel sulfate are shown in Figure 12. Water or a dilute solution of H_2SO_4 is used as dialysate. Since the weak-base anion-exchange membranes exclude the Ni^{++} ions but leak H^+ ions, a stream of pure acid can be withdrawn from the dialysate compartments and a stream of purified $NiSO_4$ solution can be withdrawn from the feed compartments.

Nishiwaki and Itoi (20) have reported a pilot-plant test for the nickel refining process in a counter-current flow dialyzer. They found that the dialysis coefficient of nickel was roughly 1/100 that of sulphuric acid. Using the same flow rate in the feed and dialysate compartment, they could obtain an acid recovery of about 75%, while more than 98% of the nickel remained in the feed liquor. They mention some of the main applications for the ion-exchange dialysis process:

- Pickling waste liquid treatment (H_2SO_4–$FeSO_4$, HCl–$FeCl_2$)
- Alumite bath adjustment (H_2SO_4–$Al_2(SO_4)_3$)

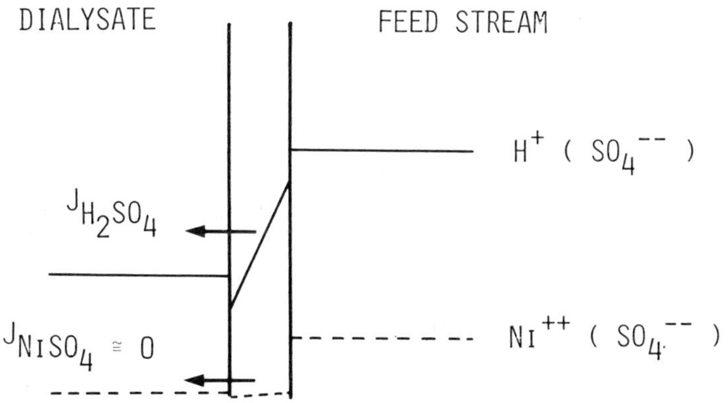

FIGURE 12.—Schematic presentation of the concentration profiles and flux directions in ion-exchange dialysis.

- Etching waste acid recovery ($HCl–AlCl_3$)
- Refining of by-product hydrochloric acid (HCl–chloromethane, etc.)
- Removal of acids from organic substances (H_2SO_4–glucose, amino acids, etc.)
- Nickel refining process (H_2SO_4–$NiSO_4$).

7. ALCOHOL REDUCTION IN BEER

A quite new industrial application of dialysis is the use of Cuprophane hollow fiber membranes to reduce the alcohol content in beer.

Recently, data have been published from a large-scale pilot-unit producing 1500 L dietetic beer per hour, whose alcohol content is reduced by 40% (21). The total membrane area is 90 m^2. The beer flows through the hollow fibers while water flows over the outside of the fibers in a counter-current direction. To prevent CO_2 release, the total system is operated at a pressure (~3 atm) above the CO_2 saturation pressure, because otherwise the diffusion would be disrupted by freed CO_2 bubbles. To prevent a diffusional loss of CO_2 from the beer, the water is presaturated with carbon dioxide. Figure 13 shows the measured degree of alcohol reduction as a function of beer flow at a constant dialysate flow of 2000 L/h in the mentioned pilot unit.

Low molecular weight aromatic matter from the beer diffuses into the dialysate along with the alcohol, producing a certain loss of other beer contents. However, it is concluded that regarding aroma, taste and stability, the dialysis process shows a clear advantage in comparison with a similar distillation process.

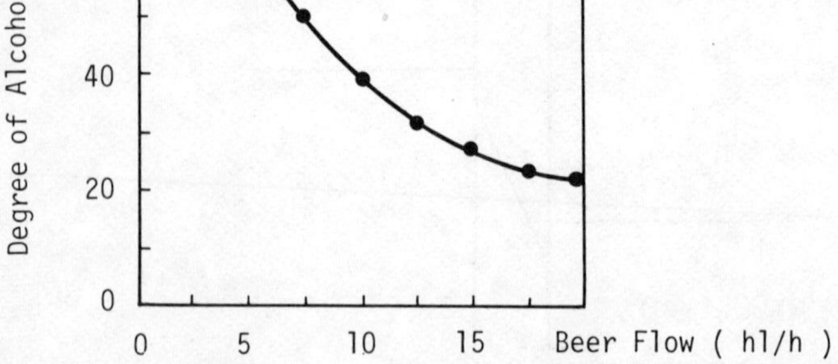

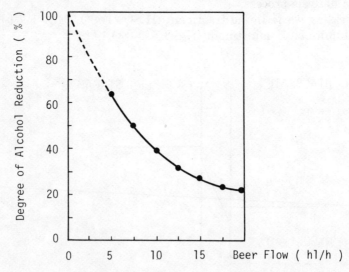

FIGURE 13.—Degree of alcohol reduction as a function of beer flow at a constant dialysate flow of 2000 L/h. Taken with permission from reference (21).

From Figure 13 it is possible to estimate the mass transfer coefficient, $\bar{k}_o$, for alcohol transport in the dialyzer as a function of beer flow using the equation for counter-current flow given in Table 1. As seen from Table 2, the calculated mass transport coefficient is independent of the beer flow rate so that the mass transfer resistance is mainly situated in the membrane and the dialysate compartment in accordance with the curves shown in Figure 8. The absolute value corresponds to 110 ml/min/1.05 m^2, which is also in reasonable agreement with the data shown in Figure 10.

TABLE 2. Calculated values of the mass transfer coefficient and the necessary membrane area using the experimental data from Figure 13 and the equation for counter-current flow given in Table 1

Alcohol reduction E	$Z = \frac{Q'}{Q''}$	Dialysate flow Q'' (L/h)	N_T	Total membrane area A (m^2)	A/E	Mass transfer coefficient $\bar{k}_o$ (cm/min)
0.22	1	2000	0.282	90	409	0.0104
0.40	0.50	4000	0.575	180	450	0.0107
0.64	0.25	8000	1.130	360	563	0.0105
0.40	1	2000	0.667	212	530	—
0.64	1	2000	1.778	566	884	—
0.64	0.50	4000	1.272	404	631	—
0.90	1	2000	9	2867	3186	—
0.90	0.50	4000	3.409	1086	1207	—
0.90	0.25	8000	2.730	870	967	—
0.90	0.10	20000	2.454	779	866	—

Basis: beer flow, Q' = 2000 L/h

For a beer flow of 2000 L/h the necessary membrane area and dialysate flow are calculated for different degrees of alcohol reduction, E, and ratio between beer flow to dialysate flow, Z. As expected, the membrane area increase is more than proportional to the alcohol reduction, especially at E-values above 0.6. Further, it is seen that the membrane area is very dependent on the Z-value at high alcohol reduction.

7.1. Comparison with a Similar Diafiltration Process

In Denmark a production plant for alcohol reduction in beer is in operation based on a similar diafiltration process. In the following an estimate of the necessary membrane area and diafiltrate volume is calculated for a plant producing 2000 L beer/h.

For a batch diafiltration process where water is added at the same rate as permeate is removed, the diafiltration flow rate, Q'', can be calculated using a differential mass balance (22):

$$Q'' = \frac{Q'}{1 - \mathcal{R}} \ln \frac{c'_{\text{In}}}{c'_{\text{Out}}} \tag{24}$$

Here $\mathcal{R}$ is the retention of alcohol and Q', c'_{In} and c'_{Out} refer to beer flow rate and alcohol concentration before and after the diafiltration process.

For the actual process in which cellulose acetate membranes are used, it is estimated that the alcohol retention is about 20% and the permeate flux around 25 L/(m²h). With these values the necessary membrane area and diafiltrate volume are calculated at different degrees of alcohol reduction, as shown in Table 3. Again the membrane area increases more than proportional to the alcohol reduction. Comparing the values given in Tables 2 and 3, it seems that the necessary membrane area and dialysate volume are roughly three to four times larger for the dialysis process than for the diafiltration process. However, due to the high pressures (~30 atm.) needed in the diafiltration process, the investment will not be that different. In addition, the operating costs will be highest for the diafiltration process so that the two processes probably are competitive.

TABLE 3. Calculated values of the necessary membrane area and diafiltrate volume, using equation 24

Alcohol reduction E	Diafiltrate flow rate Q'' (L/h)	Total membrane area A (m²)	A/E
0.22	621	25	114
0.40	1277	51	128
0.64	2554	102	159
0.90	5756	230	256

Basis: beer flow, $Q' = 2000$ L/h

REFERENCES

1. Graham, T.: 1854, "On osmotic force." Phil. Trans. Roy. Soc. London 144, pp. 177–228.
2. Kays, W.M., and London, A.L.: 1964, Compact Heat Exchangers, 2nd ed., New York, McGraw-Hill Book Co.
3. Michaels, A.S.: 1966, "Operating parameters and other performance criteria for hemodialysis and other membrane-separation devices." Trans. Am. Artif. Intern. Organs 12, pp. 387–392.
4. Colton, C.K. and Lowrie, E.G.: 1981, "Hemodialysis: Physical principles and technical considerations," in The Kidney, vol. 2, 2nd ed., Brenner, B.M. and Rector, F.C. (eds.). Philadelphia, W.B. Saunders Company.
5. Kolff, W.J., and Berk, H.T.: 1944, "The artificial kidney: A dialyser with a great area." Acta Med. Scand. 117, pp. 121–134.
6. Vertes, V., Aoyama, S., and Kolff, W.J.: 1956, "The twin coil disposable artificial kidney." Trans. Am. Soc. Artif. Intern. Organs 2, pp. 119–123.

7. Kiil, F.: 1960, "Development of a parallel-flow artificial kidney in plastics." Acta Clin. Scand. Suppl. 253, pp. 142–150.
8. Lipps, B.J., Stewart, R.D., Perkins, H.A., Holmes, G.W., McLain, E.A., Rolfs, M.R., and Oja, P.P.: 1967, "The hollow-fiber artificial kidney." Trans. Am. Soc. Artif. Intern. Organs 13, pp. 200–207.
9. Scribner, B.H., Caner, J.E., Buri, R., and Quinton, W.: 1960, "The technique of continuous hemodialysis." Trans. Am. Soc. Artif. Intern. Organs 6, pp. 88–103.
10. Lonsdale, H.K.: 1982, "The growth of membrane technology." J. Membrane Sci. 10, pp. 81–181.
11. Gotch, F.A.: 1980, "A quantitative evaluation of small and middle molecule toxicity in therapy of uremia." Dialysis and Transplantation 9, pp. 183–194.
12. Barbour, B.H., Bernstein, M., Cantor, P.A., Fischer, B.S., and Stone, W.: 1975, "Clinical use of NISR440 polycarbonate membrane for hemodialysis." Trans. Am. Soc. Artif. Intern. Organs 21, pp. 144–154.
13. Colton, C.K., Henderson, L.W., Ford, C.A., and Lysaght, M.J.: 1975, "Kinetics of hemodiafiltration. I. In vitro transport characteristics of a hollow-fiber blood ultrafilter." J. Lab. Clin. Med. 85, pp. 355–371.
14. Colton, C.K., Smith, K.A., Merrill, E.W., and Farrell, P.C.: 1971, "Permeability studies with cellulosic membranes." J. Biomed. Mater. Res. 5, pp. 459–488.
15. Klein, E., Holland, F.F., and Eberle, K.: 1979, "Comparison of experimental and calculated permeability and rejection coefficients for hemodialysis membranes." J. Membrane Sci. 5, pp. 173–188.
16. Farrell, P.C., and Babb, A.L.: 1973, "Estimation of the permeability of cellulosic membranes from solute dimensions and diffusivities." J. Biomed. Mater. Res. 7, pp. 275–300.
17. Lacey, R.E., 1972, "Membrane separation processes." Chem. Eng. 79, pp. 56–74.
18. Hwang, S-T, and Kammermeyer, K.: 1975, Membranes in Separations. New York: John Wiley and Sons.
19. Ng, P.K., and Snyder, D.D.: 1981, "Mass transport characterization of Donnan dialysis: The nickel sulfate system." J. Electrochem. Soc. 128, pp. 1714–1719.
20. Nishiwaki, T., and Itoi, S.: 1967, "Application of ion exchange membranes to dialysis process." Japan Chem. Quarterly 1, pp. 36–40.
21. Moonen, H. and Niefind, N.J.: 1982, "Alcohol reduction in beer by means of dialysis." Desalination 41, pp. 327–335.
22. Jonsson, G.: 1981, Lecture Notes in Membrane Filtration. Instituttet for Kemiindustri, Lyngby, Denmark.

SYMBOLS

Cl	clearance, equation 4
D	dialysance, equation 2
E	extraction ratio, equation 3
k_o	overall mass transfer coefficient, equation 5
$\dot{M}$	molar mass transfer rate
N_T	number of transfer units, equation 21
Z	ratio of feed to dialysate volumetric flow rates, equation 22

Superscripts

$'$	feed
$''$	dialysate or stripping solution

Subscripts

lm	log mean

BIOMEDICAL APPLICATIONS

Jerome S. Schultz

Chairman
Department of Chemical Engineering
University of Michigan
Ann Arbor, Michigan 48109

Biosensors: The need for specific *in vivo* sensors is being met by the development of fiber optic devices, which are isolated from the biological environment by appropriate membranes. Specific probes for oxygen, carbon dioxide, pH, glucose, and other metabolites have been developed by use of colorimetric and fluorometric optical systems.

Blood Oxygenation: Membrane blood oxygenators have the potential of providing a large amount of surface area in a small volume. Design considerations must include transport resistances in blood in addition to membrane characteristics. Both thin silastic and non-wettable porous membranes are of current interest.

Membrane Plasmapheresis: Removal of large molecular weight proteins from blood can be used to treat several diseases of immunological origin. The factors that influence the efficiency of membrane-based devices are the porosity of the membrane, transmembrane pressure difference, and the shear rate at the blood–membrane interface.

1. FIBER OPTIC BIOSENSORS

2. BLOOD OXYGENATION

3. MEMBRANE PLASMAPHERESIS

P. M. Bungay et al. (eds.), Synthetic Membranes: Science, Engineering and Applications, 647–665.

1. FIBER OPTIC BIOSENSORS

In recent years efforts have been directed increasingly toward the development of improved methods for "on-line" monitoring of specific metabolites or biochemicals. For example, the need for a glucose sensor has become more acute with the advent of an "artificial pancreas" concept, consisting of an implantable insulin infusion pump controlled by the blood glucose level.

The majority of approaches for the development of such sensors have been primarily focused on electrochemical principles. These devices are usually based on the measurement of a current or voltage that is related in part to the extent or rate of some biochemical enzyme reaction. The selection of an appropriate biochemical system for coupling to the electrode usually depends on the sensitivity and specificity requirements. In addition, the transducer element of the device, i.e., that containing the enzymes of other biochemical components, is usually isolated from the external environment by a membrane, with appropriate permeability characteristics for the substance to be measured.

Clark (1) was one of the first investigators to make an "enzyme-electrode" by placing a layer of glucose oxidase on an oxygen electrode. This approach was expanded greatly by Rechnitz (2).

Unfortunately, some common shortcomings of electrochemical sensors have still not been fully overcome. The measurement in enzyme-based electrode sensors usually depends on the rate of diffusion of the metabolite through a membrane and subsequent rates of reaction.

Membrane fouling and electrode inactivation may affect the sensor's calibration curve. An alternative non-electrode approach employing fiber optics was first developed at the National Institutes of Health by Goldstein and Peterson (3) for the measurement of pH in tissue and blood. An example of a fiber-optic pH sensor made by this group is shown in Figure 1. In this example, two optical fibers terminate inside a short length of permeable capillary tubing containing beads with an immobilized pH-sensitive indicator dye. The permeable tubing is a porous material (e.g., hollow dialysis fiber) to allow for diffusion and equilibration of charged ions between the interior and exterior aqueous environments. One optical fiber is connected to the light

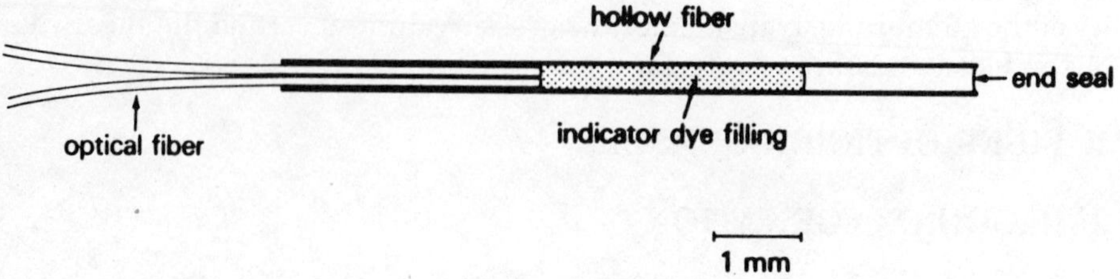

FIGURE 1.—Construction of pH probe (4).

source, and the other leads to a detector. Changes in pH result in spectral alterations of the dye, shown in Figure 2. By measuring light absorption at two wavelengths, e.g., 480 nm and 560 nm, one can estimate the pH within the transducer.

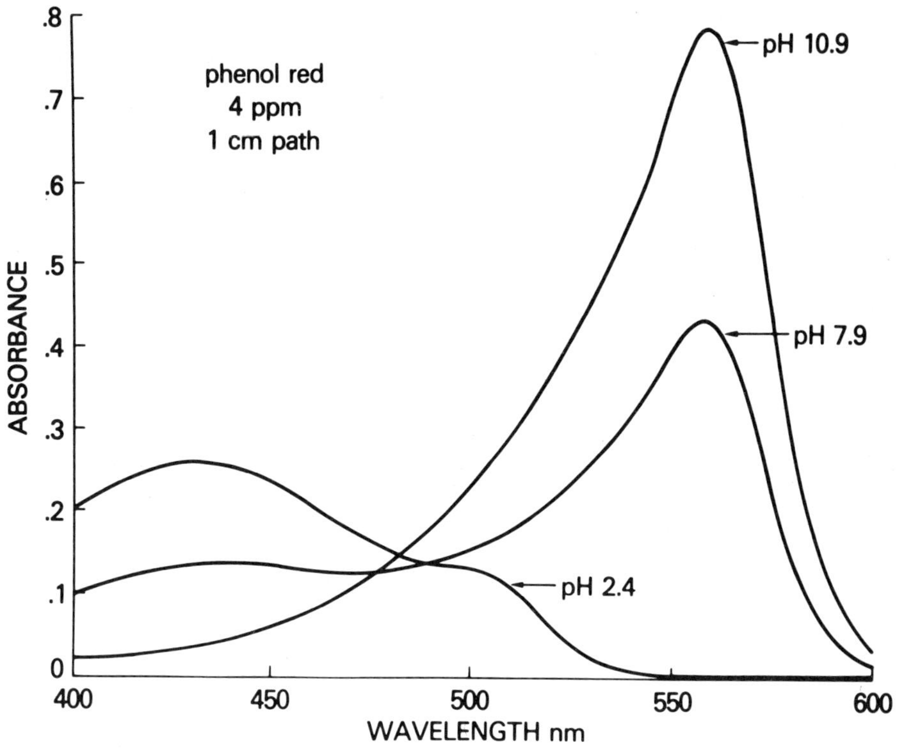

FIGURE 2.—Optical absorption of phenol red with varying pH (4).

Peterson and Goldstein (4) have also fabricated a CO_2 sensor by substituting a polymer membrane, e.g., silicone rubber, for the envelope material. Then only compounds soluble in the polymer can enter the transducer, and because CO_2 is one of the few gases that will affect the internal pH, color changes will be related to external CO_2 concentration changes.

Other schemes for circumventing the problems associated with electrochemical sensors are being explored. A new approach[1] taken by Schultz and coworkers (5, 6) uses a detection technique based on immunoassay principles, that is, reversible competitive binding to a receptor between the meta-

This project is supported by the National Institutes of Health, grant numbers RO1 AM 26858 and GM 29470.

bolite of interest and a fluorescently labeled ligand. The relative amounts of binding between the labeled ligand and the receptor are altered in the presence of the metabolite. This alteration of the labeled ligand–receptor reaction is monitored by a miniaturized optical fiber-based fluorometer placed to detect the fluorescence, thus giving a measurement of the metabolite concentration.

The advantages of this type of sensor, called an "affinity sensor" because it is based on the competition of analytes and fluorescent analogs for receptor sites, are as follows:

1. The working mechanism of affinity sensors is based on equilibrium binding rather than rate processes; therefore, the measurement is not affected by fouling of the membrane that separates the reactor system from the fluid being assayed, which is a problem with current electrochemical sensors.

2. The approach can be used for a wide variety of metabolites, as long as suitable receptors and competing ligands are provided.

3. The sensor can be miniaturized, enhancing its potential for implantation.

4. By the selection of appropriate receptors from the wide range of antibodies, binding proteins, etc., a high degree of selectivity is achievable.

A diagrammatic view of the transducer element of a glucose affinity sensor is shown in Figure 3. Concanavalin A (Con A) is a lectin with specific

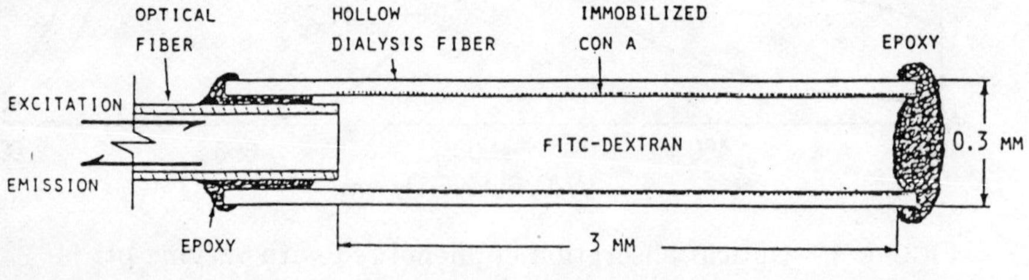

FIGURE 3.—Schematic diagram of the transducer element of a glucose affinity sensor. Con A is immobilized on the interior surface of a Cordis Dow (C-Dak) hollow fiber by periodate oxidation. The fiber is charged with 10 μg/ml FITC-dextran (70,000 MW) and is sealed with epoxy cement. The optical components include a bidirectional optical coupler (Canstar type TC3), a light source (75-W Xenon lamp), a light detector (FW 130, ITT), and a single optical fiber (SDF Corning, 0.1-mm core, 0.14-mm O.D., and 0.3 N.A.), which are connected to the three ports of the coupler. When excitation light from the source is directed into the hollow fiber by the optical coupler and optical fiber, the free FITC-dextran in the illumination field of the optical fiber is excited an emits fluorescent light. A portion of the emitted light enters the same optical fiber and is transmitted from the other port of the optical coupler into the light detector. A barrier filter, mounted in front of the detector, blocks the backscattered source radiation, allowing only fluorescence detection.

binding characteristics for few carbohydrates; among them, only glucose is present in significant levels in blood. This lectin is immobilized on the interior surface of a closed length dialysis hollow fiber. The hollow fiber also contains dextran labeled with fluorescein isothiocynate (FITC-dextran). The selected molecular weight of the dextran (70,000 daltons) is much higher than the molecular weight cutoff of the hollow fiber (5,000 daltons); thus, dextran is trapped within the hollow fiber. Exposure of the sensor to glucose solutions results in the diffusion of glucose into the hollow fiber, and the displacement of FITC-dextran from Con A sites. The change in free dextran concentration within the sensor element is measured optically by means of a single optical fiber and gives a quantitative indication of glucose in the solution.

The optical system associated with the sensor consists of a single optical fiber for transmitting the excitation light into the lumen of hollow fiber sensor element, as well as for directing some of the fluorescent emission from the labeled analog to a light detector. This configuration greatly simplifies the fabrication and miniaturization of the transducer element of the sensor. Another feature of the optical system is the capability to distinguish between the free FITC-dextran in the lumen and the FITC-dextran bound to the immobilized receptor sites on the wall. The numerical aperture of the single optical fiber is an important parameter in this regard. Smaller numerical apertures are favored to confine the view field of the optical fiber to an axial volume inside the hollow fiber and to minimize the extent of emitted light from the bound FITC-dextran entering the optical fiber.

An important consideration in fabricating an affinity sensor is the immobilization of receptor sites to the inner wall surface of a hollow fiber. Our approach involves periodate oxidation of the cellulose to yield aldehyde groups. After incubation with hexane–diamine to provide a spacer group, Con A can be coupled to the fiber with glutaraldehyde.

Typical calibration curves for the glucose sensor are shown in Figure 4. The response shows saturation behavior because at high sugar concentrations all the fluorescent dextran is displaced from the wall Con A sites, and no further increase in fluorescence can occur. The useful range for measuring glucose in this sensor extends up to about 4 mg/ml. However, the sensitivity range can be altered somewhat by changing the amount of Con A immobilized, either by using competing fluorochromes other than FITC-dextran, or by using receptor sites other than Con A.

The calibration curves for other sugars indicate the potential for changing the operational range of the sensor.

Human plasma samples with various levels of glucose (1–6 mg/ml) were tested to see whether any substance in blood might interfere with glucose measurement. The agreement between plasma glucose values measured by the sensor and values obtained by chemical analysis was very good (Figure 5), giving a correlation coefficient of 0.98.

The time required for the response of the sensor to reach 90% of its ultimate value following a step change in sugar concentration (average re-

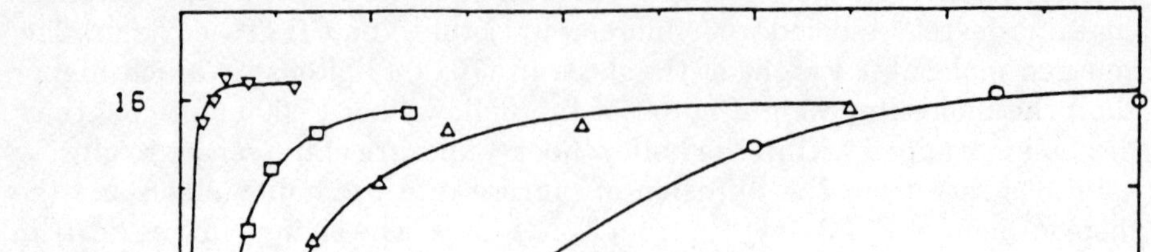

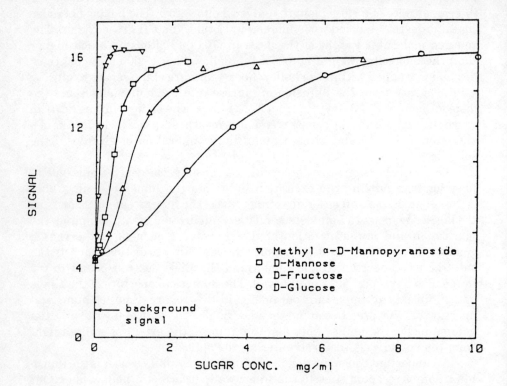

FIGURE 4.—Calibration curves of the affinity glucose sensor for four sugars dissolved in phosphate–saline buffer (pH = 7.2). The units of the ordinate are arbitrary numbers representing the fluorescence intensity measured by the photomultiplier tube. The background signal corresponds to the leakage level of excitation light into the light detector.

sponse time) is about 10 minutes. The likely limiting step in response time is diffusional transport through the hollow fiber membrane and within the lumen, suggesting that the response time may be improved by reducing the diameter of the follow fiber.

The feasibility of the optical affinity sensor concept is demonstrated: 1) the sensor's response to plasma glucose is specific and is in the range suitable for diabetic patients, 2) the sensor can be miniaturized, and 3) the response time of the sensor is within the range of blood glucose fluctuations that occur in diabetics.

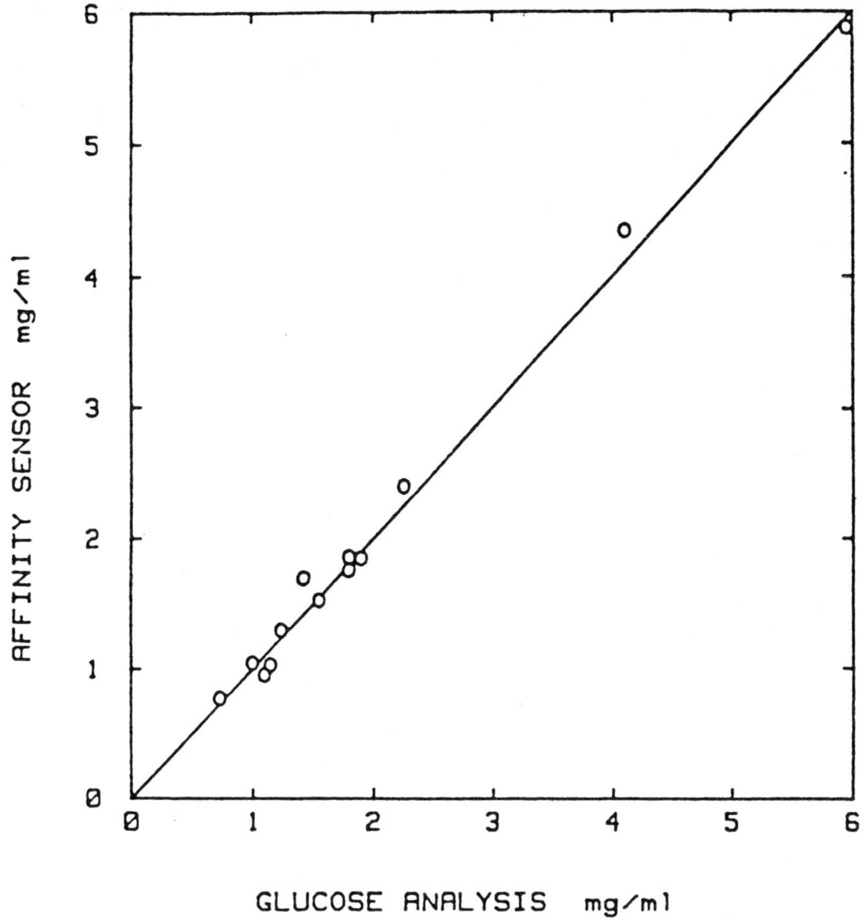

FIGURE 5.—A comparison of plasma glucose concentrations obtained by chemical analysis with concentrations measured by the sensor, using a calibration curve for glucose similar to the one in Figure 4. The pH of the plasma was adjusted to 7.2. The concentration of glucose in two of the samples (411, and 595, mg/dL) was artificially increased by the addition of glucose. The correlation coefficient between the two sets of measurements is 0.98.

2. BLOOD OXYGENATION

A variety of devices have been developed for the oxygenation of blood during surgical procedures requiring the use of a heart–lung machine. The earliest types of oxygenators were bubblers and rotating disc devices which relied on direct contact between gas and liquid. Following the introduction about 15 years ago of permeable silicone rubber membranes, many laboratories sought to develop oxygenators that were membrane-based and would avert the problems associated with direct gas–liquid contact. An advantage of membrane-based oxygenators is reduced blood degradation, due, presumably, to protein denaturation at the blood–gas interface. Another advantage is in device design and fabrication which, because of the defined blood channels, allows for precisely controlled blood-flow characteristics and, therefore, possibilities for optimal efficiency in gas exchange.

Much emphasis has been placed on oxygen transfer as the controlling factor in the design of the blood–gas exchanger system. However, examination of some of the physical properties of carbon dioxide shows that in some cases, removal of CO_2 from the blood may be a limiting factor in design.

For a 'standard 70-kg man' at rest the approximate oxygen requirement is about 250 cc/min, and the corresponding carbon dioxide production rate is about 200 cc/min. The equivalent oxygen partial pressure in venous blood is about 40 mm Hg and about 95 mm Hg for arterial blood. Use of air as the supply gas and at only one atmosphere gives a driving force for oxygenation of about $150-(40+95)/2$ or 82 mm Hg. In contrast, in venous blood the CO_2 partial pressure is about 45 mm Hg which, assuming that the exchanging air has negligible CO_2 content, constitutes the maximum driving force for CO_2 removal from blood.

The nature of blood complicates the development of analytical mathematical models for the design of a physical system. Aspects such as the particulate, cellular nature of blood as a fluid, the complexity of this chemical reaction between oxygen and hemoglobin and the non-linear characteristics of the oxyhemoglobin dissociation curve are factors that make mathematical analysis intractable. A number of sophisticated computer programs have been developed by various investigators to solve the governing non-linear diffusion–reaction equations in flowing blood. However, by postulating a few simplifying assumptions, the analytical solution of many physical situations becomes possible, and insight into the effects of various independent parameters can be gained.

If the rates of oxygen–hemoglobin reactions are relatively fast with respect to diffusional processes, the reactions may be regarded as being at equilibrium. Finally, if the particulate nature of blood is ignored, blood may be treated as if it were a homogeneous solution.

There are two limits in which simplifications of the curved oxygen binding curve make the solution of the oxygenation diffusion–reaction equations more tractable. In Figure 6, the dissociation of oxyhemoglobin is shown as a linear function of oxygen partial pressure, up to a somewhat arbitrary value of full saturation. Figure 7 shows the result of the assumption that there is a critical oxygen partial pressure below which hemoglobin has a constant value of partial saturation, and above which, it is fully saturated; this has been called a step isotherm. Mikic et al (7) have compared the calculated degree of oxygenation of hemoglobin solutions flowing through permeable

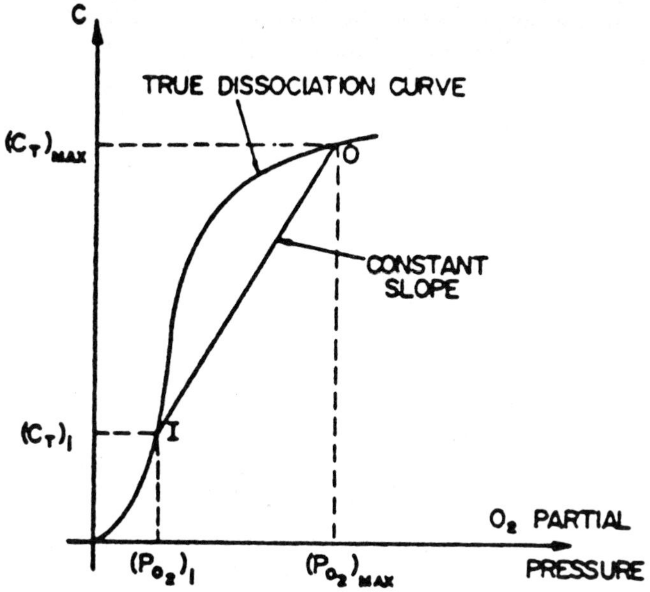

FIGURE 6.—Modeling of dissociation curve: Constant slope approach (7).

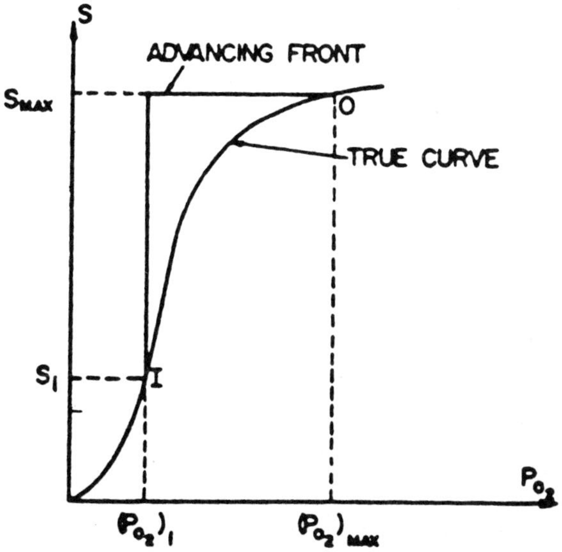

FIGURE 7.—Modeling of dissociation curve: Advancing front approach (7).

tubes for the linear, step and true isotherms. These results (Figure 8) demonstrate that calculations based on the two upper and lower limits bracket the behavior of the true oxyhemoglobin system.

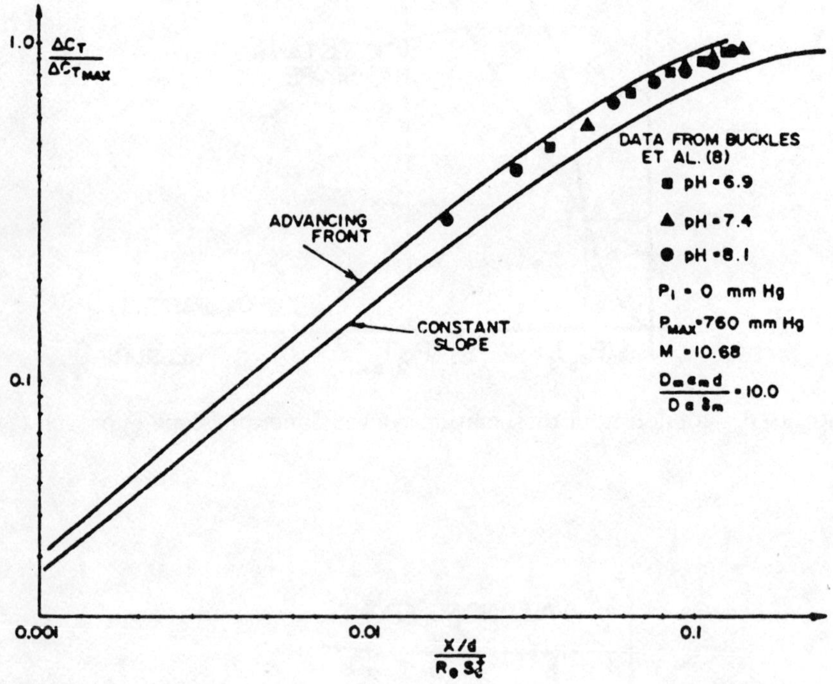

FIGURE 8.—Comparison of experimentally measured oxygen transfer with calculated values of the upper and lower bounds based on the advancing front and constant slope approximations to oxyhemoglobin dissociated curve (7).

We believe the advancing front approach of most interest. Figure 9 illustrates hypothetical concentration profiles for a flat layer of blood being oxygenated from a single surface over an increment in time, 1 to 2 (8). As the front of oxyhemoglobin advances into the film over this time increment, its rate of movement is determined by how fast oxygen can diffuse to the front.

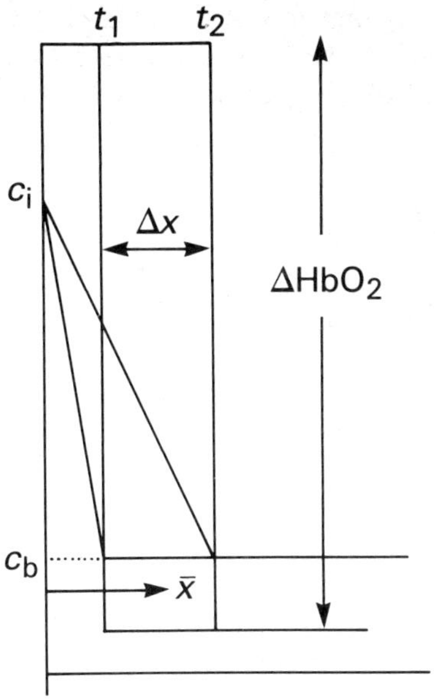

$$\frac{D(c_i - c_b)\,\Delta t}{\bar{x}} = \Delta HbO_2\,\Delta x$$

$$x = \left(\frac{2D\Delta c\ t}{HbO_2}\right)^{1/2}$$

FIGURE 9.—One-dimensional advancing front approximation of blood oxygenation.

Assuming the oxygen concentration profile to be linear, the oxygen diffusion rate is given by

$$J = (D/x)(c_{\mathrm{int}} - c_b) \tag{1}$$

and the rate of front migration is

$$J_M = \Delta HbO_2 \frac{dx}{dt} \tag{2}$$

Because of the rate of front movement is limited by the rate of oxygen transfer and $J \equiv J_M$, these equations can be solved to obtain the front position, or, equivalently, the rate of oxygen uptake as a function of time:

$$x = \left(\frac{d\,(c_{int} - c_b)\,t}{\Delta HbO_2} \right)^{1/2} \tag{3}$$

This approach, called a 'zero order' approximation by E. N. Lightfoot (9), can be applied to blood flowing between two parallel, flat membrane sheets (Figure 10). Here one must account for the parabolic laminar blood flow distribution as well as the presence of a membrane on both sides of the film of blood. An equation for the thickness of the unoxygenated blood as a function of the dimensionless length is as follows:

$$\frac{D}{Pl}\left(\frac{2}{3} - \xi + \frac{1}{3}\xi^3\right) + \frac{5}{12} - \xi + \frac{1}{2}\xi^2 + \frac{1}{3}\xi^3 - \frac{1}{4}\xi^4 = \frac{D\,(c_{int} - c_b)\,z}{(2l)^2\,\bar{u}\,\Delta HbO_2} \tag{4}$$

Lightfoot (9) gave the fraction of blood oxygenated as a function of the position of the front.

$$f = 1 - \frac{3}{2}\xi + \frac{1}{2}\xi^3 \tag{5}$$

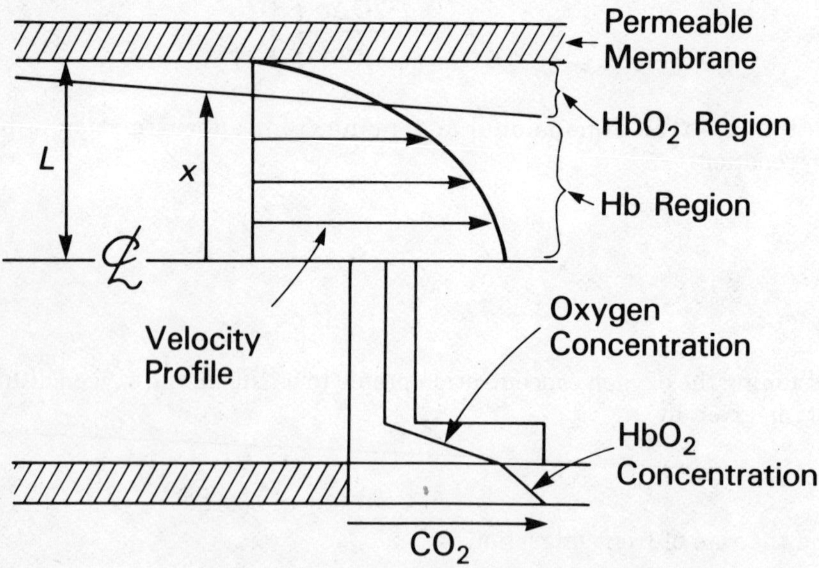

FIGURE 10.—Advancing front approximation for a plane parallel flow membrane oxygenator. Representative profiles for concentration, velocity and front position.

The results of this analysis are illustrated in Figure 11, where the parameter $(D/L)/P$ is the ratio of the oxygen permeability of the film of blood to the oxygen permeability of the polymer membrane. As this ratio exceeds a value of about one, the influence of the membrane is dominant in the oxygenation process; below a value of one, it is diffusion of oxygen in the film of blood that is limiting.

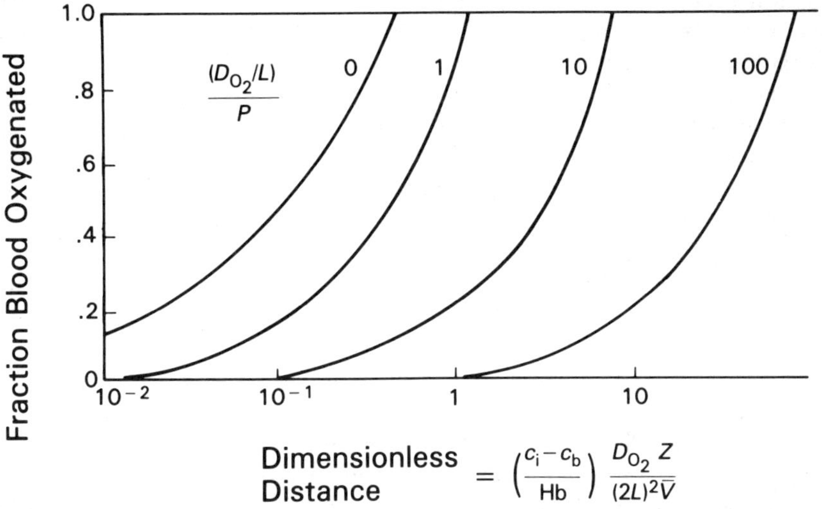

FIGURE 11.—Advancing front approximation for a plane parallel flow membrane oxygenator. Influence of the relative permeabilities of the blood and membrane films. $c_i = c_{int}$.

It is notable that the rate of oxygenation is much higher at the entry region of the device, a region where the distance that oxygen must diffuse to react with hemoglobin is shorter. A variety of oxygenation designs have been developed to capitalize on this behavior by producing many entry regions, either by mechanical mixing or by introduction of secondary flow patterns to improve oxygenator performance.

3. MEMBRANE PLASMAPHERESIS

Plasmapheresis is a process that allows the cellular elements in a person's blood to be separated from the fluid and soluble constituents, and the cellular elements to be returned to the donor. This type of process can be performed on a person for plasma donation more frequently than whole blood donation because the normal physiological rate of plasma protein regeneration is on the order of days, whereas that of erythrocytes is on the order of months.

Among the many applications of plasmapheresis is its use in blood banking. Plasma is separated from cells and then fractionated into its various constituents such as albumin, clotting factors, immunoglobulins, etc., and stored to better meet specific needs of different patients. The application of plasmapheresis as a therapeutic process often involves treatment of diseases characterized by the accumulation of deleterious immune complexes or other harmful macromolecules. By first separating out the fragile cellular elements from whole blood, one can perform a variety of fractionation procedures on the plasma to remove harmful elements.

Sterile centrifugation remains the technique in common use to separate cells from blood prior to further processing. However, advances in approaches based on membranes offer promise for on-line plasmapheresis. The earliest attempts to separate out plasma proteins using ultrafiltration membranes were of limited effectiveness because of problems with concentration polarization of the proteins at the membrane interface at high flux rates. Subsequent studies used macroporous membranes in pressurized stirred chambers, but encountered problems with hemolysis and plugging of the membrane by cells. It was the introduction of the "cross-flow" concept based on the work of Forstrom et al. (10) that led to development of practical membrane-based plasmapheresis devices. The term "cross-flow" means that the primary direction of blood flow is parallel to the membrane surface.

The system analyzed by Forstrom's group (10) is diagrammed in Figure 12. The bulk motion of the fluid parallel to the surface of the membrane produces a velocity profile near the surface that is approximately linear. The fluid shear rate close to the membrane surface is given by the slope of this line.

$$S = (\text{shear rate}) = \Delta u / \Delta x \quad (6)$$

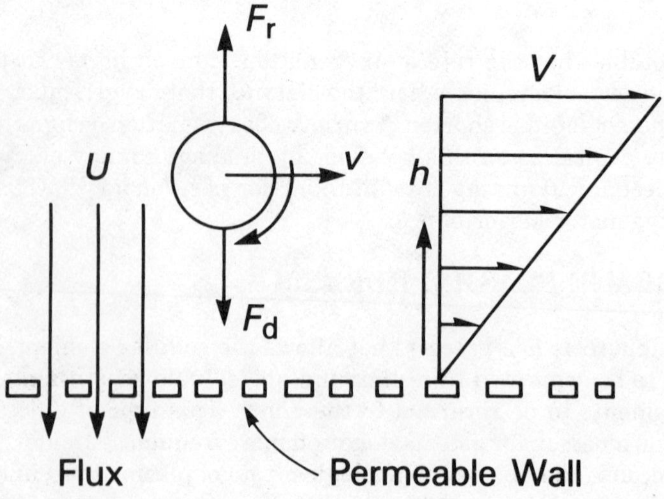

FIGURE 12.—Hydrodynamic forces acting on a particle in a suspension undergoing cross flow near the surface of a filtering membrane.

Particles near the membrane are exposed to a higher velocity on their top side, as compared with the bottom side closer to the membrane. This relative imbalance in drag forces causes deformation and rotation of particles, which in turn experience a net force, F_r, directing the particles away from the surface. Thus near the membrane surface a "skimming" layer of plasma is created that is free of particles.

The membrane is, of course, porous and when a pressure gradient is established to filter fluid through the surface, a bulk fluid velocity component is created, v(cm/sec), directed toward the membrane surface. This bulk velocity component produces another drag force on the particles, (F_d), that tends to move particles toward the surface.

In plasmapheresis, it is desirable to maximize the rate of plasma removal, v, without causing the particulate components of the blood to impact on the membrane surface. Forstrom's group (10) showed that the critical dimensionless group in this situation is

$$\frac{v \, \lambda \, \nu^{1/2}}{R^2 \, S^{3/2}} \tag{7}$$

where v is the bulk velocity normal to the membrane, λ is a correction factor for particle concentration, ν is the fluid kinematic viscosity, R is the particle radius, and S is the fluid shear rate near the surface.

The measured relationship between v_{cr} (the critical product flow velocity) and the other factors is shown in Figure 13, from which it is deduced that red blood cell deposition on the membrane occurs when

$$v_{cr} > 0.15 \, R^2 \, S^{3/2}/\lambda \, \mu^{1/2} \tag{8}$$

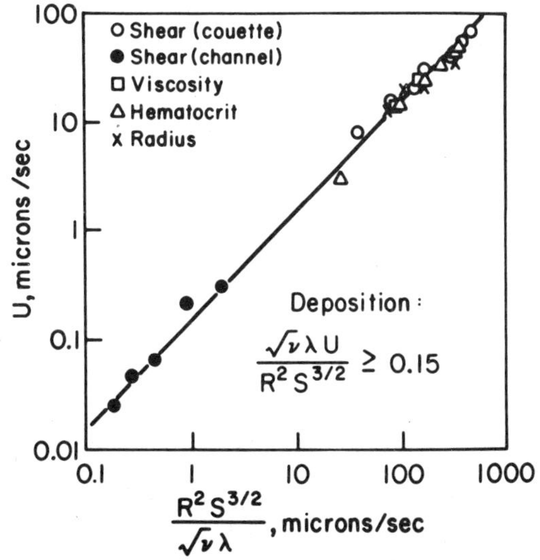

FIGURE 13.—Filtration velocity at the onset of red cell deposition versus flow parameter (10).

As seen in Figure 14, higher hematocrits produce a large increase in the correction factor λ, drastically reducing the rate of plasma production.

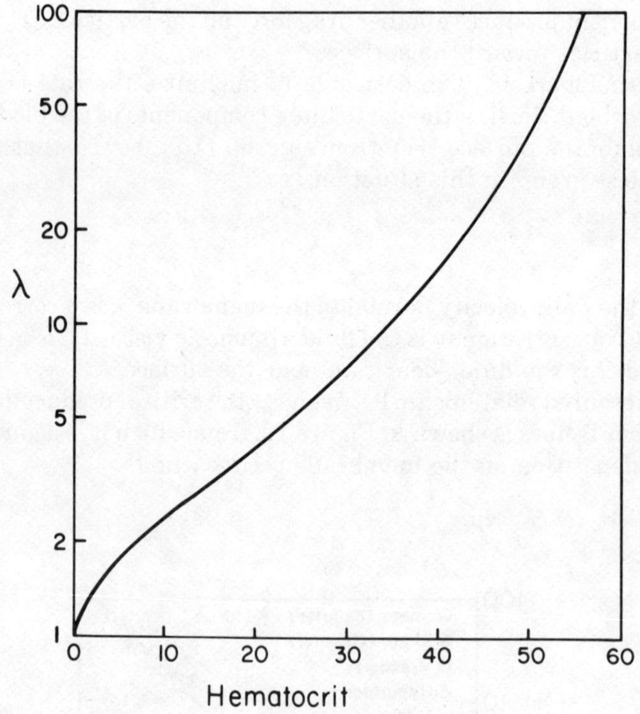

FIGURE 14.—Concentration factor of deposition parameter versus hematocrit (10).

An engineering evaluation of the "cross-flow" concept done by Castino et al. (11), using flat membranes in a channel, showed trends that agree qualitatively with theoretical expectations. They showed that, at low pressure, flux or productivity increases with wall shear rates and with transmembrane pressure (Figure 15). They also showed that plasma flux decreases with increase in hematocrit (Figure 16). And, as a result of plasmapheresis,

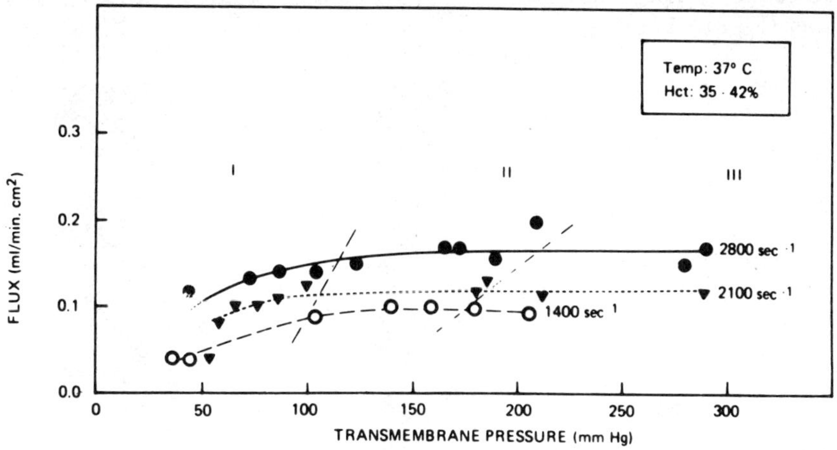

FIGURE 15.—Plasma flux as a function of transmembrane pressure at different shear rates (11).

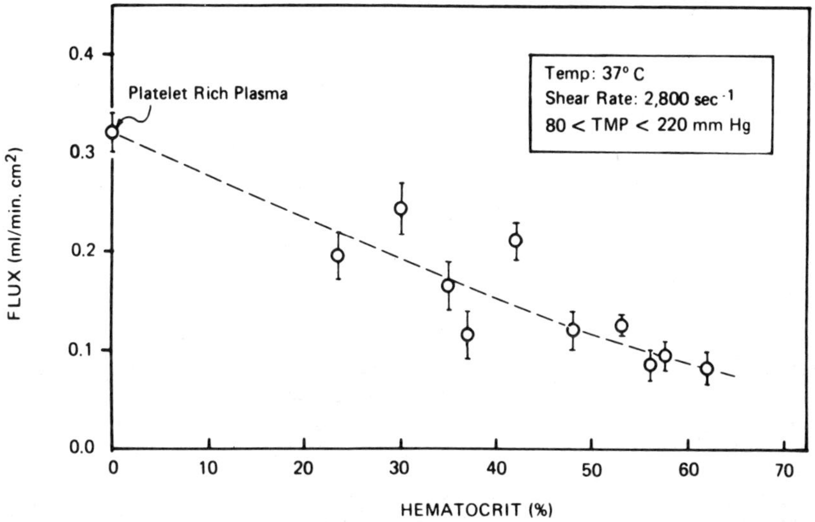

FIGURE 16.—Plasma flux as a function of hematocrit at constant shear rate (11).

no detectable change was found in the composition of the proteins in the product (Figure 17).

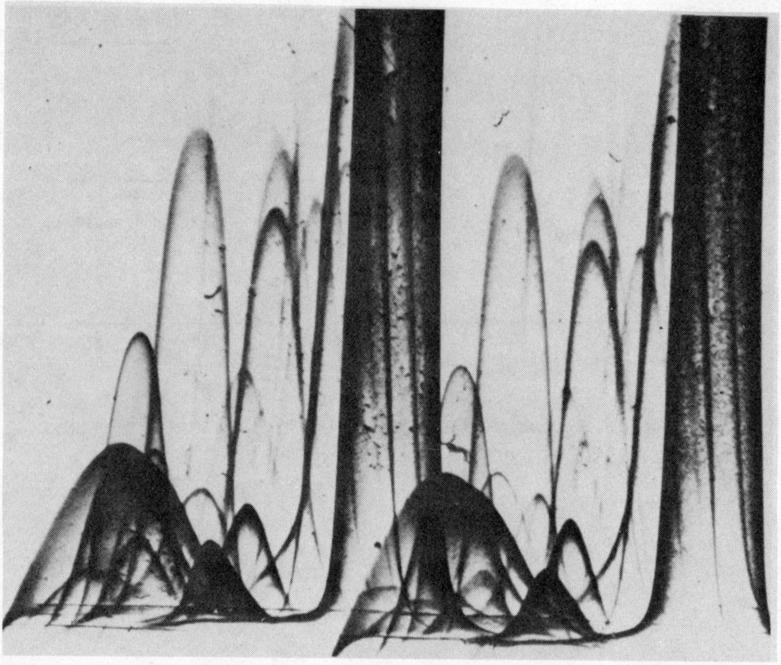

FIGURE 17.—Bidimensional immunoelectrophoresis of plasma proteins in the filtrate (left) and blood (right) phases.

It should be noted that because of better conduit stability under variations in transmembrane and axial pressures, some commercial plasmapheresis systems now use a hollow fiber "shell and tube" configuration where blood flows in the lumen of the fibers and the plasma product is collected from the shell side (12).

REFERENCES

1. Clark, L. C. and Lyons, C.: 1962, "Electrode systems for continuous monitoring in cardiovascular surgery." Ann. N. Y. Acad. Science 102, pp. 29–45.
2. Rechnitz, G. A.: 1982, "Bioanalysis with potentiometric membrane electrodes." Analyt. Chem. 54, p. 1194A.
3. Goldstein, S. R. and Peterson, J. I.: 1977, "A miniature fiber optic pH sensor suitable for in vivo application." In Advances in Bioengineering, Amer. Soc. Mech. Eng., New York, p. 81.
4. Peterson, J. I. and Goldstein, S. R.: 1982, "A miniature fiber optic pH sensor potentially suitable for glucose measurements." Diabetes Care 5, pp. 272–274.
5. Schultz, J. S. and Sims, G.: 1979, "Affinity sensors for individual metabolites." Biotech. Bioeng. Symp. 9, pp. 65–71.

6. Schultz, J. S., Mansouri, S., and Goldstein, I. J.: 1982, "Affinity glucose sensor." Diabetes Care 5, pp. 245–253.
7. Mikic, B. B., Benn, J. A., and Drinker, T. A.: 1972, "Upper and lower bounds on oxygen transfer rates. Theoretical considerations." Ann. Biomed. 1, pp. 212–220.
8. Marx, T. I., Baldwin, B. R., and Miller, D. R.: 1962, "Factors influencing oxygen uptake by blood in membrane oxygenators." Ann. Surg. 156, p. 204.
9. Lightfoot, E. N.: 1968, "Low order approximation for membrane blood oxygenators." AIChE J. 14, p. 669.
10. Forstrom, R. S., Bartel, T. K., Blackshear, Jr., P. L. and Wood, T.: 1975, "Formed elements deposition onto filtering walls." Trans. Am. Soc. Artif. Intern. Organs 21, p. 602.
11. Castino, F., Friedman, L.I., Solomon, B. A., Colton, C. K., and Lysaght, M.: 1978, "The filtration of plasma from whole blood: A novel approach to clinical detoxification." In Artificial Kidney, Artificial Liver, and Artificial Cells, Chang, T. M. S., Ed., New York, Plenum Press.
12. Solomon, B. A.: 1981, "Membrane separations: Technical principles and uses." Trans. Am. Artif. Intern. Organs 27, p. 345.

SYMBOLS

F_d	hydrodynamic drag on particle
F_r	hydrodynamic repulsive force on particle
x	"front" position
l	½ blood film thickness
HbO_2	oxygen capacity of hemoglobin under experimental conditions
ξ	(x/l) dimensionless front position
R	particle radius
S	shear rate
z	distance from entry of flat plate oxygenator
u	blood velocity in oxygenator parallel to membrane
v	velocity perpendicular to membrane
v_{cr}	critical velocity for impaction
λ	correction factor for particle concentration
ν	viscosity

MEMBRANES AND MEMBRANE PROCESSES IN BIOTECHNOLOGY

Enrico Drioli

Istituto di Principi di Ingegneria Chimica
Università degli Studi di Napoli
Naples, Italy

The relationship existing between enzyme processes and membrane processes and the important contribution that membrane science and membrane technology gave and will give to the development of enzyme engineering and biotechnology in general, are discussed. The basic properties of membrane separation processes (athermal, gentle, no additives required, etc.) are ideal for the treatment of bioactive compound solutions. Membrane processes contribute to various aspects of biotechnology, and particularly to: (1) improvements in methods for recovery and reuse of enzymes or whole cells, (2) developments in methods for their immobilization, and (3) development of enzyme membrane reactors. On the one hand, completely new enzymatic processes will be possible if based on membrane systems, and on the other, important problems of membrane separation processes (fouling, for example) might be solved by taking advantage of enzyme properties.

1. INTRODUCTION

2. ENZYME MEMBRANES

3. MEMBRANE PROCESSES IN BIOTECHNOLOGY

P. M. Bungay et al. (eds.), Synthetic Membranes: Science, Engineering and Applications, 667–682.

1. INTRODUCTION

Membrane science and membrane technology can contribute to the further development of biotechnology by offering solutions (and models) to some of the major problems of this growing field. An analysis of membrane systems in enzyme engineering, a very representative area of biotechnology, will be used in this chapter to confirm this assumption.

During the last few years significant developments have been obtained in what is generally termed "enzyme engineering" or "enzyme technology." This field covers production, separation, purification, use and recovery of enzymes[1]. More particularly, the recent developments in enzyme technology include:

a) improved methodology for inducing microorganisms to favor the production of selected enzymes;
b) improved methods for purification of enzymes;
c) methods for the immobilization of enzymes or whole cells on solid supports;
d) methods for the continuous use of enzymes in flow reactors; and
e) a technique for solid phase peptide synthesis.

Specific fields in which membranes might contribute are:

1. immobilization of biocatalysts;
2. separation, purification and concentration of bioactive compounds; and
3. development of new production techniques based on membrane reactors.

Answers offered to those points by membrane systems will be analyzed. The immobilization of enzymes in polymeric membranes will be discussed in the first part of this chapter. The second part contains an analysis of separation techniques and membrane reactors.

2. ENZYME MEMBRANES

Enzymes produced by microorganisms such as fungi and bacteria have been used for a long time in fermentation processes for the production of pharmaceuticals, beverages, food, etc. However, the industrial potential of these biocatalysts and of the correlate bioconversions appears today much

[1] Enzymes are globular proteins whose main function is to catalyze the biological reactions involved in an organism's metabolism. They are usually found within cells in a colony-organized assembly where they synergistically work and catalyze a specific reaction, both producing the energy necessary for surviving and regulating the metabolic state of the living cell, and preventing alterations. The interest in these biomolecules lies in their particular catalytic properties. Although they act with respect to a particular reaction in the same way as synthetic catalysts, i.e., reducing the energy of activation of the reaction, they achieve their catalysis in a very efficient way and with a selectivity well beyond that of synthetic catalysts.

wider, particularly because of enzyme properties such as the ability to promote reactions under mild conditions that limit side reactions. Organic reactions can be promoted which would otherwise be either impossible or too slow, or which would provide too low a yield. Also, typically the products of the catalyzed reactions are physiologically active substances.

Enzymatic reactions are useful in modern industry because they are low in energy consumption, safe with respect to sanitation, less polluting, and they conserve material. However, there are limitations to the traditional use of enzymatic reactions in industry because of:

a) low productivity of equipment per unit time;
b) high cost of purified enzymes;
c) difficulties in the recovery and reuse of used enzymes or microorganism cells; and
d) problems in maintaining uniform product quality.

Significant advantages exist when enzymes or whole microbial cells are immobilized. Some of them are (1,2):

a) cost of enzyme supply is cut;
b) plants can be built compactly;
c) a more rational process design is possible;
d) quick and smooth operation is possible;
e) utility and labor costs can be saved;
f) productivity of equipment per unit time is large and side reactions are limited because of short reaction time.

These are some of the reasons for the large efforts today devoted to the development of new immobilization techniques all over the world.

Immobilized enzymes are defined as enzymes physically confined or localized in a certain defined region of space, which retain their catalytic activities, and which can be used repeatedly and continuously. Traditional immobilization techniques today include:

a) trapping in a gel;
b) encapsulation in a membrane shell;
c) adsorption on a surface;
d) covalent bonding to solid supports;
e) crosslinking inside a support;
f) co-crosslinking with a carrier protein.

In Japan, a leader in the development of enzyme engineering, more than 70 enzymes have been already immobilized, but only few of them (about ten) represent the majority of all the existing applications; some hydrolases (α-amylase, β-galactosidase, invertase, pepsin, trypsin, α-chymotrypsin and papaine), two oxido-reductases (glucose oxidase, catalase) and one isomerase (glucose isomerase) are among the most traditional examples (1).

Because of the high cost of enzyme purification, a growing interest exists today also for the immobilization of whole microorganisms. None of the systems reported up to 1979 was based on membrane technology. Figure 1 contains a schematic summary of the reported (3) immobilization methods.

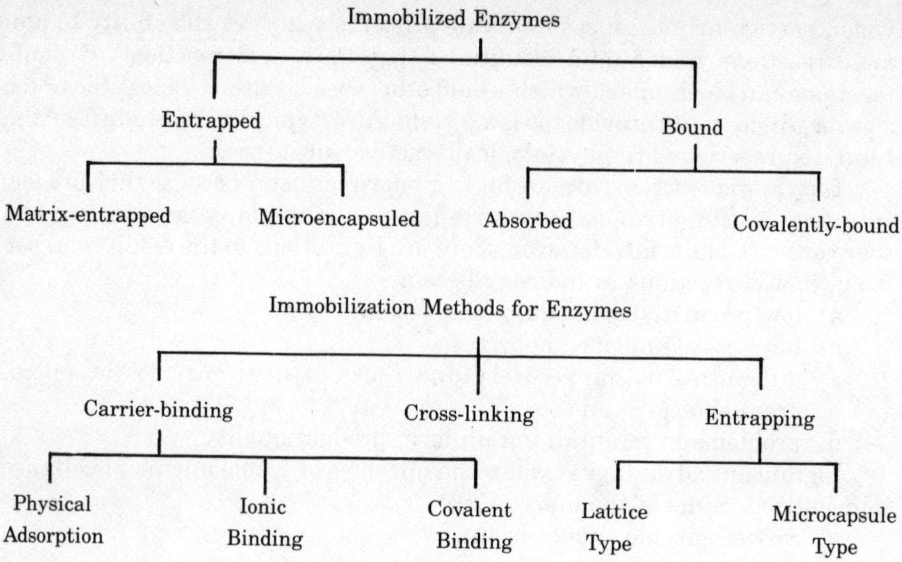

FIGURE 1.—Enzyme immobilization.

The relatively limited number of industrial applications (see Table 1), officially announced in the same country, in contrast to the large potential of immobilized enzyme systems, might be attributed to the fact that the carriers generally used for immobilization are expensive and the efficiency of the immobilization technique proposed is low. Moreover, for continuous reaction, complicated equipment has been generally proposed. In biological systems enzymes normally operate bound to membranes or in a microenvironment delimited by membranes. This fact suggested to various researchers that immobilization of enzymes should be studied in artificial membranes. Many studies have been carried out using various means for attaching enzymes to membrane-type carriers, for example, covalent bonding by glutaraldehyde (the most commonly used agent), ionic bonding, or entrapment methods. Most of the enzyme membranes reported in the literature are listed in Table 2 or referred to in (5).

Some of these methods of immobilization may be used both for enzymes and for microbial cells. A great number of different types of membrane materials have been reported: synthetic polymers such as polyacrylamide gel, polyvinylalcohol and PTFE; or biopolymers such as proteins—albumin, collagen and gelatin. The prospects exist for a considerably wider choice of suitable synthetic and natural substances. The selection of membrane material is largely determined by the nature of the enzyme being complexed, the substrate to be treated and the reaction environment to be encountered.

TABLE 1. Industrial application of immobilized enzymes and whole cells in Japan (1)

Enzymes or Cells	Carrier	Methods	Company	Year
Aminocyclase	DEAE-Sephadex	Adsorption	Tanabe Seiyaku	1969
E. Coli (aspartase)	Polyacrylamide	Entrapment	Tanabe Seiyaku	1973
Brev. Ammoniagenes (fumarase)	Polyacrylamide	Entrapment	Tanabe Seiyaku	1974
Aspartase	Duolite A7	Adsorption	Kyowa Hakko	1975
Glucose isomerase	Duolite A7	Adsorption	Kyowa Hakko	1976
Glucose isomerase	Amberlite IRA 904	Adsorption	Mitsubishi Kasei	1976
Streptomyces sp. (glucose isomerase)	Modified anion exchange resin	Adsorption	Denki Kagaku	1976
Penicillin–acylase	Celite	Adsorption	Toyo Zojo	1976

A very simple and rapid immobilization method for enzymes or microbial cells uses membranes produced by cross-linking albumin (6). The enzyme membrane can be obtained by mixing albumin solution and glutaraldehyde with a bacterial suspension. The co-crosslinking is carried out by drying the membrane surface with continuous warm ventilation. Such membranes can be useful for microorganisms without high proteolytic activity; otherwise, pretreatment of microorganisms to destroy proteolytic activity would be necessary.

Another simple technique involves the preparation of tanned protein membranes (7). Unlike the previous technique of bonding enzymes to protein materials, the whole cells are not previously diazotized, nor are the protein materials or whole cells otherwise treated chemically to produce covalent linkages. It has been found that bonding occurs directly. The tanning of the enzyme membrane is achieved using either acid or alkaline reagents, depending on the enzyme.

A different technique consists of entrapping charged enzymes in porous polymeric membranes (8). An enzyme is dispersed in an appropriate buffer solution to give an electrically charged colloidal system. Then electrophoresis is carried out. The enzyme particles adhere to the surface of a suitable carrier to form a high-density layer of enzyme. Materials such as PTFE, polyvinylidene fluoride, polyvinyl chloride, and nylon have been used. This method does not require any chemical treatment.

In most of these immobilization procedures, however, the contributions of recent progress in membrane technology have been very limited. The preparation of enzyme membranes on a large scale for industrial processes, in which selective mass transfer across the artificial membranes is combined with specific chemical reactions, would require low membrane cost and standard preparation procedures. Two other immobilization procedures, which have been recently studied in our laboratory, might accomplish those requirements.

Gelled enzyme membranes, involving labile immobilization at the membrane–solution interface, can result from concentration polarization phenomena. Both in batch unstirred ultrafiltration processes and in ultrafiltration processes with continuous recirculation of the substrate solution along the membrane, an appropriate amount of enzyme can be totally or partially immobilized in "gel" form on the pressurized face of the membrane, depending on the detailed fluid dynamics (9, 10).

Such a dynamically formed gelled enzyme membrane formation technique for acid phosphatase, urease, β-galactosidase, malic enzymes, etc., has been studied. From the experimental results it appears that enzymes forming a gel layer on an ultrafiltration membrane retain their activity and their stability is increased.

TABLE 2. Enzyme membranes (4)

Carrier and reagent	Method	Enzyme
Cellulose membranes		
Ion exchange cellulose	carrier binding	glucoamylase
DEAE–cellulose sheet	(ionic binding)	lactase
CM–cellulose sheet + carbodiimide	carrier binding (peptide binding)	peroxidase
Cellulose sheet + glutaraldehyde	carrier binding (carrier cross-linking)	peroxidase hexokinase
Nitrocellulose membrane	entrapping (lattice)	glucose oxidase peroxidase glutamate pyruvate transaminase aminoacylase
Cellulose + polyvinylidine fluoride	entrapped by pressure	alcohol dehydrogenase
Polyacrylamide gel	entrapping (lattice)	glucose oxidase catalase trypsin urease
Polyvinyl alcohol	entrapping (lattice)	invertase urease

TABLE 2. Enzyme membranes (4)—Continued

Carrier and reagent	Method	Enzyme
Polyethyleneglycol dimethacrylate	entrapping (lattice)	glucose oxidase catalase trypsin urease
Collodion membrane + bisdiazobensidine	carrier binding (carrier cross-linking)	papain
+ bensidine-2,2[1]-disulphonic acid	carrier binding (carrier cross-linking)	papain
Protein membranes		
Fibrin membrane + glutaraldehyde	carrier binding (carrier cross-linking)	trypsin
hardening	entrapping (lattice)	asparaginase
Gelatin membrane	entrapping (lattice)	urease
Protein–silicon membrane + glutaraldehyde	carrier binding (carrier cross-linking)	catalase carbonic anhydrase
Collagen membrane + hardening	entrapping (lattice)	catalase lipase lactase papain urease
Swollen collagen or zein membrane	entrapping (lattice)	lysosyme invertase
Collagen	carrier binding (peptide binding)	hexokinase + glucose 6-phosphatase hexokinase + glucose oxidase
Zein	carrier binding (peptide binding)	DNase
Tanned protein membrane	carrier binding (multiple binding)	glucose isomerase
Albumin membrane + glutaraldehyde	carrier binding (co-crosslinking)	-gole dosidase
Starch–acrylonitrile membrane + saponifying alkali-agent	entrapping or (lattice) carrier binding (covalent binding)	amylase amylase
PTFE-membrane or fluorinated ethylene–propylene resin (FEP)	electrically entrapped (lattice)	catalase glucoamylase

The method has been carried out using flat membranes and tubular membranes in a continuous recirculating system. The polymeric material forming the supporting membrane does not significantly influence the enzyme stability. The ultrafiltration membrane cut-off and the membrane morphology, on the contrary, appear to be the controlling factors. The technique has been shown to be useful also for studying the kinetic behavior of immobilized allosteric enzymes. The more traditional techniques, in fact, lead to a decrease of the specific activity or to a deactivation of these enzymes, which are particularly sensitive to conformational transitions induced by specific ligands or environmental constraints. With this technique the enzyme should be immobilized without significant changes in the enzyme microenvironment; moreover, the situation is particularly favorable because of the high enzyme concentration in the gel.

The possibility of using traditional ultrafiltration (UF) and reverse osmosis (RO) membranes filled with biocatalyst might represent a significant improvement in the development of membrane technology and enzyme engineering (9). However, the possibility of preparing industrial level UF and RO membranes filled with enzymes or whole cells, using traditional techniques, has been limited by the need for non-aqueous solvents in the casting solutions and high temperature annealing, which destroys the catalytic properties. The recent isolation of "*Caldariella acidophila*" (11), an extreme thermophile growing optimally at 87°C and whose enzymes are generally stable to protein denaturating agents, offers an interesting opportunity for using the phase inversion technique for the preparation of UF and RO membranes filled with whole cells as the enzyme source. *C. acidophila* contains enzymes of industrial interest, such as β-galactosidase and malic enzymes, for example.

Artificial membranes filled with *C. acidophila* have been prepared by several methods. Cellulose acetate and polysulfone were used to obtain asymmetric membranes by the phase inversion technique (11). Albumin and glutaraldehyde were used for cell immobilization in membranes by the co-crosslinking method. And finally, a hydrophilic polyisocyanate was used to prepare porous polyurethane structured foams in thin films[2]. The physico-chemical properties of trapped-cell β-galactosidase activity in the above models were similar to those shown by the enzyme in the free cells. At the optimal pH, the β-galactosidase exhibited maximal activity at about 100°C and appeared stable for up to 24 hours at room temperature and a pH of 3–8. Incubation of trapped-cell β-galactosidase for up to 24 hours at room temperature with organic solvents did not cause any loss of activity. After 8-9 months of wet storage at 4°C, no decrease of enzymatic activity was observed.

[2]The use of liquid polyurethane prepolymer, which contains at least two free isocyanate groups per prepolymer molecule, has been recently suggested as an immobilizing agent. Hydrophilic porous films and tubes can be prepared with this material, containing immobilized biologically active compounds (12). Physical entrapment and specific reaction, between free isocyanate groups and protein amino groups, for example, contribute to the "immobilization."

Cell entrapment imparted a significant increase in enzymatic activity in comparison with intact free cells. This effect may be a consequence of cytoplasmic membrane permeabilization of the microorganism caused by the entrapment procedures. The increase in enzymatic activity was 35-fold greater for the polyurethane system than in the membrane preparation. All the membranes had a flat sheet configuration and were tested in standard ultrafiltration systems. The permeate flow rate, β-galactosidase degree of conversion, and stability of the system were studied in the range of 70–85°C.

3. MEMBRANE PROCESSES IN BIOTECHNOLOGY

The significant progress obtained during the last few years in biochemical research and in genetic engineering has created many new opportunities for the field of biotechnology. Further development of biotechnological processes, however, particularly their realization on the industrial scale, requires: a) the availability of new separation techniques useful for extraction, purification and concentration of bioactive compounds from fermentation broths or from other liquid phases with large yields and without inducing degradation in the chemicals of interest; b) the development and optimization of new chemical reactors specific for enzymatic reactions and, in particular, for continuous operation.

The basic properties of membrane technology appear very attractive for solving many of the problems in these two areas. Cross-flow microfiltration (MF), ultrafiltration (UF), and reverse osmosis (RO), now considered standard unit operations in chemical engineering, are athermal, gentle and non-destructive. These techniques do not require additional chemicals and energy consumption is, in principle, lower than for any other competitive separation processes.

The industrial potential of those technologies has already been largely proved in areas that interface with biotechnology. As an example, more than 70,000 m^2 of ultrafiltration membranes had already been successfully installed in 1981 in the dairy industry for whey treatment.

Processing of enzymes, continuous membrane fermentation, pyrogen and cell debris removal, continuous cell culture and interferon production are some of the many possible applications for membranes in this area.

Enzymes are typically produced by fermentation (microbial enzymes) or extraction (animal or vegetable enzymes) in aqueous media. The utility of UF has been demonstrated for concentrating enzymes without denaturation and with recovery efficiencies higher than 95% (13).

A traditional fermentation (see example flow diagram in Figure 2) generally involves many process steps that are arguably inefficient and uneconomical in terms of utilization of raw materials, recovery of product, and energy consumption.

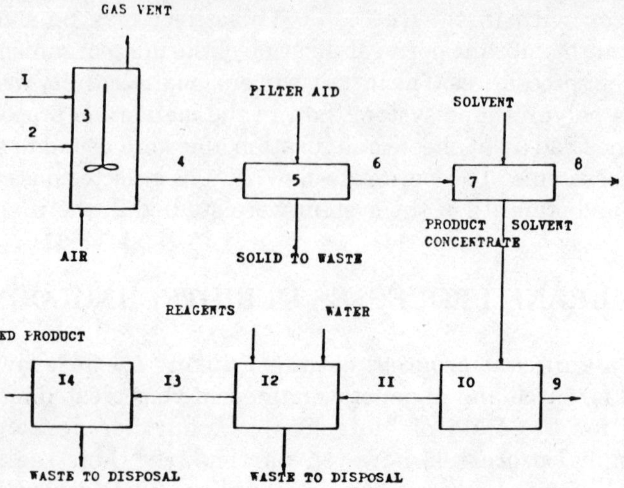

1 NUTRIENT BROTH
2 CELL CULTURE
3 FERMENTATOR BATCH
4 SPENT WHOLE BROTH
5 FILTRATION
6 CLARIFIED BROTH
7 EXTRACTION
8 EXTRACTED BROTH TO WASTE
9 SOLVENT TO RECOVERY
10 CRUDE PRODUCT EVAPORATIVE CONCENTRATION
11 CRUDE PRODUCT
12 ION EXCHANGE OR ADSORPTION PURIFICATION
13 PURE PRODUCT IN SOLUTION
14 CRYSTALLIZATION OR SPRAY/FREEZE DRYING

FIGURE 2.—Fermentation flow diagram.

Continuous cross-flow membrane MF and UF (Figure 3), when correctly introduced, have been shown to improve this process significantly. The possibility of treating liquids containing suspended solids, broadens the scope of membrane filtration. Commercially available capillary and tubular membranes having pore sizes in the range of MF and UF, have been used for the concentration of broths of bacteria, yeasts and moulds and for sterile filtration of solutions of proteins, enzymes, vaccines and amino acids.

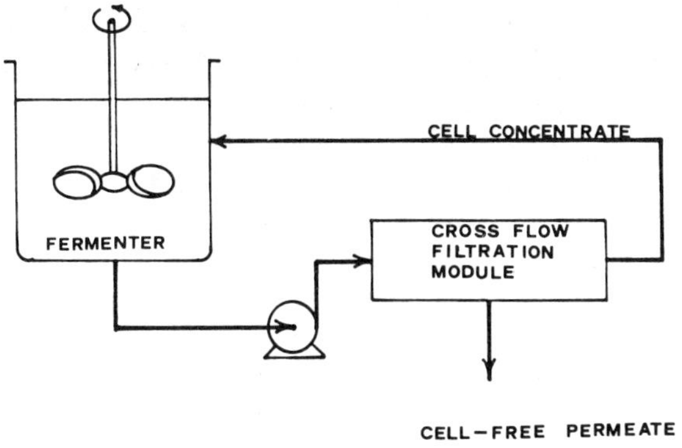

FIGURE 3.—Schematic of continuous feed/bleed cross-flow filtration of fermentation broth.

Fouling problems have been solved by a periodical backflushing of the membrane. Fluids containing upwards of 50% to 60% suspended solids by volume have been pumped through a well-designed membrane module, easily achieving very high liquid-phase recoveries from whole fermentation broths. When ultrafiltration membranes are used on the broths (or on the permeate from a microfiltration stage), retention of the solubilized macromolecular solutes, as well as particulate and colloidal material, will be accomplished, giving a filtrate containing only relatively low molecular weight solutes. The use of membrane separations (such as cross-flow MF or UF or dialysis) as a component of continuous fermentation systems is of growing interest. In this case it is possible to improve the productivity of product-inhibited fermentations, for example, by the continuous removal of the low molecular weight products. The enzymatic degradation of cellulose to alcohol, where the microorganisms are inhibited at an alcohol concentration higher than 12%, might be improved by the use of this concept.

Recently, a similar concept has been used by Degussa for the production of L-amino acids. In this case, L-amino acids are obtained by biocatalytic division of synthetically produced acetyl D, L-amino acids by means of enzymes. Unlike the previous type of fixed-bed reactor with carrier-located acylase, the new approach employs the enzyme in soluble form, and uses a membrane for separating the enzyme from the reagent solution. This avoids losses at the immobilizing stage and reduces enzyme consumption. Other advantages are that the enzyme content can be continually adjusted and the product solution is obtained free from pyrogens. The continuous removal of toxic products allows significant increases in cell mass and total product yield in batch fermentations for growth of bacteria. Hollow fiber systems have been used for increasing total cell densities up to 10^{15} cells/liter (14). Production efficiency has also been improved by 100%.

A new method for the production of interferon has been reported (14), where fibroblast cells attached to beads were placed in hollow fiber systems. The interferon production was greatly enhanced in the hollow fiber system (total yield 30 times higher than that of classical monolayer cells) due to removal of inhibitors through the membrane. The growth of hybridomas producing monoclonal antibodies, and of various animal cells, has also been tested.

When MF or UF is used in the treatment of the fermentation broth, the filtrate might in addition be easily treated by RO if the concentration of low molecular weight compounds present in the filtrate is of interest. The MF and UF processes can be considered as pretreatment for the RO step. The concentration and purification of antibiotics, by sequential UF and RO, is an example. The removal of antigenic contaminants present in biological mixtures via the combined use of immunocomplexation and UF has also been suggested (15).

Asymmetric membranes with an effective pore diameter in the range of several hundreds of Angstroms are commercially available and they are promising for use in the removal of viruses, and cell membrane fragments from plasma without retention of normal plasma protein. These membranes might also be used for the preparation of "functional membranes" by a chemical coupling or a physical immobilization of immunoreagent to the membrane surface.

Ultrafiltration, using hollow fiber membranes with a molecular weight cut off of 30,000 or lower, is often used to produce concentrated purified albumin. In a typical system, a total processing time of 18 hours is necessary to treat 2,000 liters of plasma containing 3% albumin and 20% alcohol. In a three-stage process, 214 liters of concentrate, 28% albumin and an alcohol content lower than 0.25% can be obtained. The scheme for the process is shown in Figure 4 (16).

In the previous examples, the membranes have been considered generally as semipermeable barriers for the separation of small molecules from bigger ones. When a chemical reaction takes place in the bulk solution or in the membrane itself, the system may be identified as a true membrane reactor. A classical example is a stirred-tank enzymatic reactor connected by a continuous recirculation loop to an ultrafiltration or dialysis unit. Such a system, when well designed, permits the continuous removal of the reaction products from the bulk solution without loss of enzyme (or the insoluble or macromolecular substrate).

The efficiency of enzymatic reactions that are product-inhibited can be enhanced, and quantitative conversion of substrate into products can be obtained. Highly efficient enzyme membrane reactors can be also produced by immobilizing enzymes in membranes or in hollow fibers. For example, enzymes can be confined in the porous support matrix of an asymmetric capillary membrane, while substrate-containing solution flows through the

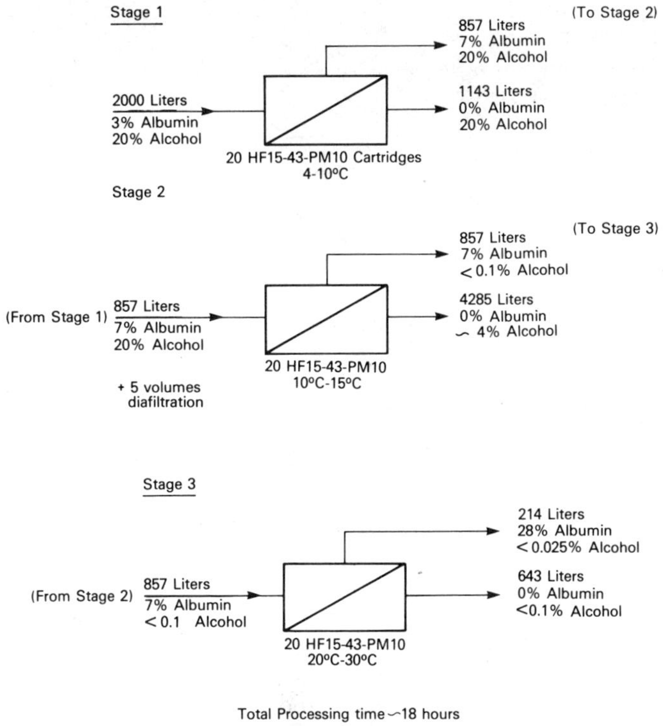

FIGURE 4.—Concentrating and purifying albumin by ultrafiltration.

fiber lumen. The dense skin layer at the lumen wall should be impermeable to the enzyme molecules, but permeable to the product and substrate molecules. The latter diffuse through the inner wall of the fiber to the enzymes into the spongy part, where the conversion takes place. Figure 5 shows a scheme for such a system. Applied transmembrane pressure and axial flow rate are parameters that contribute to the control of the reactor performance. Modelling of the above reactors has been carried out introducing various approximations. For the case, for example, of a dynamically formed gelled enzyme membrane, an analytical model was developed based on the following assumptions:

1. plug flow occurs in the reactor;
2. the enzyme is completely gelled on the UF membrane in an artificial layer of constant thickness, S_E;
3. the substrate and concentration of products upstream and downstream from the composite membrane are uniform everywhere, assuming (a) zero membrane rejection for the substrate and for the reaction products and (b) negligible back-diffusion of enzyme from the gel (checked experimentally);
4. the reaction rate in the gel follows the classical Michaelis–Menten equation, which is generally valid when the enzyme is in free solution.

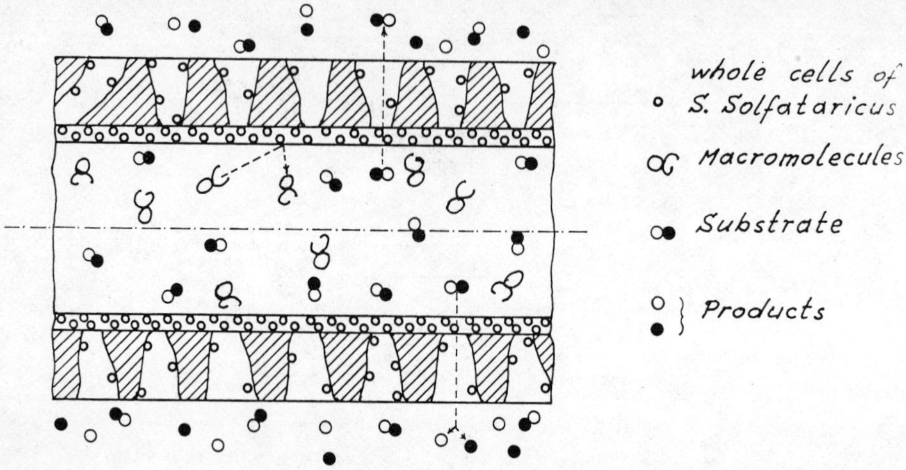

FIGURE 5.—Hollow-fiber membrane reactor.

The validity of this model has been tested by comparing different sets of experimental data with the analytical predictions (9, 10). Experimental results obtained with urease and β-glucosidase have been presented. The maximum deviation between theory and experiments is 10%.

In the continuous recirculation process the substrate concentration in the reservoir varies with time t because the reaction also takes place in the reservoir (see assumption 2, below). The analytical model developed for describing this process is based on the same assumptions made for the unstirred batch reactor; additional assumptions must, however, be introduced, specifically,

1. perfect mixing occurs in the reservoir, and
2. part of the enzyme, initially gelled on the UF membrane, is partially redissolved, because of the continuous recirculation of the solution.

The enzymatically catalyzed reaction also takes place, therefore, in the reservoir and in the fluid volume of the pipes and connectors. In general, the agreement between the experimental results and the predictions of the simple analytical model of the two classes of enzymatic heterogeneous reactors is satisfactory (17).

A detailed study of reactors with enzymes within the lumen of hollow fibers was presented by Rony (18). Immobilization was achieved preparing "cylindrical microcapsules" by filling the core of hollow fibres of suitable molecular weight cutoff with an enzyme solution, by capillary action for instance, and then sealing both ends. Bundles of such charged fibres were then assembled in a "tube-and-shell" reactor configuration. The availability of hollow fibre membranes adequately permeable to substrates and products, and the control of fluid flow all around the fibres in the bundle in order to assure uniform flow distribution and to avoid stagnation, and consequently reduce mass transfer diffusional resistances, appear to limit the applications of these bioreactors. The technique does, however, offer several advantages. Enzyme proteins can be easily retained within the core of the fibres with no deactivation due to coupling agents or to shear stresses, and the enzyme solution can be easily recovered and/or recycled.

Hollow membrane reactors with covalently bound enzymes have been proposed as extracorporeal blood detoxifiers or as devices to reduce arginine and asparagine content in blood of leukemic patients. Asparaginase and arginase have then been immobilized in commercial Spiraflo hemofilters in order to support anti-leukemia therapies: asparagine or arginine in the blood continuously circulating through such enzymatic membrane devices are enzymatically transformed and their concentration reduced both in plasma and in red blood cells.

When symmetric membranes are used or when enzymes are fed to the spongy part of asymmetric membranes, enzymes can generally penetrate the membrane wall. Enzyme immobilization thus results either in a uniform fixation of enzymes throughout the membrane wall, or in the formation of a carrier-enzyme insoluble network in the sponge of the membrane. Mass transfer through this solid phase must therefore be taken into account. A theoretical model neglecting radial convective transport and dense layer existence in asymmetric membranes is available in the literature (19). The reacting solution is still assumed to be fed to the core of hollow fibres. Steady state and isothermal condition and laminar flow pattern of reactants solution are assumed to hold; moreover, the enzyme is assumed uniformly distributed and the membrane wall curvature is neglected.

Intensive studies are in progress for better understanding and control of the various possible enzyme membrane reactors described. More practical applications in biomedical systems and in various industrial processes can be expected in the very near future.

REFERENCES

1. Thomas, D.: "Production of Biological Catalysts, Stabilization and Exploitation." Comm. of the EUR Comm., EUR 6079 EN.
2. Klibanov, A.M.: 1983, "Immobilized enzymes and cells as practical catalysts." Science 219, p. 722.
3. Wingard, L.B. (ed.): 1972, "Enzyme Engineering." New York, J. Wiley.
4. Chibata, I.: 1978, "Immobilized Enzymes." Research and Development Kodanska Scientific Books. New York, J. Wiley.
5. Johnson, J.C.: 1979, "Immobilized Enzymes: Preparation and Engineering." Chem. Tech. Rev. 133, Noyes Data Company.
6. Thomas, D., Boundillon, C., Broun G., and Kervenez, Y.P.: 1974, "Kinetic behavior of enzymes immobilized in artificial membranes: Inhibition and reversibility effects." Biochemistry 13, p. 299.
7. Wieth, W.R., Wang, S.S., and Saini, R.: 1976, U.S. Patent No. 3,972,776 (August 3).
8. Mlyauchi, T. and Furusaki, S.: 1977, U.S. Patent No. 4,051,011 (September 27).
9. Drioli, E. and Scardi, V.: 1976, "Ultrafiltration processing with enzyme-gel composite membranes." J. Membrane Sci. 1, p. 237.

10. Capobianco, C., Drioli, E., and Ragosta G.: 1977, "Heterogeneous ultrafiltration enzymatic reactors." J. Solid Phase Biochem. 2, p. 315.
11. Drioli, E., Iorio, G., DeRosa, M., Gambacorta, A., and Nicolaus, B.: 1982, "High temperature immobilized-cell ultrafiltration reactors." J. Membrane Sci. 11, p. 365.
12. Hartdegen, F.J., Swann, W.E.: 1980, "Immobilization of biological material with polyurethane polymers." U.S. Patent No. 4,237,229 (December 2).
13. Porter, M.C.: 1972, "Applications of membranes to enzyme isolation and purification." Biotech. Bioeng. Symp. 3, p. 115.
14. Tutujian, R.S.: 1982, Proc. World Filtration Congress III, p. 519.
15. Michaels, A.S.: 1980, "Membrane technology and biotechnology." Desalination 35, p. 329.
16. Short, J.L.: "UF, A Valuable Processing Technique for the Pharmaceutical Industry." ROMICON, Inc., Woburn, Massachusetts.
17. Drioli, E., Mendia, M., and Molinari, R.: 1978, Desalination 24, p. 193.
18. Rony, P.R.: 1972, "Hollow fiber enzyme reactors." J.A.C.S. 94, p. 8247.
19. Horvath, C., Shendalman, L.H., and Light, R.T.: 1973, "Open tubular heterogeneous enzyme reactors: Analysis of a theoretical model." Chem. Eng. Sci. 28, p. 375.

BIOLOGICAL MEMBRANE CONCEPTS

William R. Galey

Department of Physiology
University of New Mexico
Albuquerque, New Mexico 87131

This chapter provides an introduction to the composition, structure and transport mechanisms of biological cellular and epithelial membranes. The type of lipids found in cell membranes, their chemical properties and their arrangement into a lipid bilayer are discussed as are the classes of membrane associated proteins and their characteristics. The third major component of biological membranes, carbohydrate, is also discussed as is its apparent functional role.

The transport characteristics of the cell membrane are surveyed including its diffusive permeability, carrier-mediated transport and active transport systems.

Tissue membranes composed of sheets of cells are briefly discussed as are some of their transport properties.

A brief survey of the present and future prospects for the field of biosynthetic membranes and their utilization conclude the discussion.

P. M. Bungay et al. (eds.), Synthetic Membranes: Science, Engineering and Applications, 683–699.

1. INTRODUCTION

When one speaks of biological membranes, two structures very different in architecture and function come to mind. The first is the plasma, or cell membrane, which defines the boundary between the cell and its environment. This membrane quite obviously must provide means for acquiring nutrients for the cell, disposing of waste material and maintaining the proper environment for the multitude of biochemical reactions that must go on within the cell cytoplasm. The second membrane type is not a cellular organelle but a composite of cells that delineate the boundary between spaces within the organism. Although there are several types of membranes of this sort, defined by their particular structure and cell composition, one of the most important and certainly most studied of these tissue membranes is the epithelial membrane. Epithelial membranes are found as the barriers defining the lumen of the intestinal tract of higher organisms, secretory glands and the kidney nephron, for example, as well as the skin of most animals. Such membranes play the important role of mediating solute and water transport into and out of these organs or glands as well as in some cases providing structural integrity to the organ or tissue.

There are, of course, other types of membranes found in the bodies of man and other living organisms including those of endothelial membranes, and intracellular organelles (golgi apparatus, lysosomes, mitochondria, etc.). However, this chapter focuses on the membranes of the two former types: plasma membranes and epithelial membranes.

The goal of this chapter is to provide some understanding of the structure of these membrane types, some of their physical and biological properties and most particularly the mechanisms utilized by such membranes to accomplish their biological transport functions. Suggestions as to the future of biomembranology and how it relates to that of synthetic membranes are also briefly outlined.

2. THE CELL OR PLASMA MEMBRANE

The barrier between the cell and its environment plays many roles in the life of a cell. Not only does it mediate the uptake of nutrients and essential minerals, and the elimination of waste and cellular products, it also plays important roles in intercellular communication, information transfer and cell volume regulation and it acts as a platform for the activity of numerous intracellular enzymes.

The chemical composition of this ubiquitous and all-important structure of every cell, the membrane, includes three major chemical species: 1) lipid, 2) protein, and 3) carbohydrate.

2.1. Lipids

Lipids make up the structural backbone of membranes by forming a continuous bimolecular sheet of molecules. This sheet is composed of lipids that are amphipathic, meaning that one end of the molecule is by its chemical nature lipophilic (or hydrophobic) whereas the other part of the molecule is polar (hydrophilic) in nature. This molecular structure common to all membrane lipids leads to the formation, in an aqueous environment, of a lipid bilayer sheet arranged with the hydrophilic, polar portions of the lipid molecules at the water/membrane interfaces and the lipophilic portions of the molecules buried away from water (see Figure 1).

The lipids found in biological membranes are primarily the phosphatidyl glycerides also known as phospholipids. The general structure of this class of

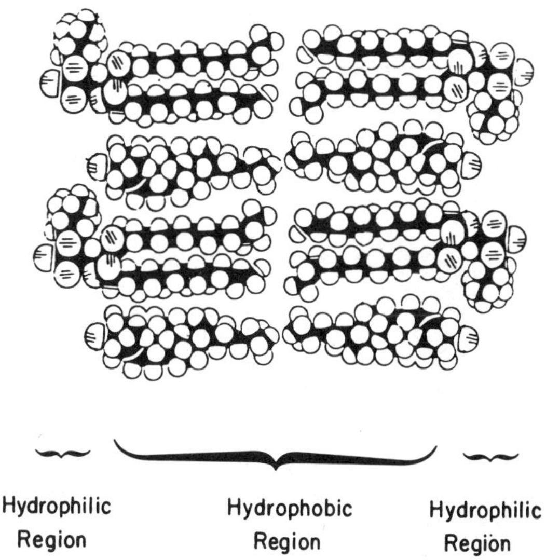

FIGURE 1.—Structure of lipid bilayer

molecules is seen in Figure 2. The polar portion of the molecules is composed of the glycerol backbone, phosphate group, and the attached base or sugar molecule. The lipophilic portion of the molecule is composed of the long fatty acid hydrocarbon chains usually 16 to 21 carbon atoms in length. The fatty acid attached at the center carbon atom of the glycerol molecule is usually unsaturated. The bend or kink in the hydrocarbon chain produced by the double bond disrupts molecular alignment and leads to a more fluid lipid structure. Of course, not all of the lipids in biological membranes are phospholipids. Some other common lipids are the sphingolipids and cardiolipins and cholesterol (Figure 2). Cholesterol is found in the membranes of higher animal organisms but not in those of bacteria or plants. The bulky steroid structure of cholesterol also mediates cell membrane fluidity by inserting between the more linear fatty acid chains of phospholipids. This results in a stabilization of the membrane in the hydrophobic region of the lipids near to their polar moiety, because the hydrocarbon steroid structure interacts avidly with neighboring lipids through hydrophobic interactions. On the other hand, the shorter length of the cholesterol molecule allows the fatty-acid chain tails more room to flex, thus fluidizing to some extent the lipid bilayer.

2.2. Proteins

Interspersed in and on this two-dimensional sea of lipid are the membrane proteins, the second major component of membranes. The proteins make up from 20 to 80% of the biomembrane mass, depending on the tissue source, and are responsible for the *specific* transport systems and enzymatic activities associated with the membrane.

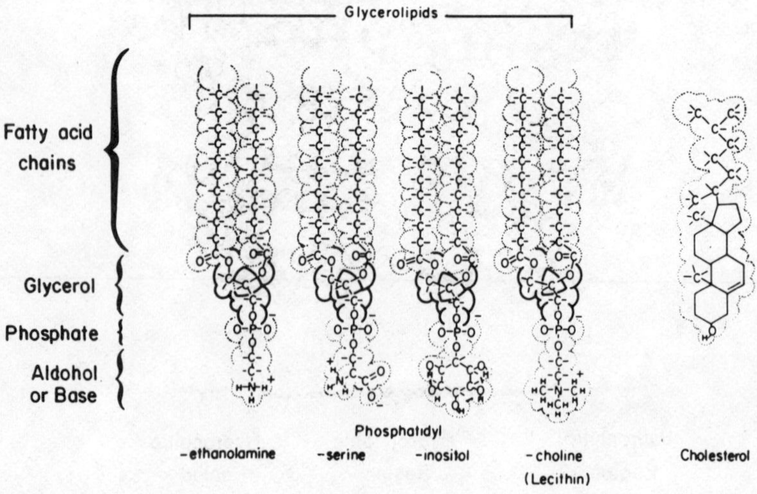

FIGURE 2.—Structures of the principal cell membrane lipids

The membrane proteins may be conveniently divided into two groups based on the nature of their association with the membrane. The peripheral (also called extrinsic) proteins are usually attached to the membrane through ionic and other polar bonds to the polar moieties of the lipids making up the bilayer surfaces. In addition they may also interact with hydrophilic surface portions of proteins embedded within the lipid bilayer. The peripheral proteins are usually relatively easily removed from the membrane by changing the pH or ionic strength of the cellular environment and are usually fairly small ($<5 \times 10^4$ dalton MW) and are water soluble. In general, a peripheral protein is associated with one membrane surface or the other (outside or inside of cell), but not to both sides of a membrane. It may be associated with particular enzymatic activity of the membrane surface, or may act as a structural element involved in giving the cell its particular shape.

The second class of proteins are the much larger integral (or intrinsic) proteins which range from 5×10^4 to 1.5×10^5 daltons in molecular weight and are much more tightly bound to the membrane. These proteins generally make up about 80% of the total membrane protein mass and are bound to the membrane through hydrophobic interactions (1). As can be seen in Figure 3, integral proteins may project deep into the lipid bilayer from either side or extend completely across it. The proteins appear to be inserted into the hydrophobic matrix during their synthesis and maintain that orientation throughout their existence. They are always associated with one surface or the other of the membrane or traverse the complete bilayer and are thereby accessible from either surface. Studies to date show that proteins are uniquely

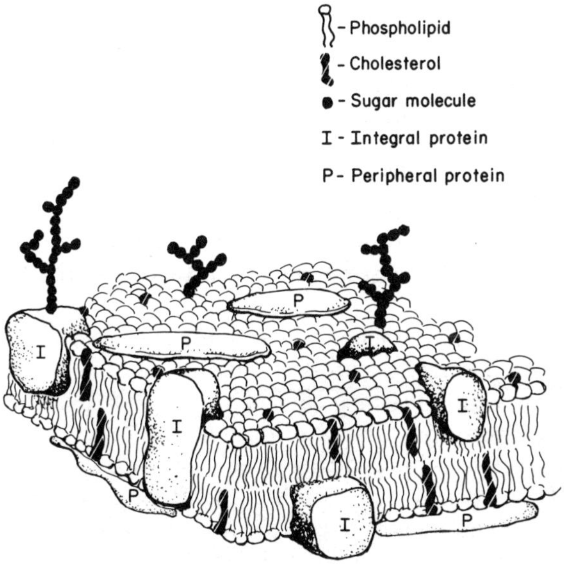

FIGURE 3.—A schematic representation of a cell membrane

associated with the inner or outer surfaces, and are not able to move from one surface of the membranes to the other. This property is dictated by the polar regions of the molecules which energetically anchor them to the surface. No proteins have been found which are not exposed to one surface or the other of the membrane. Proteins of the "intrinsic" class are responsible for the carrier-mediated, and active transport systems, as well as the channels or pores of the biological membrane.

2.3. Carbohydrates

The carbohydrates, the third membrane component, make up from only 5 to 10% of the membrane mass. They are found to be associated only with the outer (non-cytoplasmic) surface of the membrane and generally as polymers of substituted galactose molecules covalently attached to a membrane protein or lipid molecule in the bilayer membrane. The carbohydrates play little or no role in the control of membrane permeability but are important in cell–cell interactions and the immunological properties of the cell membranes. Further, the carbohydrates of cell membranes have numerous carboxylic acid groups that confer a negative surface charge on cells at biological pH's. As mentioned above, the carbohydrates are composed of branched chains of substituted simple sugars, which are covalently attached to specific proteins or lipids in the outer leaflet of the membrane bilayer. The structures of these carbohydrate chains are now under investigation though much is yet to be learned about their synthesis and function.

2.4. Additional Structural and Compositional Characteristics

Several other structural and compositional features of cellular membranes should be remembered.

First, the membrane is very thin when compared with most synthetic membranes. The total thickness of plasma membranes is between 8 and 10 nm. The basic structure contributing to this thickness is the lipid bilayer and the proteins that are associated with the polar groups of the membrane surface, as well as those hydrophobically attached to the lipids of the membrane matrix.

The second important aspect of cell membranes is that they are fluid. In fact, the plasma membrane can be thought of as a two-dimensional liquid with a viscosity some 10 times that of water, thus being rather similar in fluidity to many vegetable oils. The liquid nature of the membrane allows the lipids and integral proteins to diffuse freely across the cell surface unless they are anchored by attachment to other membrane constituents or cellular structural elements. To say that cell membranes are fluid does not mean that there is a rapid exchange of lipids between the two monolayer halves of the bilayer. While the lateral diffusion coefficient for lipids at 37°C is about

1×10^{-8} cm^2/sec, an average lipid flip-flops from one side of the bilayer to the other about 10^{10} times more slowly (2,3). The relatively slow rate of exchange between the juxtaposed monolayers allows for an asymmetric distribution of the various phospholipids to be maintained.

This asymmetry is a third important property of cell membranes. The various phospholipids are asymmetrically distributed between the two bilayers of the membrane in many if not all cells. Although the total amount of phospholipid appears to be the same on both sides of the membrane, there is a variable fraction of that lipid which is phosphatidylcholine (with choline as the alcohol, see Fig. 2), phosphatidylethanolamine (ethanolamine being the alcohol), or phosphatidylserine, depending on whether the monolayer exists on the extracellular or intracellular half of the membrane bilayer (4). Further, the proportions of the individual phospholipids on the outside or inside of the membrane vary from one cell type to another but are relatively constant within a particular cell type (i.e., blood, liver, brain, etc.) even between different species. The mechanism by which this phospholipid asymmetry arises is not clear, but is partially or fully maintained by the large thermodynamic barrier presented to the movement of the hydrophilic moiety of the molecule in flipping from one side of the hydrophobic region to the other. Further evidence is now accumulating that the phospholipid asymmetry is associated with membrane-associated enzymatic activities, which may be linked to membrane receptors of extracellular signals such as hormones (5). Another type of lipid asymmetry is seen in that rather specific "domains" or islands of particular lipids are found to be associated with certain integral proteins, which depend on these lipid associations for their activity (6).

Asymmetry is not only associated with lipids and carbohydrates, but also with the intrinsic and extrinsic membrane proteins. The extrinsic proteins are found to be associated with one membrane surface or the other but never with both. Quite certainly, the hydrophobic nature of the membrane prevents their movement to the other side of the membrane. The intrinsic proteins are also oriented in a rather specific manner. While some of these molecules span the membrane, they are always oriented in a certain direction. Their tumbling within the membrane matrix appears to be prevented by hydrophilic portion(s) of the molecule, which anchor them on each side and thus prevent molecular reorientation. Such asymmetrical orientation of molecules is clearly important in the vectorial transport of molecules by means of membrane protein-mediated "carrier" transport systems.

A final property of cellular or plasma membranes to be remembered, especially in reference to synthetic membranes, is that they are noncovalent assemblies of the constituents. In fact, the strongest organizing forces are those of hydrophobic interactions among the lipids and between lipids and intrinsic proteins. The practical consequence of such relatively weak forces acting to hold the membrane together is that it is mechanically much more fragile than most synthetic membranes and is unable to withstand significant mechanical or chemical perturbation.

3. TRANSPORT ACROSS PLASMA MEMBRANES

3.1. Physical Transport Mechanisms

Being a lipid bilayer, the cell membrane shows relatively high diffusional permeabilities to lipophilic solutes. Hence, the high membrane : water partition coefficient of lipophilic molecules facilitates the dissolution of such solutes in the membrane and accounts for this high permeability. Surprisingly, the permeability of the membranes to small hydrophilic solutes and water is anomalously high from what one would expect of a continuous lipid bilayer and the low permeability of the membrane to larger hydrophilic solutes. These high permeabilities have been attributed to small pores or channels across the membrane, although carrier-mediated processes have been suggested for all such solutes except water. It now appears that water-filled channels arise from certain integral membrane proteins and are responsible for at least the permeation of water.

In some cells, such as nerve and muscle, there are channels that are selective for K^+, Na^+ or Ca^{++} and open, allowing ion diffusion, or closed, preventing penetration, depending on the transmembrane electrical potential difference.

Although some molecules of cellular importance are lipophilic, many of the cellular nutrients such as sugars and amino acids are quite hydrophilic. Hence, for cells to survive, they must have special "carriers" to facilitate transport into the cells. However, a detailed consideration of these unique biological carriers should be preceded by a short discussion of osmosis and osmotic pressure in living cells.

Because cells contain a large number of impermeant solutes, mainly proteins, a significant colloidal osmotic pressure exists across plasma membranes. Osmosis, swelling and eventual lysis (breaking) of animal cells is avoided by a steady state asymmetric active transport of Na ions from the cell.

The magnitude of the outwardly directed flux of Na ions is such that the internal osmotic pressure is balanced and the cell volume is maintained. If this pump is inoperative, whether it be due to the lack of cellular energy (in the form of ATP) to drive the pump, or the inhibition of the pump itself, animal cells will swell and burst. On the other hand, most plants and bacteria have a rigid cell wall which lies on the outside of the cell membrane and prevents the cells from swelling to the point of lysis.

3.2. Biological Transport Mechanisms

As already noted, cells must provide a special means of transporting sugars, amino acids and other solutes necessary for their growth and metabolism. These mechanisms, known as carrier-mediated transport systems, are divided into "passive carriers," which utilize existing transmembrane activity gradients of solutes to drive the transport process, and "active transport" systems, which utilize cellular energy directly in the form of ATP and are capable of moving solutes against an electrochemical gradient.

Carrier-mediated transport First, let us look at the passive carrier mechanisms. These mechanisms have a number of properties that distinguish them as being carrier mediated. First, the rate of solute transport across the membrane is faster than it would be for the diffusion of the same solute. Second, carrier transport exhibits saturation kinetics of the type seen in site-mediated processes. Third, carriers show selectivity, transporting only a single specific molecule or a group of structurally very similar solutes. Fourth, they can be run in reverse. If the concentration (activity) gradient for the solute or solutes is reversed, so too will be the direction of solute transport. Finally, the carriers can be inhibited by rather specific carrier poisons such as phloretin.

The simplest carrier mechanism is known as "facilitated diffusion" and transports a solute down its own concentration gradient. Contrary to the implication of the name, the movement of the transported solute is not by diffusion and resembles diffusion only in that the net flux of solute is down its own activity gradient. As depicted in the schematic model, Figure 4a, a proteinaceous carrier of this type probably alternates between two or more conformational states to alternately expose the carrier site to the environments on the two sides of the membrane, and in the process carries the transported solute across the lipophilic barrier. For example, such a carrier exists for glucose in human red blood cells.

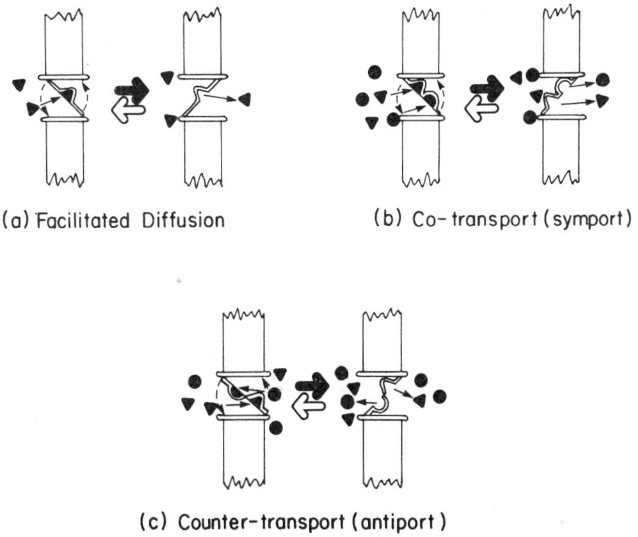

FIGURE 4.—Schematic representations of the three classes of carrier-mediated transport.

The second type of carrier process, one in which two solutes move in the same direction, is called "co-transport" or "symport". This type of carrier moves one solute across the membrane at the expense of the energy gradient provided by a second solute. This process requires both of the solutes to be present on the "feed" side of the membrane but requires that only one of them have a concentration greater on the feed side than on the product side. The second solute may actually be moved up its concentration gradient provided that the first has a sufficiently large "driving" gradient to accomplish the "uphill" transport (Figure 4b). Such carriers are very common in animal cells. For example, cells lining the intestinal wall or the kidney tubule very commonly utilize the gradient of sodium ions created by the active transport of sodium from the cells, to drive the intracellular accumulation of specific classes of nutrients such as simple sugars and amino acids.

The third type of carrier-mediated transport is "counter-transport" or "antiport," in which the two transported solutes are moved in opposite directions across the membrane, as can be seen in Figure 4c. One solute again provides the thermodynamic driving force for the second. It makes no difference which solute contributes the gradient, but the process requires at least one of the two solutes to be moving down an activity gradient sufficient to drive the movement of the other solute. Such a carrier is found in the red blood cells of most animals and exchanges Cl^- for HCO_3^- across the cell membrane.

Without going further into the nature of these processes, let us say that although the exact mechanisms by which these carriers function are not known, it is now generally agreed that they are integral proteins that act through conformational changes within the molecule rather than flip-flopping or floating from one surface of the membrane through the lipid matrix to the second side of the membrane. Because the mechanisms are not understood, note carefully that the diagrams in Figure 4 are only schematic representations and should not be considered as implying any specific mechanism.

Active transport systems Besides the aforementioned "passive" biological carriers that utilize existing solute gradients to provide the energy for transport, cells also possess "active" transport systems. To be considered active, a transport process must both: 1) be capable of moving the solute (solutes) up its (their) activity gradient(s) and, 2) utilize cellular metabolic energy (usually ATP) directly. It was, for many years, thought that a multitude of such active transport systems existed, each one specific for a particular solute. It has, however, become clear within the past five years that only a few active transport systems exist, and most of those previously thought to be active are co-transport or counter transport systems driven by gradients established by the few active transport systems. Though a number of active transport systems have been identified (i.e., Ca^{++}, H^+, K^+, etc.), the most ubiquitous (existing in essentially every cell) and certainly the most studied and best understood of the active transport systems is the "sodium and potassium pump" or Na^+, K^+ ATPase, which transports Na^+ out of cells and K^+ into cells. It is now generally accepted that 3 sodium and 2 potassium ions are transported in opposite directions for each ATP (adenosine triphosphate) molecule hydrolyzed to ADP (adenosine diphosphate) and Pi (inorganic phosphate). The large 10:1 gradient of sodium established by this pump then provides a gradient to drive many of the previously mentioned carrier transport processes for other solutes. Further, the efflux of sodium being greater than the influx of potassium acts to balance the osmotic pressure across the membrane. Finally, the pump establishes ionic gradients, which when coupled to voltage-dependent selective permeability, will lead to stationary transmembrane electrical potentials and action potentials in nerve and muscle cells.

While it is beyond the scope of this chapter to discuss these bioelectric membrane phenomena, mention should be made of the voltage and time-dependent ionic channels. These transmembrane channels are rather selective for specific ions and "open" to allow ion movement only at a given membrane potential. It is through the sequential opening and closing of these channels for Na^+, K^+ and to some extent Ca^{++}, that the electrical signals (action potentials) are created and carried along a nerve membrane.

It should be emphasized that the sodium, potassium pump must be continuously active to maintain the proper intracellular ionic and osmotic environment. This is necessary because sodium and potassium gradients across the cell membrane are continuously being dissipated. The two primary mechanisms acting to reduce the gradients are the nonspecific leak of both ions down their thermodynamic gradients; and the sodium- (or potassium-) dependent carrier-mediated transport of other solutes. Further, in "electrically active cells" the movements of sodium and potassium are also influenced by "action potential" associated changes in membrane permeability.

4. EPITHELIAL MEMBRANES

There are numerous places in multicellular organisms, particularly animals, where it is necessary to limit or mediate selective solute and/or solvent flow across the interface between organs or within the organs themselves. In these instances, one finds that cells are arranged in a manner so as to form a continuous tissue sheet. The most common of these membranes, the epithelial membrane, consists of individual cells sealed together at one end to present a solid barrier to the movement of solutes and water. The side of the membrane at which the cells are joined is called the "apical" or "mucosal" side and is generally the side of the membrane *from* which the solutes and solvent are transported. The other side of the cell sheet is called the "basal", "basolateral" or "serosal" side of the membrane and is the side to which the transport generally takes place.

As shown schematically in Figure 5, the plasma membranes on the serosal side of the tight junctions (seals between cells) form the sides of open spaces known as lateral intercellular spaces (l.i.s.). These spaces are very important to the function of epithelial membranes for they allow the develop-

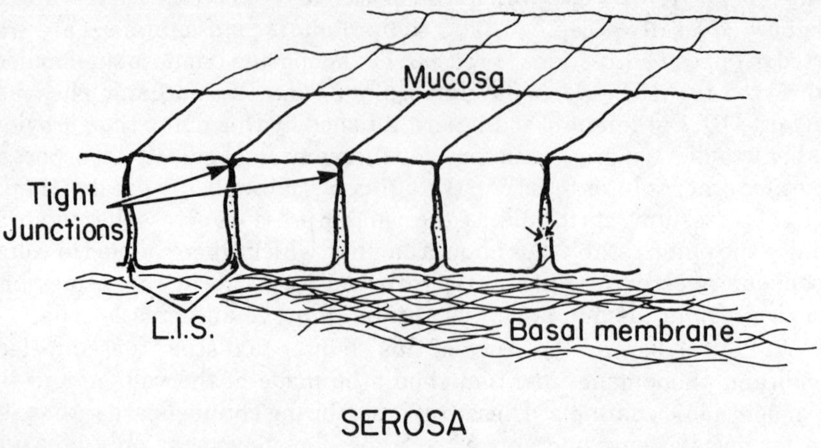

FIGURE 5.—A schematic representation of an epithelial membrane

ment of osmotic gradients along the sides of the cells. These gradients then cause the osmosis of water from the epithelial cells, which then move water across their mucosal membranes in response to the increase in intracellular osmotic pressure. The net effect is the movement of water as well as solutes across the epithelial membrane. Often the transported solution obtains isotonicity (identical osmotic pressure) with body fluids by the time it leaves the basal surface of the l.i.s. This osmotic gradient within the l.i.s. is created by the active transport of Na^+ into the l.i.s. and anions (usually Cl^-) moving into the space down their electrochemical gradient, thus creating a hypertonic solution. The rates of passive Cl^- or other anion movement into the space and osmosis of water relative to the rate of Na^+ transport determine respectively whether there will be an electrical potential across the membrane and whether the solution exiting the l.i.s. will be isotonic with other body fluids surrounding the membrane. It should not be considered that Na^+ and Cl^- (or other anions) are the only solutes that are transported into the intercellular space or across the basal membrane. In fact, a large number of solutes (particularly amino acids and sugars) are often transported across these membranes after being brought into the epithelial cells by co-transport with sodium.

The leakiness of the tight junctions in a particular tissue also greatly affects the transport of solutes and fluid. For instance, a tissue with leaky junctions between cells maintains a much lower membrane potential because the electrical conductance of the membrane is high. Furthermore, the osmotic bulk flow of water through the "leaky" junctions carries significantly more solute by the process known as solvent drag (i.e., less solute is filtered out of the solution; Figure 6).

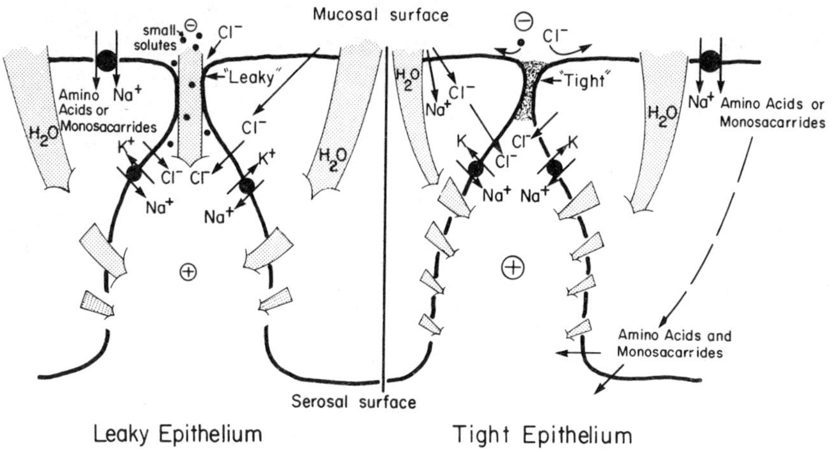

FIGURE 6.—The nature of solute and water flows across "tight" and "leaky" epithelial membranes

On the other hand, an epithelial tissue with less leaky, tight junctions, although not so effective at solvent drag, will be more efficient in the removal of $Na^+$and Cl^- because little chance exists for the leak of the ions back to the mucosal solution and a significant electrical gradient is established for anionic movement.

As should be recognized from the above discussion, epithelial membranes are not all the same, and their transport properties vary vastly from one tissue to another. Further, it should be recognized that many of the complexities of transport, of interest to individuals working with synthetic membranes—i.e., concentration polarization, filtration, electrical potentials and many others—are of interest to the biomembranologist as well.

5. THE FUNCTIONS OF BIOLOGICAL MEMBRANES

Before we conclude this brief discussion of biological membranes, it may be useful to review some of the functions of biological membrane transport in living organisms. Briefly summarized, membranes take up nutrients, eliminate waste, control environmental chemistry, regulate metabolism and act as transducers of chemicals, electricity, temperature and light energy into other energy forms. Further, they provide the barriers between the living and the inanimate world.

Just as the plasma membrane functions to regulate transport at the cellular level, epithelial membranes carry out the same functions at the organismic level in higher animals. Listed below are a few examples of these parallel functions.

Plasma membranes control the *nutrition* of cells by providing for and limiting the uptake of cellular metabolites for structural growth, division and energy production. In higher animals, the epithelial membrane lining the intestine provides the same function for the whole organism, specifically taking up the nutrients such as amino acids, sugars, ions, co-factors and water.

The plasma membrane mediates the *selective excretion* of waste materials, mainly CO_2 and H_2O, as well as certain unusable metabolic products while retaining the required nutrients and the proper amount of salts and H_2O. The kidney, through its complexly regulated epithelial membrane-mediated filtration, reabsorption and secretion mechanisms, fulfills the same selective excretion of wastes from the whole organism.

The cell membrane acts to provide the *barrier* between the activities of the cell and the environment. The skin, another epithelial membrane, provides the same barrier between the organism and the inanimate world.

The cell membrane provides for some cells a means of *storing* and *secreting* particular cellular products such as hormones and secretory proteins. Epithelial glands such as the exocrine pancreatic glands, gastric glands and sweat glands provide a coordination and organization to these secretions for the whole organism.

Many biological membrane transport functions are of cellular origin and may be modified or coordinated by epithelial organization for the whole organism, but some functions are exclusively of cellular or subcellular membrane origin and primary control. Of particular importance are 1) the capture and/or transformation of chemical or light energy into chemical energy (ATP) by mitochondria, chloroplasts and certain bacteria; 2) the development of potential energy gradients for driving carrier-mediated solute transport; and 3) the development of electrical potentials as a means of conducting information along the nerve or muscle membrane.

Caveat—Having briefly discussed the nature of the plasma membrane and some of the structural and transport properties of epithelial membranes as an example of the kinds of "membrane tissues" that exist in living organisms, one must quickly acknowledge some oversimplifications of this large and complex subject as a means of giving a perspective to the nature of biological membranes.

6. PRESENT AND FUTURE BIOMEMBRANOLOGY

What is the thrust of biomembrane research? In broad terms this is not difficult, because it is essentially the same as the thrust of all research in the natural sciences. The goal has been to understand the three areas of biomembrane composition, structure and function and how they each relate to one another. Having accomplished this, the implicit goal, though not often explicitly stated, is to utilize this body of knowledge to modify the system to bring about certain desired effects, whether they be the cure of a membrane-related disease or for the production of a more useful efficient industrial process. Until the facts and interrelationships are learned, which will surely take the rest of our lives, I believe physical membrane scientists should carefully follow the developments in biomembranology, for two important reasons. First, biological membranes show how nature has solved some of the same problems that synthetic membrane scientists face; and second, because most biological scientists tend to be less aware and/or less interested in utilitarian applications, the more physically oriented membranologist may see opportunities for the application of synthetic membranes to biological problems or the application of new biomembranology to particular physical or engineering problems.

With these thoughts in mind, let me sketch a few ideas that have recently or, I believe, will soon become feasible.

6.1. Prospects

The view of the future biomembranology, though not clear, is improving. Not only will the application of synthetic membranes to problems in biology continue and grow, but there will be a significant growth in the use of what might be called "biosynthetic" membranes. Further, I perceive the development of two types of such biosynthetic membranes.

One, a synthetic membrane that will incorporate "biological carriers" or enzyme systems, could be used in the purification and separation of certain molecules from solutions. The very high selectivity of the carrier systems from biological membranes can provide single-step separation of molecules differing only slightly in structure. For instance, the D isomer of a molecule may be separated from its L isomer. Other uses of such membranes might be for the synthesis or degradation of compounds. Single enzymes or even a series of enzyme complexes could be arranged on or within the membrane such that the product or products of one process may be acted upon by an adjacent enzyme or carrier. Although a few such membrane enzyme systems have been developed (8), with the recent progress of molecular engineering in the large-scale production of specific proteins, it will be possible to program microorganisms to produce the biological carriers or enzymes that have been either too costly or difficult to isolate. In the future, the chemical synthesis of a number of biologically relevant chemicals and pharmaceuticals may use membrane reactor systems utilizing these proteins.

Another application of bio-synthetic membranes is likely to be in the area of energy transduction. For instance, I believe that a practical membrane solar energy transducer will be developed in which some light-transducing biomolecules such as chlorophyll or bacterial rodopsin will be incorporated into synthetic membranes for the transduction of light energy into a more immediately useful form.

A second area that is sure to develop is the use of membrane-immobilized antibodies. Membrane stabilized antibodies could be used to remove specific molecules from a reaction broth or from a product mixture. In this way a one-step purification or elimination of some reaction inhibitor could be accomplished. A second use of such antibodies, already being attempted, is the attachment of tissue-specific antibodies to membrane vesicles containing a drug or diagnostic agent so as to target the vesicle to diseased tissues. Despite many problems, I'm confident that these approaches will be found useful in future medical diagnosis and treatment.

Further, membrane vesicles will be used in medicine to provide enzymes to cells in which they are deficient or to carry specific genes to be incorporated into plant or animal cells for specific protein production. Some pioneering work on extracellular enzyme replacement has already been done by Dr. Thomas Chang and numerous other investigators and the area promises further development (8). Recent years have seen the rapid development and refinement of kidney dialysis systems. I think that this area will continue to develop with organ replacement and supplementation being possible, for lung, liver or pancreas, using synthetic membrane systems.

Finally, if the problems of gas permeation and gas selectivities of synthetic membranes can be overcome it may be possible to make synthetic membrane gills a practical apparatus for scuba divers and underwater habitation.

6.2. Problems

In our attempts to utilize the very specific and very efficient biological membrane constituents in synthetic systems, we are faced with three major problems. The first, probably the most important, may limit the use of these biological wonders much more than we would like. This is the problem of the fragility or unstable nature of biological molecules and the limits on the conditions under which they are active. In most cases the conditions for incorporating biological membrane proteins into synthetic membranes will necessitate very gentle chemical and physical procedures. Furthermore, the environmental conditions under which these new systems must operate are generally within narrow ranges of pH, ionic strength and ionic composition, temperature and pressure. The solutions to these problems may not come easily.

The two other problems, while of concern, do not so severely limit the options. One is that biological membrane constituents are "designed" to work in very thin membranes. This means that in many cases it may be necessary to develop membranes on the order of 10 nm which will have mechanical and chemical strength necessary to remain intact under the specified conditions in which we wish to employ them. The other problem is that because biological molecules have a unique orientation in the membrane, it will often be necessary to stabilize the biological molecules within the synthetic membrane so as to provide the desired asymmetry of function.

Clearly biological membranes hold many promises for application and a sufficient number of problems for investigation by current and future membranologists.

REFERENCES

1. Tanford, C.: 1978, "The hydrophobic effect and the organization of living matter." Science 200, pp 1012–1018.
2. Kornberg, R.D., McConnell, H.M.: 1971, "Lateral diffusion of phospholipids in a vesicle membrane." Proc. Nat'l. Acad. Sci. USA 68, pp 2564–2568.
3. Thompson, T.E., Huang, C.: 1980, "Dynamics of lipids in biomembranes." In Membrane Physiology. Eds. Andreoli, T.E., Hoffman, J.F. and Fanestil, D.D. Plenum: New York, pp 27–48.
4. Rothman, J., Lenard, J.: 1977, Membrane Asymmetry 195, pp 743–753.
5. Hirata, E., Axelrod, J.: 1980, "Phospholipid methylation and biological signal transmission." Science 209, pp 1082–1090.
6. Karnovsky, M.J.: 1979, "Lipid domains in biological membranes." Am. J. Path. 97, pp 212–221.
7. Allen, D.: 1982, "Inositol lipids and membrane function in erythrocytes." Cell Calcium 3, pp 451–465.
8. Chang, TMS.: 1972, Artificial Cells. Charles C. Thomas Pub. Springfield, Illinois.

SELECTED BIBLIOGRAPHY—GENERAL REFERENCES

1) Alberts, B., Bray, D., Lewis J., Raff, M., Roberts, K., Watson J.D.: 1983, "Molecular biology of the cell." Chapter 6: The Plasma Membrane. Garland Publishing, New York.
2) Andreoli, T.E., Hoffman, J.F., Fanestil, D.D.: 1980, Membrane Physiology. Plenum Medical Publishers, New York.
3) Weissman, G., Clairborne, R.: 1975, Cell Membranes. H.P. Publishing Co. New York.
4) Stryer, L. 1981, "Introduction to biological membranes," in Biochemistry. W. H. Freeman and Co., San Francisco.

HOW TO BRIDGE THE GAP BETWEEN MEMBRANE BIOLOGY AND POLYMER SCIENCE
Oriented Polymers as Models for Biomembranes and Cells*

Helmut Ringsdorf and Brigitta Schmidt

Institute for Organic Chemistry
University of Mainz
West Germany

*Extended abstract of a lecture presented at the NATO Advanced Study Institute on Synthetic Membranes in Alcabideche, Portugal, 1983. The lecture was based on a chapter to appear in Advances in Polymer Science, 1984.

P. M. Bungay et al. (eds.), Synthetic Membranes: Science, Engineering and Applications, 701–712.

Can polymer chemists contribute to the understanding or even mimicking of cell membrane functions and cell–cell interactions? Fascinated by the specificity and efficiency of, for example, the destruction of tumor cells by lymphocytes (1) and having in mind what biochemical analyses tell us about membrane composition, we may try to "synthesize" membrane and cell models. The commonly used model systems, such as planar lipid monolayers at the gas–water interface, bimolecular lipid membranes and spherical liposomes, are much less stable than natural membrane systems (Figure 1).

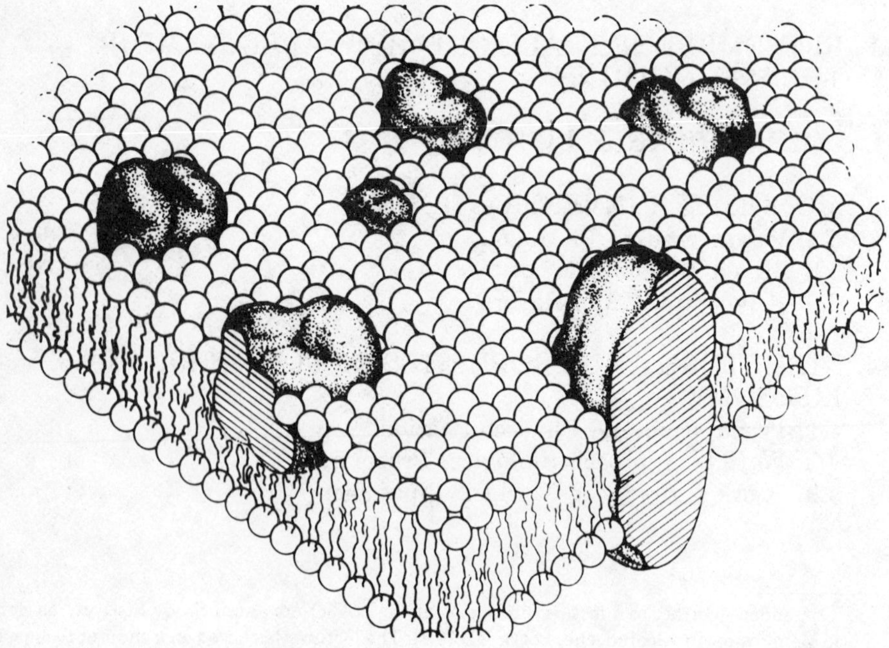

FIGURE 1.—Fluid mosaic model of biomembranes (2).

In an attempt to increase the stability of biomembrane models (3) a variety of polymerizable lipids were synthesized. Possible preparations of polymeric model membranes are demonstrated in Figure 2 (4,5).

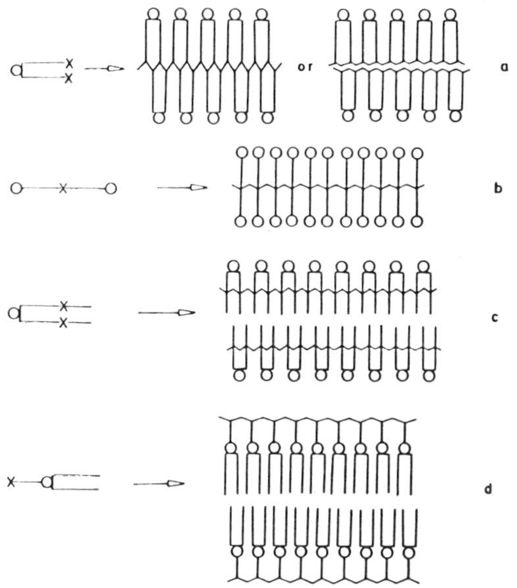

FIGURE 2.—Schematic representation of different methods to stabilize membrane models by polymerization of reactive units (X) at different locations in the lipid (4,5).

Dienes (5,6,11), diacetylenes (4,5,7,8,11) and methacrylates (6,8–10) have been used as polymerizable units. The analogues of natural phospholipids such as lecithin and cephalin have also been prepared (7,11). These compounds can be oriented in monolayers at the gas–water interface. Polymerization can be achieved by UV-irradiation of the monomeric films (11), preserving the membrane-like orientation of the molecules. The polymeric films are either very rigid or flexible depending upon the nature of the polymerizable group.

From the synthetic lipids, liposomes can be formed by ultrasonication of the aqueous lipid dispersions above the phase transition temperature (12,13). These monomeric liposomes are transformed into polymeric liposomes by UV-irradiation or by self-condensation of long-chain amino acids (Figure 3; reference 14).

The preservation of the spherical shape of liposomes after polymerization can be visualized by scanning electron microscopy. The stability of the polymeric liposomes has been found to be much higher than that of monomeric liposomes. The increased stability, e.g., toward detergents such as sodium

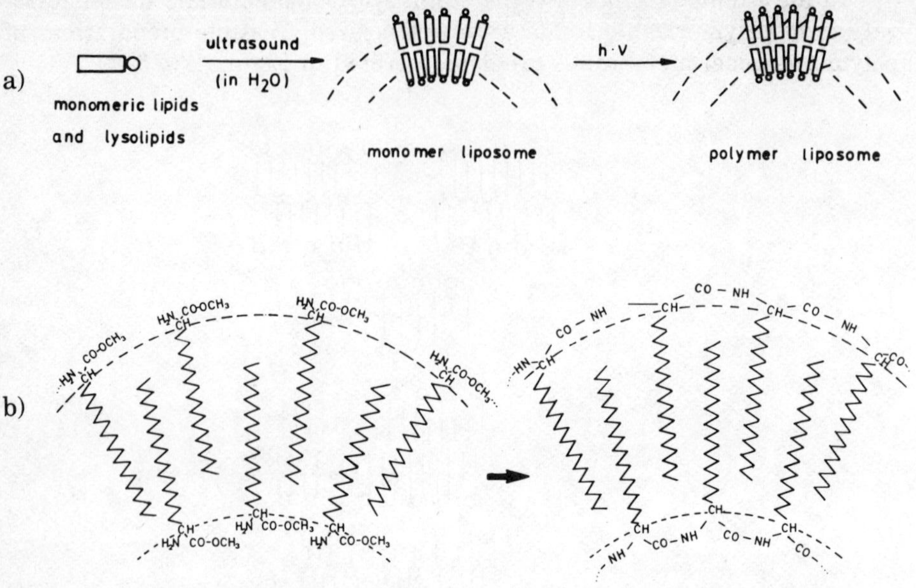

FIGURE 3.—Formations of polymeric vesicles from polymerizable lipids a) by UV-irradiation and b) by self condensation of long-chain amino acids.

dodecylsulphate (SDS), can be shown using 6-carboxyfluorescein (6-CF) as a marker for the membrane permeability. This technique is explained schematically in Figure 4.

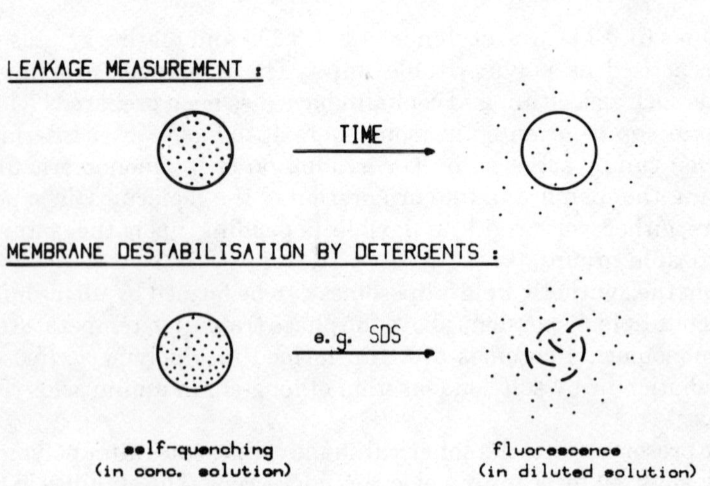

FIGURE 4.—Use of 6-carboxyfluorescein (6-CF) as marker for membrane permeability of liposomes.

Liposomes of a dienoyl-lecithin are more sensitive toward SDS-treatment than those of the saturated analogue, DPPC. In contrast to this, the polymerized vesicles show only a very small increase in 6-CF permeability up to quite high detergent concentrations (Figure 5).

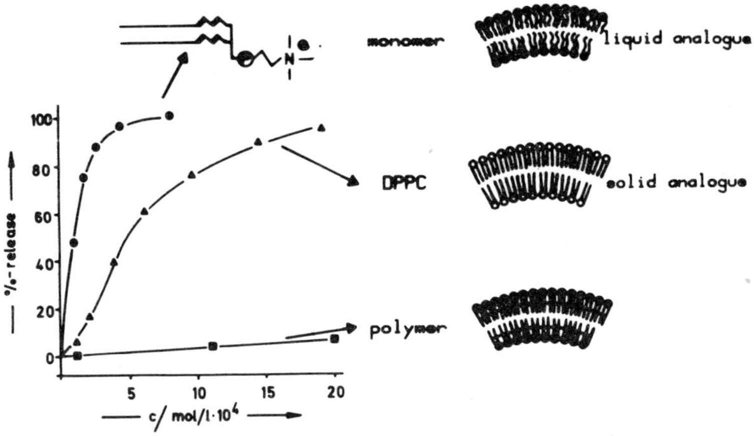

FIGURE 5.—Demonstration of the enhanced stability of polymeric liposomes in the presence of SDS compared to liposomes prepared from the monomer and from a saturated analogue (DPPC).

One of the important properties of cell membranes is fluidity. Whether polymerized model membranes systems are too rigid for showing a phase transition depends on the type and the position of the polymerizable group in the lipid used for the preparation of the membrane. In the case of polydiacetylene lipids, the lack of phase transition is caused by the formation of a rigid fully conjugated polymer backbone (15), which is shown in the following differential scanning calorimetry (DSC)-diagrams in Figure 6.

The phase transition of polymeric liposomes is retained if the polymer chain is more flexible or is located on the hydrophilic surface of the vesicles (Figure 7).

Mixed and phase-separated monolayers from natural and polymerizable lipids (16) can be prepared as shown in Figure 8.

The phase separation in mixed liposomes prepared from polymerizable lipids and saturated lecithins (e.g. dimyristoyl-lecithin, DML) can be demonstrated by freeze–fracture electron microscopy (17). The domains of DML can be identified by their typical "ripple" structure (Figure 9).

Partially polymerized phase-separated liposomes can be used for the incorporation of proteins or the selective opening of membrane compartments in order to release entrapped substances (Figure 10).

To "uncork" a partially polymerized vesicle, its membrane has to contain destabilizable areas that can be opened by variation of pH, temperature increase, photochemical destabilization or enzymatic processes (Figure 11).

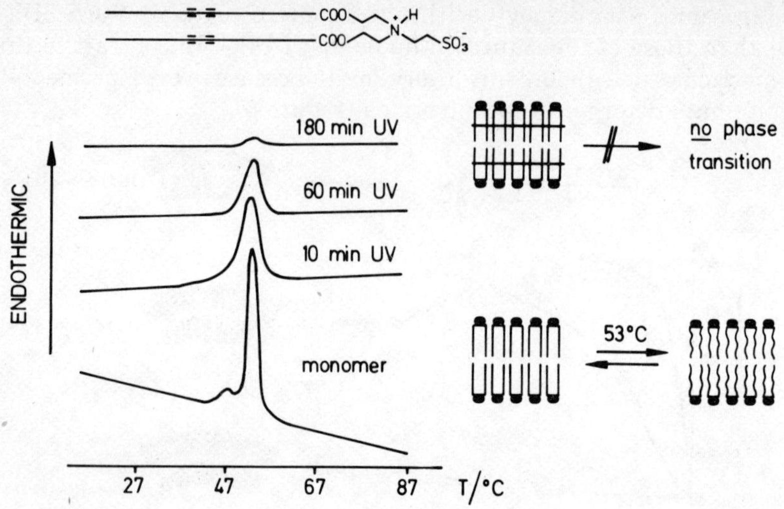

FIGURE 6.—Influence of the polymerization on the phase transition of liposomes prepared from lipids with diacetylene units in the hydrophobic chain.

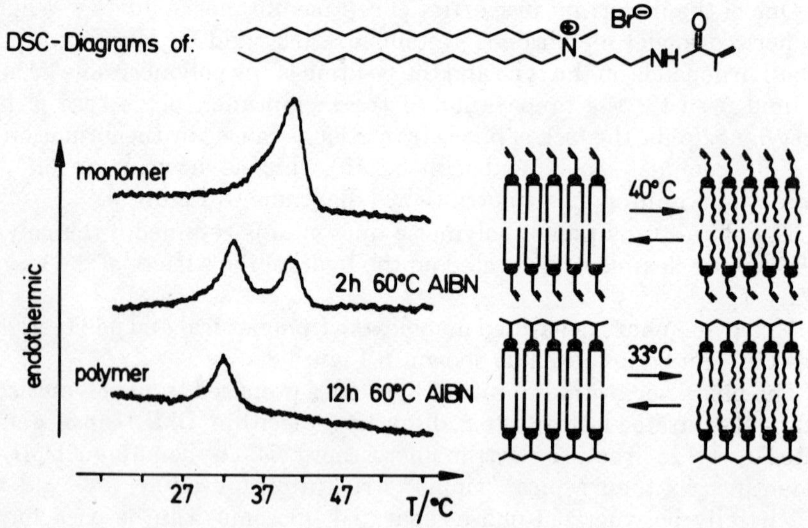

FIGURE 7.—Change of the phase transition temperature during polymerization of liposomes prepared from lipids with methacryloyl units in the head group. Polymerization is carried out with azobisisobutyronitrile (AIBN) at 60°C.

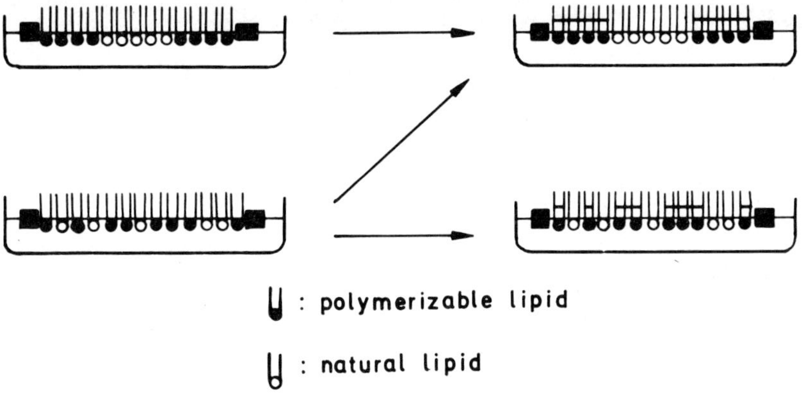

FIGURE 8.—Mixed monolayers of natural and polymerizable lipids.

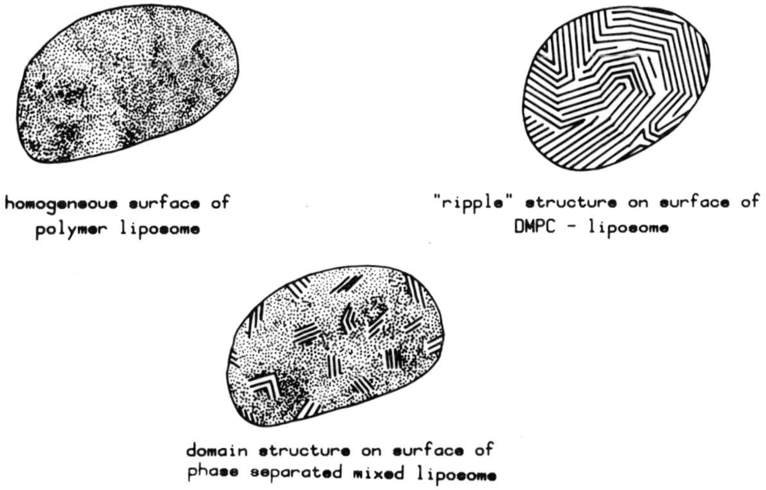

FIGURE 9.—Schematic representation of a phase-separated liposome consisting of the homogeneous surface of polymer lipids and domains with the ripple structure of the lecithin in comparison with the pure systems.

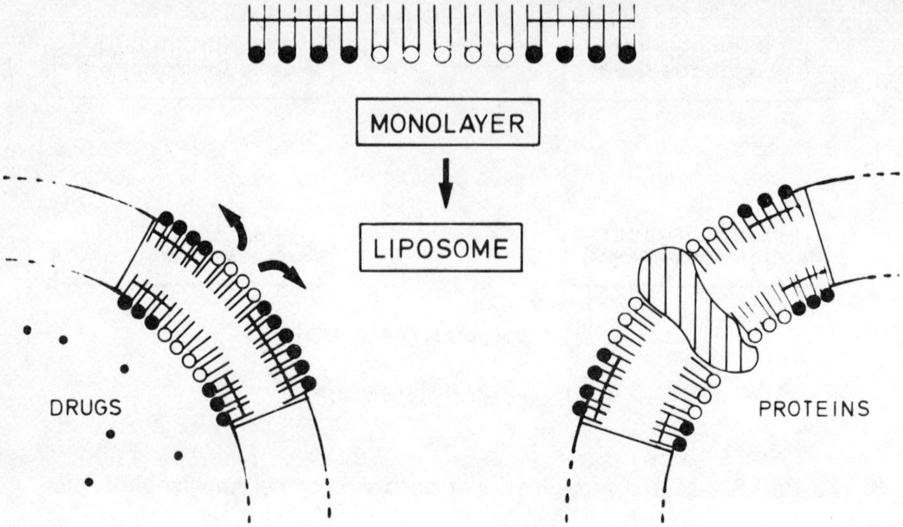

FIGURE 10.—Application of mixed polymeric membranes.

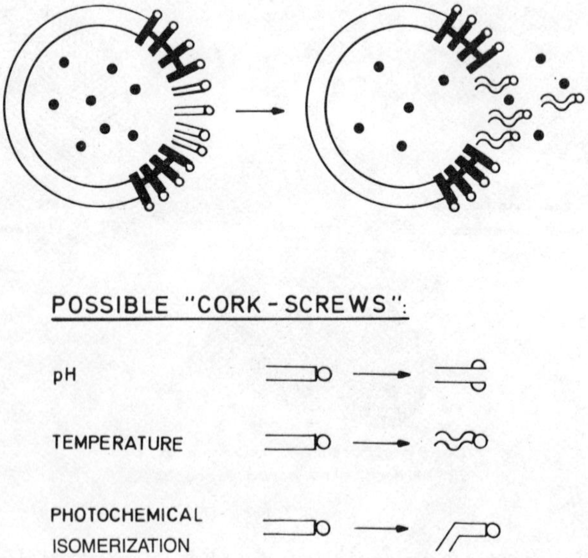

FIGURE 11.—"Uncorking" of mixed polymeric liposomes.

After the realization of these first steps, more complex mixed systems have been studied. Thus, an ATP-synthetase enzyme was incorporated into polymerizable liposomes (Figure 12). The enzyme was still active after polymerization (18) and its activity did not decrease after several weeks, due to the high stability of the proteoliposomes. In addition, the chromoprotein bacteriorhodopsin has been incorporated into liposomes of fully synthetic, polymerizable lipids (19). Such light-driven polymer-embedded proton pumps show the expected long-term stability over months.

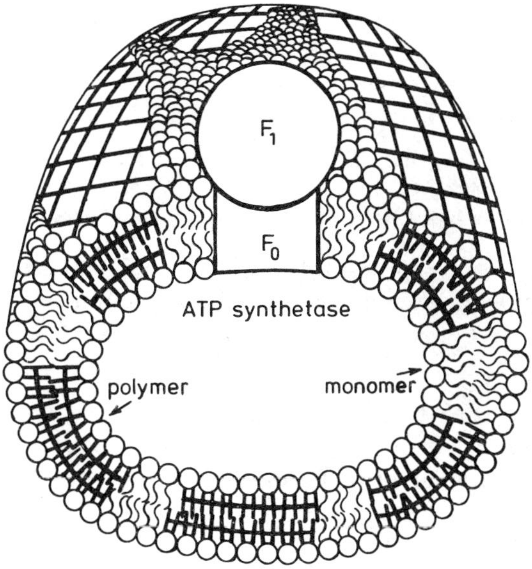

FIGURE 12.—Schematic representation of a partially polymerized proteoliposome incorporating ATP synthetase (18).

An important feature of biological membranes, surface recognition, could be modelled with polymerized liposomes bearing sugar moieties. These moieties are recognized by sugar-specific proteins such as Concanavalin A (20).

Membrane models, though oversimplified, are of high potential value for the understanding of important biological processes such as intercellular communication, cell uptake of extracellular material, and transformation of external impulses into intracellular effects.

As shown schematically in Figure 13, four different synthetic approaches to stable biomembrane and cell models via polymerizable phospholipids can be discussed (8).

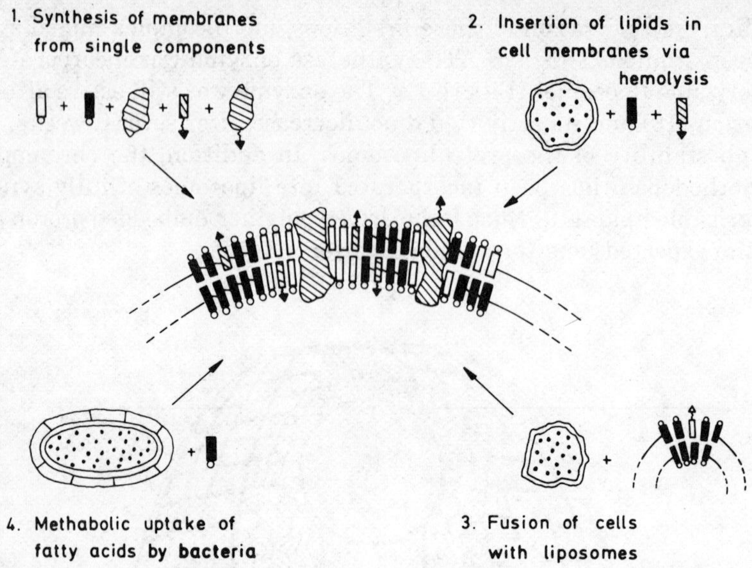

FIGURE 13.—Schematic representation of different synthetic ways to build up highly stable biomembranes and cell models by partial polymerization of the cell membrane.

As a first approach to prepare biomembrane models (demonstrated in scheme 3 of Figure 13), fusion experiments have been carried out. The fusion of giant liposomes indicated in Figure 14 has already been accomplished (21,22).

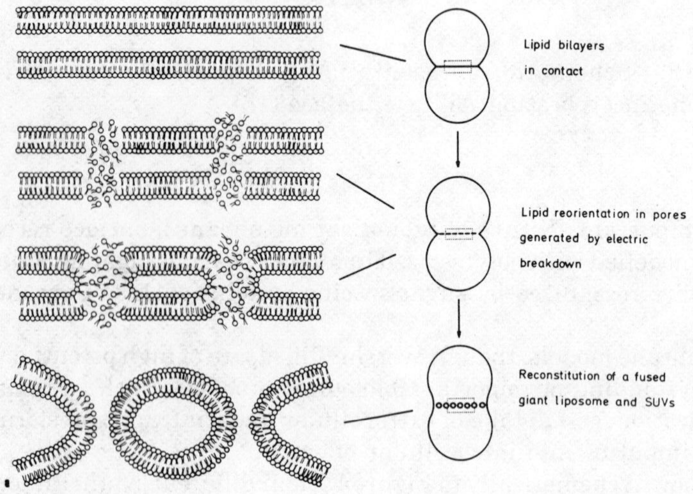

FIGURE 14.—Fusion of giant liposomes with the formation of small unilamellar vesicles.

Recently the biosynthetic incorporation of diacetylene fatty acids into biomembranes of *Acholeplasma Laidlawii* cells (scheme 4 in Figure 13), and their polymerization by irradiation with UV-light, have been realized (23).

The investigation of pure and mixed oriented polymeric membranes may provide a useful approach to stabilize cell models for the investigation and characterization of cell properties and for the simulation of biological interactions. However, the scope of possible applications may be even wider—e.g., as drug carriers, catalytic and biomimetic devices or as energy transfer systems.

In addition, polymerized films of defined structure may be used for separation purposes, as already discussed several years ago for monomeric Langmuir–Blodgett (LB) multilayers (24,25). Now, polymerizable LB multilayers are being investigated as hyperfiltration (26) and gas separation composite membranes (27–29) because they combine extreme thinness with high mechanical and chemical stability. Such composite membranes might be achieved by alternative techniques, too (30–32).

REFERENCES

1. Old L.J.: 1977, "Cancer immunology." Sci. Am. 236, p. 62.
2. Singer S.J., Nicolson G.L.: 1972, "The fluid mosaic model of the structure of cell membranes." Science 175, p. 720.
3. Weisman G. (ed.): 1975, "Cell Membranes, Biochemistry, Cell Biology and Pathology." HP Publishing Co., New York.
4. Hub H., Hupfer B., Koch H., Ringsdorf H.: 1980, "Polymerizable phospholipid analogues—new stable biomembrane and cell models." Angew. Chem. Int. Ed. Engl. 19, p. 938.
5. Akimoto A., Dorn K., Gros L., Ringsdorf H., Schupp H.: 1981, "Polymer model membranes." Angew. Chem. Int. Ed. Engl. 20, p. 90.
6. Dorn K., Klingbiel R.T., Specht D.P., Tyminski P.N., Ringsdorf H., O'Brien D.F.: 1984, "Permeability characteristics of polymeric bilayer membranes from methacryloyl and butadiene lipids." J. Am. Chem. Soc. 106, p. 1627.
7. Johnston D.S., Sanghera S., Chapman, D.: 1980, "The formation of polymeric model biomembranes from diacetylenic fatty acids and phospholipids." Biochim. Biophys. Acta 602, p. 213.
8. Gros L., Ringsdorf H., Schupp H.: 1981, "Polymeric antitumor agents on a molecular and on a cellular level?" Angew. Chem. Int. Ed. Engl. 20, p. 305;
   1981, "Polyreactions in oriented systems, 21: Polymeric phospholipid monolayers." Makromol. Chem. 182, p. 247.
9. Regen S.L., Czech B.: 1980, "Polymerized vesicles." J. Am. Chem. Soc. 102, p. 6638.
   1982, "Polymerized phosphatidylcholine vesicles. Synthesis and characterization." J. Am. Chem. Soc. 104, p. 791.
10. Tundo P., Kippenberger D.G., Klahn P.H., Prieto N.E.: 1982, "Functionally polymerized surfactant vesicles. Synthesis and characterization." J. Am. Chem. Soc. 104, p. 456.
11. Hupfer B., Ringsdorf H.: 1983, "Spreading and polymerization behavior of diacetylenic phospholipids at the gas–water interface." Chem. Phys. Lip. 33, p. 263;
    Hupfer B., Ringsdorf H., Schupp H.: 1983, "Liposomes from polymerizable phospholipids." Chem. Phys. Lip. 32, p. 271.
12. Büschl R., Folda T., Ringsdorf H.: 1984, "Polymeric monolayers and liposomes as models for biomembranes." Makromol. Chem., Suppl. 6, p. 245.
13. Bader H., Ringsdorf H., Schmidt B.: 1984, "Water-soluble polymers in medicine." Angew. Makromol. Chem. 123/124, p. 457.
14. Folda T., Gros L., Ringsdorf H.: 1982, "Polyreactions in oriented polypeptides in monolayers and liposomes." Makromol. Chem. Rapid Commun. 3, p. 167.

15. Tieke B., Wegner G., Naegele D., Ringsdorf H.: 1976, "Polymerization von Tricosa–10, 12 diin-1-säure in Multischichten." Angew. Chem. 88, p. 805.
16. Büschl R., Hupfer B., Ringsdorf H.: "Polyreactions in oriented systems, 30: Mixed monolayers and liposomes from natural and polymerizable lipids." Makromol. Chem. Rapid Commun. 3, p. 589.
17. Gaub H., Sackmann E., Büschl R., Ringsdorf H.: 1984, "Lateral diffusion and phase separation in two-dimensional solutions of polymerized butadiene lipids in dimyristoylphosphatidylcholine bilayers." Biophys. J. 45, p. 725.
18. Wagner N., Dose K., Koch H., Ringsdorf H.: 1980, "Incorporation of ATP synthetase into long-term stable liposomes of a polymerizable synthetic sulfolipid." FEBS Lett. 132, p. 313.
19. Pabst R., Ringsdorf H., Koch H., Dose K.: 1983, "Light-driven proton transport of bacteriorhodopsin incorporated into long-term stable liposomes of a polymerizable sulfolipid." FEBS Lett. 154, p. 5.
20. Bader H., Ringsdorf H., Skura J.: 1981, "Liposomes from polymerizable glycolipids." Angew. Chem. 93, p. 109;
1981, "Polymer model membranes." Angew. Chem. Int. Ed. Engl. 20, p. 91.
21. Zimmermann U., Scheurich P., Pilwat G., Benz K.: 1981, "Cells with manipulated functions: New perspectives for cell biology, medicine, and technology." Angew. Chem. Int. Ed. Engl. 20, p. 325.
22. Büschl R., Ringsdorf H., Zimmermann U.: 1982, "Electric field-induced fusion of large liposomes from natural and polymerizable lipids." FEBS Lett. 150, p. 38.
23. Leaver J., Alonsu A., Durrani A.A., Chapman D.: 1983, "The biosynthetic incorporation of diacetylenic fatty acid into the biomembranes of *Acholeplasma Laidlawil* A cells and polymerization of the biomembranes by irradiation with ultraviolet light." Biochim. Biophys. Acta 727, p. 327.
24. Rose G.D., Quinn J.A.: 1968, "Composite membranes: The permeation of gases through deposited monolayers." Science 159, p. 636.
25. Miller M.L., Sutherland J., Schmitt T.: 1968, "Transfer of monolayers to gel-cellulose." J. Colloid & Interface Sci. 28, p. 28.
26. Heckmann K., Strobl G., Bauer S.: 1983, "Hyperfiltration through cross-linked monolayers." Thin Solid Films 99, p. 265.
27. Albrecht O., Laschewsky A., Ringsdorf H.: 1984, "Polymerizable built-up multilayers on polymer supports." Macromolecules 17, p. 937.
28. Albrecht O., Laschewsky A., Ringsdorf H.: 1984, "Investigations of polymerizable multilayers as gas separation membranes." J. Membrane Sci., in press.
29. Lando J.B.: 1983, Personal communication.
30. Langmuir I., Schaefer V.: 1938, "Activities of urease and pepsin monolayers." J. Am. Chem. Soc. 60, p. 1351.
31. Regen S.L., Kirszenstejn B., Singh A.: 1983, "Communications to the Editor: Polymer-supported membranes: A new approach for modifying polymer surfaces." Macromolecules 16, p. 335.
32. Albrecht O., Laschewsky A.: 1984, "Communications to the Editor: Polymer-supported membranes." Macromolecules 17, p. 1292.

AUTHORS

Dr. Philippe Aptel
Laboratoire de Génie Chimique (CNRS L.A. 192)
Université Paul Sabatier
118, route de Narbonne
31062 Toulouse Cedex, FRANCE

Dr. Richard W. Baker, President
Membrane Technology and Research, Inc.
1030 Hamilton Court
Menlo Park, California 94025, U.S.A.

Dr. D. Bargeman
Department of Chemical Technology
Technical University of Twente
P.O. Box 217
7500 AE Enschede, THE NETHERLANDS

Dr. Peter M. Bungay
Biomedical Engineering and Instrumentation Branch
National Institutes of Health
Building 13, Room 3W13
Bethesda, Maryland 20892, U.S.A.

Dr. Michael J. Clifton
Laboratoire de Génie Chimique (CNRS L.A. 192)
Université Paul Sabatier
118, route de Narbonne
31062 Toulouse Cedex, FRANCE

Professor Enrico Drioli
Università degli Studi di Napoli
Istituto di Principi di Ingegneria Chimica
Piazzale Tecchio
80125 Napoli, ITALIA

Professor William R. Galey
Department of Physiology
University of New Mexico
Albuquerque, New Mexico 87131 U.S.A.

Professor Gunnar Jonsson
Instituttet for Kemiindustri
DtH Building 227
Technical University of Denmark
DK-2800 Lyngby, DENMARK

Dr. Harold K. Lonsdale
Bend Research, Inc.
64550 Research Road
Bend, Oregon 97701, U.S.A.

Professor Patrick Meares, Head
Department of Chemistry
University of Aberdeen
Meston Walk
Old Aberdeen, AB9, 2UE, SCOTLAND (U.K.)

Professor Jean Néel
Laboratoire de Chimie-Physique Macromoléculaire (CNRS E.R.A. 23)
Ecole Nationale Supérieure des Industries Chimiques
1, rue Granville
54042 Nancy Cedex, FRANCE

Dr. Geoffrey S. Park
Department of Applied Chemistry
University of Wales
Institute of Science and Technology
Cardiff, CF1 3NU, WALES (U.K.)

Dr. Maria Norberta de Pinho,
Departamento de Engenharia Química
Instituto Superior Técnico
Avenida Rovisco Pais
1096 Lisboa Codex, PORTUGAL

Dr. Mark C. Porter
Consultant
3449 Byron Court
Pleasanton, California 94566, U.S.A.

Professor Dr. Robert Rautenbach
Institut für Verfahrenstechnik
Rheinisch-Westfälische Technische
Hochschule Aachen
Turmstrasse 46
D–5100 Aachen, WEST GERMANY

Professor Dr. Helmut Ringsdorf
Institut für Organische Chemie
der Johannes Gutenberg Universität
J.J.-Becher-Weg 18–20
Postfach 3980
D–6500 Mainz 1, WEST GERMANY

Dr. Lynda M. Sanders
Syntex Research Corporation
Division of Syntex (U.S.A.) Inc.
3401 Hillview Avenue
Palo Alto, California 94304, U.S.A.

Dr. Brigitta Schmidt
Institut für Organische Chemie
der Johannes Gutenberg Universität
J.J.-Becher-Weg 18–20
Postfach 3980
D–6500 Mainz 1, WEST GERMANY

Professor Jerome S. Schultz, Chairman
Department of Chemical Engineering
University of Michigan
Ann Arbor, Michigan 48109, U.S.A.

Professor C.A. Smolders
Department of Chemical Technology
Technical University of Twente
P.O. Box 217
7500 AE Enschede, THE NETHERLANDS

Dr. H. Strathmann
Fraunhofer-Institut für Grenzflächen-und
Bioverfahrenstechnik
Institutszentrum Stuttgart
Nobelstrasse 12
D–7000 Stuttgart 80, WEST GERMANY

Dr. William J. Ward III
General Electric Research and Development
Center
P.O. Box 8, K1/4B15
Schenectady, New York 12301, U.S.A.

Dr. J. G. Wijmans
Department of Chemical Technology
Technical University of Twente
P.O. Box 217
7500 AE Enschede, THE NETHERLANDS

PARTICIPANTS

North Atlantic Treaty Organization Advanced Study Institute on Synthetic Membranes, held at the Hotel Sintra Estoril, Alcabideche, Portugal, June 26 - July 8, 1983.

Belgium

Dr. Serge Joris
Laboratoire Central
Solvay & Cie
rue de Ransbeek, 310
1120 Bruxelles

Brazil

Dr. Ronaldo Nobrega
Programa de Engenharia Química
COPPE, Universidade Federal do Rio de Janeiro
Cidade Universitaria C.T. Bloco G - C.P. 68502
21944 - Rio de Janeiro, RJ

Canada

Dr. Abdul-Fattah Asfour
Department of Chemical Engineering
University of Windsor
Windsor, Ontario N9B 3P4

Prof. Louis-Phillippe Blanchard
Department of Chemical Engineering
Laval University
Ste. Foy, Québec G1K 7P4

Denmark

Mr. Peter Bo
National Laboratory Risoe
Roskilde

Mr. Per Michael Christensen
The Technical University of Denmark
Instituttet for Kemiindustri
Building 227, Room 102
DK–2800, Lyngby

France

M. René Angleraud
Centre de Recherches de Saint-Fons
Rhône-Poulenc
B.P. 62
69190 Saint-Fons

Dr. Pascal Belleville
Département Génie Chimique
Institut Universitaire de Technologie Saint-Nazaire
B.P. 420
44606 Saint-Nazaire

Dr. Jean-Pierre Brun
Laboratoire de Thermodynamique et d'Electrochimie des Matériaux
Université Paris - Val de Marne
Av. du Gl de Gaulle
94010 Creteil Cedex

Dr. André Larbot
Laboratoire Professeur Cot
Ecole Nationale Supérieure de Chimie
rue de l'Ecole Normale
34075 Montpellier Cedex

M. André Le Jeune
Commissariat à l'Energie Atomique
CEN/Saclay-IRDI/DGI
91191 Gif-sur-Yvette Cedex

M. Jean-Marie Martinet
Centre d'Etudes Nucléaires de la Vallée du Rhône, DGI
Commissariat à l'Energie Atomique
B.P. 171
30205 Bagnols-s-Ceze

Professeur Eric Sélégny
Laboratoire de Chimie Macromoléculaire
Faculté des Sciences
Université de Rouen
76130 Mont-Saint-Aignan

Germany

Dr. C. Thomas Badenhop
Seitz-Filter-Werke
Postfach 889
D–6550 Bad Kreuznach

Dr. Horst Chmiel
Fraunhofer - IGB
Nobelstrasse 12
D–7000 Stuttgart 80

Ms. Sabine Englich
Helmholtz-Institut
für Biomedizinische Technik
RWTH Aachen
Goethestrasse 27/29
D–5100 Aachen

Dr. Herbert Harttig
Gambro Dialysatoren KG
Postfach 1323
D–7459 Hechingen

Mr. Klaus Kimmerle
Fraunhofer-IGB, Stuttgart
Rosentalstrasse 56
D–7000 Stuttgart 80

Dr. Hans-Dieter Lehmann
Gambro Dialysatoren KG
Postfach 1323
D–7450 Hechingen

Dr. Klaus-Viktor Peinemann
GKSS-Forschungszentrum
Max-Planck-Strasse
D–2054 Geesthacht

Dr. Achinto Sen-Gupta
Unilever Forschungs GmbH
Behringstrasse 154
D–2000 Hamburg

Mr. Kurt Tippmer
Firma Carl Still GmbH
Kaiserwall 21
Postfach 1480
D–4350 Recklinghausen

Prof. Dr. Ing. Udo Werner
Lehrstuhl für Mechanische
Verfahrenstechnik
Universität Dortmund
Postfach 500500
D–4600 Dortmund 50

Greece

Dr. Kostas A. Matis
Laboratory of General and Inorganic Chemical Technology
University of Thessaloniki
Thessaloniki

Israel

Dr. Esther Zeigerson
Research and Development Authority
P.O. Box 1025
Beer-Sheva 84110

Italy

Ms. Catia Bastioli
Istituto Guido Donegani
Via G. Fauser, 4
Centro Richerche Novara
28100 Novara

Dr. Aldo Bottino
Istituto di Chimica Industriale
Università di Genova
C.so Europa, 30
16132 Genova

Eng. Gerardo Catapano
Istituto di Principi di Ingegneria Chimica
Università degli Studi di Napoli
Piazzale Tecchio
80125 Napoli

Dr. Amalia Imperato
Istituto di Chimica Industriale
Università di Genova
C.so Europa, 30
16132 Genova

Eng. Raffaele Molinari
Istituto di Principi de Ingegneria Chimica
Università degli Studi di Napoli
Piazzale Tecchio
80125 Napoli

Assoc. Prof. Amabile Penati
Dipartimento di Chimica Industriale
e Ingegneria Chimica
Politecnico di Milano
Piazza Leonardo da Vinci 32
20133 Milano

Eng. Alfredo Punzo
Istituto di Principi di Ingegneria Chimica
Università degli Studi di Napoli
Piazzale Tecchio
80125 Napoli

Japan

Dr. Masuo Aizawa
Institute of Materials Science
University of Tsukuba
Sakura-mura, Ibaraki 305

Netherlands

Dr. Jun Kong Liou
Gr. v. Prinstererstr. 12
6702 CR Wageningen

Dr. Hans Stapersma
Kon./Shell Lab. Amsterdam
Badhuisweg 3
Amsterdam

Mr. Wim Wes
N.V. Nederlandse Gasunie
Laan Corpus den Hoorn 102
Groningen

Prof.ir. Johannes A. Wesselingh
Laboratorium voor Chemische Technologie
Technische Hogeschool Delft
Julianalaan 136
2628 BL Delft

Portugal

Prof. Manuel Alves da Silva
Departamento de Química
Faculdade de Ciências e Technologia
Universidade de Coimbra
3000 Coimbra

Ms. Ana Maria Brites Nunes
Departamento de Engenharia Química (Secção 41)
Instituto Superior Técnico
Av. Rovisco Pais
1096 Lisboa Codex

Mr. Antonio Carlos Lopes da Conceicão
Dept. de Engenharia Química, Secção Fen. Transf. Apl.
Instituto Superior Técnico
1096 Lisboa Codex

Ms. Maria E.M. Da Cruz
Laboratório Nacional de Engenharia e Tecnologia Industrial
Estrada das Palmeiras
Queluz de Baixo
2745 Queluz

Dr. Teresa Maria Fonseca de Moura
Faculdade de Ciências e Tecnologia
Universidade Nova de Lisboa
Rua Prof. Luiz Reis Santos BL3/1H
1600 Lisboa

Dr. César Sequeira
Laboratório de Electroquímica
Instituto Superior Técnico
Av. Rovisco Pais
1000 Lisboa

Spain

Prof. Dr. Juan I. Arribas
Dpto. de Termologia
Facultad de Ciencias
Universidad de Valladolid
Valladolid

Dr. Javier Garrido
Facultad de Fisicas
Universidad de Valencia
Burjasot (Valencia)

Dr. Ing. Gus Reimers
Arturo Sorio 320
Madrid 33

Dr. Benjamin Seoane
Dpto. de Termologia
Fac. de Ciencias Fisicas
Universidad Complutense de Madrid
Madrid

Switzerland

Dr. Heinz Pfenninger
Ciba-Geigy AG
K–667.306
4002 Basel

Turkey

Dr. Inci Eroğlu
Department of Chemical Engineering
Middle East Technical University (ODTÜ)
Ankara

Dr. Erhan Pişkin
Chemical Engineering Department
Hacettepe University
P.K. 716, Kizilay
Ankara

Dr. Mehmet Şahan
Department of Chemistry
Hacettepe University
Beytepe, Ankara

Dr. Meral Yücel
Department of Biological Sciences
Middle East Technical University (ODTÜ)
Ankara

United States

Mr. Ingo Blume
Membrane Technology and Research, Inc.
1030 Hamilton Court
Menlo Park, CA 94025

Professor James C.W. Chien
Polymer Science and Engineering
University of Massachusetts
Graduate Research Center, Room 701
Amherst, MA 01003

Dr. Gary Dirks
Research & Development Department
The Standard Oil Company (Ohio)
4440 Warrensville Center Road
Cleveland, OH 44128

Mr. Randy Duran
Dept. of Macromolecular Science
Case Western Reserve University
606 Olin Building
Cleveland, OH 44106

Assoc. Prof. Alan Grodzinsky
Dept. of Electrical Engineering
and Computer Science
Massachusetts Institute of Technology
Room 38–377
Cambridge, MA 02139

Asst. Prof. Douglas R. Lloyd
Department of Chemical Engineering
The University of Texas at Austin
Austin, TX 78712

Dr. Henri J.R. Maget
Consultant
6455 La Jolla Boulevard(I–208)
La Jolla, CA 92037

Professor F.P. McCandless
Department of Chemical Engineering
Montana State University
Bozeman, MT 59717

Mr. Gerald Minkowitz
School of Engineering and Applied Sciences
Columbia University
New York, NY 10027

Prof. Keith Pannell
Department of Chemistry
University of Texas at El Paso
El Paso, TX 79968

Mr. Tom Stanoch
Membrane Development Venture
Albany International Corporation
921 Providence Highway
Norwood, MA 02062

Asst. Prof. James Watters
Chemical and Environmental Engineering
Department
University of Louisville
Louisville, KY 40292

United Kingdom

Dr. Timothy deV. Naylor
British Petroleum Research Centre
Chertsey Road
Sunbury, TW16 7 ENGLAND

Dr. David E. Griffiths
Department of Chemistry and
Molecular Sciences
University of Warwick
Coventry, CV47AL, ENGLAND

INDEX

A

B

D

E

F

G

H

I

X

Y

Z